Theo de Jong
Gerhard Pfister

Local Analytic Geometry

Advanced Lectures in Mathematics

Introduction to Markov Chains
Ehrhard Behrends

Einführung in die Symplektische Geometrie
Rolf Berndt

Wavelets – Eine Einführung
Christian Blatter

Local Analytic Geometry
Theo de Jong, Gerhard Pfister

Dirac-Operatoren in der Riemannschen Geometrie
Thomas Friedrich

Hypergeometric Summation
Wolfram Koepf

The Steiner Tree Problem
Hans-Jürgen Prömel, Angelika Steger

The Basic Theory of Power Series
Jesús M. Ruiz

vieweg

Theo de Jong
Gerhard Pfister

Local Analytic Geometry

Basic Theory and Applications

vieweg

Dr. Theo de Jong
Universität des Saarlandes
Fachbereich Mathematik
Postfach 15 11 50
D-66041 Saarbrücken, Germany
dejong@math.uni-sb.de

Prof. Dr. Gerhard Pfister
Universität Kaiserslautern
Fachbereich Mathematik
Erwin-Schrödinger-Str.
D-67663 Kaiserslautern, Germany
pfister@mathematik.uni-kl.de

Die Deutsche Bibliothek - CIP-Cataloguing-in-Publication-Data
A catalogue record for this publication is available from
Die Deutsche Bibliothek.

Originally published by Friedr. Vieweg & Sohn Verlagsgesellschaft mbH,Braunschweig/Wiesbaden in 2000.

Vieweg is a company in the specialist publishing group BertelsmannSpringer.

www.vieweg.de

Cover design: Ulrike Weigel, www.CorporateDesignGroup.de

Printed on acid-free paper

ISBN 978-3-528-03137-4 ISBN 978-3-322-90159-0 (eBook)
DOI 10.1007/978-3-322-90159-0

Voor Jeannette, Nils en Noah

Für Marlis

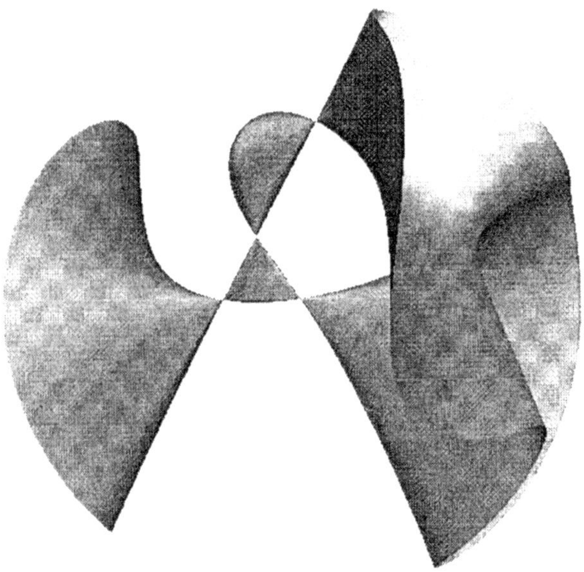

Preface

Algebraic geometry is, loosely speaking, concerned with the study of zero sets of polynomials (over an algebraically closed field). As one often reads in prefaces of introductory books on algebraic geometry, it is not so easy to develop the basics of algebraic geometry without a proper knowledge of commutative algebra. On the other hand, the commutative algebra one needs is quite difficult to understand without the geometric motivation from which it has often developed.

Local analytic geometry is concerned with germs of zero sets of analytic functions, that is, the study of such sets in the neighborhood of a point. It is not too big a surprise that the basic theory of local analytic geometry is, in many respects, similar to the basic theory of algebraic geometry. It would, therefore, appear to be a sensible idea to develop the two theories simultaneously. This, in fact, is not what we will do in this book, as the "commutative algebra" one needs in local analytic geometry is somewhat more difficult: one has to cope with convergence questions. The most prominent and important example is the substitution of division with remainder. Its substitution in local analytic geometry is called *the Weierstraß Division Theorem.*

The above remarks motivated us to organize the first four chapters of this book as follows. In Chapter 1 we discuss the algebra we need. Here, we assume the reader attended courses on linear algebra and abstract algebra, including some Galois theory. Probably the reader could just start with Chapter 2, referring to Chapter 1 only when needed. Chapter 2 deals with the basics of affine algebraic geometry, up to say, Hilbert's Nullstellensatz and decomposition into irreducible components. In Chapter 3 we tackle the corresponding basics for local analytic geometry. Here it is assumed that the reader has knowledge of the theory of holomorphic functions in one variable. Although Chapter 4 is written in the language of local analytic geometry, most statements and proofs make sense for affine algebraic geometry, too.

As an application of the general theory, in Chapter 5 we study the "simplest" germs of local analytic spaces: plane curve singularities. Topics here are Puiseux expansion, semigroups of curves, and resolutions of plane curve singularities.

Many of the topics in the rest of the book cannot usually be found in books on local analytic geometry. The principle of conservation of number in Chapter 6 is such an example. This is about interpretation of invariants in "families". The simplest example of this is the intersection number of two plane curve singularities. This intersection number can be interpreted as the number of intersection points appearing after slightly "perturbing" the singularities. To put the proofs in their proper context, we discuss and prove the basic coherence theorems of Oka and Cartan, and the finite mapping theorem of Grauert and Remmert. Next in line is the theory of standard bases in a power series ring, due to Hironaka and Grauert. The corresponding notion in a polynomial ring is called Gröbner basis, and has received much attention lately because of its applications in computer algebra. We refer to the computer algebra system SINGULAR [Singular 2000].[1] One of the main ideas is that standard bases allow a well-defined representative of an element f modulo an ideal I, called the normal form of f. In particular, this normal form is zero

[1] Our advise to the reader is to compute many examples of the theory in this book by means of a computer algebra system, which can calculate in localizations of polynomials rings.

if and only if f is an element of I.

Chapter 8 is devoted to approximation theorems. Most famous, and the easiest to prove, is the Artin Approximation Theorem. It states that if one has a formal solution (that is, without regarding the convergence) of a set of analytic equations, then there exists an analytic, that is, convergent solution to these equations. Further, a formal solution can be arbitrarily approximated by an analytic one. For applications it is sometimes necessary that some solutions do not depend on some of the variables. This so-called nested approximation theorem does not hold in general, however. Nevertheless, it does hold under stronger assumptions. This is called Grauert's Approximation Theorem.

In Chapter 9 we give the classification of simple hypersurface singularities in all its details. One of the important tools here is the finite determinacy theorem. The fact that an isolated hypersurface singularity is finitely determined is quite easy to prove: one can prove it by an application of Newton's Lemma, which is a "souped-up" version of the implicit function theorem. For a sharper statement we use the Artin Approximation Theorem. Finally, in Chapter 10, we give, as application of Grauert's Approximation Theorem, a proof of the existence of a semi-universal deformation of an isolated singularity.

Some proofs in this book are new, or at least we have not seen them in this form in the literature. We mention:

- The proof of the convergence of the Puiseux expansion by using Newton's Lemma.
- The proof of the inversion theorem for Puiseux pairs.
- Direct construction of factors of a power series in two variables, without using the Puiseux expansion.
- Certain parts of the proof of Grauert's Approximation Theorem, in particular the proof of Cartan's Lemma by using standard bases.
- The fact that the resolution graph determines the semigroup of the curve.
- Parts of the proof of the Mather-Yau Theorem.

Assuming the reader has the knowledge mentioned above, it was our intention to give either full proofs of all the statements in the book, or to put them into exercises with sufficient hints. Except for the descriptive third section of Chapter 8, we hope to have succeeded in this goal.

Acknowledgment: Various people read parts of the manuscript and gave us useful comments. We mention Holger Cröni, Anne Frühbis Krüger, Tobias Hirsch, Rainer Kaenders and Dorin Popescu. We thank Thomas Keilen, who also made some of the pictures for us. We thank Pauline Bitsch for typing a large part of the manuscript. In particular we thank Olaf Bachmann for helping us with LaTeX, and explaining us the use of the cvs system.

Last but not least, we thank our wives Jeannette and Marlis for being so patient with us during the three years we were writing this book.

Introductory Guide

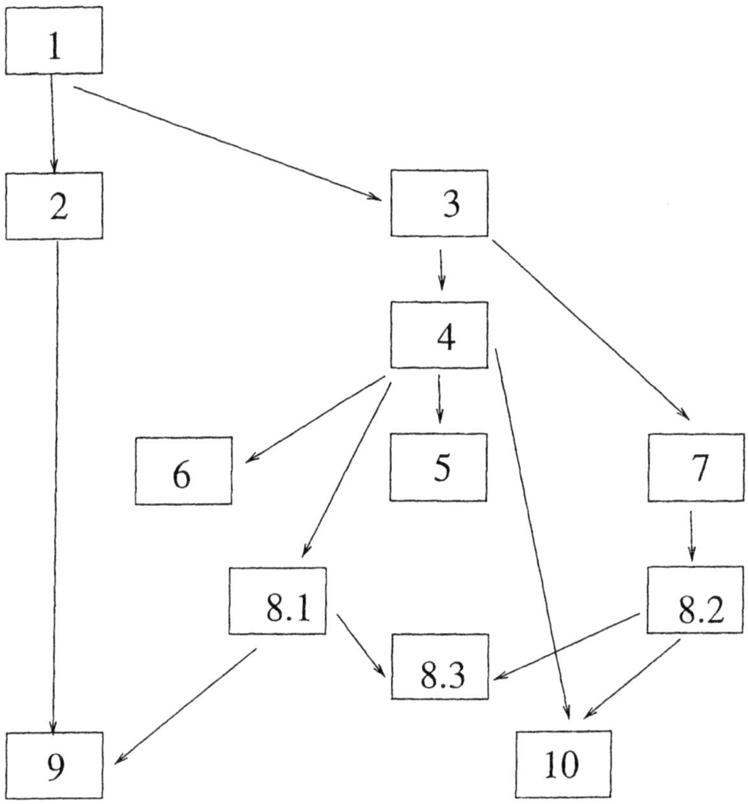

Contents

1 Algebra

In the first chapter of the book we collect some facts from algebra which we will use in this book. We assume that the reader is familiar with the concepts of *ring*, *ideal in a ring*, *principal ideal domain*, *prime ideal*, *unique factorization domain*, etc. Throughout this book, a ring will mean a commutative ring with 1. All homomorphisms of rings always take 1 to 1. The first topic we discuss in this chapter is Noetherian rings, that is, rings in which every ideal is finitely generated. The main thing to prove here is the Hilbert Basis Theorem. This says that the polynomial ring $K[x_1, \ldots, x_n]$, with K a field, is Noetherian. Then we will discuss the basic notions of the theory of modules over a ring, a notion which is analogous to vector spaces over a field. In the third section we will discuss local rings, that is, rings which have a unique maximal ideal. They play a very important role in this book. For us, the most important example will appear in the third chapter, the power series ring in n variables $\mathbb{C}\{x_1, \ldots, x_n\}$. Localization of rings will be defined. Geometric interpretations of localization will be deferred to Chapter two. In section four we discuss primary decomposition of ideals in Noetherian rings. Geometrically, this corresponds to the decomposition of an algebraic or analytic set into irreducible components: these notions will also be discussed in chapters two and three. Algebraically, it can be seen as a generalization of unique factorization in $\mathbb{Z}$, or in $K[x_1, \ldots, x_n]$. In the final section of this chapter we treat ring extensions. The notions of finite and algebraic extensions of fields here correspond to finite and integral extensions of rings. A reduced ring is called normal (or integrally closed), if every element of the total quotient ring $Q(R)$ which is integral over R is already in R. The most important examples are $K[x_1, \ldots, x_n]$, and, as we will see in Chapter three, the power series rings $\mathbb{C}\{x_1, \ldots, x_n\}$. We will prove the Cayley-Hamilton Theorem, and the very important theorem of the finiteness of integral closure, which is due to Dedekind and Kronecker. It says that, under some weak hypothesis, the normalization of a ring R is finitely generated as an R-module. This is a nontrivial statement, and it is only because of this proof that we need some Galois theory as prerequisite.

1.1 Noetherian Rings

Definition 1.1.1. A ring R is called *Noetherian* if every ideal in R is finitely generated.

This means that for every ideal $I \subset R$ there exist finitely many elements $f_1, \ldots, f_s$ in I such that $I = (f_1, \ldots, f_s) = \{\sum_{\nu=1}^{s} a_\nu f_\nu \colon a_\nu \in R\}$. In particular, principal ideal domains (domains in which every ideal can be generated by *one* element, for example fields) are Noetherian. But the Noetherian condition is much less restrictive, and almost all rings we consider in this book are Noetherian. Before giving some more examples we give the following characterizations of Noetherian rings.

Lemma 1.1.2. *Let R be a ring. Then the following conditions are equivalent:*

(1) R is Noetherian.

(2) Any chain of ideals in R

$$I_1 \subset I_2 \subset \cdots \subset I_k \subset \cdots$$

becomes stationary, that is, there exists an $n > 0$ such that $I_k = I_n$ for all $k \geq n$.

(3) Every nonempty set of ideals in R has a maximal element with respect to inclusion.

Proof.
(1) $\Longrightarrow$ (2) It is easily checked that the union $\cup_{k=1}^{\infty} I_k$ is an ideal again, which by assumption is finitely generated, say by $a_1, \ldots, a_s$. For n big enough, one has $a_i \in I_n$ for $i = 1, \ldots, s$. Then $I_k = I_n$ for all $k \geq n$.

(2) $\Longrightarrow$ (1) Suppose that there exists an ideal I in R which is not finitely generated. Then we can inductively define elements as follows:

- Take $a_1 \in I$.
- Given $a_1, \ldots, a_{k-1}$ take an element $a_k \in I \setminus (a_1, \ldots, a_{k-1})$.

But then the chain $(a_1) \subset (a_1, a_2) \subset \cdots$ is not stationary, which is a contradiction.

(2) $\Longrightarrow$ (3) If (3) were false, then there would exist a set of ideals such that for any ideal I_k in this set we can find an ideal I_{k+1} with $I_k \subsetneq I_{k+1}$. This would lead to a chain of ideals which is not stationary, a contradiction.

(3) $\Longrightarrow$ (2) Apply (3) to the set of ideals I_k for $k \geq 1$. If I_n is a maximal element, then $I_k = I_n$ for all $k \geq n$. □

Theorem 1.1.3 (Hilbert's Basis Theorem). *Let R be a Noetherian ring. Then the polynomial ring $R[x]$ is also Noetherian.*

Proof. By contradiction. Let $I \subset R[x]$ be an ideal in $R[x]$ which is not finitely generated. We then inductively define a sequence of polynomials $f_1, f_2, \ldots$ in I by:

- Let $f_1 \in I$ be a polynomial of minimal degree.
- Given $f_1, \ldots, f_{k-1}$, let f_k be a polynomial of minimal degree in $I \setminus (f_1, \ldots, f_{k-1})$.

Let a_k be the leading coefficient of f_k. Then we claim that for all $k \geq 2$:

$$(a_1, \ldots, a_{k-1}) \subsetneq (a_1, \ldots, a_k)$$

which would give us a chain of ideals in R which does not stabilize, in contradiction to the assumption that R is Noetherian. Suppose on the contrary that $a_k \in (a_1, \ldots, a_{k-1})$ for some k. We then can write $a_k = b_1 a_1 + \ldots + b_{k-1} a_{k-1}$ for some $b_i \in R$. The polynomial

$$f_k - b_1 x^{\deg(f_k) - \deg(f_1)} f_1 - \cdots - b_{k-1} x^{\deg(f_k) - \deg(f_{k-1})} f_{k-1}$$

is in $I \setminus (f_1, \ldots, f_{k-1})$, but has degree smaller than f_k, in contradiction to the choice of f_k. □

The following corollary gives us the first class of examples of Noetherian rings using the fact that fields are Noetherian.

Corollary 1.1.4. *Let R be a Noetherian ring, and $I \subset R[x_1,\dots,x_n]$ be an ideal. Then the quotient ring $R[x_1,\dots,x_n]/I$ is Noetherian.*

Proof. Using induction and Theorem 1.1.3 we obtain that $R[x_1,\dots,x_n]$ is Noetherian. The rest is given as Exercise 1.1.20. □

Definition 1.1.5.

(1) Let R be a ring. An *R–algebra* S is a ring S together with a ring homomorphism $\varphi : R \longrightarrow S$. Via this ring homomorphism we can define multiplication with scalars $r \cdot s := \varphi(r)s$. Alternatively, one can define an R–algebra S to be a ring S which is at the same time an R–module, with the condition that $r \cdot (st) = (rs) \cdot t$ for $r,s \in R$ and $t \in S$.

(2) Let R be a ring and $S = R[x_1,\dots,x_n]/I$ be the quotient of the polynomial ring by an ideal I. Then the R–algebra S is called a *ring of finite type* over R or, alternatively, an *affine algebra* over R.

From Corollary 1.1.4 we obtain that affine algebras over a Noetherian ring (or in another language, rings of finite type over a Noetherian ring) are Noetherian. Our most important case is the case of affine algebras over a field (which is Noetherian).

Now we will investigate another special class of Noetherian rings.

Definition 1.1.6. Let K be a field and R be a K–algebra. R is called an *Artinian K–algebra* if R, considered as K–vector space, has $\dim_K(R) < \infty$.

Lemma 1.1.7. *Let R be an Artinian K–algebra. Then R is Noetherian.*

Proof. Let $I \subset R$ be an ideal, then I is a finite-dimensional K–vector space because R is a finite-dimensional K–vector space. Any K–basis of I, in particular, generates I as ideal and, therefore, I is finitely generated. □

Lemma 1.1.8. *Let R be an Artinian K–algebra. Then any chain of ideals in R*

$$I_1 \supset I_2 \supset \dots \supset I_k \supset \dots$$

becomes stationary, that is, there exists an $n > 0$ such that $I_k = I_n$ for all $k \geq n$.[1]

Proof. A chain of ideals is especially a chain of K–vector spaces. Such a chain becomes stationary because R is a finite-dimensional K–vector space. □

Examples 1.1.9. Let K be a field.

(1) $K[x]/(x^n), n \geq 1$, is an Artinian K–algebra.

(2) Let A be an Artinian K–algebra and $I \subset A$ be an ideal. Then A/I is an Artinian K–algebra.

At the end of this section, we introduce the notion of graded rings and prove some other useful tools.

[1] This is the usual way to define an Artinian ring. We need later on only Artinian K–algebras in the sense of 1.1.6 which makes proofs a little simpler. Note that $\mathbb{C}[x]/(x^2)$ is a $\mathbb{Q}$–algebra which is an Artinian ring, but not an Artinian $\mathbb{Q}$–algebra in the sense of our definition.

Definition 1.1.10. Let G be an abelian group. A *G-graded ring* is a ring R together with a direct sum decomposition

$$R = \bigoplus_{g\in G} R_g, \tag{1.1}$$

such that

$$R_g \cdot R_h \subset R_{g+h} \quad \text{for all } g,h \in G. \tag{1.2}$$

The elements of R_g are called *homogeneous of degree g*. A $\mathbb{Z}$–graded ring is simply called a *graded ring.*

Examples 1.1.11.

(1) Let S be a ring (for example a field), and $R := S[x_1,\dots,x_n]$ be the polynomial ring in n variables with coefficients in S. Then R is a graded ring, by defining

$$R_d := \left\{ \sum_{\nu_1+\dots+\nu_n=d} c_{\nu_1,\dots,\nu_n} x_1^{\nu_1}\dots x_n^{\nu_n}; \quad c_{\nu_1,\dots,\nu_n} \in S \right\}, \quad \text{and } R_d = (0) \text{ for } d < 0.$$

An element of R_d is called a homogeneous polynomial of degree d. Consider $f \in R$. Then we can write

$$f = f_0 + f_1 + \dots + f_d, \quad f_i \in R_i,\ f_d \neq 0$$

in a unique way. We call $d =: \deg(f)$ the degree of the polynomial f.

(2) More generally, consider the polynomial ring $R := S[x_1,\dots,x_n]$ as above and let $\lambda = (\lambda_1,\dots,\lambda_n) \in \mathbb{N}^n$. We call $f \in S[x_1,\dots,x_n]$ quasi-homogeneous of degree d with respect to the weight λ if f can be written as

$$f = \sum_{\nu_1\lambda_1+\dots+\nu_n\lambda_n=d} c_{\nu_1,\dots,\nu_n} x_1^{\nu_1}\dots x_n^{\nu_n}.$$

Again, $R = \oplus_{d\geq 0} R_d$, where R_d consists of all quasi-homogeneous polynomials of degree d.

Let $f \in R$, $f = f_0 + \dots + f_d$, $f_i \in R_i$, $f_d \neq 0$. If we wish to distinguish between the usual degree and the quasi-homogeneous degree, we call $d =:$ w-deg(f) (the weighted degree).

Theorem 1.1.12 (Chinese Remainder Theorem). *Let R be a ring and $I_1,\dots,I_m$ be ideals in R. Assume that*

(1) $\bigcap_{\nu=1}^{m} I_\nu = (0)$,

(2) $I_\nu + I_\mu = R$ *for* $\nu \neq \mu$.

Then the canonical map

$$\begin{aligned} R &\longrightarrow \bigoplus_{\nu=1}^{m} R/I_\nu, \\ x &\mapsto (x + I_1,\dots,x + I_m) \end{aligned}$$

is an isomorphism of rings.

Proof. The first condition implies that the map is injective. To prove the surjectivity it is enough to see that for all j there exists an $x \in R$ such that $x \in I_\nu$ for all $\nu \neq j$ and $x \equiv 1 \mod I_j$. Without loss of generality, it suffices to prove this for $j = 1$. By assumption we have $I_\nu + I_1 = R$ for all $\nu \neq 1$. In particular we can find $x_\nu \in I_\nu$ and $y_\nu \in I_1$ for $\nu = 2, \dots, m$ such that $x_\nu + y_\nu = 1$. We define

$$x := \prod_{\nu=2}^{m} x_\nu = \prod_{\nu=2}^{m} (1 - y_\nu).$$

As $x_\nu \in I_\nu$ for $\nu = 2, \dots, m$, it follows that $x \in I_\nu$ for $\nu = 2, \dots, m$. As $y_\nu \in I_1$, it follows that $x \equiv 1 \mod I_1$. □

Finally, we prove a useful Lemma which will be needed in Chapters 4 and 6.

Lemma 1.1.13 (Prime Avoidance). *Let R be a Noetherian ring and $\mathfrak{p}_1, \dots, \mathfrak{p}_n$ be ideals in R. Suppose that $\mathfrak{p}_1, \dots, \mathfrak{p}_{n-1}$ are prime ideals. Let $I \subset \cup_{i=1}^n \mathfrak{p}_i$ be an ideal. Then there exists k such that $I \subset \mathfrak{p}_k$.*

To put it differently, let I be an ideal such that $I \not\subset \mathfrak{p}_i$ for all i. Then $I \not\subset \cup_{i=1}^n \mathfrak{p}_i$. This means that we can find $f \in I$ with $f \notin \mathfrak{p}_i$ for all i, whence the name prime avoidance.

Proof. The proof is done by induction on n: the case $n = 1$ is obvious. So it suffices to show that $I \subset \cup_{j \neq i} \mathfrak{p}_j$ for some i. This we will show by contradiction. So suppose that $I \not\subset \cup_{j \neq i} \mathfrak{p}_j$ for all i. Therefore, we can choose $x_1, \dots, x_n \in I$ such that $x_i \notin \cup_{j \neq i} \mathfrak{p}_j$. Because of the fact that $x_i \in I \subset \cup_{j=1}^n \mathfrak{p}_j$ we get $x_i \in \mathfrak{p}_i$.

Now we consider $x_1 + x_2 \dots x_n \in I$. In particular, there exists a k such that $x_1 + x_2 \dots x_n \in \mathfrak{p}_k$.

If $k = 1$ then $x_1 \in \mathfrak{p}_1$ and, therefore, $x_2 \dots x_n \in \mathfrak{p}_1$. This implies $x_\nu \in \mathfrak{p}_1$ for some $\nu > 1$ which is not possible by the choice of x_ν. If $k > 1$ then $x_2 \dots x_n \in \mathfrak{p}_k$ because $x_k \in \mathfrak{p}_k$ and, therefore, $x_1 \in \mathfrak{p}_k$ which is again not possible. Altogether we obtain a contradiction to the assumption that $I \not\subset \cup_{j \neq i} \mathfrak{p}_j$ for all i. This shows the Lemma. □

Exercises

1.1.14. Let $I, J \subset R$ be ideals. The product $I \cdot J \subset R$ is the ideal in R generated by the elements $x \cdot y$, where x runs through I and y runs through J. Show that $I \cdot J \subset I \cap J$.

1.1.15.

(1) Let $I, J \subset R$ be ideals. Show that $I : J := \{x \in R : xJ \subset I\}$ is an ideal in R, and that $I \subset I : J$.

(2) Let $a \in R$. Prove that $(I \cap J) : a = (I : a) \cap (J : a)$.

(3) Let $I : f^\infty := \{x \in R : xf^s \in I \text{ for some } s\}$. If R is Noetherian, show that $I : f^\infty = I : f^k$ for some $k \in \mathbb{N}$.

(4) Let $I \subset R$ be an ideal, and $x,y \in R$. Prove that $(I : x) : y = I : xy$.

1.1.16. Let $I \subset R$ be an ideal. The radical $\sqrt{I}$ of I is defined by:

$$\sqrt{I} = \{f \in R : f^s \in I \text{ for some } s \in \mathbb{N}\}.$$

An ideal is called *radical* if and only if $\sqrt{I} = I$.

(1) Prove that $\sqrt{I}$ is an ideal.

(2) Prove that $\sqrt{I + J} = \sqrt{\sqrt{I} + \sqrt{J}}$.

(3) Find ideals I,J for which $\sqrt{I+J} \supsetneqq \sqrt{I} + \sqrt{J}$.

(4) Prove that $\sqrt{I \cap J} = \sqrt{I} \cap \sqrt{J}$. Generalize to finitely many ideals.

1.1.17. Let R be a ring. An element $x \in R$ is called *nilpotent* if there exists an $n \in \mathbb{N}$ with $x^n = 0$. Show that the set of nilpotent elements is an ideal N.

The ring $R_{red} := R/N$ is called the *reduction* of R. A ring is called *reduced* if it has no nonzero nilpotent elements. Show that R_{red} is reduced.

1.1.18. Let R be a G–graded ring. An ideal $I \subset R$ is called homogeneous if $I = \sum_{g \in G}(I \cap R_g)$. Prove the following.

(1) I is homogeneous if and only if it can be generated by homogeneous elements.

(2) Sums, products and intersections of homogeneous ideals are homogeneous.

1.1.19. An abelian group G is called ordered, if there is a total ordering $<$ on G satisfying:

$$g < h \Longrightarrow a + g < a + h \text{ for all } a \in G.$$

For example, the group $\mathbb{Z}$ with its natural ordering is an ordered group. Let G be an ordered group and R be a G–graded ring.

(1) Prove that a *homogeneous* ideal I is prime if and only if for all *homogeneous* elements $a,b \in R$ with $ab \in I$ it follows that $a \in I$ or $b \in I$.

(2) Prove that the radical of a homogeneous ideal is homogeneous.

1.1.20. Let R be a Noetherian ring and $I \subset R$ be an ideal. Prove that R/I is Noetherian.

1.1.21. Consider two ideals $I \supset J$ in a Noetherian ring R. Suppose that for all $f \in I$ there exists a k (depending on f) with $f^k \in J$. Prove that some power of I is contained in J. In particular $\sqrt{I}^k \subset I$ for some k.

1.2 Modules

The notion of a module over a ring is the analog of the notion of vector space over a field, and is obtained by simply copying the definition. Here it is.

Definition 1.2.1. Let R be a ring. An *R–module* M is a set M with two operations:

(1) There is an *addition* $+ : M \longrightarrow M$:

$$x,y \in M \mapsto x + y \in M$$

(2) We have *multiplication* with scalars $\cdot R \times M \longrightarrow M$:

$$x \in M, \ a \in R \mapsto a \cdot x = ax \in M.$$

These should satisfy the following conditions.

- The set M is an abelian group with respect to the addition $+$.
- $a(x + y) = ax + ay$
  $(ab)x = a(bx)$
  $(a + b)x = ax + bx$
  $1 \cdot x = x$
  This holds for all $x,y \in M$ and $a,b \in R$.

As in the theory of rings, groups, and vector spaces, we want to consider modules up to *isomorphism*. The definition should not surprise the reader:

Definition 1.2.2. Let M and N be R–modules, $\varphi : M \to N$ a map.

- φ is called a *homomorphism* of R–modules (or simply homomorphism) if
  - $\varphi(x + y) = \varphi(x) + \varphi(y)$
  - $\varphi(ax) = a\varphi(x)$

  for all x,$y \in M$ and all $a \in R$.
- φ is called an *isomorphism* if and only if φ is a bijective homomorphism.
- Two modules M and N are called *isomorphic*, notation $M \cong N$, if and only if there exists an isomorphism $\varphi : M \to N$.

From the definition it is not immediate that isomorphy is an equivalence relation. Of course, the fact that M is isomorphic to itself is trivial, as is the fact that the isomorphism property is transitive. To prove that it is symmetric too, one has to show that the inverse $\varphi^{-1} : N \to M$ of a bijective homomorphism $\varphi : M \to N$ is a homomorphism again. This will be left to the reader as an exercise.

This property distinguishes algebra from for example topology. A bijective continuous map between topological space need not be a homeomorphism. To give a very trivial example, the identity map from a set with more than one element to itself, the first equipped with the discrete topology, and the second equipped with the trivial topology, is a bijective continuous map which is not a homeomorphism.

We now give some examples of modules.

Examples 1.2.3.

(1) A K–module, for a field K, is just a K–vector space.

(2) A $\mathbb{Z}$–module is an abelian group.

(3) An ideal I of R is an R–module. In particular, R itself is an R–module.

(4) Consider the polynomial ring $K[X]$, and let M be a $K[X]$–module. Then in particular M is a K–module, that is, a K–vector space. We also get a map (multiplication with X):

$$X : M \longrightarrow M$$

which one checks to be K–linear. In fact, the data M is a $K[X]$–module, and M is a K–vector space together with a linear map, are equivalent. Classifying $K[X]$–modules up to isomorphism is the same as classifying endomorphisms of K–vector spaces up to conjugacy, see Exercise 1.2.30.

We collect some more definitions.

Definition 1.2.4. Let R be a ring, M and N be R–modules.

(1) Let $\emptyset \neq A \subset M$. Then A is called a *submodule* of M if for all $x,y \in A$ and $a \in R$ we have $x + y$ and $ax \in A$. Then A is in a natural way an R–module, where addition and scalar multiplication are induced by those in M. One shows without difficulty that the image $\varphi(M)$ of a homomorphism $\varphi : M \longrightarrow N$ is in a natural way a submodule of N.

Let $N \subset M$ be a submodule. Then the *quotient module* M/N is the set of equivalence classes $\overline{x} := x + N$ where x runs through M. It is in a natural way an R–module via

- $(x + N) + (y + N) := (x + y) + N$,
- $a(x + N) := ax + N$.

Here $x,y \in M$ and $a \in R$. One checks that addition and scalar multiplication are well-defined.

(2) Let $\varphi : M \longrightarrow N$ be a homomorphism of R–modules. We define

- the *kernel* of φ, notation $\text{Ker}(\varphi)$, as
$$\text{Ker}(\varphi) := \{x \in M : \varphi(x) = 0\}.$$
One checks, exactly as in linear algebra, that $\text{Ker}(\varphi) = \{0\}$ if and only if φ is injective. Moreover, $\text{Ker}(\varphi)$ is a submodule of M,
- the *cokernel* of φ, notation $\text{Coker}(\varphi)$, to be the quotient module $N/\varphi(M)$.

(3) The *direct sum* $M \oplus N$ of M and N is defined to be the direct sum of abelian groups $M \oplus N = \{(m,n), m \in M, n \in N\}$, with obvious addition, and with scalar multiplication defined by $a(m,n) := (am,an)$. It is sometimes also called the *direct product* $M \times N$. Similarly, one defines the *direct sum* $\oplus_{i \in I} M_i$ of R–modules. Here I is any index set. As a set:
$$\bigoplus_{i \in I} M_i = \{(m_i)_{i \in I} \text{ with only finitely many } m_i \neq 0\}$$
and we have the obvious addition and scalar multiplication. One also has the *direct product*
$$\prod_{i \in I} M_i = \{(m_i)_{i \in I}\}.$$
with obvious addition and scalar multiplication. Here it is allowed that infinitely many m_i are nonzero.

(4) An R–module is called a *free* module if it is isomorphic to a direct sum $\oplus_{i \in I} R$ of copies of R. Often we will have the case that I is a finite set of cardinality t. This free module is called the free module of *rank* t, denoted by R^t. The fact that the rank is well-defined is not trivial and will be shown in the exercises.

(5) Let M be an R–module, $m_1, \ldots ,m_t$ be elements of M. The submodule $(m_1, \ldots ,m_t)$ generated by $m_1, \ldots ,m_t$ is by definition the submodule of M given by
$$\{a_1 m_1 + \ldots + a_t m_t : a_1, \ldots ,a_t \in R\}.$$
Similarly one defines the submodule generated by an arbitrary number of elements of M.

(6) An R–module M is called *finitely generated* if there exist finitely many elements $m_1,\dots,m_t$ of M such that $M = (m_1,\dots,m_t)$. If this is the case, the elements $m_1,\dots,m_t$ are called generators of M. Finite generation of a module is equivalent to having a *surjective* homomorphism

$$\varphi : R^t \longrightarrow M,$$
$$(a_1,\dots,a_t) \mapsto a_1m_1 + \ldots + a_tm_t.$$

(7) A set $\{m_1,\dots,m_t\}$ is called a *basis* of M if every $m \in M$ can be written as an R–linear combination of the elements $m_1,\dots,m_t$ in a *unique way*, that is, for all $m \in M$ there exist uniquely defined $a_1,\dots,a_t \in R$ such that $m = a_1m_1+\ldots+a_tm_t$. One can show (see the exercises) that a module has a basis if and only if it is isomorphic to a finitely generated free module. Similarly an infinite set is called a basis of M if every element of M can be written as an R–linear combination of finitely many elements of the set in a unique way.

(8) The *torsion module* of M is defined by:

$$\mathrm{Tors}(M) := \{m \in M : \text{there exists a nonzerodivisor } a \in R \text{ with } a \cdot m = 0\}.$$

So in case R is an integral domain, we have to find a nonzero element $a \in R$ with $am = 0$. A module M is called *torsion free* if $\mathrm{Tors}(M) = 0$. M is called a torsion module if $\mathrm{Tors}(M) = M$.

(9) The set $\mathrm{Hom}_R(M,N)$ of all homomorphisms from M to N is an R–module via the operations

$$(\varphi + \psi)(m) := \varphi(m) + \psi(m).$$
$$(a\varphi)(m) := a\varphi(m),$$

for all $m \in M$ and $a \in R$.

(10) Let $m \in M$. The *annihilator* of m is defined by $\mathrm{Ann}(m) := \{r \in R : r \cdot m = 0\}$. The annihilator of M by definition is $\mathrm{Ann}(M) := \{r \in R : r \cdot m = 0 \text{ for all } m \in M\}$. The annihilator is obviously an ideal.

Despite the obvious similarity with vector spaces, the general theory of modules over a ring is much more difficult and much richer. The inability of to divide is just too much an obstacle. For example, we know from the theory of vector spaces, that any finitely generated vector space over a field K is isomorphic to K^t for some t. This is certainly not true for rings, as the following trivial example shows:

Example 1.2.5.

(1) The $\mathbb{Z}$–module (that is, abelian group) $\mathbb{Z}/(2)$ is not free. Indeed a free group either has 1 or an infinite number of elements.

(2) More generally, any module M whose torsion submodule $\mathrm{Tors}(M)$ is nontrivial, is not a free module. Indeed, suppose the converse, that is, let $\{m_i\}_{i\in I}$ be a basis of M. Let $m \in \mathrm{Tors}(M)$, $m \neq 0$ and assume

$$m = a_1m_1 + \ldots + a_tm_t.$$

As $m \neq 0$ we have $a_i \neq 0$ for some i. Because $m \in \text{Tors}(M)$ there exists a nonzerodivisor $a \in R$ with $am = 0$. We obtain

$$a \cdot m = aa_1m_1 + \ldots + aa_tm_t = 0.$$

By definition of basis $aa_i = 0$, which is a contradiction to the fact that a is a nonzerodivisor.

One could ask the question whether at least finitely generated torsion free modules are free. But this is also much too much to ask for, as the following simple lemma shows:

Lemma 1.2.6. *Let $I \subset R$ be an ideal, with R an integral domain. Then I is a free module if and only if I is principal.*

Proof. The case where $I = (0)$ is trivial. Otherwise, suppose that I is principal, that is, $I = (f)$ for some $f \in R$. Then the map: $R \longrightarrow I$ given by $a \mapsto af$ gives an isomorphism of R–modules. It is injective because R is supposed to be an integral domain. To prove the converse, let I be a free R-module. If the rank is one, let $\varphi : R \longrightarrow I$ be an isomorphism. It is then immediate that I is the ideal generated by $\varphi(1)$. We are finished if we show that an ideal cannot be isomorphic to a free module of rank at least two. For this, it suffices to show that there does not exist an injective map $\alpha : R^n \longrightarrow I$, as soon as $n \geq 2$. Suppose the converse. Without loss of generality we may assume $n = 2$. Let $f = \alpha(1,0)$, and $g = \alpha(0,1)$. Since α is injective, both f and g are nonzero. But then $\alpha(g, -f) = gf - fg = 0$, a contradiction. □

So for example the ideal (x,y) in the polynomial ring $K[x,y]$ is not a free $K[x,y]$–module. From this lemma one might hope that at least for *principal* ideal domains (that is, domains where every ideal can be generated by one element), a finitely generated torsion free module is free. This is indeed the case, in fact one has the following classification theorem:

Theorem 1.2.7. *Let R be a principal ideal domain, and M be a finitely generated R–module. Then*

$$M \cong R^s \oplus \text{Tors}(M).$$

The torsion module has the following form:

$$\text{Tors}(M) \cong R/(a_1) \oplus \cdots \oplus R/(a_t) \text{ with } (a_1) \supset (a_2) \supset \ldots \supset (a_t) \neq (0).$$

For a proof, we refer to Lang, [Lang 1965], Chapter XV, §2. At one point in the book we will need the statement that a finitely generated torsion free module over the power series ring in one variable is free. This is a special case of the theorem, for which an easier proof is available. We refer to 1.3.9.

Of course, the finitely generated free modules R^p with $p \in \mathbb{N}$ are the simplest R–modules. Just as in linear algebra, maps between such modules can be given by matrices. Consider R^p with basis $\{e_1, \dots, e_p\}$, and R^q with basis $\{f_1, \dots, f_q\}$. Let $\varphi : R^p \longrightarrow R^q$ be an R–module homomorphism. The matrix of φ (with respect to the chosen bases) is the matrix (a_{ij}) with entries in R given by $\varphi(e_i) = \sum_{j=1}^{q} a_{ji} f_j$. All the results in linear algebra which do not use division in their proofs carry over to the general case of rings. In particular, the determinant of a map between free modules of the same rank is defined. On the other hand, because in a general ring we cannot divide, Cramer's rule does not carry over. But at least we can save the following statement.

Theorem 1.2.8 (Cramer's Rule). *Let a ring R, an $n \times n$ matrix $A = (a_{ij})$ with coefficients in R and vectors $(x_1, \dots, x_n)$ and $(b_1, \dots, b_n)$ with entries in R be given. For all j let A_j be the matrix obtained from A by replacing its j-th column[2] by $(b_1, \dots, b_n)^t$. Consider the system of equations*

$$\sum_{j=1}^{n} a_{ij} x_j = b_i \qquad i = 1, \dots, n.$$

Then

$$\det(A) x_j = \det(A_j) \quad \text{for } j = 1, \dots, n..$$

For a proof, look in any textbook on linear algebra, and forget to do the division at the end of the proof, or do the proof yourself, Exercise 1.2.28.

Definition 1.2.9. Let A be an $n \times n$ matrix with entries in R. The *adjoint matrix*

$$A^{ad} = (a_{ij}^{ad})$$

is defined by

$$a_{ij}^{ad} = (-1)^{i+j} \det(A^{ij})$$

where A^{ij} is the matrix obtained from A by deleting the i–th column and j–th row.

For a matrix A with adjoint matrix A^{ad} we have the following theorem, which follows from Cramer's rule:

Theorem 1.2.10.

$$A^{ad} A = A A^{ad} = \det(A) \cdot Id_n,$$

where Id_n denotes the $n \times n$ unit matrix.

So, up to a factor (the determinant), the adjoint matrix describes "the inverse matrix".

Definition 1.2.11. A (finite or infinite) sequence of R–modules and homomorphisms

$$\cdots \longrightarrow M_{k-1} \xrightarrow{\alpha_{k-1}} M_k \xrightarrow{\alpha_k} M_{k+1} \longrightarrow \cdots$$

is called a *complex* if $\operatorname{Ker}(\alpha_k) \supset \operatorname{Im}(\alpha_{k-1})$. It is called *exact* at M_k if $\operatorname{Ker}(\alpha_k) = \operatorname{Im}(\alpha_{k-1})$. It is called *exact* if it is exact at all M_k. In particular we have:

[2] Given a matrix A we denote by A^t the transposed matrix.

- $0 \longrightarrow M \xrightarrow{\alpha} N$ *is exact* $\Longleftrightarrow$ α *is injective,*
- $M \xrightarrow{\alpha} N \longrightarrow 0$ *is exact* $\Longleftrightarrow$ α *is surjective,*
- $0 \longrightarrow M \xrightarrow{\alpha} N \longrightarrow 0$ *is exact* $\Longleftrightarrow$ α *is an isomorphism.*

An exact sequence of type

$$0 \longrightarrow M_1 \xrightarrow{\alpha} M_2 \xrightarrow{\beta} M_3 \longrightarrow 0$$

is called a *short exact sequence.* This is equivalent to saying that α is injective, β is surjective, and β induces an isomorphism $M_2/\operatorname{Im}(\alpha) \cong M_3$.

Lemma 1.2.12 (Snake Lemma). *Consider a commutative diagram*[3]

$$\begin{array}{ccccccccc} 0 & \longrightarrow & M_1 & \xrightarrow{f_1} & M_2 & \xrightarrow{f_2} & M_3 & \longrightarrow & 0 \\ & & h_1\downarrow & & h_2\downarrow & & h_3\downarrow & & \\ 0 & \longrightarrow & N_1 & \xrightarrow{g_1} & N_2 & \xrightarrow{g_2} & N_3 & \longrightarrow & 0 \end{array}$$

of R–modules, with exact rows. Then the sequence

$$0 \to \operatorname{Ker}(h_1) \xrightarrow{f_1'} \operatorname{Ker}(h_2) \xrightarrow{f_2'} \operatorname{Ker}(h_3) \xrightarrow{\delta} \operatorname{Coker}(h_1) \xrightarrow{\overline{g}_1} \operatorname{Coker}(h_2) \xrightarrow{\overline{g}_2} \operatorname{Coker}(h_3) \to 0$$

is exact.

In this lemma f_1' and f_2' are restrictions of f_1 and f_2, and $\overline{g}_1$ and $\overline{g}_2$ are induced by g_1 and g_2. The map δ is defined as follows. Let $x \in \operatorname{Ker}(h_3)$, that is, $x \in M_3$ with $h_3(x) = 0$. As the upper row is exact, it follows that f_2 is surjective, hence we can take $y \in M_2$ with $f_2(y) = x$. Because of the commutativity of the diagram we have $g_2h_2(y) = h_3f_2(y) = h_3(x) = 0$. Therefore, $h_2(y) \in \operatorname{Ker}(g_2)$. Since the lower row is exact, we have $\operatorname{Ker}(g_2) = \operatorname{Im}(g_1)$. Hence we can find $w \in N_1$ with $g_1(w) = h_2(y)$. Now we put $\delta(x)$ to be the class of w in $\operatorname{Coker}(h_1)$. We have to prove that δ is well-defined and that this sequence is exact. The proof of the Snake Lemma is left as Exercise 1.2.27.

Definition 1.2.13. Let R be a ring, and M be an R–module. Then M is called *Noetherian* if every submodule $N \subset M$ is finitely generated.

This just means that we can find finitely many elements $m_1, \ldots, m_t \in N$ such that

$$N = R \cdot m_1 + \ldots + R \cdot m_t = (m_1, \ldots, m_t).$$

Note that a ring R is Noetherian if and only if R, considered as an R–module, is Noetherian. The following lemma is left as Exercise 1.2.29.

Lemma 1.2.14.

- *Submodules and quotient modules of Noetherian modules are Noetherian.*

---

[3] A diagram $\begin{array}{ccc} M_1 & \xrightarrow{f} & M_2 \\ h_1\downarrow & & h_2\downarrow \\ N_1 & \xrightarrow{g} & N_2 \end{array}$ is called *commutative* if $h_2 f = g h_1$, that is it does not matter which route one takes along the diagram.

- *Suppose that $N \subset M$ are R-modules. Then*

$$M \text{ is Noetherian} \iff N \text{ and } M/N \text{ are Noetherian.}$$

We could rephrase this by saying that, in a short exact sequence of modules, if any two are Noetherian, then the third is Noetherian.

Noetherian modules over a Noetherian ring are easy to construct:

Lemma 1.2.15. *Let R be a Noetherian ring, and M be a finitely generated R-module. Then M is a Noetherian R-module.*

Proof. As M is a quotient of a free module, we only have to prove the lemma for $M = R^t$, by the previous lemma. This will be done by induction on t. The case $t = 1$ is the assumption. Apply the previous lemma to the exact sequence

$$0 \to R \xrightarrow{\begin{pmatrix} 1 \\ 0 \\ \vdots \\ 0 \end{pmatrix}} R^t \xrightarrow{\begin{pmatrix} 0 & 1 & 0 & \dots & 0 \\ \vdots & \vdots & \vdots & & \vdots \\ 0 & 0 & 0 & \dots & 1 \end{pmatrix}} R^{t-1} \to 0.$$

□

Definition 1.2.16. (1) The *tensor product* $M \otimes_R N$ is an R-module with the following universal property:

There exists a canonical bilinear map $\lambda : M \times N \longrightarrow M \otimes_R N$. If $\varphi : M \times N \longrightarrow P$ is any bilinear map, then there exists a unique homomorphism $h : M \otimes_R N \longrightarrow P$ such that $h \circ \lambda = \varphi$.

$$\begin{array}{ccc} M \times N & \xrightarrow{\lambda} & M \otimes_R N \\ & {\scriptstyle\varphi}\searrow & \downarrow{\scriptstyle h} \\ & & P. \end{array}$$

One can give a direct construction of the tensor product as follows. $M \otimes_R N$ is the quotient of the free R-module generated by $\{m \otimes n : m \in M,\ n \in N\}$ divided out by the submodule of the tensor product relations generated by

$$\{(am + bn) \otimes (cp + dq) - ac \cdot m \otimes p - ad \cdot m \otimes q - bc \cdot n \otimes p - bd \cdot n \otimes q \mid a,b,c,d \in R,$$
$$m,n \in M,\ p,q \in N\}.$$

The bilinear form $\lambda : M \times N \longrightarrow M \otimes_R N$ is defined by $\lambda(m,n) = m \otimes n$ and has the property mentioned above (Exercise 1.2.45).

(2) Let S be an R-algebra, then $M \otimes_R S$ is an S-module with the multiplication $s \cdot m \otimes s' := m \otimes (ss')$ (Exercise 1.2.44). For example, if V is a real vector space, then $V \otimes_{\mathbb{R}} \mathbb{C}$ is a $\mathbb{C}$-vector space, called the complexification of V.

(3) Let S,T be R-algebras, then $S \otimes_R T$ is an R-algebra with the multiplication $(s \otimes t) \cdot (s' \otimes t') := (ss') \otimes (tt')$. The simplest example here is the case that $R = k$ is a field, $S = k[X]$ and $T = k[Y]$. Then $S \otimes_R T = k[X,Y]$.

Let $i_1 : S \longrightarrow S \otimes_R T$, $i_2 : T \longrightarrow S \otimes_R T$ be defined by $i_1(s) = s \otimes 1$, $i_2(t) = 1 \otimes t$, then

$$\begin{array}{ccc} S\otimes_R T & \stackrel{i_1}{\longleftarrow} & S \\ {\scriptstyle i_2}\uparrow & & \uparrow \\ T & \longleftarrow & R \end{array}$$

is commutative, and universal (cf. Exercise 1.2.45), that is: let A be an R–algebra and $j_1 : S \longrightarrow A$, $j_2 : T \longrightarrow A$ be homomorphisms such that

$$\begin{array}{ccc} A & \stackrel{j_1}{\longleftarrow} & S \\ {\scriptstyle j_2}\uparrow & & \uparrow \\ T & \longleftarrow & R \end{array}$$

is commutative, then there exists a unique homomorphism $h : S \otimes_R T \longrightarrow A$ such that $h \circ i_\nu = j_\nu$, $\nu = 1,2$.

Definition 1.2.17. Let R be a G–graded ring. An R–module M is called *G-graded* if there is a decomposition

$$M = \oplus_{g\in G} M_g$$

such that

$$R_g M_h \subset M_{g+h}$$

for all $g,h \in G$. In particular a G–graded ring is a G–graded module over itself. A $\mathbb{Z}$–graded module is called a *graded module*. A homomorphism $\varphi : M \longrightarrow N$ is called *homogeneous* if $\varphi(M_g) \subset N_g$ for all $g \in G$. More generally, one says that $\varphi : M \longrightarrow N$ is homogeneous of degree h if $\varphi(M_g) \subset N_{g+h}$ for all $g \in G$. For a G–graded module M and $g \in G$ we define the G–graded module $M(g)$. It has M as the underlying module, but has different grading: $M(g)_h := M_{g+h}$. So a homogeneous homomorphism $\varphi : M \longrightarrow N$ of degree g induces a homogeneous homomorphism $\varphi : M \longrightarrow N(g)$ (or $\varphi : M(-g) \longrightarrow N$).

Definition 1.2.18. Let R be a ring and $\varphi : R^n \longrightarrow R^m$ be an R–module homomorphism. Assume that φ is defined by the matrix A for some choice of bases in R^n and R^m. Then for all $1 \le t \le \min\{n,m\}$ the ideal generated by the t–minors of A does not depend on A and is called the t-th *Fitting ideal* $I_t(\varphi)$ of φ.

For convenience we define $I_t(\varphi) = R$ if $t \le 0$.

Remark 1.2.19.

(1) The proof that $I_t(\varphi)$ is independent on the choice of the bases is left as Exercise 1.2.47.

(2) Let $R^n \stackrel{\varphi}{\longrightarrow} R^m \longrightarrow M \longrightarrow 0$ and $R^s \stackrel{\psi}{\longrightarrow} R^t \longrightarrow M \longrightarrow 0$ be exact. Then $I_{m-\nu}(\varphi) = I_{t-\nu}(\psi)$ for all ν. The proof uses some facts on so-called local rings, to be treated in the next section. The proof is left as Exercise 1.3.28.

(3) If S is an R–algebra, then $I_\nu(\varphi \otimes id_S) = I_\nu(\varphi)S$. This is left as Exercise 1.2.47.

Exercises

1.2.20. Prove that the ring of integers $\mathbb{Z}$ and the polynomial ring in one variable $K[x]$ over a field K are principal ideal domains.
(Hint: Use division with remainder.)

1.2.21.

(1) Let $\varphi : R^p \to R^s$ be an *isomorphism* between free R-modules. Show that $p = s$.
(Hint: Suppose not, for example $p > s$. Let $\psi : R^s \to R^p$ be the inverse of φ. Extend φ resp. ψ to maps

$$\varphi' : R^p \to R^s \oplus R^{p-s}$$

resp.

$$\psi' : R^s \oplus R^{p-s} \to R^p$$

in the obvious way. Derive a contradiction by looking at determinants.)

(2) Conclude that the rank of a free module is well-defined.

1.2.22. Let M be an R-module, and $I = \mathrm{Ann}(M) = \{x \in R : \ x \cdot M = 0\}$. Then M is in a natural way an R/I-module. Prove this.

1.2.23. Let R be a subring of S. Show that one can, in a natural way, view S as an R-module.

1.2.24. Prove that $\mathrm{Tors}(M)$ is a submodule of M.

1.2.25. Consider an exact sequence

$$\cdots \longrightarrow M_{k-1} \xrightarrow{\alpha_{k-1}} M_k \xrightarrow{\alpha_k} M_{k+1} \longrightarrow \cdots.$$

Put $N_k = \mathrm{Im}(\alpha_k) = \mathrm{Ker}(\alpha_{k+1})$. Show that for all k one has short exact sequences:

$$0 \longrightarrow N_{k-1} \longrightarrow M_k \longrightarrow N_k \longrightarrow 0.$$

(So any exact sequence can be split into short exact sequences.)

1.2.26. Let K be a field, and consider an exact sequence

$$0 \longrightarrow M_1 \longrightarrow M_2 \longrightarrow \cdots \longrightarrow M_s \longrightarrow 0$$

of finite-dimensional K-vector spaces. Prove that $\sum_{k=1}^{s}(-1)^k \dim_K(M_k) = 0$.

1.2.27. Prove the Snake Lemma 1.2.12.

1.2.28. Prove Cramer's rule 1.2.8.

1.2.29. Prove Lemma 1.2.14.

1.2.30.

(1) Let M be a $K[X]$-module, which is finite-dimensional as a K-vector space. Show that M is a torsion module.

(2) Prove the theorem on Jordan normal form of matrices over an algebraically closed field, using the classification of finitely generated modules over a principal ideal domain, see Theorem 1.2.7.

1.2.31. Let K be a field, S be a K-algebra, and M be an S-module. Suppose that $\mathrm{Ann}(M) = (0)$ and that $\dim_K(M) < \infty$. Prove that $\dim_K(S) < \infty$.
(Hint: Show that one can embed S into $\mathrm{Hom}_S(M,M)$.)

1.2.32. Let N be an R–module. Prove that the map $\phi : \mathrm{Hom}_R(R,N) \longrightarrow N$, $\phi(\varphi) = \varphi(1)$, is an isomorphism.

1.2.33. Let M,M',N,N' be R–modules and $\alpha : M' \longrightarrow M$, $\beta : N \longrightarrow N'$ be homomorphisms. Prove that $\varphi \mapsto \beta \circ \varphi \circ \alpha$ defines a homomorphism $\mathrm{Hom}_R(M,N) \longrightarrow \mathrm{Hom}_R(M',N')$.

1.2.34. Let M,N be R–modules and $M = \bigoplus_{i=1}^{n} M_i$. Prove that

$$\mathrm{Hom}(M,N) \cong \bigoplus_{i=1}^{n} \mathrm{Hom}(M_i,N)$$

$$\mathrm{Hom}(N,M) \cong \bigoplus_{i=1}^{n} \mathrm{Hom}(N,M_i).$$

1.2.35. Consider an exact sequence

$$\cdots \longrightarrow M_{k+1} \longrightarrow M_k \longrightarrow M_{k-1} \longrightarrow \cdots$$

of R–modules. Let $x \in R$ be a nonzero divisor of M_k for all k. Prove that the induced sequence

$$\cdots M_{k+1}/xM_{k+1} \longrightarrow M_k/xM_k \longrightarrow M_{k-1}/xM_{k-1} \longrightarrow \cdots$$

is also exact.
(Hint: Split the exact sequence in many short exact sequences and use the Snake Lemma.)

1.2.36. Let $0 \longrightarrow M \longrightarrow N \longrightarrow L$ be a complex of R–modules. Prove that the complex is exact if and only if for all R–modules A

$$0 \longrightarrow \mathrm{Hom}_R(A,M) \longrightarrow \mathrm{Hom}_R(A,N) \longrightarrow \mathrm{Hom}_R(A,L)$$

is exact.

1.2.37. Let $M \longrightarrow N \longrightarrow L \longrightarrow 0$ be a complex of R–modules. Prove that the complex is exact if and only if for all R–modules A

$$0 \longrightarrow \mathrm{Hom}_R(L,A) \longrightarrow \mathrm{Hom}_R(N,A) \longrightarrow \mathrm{Hom}_R(M,A)$$

is exact.

1.2.38. Let M,M',N,N' be R–modules and $\alpha : M' \longrightarrow M$, $\beta : N' \longrightarrow N$ be homomorphisms. Prove that $m' \otimes n' \mapsto \alpha(m') \otimes \beta(n')$ defines a homomorphism $M' \otimes_R N' \longrightarrow M \otimes_R N$.

1.2.39. Prove that $M \otimes_R N \cong N \otimes_R M$.

1.2.40. Prove that the canonical map $\phi : \mathrm{Hom}_R(M \otimes_R N,P) \longrightarrow \mathrm{Hom}_R(M, \mathrm{Hom}_R(N,P))$ defined by $\phi(\lambda)(m)(n) := \lambda(m \otimes n)$ is an isomorphism.

1.2.41. Let M,N be R–modules and $M = \bigoplus_{i=1}^{n} M_i$. Prove that $M \otimes_R N \cong \bigoplus_{i=1}^{n} (M_i \otimes_R N)$.

1.2.42. Prove that $M \otimes_R R \cong M$.

1.2.43. Let $M \longrightarrow N \longrightarrow L \longrightarrow 0$ be an exact sequence of R–modules and A be an R–module. Prove that

$$M \otimes_R A \longrightarrow N \otimes_R A \longrightarrow L \otimes_R A \longrightarrow 0$$

is exact.
(Hint: Use Exercises 1.2.37 and 1.2.40.)

1.2.44. Let M be an R–module and S be an R–algebra. Prove that with the multiplication $s \cdot m \otimes s' = m \otimes ss'$, $M \otimes_R S$ is an S–module.

1.2.45. Prove the universal properties mentioned in Definition 1.2.16.

1.2.46. Let V and W be vector spaces over K. For $\varphi \in V^*$ and $\sum v_i \otimes w_i \in V \otimes_K W$, consider $\sum \varphi(v_i)w_i \in W$. Show that we get a well defined map $V^* \longrightarrow W$. Show that $V \otimes_K W \cong \mathrm{Hom}(V^*, W)$.

1.2.47. Prove (1) and (3) of Remark 1.2.19.
(Hint: It is enough to consider the case of two matrices A and B, B obtained from A by one column operation, and to prove that $I_t(A) = I_t(B)$.)

1.2.48. Let M,N be R–modules and $0 \longrightarrow M \xrightarrow{\varphi} N \xrightarrow{\psi} R^s \longrightarrow 0$ be an exact sequence. Prove that $N \cong M \oplus R^s$, more precisely

$$\begin{array}{ccccccccc} 0 & \longrightarrow & M & \xrightarrow{\varphi} & N & \xrightarrow{\psi} & R^s & \longrightarrow & 0 \\ & & \| & & \uparrow & & \| & & \\ 0 & \longrightarrow & M & \xrightarrow[i]{} & M \oplus R^s & \xrightarrow[\pi]{} & R^s & \longrightarrow & 0 \end{array}$$

commutes, i,π the canonical injection resp. projection. In this case one says that M is a direct summand of N.

1.3 Local Rings and Localization

Definition 1.3.1. A ring R is called *local* if it has a unique maximal ideal $\mathfrak{m}$. One often says that $(R,\mathfrak{m})$ is a *local ring*, to indicate that $\mathfrak{m}$ is its unique maximal ideal.

Later on in this section we will see how to construct local rings from general rings by means of "localization". In this book, the local rings we are dealing with most frequently are power series rings and their quotients, see Section 3.1. Here we define those rings for the case of one variable.

Definition 1.3.2. The *formal power series ring* $\mathbb{C}[[x]]$ in one variable consists of elements $\sum_{k=0}^{\infty} a_k x^k$, with $a_k \in \mathbb{C}$. We have the obvious addition and multiplication. The convergent power series ring $\mathbb{C}\{x\}$ in one variable consists of elements $\sum_{k=0}^{\infty} a_k x^k$, with $a_k \in \mathbb{C}$, that converge in some open neighborhood of 0 in $\mathbb{C}$.

Note that in the formal power series ring as well as in the convergent power series ring, the elements which do not lie in (x), that is, those with nonzero constant term, are units. Therefore, the fact that the power series rings are local rings follows from the following lemma.

Lemma 1.3.3. *Let R be a ring, and $\mathfrak{m} \subset R$ be an ideal. Then R is local, with maximal ideal $\mathfrak{m}$, if and only if $R \setminus \mathfrak{m}$ is the set of units of R.*

Proof. Suppose R is local with maximal ideal $\mathfrak{m}$. Take an element $a \in R$. If $(a) \neq R$, then (a) is contained in a maximal ideal, which by definition of a local ring must be $\mathfrak{m}$. Therefore, if $a \notin \mathfrak{m}$, then $(a) = R$, that is, a is a unit. Moreover, if $a \in \mathfrak{m}$, then a is not a unit, as otherwise $(a) = R$.

Suppose, on the other hand, that $R \setminus \mathfrak{m}$ is the set of units. Take an ideal $I \neq R$. Then the ideal I does not contain units, and therefore must be contained in $\mathfrak{m}$. This shows that $\mathfrak{m}$ is the unique maximal ideal of R. □

Theorem 1.3.4 (Nakayama's Lemma). *Let* $(R,\mathfrak{m})$ *be a local ring and* M *be a* finitely generated R*-module with* $\mathfrak{m}\cdot M = M$. *Then* $M = 0$.

Proof. By contradiction. Let $m_1,\dots,m_t$ be a *minimal* number of generators for M. As $M = \mathfrak{m}\cdot M$ there exist $a_1,\dots,a_t \in \mathfrak{m}$ such that $m_t = a_1m_1 + \dots + a_tm_t$. Therefore

$$(1-a_t)m_t = a_1m_1 + \dots + a_{t-1}m_{t-1}.$$

Now $1 - a_t \notin \mathfrak{m}$, and as R is a local ring it follows that $1 - a_t$ is a unit. Therefore $m_t \in (m_1,\dots,m_{t-1})$, in contradiction to the minimality of t. □

For more a general version of Nakayama's Lemma, we refer to Exercise 1.3.17. A corollary of this important lemma is the following corollary.

Corollary 1.3.5 (Krull's Intersection Theorem). *Let* $(R,\mathfrak{m})$ *be a Noetherian local ring, and* M *be a finitely generated* R*-module. Then* $\cap_{k\in\mathbb{N}}\mathfrak{m}^kM = (0)$.

Proof. We will prove this theorem for the case $M = R$, the general case is similar, see Exercise 1.3.24. Put $I = \cap_{k\in\mathbb{N}}\mathfrak{m}^k$. Then I is an ideal (check this), finitely generated because R is Noetherian. We will show that $\mathfrak{m}\cdot I = I$, so that $I = (0)$ by Nakayama's Lemma. For this purpose consider the following set of ideals

$$\mathscr{A} := \{J \text{ ideal in } R : J \cap I = \mathfrak{m}I\}.$$

Our ring R is Noetherian, so that this set has a maximal element which we call J. Obviously $\mathfrak{m}I \subset J$. Note that it suffices to prove that $\mathfrak{m}^k \subset J$ for certain k. Indeed, as obviously $I \subset \mathfrak{m}^k$, it follows $I \subset \mathfrak{m}^k \subset J$, hence $I \subset J \cap I = \mathfrak{m}I$, and we can apply Nakayama. As $\mathfrak{m}$ is finitely generated, it suffices to find for each $f \in \mathfrak{m}$ an α with $f^\alpha \in J$. This we will do by showing $J + (f^\alpha) = J$ by using the maximality property of J. Thus it remains to show that $(J+(f^\alpha))\cap I = \mathfrak{m}I$. The inclusion $\supset$ is obvious. As R is Noetherian, the chain $J : f \subset J : f^2 \subset \cdots$ stabilizes. Hence there exists an α with $J : f^\alpha = J : f^{\alpha+1}$, and this is the α we need. Now let $x \in (J + (f^\alpha)) \cap I$, say $x = y + af^\alpha \in I$ for $y \in J$. It follows that $f^{\alpha+1}a = fx - fy \in \mathfrak{m}I + J = J$, as $\mathfrak{m}I \subset J$. Hence $a \in J : f^\alpha = J : f^{\alpha+1}$. Thus $f^\alpha a \in J$, so that $x \in J$ follows. As $x \in I$ it follows $x \in J \cap I = \mathfrak{m}I$. This holds for all $x \in (J + (f^\alpha)) \cap I$ so that $(J + (f^\alpha)) \cap I \subset \mathfrak{m}I$ follows. This is what we had to show. □

Another corollary is the following lemma on the number of generators of a module over a local ring. The proof is left as Exercise 1.3.18.

Corollary 1.3.6. *Let* M *be a finitely generated module over a local ring* $(R,\mathfrak{m})$.

(1) Let $f_1,\dots,f_s \in M$ *such that the classes of the* f_i *generate* $M/\mathfrak{m}M$ *as* $R/\mathfrak{m}$*-vector space. Then* $f_1,\dots,f_s$ *generate* M.

(2) The minimal number of generators of M *is equal to* $\dim_{R/\mathfrak{m}}(M/\mathfrak{m}M)$.

Definition 1.3.7. Let R be a ring, and M be an R-module. A *presentation of* M is a short exact sequence

$$G \xrightarrow{\alpha} F \xrightarrow{\beta} M \longrightarrow 0 \tag{1.3}$$

of R-modules, in which F and G are free R-modules. In case R is a local ring, and M is finitely generated, a presentation as in 1.3 is called *minimal* if the rank of F is equal to the minimal number of generators of M, that is, equal to $\dim_{R/\mathfrak{m}}(M/\mathfrak{m}M)$, see 1.3.6. This is equivalent to the statement $\mathrm{Im}(\alpha) \subset \mathfrak{m}\cdot F$.

Proposition 1.3.8. *Let R be a local ring and $R^n \xrightarrow{\varphi} R^m \longrightarrow M \longrightarrow 0$ be a presentation of M. The following conditions are equivalent:*

(1) M is a free module of rank $m-r$,

(2) $Im(\varphi)$ is a free module of rank r and a direct summand of R^m,

(3) $I_r(\varphi) = R$ and $I_{r+1}(\varphi) = (0)$.

Proof.
(3) $\Rightarrow$ (2) The fact that $I_r(\varphi) = R$ and R is local implies that for a matrix A representing φ there is an r–minor which is not in the maximal ideal of R, hence a unit. By renumbering we may assume that this minor is obtained by taking the determinant of the first r rows and columns. We write $A = \left(\begin{smallmatrix} B & * \\ * & * \end{smallmatrix}\right)$ with B an $r \times r$–matrix. Then $\frac{1}{\det B}\begin{pmatrix} B^{ad} & 0 \\ 0 & Id_{n-r} \end{pmatrix}$ gives an automorphism of R^n and we multiply A from the right with this matrix. So we may assume that A is of type $\left(\begin{smallmatrix} Id_r & * \\ * & * \end{smallmatrix}\right)$. As in linear algebra, we can now do row and column operations, corresponding to a change of basis in target and source. We may then assume that A is of type $\left(\begin{smallmatrix} Id_r & 0 \\ 0 & * \end{smallmatrix}\right)$. As by assumption $I_{r+1}(\varphi) = 0$, it then follows that A is of type $\left(\begin{smallmatrix} Id_r & 0 \\ 0 & 0 \end{smallmatrix}\right)$. This immediately implies (2).

(2) $\Rightarrow$ (1) is trivial, using Exercise 1.3.26.

(1) $\Rightarrow$ (3) By Exercise 1.2.48 and Exercise 1.3.26 we obtain that $R^m \cong M \oplus \mathrm{Im}(\varphi)$, and $\mathrm{Im}(\varphi)$ is free of rank r. Similarly, $R^n \cong \mathrm{Ker}(\varphi) \oplus \mathrm{Im}(\varphi)$ and $\mathrm{Ker}(\varphi)$ is free of rank $n-r$. So we may assume that φ is given by the matrix $\left(\begin{smallmatrix} Id_r & 0 \\ 0 & 0 \end{smallmatrix}\right)$, see Remark 1.2.19. This implies (3). □

The following theorem is a special case of the classification of finitely generated modules over principal ideal domains.

Theorem 1.3.9. *Let M be a finitely generated torsion free $\mathbb{C}\{x\}$–module. Then M is free.*

As a finitely generated vector space is free, the theorem immediately follows from the following lemma.

Lemma 1.3.10. *Let $(R,\mathfrak{m})$ be a local ring, M a finitely generated R–module, and $x \in \mathfrak{m}$ a nonzero divisor of M. Then*

$$M \text{ is a free } R\text{–module} \iff M/xM \text{ is a free } R/(x)\text{–module.}$$

Proof. The implication $\Longrightarrow$ is trivial. Suppose, therefore, that M/xM is a free $R/(x)$-module. Take a map $\alpha : R^n = F \to M$, sending each basis element to a generator of M. We might take n to be the minimal number of generators of M. It follows that the map $F/xF \to M/xM$ is also surjective. By 1.3.6 $M/\mathfrak{m}M \cong (R/\mathfrak{m})^n$. As we supposed that M/xM is a free $R/(x)$–module the map α induces an isomorphism $(R/(x))^n \cong M/xM$. Let K be the kernel of the map α. We obtain a commutative diagram:

$$\begin{array}{ccccccccc}
0 & \longrightarrow & K & \longrightarrow & F & \longrightarrow & M & \longrightarrow & 0 \\
 & & \downarrow{\scriptstyle \cdot x} & & \downarrow{\scriptstyle \cdot x} & & \downarrow{\scriptstyle \cdot x} & & \\
0 & \longrightarrow & K & \longrightarrow & F & \longrightarrow & M & \longrightarrow & 0
\end{array}$$

where $\cdot x$ denotes multiplication by x. By assumption, x is not a zero divisor of M, so that the kernel of $\cdot x : M \to M$ is zero. It follows from the Snake Lemma that we have an exact sequence:

$$0 \longrightarrow K/xK \longrightarrow F/xF \longrightarrow M/xM \longrightarrow 0.$$

From the remark above, we have that $F/xF \to M/xM$ is an isomorphism. It follows that $K/xK = 0$. From Nakayama, in the version of Exercise 1.3.17 it follows that $K = 0$ or, equivalently, that $M \cong F$ is free. □

There is a procedure to construct local rings from a given ring, called *localization.* This is a special case of forming rings (and modules) of fractions, which is itself a generalization of defining the quotient field of an integral domain. In order to define rings of fractions, we have to introduce the concept of a *multiplicatively closed set.*

Definition 1.3.11. Let R be a ring. A set $S \subset R$ is called *multiplicatively closed* if and only if it satisfies the following conditions.

- $1 \in S$,
- If $f,g \in S$, then $f \cdot g \in S$.

Definition 1.3.12. Let R be a ring and S be a multiplicatively closed set. We will define the *ring of fractions* R_S, also denoted by $S^{-1}R$. As a set it is the following:

$$R_S = \{\tfrac{r}{s} : r \in R, s \in S\}$$

and satisfies the following properties:

- $$\frac{r_1}{s_1} = \frac{r_2}{s_2} \in R_S$$

  if and only if there exists an $s \in S$ such that $s(r_1s_2 - r_2s_1) = 0$.
- Multiplication:

  $$\frac{r_1}{s_1} \cdot \frac{r_2}{s_2} := \frac{r_1r_2}{s_1s_2}.$$
- Addition:

  $$\frac{r_1}{s_1} + \frac{r_2}{s_2} := \frac{r_1s_2 + r_2s_1}{s_1s_2}.$$

Similarly, for any R–module M, and any multiplicatively closed set $S \subset R$, one can define the R_S–module M_S, also denoted $S^{-1}M$. In particular this can be done for an ideal $I \subset R$. Its definition is straightforward, and is left to the reader.

Examples 1.3.13.

(1) Let R be an integral domain, and $S = R \setminus \{0\}$. Then R_S is the quotient field $Q(R)$ of R. It is indeed a field, as $Q(R) = \{\frac{a}{b} : a,b \in R, b \neq 0\}$. So if $\frac{a}{b} \neq 0$, then $a \neq 0$, and $\frac{b}{a} \in Q(R)$.

(2) More generally, for a ring R, one can take as multiplicatively closed set S the set of all nonzero divisors. The resulting ring R_S is called the *total quotient ring*, or *total ring of fractions* of R, notation $Q(R)$. We will prove in Proposition 1.4.27 that for a reduced Noetherian ring $Q(R)$ is a direct sum of fields.

(3) Consider the multiplicatively closed set $S = \{1,f,f^2,f^3,\dots\}$ for $f \in R$. In this case one usually writes R_f instead of R_S.

(4) Let $\mathfrak{p}$ be a prime ideal of R. According to the definition of a prime ideal the set $R \setminus \mathfrak{p}$ is a multiplicatively closed set. The resulting ring of fractions is denoted by $R_\mathfrak{p}$ (instead of the "correct" $R_{R\setminus\mathfrak{p}}$), and is called the localization of R in $\mathfrak{p}$. So $R_\mathfrak{p} = \{\frac{a}{b} : a \in R,\, b \notin \mathfrak{p}\}$.

At first it might be surprising that the following lemma holds generally for prime ideals, and not only for maximal ideals. Nevertheless, the proof is easy!

Lemma 1.3.14. *Let R be a ring and $\mathfrak{p} \subset R$ be a prime ideal. Then $R_\mathfrak{p}$ is a local ring, with maximal ideal $\mathfrak{p}R_\mathfrak{p} = \{\frac{a}{b} : a \in \mathfrak{p},\, b \notin \mathfrak{p}\}$.*

Proof. By 1.3.3, it suffices to show that the elements of $\mathfrak{p}R_\mathfrak{p}$ are the nonunits of $R_\mathfrak{p}$. An element in $R_\mathfrak{p} \setminus \mathfrak{p}R_\mathfrak{p}$ is of type $\frac{a}{b}$, where $a,b \notin \mathfrak{p}$. But then $\frac{b}{a} \in R_\mathfrak{p}$, so that $\frac{a}{b}$ is a unit. Of course, every element $\frac{a}{b}$, with $a \in \mathfrak{p}$ is not a unit in $R_\mathfrak{p}$. □

Theorem 1.3.15. *Let R be a ring, S a multiplicatively closed set. There is a one-one correspondence:*

$$\text{prime ideals } \mathfrak{p} \subset R \text{ with } \mathfrak{p} \cap S = \emptyset \quad \longleftrightarrow \quad \text{prime ideals } \mathfrak{p}_S \subset R_S.$$

The correspondence is given by the following rules. Given a prime ideal $\mathfrak{p}_S \subset R_S$, the ideal $\mathfrak{p} := \mathfrak{p}_S \cap R$ is a prime ideal in R. Given a prime ideal $\mathfrak{p}$ in R whose intersection with S is empty, its localization $\mathfrak{p}_S$ is a prime ideal in R_S.

Proof. Suppose $\mathfrak{p}$ is a prime ideal in R, with $\mathfrak{p} \cap S = \emptyset$. Then the inclusion $\mathfrak{p}_S \subsetneq R_S$ is strict. This is because otherwise $1 = \frac{a}{s}$, with $a \in \mathfrak{p}$, $s \in S$. Hence there would exist a $t \in S$ such that $t(s-a) = 0$, that is $ta = ts$. But $ts \in S$, and $ta \in \mathfrak{p}$, which by assumption cannot be the case. Now suppose $\frac{a}{s}\frac{b}{t} \in \mathfrak{p}_S$. It follows that $ab \in \mathfrak{p}$, and therefore either $a \in \mathfrak{p}$ or $b \in \mathfrak{p}$. Therefore either $\frac{a}{s} \in \mathfrak{p}_S$ or $\frac{b}{t} \in \mathfrak{p}_S$. The converse is left to the reader. □

Exercises

1.3.16. Prove that $IR_f \cap R = I : f^\infty$ (see 1.1.15).

1.3.17 (Nakayama's Lemma). Let R be a ring, $I \subset R$ be an ideal which is contained in all the maximal ideals of R, and M be a finitely generated R–module.

(1) Suppose that $M = I \cdot M$. Prove that $M = 0$.

(2) Let N be a submodule of M such that $M = N + I \cdot M$. Prove that $M = N$.

1.3.18. Prove Corollary 1.3.6.
(Hint: Let $N = (f_1, \dots, f_s)$ and apply Nakayama's Lemma.)

1.3.19. Let $(R,\mathfrak{m})$ be a Noetherian local ring, and J an ideal in R.

(1) Suppose $\mathfrak{m}J = (0)$.

(a) Prove that J is a finite-dimensional $R/\mathfrak{m}$–vector space.

(b) Prove that there exists an $n \in \mathbb{N}$ with $J \cap \mathfrak{m}^n = (0)$.

(2) Show that that there exists an n with $J \cap \mathfrak{m}^n \subset \mathfrak{m}J$.

(3) Prove that for such an n we have $J/\mathfrak{m}J \cong J + \mathfrak{m}^n/\mathfrak{m}J + \mathfrak{m}^n$.

1.3.20. Let K be a field, and R be a local K-algebra, with maximal ideal $\mathfrak{m}$.

(1) Suppose that $\dim_K R = n < \infty$. Show that $\mathfrak{m}^n = 0$.
(Hint: Show that the sequence of ideals $\mathfrak{m} \supset \mathfrak{m}^2 \supset \cdots$ stabilizes and use Nakayama's Lemma.)

(2) Let $I \subset R$ be an ideal such that $\dim_K(R/I) = n < \infty$. Show that $\mathfrak{m}^n \subset I$.

1.3.21. Let M be an R-module. Prove that the following conditions are equivalent.

(1) $M = 0$.

(2) $M_\mathfrak{p} = 0$ for all prime ideals $\mathfrak{p} \subset R$.

(3) $M_\mathfrak{m} = 0$ for all maximal ideals $\mathfrak{m} \subset R$.

(Hint: For (3) $\Longrightarrow$ (1) show that for all $m \in M$ the ideal $\mathrm{Ann}(m)$ is not contained in any maximal ideal of R.)

1.3.22. Let R be a ring.

(1) Let $S \subset R$ a multiplicatively closed subset of R and $M \longrightarrow N \longrightarrow P$ be an exact sequence of R-modules. Prove that the induced sequence $M_S \longrightarrow N_S \longrightarrow P_S$ is exact. One phrases this by saying that localization is an exact functor.

(2) Let $\cdots \longrightarrow M_{i-1} \longrightarrow M_i \longrightarrow M_{i+1} \longrightarrow \cdots$ be a complex. Prove that the complex is exact if and only if for all prime ideals $\mathfrak{p} \subset R$ the induced complex $\cdots \longrightarrow M_{i-1\,\mathfrak{p}} \longrightarrow M_{i\mathfrak{p}} \longrightarrow M_{i+1\,\mathfrak{p}} \longrightarrow \cdots$ is exact.

(3) Show that $R_S \otimes_R M \cong M_S$.

(4) Show that $(M \otimes_R N)_S \cong M_S \otimes_{R_S} N_S$.

1.3.23. Let R be a ring, and $\mathfrak{p} \subset R$ be a prime ideal. We define the n-th *symbolic power* $\mathfrak{p}^{(n)}$ of $\mathfrak{p}$ by

$$\mathfrak{p}^{(n)} := \mathfrak{p}^n R_\mathfrak{p} \cap R.$$

We obviously have $\mathfrak{p}^n \subset \mathfrak{p}^{(n)}$.

(1) Let $R = k[x,y,z]/(z^2 - xy)$, and $\mathfrak{p} = (x,z)$. Show that $\mathfrak{p}$ is prime, and that $\mathfrak{p}^{(2)} = (z^2,x)$. In particular $\mathfrak{p}^2 \neq \mathfrak{p}^{(2)}$.

(2) Let $f,g \in R$ be given such that $f \notin \mathfrak{p}$ and $f \cdot g \in \mathfrak{p}^{(n)}$. Show that $g \in \mathfrak{p}^{(n)}$.

1.3.24. Prove Krull's Intersection Theorem 1.3.5 for modules.

1.3.25. Let R be a ring. The *Jacobson radical* $J(R)$ of R by definition is the intersection of all maximal ideals of R. Show that $x \in J(R)$ implies that $1 + x$ is a unit.

1.3.26. Let R be a local ring and $R^n = R^s \oplus N$. Prove that N is free of rank $n - s$.
(Hint: Use 1.3.6.) This statement does not hold for general (nonlocal) rings.

1.3.27. Let R be a ring and $\mathfrak{p}_1, \ldots, \mathfrak{p}_m$ be prime ideals. Let $(0) \neq M$ be a finitely generated R-module such that $M_{\mathfrak{p}_j} \neq (0)$ for all j. Prove that there exists an $x \in M$ such that $x \notin \mathfrak{p}_j M_{\mathfrak{p}_j}$ for all j.
(Hint: Use the idea of the proof of prime avoidance, see 1.1.13: Use induction on m and assume, therefore, $y \notin \mathfrak{p}_j M_{\mathfrak{p}_j}$ for $j = 1, \ldots, m-1$ and a suitable y. Assume the worst case, that is, $y \in \mathfrak{p}_m M_{\mathfrak{p}_m}$. You may also assume that $\bigcap_{j=1}^{m-1} \mathfrak{p}_j \not\subset \mathfrak{p}_m$ and choose $r \in \bigcap_{j=1}^{m-1} \mathfrak{p}_j \setminus \mathfrak{p}_m$. Let $M = (y, y_2, \ldots, y_n)$ and $M' = (y, ry_2, \ldots, ry_n)$ then $M'_{\mathfrak{p}_m} = M_{\mathfrak{p}_m}$ and $y + ry_i \notin \mathfrak{p}_j M_{\mathfrak{p}_j}$ for $i \geq 2$ and $j = 1, \ldots, m-1$. One of these elements is the element we are looking for because, otherwise, $y, y_2, \ldots, y_n \in \mathfrak{p}_m M_{\mathfrak{p}_m}$, that is, $M_{\mathfrak{p}_m} = (0)$.)

1.3.28. Prove (2) of Remark 1.2.19.
(Hint: Localize by prime ideals to reduce the assertion to the case of local rings. Let $\mathfrak{m}$ be the maximal ideal of R and $R^{n_1} \xrightarrow{\varphi_1} R^{m_1} \longrightarrow M \longrightarrow 0$, $R^{n_2} \xrightarrow{\varphi_2} R^{m_2} \longrightarrow M \longrightarrow 0$ be two presentations of M. Show that we may assume that $\mathrm{Ker}(\varphi_1) \subset \mathfrak{m}R^{n_1}$ and $\mathrm{Im}(\varphi_1) \subset \mathfrak{m}R^{m_1}$. Suppose first that $\mathrm{Ker}(\varphi_i) \subset \mathfrak{m}R^{n_i}$ and $\mathrm{Im}(\varphi_i) \subset \mathfrak{m}R^{m_i}$. Prove that there exist isomorphisms λ,μ such that

$$\begin{array}{ccccc} R^{n_1} & \xrightarrow{\varphi_1} & R^{m_1} & \longrightarrow & M \\ \downarrow{\scriptstyle\lambda} & & \downarrow{\scriptstyle\mu} & & \| \\ R^{n_2} & \xrightarrow{\varphi_2} & R^{m_2} & \longrightarrow & M \end{array}$$

commutes. Use Remark 1.2.19 (1) for this case. Now show that for a suitable choice of basis in R^{n_2} and R^{m_2} the map φ_2 has a matrix $\left(\begin{smallmatrix} \overline{\varphi}_2 & 0 & 0 \\ 0 & Id & 0 \end{smallmatrix}\right)$ with $\overline{\varphi}_2 : R^{n_1} \longrightarrow R^{m_1}$ and Id the identity matrix.)

1.4 Primary Decomposition

It is well-known that in $\mathbb{Z}$ every number is a product of prime numbers. More precisely, let $a \in \mathbb{Z}$ and $a \neq \pm1,0$, then

$$a = \mathrm{sign}(a) \prod p_1^{\nu_1} \cdots p_r^{\nu_r}$$

with pairwise different prime numbers p_i. The p_i and the ν_i are uniquely determined (up to permutation). It is easily shown that equivalently

$$(a) = (p_1)^{\nu_1} \cap \cdots \cap (p_r)^{\nu_r}$$

holds. This property can be generalized in the following manner:

Definition 1.4.1. Let R be an integral domain.

(1) An element $0 \neq p \in R$ with $(p) \neq R$ is called *prime* if $p \mid gh$ implies that $p \mid g$ or $p \mid h$. An element f in R is called *irreducible* if $f = gh$ implies that g or h is a unit.

(2) R is called a *unique factorization domain*, or a *factorial ring*, if every element different from zero and not being a unit is a product of finitely many primes.

(3) Let R be a unique factorization domain and $a_1, \ldots, a_n \in R$ elements different from zero. An element $d \in R$ is called a *common divisor* of $a_1, \ldots, a_n$ if $d \mid a_i$ for $i = 1, \ldots, n$. An element $d \in R$ is called a *greatest common divisor* , $d = \gcd(a_1, \ldots, a_n)$, if d is a common divisor of $a_1, \ldots, a_n$ and every common divisor of $a_1, \ldots, a_n$ divides d.

Remarks 1.4.2.

(1) It follows directly from the definition that a prime element is irreducible. The converse is not true in general. For example, in the ring $\mathbb{Z}[\sqrt{-5}]$ the element 2 is irreducible, but not prime as $2 \mid 6$, and $6 = (1 + \sqrt{-5})(1 - \sqrt{-5})$. However, it can be shown that 2 does not divide $1 + \sqrt{-5}$ and $1 - \sqrt{-5}$. To give a different example, consider the ring $\mathbb{C}[x,y,z]/(z^2 - xy)$. Then we have $z \mid xy$, but z does not divide x or y.

The converse does hold however in a unique factorization domain.

(2) One could, alternatively, define a unique factorization domain as a domain in which each nonzero element which is not a unit has (up to units and renumbering) a unique decomposition into irreducibles. This will be shown in the exercises.

(3) The existence of a greatest common divisor will be proved in Exercise 1.4.37.

Example 1.4.3. Any principal ideal domain, especially $\mathbb{Z}$ and the polynomial ring $K[x]$ over a field K in one variable, is a unique factorization domain. This will be proved in Exercise 1.4.35. Moreover, it follows from the Lemma of Gauß, to be proved next, that the polynomial ring $K[x_1, \ldots, x_n]$ over any field K is a unique factorization domain.

Lemma 1.4.4 (Lemma of Gauß). *Let R be a unique factorization domain. Then $R[x]$ is a unique factorization domain.*

One of the main steps in the proof of this Theorem is the following lemma.

Lemma 1.4.5. *Let R be a unique factorization domain, $K = Q(R)$ its quotient field, and $f \in R[x]$ irreducible. Then f is also irreducible considered as element of $K[x]$.*

Proof. Suppose the converse. Let $f = \overline{g} \cdot \overline{h}$, $\overline{g},\overline{h} \in K[x]$ and $\deg(\overline{h}) < \deg(f)$, $\deg(\overline{g}) < \deg(f)$. We choose $c \in R$ such that $c\overline{g} =: g$, $c\overline{h} =: h$ are in $R[x]$ and obtain $c^2 f = gh$. Now c^2 is not a unit in R. This is because f is by assumption irreducible in $R[x]$. So we can write $c^2 = p_1 \cdots p_s$, $p_i \in R$ prime. Exercise 1.4.39 implies $p_1 | g$ or $p_1 | h$. So we can find g' and h' in $R[x]$ with $p_2 \cdots p_s f = g'h'$. Repeating this argument, shows that $f = g_0 \cdot h_0$, $\deg(g_0) = \deg(g)$, $\deg(h_0) = \deg(h)$. This is a contradiction to the fact that f is irreducible in $R[x]$. □

Proof of the Lemma of Gauß 1.4.4. Let $K = Q(R)$ be the field of fractions of R. Let $f \in R[x]$. We have to prove that $f = f_1 \cdots f_n$, $f_i \in R[x]$ prime. We can write $f = \sum_{k=0}^{m} a_k x^k$. Define $d = \gcd(a_0, \ldots, a_m)$. Then $f = d \cdot \sum_{k=0}^{m} \frac{a_k}{d} x^k$. We know, because R is a unique factorization domain, that d is a unit or a product of finitely many primes. Because of Exercise 1.4.39 they are also prime in $R[x]$. Therefore, we may assume that $d = 1$.

Let $f = f_1 \cdots f_n$, $f_i \in R[x]$ be irreducible. Such a decomposition into finitely many polynomials exists because $\deg(f) = \sum \deg(f_i)$. Therefore, we may assume that f is irreducible and we have to prove that f is prime. From Lemma 1.4.5 it follows that f is irreducible in $K[x]$. From Exercise 1.4.33 it follows that f is prime in $K[x]$.

Now we are ready to prove the theorem. Let $f \mid g \cdot h$, $g,h \in R[x]$. Then $f \mid g$ or $f \mid h$ in $K[x]$. Without loss of generality, suppose that $f \mid h$. Then it follows that $f \cdot \overline{b} = h$ for a suitable $\overline{b} \in K[x]$. Then we choose $c \in R$ such that $c \cdot \overline{b} =: b \in R[x]$ and obtain $fb = c \cdot h$. Now $f = \sum_{k=0}^{m} a_k x^k$ and $\gcd(a_0, \ldots, a_m) = 1$ implies, using 1.4.39 that $c \mid b$, that is, $\overline{b} \in R[x]$. □

Corollary 1.4.6. *Let R be a field or $\mathbb{Z}$. Then $R[x_1, \ldots, x_n]$ is a unique factorization domain.*

In the rest of the section, we discuss a generalization of this result for an arbitrary Noetherian ring. From now on we fix a Noetherian ring R. First we need to find a substitute of the notion of prime power for a general ring.

Definition 1.4.7. An ideal $\mathfrak{q} \subsetneq R$ is called a *primary ideal* if $ab \in \mathfrak{q}$, $a \notin \mathfrak{q}$ implies $b^n \in \mathfrak{q}$ for some $n \in \mathbb{N}$.

An ideal $\mathfrak{p}$ is prime, if $R/\mathfrak{p}$ is a domain, that is, has no zero divisors. Similarly, an ideal I is primary, if every zerodivisor of R/I is nilpotent. An ideal I is *radical* , exactly if R/I does not have any *nilpotent* elements, except for the zero element. This means exactly that R/I is a *reduced ring.*

Lemma 1.4.8.

(1) Let $\mathfrak{m}$ be a maximal ideal of R. Then $\mathfrak{m}$ is a prime ideal.

(2) Let $\mathfrak{q}$ be an ideal in R such that $\mathfrak{m} := \sqrt{\mathfrak{q}}$ is a maximal ideal. Then $\mathfrak{q}$ is a primary ideal.

(3) Let $\mathfrak{q} \subset R$ be a primary ideal. Then $\sqrt{\mathfrak{q}}$ is prime.

(4) Let $\mathfrak{q} \subset R$ be a primary ideal and $d \notin \mathfrak{q}$ be an element of R. Then $\mathfrak{q} : d$ is a primary ideal.

Proof.
(1) Let $ab \in \mathfrak{m}$ and $a \notin \mathfrak{m}$. Then $(a) + \mathfrak{m} = R$ because of the maximality of $\mathfrak{m}$. Let $1 = r \cdot a + m$ for some $r \in R$, $m \in \mathfrak{m}$, then $b = r \cdot a \cdot b + mb \in \mathfrak{m}$ because $a \cdot b$ and $m \cdot b \in \mathfrak{m}$. This implies that $\mathfrak{m}$ is a prime ideal.

(2) Now let $a \cdot b \in \mathfrak{q}$ with $a \notin \mathfrak{q}$. If $b^s \notin \mathfrak{q}$ for all $s \in \mathbb{N}$, that is, $b \notin \mathfrak{m}$, then again $(b) + \mathfrak{m} = R$. Thus $1 = rb + m$ for some $r \in R$ and $m \in \mathfrak{m}$. Hence $a = rba + m \cdot a$. Iterating this we obtain $a = r(1 + m + m^2 + \cdots + m^k)ba + m^{k+1}a$ for all $k > 0$. As $\sqrt{\mathfrak{q}} = \mathfrak{m}$, we have $m^{k+1} \in \mathfrak{q}$ for some k (R is Noetherian). This is in contradiction to $a \notin \mathfrak{q}$. This implies $b \in \mathfrak{m}$ and thus $b^s \in \mathfrak{q}$ for some s.

(3) Let $ab \in \sqrt{\mathfrak{q}}$ and $a \notin \sqrt{\mathfrak{q}}$. Then $(ab)^m \in \mathfrak{q}$ for some m but $a^m \notin \mathfrak{q}$. This implies that for some s, $(b^m)^s \in \mathfrak{q}$ and, therefore, $b \in \sqrt{\mathfrak{q}}$.

(4) Exercise 1.4.42. □

Remarks 1.4.9.

(1) It is not true that any power of a prime ideal is primary. For example, the ideal $\mathfrak{p} = (x,z)$ is prime in $R := K[x,y,z]/(xy - z^2)$, because $R/\mathfrak{p} \cong K[y]$ does not have zero divisors. However, $\mathfrak{p}^2$ is not primary. To see this, note that $z^2 = xy \in \mathfrak{p}^2$, $x \notin \mathfrak{p}^2$ and no power of y is in $\mathfrak{p}^2$.

(2) It is always true from (f) is a prime ideal it follows that f is an irreducible element. The converse does not always hold, but it holds if R is a unique factorization domain. Furthermore (f^m) is primary in this case. This applies for example for rings like $K[x_1,\ldots,x_n]$ or $\mathbb{Z}[x_1,\ldots,x_n]$ by the Lemma of Gauß.

One of the main goals in this section is to show that any ideal in a Noetherian ring is a finite intersection of primary ideals. Before proving this, we need the following lemma.

Lemma 1.4.10. *Let $I \subset R$ be an ideal, and $f \in R$ such that $I : f = I : f^2$. Then $(I : f) \cap (I,f) = I$.*

Proof. The inclusion $I \subset (I : f) \cap (I,f)$ is obvious. For the other inclusion, let $g \in (I : f) \cap (I,f)$. We have to prove $g \in I$. Let $g = x + y \cdot f$, $x \in I$, $y \in R$. As $g \in I : f$ it follows that $fg = xf + y \cdot f^2 \in I$. Hence $y \cdot f^2 \in I$, so that $y \in I : f^2 = I : f$. It follows that $yf \in I$, and therefore $g \in I$. □

Theorem 1.4.11 (Primary Decomposition). *Let R be a Noetherian ring and $I \subsetneq R$ be an ideal. Then there exist finitely many* primary *ideals $\mathfrak{q}_1, \dots, \mathfrak{q}_r$ in R such that $I = \mathfrak{q}_1 \cap \dots \cap \mathfrak{q}_r$.*

Proof. Suppose the converse. Then the set $\mathscr{A}$ of ideals for which the theorem is not true, that is,

$$\mathscr{A} := \{I \text{ ideal in } R : I \text{ is not an intersection of finitely many primary ideals}\},$$

is nonempty. The ring R is Noetherian by assumption, and therefore by 1.1.2 the set $\mathscr{A}$ has a maximal element with respect to inclusion. Let I be such a maximal element. In particular, I is not primary! Therefore, there exist $a,b \in R$ such that $a \cdot b \in I$, $a \notin I$ and $b^n \notin I$ for all n. We have a chain of ideals $I : b \subset I : b^2 \subset \cdots$. As our ring R is Noetherian, we get by 1.1.2 that this chain is stationary. Hence, there exists an n with the property $I : b^n = I : b^{n+1} = I : b^{n+2} = \cdots = I : b^{2n}$. We use Lemma 1.4.10 to obtain $I = (I : b^n) \cap (I,b^n)$.

Now, since $b^n \notin I$, we have $I \subsetneq (I,b^n)$. Moreover as $a \notin I$ and $ab^n \in I$ we have $I \subsetneq (I : b^n)$. As I is maximal in $\mathscr{A}$ it follows that both $(I : b^n),(I,b^n) \notin \mathscr{A}$. Therefore, both ideals are intersections of finitely many primary ideals. But as $I = (I : b^n) \cap (I,b^n)$, it would follow that I is the intersection of finitely many primary ideals too, in contradiction to our assumption. □

Examples 1.4.12.

(1) $(xy,xz) = (x) \cap (y,z) \subset K[x,y,z]$;

(2) $(xy - s,xz - s,yz - s) = (x,y,s) \cap (x,z,s) \cap (y,z,s) \cap (x - y,y - z,x^2 - s) \subset K[x,y,z,s]$.

We now turn our attention to the problem of uniqueness of the primary decomposition.

Definition 1.4.13.

- Let $\mathfrak{q}$ be a primary ideal. The prime ideal $\mathfrak{p} = \sqrt{\mathfrak{q}}$ is called the *associated prime* of $\mathfrak{q}$. In this case one says that $\mathfrak{q}$ is a *$\mathfrak{p}$-primary ideal.*
- A primary decomposition $I = \mathfrak{q}_1 \cap \dots \cap \mathfrak{q}_r$ is called *irredundant*[4] if the following two conditions are satisfied:
  (1) None of the $\mathfrak{q}_i$ can be removed from the primary decomposition, that is for all i, the inclusion $I \subset \cap_{j \neq i} \mathfrak{q}_j$ is strict.
  (2) The associated primes $\mathfrak{p}_i := \sqrt{\mathfrak{q}_i}$ are pairwise different.

Lemma 1.4.14.

(1) Let $\mathfrak{q}$ and $\mathfrak{q}' \subset R$ be $\mathfrak{p}$-primary ideals. Then $\mathfrak{q} \cap \mathfrak{q}'$ is a $\mathfrak{p}$-primary ideal.

(2) Let R be Noetherian and $I \subsetneq R$ be an ideal in R. Then there exists an irredundant primary decomposition of I.

[4] In the literature this is often also called minimal, normal or reduced. Some authors, for example Eisenbud, [Eisenbud 1995], reserve the notion irredundant primary decomposition for a primary decomposition in which none of the terms can be dropped.

Proof.

(1) From 1.1.16, parts 2 and 4, we have $\sqrt{\mathfrak{q}\cap\mathfrak{q}'}=\mathfrak{p}$. We have to prove that $\mathfrak{q}\cap\mathfrak{q}'$ is primary. So let $ab\in\mathfrak{q}\cap\mathfrak{q}'$, and $a\notin\mathfrak{q}\cap\mathfrak{q}'$. We may assume without restriction of generality that $a\notin\mathfrak{q}$. Then it follows that $b\in\sqrt{\mathfrak{q}}=\mathfrak{p}=\sqrt{\mathfrak{q}\cap\mathfrak{q}'}$, which shows the lemma for $\mathfrak{q}\cap\mathfrak{q}'$.

(2) This follows from the first part of the lemma, as we can combine different $\mathfrak{p}$–primary ideals into one $\mathfrak{p}$–primary ideal, for every prime ideal $\mathfrak{p}$. □

It is important to note that an irredundant primary decomposition might not be unique. The simplest example is:

Example 1.4.15.

$$(x^2,xy)=(x)\cap(x,y)^2=(x)\cap(x^2,y).$$

But at least the associated prime ideals are the same, that is, (x,y) and (x), in both cases. This turns out to be the case in general.

Theorem 1.4.16 (First Uniqueness Theorem of Primary Decomposition). *Let R be a Noetherian ring and $I=\mathfrak{q}_1\cap\ldots\cap\mathfrak{q}_r$ be an irredundant primary decomposition of I. Then the prime ideals $\mathfrak{p}_1=\sqrt{\mathfrak{q}_1},\ldots,\mathfrak{p}_r=\sqrt{\mathfrak{q}_r}$ are uniquely determined by I, up to permutation.*

For the proof, we need the following lemma.

Lemma 1.4.17. *Let R be a Noetherian ring and $\mathfrak{q}$ be a primary ideal with associated prime $\mathfrak{p}$ and $I=J\cap\mathfrak{q}$ for some ideal $J\not\subset\mathfrak{q}$. Then there exists a $d\in R$ such that $I:d=\mathfrak{p}$.*

Proof.
Step 1. We first reduce to the case that $J=R$, that is, $I=\mathfrak{q}$. We choose $d\in J$ and $d\notin\mathfrak{q}$. Then it follows that

$$I:d=(J\cap\mathfrak{q}):d=(J:d)\cap(\mathfrak{q}:d)=\mathfrak{q}:d.$$

As $\mathfrak{q}:d$ is also primary by Exercise 1.4.47, and because of the property $(I:x):y=I:xy$ (Exercise 1.1.15) we reduced the problem to the case $J=R$.

Step 2. The crucial thing to note is that for any $g\in\mathfrak{p}\setminus\mathfrak{q}$ it follows that $\mathfrak{q}\subsetneq\mathfrak{q}:g$. Indeed, $g\in\mathfrak{p}\setminus\mathfrak{q}$ implies $g^m\in\mathfrak{q}$ for some m, and $g\notin\mathfrak{q}$ implies $m\geq 2$. If we choose m minimal such that $g^m\in\mathfrak{q}$, then $g^{m-1}\in\mathfrak{q}:g$, but $g^{m-1}\notin\mathfrak{q}$. Note, moreover, that by 1.4.47 $\mathfrak{q}:g$ is $\mathfrak{p}$–primary. In particular $\mathfrak{q}:g\subset\mathfrak{p}$.

Step 3. The proof of the lemma is by contradiction. So suppose that the set

$$A:=\{\mathfrak{q}\subset R\ \mathfrak{p}\text{–primary}:\mathfrak{p}\neq\mathfrak{q}:d \text{ for all } d\in R\}$$

is nonempty. As R is Noetherian, we can take a maximal element $\mathfrak{q}$ in A. In particular, $\mathfrak{p}\neq\mathfrak{q}$, so that we can take $g\in\mathfrak{p}\setminus\mathfrak{q}$ with $\mathfrak{q}\subsetneq\mathfrak{q}:g$. Now $\mathfrak{q}:g$ is $\mathfrak{p}$–primary, and by maximality of $\mathfrak{q}$ there exists a d with $(\mathfrak{q}:g):d=\mathfrak{p}$. Hence $\mathfrak{q}:(gd)=\mathfrak{p}$, which is a contradiction. □

The prime ideals one can get by taking ideal quotients are called the associated primes of I. More precisely:

Definition 1.4.18. Let $I \subsetneq R$ be an ideal. A prime ideal $\mathfrak{p}$ of R is called an *associated prime* of I if it can be written as

$$\mathfrak{p} = I : d$$

for some $d \in R$. The set of associated primes of I is denoted by $\text{Ass}(I)$.

Remarks 1.4.19.

(1) For a primary ideal $\mathfrak{q}$, we have now two definitions of an associated prime of $\mathfrak{q}$. We will prove in Theorem 1.4.20 that the two definitions coincide.

(2) Alternatively, but very similarly, one can define the set of associated primes by

$$\text{Ass}(I) := \{\mathfrak{p} : \mathfrak{p} \text{ prime}, \mathfrak{p} = \ \text{Ann}(\bar{b}), \bar{b} \in R/I.\}$$

The First Uniqueness Theorem 1.4.16 follows directly from the following theorem.

Theorem 1.4.20. *Let $I = \mathfrak{q}_1 \cap \cdots \cap \mathfrak{q}_r$ be an irredundant primary decomposition of I, and $\mathfrak{p}_i = \sqrt{\mathfrak{q}_i}$. Then*

$$\text{Ass}(I) = \{\mathfrak{p}_1, \ldots, \mathfrak{p}_r\},$$

that is, the set of associated primes is equal to the set of prime ideals which are associated to the primary ideals occurring in an irredundant primary decomposition of I.

Proof. We apply 1.4.17 with $J = \cap_{j \neq i} \mathfrak{q}_j$, and $\mathfrak{q} = \mathfrak{q}_i$. We conclude that there exist d_i such that $I : d_i = \mathfrak{p}_i$ for $i = 1, \ldots, r$. This gives the inclusion $\text{Ass}(I) \supset \{\mathfrak{p}_1, \ldots, \mathfrak{p}_r\}$. To show the other inclusion, let $\mathfrak{p} = I : d \in \text{Ass}(I)$. Then

$$\mathfrak{p} = I : d = (\mathfrak{q}_1 : d) \cap \cdots \cap (\mathfrak{q}_r : d).$$

Let $\{i_1, \ldots, i_t\}$ be the set of indices with $d \notin \mathfrak{q}_{i_j}$. This set is not empty because $d \notin I = \cap \mathfrak{q}_i$. Then

$$\mathfrak{p} = (\mathfrak{q}_{i_1} : d) \cap \cdots \cap (\mathfrak{q}_{i_t} : d).$$

By Exercise 1.4.47, $\mathfrak{q}_{i_k} : d$ is $\mathfrak{p}_{i_k}$-primary. In particular $\sqrt{\mathfrak{q}_{i_k} : d} = \sqrt{\mathfrak{q}_{i_k}} = \mathfrak{p}_{i_k}$. By Exercise 1.1.16 part 4 it follows that:

$$\mathfrak{p} = \sqrt{\mathfrak{p}} = \sqrt{\mathfrak{q}_{i_1} : d} \cap \cdots \cap \sqrt{\mathfrak{q}_{i_t} : d} = \sqrt{\mathfrak{q}_{i_1}} \cap \cdots \cap \sqrt{\mathfrak{q}_{i_t}} = \mathfrak{p}_{i_1} \cap \cdots \cap \mathfrak{p}_{i_t}$$

As $\mathfrak{p}$ is a prime ideal it follows that $\mathfrak{p} \supset \mathfrak{p}_{i_k}$ for some k. As the other inclusion is clear it follows that $\mathfrak{p} = \mathfrak{p}_{i_k}$. □

There is a second uniqueness theorem which we now want to formulate.

Definition 1.4.21.

(1) Let $\mathfrak{p} \in \text{Ass}(I)$. Then we define

$$M(I,\mathfrak{p}) = \{\mathfrak{q} : \mathfrak{q} \in \ \text{Ass}(I), \ \mathfrak{q} \subset \mathfrak{p}\}.$$

(2) Suppose that $\mathfrak{p} \in \mathrm{Ass}(I)$ with $M(I,\mathfrak{p}) = \{\mathfrak{p}\}$. Then $\mathfrak{p}$ is called a *minimal associated prime* of I. This means that there does not exist an associated prime $\mathfrak{q}$ of I with $\mathfrak{q} \subsetneq \mathfrak{p}$.

(3) If $\mathfrak{p},\mathfrak{q} \in \mathrm{Ass}(I)$, $\mathfrak{p} \subsetneq \mathfrak{q}$, then $\mathfrak{q}$ is called an *embedded prime ideal* of I. This exactly means that $M(I,\mathfrak{q}) \neq \{\mathfrak{q}\}$.

Example 1.4.22. Let $I = (x^2,xy) \subset K[x,y]$. Then $I : x = (x,y)$, $I : y = (x)$. $\mathrm{Ass}(I) = \{(x),(x,y)\}$, $M\big(I,(x,y)\big) = \mathrm{Ass}(I)$. The ideal (x,y) is an embedded prime ideal.

Theorem 1.4.23 (Second Uniqueness Theorem of Primary Decomposition). *Let R be a Noetherian ring and $I \subset R$ be an ideal, $I = \mathfrak{q}_1 \cap \cdots \cap \mathfrak{q}_r$ an irredundant primary decomposition of I. Let $\mathfrak{p}$ be an associated prime ideal of I, and put*

$$M(I,\mathfrak{p}) = \{\sqrt{\mathfrak{q}_{i_1}},\ldots,\sqrt{\mathfrak{q}_{i_s}}\}.$$

Then $\mathfrak{q}_{i_1} \cap \cdots \cap \mathfrak{q}_{i_s}$ is uniquely determined (that is, independent of the irredundant decomposition). In fact:

$$\mathfrak{q}_{i_1} \cap \cdots \cap \mathfrak{q}_{i_s} = I_{\mathfrak{p}} \cap R.$$

In particular, the primary ideals belonging to minimal associated primes are uniquely determined. For such a minimal prime ideal $\mathfrak{p}_i$ one, therefore, has $\mathfrak{q}_i = I_{\mathfrak{p}_i} \cap R$.

Proof. There is the following obvious generalization of 1.3.15:

$$\text{primary ideals } \mathfrak{q} \subset R \text{ with } \mathfrak{q} \cap S = \emptyset \quad \longleftrightarrow \quad \text{primary ideals } \mathfrak{q}_S \subset R_S.$$

The correspondence is that to a primary ideal $\mathfrak{q}$ we assign $\mathfrak{q}_S$. The ideal $\mathfrak{q}_S$ is primary if $\mathfrak{q} \cap S = \emptyset$, it is equal to R_S if $\mathfrak{q} \cap S \neq \emptyset$. On the other hand, for a primary ideal $\mathfrak{q}_S$ in R_S, we get a primary ideal $\mathfrak{q} := \mathfrak{q}_S \cap R$ in R. For a proof, see Exercise 1.4.49. Using Exercise 1.4.50 we get $I_{\mathfrak{p}} = \mathfrak{q}_{i_1\mathfrak{p}} \cap \ldots \cap \mathfrak{q}_{i_s\mathfrak{p}}$ and $I_{\mathfrak{p}} \cap R = \mathfrak{q}_{i_1} \cap \cdots \cap \mathfrak{q}_{i_s}$. This is what we had to show. □

Theorem 1.4.24. *Let R be a Noetherian ring and $\mathrm{Ass}\big((0)\big) = \{\mathfrak{p}_1,\ldots,\mathfrak{p}_k\}$. Then*

(1) $\sqrt{(0)} = \mathfrak{p}_1 \cap \cdots \cap \mathfrak{p}_k$. *In particular this means that the intersection of the minimal prime ideals is the set of nilpotent elements.*

(2) Let $\mathfrak{p} \subset R$ be a prime ideal. Then $\mathfrak{p}_i \subset \mathfrak{p}$ for some i.

(3) $\bigcup_{i=1}^{k} \mathfrak{p}_i \setminus \{0\}$ *is the set of zerodivisors of R.*

(4) If R is reduced, then the set of zerodivisors of R is equal to the union of the minimal prime ideals of R.[5]

Proof. Let $(0) = \mathfrak{q}_1 \cap \cdots \cap \mathfrak{q}_k$ be an irredundant primary decomposition. Then $\sqrt{(0)} = \mathfrak{p}_1 \cap \cdots \cap \mathfrak{p}_k$ follows, which is (1). For (2), as $\mathfrak{p}$ is a prime ideal, it follows that $\mathfrak{p}_1 \cap \cdots \cap \mathfrak{p}_k = \sqrt{(0)} \subset \mathfrak{p}$, so (2) follows.

[5] Note that minimal prime ideals of R are the minimal associated prime ideals of (0).

To prove (3), let x be a zero divisor, that is, there is a $y \in R$, $y \neq 0$ and $xy = 0$. Let $(0) = \mathfrak{q}_1 \cap \cdots \cap \mathfrak{q}_k$ be an irredundant primary decomposition and $\sqrt{\mathfrak{q}_i} = \mathfrak{p}_i$. Since $y \neq 0$ there exists an i such that $y \notin \mathfrak{q}_i$. This implies $x \in \mathfrak{p}_i$. To prove the other direction, let $x \in \mathfrak{p}_i$ for some i. If x is nilpotent, we are done. Otherwise $x^m \in \mathfrak{q}_i$ for some m and $x^m \neq 0$. In particular, $k \geq 2$.

But $\mathfrak{q}_i \cdot (\mathfrak{q}_1 \cap \cdots \cap \mathfrak{q}_{i-1} \cap \mathfrak{q}_{i+1} \cap \cdots \cap \mathfrak{q}_k) \subset \mathfrak{q}_1 \cap \cdots \cap \mathfrak{q}_k = (0)$ implies that there exists a $y \neq 0$ such that $x^m y = 0$. Let $n \geq 1$ be minimal such that $x^n y = 0$. Then $x \cdot (x^{n-1}y) = 0$ and, consequently, x is a zero divisor. The final statement follows from the third. □

As an application, we first want to study the structure of Artinian K–algebras (cf. 1.1.6).

Lemma 1.4.25. *Let R be an Artinian K-algebra and $\mathfrak{p} \subset R$ be a prime ideal. Then $\mathfrak{p}$ is a maximal ideal.*[6]

Proof. $R/\mathfrak{p}$ is an Artinian K–algebra, see 1.1.9, and an integral domain. We have to prove that $R/\mathfrak{p}$ is a field. Let $x \in R/\mathfrak{p}$, $x \neq 0$, then Lemma 1.1.8 implies that $(x^n) = (x^{n+1})$ for some n. Hence, $x^n = x^{n+1}y$ for some $y \in R/\mathfrak{p}$. Since $R/\mathfrak{p}$ is an integral domain, this implies $1 = xy$. So x is invertible, and $R/\mathfrak{p}$ is a field. □

Theorem 1.4.26. *Let R be an Artinian K-algebra. Then R is uniquely (up to isomorphism) a finite direct sum of local Artinian K-algebras.*

Proof.
Step 1. We first prove the existence. Let $(0) = \mathfrak{q}_1 \cap \cdots \cap \mathfrak{q}_n$ be an irredundant primary decomposition of the zero ideal. Then all the associated primes $\mathfrak{m}_i := \sqrt{\mathfrak{q}_i}$ are maximal (1.4.25). By the Second Uniqueness Theorem 1.4.23 the $\mathfrak{q}_i$ are uniquely determined. We claim that furthermore $\mathfrak{q}_i + \mathfrak{q}_j = R$ if $i \neq j$. Indeed, otherwise $\mathfrak{m} \supset \mathfrak{q}_i + \mathfrak{q}_j$ for some maximal ideal $\mathfrak{m}$. But then $\mathfrak{m} \supset \sqrt{\mathfrak{q}_i} = \mathfrak{m}_i$ and $\mathfrak{m} \supset \sqrt{\mathfrak{q}_j} = \mathfrak{m}_j$. This implies $\mathfrak{m} = \mathfrak{m}_i = \mathfrak{m}_j$ and, therefore, $i = j$ (the decomposition was irredundant).

Therefore, the conditions to apply the Chinese Remainder Theorem 1.1.12 are fulfilled and the canonical map

$$R \longrightarrow \bigoplus_{i=1}^{n} R/\mathfrak{q}_i$$

is an isomorphism.

Because of 1.1.9 the $R/\mathfrak{q}_i$ are Artinian K–algebras. They are local because $\mathfrak{q}_i \neq \mathfrak{q}_j$ for $i \neq j$, again because the primary decomposition was irredundant.

Step 2. We now prove the uniqueness of the decomposition. So consider $R \cong \oplus_{i=1}^m R_i$ with R_i local. In step 1 we proved that an irredundant primary decomposition of the zero ideal in an Artinian K–algebra is uniquely determined (up to permutation). We will show that if $\mathfrak{q}_i := \mathrm{Ker}(R \longrightarrow R_i)$, then $\mathfrak{q}_1 \cap \ldots \cap \mathfrak{q}_m$ is an irredundant primary decomposition of the zero ideal in R. This suffices to show the uniqueness, as $R_i \cong R/\mathfrak{q}_i$.

(1) Consider the zero ideal in R_i. As R_i has only one maximal ideal, there is only one primary ideal in an irredundant primary decomposition. This ideal has to be (0), hence (0) is a primary ideal in R_i. As $R_i = R/\mathfrak{q}_i$, this shows that the ideals $\mathfrak{q}_i$ are primary.

[6] This is a very special property, and says that an Artinian K–algebra has Krull dimension zero. For a discussion on these matters, see Section 4.1.

(2) $\cap_{i=1}^m \mathfrak{q}_i = (0)$ because $R \cong \oplus_{i=1}^m R_i$. Thus $\mathfrak{q}_1 \cap \cdots \cap \mathfrak{q}_m = (0)$ is a primary decomposition.

(3) To prove that the primary decomposition is irredundant, it suffices to show that $\sqrt{\mathfrak{q}_i} \neq \sqrt{\mathfrak{q}_j}$ for $i \neq j$. For this it suffices to show that $\mathfrak{q}_i + \mathfrak{q}_j = R$ for $i \neq j$. Now, because $R \cong \oplus_{i=1}^m R_i$ we find $x \in R$ corresponding to $(0,\ldots,0,1,0,\ldots,0)$. So $x - 1 \in \mathfrak{q}_i$ and $x \in \mathfrak{q}_j$. This implies $1 = x + (1-x) \in \mathfrak{q}_i + \mathfrak{q}_j$. □

Next we study the structure of quotient rings, as promised in the Examples 1.3.13.

Proposition 1.4.27. *Let $R = Q(R)$ be a reduced Noetherian ring and* $\mathrm{Ass}((0)) = \{\mathfrak{p}_1,\ldots,\mathfrak{p}_s\}$. *Then $R \cong \oplus_{i=1}^s R/\mathfrak{p}_i$.*

Proof. Note that $\mathfrak{p}_1,\ldots,\mathfrak{p}_s$ are minimal, because R is reduced. By prime avoidance, Lemma 1.1.13, $\mathfrak{p}_i + \mathfrak{p}_j \not\subset \cup_{\ell=1}^s \mathfrak{p}_\ell$ if $i \neq j$.

By Theorem 1.4.24 there is a nonzerodivisor in $\mathfrak{p}_i + \mathfrak{p}_j$. As $R = Q(R)$, all nonzerodivisors are units, so that $\mathfrak{p}_i + \mathfrak{p}_j = R$. As R is, moreover, reduced, we have $(0) = \mathfrak{p}_1 \cap \cdots \cap \mathfrak{p}_s$ by 1.4.24. The conditions of the Chinese Remainder Theorem (1.1.12) are satisfied and we get $R \cong \oplus_{i=1}^s R/\mathfrak{p}_i$. □

Finally, we wish to study the rank of a module.

Definition 1.4.28. Let R be a Noetherian ring and M be a finitely generated R-module. One says that M has *rank* r if $M \otimes_R Q(R)$ is a free $Q(R)$-module of rank r. We write $\mathrm{rank}(M) = r$.

Lemma 1.4.29. *Let R be a Noetherian ring and M be a finitely generated R-module. M has rank r if and only if $M_\mathfrak{p}$ is a free $R_\mathfrak{p}$-module of rank r for all $\mathfrak{p} \in \mathrm{Ass}((0))$.*

Proof. Since $R_\mathfrak{p} = Q(R)_{\mathfrak{p}Q(R)}$, see Exercise 1.4.51, we may assume that $R = Q(R)$. One implication is clear. Assume that $M_\mathfrak{p}$ is free of rank r for all prime ideals $\mathfrak{p}$. We use induction on r. The case $r = 0$ is trivial. Suppose $r > 0$. Now we choose $x \in M$ such that $x \notin \mathfrak{p}M_\mathfrak{p}$ for all prime ideals $\mathfrak{p}$ in $\mathrm{Ass}((0))$. Such an x exists because of Exercise 1.3.27. Now x is an element of a minimal system of generators of $M_\mathfrak{p}$ for all $\mathfrak{p}$, by Nakayama, that is, an element of a basis of the free module $M_\mathfrak{p}$ for all $\mathfrak{p}$. This implies that $(M/xR)_\mathfrak{p}$ is free of rank $r-1$ for all $\mathfrak{p}$. Using induction, we obtain that M/xR is free of rank $r-1$ and, therefore, by Exercise 1.2.48, $M \cong xR \oplus M/xR$ is free of rank r because $xR \cong R$ ($xR_\mathfrak{p} \cong R_\mathfrak{p}$ for all $\mathfrak{p}$). □

Remark 1.4.30. If R is an integral domain, then every finitely generated R-module M has a rank, namely $\mathrm{rank}(M) = \dim_{Q(R)} M \otimes_R Q(R)$. In general this is not true: let $R = \mathbb{C}[x,y]/(xy)$ and $M = (x)$. Then $M_{(x)} = (0)$ and $M_{(y)} = R_{(y)}$.

Proposition 1.4.31. *Let R be a Noetherian ring and $0 \longrightarrow M \xrightarrow{i} N \xrightarrow{\pi} P \longrightarrow 0$ be an exact sequence of R-modules. If two of the modules M,N,P have a rank then the third one has a rank and rank(M)+ rank(P) = rank(N).*

Proof. Because of 1.4.29, we may assume that R is local, and that $\mathfrak{m} \in \mathrm{Ass}((0))$. Because of Exercise 1.2.48 we may assume that N is always free. If N and P are free, we can use 1.3.26 to deduce that M is free. Now let N and M be free.

Suppose $M \subset \mathfrak{m}N$, then $M = (0)$. Indeed, as $\mathfrak{m} \in \mathrm{Ass}((0))$, there exists a $d \in R$ with $\mathfrak{m} = (0) : d$. It follows that $dM = d\mathfrak{m}N = (0)$. As M is free, this implies $M = (0)$.

If $M \not\subset \mathfrak{m}N$ we use induction on the rank of M. If the rank of M is 1 and m is a free generator then $i(m) \notin \mathfrak{m}N$ implies that $i(m)$ can be extended to a set of free generators of N. This implies that $N \cong M \oplus P$ and P is free because of 1.3.26. If the rank of M is bigger than 1 we can, for the same reason, write $M = M' \oplus R$, $N = N' \oplus R$ such that $i(M') \subset N'$ and the induced sequence $0 \longrightarrow M' \longrightarrow N' \longrightarrow P \longrightarrow 0$ is still exact. Then P is free by induction. □

Proposition 1.4.32. *Let R be a Noetherian ring and F, G be free R–modules of finite rank. Let $\varphi : F \longrightarrow G$ be a homomorphism. Then the following conditions are equivalent:*

(1) $\mathrm{Im}(\varphi)$ *has rank r,*

(2) $I_r(\varphi)$ contains a nonzerodivisor and $I_{r+1}(\varphi) = 0$.

Proof. (1) ⇒ (2). Assume that all elements of $I_r(\varphi)$ are zerodivisors. Then $I_r(\varphi) \subset \mathfrak{p}$ for some $\mathfrak{p} \in \mathrm{Ass}((0))$ by prime avoidance. But $\mathrm{Im}(\varphi) \otimes_R Q(R)$ is free of rank r, especially $\mathrm{Im}(\varphi) \otimes_R R_\mathfrak{p}$ is free of rank r. This implies that $I_r(\varphi)R_\mathfrak{p} = R_\mathfrak{p}$ by Proposition1.3.8, and is a contradiction to $I_r(\varphi) \subset \mathfrak{p}$.

Now assume $I_{r+1}(\varphi) \neq 0$. As $\mathrm{Im}(\varphi)$ has rank r, $I_{r+1}(\varphi)$ does not contain a nonzerodivisor. (Otherwise this would imply $I_{r+1}(\varphi)Q(R) = Q(R)$ and by Exercise 1.3.8 give that $\mathrm{Im}(\varphi)$ has rank at least $r+1$.) By prime avoidance and 1.4.24 there exists a $\mathfrak{p} \in \mathrm{Ass}((0))$ with $I_{r+1}(\varphi)R_\mathfrak{p} \subset \mathfrak{p}R_\mathfrak{p}$. But we just proved that $I_r(\varphi)R_\mathfrak{p} = R_\mathfrak{p}$ and, therefore, for a suitable choice of bases in $F \otimes_R R_\mathfrak{p}$ and $G \otimes_R R_\mathfrak{p}$ the matrix corresponding to $\varphi : F_\mathfrak{p} \longrightarrow G_\mathfrak{p}$ is $\left(\begin{smallmatrix} Id_r & 0 \\ 0 & H \end{smallmatrix}\right)$ with $H \neq 0$ because $I_{r+1}(\varphi) \neq 0$. Now we can choose $x \in \mathfrak{p} \setminus \{0\}$ such that $x\mathfrak{p} = (0)$, then $xH = 0$. This implies that $\mathrm{Im}(\varphi) \otimes_R R_\mathfrak{p}$ has torsion elements, and therefore is not free which is a contradiction.

For (2) ⇒ (1), the existence of a nonzerodivisor, implies $I_r(\varphi)Q(R) = Q(R)$. Together with Proposition 1.3.8 we get that $\mathrm{Im}(\varphi)$ has rank r. □

Exercises

1.4.33. Let R be a principal ideal domain. Prove that $f \in R$ is prime if and only if it is irreducible.
(Hint: Irreducible elements in a principal ideal domain generate maximal ideals. Maximal ideals are prime ideals.)

1.4.34. Prove that $y^2 - x^2(x+1) \in \mathbb{C}[x,y]$ is irreducible.

1.4.35. Use Exercise 1.4.33 to prove that principal ideal domains are unique factorization domains.
(Hint: Use the property that a principal ideal domain is Noetherian to prove that every $f \neq 0$ not being a unit is a finite product of irreducible elements.)

1.4.36. Let R be a unique factorization domain. Prove that R has the unique factorization property: let $p_1, \ldots, p_n$, $q_1, \ldots, q_m$ be primes in R and $\prod_{i=1}^{n} p_i = \prod_{i=1}^{m} q_i$.
Prove:

(1) $n = m$.

(2) There exist a permutation $\pi \in S_n$ and units $u_1, \ldots, u_n \in R$ such that $p_i = u_i q_{\pi(i)}$.

1.4.37. Let R be a unique factorization domain and $a_1, \ldots, a_n \in R$ be elements different from zero. Prove that $a_1, \ldots, a_n$ have a greatest common divisor.

1.4.38. Recall the Euclidean algorithm and prove that it computes in $\mathbb{Z}$ and $K[x]$ the greatest common divisor of two elements.

1.4.39. Let R be a unique factorization domain and $p \in R$ prime. Let $f,g \in R[x]$ be two polynomials and assume $p|f \cdot g$. Prove that $p|f$ or $p|g$, that is, p is also prime in $R[x]$.

1.4.40. Let $a,b \in \mathbb{Z}$. Prove that $(a) \cdot (b) = (a) \cap (b)$ iff $\gcd(a,b) = 1$.

1.4.41. Let R be a unique factorization domain and $f \in R$ be an irreducible element. Prove that (f) is a prime ideal and $(f)^m$ is primary for $m \geq 1$.

1.4.42. Prove (4) of Lemma 1.4.8.

1.4.43. Let $I \subset R$ be an ideal and $\mathfrak{q} \supset I$ be a prime (respectively primary) ideal. Prove that $\overline{\mathfrak{q}} = \{\overline{f} : f \in \mathfrak{q}\} \subset R/I$ is prime (respectively primary).

1.4.44. Prove that $(x_{i_1},\dots,x_{i_k}) \subset K[x_1,\dots,x_n]$ is a prime ideal.

1.4.45. Let $\mathfrak{q}$ be a prime (respectively primary) ideal in a Noetherian ring R. Let $I \cdot J \subset \mathfrak{q}$ and $I \not\subset \mathfrak{q}$ for two ideals $I,J \subset R$. Prove that J (respectively J^s for some s) is contained in $\mathfrak{q}$.

1.4.46. Let $I = (x^2yz^3,xyw,zw^2) \subset K[x,y,z,w]$. Compute the associated primes of I.

1.4.47. Let $\mathfrak{q}$ be a primary ideal and $\mathfrak{p}$ its associated prime ideal. Let $x \in R$, then

$$\mathfrak{q} : x = \begin{cases} R \text{ if } x \in \mathfrak{q} \\ \mathfrak{q} \text{ if } x \notin \mathfrak{p} \\ \text{primary with associated prime } \mathfrak{p} \text{ if } x \notin \mathfrak{q}. \end{cases}$$

1.4.48. Prove that

(1) $\sqrt{(I,f\cdot g)} = \sqrt{(I,f)} \cap \sqrt{(I,g)}$.

(2) $\sqrt{I} = \sqrt{IR_f \cap R} \cap \sqrt{(I,f)}$.

1.4.49. Formulate and prove a generalization of 1.3.15 to the case of primary ideals.

1.4.50. Let R be a Noetherian ring, $\mathfrak{p} \subset R$ be a prime ideal.

(1) Let I,J be ideals in R. Prove that $(I \cap J)R_\mathfrak{p} = IR_\mathfrak{p} \cap JR_\mathfrak{p}$.

(2) Let $\mathfrak{q}$ be a primary ideal. Show that

   (a) $\mathfrak{q}R_\mathfrak{p} = R_\mathfrak{p} \iff \mathfrak{q} \not\subset \mathfrak{p}$.

   (b) Suppose that $\mathfrak{q} \subset \mathfrak{p}$. Then $\mathfrak{q}R_\mathfrak{p}$ is primary, and $\mathfrak{q}R_\mathfrak{p} \cap R = \mathfrak{q}$.

(3) Let $\mathfrak{q},\mathfrak{q}' \subset \mathfrak{p}$ be two primary ideals with different associated primes. Prove that the associated primes of $\mathfrak{q}R_\mathfrak{p}$ and $\mathfrak{q}'R_\mathfrak{p}$ are different.

(4) Let $I = \mathfrak{q}_1 \cap \dots \cap \mathfrak{q}_s \cap \mathfrak{q}_{s+1} \cap \dots \cap \mathfrak{q}_t$ be an irredundant primary decomposition of I. Suppose $\mathfrak{q}_1,\dots,\mathfrak{q}_s \subset \mathfrak{p}$, and $\mathfrak{q}_{s+1},\dots,\mathfrak{q}_t \not\subset \mathfrak{p}$ Show that

$$(\mathfrak{q}_1 \cap \dots \cap \mathfrak{q}_{s-1})R_\mathfrak{p} \not\subset \mathfrak{q}_s R_\mathfrak{p}.$$

Deduce that

$$IR_\mathfrak{p} = \mathfrak{q}_1 R_\mathfrak{p} \cap \dots \cap \mathfrak{q}_s R_\mathfrak{p}$$

is an irredundant primary decomposition of $IR_\mathfrak{p}$.

(5) With the assumptions of (4), let $\mathfrak{q} \in \operatorname{Ass}(I)$ and $\mathfrak{q} \subset \mathfrak{p}$ be a prime ideal. Suppose that $IR_\mathfrak{p}$ is prime and prove $IR_\mathfrak{p} = \mathfrak{q}R_\mathfrak{p}$.

1.4.51. Let R be a Noetherian ring and $\mathfrak{p} \in \operatorname{Ass}((0))$. Prove that $R_\mathfrak{p} \cong Q(R)_{\mathfrak{p}Q(R)}$.

1.4.52. Let R be a Noetherian ring and M be a finitely generated R–module, $N \subset M$ a submodule. Define $\sqrt[M]{N} = \{x \in R : x^q M \subset N \text{ for some } q\}$, $N : M = \{x \in R : xM \subset N\}$ and $\text{Ann}(M/N) = \{x \in R : xM/N = 0\}$. An element $x \in R$ is called a zerodivisor in M/N if there exists a nonzero $m \in M/N$ such that $xm = 0$. $N \subset M$ is primary in M if $N \neq M$ and every zerodivisor in M/N is nilpotent, that is, x being a zerodivisor in M/N implies that $x^\rho \in \text{Ann}(M/N)$ for some ρ.
Prove the following statements.

(1) $\sqrt[M]{N} = \sqrt{\text{Ann}(M/N)}$.

(2) N being primary in M implies that $N : M$ is a primary ideal (call $\sqrt{N : M}$ the associated prime ideal to N).

(3) N has an irredundant primary decomposition and the associated prime ideals are uniquely determined.

(4) If $\mathfrak{p}$ is an associated prime to N, then $\mathfrak{p} = N : (m)$ for some $m \in M$.

(5) Let $\mathfrak{p}_1, \ldots, \mathfrak{p}_s$ be the set of associated primes of N, then the zerodivisors of M/N are $\bigcup_{i=1}^{s} \mathfrak{p}_i \setminus \{0\}$.

1.4.53.

(a) In 1.4.24 we proved that the set of nilpotent elements in a Noetherian ring is the intersection of all (minimal) prime ideals in the ring. Prove that this holds in fact for all rings (Noetherian or not) by an application of Zorn's Lemma.

(b) Let R be a ring and $I \subset R$ be an ideal. Prove that $\sqrt{I} = \bigcap_{\substack{\mathfrak{p} \supset I \\ \mathfrak{p} \text{ prime}}} \mathfrak{p}$.

1.4.54. Let $(R,\mathfrak{m})$ be a Noetherian local ring and M be an R–module. Suppose that for all prime ideals $\mathfrak{p} \neq \mathfrak{m}$ the localization $M_\mathfrak{p}$ is the zero module. Show that there exists a k such the $\mathfrak{m}^k \cdot M = 0$.
(Hint: Look at a primary decomposition of $\text{Ann}(M)$.)

1.4.55. Let $(R,\mathfrak{m})$ be a Noetherian local ring. Suppose that all $x \in \mathfrak{m} \setminus \mathfrak{m}^2$ are zerodivisors. Show that all $x \in \mathfrak{m}$ are zerodivisors.
(Hint: Use 1.4.24.)

1.5 Finite and Integral Extensions

Finite and integral extensions of rings are the ring counterparts of finite and algebraic extensions of fields, which you (hopefully) know from a course in Galois theory. First we give the relevant definitions.

Definition 1.5.1. Let $R \subset S$ be rings.

- $R \subset S$ is called a *finite extension* if S, considered as an R–module, is finitely generated.

- An element $\alpha \in S$ is called *integral* over R if and only if α satisfies an equation:

$$\alpha^n + a_1\alpha^{n-1} + \ldots + a_{n-1}\alpha + a_n = 0$$

  with all $a_i \in R$. Such an equation is called an *integral equation* for α. The extension $R \subset S$ is called *integral* if every element $x \in S$ is integral over R.

- On the other hand, R is called *integrally closed* in S, if every element in S which is integral over R already belongs to R.
- R is called *integrally closed* if R is integrally closed in $Q(R)$, the total quotient ring of R.
- R is called *normal* if R is reduced and integrally closed.

We obtain the first lemma by copying the proofs of the corresponding statements in field theory. As usual we denote by $R[\alpha] \subset S$ the smallest ring in S containing R and α.

Lemma 1.5.2. *Let $R \subset S$ and $\alpha \in S$.*

(1) Let $R \subset S \subset T$ be ring extensions, with $R \subset S$ and $S \subset T$ finite. Then $R \subset T$ is finite.

(2) α is integral over $R \iff R[\alpha]$ is a finitely generated R–module.
In particular, every finite extension $R \subset S$ is an integral extension.

(3) Let $R \subset S \subset T$ be ring extensions, with $R \subset S$ and $S \subset T$ integral. Then $R \subset T$ is an integral extension.

(4) Let S be a finitely generated R–algebra. Then S is finitely generated as an R–module if and only if every element of S is integral over R.

Proof.
(1) If $u_1, \dots, u_s$ are generators of S as R–module and $v_1, \dots, v_t$ are generators of T as S–module, then one proves directly that the products $u_i v_j$ generate T as an R–module.

(2) Let α be integral, with integral equation $p(\alpha) = 0$, for some monic[7] polynomial $p \in R[X]$ of degree n, and $g(\alpha) \in R[\alpha]$ for some polynomial $g \in R[X]$. Then we can perform division with remainder to write $g = qp + r$, where $\deg(r) < n$ (cf. Theorem 2.1.5). Plugging in α, we see that $g(\alpha) = r(\alpha)$. This shows that the elements $1, \alpha, \dots, \alpha^{n-1}$ generate $R[\alpha]$ as R–module. On the other hand, if we have finitely many generators $q_1, \dots, q_t$ of $R[\alpha]$ as R–module, and if we put $n-1$ to be maximum of the degrees with respect to α occurring in the q_i, it follows that the element α^n is an R–linear combination of the q_i. This gives an integral equation for α.

(3) We give here a proof for the case that R is Noetherian. For the general case, see Exercise 1.5.28. Let $\alpha \in T$. By hypothesis, it satisfies an integral equation:

$$\alpha^n + a_1 \alpha^{n-1} + \ldots + a_{n-1}\alpha + a_n = 0$$

with all $a_i \in S$. So in fact α is integral over $R[a_1, \dots, a_n]$. Therefore $R[a_1, \dots, a_n][\alpha]$ is a finitely generated $R[a_1, \dots, a_n]$-module, by part (2). As all elements a_i are integral over R by hypothesis, it follows from the second part and induction that $R[a_1, \dots, a_n]$ is a finitely generated R–module. Therefore, $R[\alpha]$ is a submodule of $R[a_1, \dots, a_n, \alpha]$ which is also a finitely generated R–module by the first part. As R is Noetherian, it follows that $R[\alpha]$ is finitely generated. Now use (2) again, to conclude that α is integral.

(4) This follows from part (2). □

[7] This means that the leading coefficient of p is 1.

Definition 1.5.3. The *integral closure* of R in S is the set of all elements in S which are integral over R. In the case that S is the *total quotient ring* of R, and R is reduced, we simply call it the *normalization* of R. We denote the normalization by $\widetilde{R}$.

To use the standard proof from field theory that sum, difference and products of integral elements are integral again we need the assumption that R is Noetherian. The statement holds without this condition however, see Exercise 1.5.29.

Lemma 1.5.4. *Let $R \subset S$ be rings. Then the integral closure of R in S is a ring. The integral closure of R in S is integrally closed in S. In particular, the normalization of a ring is normal.*

Proof. We give the proof here for the case that R is Noetherian. We have to show that if α and β are integral, then $\alpha - \beta$ and $\alpha\beta$ are integral. But those elements are in $R[\alpha,\beta]$, which is a finitely generated R-module by 1.5.2. As we supposed R to be Noetherian, it follows that $R[\alpha\beta]$ and $R[\alpha - \beta]$ are finitely generated R–modules. By the second part of 1.5.2 it follows that $\alpha - \beta$ and $\alpha\beta$ are integral. The other statement follows from the third part of 1.5.2. □

To give examples of normal rings, we prove the following theorem:

Theorem 1.5.5. *Let R be a unique factorization domain. Then R is normal.*

Proof. Suppose $\frac{r}{s} \in Q(R)$ is in lowest terms, and is integral over R. So $\frac{r}{s}$ satisfies an equation:

$$\left(\tfrac{r}{s}\right)^n + a_1 \left(\tfrac{r}{s}\right)^{n-1} + \ldots + a_n = 0$$

with $a_i \in R$. We multiply the equation with s^n and conclude:

$$r^n = -s(a_1 r^{n-1} + \ldots + a_n s^{n-1})$$

Hence s divides r^n, and thus s divides r. As we assumed that they have no common factor, it follows that s is a unit in R, and so $\frac{r}{s} \in R$. □

In particular, polynomial rings over fields are normal! The converse of the theorem does not hold, see Exercise 1.5.33.

Examples 1.5.6.

(1) Consider the ring $R := \mathbb{C}[x,y]/(y^2 - x^3)$. This ring is not normal. Indeed the element $t := \frac{y}{x}$ satisfies $t^2 = \frac{y^2}{x^2} = \frac{x^3}{x^2} = x$. Therefore, t is integral, as it satisfies the integral equation $X^2 - x = 0$. The normalization $\widetilde{R}$ is in fact $\mathbb{C}[x,y,t]/(y^2 - x^3, x - t^2, y - t^3) \cong \mathbb{C}[t]$. To show that $t \notin R$, consider the map $\varphi : \mathbb{C}[x,y] \longrightarrow \mathbb{C}[t]$, which sends x to t^2 and y to t^3. Obviously $y^2 - x^3$ is in the kernel of φ, and in fact the kernel is generated by $y^2 - x^3$. Indeed, let $g \in \operatorname{Ker}(\varphi)$. Then write $g = q(y^2 - x^3) + r$, where $r = r_0 + r_1 y$, $r_i \in \mathbb{C}[x]$. Then $g(t^2,t^3) = r_0(t^2) + r_1(t^2) \cdot t^3 = 0$. The first term only has terms in even degree in t, whereas the second has only odd terms in t. From this it follows that $r_1 = r_2 = 0$. Hence $\mathbb{C}[x,y]/(y^2 - x^3) = \mathbb{C}[t^2,t^3]$, and $t = \frac{y}{x}$ is not an element of R.

(2) Consider the ring $R := \mathbb{C}[x,y]/(x \cdot y)$. This ring is not normal. Namely, $x+y$ is not a zerodivisor in R because $x + y \notin (x) \cup (y)$ (Theorem 1.4.24). Consider $u = \frac{x}{x+y}$. Then $u^2 = u$ because $x^2(x+y) = x^3 = x(x^2+y^2) = x(x+y)^2$ in R. But it can easily be checked that $u \notin R$.

(3) Consider the ring $R := \mathbb{C}[x]/(x^2)$. Then $R = Q(R)$ and therefore R is integrally closed. However, R is not normal, as R is not reduced.

In the second example we used quite essentially that R is not an integral domain. In fact, the following theorem holds.

Theorem 1.5.7. *Let R be a Noetherian normal local ring. Then R is an integral domain.*

Proof. R is normal and, therefore, reduced. Let $(0) = \mathfrak{p}_1 \cap \cdots \cap \mathfrak{p}_r$ be an irredundant prime decomposition. We have to prove that $r = 1$. Assuming $r \geq 2$, we can find $g \in \mathfrak{p}_2 \cap \cdots \cap \mathfrak{p}_r$, $g \notin \mathfrak{p}_1$ (the decomposition is irredundant) and $f \in \mathfrak{p}_1$, $f \notin \mathfrak{p}_2 \cup \cdots \cup \mathfrak{p}_r$ (Prime avoidance 1.1.13). By the choice of f and g we have that $f + g \notin \mathfrak{p}_i$ for $i = 1, \ldots, r$. This implies that $f + g$ is not a zerodivisor (Theorem 1.4.24). On the other hand, $f \cdot g = 0$ because $f \cdot g \in \mathfrak{p}_1(\mathfrak{p}_2 \cap \cdots \cap \mathfrak{p}_r) \subset \mathfrak{p}_1 \cap \cdots \cap \mathfrak{p}_r = (0)$. Consider $u := \frac{f}{f+g}$, then $u^2 = u$ because $f^2(f+g) = f^3 = f(f^2+g^2) = f(f+g)^2$. This implies $u \in R$ because R is normal. Note that $u \neq 1$, as otherwise $\frac{f}{f+g} = 1$, and therefore $g = 0$. Now $u(1-u) = 0$, hence u is a zerodivisor in R. The element u is therefore contained in the maximal ideal of R. Hence $1 - u$ is a unit because R is local. This implies $u = 0$ and, therefore, $f = 0$ which is a contradiction to the choice of f. □

The assumption on R to be local in the theorem can be weakened to the assumption that R has no idempotents different from 1.[8] The last part of the proof just showed that a local ring does not contain idempotents different from 1.

For more subtle arguments concerning integral elements given later, we need the following important theorem:

Theorem 1.5.8 (Cayley-Hamilton). *Let $R \subset S$, $\alpha \in S$. Let M be an $R[\alpha]$-module, which is finitely generated as an R–module. Suppose moreover that M is a* faithful[9] *$R[\alpha]$-module. Then α is integral over R.*

Proof. Let $u_1, \ldots, u_t$ be generators of M as an R–module. Then for each i the element αu_i is defined and is an element of M. As the u_i's generate M there exist elements $\varphi_{ij} \in R$, such that $\alpha u_i = \sum_{j=1}^{t} \varphi_{ij} u_j$. In matrix notation:

$$\Big(\alpha Id - (\varphi_{ij})\Big) \begin{pmatrix} u_1 \\ \vdots \\ u_t \end{pmatrix} = 0, \quad Id \text{ the unit matrix.}$$

We multiply on the left with the adjoint (see 1.2.9) of $\alpha Id - (\varphi_{ij})$ to conclude that $\det(\alpha Id - (\varphi_{ij}))u_k = 0$ for $k = 1, \ldots, t$. This is Cramer's rule, see 1.2.8. As we assumed M to be faithful, it follows that $\det\big(\alpha Id - (\varphi_{ij})\big) = 0$. Expand the determinant to find an integral equation for α. □

[8] An element x in a ring is called idempotent if $x^2 = x$.

[9] An R–module M is called faithful if given $a \in R$ such that $a \cdot m = 0$ for all $m \in M$ it follows that $a = 0$.

Remarks 1.5.9.

(1) Note that the ring S is "dummy" in some sense. We just need that our element lives in some ring. To see why this theorem is called the Cayley-Hamilton Theorem, we refer to Exercise 1.5.30.

(2) If moreover α sends M into $I \cdot M$ for some ideal I, then α satisfies an integral equation of type

$$\alpha^n + a_1\alpha^{n-1} + \cdots + a_n = 0, \quad a_j \in I.$$

This motivates the following definition.

Definition 1.5.10. Let $R \subset S$ be a ring extension, and $I \subset R$ be an ideal. An element x is called *integral over I* if it satisfies an equation of type[10]

$$x^n + a_1x^{n-1} + \ldots + a_n = 0, \quad a_j \in I.$$

Lemma 1.5.11. *Let $R \subset S$ be rings, $I \subset R$ be an ideal, and $x \in S$. The following conditions are equivalent.*

(1) x is integral over I.

(2) The ring $S' := R[x]$ is a finitely generated R-module, and $x \in \sqrt{I \cdot S'}$.

(3) $R[x]$ is contained in a subring S' of S, where S' is a finitely generated R-module, and $x \in \sqrt{I \cdot S'}$.

Proof. Of course, (2) $\Longrightarrow$ (3) is trivial. The equivalence (1) $\Longleftrightarrow$ (2) can be proved as in 1.5.2. The implication (3) $\Longrightarrow$ (1) uses the Cayley-Hamilton Theorem. Indeed, take an $m \in \mathbb{N}$ such that $x^m \in I \cdot S'$. Therefore, multiplication with x^m sends S' to $I \cdot S'$. It follows from the proof of the Cayley-Hamilton Theorem that x^m satisfies an integral equation with coefficients in I (S' is a faithful $R[x^m]$-module because $1 \in S'$). □

Corollary 1.5.12. *Let $R \subset S$ be rings, and $I \subset R$. Let $\overline{R}$ be the integral closure of R in S. Then $\sqrt{I\overline{R}}$ is the set of all elements of S which are integral over I.*

Proof. If $x \in S$ and is integral over I, then by the second part of the previous lemma we get $x \in \sqrt{IR[x]} \subset \sqrt{I\overline{R}}$. Conversely, if $x \in \sqrt{I\overline{R}}$ then $x^m \in I\overline{R}$ for certain m. Hence $x^m \in I \cdot R[x_1, \ldots, x_n]$ for certain $x_i \in \overline{R}$, so that from (3) $\Longrightarrow$ (1) of the previous lemma it follows that x is integral over I. □

Next we will give a criterion for normality.

Theorem 1.5.13 (Criterion for Normality). *Let R be a Noetherian reduced ring. Let $I = \sqrt{I} \subset R$ be a radical ideal with the following properties:*

(1) I contains a nonzerodivisor of R.

(2) Let $\mathfrak{p}$ be a prime ideal of R such that $R_\mathfrak{p}$ is not normal. Then $\mathfrak{p} \supset I$.

[10] There is a related notion of *integrally dependence on I* by using the condition $a_j \in I^j$. This notion will not be used in this book.

Then R is normal if and only if the canonical inclusion $R \subset \mathrm{Hom}_R(I,I)$ is an equality.

To prove the theorem we need the following lemma.

Lemma 1.5.14. *Let R be a Noetherian reduced ring, $\widetilde{R}$ its normalization. Let $I \subset R$ be an ideal containing a nonzerodivisor of R. Then*

(1) $R \subset \mathrm{Hom}_R(I,I) \subset \widetilde{R}$.

(2) If, moreover, $I = \sqrt{I}$, that is, I is radical, then $\mathrm{Hom}_R(I,I) = \widetilde{R} \cap \mathrm{Hom}_R(I,R)$.

Proof.
(1) The canonical map $R \longrightarrow \mathrm{Hom}_R(I,I)$ is defined by the multiplication: to $y \in R$ we associate the map $\varphi_y : I \longrightarrow I$ defined by $\varphi_y(x) = yx$. The injectivity of the canonical map follows from the assumption that I contains a nonzerodivisor.

Now let $a \in I$ be a nonzerodivisor and $\varphi \in \mathrm{Hom}_R(I,I)$. Then for every $b \in I$ we have $\varphi(b) = b \cdot \frac{\varphi(a)}{a}$ because $a\varphi(b) = \varphi(ab) = b\varphi(a)$. Especially, $\frac{\varphi(a)}{a} \in Q(R)$ is independent of the choice of the nonzerodivisor. The map $\varphi \mapsto \frac{\varphi(a)}{a}$ gives the inclusion $\mathrm{Hom}_R(I,I) \subset Q(R)$. In particular, we obtain an identification

$$\mathrm{Hom}_R(I,I) = \{x \in Q(R) : xI \subset I\}.$$

We claim that this module is already contained in the normalization $\widetilde{R}$ of R. Let $x \in Q(R)$ and $xI \subset I$ then I can be considered as an $R[x]$–module.

As an ideal containing a nonzerodivisor it is a faithful $R[x]$–module and we can apply the Cayley-Hamilton Theorem 1.5.8 to obtain that x is integral over R. This implies that $\mathrm{Hom}_R(I,I) \subset \widetilde{R}$.

(2) We just proved that $\mathrm{Hom}_R(I,I) \subset \widetilde{R}$. Moreover, the inclusion $\mathrm{Hom}_R(I,I) \subset \mathrm{Hom}_R(I,R)$ is obvious. So we get $\mathrm{Hom}_R(I,I) \subset \widetilde{R} \cap \mathrm{Hom}_R(I,R)$. As above, note that $\mathrm{Hom}_R(I,R) = \{x \in Q(R) : xI \subset R\}$.

To prove the converse inclusion, let $z \in \widetilde{R} \cap \mathrm{Hom}_R(I,R)$. We have to prove that $zI \subset I$. Let $z^n + a_{n-1}z^{n-1} + \cdots + a_0 = 0$ for suitable $a_0,\ldots,a_{n-1} \in R$. Let $f \in I$, then by multiplying the above equation with f^n we obtain

$$(zf)^n + a_{n-1}f(zf)^{n-1} + \cdots + a_0 f^n = 0.$$

This implies $(zf)^n \in I$ because $zf \in R$ and $f \in I$. But $I = \sqrt{I}$ and, therefore, $zf \in I$. This implies $zI \subset I$ and proves the second part of the lemma. □

Proof of Theorem 1.5.13. Let R be normal, that is, $R = \widetilde{R}$. Then Lemma 1.5.14 (1) implies that $R = \mathrm{Hom}_R(I,I)$.

To prove the converse, assume that $R = \mathrm{Hom}_R(I,I)$. Suppose that the theorem is wrong, that is, $R \neq \widetilde{R}$. Let $h \in \widetilde{R} \setminus R$. Consider the ideal

$$J := \{u \in R : hu \in R\}$$

in R. We claim that $I \subset \sqrt{J}$. As $\sqrt{J} = \cap_{\mathfrak{p} \supset J}\mathfrak{p}$ is the intersection of all prime ideals containing J, by Exercise 1.4.53, it suffices to show that $\mathfrak{p} \supset I$ for all $\mathfrak{p} \supset J$. So let $\mathfrak{p}$ be a prime ideal with $\mathfrak{p} \supset J$. Now $h \notin R_{\mathfrak{p}}$. Indeed, otherwise $ah \in R$ for some $a \notin \mathfrak{p}$. But then also $a \notin J$, in contradiction to the definition of J. So h gives an element of $Q(R_{\mathfrak{p}})$ which is integral over $R_{\mathfrak{p}}$ (use the same integral equation) and, therefore, $R_{\mathfrak{p}}$ is not normal. This implies $\mathfrak{p} \supset I$ by the second assumption, and proves the claim.

As our ring is Noetherian, we can now choose $d \in \mathbb{N}$ such that $I^d \subset J$. This implies $hI^d \subset hJ \subset R$. As $h \notin R$, we have $d > 0$. Let d be minimal with $hI^d \subset R$. Thus there exists an $x \in I^{d-1}$ such that $hx \notin R$. But $hx \in \widetilde{R}$ and $hxI \subset hI^d \subset R$ implies $hx \in \widetilde{R} \cap \mathrm{Hom}_R(I,R) = \mathrm{Hom}_R(I,I) = R$. This is a contradiction to the choice of x. Therefore, our original assumption $h \notin R$ is wrong. This proves the theorem. □

Remark 1.5.15. We will see later that for rings of finite type over a field (cf. Chapter 2) or analytic algebras (cf. Chapter 3) the ideal I describing the singular locus (which can be computed using the Jacobian Criterion, cf. Chapter 4) has the properties required in Theorem 1.5.13.

Example 1.5.16 (Continuation of Example 1.5.6).

(1) Let $R = \mathbb{C}[t^2,t^3]$, $I = (t^2,t^3)$. Then
$\mathrm{Hom}_R(I,I) = \{x \in Q(\mathbb{C}[t^2,t^3]) = Q(\mathbb{C}[t]) : xI \subset I\} = \mathbb{C}[t]$.

(2) Let $R = \mathbb{C}[x,y]/(x \cdot y)$, $I = (x,y)$ and $u = \frac{x}{x+y}$ then $R[u] = \mathrm{Hom}_R(I,I)$ because $R[u] \subset \mathrm{Hom}(I,I)$ and $R[u] = R[u]/(u) \times R[u]/(1-u)$ is the product of the two normal rings $R[u]/(u) \cong \mathbb{C}[y]$ and $R[u]/(1-u) = \mathbb{C}[x]$.

We now come to the very important theorem of the finiteness of the normalization. This theorem, due to Kronecker and Noether, is of utmost importance for the development of the theory in this book, and will be used at several points. For example, it is used in the proof of the Nullstellensatz in local analytic geometry, see 3.4.4, and in the proof of the Local Parametrization Theorem, see 3.4.14, which is also very important in our development of the theory of local analytic geometry. It is also used in Theorem 4.4.8, in order to prove that the "normalization of a space" always exists. Before stating the theorem of the finiteness of normalization, we recall some facts concerning the discriminant of a polynomial in one variable.

Definition 1.5.17. Let k be a field (not necessarily algebraically closed), and $f = x^r + a_1x^{r-1} + \ldots + a_r$ be a polynomial with coefficients in k. Consider a field extension $k \subset K$ in which f splits:

$$f = (x - \alpha_1) \cdots (x - \alpha_r).$$

Then the *discriminant* $\Delta = \Delta_x(f)$ with respect to the variable x is defined by:

$$\Delta = \prod_{i \neq j} (\alpha_i - \alpha_j).$$

The discriminant therefore is zero exactly if f has a multiple root. It follows from the Theorem on Symmetric Functions, and is a standard fact in field theory, that the discriminant in fact is a polynomial in the coefficients $a_1, \ldots, a_r$ of f.

Lemma 1.5.18. *Let R be an normal integral domain, K be its quotients field, and $L \supset K$ be a field extension. Let $x \in L$ be integral over R. In particular, x is algebraic over K. Then the minimal polynomial $f(X) \in K[X]$ of x has coefficients in R.*

Proof. Consider a complete set of conjugates x_i of x in an algebraic closure of L. This means that we can write

$$f(X) = \prod (X - x_i).$$

(Each x_i is counted p^e times, where p^e is the degree of inseparability. In particular x_i is counted once, if the characteristic is zero.) By a standard result in field theory, there exists a K–isomorphism $\sigma_i : K(x) \cong K(x_i)$, with $\sigma_i(x) = x_i$ for all i. It follows that also x_i is integral over R. Indeed, apply the isomorphism σ_i to an integral equation of x

$$x^n + a_1 x^{n-1} + \dots, + a_n = 0$$

to obtain the integral equation

$$x_i^n + a_1 x_i^{n-1} + \dots, + a_n = 0.$$

The coefficients of $f(X)$ are the elementary symmetric functions of the x_i. As sums and products of integral elements are integral, it follows that the coefficients of f are integral over R. As they are moreover in K, and R is normal by assumption, it follows that the coefficients are in R. □

Theorem 1.5.19 (Finiteness of the Normalization). *We consider:*

- *a finite ring extension $R \subset S$ of integral domains with R a* normal *ring;*
- *K, the quotient field of R and L the quotient field of S;*
- *$\alpha \in S$ a primitive element of the field extension $K \subset L$, that is $L = K(\alpha)$. We suppose that α is separable;* [11]
- *$P(X) \in K[X]$ the minimal polynomial of α;*
- *$0 \neq \Delta$ the discriminant of $P(X)$.*

Then we have the following statements.

(1) For all $\beta \in S$, the minimal polynomial of β has coefficients in R. In particular $P(X) \in R[X]$ and $\Delta \in R$.

(2) Let a polynomial $q \in R[X]$ with $q(\alpha) = 0$ be given. Then $q \in (P)$.

(3) Every element f in S is in the R–module generated by

$$\frac{1}{\Delta}, \dots, \frac{\alpha^{r-1}}{\Delta}.$$

To put it in another way, for any $f \in S$ there exists a polynomial $q \in R[X]$ with $\deg(q) < \deg(P)$ *such that $\Delta f = q(\alpha)$.*

(4) Let $\widetilde{S}$ be the normalization of S. Then $\widetilde{S} \subset \frac{1}{\Delta} R[\alpha]$. In particular $\widetilde{S}$ is a finitely generated S–module.

(5) $S_\Delta = R_\Delta[\alpha]$ is normal.

[11] Recall that separable means that a minimal polynomial for α has no multiple roots.

Proof. For the proof we need some Galois theory. Let r be the degree of P.

(1) The first statement is a consequence of Lemma 1.5.18.

(2) Note that a minimal polynomial P is monic. We perform division with remainder. Thus $q = HP + B$, with degree of B smaller than r. If B is nonzero, then we divide B by its leading coefficient. We get a monic polynomial B' in $K[X]$ with $B'(\alpha) = 0$. This is a contradiction to the fact that P is the minimal polynomial of α. Thus $B = 0$, and $q \in (P)$.

(3) We consider the splitting field M of P. The element α has r (the degree of P) conjugates $\alpha_1, \ldots, \alpha_r$ in M, that is, the $\alpha = \alpha_1, \alpha_2, \ldots, \alpha_r$ are the (pairwise different) zeros of P in M. As $P(\alpha_i) = 0$, it follows that the α_i are integral over R. Now consider an element $f \in S \subset L$. It is an easy fact from field theory that we can write:

$$f = q_0 \cdot 1 + q_1 \cdot \alpha + \ldots + q_{r-1} \cdot \alpha^{r-1} \tag{1.4}$$

with $q_i \in K$ for all[12] i. We have to prove that $\Delta \cdot q_i \in R$ for all i. Now define

$$f_1 := q_0 \cdot 1 + q_1 \cdot \alpha_1 + \ldots + q_{r-1} \cdot \alpha_1^{r-1}$$

$$\vdots$$

$$f_r := q_0 \cdot 1 + q_1 \cdot \alpha_r + \ldots + q_{r-1} \cdot \alpha_r^{r-1}.$$

We need that the f_i are also integral over R. For this we need some Galois theory. There exist automorphisms σ_i of M over K mapping α to α_i. If we let σ_i operate on equation (1.4) it follows that $f_i = \sigma_i(f)$. If we now let σ_i operate on an integral equation for f, it follows that the f_i are also integral over R. We can view the previous equations as a system of linear equations:

$$A \begin{pmatrix} q_0 \\ \vdots \\ q_{r-1} \end{pmatrix} = \begin{pmatrix} f_1 \\ \vdots \\ f_r \end{pmatrix}; \quad A = \begin{pmatrix} 1 & \alpha_1 & \ldots & \alpha_1^{r-1} \\ \vdots & \vdots & & \vdots \\ 1 & \alpha_r & \ldots & \alpha_r^{r-1} \end{pmatrix},$$

to which we apply our knowledge of linear algebra. First of all, the determinant $D := \det(A)$, as a sum and product of integral elements itself is integral. But we know more: the matrix A is the well-known Vandermonde matrix and its determinant can be computed to be $D = \prod_{i>j}(\alpha_i - \alpha_j)$. Furthermore, by definition of the discriminant $\Delta = \pm D^2$. It follows from Cramer's rule 1.2.8 that

$$D \cdot q_\nu = P_\nu(f_i, \alpha_j)$$

for all ν for some polynomial P_ν with *integer* coefficients. It follows that Dq_ν is integral over R. As D also was integral we conclude that $\Delta q_\nu = DDq_\nu$ is integral over R. Now $\Delta \in R \subset K$, and the q_ν are in K, so that $\Delta \cdot q_\nu$ is an element of K, the quotient ring of R. We can now use our assumption that R is *normal* to conclude that $\Delta \cdot q_\nu \in R$ for all ν. This is what we had to prove.

(4) Let $\beta \in \widetilde{S}$. Then $S[\beta]$ is a finitely generated R–module with quotient field L and, therefore, by (3) of the theorem (applied to $R \subset S[\beta]$) included in $R_\Delta[\alpha]$. This implies that $\widetilde{S} \subset R_\Delta[\alpha]$ and is, therefore, finitely generated.

[12] Therefore, it is trivially true that for all f there exists a Δ_f with $\Delta_f \cdot f \in R[\alpha]$. The point of the theorem is, that we can take a universal Δ which works for all f.

(5) Note that $S_\Delta = R_\Delta[\alpha]$ and, because of (4), $\widetilde{S}_\Delta = R_\Delta[\alpha]$. □

Theorem 1.5.20 (Splitting of Normalization). *Let R be a reduced Noetherian ring and $(0) = \mathfrak{p}_1 \cap \cdots \cap \mathfrak{p}_s$ be an irredundant primary decomposition. Then there is a canonical isomorphism between the normalization $\widetilde{R}$ and the direct sum of the normalizations $\widetilde{R/\mathfrak{p}_i}$:*

$$\widetilde{R} \cong \bigoplus_{i=1}^{s} \widetilde{R/\mathfrak{p}_i}.$$

Proof. We have the following statements that are easy to prove, see Exercise 1.5.27.

(1) $\mathfrak{p}_i Q(R)$ is a prime ideal in $Q(R)$.

(2) $\cap_{i=1}^{s} \mathfrak{p}_i Q(R) = (0)$.

(3) $Q(R)/\mathfrak{p}_i Q(R) = Q(R/\mathfrak{p}_i)$.

Now we use Proposition 1.4.27 and obtain

$$Q(R) \cong \bigoplus_{i=1}^{s} Q(R)/\mathfrak{p}_i Q(R) \cong \bigoplus_{i=1}^{s} Q(R/\mathfrak{p}_i).$$

The second isomorphism is the third statement from above. Let $u_i \in Q(R)$ correspond to $(0,\ldots,0,1,0,\ldots,0)$, with 1 at the i–th place. It follows immediately from the isomorphism that $u_i^2 = u_i$. So $u_i \in \widetilde{R}$. Note, moreover, that $u_i \in \cap_{j \neq i} \mathfrak{p}_j Q(R)$, $u_i - 1 \in \mathfrak{p}_i Q(R)$ for $i = 1,\ldots,s$. Thus we get $u_i \in \cap_{j \neq i}(\mathfrak{p}_j Q(R) \cap \widetilde{R})$ and $u_i - 1 \in \mathfrak{p}_i Q(R) \cap \widetilde{R}$. The conditions of the Chinese Remainder Theorem are satisfied. We get an isomorphism

$$\widetilde{R} \cong \bigoplus_{i=1}^{s} \widetilde{R}/(\mathfrak{p}_i Q(R) \cap \widetilde{R}).$$

The theorem now follows because the canonical map

$$R/\mathfrak{p}_i \longrightarrow \widetilde{R}/(\mathfrak{p}_i Q(R) \cap \widetilde{R})$$

induces an isomorphism $\widetilde{R/\mathfrak{p}_i} \cong \widetilde{R}/(\mathfrak{p}_i Q(R) \cap \widetilde{R})$, see Exercise 1.5.27. □

Remark 1.5.21.

(1) Note that we reproved that a Noetherian normal local ring is an integral domain.

(2) Let K be a field, and $S = K[x_1,\ldots,x_n]/\mathfrak{p}$ with $\mathfrak{p}$ a prime ideal. We will prove in 2.2.9 (Noether normalization) that there exists $R \subset S$ satisfying the assumptions of the Theorem 1.5.19. This implies that the normalization $\widetilde{S}$ is a finitely generated S–module. In general this is not necessarily true. A similar statement holds for so-called analytic algebras, which we will consider in the third chapter.

Theorem 1.5.22 (Lying-Over Theorem). *Let R be a ring, $R \subset S$ be an integral extension, and $\mathfrak{p} \subset R$ be a prime ideal. Then there exists a prime ideal $\mathfrak{q}$ in S with $\mathfrak{q} \cap R = \mathfrak{p}$.*

For the proof, we need the following two lemmas.

Lemma 1.5.23. *Let $R \subset S$ be an integral extension of rings, and $\mathfrak{p} \subset R$ be a prime ideal. Then $\mathfrak{p}S \cap R = \mathfrak{p}$. In particular $\mathfrak{p}S \neq S$, so that there is an inclusion $R/\mathfrak{p} \hookrightarrow S/\mathfrak{p}S$.*

Proof. Let $x \in \mathfrak{p}S \cap R$. Let $S' \subset S$ be a subring such that $R \subset S'$ is a finite extension and $x \in \mathfrak{p}S'$. We can find such an S' as follows. Write $x = \sum_{i=1}^{n} x_i s_i$ with $s_i \in S$ and $x_i \in \mathfrak{p}$. Then $S' := R[s_1, \ldots, s_n]$. It follows from 1.5.11 that x is integral over $\mathfrak{p}$. Let

$$x^n + a_1 x^{n-1} + \ldots + a_n = 0,$$

with $a_i \in \mathfrak{p}$. Now from $x \in R$ it then follows $x^n \in \mathfrak{p}$. The ideal $\mathfrak{p}$ is prime, so that $x \in \mathfrak{p}$. Thus $\mathfrak{p}S \cap R = \mathfrak{p}$. □

The following lemma is a generalization of the fact that every ideal is contained in a maximal ideal, corresponding to the case $S = \{1\}$ in the lemma.

Lemma 1.5.24. *Let R be a ring, and S be a multiplicatively closed subset of R. Let I be an ideal of R, with $I \cap S = \emptyset$. Then there exists a prime ideal $\mathfrak{p}$ of R with $I \subset \mathfrak{p}$ and $\mathfrak{p} \cap S = \emptyset$.*

Proof. As $I \cap S = \emptyset$, we get $IR_S \neq R_S$. Let $\mathfrak{m} \subset R_S$ be a maximal ideal with $IR_S \subset \mathfrak{m}$. Then $I \subset \mathfrak{m} \cap R =: \mathfrak{p}$. □

Proof of the Lying-Over Theorem 1.5.22. By 1.5.23 we get $\mathfrak{p}S \cap N = \emptyset$, for the multiplicatively closed subset $N = R \setminus \mathfrak{p}$. By 1.5.24 there exists a prime ideal $\mathfrak{P}$ in S with $\mathfrak{p}S \subset \mathfrak{P}$ and $\mathfrak{P} \cap N = \emptyset$. Therefore, $\mathfrak{P} \cap R \supset \mathfrak{p}$ follows from the first, and $\mathfrak{P} \cap R \subset \mathfrak{p}$ follows from the second condition. This shows that $\mathfrak{P} \cap R = \mathfrak{p}$. □

Corollary 1.5.25 (Going-Up Theorem). *Let R be a ring, $R \subset S$ be an integral extension, and $\mathfrak{q} \subset \mathfrak{p} \subset R$ be prime ideals. Let $\mathfrak{Q}$ be a prime ideal in S lying over $\mathfrak{q}$, that is $\mathfrak{Q} \cap R = \mathfrak{q}$. Then there exists a prime ideal $\mathfrak{P}$ in S with $\mathfrak{P} \cap R = \mathfrak{p}$ and $\mathfrak{Q} \subset \mathfrak{P}$.*

Proof. Apply the Lying-Over Theorem to the integral extension $R/\mathfrak{q} \subset S/\mathfrak{Q}$. □

Theorem 1.5.26 (Going-Down Theorem). *Let $R \subset S$ be an integral extension of integral domains, and suppose that R is* normal. *Let $\mathfrak{p} \subset \mathfrak{q}$ be two prime ideals in R. Let $\mathfrak{Q}$ be a prime ideal in S with $\mathfrak{Q} \cap R = \mathfrak{q}$. Then there exists a prime ideal $\mathfrak{P}$ in S with $\mathfrak{P} \subset \mathfrak{Q}$ and $\mathfrak{P} \cap R = \mathfrak{p}$.*

Proof. We have multiplicatively closed sets $N' := R \setminus \mathfrak{p}$ and $N'' := S \setminus \mathfrak{Q}$. Therefore, $N := N' \cdot N''$ is a multiplicatively closed subset of S.
Claim: We have $\mathfrak{p}S \cap N = \emptyset$.

Otherwise, we have an element $x \in \mathfrak{p}S \cap N$. By 1.5.11 x is integral over $\mathfrak{p}$. We will show that the *minimal* polynomial $P(X) = X^n + a_1 X^{n-1} + \ldots + a_n$ of x over $Q(R)$ has coefficients in $\mathfrak{p}$. By Lemma 1.5.18 the a_i are elements of R. Let $x_1, \ldots, x_n$ be the zeros of $P(X)$. Then $x_1, \ldots, x_n$ are also integral over $\mathfrak{p}$, as $P(X)$ divides any polynomial which gives an integral equation for x over $\mathfrak{p}$. By Corollary 1.5.12 it follows that the elementary symmetric functions $a_1, \ldots, a_n$ of the $x_1, \ldots, x_n$ are integral over $\mathfrak{p}$. As $\mathfrak{p}$ is prime, it follows that $a_1, \ldots, a_n \in \mathfrak{p}$.

Moreover, as $x \in N$, we can write $x = u \cdot v$, with $u \in N'$, and $v \in N''$. Consider the element $v = \frac{x}{u}$. It is in S, and satisfies the equation

$$v^n + \frac{a_1}{u} v^{n-1} + \ldots + \frac{a_n}{u^n} = 0,$$

which is a minimal equation for v. By Lemma 1.5.18, the $\frac{a_i}{u^i}$ are elements of R. So we can write

$$\frac{a_i}{u^i} = b_i,$$

or, what is the same $a_i = b_i u^i$. Now $a_i \in \mathfrak{p}$, and, therefore, as $u^i \notin \mathfrak{p}$, it follows that $b_i \in \mathfrak{p}$. This shows that the element v is integral over $\mathfrak{p}$. Therefore, $v \in \sqrt{\mathfrak{p}S} \subset \sqrt{\mathfrak{q}S} \subset \mathfrak{Q}$. This is a contradiction to $v \in N'' = S \setminus \mathfrak{Q}$, and therefore proves the claim.

By 1.5.24 there exists a prime ideal $\mathfrak{P} \subset S$ with $\mathfrak{P} \cap N = \emptyset$ and $\mathfrak{p}S \subset \mathfrak{P}$. In particular $\mathfrak{P} \cap N'' = \emptyset$, so that $\mathfrak{P} \subset \mathfrak{Q}$. As $\mathfrak{P} \cap N' = \emptyset$, it follows that $\mathfrak{P} \cap R \subset \mathfrak{p}$, and therefore $\mathfrak{p} = \mathfrak{P} \cap R$. □

The condition that R is normal in the Going-Down Theorem is necessary. See Example 2.3.16.

Exercises

1.5.27. Show the unproved statements in the proof of Theorem 1.5.20.
(Hint: To prove that $\widetilde{R/\mathfrak{p}_i} \cong \widetilde{R}/(\mathfrak{p}_i Q(R) \cap \widetilde{R})$ consider the canonical inclusions

$$R \subset \bigoplus R/\mathfrak{p}_i \subset \bigoplus \widetilde{R}/(\mathfrak{p}_i Q(R) \cap \widetilde{R}) \subset \bigoplus Q(R/\mathfrak{p}_i).)$$

1.5.28.

(1) Let $R \subset S$ be a ring extension, and $\alpha \in S$. Here R is not necessarily Noetherian. Use the Cayley-Hamilton Theorem to prove that the following conditions are equivalent.

  (a) α is integral over R.

  (b) $R[\alpha]$ is a finitely generated R–module.

  (c) $R[\alpha]$ is contained in a ring $T \subset S$, such that T is a finitely generated R–module.

  (d) There exists a faithful $R[\alpha]$–module M which is finitely generated as an R–module.

(2) Prove that 1.5.2 (3) holds without the hypothesis that R is Noetherian.

1.5.29. Use the Cayley-Hamilton Theorem to prove that the Noetherian hypothesis in 1.5.4 is superfluous.

1.5.30.

(1) Let K be a field, V be a finite-dimensional vector space, and $\varphi : V \to V$ be a linear map. Interprete 1.5.8 as the Theorem of Cayley-Hamilton you know from linear algebra. Note that we could just as well take any ring R instead of the coefficient field K, and any finitely generated module M instead of V.

(2) Let R be a ring, M be a finitely generated R–module, and $\varphi : M \to M$ be a *surjective* R–module homomorphism. Prove that φ is an isomorphism.
(Hint: Consider M as $R[X]$–module, via $X \cdot m := \varphi(m)$, and extend by linearity. Then $(X) \cdot M = M$. Consider M as $R[X,Y]$–module, via $Y \cdot m = Id(m) = m$. Then apply Remark 1.5.9 (2) to deduce that Y integral over X.)

1.5.31. Let R be a reduced ring, and $S \subset R$ be a multiplicatively closed set. Show that $\widetilde{R_S} = \widetilde{R}_S$. Informally speaking, one says that taking integral closure commutes with localization.

1.5.32. Let R be a reduced ring. Prove that the following conditions are equivalent.

(1) R is normal.

(2) $R_\mathfrak{p}$ is normal for all prime ideals $\mathfrak{p}$ in R.

(3) $R_\mathfrak{m}$ is normal for all maximal ideals $\mathfrak{m}$ in R.

1.5.33. Prove that the ring $R = k[x,y,z]/(z^2 - xy)$ is normal.
(Hint: Localize in the prime ideals (x,z) and in (y,z), and show that those localizations are normal.) Show however, that R is not a unique factorization domain. This shows that the converse of 1.5.5 in general does not hold.

1.5.34. Prove Nakayama's Lemma using 1.5.8.

1.5.35. Let $f = x_n^s + a_1 x_n^{s-1} + a_2 x_n^{s-2} + \ldots + a_s$ with $a_i \in K[x_1, \ldots ,x_{n-1}]$. Consider the discriminant $\Delta \in K[x_1, \ldots ,x_{n-1}]$ of f with respect to x_n. Suppose that f, viewed as an element of $K[x_1, \ldots ,x_n]$, is square-free.[13] Prove that $\Delta \neq 0$.

1.5.36. To illustrate the Going-Up Theorem, let $R = K[x,y] \subset S = K[x,y,z]/(z^2 - xy)$.

(1) Show that S is an integral extension of R.

(2) Find a prime ideal $\mathfrak{q}$ in S with $\mathfrak{q} \cap R = (x)$. The ideal (x) is of course prime in R.

(3) Find a prime ideal $\mathfrak{q}$ in S with $\mathfrak{q} \cap R = (y^2 - x^3) \subset K[x,y]$.

[13] A polynomial is by definition called square-free, if it does not have multiple factors.

2 Affine Algebraic Geometry

The purpose of this chapter is to introduce the reader to the basic definitions, ideas and notations of affine algebraic geometry. Loosely speaking, an affine algebraic set is the zero set of finitely many polynomials in an affine space K^n, where K is an algebraically closed field.[1] Some of the results of this chapter are true and proved over arbitrary algebraically closed fields. This is the reason why we do not consider only the case $K = \mathbb{C}$ as we will do from Chapter 3 onwards. Admittedly, our main interest in this book is the behavior of such zero sets in the neighborhood of a point. We will discuss and study the algebraic case first, however, simply because the proofs of the statements in this case usually are easier. This is because one does not have to bother one-selves with convergence questions. Having grasped the main geometric and algebraic ideas in the affine case, the reader is hopefully ready to tackle the local case in Chapter 3. A second reason for doing the affine case, too, is that we need to do some algebraic geometry, in order to classify hypersurface singularities, see Chapter 9.

Again for reasons of simplicity, we first study the zero set $V(f)$ of a single polynomial f. These zero sets are called affine hypersurfaces. The main result in this section is Study's Lemma. It says that if f and g are polynomials, with f reduced and $V(f)$ a subset of $V(g)$, then f divides g. We give two proofs, whose ideas permeate large parts of the first part of the book. Namely, we consider a "general" linear projection of the hypersurface $V(f)$ on a hyperplane. This is called Noether normalization. One shows that the projection is surjective (this is quite trivial), and the proof is completed by using some elementary algebra, in particular, division with remainder. A second proof of Study's Lemma is given. Here it is not only proved that the projection is surjective, but also the number of points in the fiber are studied.

Study's Lemma allows us to interpret certain algebraic operations and statements geometrically. We mention two of them.

The first is the decomposition of a hypersurface in irreducible components. A hypersurface is called reducible, if it can be written in a nontrivial way as the union of two (or more) hypersurfaces. Now, if f and g are polynomials, the zero set of $f \cdot g$, obviously, is the union of the zero sets of f and g. Study's Lemma allows us to conclude that for the zero set of an irreducible polynomial f such a decomposition is not possible.

The second is the interpretation of the K-algebra $K[x_1, \ldots, x_n]/(f)$. On a hypersurface $V = V(f)$, where f is a polynomial, it is natural to consider functions which are restrictions of polynomials on the ambient space K^n. This set of functions is in a natural way a K-algebra. If f is reduced, this algebra can be identified with the quotient $K[x_1, \ldots, x_n]/(f) := K[V]$. In the proof, essential use of Study's Lemma is made.

In the second section, we will study affine algebraic sets in general. That is, we study zero sets $V(I) := \{a \in K^n : f(a) = 0 \text{ for all } f \in I\}$ of ideals I in the polynomial ring $K[x_1, \ldots, x_n]$. Hilbert's Basis Theorem allows us to consider zero sets of only finitely many polynomials. The main discussion is on a generalization of Study's Lemma, called Hilbert's Nullstellensatz. A weaker form is given by the Weak Nullstellensatz. It says that

[1] Just keep in mind the case $K = \mathbb{C}$.

if I is a proper ideal, then $V(I) \neq \emptyset$. This statement is in fact equivalent to the structure of maximal ideals in the polynomial ring: every maximal ideal in $K[x_1, \dots, x_n]$ is of type $(x_1 - a_1, \dots, x_n - a_n)$, for some $(a_1, \dots, a_n) \in K^n$. So we have a one-one correspondence between maximal ideals and points. The proof of the Weak Nullstellensatz is a consequence of the Projection Theorem, which is a statement on how one can describe, under certain circumstances, the image of an affine space under projection. To prove Hilbert's Nullstellensatz, we again use the Projection Theorem and Noether normalization. This means that we study a general projection $\pi : X = V(I) \longrightarrow K^r$, where r is determined by X. Morally speaking, r is equal to the dimension of X, a property of X which is not defined at this point of our study. The proof of the fact that the map π for a Noether normalization then is surjective, is done by means of the Projection Theorem. Hilbert's Nullstellensatz itself is an easy consequence of the surjectivity of π, and is an immediate generalization of the second part of the proof of Study's Lemma. A shorter proof of the Nullstellensatz out of the weak Nullstellensatz is by using the so-called Rabinowitch trick. This is done in the exercises. We presented the other proof however, as a substitute for the Rabinowitch trick in the local case does not, to our knowledge, exist. We want to have similar proofs for both the affine and the analytic case.

In Section 2.3, we consider regular maps between affine algebraic sets. These maps are restrictions of polynomial maps. In particular, we define isomorphisms between affine algebraic sets. If $\varphi : V \longrightarrow W$ is a regular map between affine algebraic varieties, we get, by composition, a K-algebra homomorphism $\varphi^* : K[W] \longrightarrow K[V]$. (Arrows reverse!) Again geometric statements about maps between V and W are interpreted as algebraic statements on the map $\varphi^* : K[W] \longrightarrow K[V]$. For example, it is shown that φ is injective, if and only if φ^* is surjective.

2.1 Affine Hypersurfaces

Definition 2.1.1. Let K be a field, $f \in K[x_1, \dots, x_n]$ be a polynomial. The *zero set* $V(f) \subset K^n$ of f is defined by:

$$V(f) := \{(a_1, \dots, a_n) \in K^n : f(a_1, \dots, a_n) = 0\}.$$

A set $V \subset K^n$ is called an *affine hypersurface* if there exists a nonzero $f \in K[x_1, \dots, x_n]$, such that $V = V(f)$.

We give some examples. For the examples, we use $K = \mathbb{R}$ for two reasons: real pictures are easier to draw; $\mathbb{R}$ is not algebraically closed and almost everything which fails over a field which is not algebraically closed fails already over $\mathbb{R}$.

Examples 2.1.2.

(1) $f = x^2 + y^2 - 1 \in \mathbb{R}[x,y]$. The set $V(f)$ is a circle in the plane.

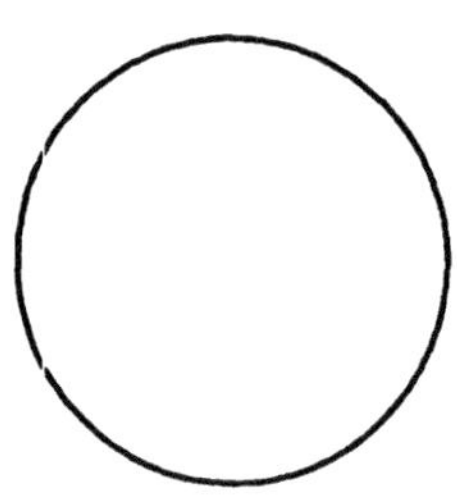

(2) $f = x^2 - yz \in \mathbb{R}[x,y,z]$. The set $V(f)$ is a cone in three-dimensional space.

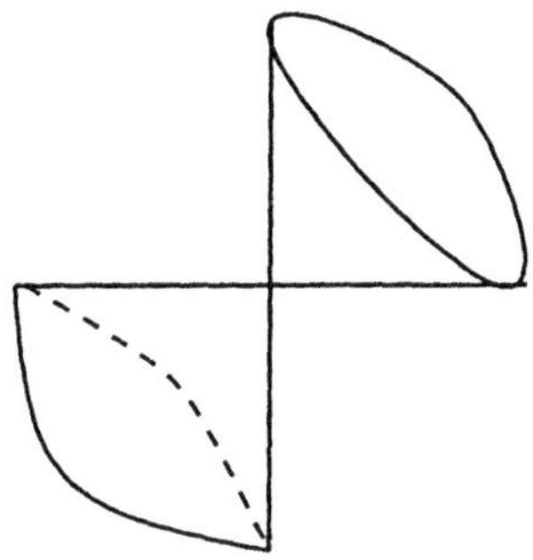

As is well-known, see 1.4.4, the polynomial ring $K[x_1, \dots, x_n]$ is factorial. It follows easily from the definition, see Exercise 2.1.18, that $f|g$ (or what is the same $(f) \supset (g)$) implies that there is an inclusion of zero sets $V(f) \subset V(g)$. The converse does not hold, as the following simple examples shows.

Examples 2.1.3.

(1) Take $f = x^2 + 1 \in \mathbb{R}[x]$, $g = x \in \mathbb{R}[x]$. Then $\emptyset = V(f) \subset V(g) = \{0\}$. But as the degree of f is bigger than the degree of g, f cannot divide g.

(2) Let f be a nonconstant polynomial. Then $V(f^2) = V(f)$, so in particular $V(f^2) \subset V(f)$. But obviously f^2 does not divide f.

The phenomena occurring in the first example is typical for fields which are *not algebraically closed.* In fact, one has the following theorem.

Theorem 2.1.4 (Study's Lemma). *Let K be an algebraically closed field, $f,g \in K[x_1, \dots, x_n]$. Suppose that f is* irreducible, *and that $V(f) \subset V(g)$. Then $f|g$.*

There is a proof of Study's Lemma using so-called resultants, see for example, [Brieskorn-Knörrer 1986], page 192. We present two other proofs of Study's Lemma here. Both proofs are based on the following two facts.

- Division with remainder for polynomials.
- Noether normalization for hypersurfaces.

We will first discuss the well-known Division with Remainder Theorem.

Theorem 2.1.5 (Division with Remainder). *Let R be a ring and $f \in R[x]$ be a polynomial. Suppose that the leading coefficient of f is a unit in R. Then for all $g \in R[x]$ there exist $q,r \in R[x]$, with* $\deg(r) < \deg(f)$ *such that:*

$$g = qf + r.$$

Moreover, q and r are uniquely determined by g and f.

Proof. Without loss of generality, we may assume that f is monic, that is, the leading coefficient of f is equal to 1.
Step 1: Existence: The proof is by induction on $\deg(g)$. If $\deg(g) < \deg(f)$ then one can take $q = 0$ and $r = g$. Otherwise, write

$$f = x^d + b_{d-1}x^{d-1} + \ldots + b_0; \quad g = a_m x^m + a_{m-1}x^{m-1} + \ldots + a_0,$$

with $m \geq d$. Consider

$$\widetilde{g} = g - a_m x^{m-d} f = (a_{m-1} - a_m b_{d-1})x^{m-1} + \ldots$$

The degree of $\widetilde{g}$ is smaller than the degree of g. By induction $\widetilde{g} = \widetilde{q}f + r$ with $\deg(r) < \deg(f)$. With $q = \widetilde{q} + a_m x^{m-d}$ the equality $g = qf + r$ holds.

Step 2: Uniqueness. Suppose $g = q_1 f + r_1 = q_2 f + r_2$ with $\deg(r_1), \deg(r_2) < \deg(f)$. It suffices to show that $q_1 = q_2$, as $r_1 = r_2$ follows immediately from it. Subtracting we get:

$$(q_1 - q_2)f = (r_2 - r_1)$$

As the degree of r_1 and r_2 is smaller than the degree of f, all the terms on the left hand side of degree at least d have to vanish. Because f is *monic*, and looking at the term of highest degree, it follows that q_1 and q_2 have the same degree and the same leading coefficient. Subtraction of this highest degree term from both q_1 and q_2 gives us q_1' and q_2' with $(q_1' - q_2')f = (r_2 - r_1)$. By (descending) induction we have that $q_1' = q_2'$. Hence $q_1 = q_2$. □

Lemma 2.1.6 (Noether Normalization for Hypersurfaces). *Let K be a field with an infinite number of elements, and let $f \in K[x_1, \ldots, x_n]$ be a polynomial of degree s. Then after a general linear change of coordinates, one has:*

$$f = c(x_n^s + a_1 x_n^{s-1} + \ldots + a_s) \tag{2.1}$$

with $c \in K$ a nonzero constant, and the a_i are elements of $K[x_1, \ldots, x_{n-1}]$.

Proof. We write

$$f = f_s + f_{s-1} + \ldots + f_0$$

where each f_i is homogeneous of degree i. The polynomial f_s is nonzero, so by Exercise 2.1.19 for general $(b_1, \ldots, b_n) \in K^n$ the value $c := f_s(b_1, \ldots, b_n) \neq 0$. After renumbering the coordinates, we may furthermore assume that $b_n \neq 0$. Take new coordinates $y_1, \ldots, y_n$ related to $x_1, \ldots, x_n$ by the formula's:

$$x_i = b_i y_n + y_i \text{ for } 1 \leq i \leq n-1, \quad x_n = b_n y_n$$

Then $g(y_1, \ldots, y_n) := f(x_1(y_1, \ldots, y_n), \ldots, x_n(y_1, \ldots, y_n))$ is such that $g(0, \ldots, 0, y_n) = f_s(b_1, \ldots, b_n)y_n^s + f_{s-1}(b_1, \ldots, b_n)y_n^{s-1} + \ldots$, with $f_s(b_1, \ldots, b_n) = c \neq 0$. Hence $g = c(y_n^s + a_1 y_n^{s-1} + \ldots + a_s)$, for certain $a_i \in K[y_1, \ldots, y_{n-1}]$. This shows the lemma. □

The Noether normalization for hypersurfaces can be reformulated in a more algebraic manner. The algebraic formulation is the one which is going to be generalized later.

Corollary 2.1.7 (Noether Normalization for Hypersurfaces). *For a general coordinate system,*

$$K[x_1, \ldots, x_n]/(f)$$

is a finitely generated $K[x_1, \ldots, x_{n-1}]$-module.[2]

[2] In this case it is even a free module.

Proof. After a general change of coordinates we may assume that f is of type (2.1). We claim that $K[x_1, \dots, x_n]/(f)$ is as $K[x_1, \dots, x_{n-1}]$-module generated by $\{1, x_n, \dots, x_n^{s-1}\}$. Consider a $g \in K[x_1, \dots, x_n]$. We do division with remainder

$$g = qf + r$$

where r is a polynomial in x_n of degree less than s with coefficients in $K[x_1, \dots, x_{n-1}]$. Hence, modulo f, the element g is indeed in the module generated by $\{1, x_n, \dots, x_n^{s-1}\}$. □

Examples 2.1.8.

(1) Consider $f = xy$. Then obviously $\mathbb{C}[x] \subset \mathbb{C}[x,y]/(xy)$, but $\mathbb{C}[x,y]/(xy)$ is not a finitely generated $\mathbb{C}[x]$-module. Indeed, if it was finitely generated, say by the elements $g_1, \dots, g_s \in \mathbb{C}[x,y]$, then let d be the maximum of the degree in y in the polynomials g_i. Then y^{d+1} is not in the module generated by the $g_1, \dots, g_s$. The module $\mathbb{C}[x,y]/(xy)$ is as $\mathbb{C}[x]$-module generated by $\{1, y, y^2, \dots\}$.

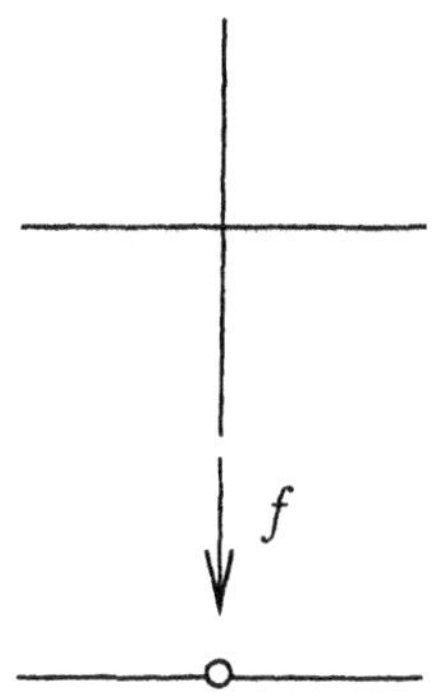

(2) Consider $f = xy - 1$. Then $\mathbb{C}[x] \subset \mathbb{C}[x,y]/(xy-1)$. But again $\mathbb{C}[x,y]/(xy-1) = \mathbb{C}[x,x^{-1}]$ is not a finitely generated $\mathbb{C}[x]$-module. It is generated by $\{1, y, y^2, \dots\} = \{1, x^{-1}, x^{-2}, \dots\}$.

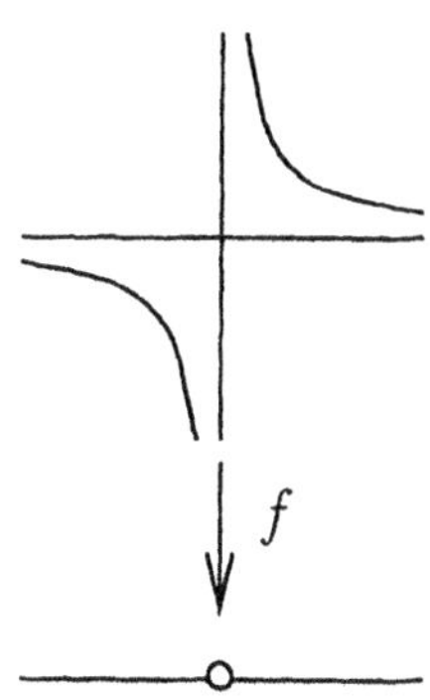

These two examples are quite typical. In both cases, we considered the inclusion of rings $\mathbb{C}[x] \subset \mathbb{C}[x]/(f)$. Consider the projection $\pi : X = V(f) \longrightarrow \mathbb{C}$ which sends (a,b) to a. In the first example, this map is surjective, but the fiber $\pi^{-1}(0)$ consists of infinitely many points, in fact the whole y–axis. In the second example, the map π is not surjective: $\pi^{-1}(0) = \emptyset$. But, if f is as in formula (2.1), it turns out that the projection $\pi : V(f) \longrightarrow \mathbb{C}^{n-1}$ on the first $n-1$ coordinates is surjective, and has only finitely many points in each fiber. This fact, which is very easy to prove, is crucial for our first proof of Study's Lemma.

Lemma 2.1.9. *Let K be an algebraically closed field, and $f = x_n^s + a_1 x_n^{s-1} + \ldots + a_s$, with $a_i \in K[x_1, \ldots, x_{n-1}]$. Consider the projection*

$$\pi : V(f) \longrightarrow K^{n-1}; \quad (p_1, \ldots, p_n) \mapsto (p_1, \ldots, p_{n-1}).$$

Then π is surjective.

Proof. This is quite obvious. For, let $p = (p_1, \ldots, p_{n-1}) \in K^{n-1}$. Because K is algebraically closed, the equation

$$X^s + a_1(p) X^{s-1} + \ldots + a_s(p) = 0,$$

has a solution p_n. Then $(p_1, \ldots, p_n) \in V(f)$. □

First proof of Study's Lemma 2.1.4. Let $g \in K[x_1, \ldots, x_n]$ with $V(f) \subset V(g)$. We have to show that $f \mid g$. By the Noether Normalization Lemma for Hypersurfaces, we may suppose that f is of type

$$f = x_n^s + a_1 x_n^{s-1} + \ldots + a_s.$$

By 2.1.7 $K[x_1, \ldots, x_n]/(f)$ is a finitely generated $K[x_1, \ldots, x_{n-1}]$-module. Consider the class $\overline{g}$ of g in $K[x_1, \ldots, x_n]/(f)$. Hence, by 1.5.2, $\overline{g}$ is integral over $K[x_1, \ldots, x_{n-1}]$, that is, it satisfies an equation of type

$$\overline{g}^k + b_1 \overline{g}^{k-1} + \ldots + b_k = 0,$$

with $b_i \in K[x_1, \ldots, x_{n-1}]$. This equation is to be considered in the ring $K[x_1, \ldots, x_n]/(f)$. Therefore

$$g^k + b_1 g^{k-1} + \ldots + b_k \in (f).$$

We take k minimal with this property. Take a point $p \in V(f)$, so that $f(p) = 0$. From $V(f) \subset V(g)$ it follows that $g(p) = 0$. Therefore $b_k(p) = 0$. But b_k only depends on the first $n-1$ coordinates. As $\pi : V(f) \longrightarrow K^{n-1}$ is surjective by the previous lemma, it follows that $b_k(p_1, \ldots, p_{n-1}) = 0$ for all $(p_1, \ldots, p_{n-1}) \in K^{n-1}$. Hence b_k is the zero polynomial. Therefore

$$g \cdot (g^{k-1} + b_1 g^{k-2} + \ldots + b_{k-1}) \in (f).$$

As f is irreducible, either $f \mid g$ or $f \mid g^{k-1} + b_1 g^{k-2} + \ldots + b_{k-1}$. But the second possibility is excluded by our choice of k to be minimal. □

The second proof of Study's Lemma we present looks somewhat more complicated. It gives, however, also some more information on the fibers of the map $\pi : V(f) \longrightarrow K^{n-1}$. Moreover, the condition irreducible in this proof can be somewhat weakened: we only need f to be square-free.

Lemma 2.1.10. *Let K be an algebraically closed field. Consider a square-free polynomial $f \in K[x_1 \dots, x_n]$ of the form*

$$f = x_n^s + a_1 x_n^{s-1} + a_2 x_n^{s-2} + \ldots + a_s; \quad a_i \in K[x_1, \dots, x_{n-1}].$$

Let $\Delta \in K[x_1, \dots, x_{n-1}]$ be the discriminant of f (with respect to x_n), see 1.5.17. Consider the projection

$$\pi : V(f) \longrightarrow K^{n-1},$$
$$(p_1, \dots, p_n) \mapsto (p_1, \dots, p_{n-1}) =: q.$$

Then $\pi^{-1}(q)$ has s, the degree of f, elements exactly when $\Delta(q) \neq 0$.

Proof. Consider $q = (p_1, \dots, p_{n-1}) \in K^{n-1}$. Look at the polynomial

$$x_n^s + a_1(q) x_n^{s-1} + a_2(q) x_n^{s-2} + \ldots + a_s(q) \in K[x_n]. \tag{2.2}$$

Because K is algebraically closed, this polynomial has s (distinct) roots exactly when its discriminant, which is $\Delta(q)$, is nonzero. If $b_1, \dots, b_t$, $t \leq s$ are the roots of the polynomial (2.2), then the points in the fiber $\pi^{-1}(q)$ are $(p_1, \dots, p_{n-1}, b_1), \dots, (p_1, \dots, p_{n-1}, b_t)$. □

Example 2.1.11. Consider the curve C given by the equation $y^2 - x^2(x+1) = 0$. The discriminant is given by $4x^2(x+1)$, so is zero for $x = 0$ and $x = -1$. Consider the projection π from C to the x-axis.

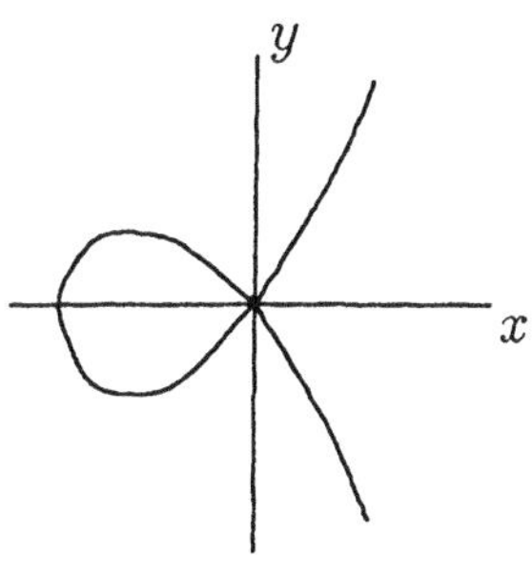

We see indeed from the picture (over $\mathbb{C}$!) that $\pi^{-1}(p)$ consists of two points for $p \neq 0, -1$, whereas for $p = 0$ or -1 it consists of one point. For other examples see Exercise 2.1.21.

Second proof of Study's Lemma. After, if necessary, performing a linear change of coordinates, we may assume, by Noether normalization for hypersurfaces, that

$$f = x_n^s + a_1 x_n^{s-1} + a_2 x_n^{s-2} + \ldots + a_s$$

with $a_i \in K[x_1, \dots, x_{n-1}]$. As f is square-free, by Exercise 2.1.20 the discriminant Δ of f is nonzero. We consider the projection $\pi : V(f) \longrightarrow K^{n-1}$ as in Lemma 2.1.10. Let $D = \{q \in K^{n-1} : \Delta(q) = 0\}$. We have shown that the inverse image of $q \notin D$ under π has exactly s points. Now do division with remainder for polynomials in the variable x_n:

$$g = qf + r$$
$$r = r_0 + r_1 x_n + \ldots + r_{s-1} x_n^{s-1}, \quad r_i \in K[x_1, \dots, x_{n-1}].$$

Let $(p_1, \dots, p_{n-1}, p_n) \in V(f)$. As by assumption $V(f) \subset V(g)$, it follows that $g(p_1, \dots, p_{n-1}, p_n) = 0$. As $g = qf + r$, it follows that $r(p_1, \dots, p_{n-1}, p_n) = 0$. Hence, if $q = (p_1 \dots, p_{n-1}) \notin D$ the polynomial

$$r_0(q) + r_1(q)x_n + \ldots + r_{s-1}(q)x_n^{s-1}$$

has s roots. But the polynomial $r(p_1, \dots, p_{n-1})$ has degree $s-1$, has s roots, so must be the zero polynomial. Therefore, its coefficients are zero. Hence

$$r_0(p_1, \dots, p_{n-1}) = \cdots = r_{s-1}(p_1, \dots, p_{n-1}) = 0.$$

As we can take $q = (p_1, \dots, p_{n-1})$ general, it follows from Exercise 2.1.19 that the polynomials, $r_0, \dots, r_{s-1} \in K[x_1, \dots, x_{n-1}]$ are zero. Hence $r = 0$. Therefore, $g = qf$, that is, $f \mid g$. □

Definition 2.1.12. An affine hypersurface $C \subset K^n$ is called *irreducible* if from $C = C_1 \cup C_2$, with C_1 and C_2 affine hypersurfaces, it follows that either $C_1 = C$ or $C_2 = C$. A hypersurface is called *reducible* if it is not irreducible.

An *irreducible component* of an affine hypersurface C is an irreducible affine hypersurface contained in C.

A consequence of Study's Lemma is that one can characterize irreducible hypersurfaces by its defining equation.

Theorem 2.1.13. *Let K be an algebraically closed field. Let $C = V(f)$ be an affine hypersurface. Then the following conditions are equivalent.*

(1) C is irreducible.

(2) There exists an irreducible $g \in K[x_1, \dots, x_n]$, and a $k \in \mathbb{N}$ with $f = g^k$.

It follows that that for all affine hypersurfaces C we have a decomposition

$$C = C_1 \cup \cdots \cup C_s,$$

where C_i are the irreducible components of C, and $C_i \neq C_j$ for $i \neq j$.

Proof. Suppose first that C is irreducible. Suppose $f = f_1^{n_1} \cdots f_s^{n_s}$. Then $V(f) = V(f_1 \cdots f_s)$. So we may assume that f is square-free. We have to prove that f is irreducible under this assumption. Suppose the converse. Then we can write $f = f_1 \cdot f_2$, where f_1 and f_2 do not have common factors and are not constant. In particular $f_1 \nmid f_2$ and $f_2 \nmid f_1$. From Study's Lemma it follows that $V(f_1) \not\subset V(f_2)$ and $V(f_2) \not\subset V(f_1)$. Therefore $C \neq V(f_1)$ and $C \neq V(f_2)$. But as $f = f_1 \cdot f_2$ it follows that $C = V(f_1) \cup V(f_2)$. This is a contradiction to the assumption that C is irreducible.

To prove the converse, suppose $C = C_1 \cup C_2$, $C = V(f)$, $C_1 = V(f_1)$ and $C_2 = V(f_2)$, and both C_1 and C_2 are not equal to C. Let h be irreducible and $h \mid f_1$. Then $V(h) \subset V(f_1) \subsetneqq V(f)$ (Exercise 2.1.18). Study's Lemma implies $h \mid f$. and thus $h \mid g^k$. As the polynomial ring is a unique factorization domain, and g is irreducible, it follows that $h = g \cdot c$ for some $c \in K$. This implies $V(h) = V(g) = V(f)$ which is in contradiction to the assumption $V(h) \subsetneqq V(f)$. □

Examples 2.1.14.

(1) $f(x,y) = x \cdot y$. Then $V(f)$ has two components, given by $x = 0$ and $y = 0$.

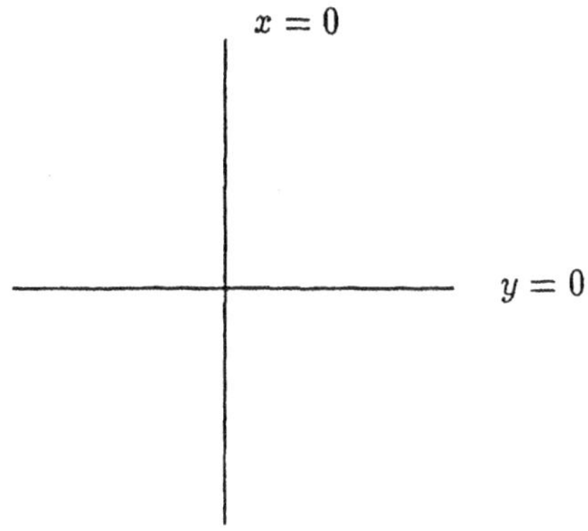

(2) As $f(x,y) = y^2 - x^2(x+1)$, see 2.1.11, is an irreducible polynomial, see Exercise 1.4.34, the affine hypersurface $V(f)$ is irreducible.

We now consider polynomial functions on affine hypersurfaces.

Definition 2.1.15. Let $C = V(f) \subset K^n$ be an affine hypersurface. A function $g : C \longrightarrow K$ is called *regular*, if g is the restriction of a polynomial map $g : K^n \longrightarrow K$. With the obvious addition and multiplication the regular functions on C form a ring . It is in fact a K-algebra. The *ring of regular functions* of C we denote by $K[C]$.

A polynomial map $g : K^n \longrightarrow K$ is just given by a polynomial $g \in K[x_1, \dots, x_n]$. Let such a g be given. The restrictions g and $g + \alpha f$ for any $\alpha \in K[x_1, \dots, x_n]$ define the *same* function on $C = V(f)$. Using Study's Lemma, we have the converse, as soon as we assume that the hypersurface C is defined by a square-free f.

Theorem 2.1.16. *Let K be an algebraically closed field. Let $f \in K[x_1, \dots x_n]$ be square-free. Then the ring of regular functions on $C = V(f)$ is equal to $K[x_1, \dots, x_n]/(f)$.*

Proof. Let $g_1, g_2 \in K[x_1, \dots x_n]$ be polynomial functions such that the functions $g_{1|C}$ and $g_{2|C}$ are equal. We have to show that the classes of g_1 and g_2 in $K[x_1, \dots, x_n]/(f)$ are equal. By assumption $g_1 - g_2$ vanish on C, which means that $V(f) \subset V(g_1 - g_2)$. From Study's Lemma it follows that $f | g_1 - g_2$, or, what is the same $g_1 - g_2 \in (f)$. This shows the theorem. □

The assumption that f is square-free is essential. For example, let $f = x^2 \in K[x,y]$, and $C = V(f)$. Then x and 0 define the same function on $V(f)$, but $x \neq 0 \in K[x,y]/(x^2)$.

Exercises

In these exercises, K is an algebraically closed field.

2.1.17. Deduce Study's Lemma for square-free f from Study's Lemma for irreducible f.

2.1.18. Let $f,g \in K[x_1, \dots, x_n]$ with $f \mid g$. Show that $V(f) \subset V(g)$.

2.1.19. Let K be a field with an infinite number of elements. Let $f \in K[x_1, \dots, x_n]$ be a nonconstant polynomial. Prove, by induction on n, that for "general" points $(a_1, \dots, a_n) \in K^n$ we have $f(a_1, \dots, a_n) \neq 0$. Specify what we mean by general.

2.1.20. Let $f = x_n^s + a_1 x_n^{s-1} + a_2 x_n^{s-2} + \ldots + a_s$, with $a_i \in K[x_1, \ldots, x_{n-1}]$. Suppose that f is square-free. Prove that the discriminant of f is nonzero.

2.1.21.

(1) Consider the curve defined by $y^2 - xy + x^2 = 0$, and the projection π on the x-axis. Compute for which points on the x-axis the inverse image of π consists of less than two points.

(2) Consider the cone $x^2 - yz$, see example 2.1.2., and the projection π on the y,z-plane. Compute for which points on the y,z-plane the inverse image of π consists of less than two points. What about projection on the x,y-plane?

2.1.22. Prove, using only the statement of Study's Lemma, that $V(f) \neq \emptyset$ for any nonunit $f \in K[x_1, \ldots, x_n]$.

2.2 Affine Varieties

In this rest of this chapter we assume, unless said otherwise explicitly that K is algebraically closed. In this section, we will, more generally, consider zero sets of any set of elements in $K[x_1, \ldots, x_n]$.

Definition 2.2.1. Let $F \subset K[x_1, \ldots, x_n]$ be a subset. We put

$$V(F) := \{p \in K^n : f(p) = 0 \text{ for all } f \in F\}.$$

and call it the *zero set* of F. A subset V of K^n is called an *(affine) algebraic set*, or an *affine variety* if there exists a subset $F \subset K[x_1, \ldots, x_n]$ with $V = V(F)$. Conversely, for any set $A \subset K^n$ we define

$$\mathscr{I}(A) = \{f \in K[x_1, \ldots, x_n] : f(p) = 0 \text{ for all } p \in A\},$$

and call it the *ideal of (functions vanishing on) A*.

Elementary properties of these notions are gathered in the following lemma.

Lemma 2.2.2. *Let V,W be algebraic subsets of K^n, A be a subset of K^n, I,J,I_α be ideals of $K[x_1, \ldots, x_n]$, F be a subset of $K[x_1, \ldots, x_n]$, and B be an index set.*

(1) $\mathscr{I}(A)$ is an ideal. Even more, $\mathscr{I}(A)$ is a radical *ideal.*

(2) $V \subset W \implies \mathscr{I}(V) \supset \mathscr{I}(W)$.

(3) $I \subset J \implies V(I) \supset V(J)$.

(4) Let I be the ideal generated by $F \subset K[x_1, \ldots, x_n]$. Then $V(F) = V(I)$.

(5) $\emptyset$ and K^n are algebraic sets.

(6) $V(I \cdot J) = V(I \cap J) = V(I) \cup V(J)$. In particular, (induction) finite unions of algebraic sets are algebraic.

(7) $V(\sum_{\alpha \in B} I_\alpha) = \bigcap_{\alpha \in B} V(I_\alpha)$. In particular, arbitrary intersections of algebraic sets are algebraic sets.

(8) $\mathscr{I}(V \cup W) = \mathscr{I}(V) \cap \mathscr{I}(W)$.

(9) $\sqrt{I} \subset \mathscr{I}(V(I))$.

Proof.
(1) The fact that $\mathscr{I}(A)$ is an ideal is easy. Moreover, if $f^m \in \mathscr{I}(A)$, then $f^m(p) = 0$ for all $p \in A$. Hence $f(p) = 0$ for all $p \in A$, that is $f \in \mathscr{I}(A)$.

(2) and (3) are trivial.

(4) As $F \subset I$, it follows that $V(F) \supset V(I)$. Suppose on the other hand that $a \in V(F)$, and let $f = \sum_{i=1}^{n} a_i f_i$, $f_i \in F$ be an element of I. As $a \in V(F)$, we have that $f_i(a) = 0$ for all i. Hence $f(a) = 0$. As f is arbitrary it follows that $a \in V(I)$.

(5) $\emptyset = V(\{1\})$, hence is algebraic. Furthermore, $K^n = V(\{0\})$, hence is algebraic too.

(6) We prove three inclusions.

(a) $V(I) \cup V(J) \subset V(I \cap J)$: let a in $V(I) \cup V(J)$. Therefore, either $f(a) = 0$ for all $f \in I$ or $f(a) = 0$ for all $f \in J$. In particular, if $f \in I \cap J$, it follows that $f(a) = 0$.

(b) $V(I \cap J) \subset V(I \cdot J)$: this follows from the inclusion of ideals $I \cdot J \subset I \cap J$.

(c) $V(I \cdot J) \subset V(I) \cup V(J)$: let $a \in V(I \cdot J)$. In particular $f(a) \cdot g(a) = 0$ for all $f \in I$ and $g \in J$. Suppose $a \notin V(J)$. Then we must show $a \in V(I)$. As $a \notin V(J)$ there exists a $g \in J$ with $g(a) \neq 0$. But as $f(a)g(a) = 0$ for all $f \in I$, it follows that $f(a) = 0$ for all $f \in I$. This gives $a \in V(I)$.

(7) As the ideal generated by $\cup_{\alpha \in B} I_\alpha$ is exactly $\sum_{\alpha \in B} I_\alpha$, it follows that $V(\cup_{\alpha \in B} I_\alpha) = V(\sum_{\alpha \in B} I_\alpha)$. The equality $V(\cup_{\alpha \in B} I_\alpha) = \cap_{\alpha \in B} V(I_\alpha)$ is easy, and therefore left to the reader.

(8) If f vanishes on $V \cup W$, it must vanish on both. The inclusion $\mathscr{I}(V \cup W) \subset \mathscr{I}(V) \cap \mathscr{I}(W)$ follows. Suppose on the other hand that $f \in \mathscr{I}(V) \cap \mathscr{I}(W)$, and let $a \in V \cup W$. Then either $a \in V$ or $a \in W$. In any case, for $f \in \mathscr{I}(V) \cap \mathscr{I}(W)$ it follows that $f(a) = 0$.

(9) The inclusion $I \subset \mathscr{I}(V(I))$ is a tautology. As $\mathscr{I}(V(I))$ is a radical ideal by the first part of the lemma, it follows that $\sqrt{I} \subset \mathscr{I}(V(I))$. □

Examples 2.2.3.

(1) Consider $I = (xy,xz,yz)$. It is not too difficult to show, see Exercise 2.2.30, that $I = (x,y) \cap (x,z) \cap (y,z)$. Hence $X := V(I)$ consist of the union of the coordinate axes.

On the other hand, X is also the zero set of the ideal $I' := (x,y) \cdot (x,z) \cdot (y,z) = (x^2y, x^2z, xyz, xz^2, xy^2, yz^2, y^2z)$. This ideal is smaller than I and, therefore, is not radical. This can be seen immediately. For example, $xy \notin I'$ but $(xy)^2 = y \cdot x^2y$ is in I'. This is quite typical. Products of radical ideals in general are not radical.

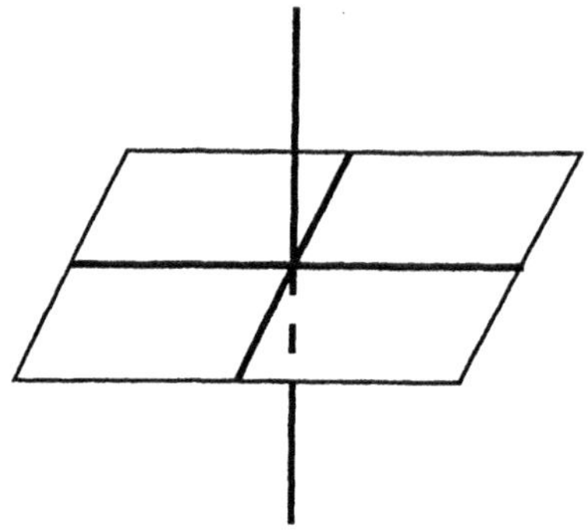

(2) Consider $I_1 = (z, y^2 - x^3)$, and $I_2 = (x, y)$, and $X_1 = V(I_1), X_2 = V(I_2)$. Then the union $X = X_1 \cup X_2$ is the zero set of the ideal $I = I_1 \cdot I_2 = \big(xz, yz, x(y^2 - x^3), y(y^2 - x^3)\big)$. Again this ideal is not radical. Indeed, $y^2 - x^3 \notin I$, but $(y^2 - x^3)^2 = y \cdot y(y^2 - x^3) - x^2 \cdot x(y^2 - x^3) \in I$, so that $y^2 - x^3 \in \sqrt{I}$.

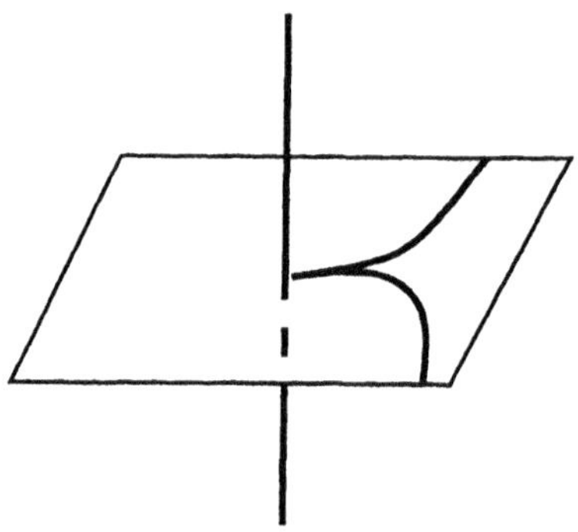

Definition 2.2.4. We define a topology on K^n by taking the closed subsets to be the algebraic subsets. That this is indeed a topology follows from Lemma 2.2.2 (5), (6) and (7). This topology is called the *Zariski topology* of K^n. More generally, for any algebraic set $V \subset K^n$, we define the *Zariski topology* on V to be the topology on V induced by the Zariski topology on K^n. So a closed subset in V is of type $V(I) \cap V$ for some ideal $I \subset K[x_1, \ldots, x_n]$.

The Zariski topology of $\mathbb{C}^n$ is much coarser than the usual (Euclidean) topology of $\mathbb{C}^n$. For example, for $n = 1$, a subset of $\mathbb{C}$ is closed in the Zariski topology when this subset is finite or equal to $\mathbb{C}$. The Zariski topology does not behave well under products, see Exercise 2.2.34.

Lemma 2.2.5. *Let $A \subset K^n$ be a subset. Then $\overline{A} = V(\mathscr{I}(A))$.*

Proof. The inclusion $A \subset V(\mathscr{I}(A))$ is a tautology. Moreover, $V(\mathscr{I}(A))$ is the smallest closed subset containing A. Indeed, let B be a closed subset with $A \subset B$. As B is closed, there exists an ideal J with $B = V(J)$. Then $J \subset \mathscr{I}(V(J))$, hence $B = V(J) \supset V(\mathscr{I}(V(J))) = V(\mathscr{I}(B)) \supset V(\mathscr{I}(A))$. The last inclusion is by combining (2) and (3) of the Lemma 2.2.2. As $\overline{A}$ is the smallest closed subset containing A, the lemma follows. $\square$

Theorem 2.2.6 (Hilbert's Nullstellensatz). *Let $I \subset K[x_1, \ldots, x_n]$ be an ideal. Then*

$$\sqrt{I} = \mathscr{I}(V(I)).$$

We will give several proofs of the Nullstellensatz in this book. For still other proofs, we refer to the book of Eisenbud [Eisenbud 1995]. Hilbert's Nullstellensatz is a generalization of Study's Lemma, see Exercise 2.2.31. Why is it called the Nullstellensatz (German for Theorem of Zeros)? Well, let $I \subsetneq K[x_1, \ldots, x_n]$. Then it follows from the Nullstellensatz that $V(I) \neq \emptyset$: indeed, otherwise $\mathscr{I}(V(I)) = \mathscr{I}(\emptyset) = K[x_1, \ldots, x_n]$, hence $\sqrt{I} = (1)$. Therefore for some n we have $1 = 1^n \in I$ and, therefore, $I = K[x_1, \ldots, x_n]$, a contradiction. Hence from the Nullstellensatz it follows:

Theorem 2.2.7 (Weak Nullstellensatz).
Let $I \subsetneq K[x_1, \ldots, x_n]$ be an ideal. Then $V(I) \neq \emptyset$.

In fact, one is able to show via the so-called Rabinowitch trick, see Exercise 2.2.32, that conversely, the weak Nullstellensatz implies the Nullstellensatz. In this reduction, both the number of variables and the number of equations grow by one! In particular we get an alternative proof of Study's Lemma, by following the lines in this proof of the Nullstellensatz, in which one would have to prove the weak Nullstellensatz for two functions. We will prove the Weak Nullstellensatz by using the Projection Theorem, which we treat now.

Theorem 2.2.8 (Projection Theorem). *Consider a nonzero ideal $I \subset K[x_1, \ldots, x_n]$. Suppose that I contains an element f_0 which is regular in x_n, that is*

$$f_0 = x_n^d + a_1 x_n^{d-1} + \ldots + a_0$$

for certain elements $a_i \in K[x_1, \ldots, x_{n-1}]$. Let $p : K^n \longrightarrow K^{n-1}$ be the map which sends $(x_1, \ldots, x_n)$ to $(x_1, \ldots, x_{n-1})$. Define moreover $I' := I \cap K[x_1, \ldots, x_{n-1}]$, and put $X' = V(I')$. Then

$$p(X) = X'.$$

Proof. The inclusion $p(X) \subset X'$ is the easy inclusion. Indeed if $a \in p(X)$, then there exists a $b \in K$ with $(a,b) \in X$. Let $f \in I'$. Then $f \in I$ and thus $f(a,b) = 0$. But f does not depend on x_n so that $f(a) = 0$.

For the inclusion $X' \subset p(X)$ we will use the Nullstellensatz for the case $n = 1$. This case is easy, as $K[x]$ is a principal ideal domain, and every f has a root, since by assumption K is algebraically closed. Assume that $a \notin p(X)$. We will show that $a \notin X'$. To prove this, it suffices to find an $g \in I'$ with $g(a) \neq 0$. This g is constructed in two steps.

Step 1. First we will show that for all $f \in K[x_1, \ldots, x_n]$ there exists a representation

$$f = g_0 + g_1 x_n + \ldots + g_{n-1} x_n^{d-1} + h_f \tag{2.3}$$

with $h_f \in I$, $g_i \in K[x_1, \ldots, x_{n-1}]$ and $g_i(a) = 0$.

To show this we look at the map

$$\varphi : K[x_1, \ldots, x_n] \longrightarrow K[x_n], \quad f \mapsto f(a, x_n).$$

Now

$$\exists b : b \in V(\varphi(I)) \iff \exists b : f(a,b) = 0 \text{ for all } f \in I \iff a \in p(X),$$

By assumption $a \notin p(X)$, so that $V(\varphi(I)) = \emptyset$. From the Nullstellensatz in one variable we get $\varphi(I) = K[x_n]$. In particular for all f we can find an $h'_f \in I$ with $\varphi(h'_f) = \varphi(f)$. With $g_f := f - h'_f$ we get a decomposition

$$f = g_f + h'_f,$$

such that $h'_f \in I$ and $g_f(a,x_n) = 0$. We divide g_f through f_0. We get

$$g_f = qf_0 + \sum_{j=0}^{d-1} g_j x_n^j.$$

If we plug in (a,x_n) we get $0 = q(a,x_n)f_0(a,x_n) + \sum_{j=0}^{d-1} g_j(a)x_n^j$. By assumption $f_0(a,x_n)$ is a polynomial in x_n of degree d. By the uniqueness statement in division with remainder it follows that $g_j(a) = 0$ for all j. Now we put $h_f := h'_f + qf_0$ and we get the representation (2.3).

Step 2. We apply (2.3) to the functions $1,x_n,\dots,x_n^{d-1}$. We get

$$\begin{array}{ccccccc} 1 & = & g_{0,0} & +\dots+ & g_{0,d-1}x_n^{d-1} & + & h_0 \\ \vdots & & & \vdots & & & \vdots \\ x_n^{d-1} & = & g_{d-1,0} & +\dots+ & g_{d-1,d-1}x_n^{d-1} & + & h_{d-1} \end{array}$$

with $g_{ij}(a) = 0$ for all i,j and $h_j \in I$ for all j. In matrix notation

$$(Id - g_{ij}) \begin{pmatrix} 1 \\ \vdots \\ x_n^{d-1} \end{pmatrix} = \begin{pmatrix} h_0 \\ \vdots \\ h_{d-1} \end{pmatrix}.$$

Let B be the adjoint matrix of $(Id - g_{ij})$. We multiply with B from the left, and it follows from Cramer's rule that

$$\det(Id - g_{ij}) \begin{pmatrix} 1 \\ \vdots \\ x_n^{d-1} \end{pmatrix} = B \begin{pmatrix} h_0 \\ \vdots \\ h_{d-1} \end{pmatrix}. \tag{2.4}$$

We define $g := \det(Id - g_{ij}) \in K[x_1,\dots,x_{n-1}]$. From the first row of (2.4) we see that $g \in I$, so $g \in I'$. As moreover $g_{ij}(a) = 0$, it follows that $g(a) = 1 \neq 0$. This was our goal. □

Proof of the weak Nullstellensatz. The proof is by induction on n. The case $I = (0)$ is obvious. Otherwise there exists an $f_0 \in I$, for which we may, by the Noether normalization for hypersurfaces 2.1.6, assume to fulfill the conditions of the Projection Theorem. Furthermore $I' \neq (1)$, as otherwise $I = (1)$. By induction $V(I') \neq \emptyset$, and thus by the Projection Theorem $p(X)$, and in particular $X = V(I)$ is nonempty. □

For the following theorem we do not need that K is algebraically closed. The statement of the following so-called Noether normalization is a direct generalization of the Noether normalization for hypersurfaces, see 2.1.7.

Theorem 2.2.9 (Noether Normalization). *Let K be a field with infinitely many elements, $A = K[x_1, \dots, x_n]/I$ a finitely generated K-algebra. Then after a general linear coordinate change, there exists a number $r \leq n$, and an inclusion*

$$K[x_1, \dots, x_r] \subset A,$$

such that A is a finitely generated *$K[x_1, \dots, x_r]$-module. If moreover $I \neq (0)$, then $r < n$.*

If $V = V(I)$, we also say that the projection on the first r coordinates

$$\pi : V \longrightarrow K^r$$

is a Noether normalization of V.

Proof. If $I = (0)$, we can take $x_1, \dots, x_r$ to be $x_1, \dots, x_n$. Otherwise the proof goes by induction on n, the case $n = 0$ being trivial. Let $g \in I$ with $g \neq 0$. By Noether normalization for hypersurfaces see 2.1.7, after a general (linear) change of coordinates, we may assume that $K[x_1, \dots, x_n]/(g)$ is a finitely generated $K[x_1, \dots, x_{n-1}]$-module. Because $g \in I$, then certainly $K[x_1, \dots, x_n]/I$ is a finitely generated $K[x_1, \dots, x_{n-1}]/(I \cap K[x_1, \dots, x_{n-1}])$-module. We apply the induction hypotheses to $I \cap K[x_1, \dots, x_{n-1}]$. After a general linear coordinate change in $x_1, \dots, x_{n-1}$ we have that

$$K[x_1, \dots, x_r] \subset K[x_1, \dots, x_{n-1}]/\big(I \cap K[x_1, \dots, x_{n-1}]\big)$$

is finitely generated. Composing with the inclusion

$$K[x_1, \dots, x_{n-1}]/\big(I \cap K[x_1, \dots, x_{n-1}]\big) \subset K[x_1, \dots, x_n]/I,$$

allows us to deduce that $K[x_1, \dots, x_r] \subset K[x_1, \dots, x_n]/I$ is finitely generated. This proves the theorem. □

Corollary 2.2.10. *Let $\pi : V \longrightarrow K^r$ be a Noether normalization. Then π is surjective.*

Proof. This follows by induction from the Projection Theorem 2.2.8. □

Corollary 2.2.11. *Hilbert's Nullstellensatz holds for prime ideals $\mathfrak{p} \subset K[x_1, \dots, x_n]$.*

Proof. As a prime ideal is radical, the Nullstellensatz says that $\mathfrak{p} = \mathscr{I}(V(\mathfrak{p}))$ for a prime ideal $\mathfrak{p}$. The inclusion $(\sqrt{\mathfrak{p}} = \mathfrak{p} \subset \mathscr{I}(V(\mathfrak{p}))$ is easy, see Lemma 2.2.2 (9). For the other inclusion, consider a Noether normalization

$$K[x_1, \dots, x_r] \hookrightarrow K[x_1, \dots, x_n]/\mathfrak{p}.$$

Now let $f \in \mathscr{I}(V(\mathfrak{p}))$ be given. By looking at its class in $K[x_1, \dots, x_n]/\mathfrak{p}$, we see that there exists an integral equation

$$f^s + a_1 f^{s-1} + \dots + a_s = 0 \in K[x_1, \dots, x_n]/\mathfrak{p}, a_i \in K[x_1, \dots, x_r].$$

We take s minimal with this property. From $\mathfrak{p} \subset \mathscr{I}(V(\mathfrak{p}))$, it follows $a_s \in \mathscr{I}(V(\mathfrak{p}))$. This implies that $a_s(p) = 0$ for all $p \in V(\mathfrak{p})$. If $p = (p_1, \dots, p_n)$ then $a_s(p) = a_s(p_1, \dots, p_r)$. Using Corollary 2.2.10 we know that $V(\mathfrak{p}) \longrightarrow K^r$ is surjective and, consequently, for all $(p_1, \dots, p_r) \in K^r$ there exists a $p = (p_1, \dots, p_r, p_{r+1}, \dots, p_n) \in V(\mathfrak{p})$ which implies $a_s(p_1, \dots, p_r) = 0$ for all $(p_1, \dots, p_r) \in K^r$. This implies $a_s \in \mathscr{I}(K^r) = 0$. Hence

$$f(f^{s-1} + a_1 f^{s-2} + \ldots + a_{s-1}) \in \mathfrak{p}.$$

Because $\mathfrak{p}$ is a prime ideal, and s was to be taken minimal, it follows that $f \in \mathfrak{p}$. This is what we had to show. □

To complete our proof of the Nullstellensatz it remains to show the following lemma.

Lemma 2.2.12. *Suppose the Nullstellensatz holds for all prime ideals $\mathfrak{p} \subset K[x_1, \ldots, x_n]$. Then the Nullstellensatz holds for all ideals $I \subset K[x_1, \ldots, x_n]$.*

Proof. Let $I = \mathfrak{q}_1 \cap \ldots \cap \mathfrak{q}_r$ be a primary decomposition of I. Let $\mathfrak{p}_i = \sqrt{\mathfrak{q}_i}$ for all i. In particular $V(\mathfrak{p}_i) = V(\mathfrak{q}_i)$. Then $V(I) = V(\mathfrak{p}_1 \cap \ldots \cap \mathfrak{p}_r) = V(\mathfrak{p}_1) \cup \ldots \cup V(\mathfrak{p}_r)$ by Lemma 2.2.2 (6). Hence by (8) of the same lemma

$$\mathscr{I}(V(I)) = \mathscr{I}(V(\mathfrak{p}_1)) \cap \ldots \cap \mathscr{I}(V(\mathfrak{p}_r)).$$

Thus $\sqrt{I} = \sqrt{\cap_{i=1}^r \mathfrak{q}_i} = \cap_{i=1}^r \sqrt{\mathfrak{q}_i} = \cap_{i=1}^r \mathfrak{p}_i = \cap_{i=1}^r \mathscr{I}(V(\mathfrak{p}_i)) = \mathscr{I}(V(I))$. □

Hilbert's Nullstellensatz can be used to give a geometric interpretation of primary decomposition for radical ideals in a polynomial ring over an algebraically closed field. For this, we need the following definition.

Definition 2.2.13. Let X be a topological space. Then X is called *irreducible* if from $X = X_1 \cup X_2$ with X_1 and X_2 closed subsets of X it follows that either $X_1 = X$ or $X_2 = X$. The space X is called *reducible* if it is not irreducible. An *irreducible decomposition of a topological space* is a decomposition

$$X = X_1 \cup \ldots \cup X_r,$$

where the X_i are irreducible, and for all $i \neq j$, X_i is not contained in X_j.

We will prove that irreducible algebraic sets with the Zariski topology correspond to zero sets of prime ideals. By an application of the Primary Decomposition Theorem it will follow that an affine algebraic set X always has an irreducible decomposition.

Example 2.2.14. Let $X = V(zx, zy) \subset K^2$, with the Zariski topology.

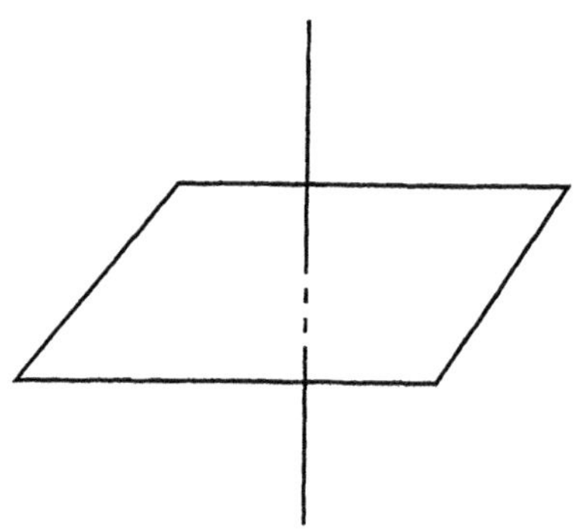

Then $X = X_1 \cup X_2$, where $X_1 = V(x,y)$ and $X_2 = V(z)$. Moreover, none of the X_i, for $i = 1,2$ is contained in the other.

Theorem 2.2.15.

(1) Let $X \subset K^n$ be an irreducible algebraic set. Then $\mathscr{I}(X)$ is a prime ideal.

(2) Conversely let $\mathfrak{p}$ be a prime ideal in $K[x_1, \ldots, x_n]$. Then $V(\mathfrak{p})$ is irreducible.

Proof.
(1) Suppose $f \cdot g \in \mathcal{I}(X)$. Then $V(f) \cup V(g) = V(f \cdot g) \supset V(\mathcal{I}(X)) = X$. The last equality is by 2.2.5. Hence

$$X = (X \cap V(f)) \cup (X \cap V(g)),$$

so that by assumed irreducibility of X either $X = X \cap V(f)$, or $X = X \cap V(g)$. So either $X \subset V(f)$, of $X \subset V(g)$, which means that either $f \in \mathcal{I}(X)$ or $g \in \mathcal{I}(X)$. This is what we had to show.

(2) Suppose $V(\mathfrak{p}) = X_1 \cup X_2$ with X_1 and X_2 closed. Then by the Nullstellensatz $\mathfrak{p} = \mathcal{I}(V(\mathfrak{p})) = \mathcal{I}(X_1) \cap \mathcal{I}(X_2)$. As $\mathfrak{p}$ is prime, either $\mathfrak{p} = \mathcal{I}(X_1)$ or $\mathfrak{p} = \mathcal{I}(X_2)$. Therefore, either $V(\mathfrak{p}) = X_1$ or $V(\mathfrak{p}) = X_2$. □

Theorem 2.2.16. *Let I be a radical ideal and let $I = \mathfrak{p}_1 \cap \ldots \cap \mathfrak{p}_r$ be an irredundant primary decomposition of I. Let $X_i = V(\mathfrak{p}_i)$. Then $X = X_1 \cup \ldots \cup X_r$ is an irreducible decomposition of X. An irreducible decomposition of X is, up to permutation of the X_i, uniquely determined.*

Proof. For the first statement, we have $X = V(I) = V(\mathfrak{p}_1) \cup \ldots \cup V(\mathfrak{p}_r)$. The $V(\mathfrak{p}_i)$ are irreducible by Theorem 2.2.15. Suppose $V(\mathfrak{p}_i) \subset V(\mathfrak{p}_j)$. Then by the Nullstellensatz $\mathfrak{p}_i = \mathcal{I}(V(\mathfrak{p}_i)) \supset \mathcal{I}(V(\mathfrak{p}_j)) = \mathfrak{p}_j$. As the primary decomposition of I is irredundant, it follows that $i = j$. This shows that the X_i are pairwise not contained in each other. Hence, $X = X_1 \cup \ldots \cup X_r$ is a decomposition of X into irreducible components.

To show that an irreducible decomposition is uniquely determined, consider an irreducible decomposition $X = X_1 \cup \ldots \cup X_r$. Let $\mathfrak{p}_i = \mathcal{I}(X_i)$. Then the $\mathfrak{p}_i$ are prime ideals and $\mathcal{I}(X) = \mathfrak{p}_1 \cap \ldots \cap \mathfrak{p}_r$. The irredundancy of the primary decomposition follows from $\mathfrak{p}_j \not\subset \mathfrak{p}_i$ for $i \neq j$, which follows from $X_i \not\subset X_j$. By the Uniqueness Theorems of a Primary Decomposition, a primary decomposition of a radical ideal is uniquely determined. Now $X_i = V(\mathcal{I}(X_i))$ and it follows that the irreducible decomposition is uniquely determined. □

Lemma 2.2.17. *Let $X \subset K^n$ be an irreducible algebraic set, $U \subset X$ an open nonempty subset. Then $\overline{U} = X$. In particular, let $h \in K[x_1, \ldots, x_n]$ be a polynomial with $h \notin \mathcal{I}(X)$. Then $\overline{X \setminus V(h)} = X$.*

Proof. As X is closed the inclusion $\overline{U} \subset X$ is trivial. Suppose the inclusion is proper. Then $\overline{U}$ is closed, and we can write $X = \overline{U} \cup \overline{X \setminus \overline{U}}$. This gives a nontrivial decomposition of X into closed subspaces, a contradiction. Hence the inclusion is an equality. □

Despite the Nullstellensatz, it is often important to consider ideals which are not radical.

Example 2.2.18. Let $I = (y^2, xy) \in K[x,y]$. Then, see 1.4.15, $I = (y) \cap (y^2, x)$ is a primary decomposition. Hence $V(I) = V(y) \cup V(x^2, y)$. Hence $V(I) = \{(x,y) : y = 0\} \cup \{(0,0)\}$. One tries to visualize this by regarding the point $(0,0)$ to be "embedded" in the line $\{(x,y) : y = 0\}$ as the following picture suggests.

This is the reason why, for an ideal I in a Noetherian ring R, the associated primes of I which are not minimal are called *embedded primes*.

The next thing we do is to give a geometric interpretation of the the ideal quotient.

Theorem 2.2.19. *Consider an ideal $I \subset K[x_1,\dots,x_n]$, and let d be an element in $K[x_1,\dots,x_n]$. Let $I = \mathfrak{q}_1 \cap \dots \cap \mathfrak{q}_r$ be an irredundant primary decomposition of I. Then*

$$V(I:d) = \bigcup_{i:d\notin \mathfrak{q}_i} V(\mathfrak{q}_i).$$

Proof. From (the easy) Exercise 1.1.15 it follows that $I : d = (\mathfrak{q}_1 : d)\cap\dots\cap(\mathfrak{q}_r : d)$ so that $V(I : d) = V(\mathfrak{q}_1 : d) \cup \dots \cup V(\mathfrak{q}_r : d)$. Now if $d \in \mathfrak{q}_i$, then $\mathfrak{q}_i : d$ is the whole polynomial ring. In particular $V(\mathfrak{q}_i : d)$ is empty. Suppose, on the other hand, that $d \notin \mathfrak{q}_i$. Then $\sqrt{\mathfrak{q}_i} = \sqrt{\mathfrak{q}_i : d}$ by 1.4.47, so it follows that $V(\mathfrak{q}_i : d) = V(\mathfrak{q}_i)$. From this the theorem follows. □

For radical ideals I the theorem means that $V(I : d)$ picks out the irreducible components of $V(I)$ on which the function d does not vanish identically. In mathematical terms, we can write $V(I : d) = \overline{V(I) \setminus V(d)}$. Here the bar means taking the closure in the Zariski topology. This will be slightly generalized in Exercise 2.2.38.

Examples 2.2.20.

(1) Let $I = (xy,xz,yz) = (x,y)\cap(x,z)\cap(y,z)$. Then I is a radical ideal. Let $d = x$. Then d vanishes on the y and z–axis so that $V(I : d)$ is the x–axis. Indeed $I : d = (y,z)$.

(2) Consider the ideal $I = (x^2,xy)$. Let $d = x$. Then $I : d = (x,y)$. So $V(I : d)$ does consist of the origin. So the statement $V(I : d) = \overline{V(I) \setminus V(d)}$ does not generalize to nonradical ideals.

The final topic in this section is the discussion of localizations. First a definition.

Definition 2.2.21. Let $V \subset K^n$ be an algebraic set

(1) A function $f : V \longrightarrow K$ is called *regular*, if f is the restriction of a polynomial map $f : K^n \longrightarrow K$. The regular functions on V form a K–algebra, which we denote by $K[V]$. It is easy to see, see Exercise 2.2.27, that $K[V]$ is isomorphic to $K[x_1,\dots,x_n]/\mathscr{I}(V)$. The ring $K[V]$ is called the *coordinate ring*, or *ring of regular functions* of V.

(2) Let V be an irreducible algebraic set. A *rational function* on V is an element of $K(V) := Q(K[V])$, the quotient field of $K[V]$.

Note that a rational function on V in general is not a genuine function on V: if $f = \frac{g}{h}$, then, in general, the value of f at a point p with $h(p) = 0$ is not defined.

Lemma 2.2.22.

(1) Let V be an irreducible algebraic set, and p a point of V. Let $\mathfrak{m}_p$ be the maximal ideal of $K[V]$ of regular functions of V vanishing at the point p. Then $K[V]_{\mathfrak{m}_p}$ are those rational functions f on V which have a well-defined value at p.

(2) More generally, let $W \subset V$ be an irreducible subset of V, and $\mathfrak{p} \subset K[V]$ induced by $\mathscr{I}(W)$. Elements of $K[V]_{\mathfrak{p}}$ are those rational functions on V which have a well defined value on a nonempty Zariski open subset of W.

(3) *Let $f \neq 0$ be an element of $K[V]$. Then elements of $K[V]_f$ have a well defined value on $V \setminus V(f)$.*

Proof. Elements f in $K[V]_{\mathfrak{m}_p}$ are of type $f = \frac{g}{h}$ with $h \notin \mathfrak{m}_p$. In particular $h(p) \neq 0$. So $f(p) = \frac{g(p)}{h(p)}$ is a well defined element of K. The other two statements are proved in a similar way. □

Now about localizations for not necessarily irreducible V. First an example.

Example 2.2.23.

Consider the coordinate axes $V(x \cdot y)$, so $K[V] = K[x,y]/(xy)$. Let $\mathfrak{p} = (x)$. Then all elements which are not divisible by x are units in $K[V]_\mathfrak{p}$. In particular y is a unit. Hence $x = 0$ in $K[V]_\mathfrak{p}$, so that $K[V]_\mathfrak{p}$ consists of the rational functions on the y–axis.

This was an example for the following theorem, whose proof is left as an exercise.

Theorem 2.2.24. *Let $X = X_1 \cup \ldots \cup X_r$ be an irreducible decomposition of an algebraic set X, and Y be an irreducible subset of X, defined by the prime ideal $\mathfrak{p}$. Then $K[X]_\mathfrak{p} \cong \oplus_{Y \subset X_i} K[X_i]_\mathfrak{p}$. In particular, if $\mathfrak{p}_i$ is the ideal of X_i then $K[X]_{\mathfrak{p}_i} \cong K(X_i)$, the ring of rational functions on X_i.*

Remark 2.2.25. Let V be an algebraic set, with coordinate ring $K[V]$. Then for any set $F \subset K[V]$ we can define the zero set $V(F)$ which is a subset of V, and for any subset $A \subset V$ we can define the ideal of functions vanishing on A. All the properties of 2.2.2 hold in this more general case, and the proofs are literally the same. Furthermore, the Nullstellensatz holds in this more general case, and follows from the ordinary Nullstellensatz. The proofs are left as Exercise 2.2.27.

Exercises

2.2.26.

(1) Let $(a_1, \ldots, a_n) \in K^n$ be a point. Show that $(x_1 - a_1, \ldots, x_n - a_n)$ is a maximal ideal of $K[x_1, \ldots, x_n]$.
(Hint: Consider the map $K[x_1, \ldots, x_n] \longrightarrow K : f \mapsto f(a_1, \ldots, a_n)$).

(2) Let $\mathfrak{m} \subset K[x_1, \ldots, x_n]$ be a maximal ideal. Prove that the following two statements are equivalent.

  (a) $V(\mathfrak{m}) \neq \emptyset$.
  (b) There exists $(a_1, \ldots, a_n) \in K^n$ such that $\mathfrak{m} = (x_1 - a_1, \ldots, x_n - a_n)$.

(3) Let $V \subset K^n$ be an algebraic set. Prove that there is a one-one correspondence,

$$\text{points of } V \longleftrightarrow \text{ maximal ideals of } K[V],$$

where $K[V] = K[x_1, \ldots, x_n]/\mathscr{I}(V)$. The point $(a_1, \ldots, a_n) \in V$ corresponds to the maximal ideal generated by the classes of $x_1 - a_1, \ldots, x_n - a_n$.

2.2.27.

(1) Show that $K[V]$ is isomorphic to $K[x_1, \ldots, x_n]/\mathscr{I}(V)$.

(2) For a subset $F \subset K[V]$, define $V(F) := \{a \in V : f(a) = 0 \text{ for all } f \in F\}$. $V(F)$ is called the zero set of F. Show that the properties of Lemma 2.2.2 also holds in this more general context. Show that the Nullstellensatz holds in the ring $K[V]$.

(3) Show that V is irreducible if and only if $K[V]$ is an integral domain.

(4) Let $a,b \in V$. Show that there exists a regular function $f \in K[V]$ with $f(a) \neq f(b)$. (Hint: Reduce to the case $V = K^n$.)

2.2.28. We proved hat $V(I+J) = V(I) \cap V(J)$. It is not true in general that $\mathscr{I}(V \cap W) = \mathscr{I}(V) + \mathscr{I}(W)$. Give a counterexample.

2.2.29.

(1) Let $I \subset K[x_1,\ldots,x_n]$ be an ideal, such that the vector space $K[x_1,\ldots,x_n]/I$ has finite dimension s. Show that the set $V(I)$ consists of *at most* s points.
(Hint: Use the fact that an Artinian algebra is the direct sum of local Artinian algebras, see 1.4.26.)

(2) Suppose moreover that I is radical. Show that $V(I)$ consists of exactly s points.

(3) Show that for a Noether normalization $\pi : X \to K^k$, with s the degree of the corresponding field extension $K(x_1,\ldots,x_k) \subset K(X) = Q(K[X])$, the number of points in *a fiber* is *at most* s.

2.2.30.

(1) Prove that $(x,y) \cap (x,z) \cap (y,z)$ is an irredundant primary decomposition of $I = (xy,xz,yz)$.

(2) Prove that $(z,y^2 - x^3) \cap (x,y) = (yz,xz,y^2 - x^3)$.

2.2.31. Prove that Study's Lemma is a consequence of Hilbert's Nullstellensatz.

2.2.32. Show that the Nullstellensatz follows from the weak Nullstellensatz by using the so-called Rabinowitch trick which goes as follows.
Take $f \in \mathscr{I}(V(I))$, and consider the polynomial ring $K[x_1,\ldots,x_n,y]$. Take the ideal J generated by I and $yf-1$. Show that $V(J) = \emptyset$. Apply the weak Nullstellensatz to conclude that $1 \in J$. Then use the fact that modulo J, the equality $y = \frac{1}{f}$ holds.

2.2.33. Consider affine varieties,

$$X = V(f_1,\ldots,f_s) \subset K^n,\ Y = V(g_1,\ldots,g_t) \subset K^m,$$

where $f_1,\ldots,f_s \in K[x_1,\ldots,x_n]$, $g_1,\ldots,g_t \in K[y_1,\ldots,y_m]$. Show that the Cartesian product $X \times Y$ is also an affine variety, as zero set of the ideal

$$(f_1,\ldots,f_s,g_1,\ldots,g_t),$$

where the f_i and g_i are interpreted as elements of $K[x_1,\ldots,x_n,y_1,\ldots,y_m]$.

2.2.34. Suppose $m,n > 0$. Prove that the Zariski topology of $K^n \times K^m$ is not equal to the product of the Zariski topologies of K^n and K^m.

2.2.35. Let V be an algebraic set. Let $\mathfrak{m}_k = (x_1,\ldots,x_k) \subset K[x_1,\ldots,x_k]$, and let $\mathfrak{m}$ be a maximal ideal of $K[V]$. A Noether normalization of $K[V]_\mathfrak{m}$, by definition is an inclusion

$$K[x_1,\ldots,x_k]_{\mathfrak{m}_k} \subset K[V]_\mathfrak{m}$$

for some k, such that $K[V]_\mathfrak{m}$ is a finitely generated $K[x_1,\ldots,x_k]_{\mathfrak{m}_k}$-module. Show that after a general coordinate change, a Noether normalization of $K[V]_\mathfrak{m}$ exists.

2.2.36.

(1) Let V be an irreducible topological space, $f : V \longrightarrow W$ be continuous. Show that $f(V)$ is irreducible.

(2) Let $V \subset W$ be topological spaces. Suppose that V is irreducible. Show that the closure $\overline{V}$ of V in W is also irreducible.

2.2.37. In this exercise we look at the Going-Up Theorem in the special case that $\mathfrak{p}$ is a maximal ideal.

(1) Let $R \subset S$ be an integral extension of integral domains. Show that R is a field if and only if S is a field.

(2) Let $R \subset S$ be an integral extension of rings. Let $\mathfrak{m}$ be a maximal ideal in R, and $\mathfrak{q} \subset S$ be a prime ideal with $\mathfrak{q} \cap R = \mathfrak{m}$. Prove that $\mathfrak{q}$ is a maximal ideal.

Give geometric interpretations of these statements.

2.2.38. Let I,J be ideals in the polynomial ring $K[x_1, \dots ,x_n]$. Suppose I is radical. Prove that $V(I : J) = \overline{V(I) \setminus V(J)}$.

2.2.39. Let $Y \subset V$ be algebraic sets. Show that $\mathscr{I}(Y) = \bigcap_{p \in Y} \mathfrak{m}_p$.

2.3 Maps between Algebraic Sets

In this section we look at the morphisms we allow between algebraic sets. As an algebraic set is defined by polynomial equations, it is a natural idea to allow between them only *polynomial mappings.* The simple case of regular functions, that is, regular maps $V \longrightarrow K$ has already been considered in 2.2.21.

Definition 2.3.1. Let $V \subset K^n$ and $W \subset K^m$ be affine varieties. A map $\varphi : V \longrightarrow W$ is called a *regular map* between the affine varieties V and W if φ is the restriction of a polynomial map $\varphi : K^n \longrightarrow K^m$. Regular maps $V \longrightarrow K$ are called *regular functions.*

This means the following: suppose $(x_1, \dots ,x_n)$ are coordinates on K^n, and $(y_1, \dots ,y_m)$ are coordinates on K^m. Then there exist functions $\varphi_1, \dots ,\varphi_m \in K[x_1, \dots ,x_n]$, such that the point with coordinates $(x_1, \dots ,x_n) \in V$ is sent to the point with coordinates $(y_1, \dots ,y_m) = (\varphi_1(x_1, \dots ,x_n), \dots ,\varphi_m(x_1, \dots ,x_n)) \in W$.

Definition 2.3.2. An *isomorphism* $\varphi : V \longrightarrow W$ is a regular map which has a two-sided inverse. Two affine varieties V,W are called *isomorphic* if there exists an isomorphism $\varphi : V \longrightarrow W$ between them. Of course, being isomorphic is an equivalence relation.

It is easy to check that the composition of two polynomial maps is a polynomial map. This in particular gives the following. Let $\varphi : V \longrightarrow W$ be a regular map. Let $f \in K[W]$, that is, $f : W \longrightarrow K$ be a regular function. Then the composition $f \circ \varphi : V \xrightarrow{\varphi} W \xrightarrow{f} K$ is a regular function too. Hence we get a map

$$\varphi^* : K[W] \longrightarrow K[V]; \quad f \mapsto \varphi \circ f$$

which one checks to be a K-algebra homomorphism. Note that if moreover $\psi : W \longrightarrow X$, we have the property $(\psi \circ \varphi)^* = \varphi^* \circ \psi^*$. So we assign to each regular map a K-algebra homomorphism. There is a converse to this.

Theorem 2.3.3. *Suppose a K-algebra homomorphism $\alpha : K[W] \longrightarrow K[V]$ is given. Then there is a unique regular map $\varphi : V \longrightarrow W$ with $\varphi^* = \alpha$.*

Proof. We write

$$K[V] = K[x_1, \dots ,x_n]/\mathscr{I}(V), \quad K[W] = K[y_1, \dots ,y_m]/\mathscr{I}(W).$$

The composition $K[y_1, \dots, y_m] \longrightarrow K[W] \xrightarrow{\alpha} K[V]$ we also denote by α. Consider the "coordinate" functions $y_1, \dots, y_m \in K[y_1, \dots, y_m]/\mathscr{I}(W)$. Then $\alpha(y_i) = \overline{\varphi}_i(x_1, \dots, x_n) \in K[V]$ for some $\overline{\varphi_i}$. Take lifts $\varphi_i \in K[x_1, \dots, x_n]$ of $\overline{\varphi_i}$. Then $\varphi := (\varphi_1, \dots, \varphi_m)$ is a polynomial map $\varphi : K^n \longrightarrow K^m$. As both α and φ^* are K–algebra maps, and they agree on y_i, it follows that they agree on all polynomials in the y_i. Hence $\alpha = \varphi^* : K[y_1, \dots, y_m] \longrightarrow K[V]$. So, uniqueness is clear. As different lifts differ by elements in $\mathscr{I}(V)$ the restriction of φ to V is well-determined. We need to show

$$\varphi(V) \subset W.$$

Take $p \in V$, and consider $\varphi(p)$. As $W = V(\mathscr{I}(W))$ by 2.2.5 it suffices to show $f(\varphi(p)) = 0$ for all $f \in \mathscr{I}(W)$. Now $f(\varphi(p)) = f \circ \varphi(p) = \varphi^*(f)(p) = \alpha(f)(p)$. As $\alpha : K[W] \longrightarrow K[V]$, it maps all functions in $\mathscr{I}(W)$ to zero. Hence $\alpha(f)$ is the zero function, in particular $\alpha(f)(p) = 0$. □

The following follows immediately.

Corollary 2.3.4. *Algebraic sets V and W are isomorphic if and only if $K[V]$ and $K[W]$ are isomorphic as K-algebras.*

As K–algebra homomorphisms between affine rings correspond to regular maps of the corresponding varieties, one would like to translate certain properties of K–algebra homomorphisms into properties of the corresponding regular map. The following theorem gives a first example of this principle.

Theorem 2.3.5. *Let algebraic sets V,W, algebraic subsets $i : X \hookrightarrow V$, $j : Y \hookrightarrow W$, and a regular map $\varphi : V \longrightarrow W$ be given. Then*

(1) $\varphi(X) \subset Y \iff \varphi^*(\mathscr{I}(Y)) \subset \mathscr{I}(X)$,

(2) the fiber $\varphi^{-1}(Y)$ is the zero set of the set $\varphi^*(\mathscr{I}(Y))$,

(3) a regular map is continuous in the Zariski topology.

Proof. (1) As $\mathscr{I}(X) \supset \mathscr{I}(V)$, we get a canonical surjection $i^* : K[V] \longrightarrow K[X]$ which is just restriction of regular maps. Similarly, we have a canonical surjection $j^* : K[W] \longrightarrow K[Y]$. Now if $\varphi^*(\mathscr{I}(Y)) \subset \mathscr{I}(X)$, we get an induced map $\overline{\varphi}^* : K[Y] \longrightarrow K[X]$, with $i^* \circ \varphi^* = \overline{\varphi}^* \circ j^*$ By the previous theorem we get a map $\overline{\varphi} : X \longrightarrow Y$, such that $j \circ \overline{\varphi} = \varphi \circ i$. This exactly means that $\varphi(X) \subset Y$. If, on the other hand φ sends X to Y, that is, we have a map $\overline{\varphi} : X \longrightarrow Y$, we get an induced map $\overline{\varphi}^* : K[Y] \longrightarrow K[X]$, so a commutative diagram

$$\begin{array}{ccc} K[W] & \xrightarrow{\varphi^*} & K[V] \\ \downarrow{\scriptstyle i^*} & & \downarrow{\scriptstyle j^*} \\ K[Y] & \xrightarrow{\varphi^*} & K[X]. \end{array}$$

It follows that $\varphi^*(\mathscr{I}(Y)) \subset \mathscr{I}(X)$.

(2) We apply the first part to $X = \varphi^{-1}(Y)$. We get

$$\varphi^*(\mathscr{I}(Y)) \subset \mathscr{I}(\varphi^{-1}(Y)),$$

so that, by taking zero sets, $V(\varphi^*(\mathscr{I}(Y))) \supset \varphi^{-1}(Y)$. To prove the other inclusion, let $p \in V(\varphi^*(\mathscr{I}(Y))$. Then $\mathfrak{m}_p = \mathscr{I}(p) \supset \mathscr{I}(V(\varphi^*(\mathscr{I}(Y))) \supset \varphi^*(\mathscr{I}(Y))$. Therefore $\varphi(p) \in Y$ follows from the first part applied to $X = \{p\}$. Hence $p \in \varphi^{-1}(Y)$ as was to be proved.

(3) From the second part it follows that inverse images of Zariski closed sets are Zariski closed. Thus, φ is continuous. □

Example 2.3.6. Let $V = V(I)$, where $I = (y^2 - x) \subset K[x,y]$. Consider the projection $\varphi : V \longrightarrow W = K$, the projection on the x–axis. Then the map φ^* is just the inclusion map $\varphi^* : K[x] \subset K[x,y]/(y^2 - x)$. We take the subset Y of $W = K$ defined by $x(x-1) = 0$, so consisting of the points 0 and 1. Hence the fiber is defined by the ideal $(x(x-1), y^2 - x))$ in K^2 so consists of three points, namely $(0,0)$,$(1,1)$ and $(1, -1)$.

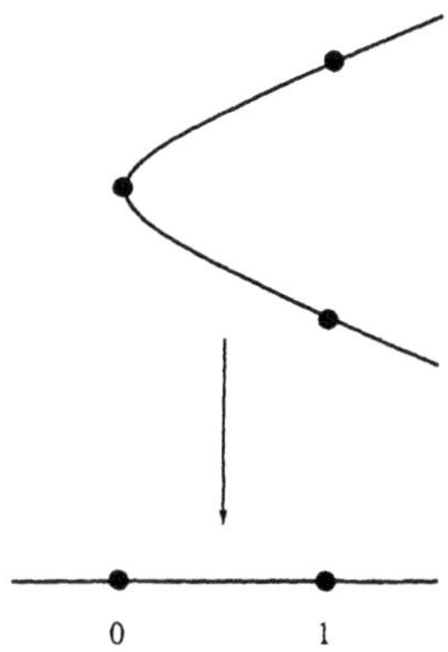

Lemma 2.3.7. *Let a regular map $\varphi : V \longrightarrow W$ be given, and let $\varphi^* : K[W] \longrightarrow K[V]$ be the induced map on coordinate rings. Then*

$$\varphi \text{ is injective } \iff \varphi^* \text{ is surjective.}$$

Proof. First suppose that φ^* is surjective. Suppose that $a,b \in V$ with $\varphi(a) = \varphi(b)$. Then for all $g \in K[W]$, it follows that $\varphi^*(g)(a) = g \circ \varphi(a) = g \circ \varphi(b) = \varphi^*(g)(b)$. Suppose $a \neq b$. Choose an $f \in K[V]$ with $f(a) \neq f(b)$. This we can do by Exercise 2.2.27. Because φ^* is surjective, there is a $g \in k[W]$ with $f = \varphi^*(g)$. This is a contradiction.

For the converse, suppose φ is injective. Hence V can be viewed as a subset of W: $V \subset W \subset K^n$. Hence $\mathscr{I}(V) \supset \mathscr{I}(W)$, so that $K[W] = K[x_1, \ldots, x_n]/\mathscr{I}(W) \longrightarrow k[V] = K[x_1, \ldots, x_n]/\mathscr{I}(V)$ is surjective. □

The corresponding statement with injective and surjective interchanged, is not true, by the following example which we have seen before.

Example 2.3.8. Consider $V = \{(x,y) \in K^2 : xy - 1 = 0\}$, let W be the x–axis, and $\varphi : V \longrightarrow W$ be the projection. Then φ^* is the inclusion $K[W] = K[x] \subset K[V] = K[x,y]/(xy - 1) = K[x,x^{-1}]$. However, φ is not surjective: it misses the point 0.

Note that in this example, $K[V]$ is not a finitely generated $K[W]$–module. If we insist however on this, the converse holds. The following therefore is a generalization of the fact that for a Noether normalization $\pi : V \longrightarrow K^r$, the map π is surjective, and that the fibers consist of finitely many points. The proof is left as Exercise 2.3.17.

Theorem 2.3.9. *Let $\varphi : V \longrightarrow W$ be a regular map, with $\varphi^* : K[W] \longrightarrow K[V]$ injective, and such that $K[V]$ is a finitely generated $K[W]$-module. Then φ is surjective, and the fibers consist of finitely many points.*

Note that although in Example 2.3.8, the map φ is not surjective, it is at least "almost" surjective. To make this more precise, we need to understand localizations of affine rings.

Lemma 2.3.10. *Let $V \subset K^n$ be an algebraic set, with coordinate $K[V]$. Take $0 \neq f \in K[V]$. Then $V \setminus V(f) \subset V$ is homeomorphic in the Zariski topology to an algebraic set X, whose coordinate ring $K[X]$ is the localization $K[V]_f$.*

Proof. Let $K[V] = K[x_1, \dots, x_n]/I$. Take a new coordinate y, and consider the set $X := V(I, y \cdot f - 1) \subset K^{n+1}$. The coordinate ring of X is $K[x_1, \dots, x_n, y]/(I, y \cdot f - 1)$ which is as K-algebra, by sending y to $\frac{1}{f}$, isomorphic to $K[V]_f$. The projection $\pi : X \longrightarrow K^n$ on the first n coordinates is easily seen to be a one to one map of X onto its image $V \setminus V(f)$. The map $\pi : X \longrightarrow V \setminus V(f)$ is regular, hence continuous. To show that π is a homeomorphism, we have to show that a closed subset in X is sent to a closed subset in $V \setminus V(f)$. It suffices to show this for irreducible sets Y in X, corresponding to a prime ideal $\mathfrak{p}_f$ in $K[V]_f$. By 1.3.15 $\mathfrak{p}_f \cap K[V] = \mathfrak{p}$ is a prime ideal in $K[V]$, and conversely a prime ideal in $K[V]$ with $f \notin \mathfrak{p}$ extends to a prime ideal in $K[V]_f$. We claim that the image of $V(\mathfrak{p}_f)$ is equal to $V(\mathfrak{p}) \setminus V(f)$, which would prove the lemma. Indeed, let $a \in V(\mathfrak{p}) \setminus V(f)$. Then for all $g \in \mathfrak{p}$, and all $k \in \mathbb{N}$ we have $\frac{g}{f^k}(a) = 0$, as $g(a) = 0$, and $f(a) \neq 0$. As $\mathfrak{p}_f$ consists of such elements it follows that $a \in V(\mathfrak{p}_f)$. On the other hand, if $a \in V(\mathfrak{p}_f)$, then $\frac{g}{f^k}(a) = 0$ for all $g \in \mathfrak{p}$ and $k \in \mathbb{N}$. Hence $g(a) = 0$, and therefore $a \in V(\mathfrak{p})$. This proves the lemma. □

Definition 2.3.11. Consider irreducible affine varieties, and $\varphi : V \longrightarrow W$ a regular map. Then φ is called *dominant* if $\varphi(V)$ contains a nonempty Zariski open subset U. So $\emptyset \neq U \subset \varphi(V)$.

Theorem 2.3.12. *Let V, W be irreducible affine varieties, and $\varphi : V \longrightarrow W$ be a regular map. Then the following conditions are equivalent:*

(1) $\varphi^* : K[W] \longrightarrow K[V]$ is injective.

(2) φ is dominant.

(3) $\overline{\varphi(V)} = W$.

Proof.
The implication (2) $\Longrightarrow$ (3) is easy, see Exercise 2.2.36.
(3) $\Longrightarrow$ (1). Take $f, g \in K[W]$. Then we have the following two equivalences.

$$f = g \iff f_{|\varphi(V)} = g_{|\varphi(V)} \iff f \circ \varphi = g \circ \varphi.$$

The second equivalence is clear. For the first, the only not obvious implication is $\Longleftarrow$. As regular functions are continuous, it follows that $(f - g)^{-1}(0)$ is closed. By assumption, this set contains $\varphi(V)$, and therefore contains $\overline{\varphi(V)} = W$. Hence $f = g$ on the whole of W.

The implication (1) $\Longrightarrow$ (2) is the "difficult" case. We have to prove that $W \smallsetminus V(f) \subset \varphi(V)$ for some $0 \neq f \in K[W]$. The idea is to find a nonzero $f \in K[W] \subset K[V]$ and an r such that there exists an inclusion $K[W]_f[x_1, \dots ,x_r] \subset K[V]_f$, such that $K[V]_f$ is a finitely generated $K[W]_f[x_1, \dots ,x_r]$-module. Indeed, this inclusion gives

- a finite map, hence surjective map $\widetilde{\varphi} : V \setminus V(f) \longrightarrow \Big(W \smallsetminus V(f)\Big) \times K^r$,
- a projection map $\pi : \Big(W \smallsetminus V(f)\Big) \times K^r \longrightarrow W \smallsetminus V(f)$, which is obviously surjective,

such that $\varphi = \pi \circ \widetilde{\varphi}$. Hence, the image of φ would contain $W \smallsetminus V(f)$, which is a nonempty Zariski open subset of W.

Now we will find the f and r. For simplicity we write $R = K[W]$, and $S = K[V]$. By assumption we have an inclusion $R \subset S$. Consider the following diagram:

$$\begin{array}{ccc} R & \hookrightarrow & S \\ \downarrow & & \downarrow \\ E = Q(R) & \hookrightarrow & S_{R \smallsetminus \{0\}} \subset Q(S). \end{array}$$

Obviously, $S_{R\smallsetminus\{0\}}$ is a finitely generated E–algebra, and E is a field. In this situation we can apply the Noether Normalization Theorem, see Theorem 2.2.9. There exist $x_1, \dots ,x_r \in S_{R\smallsetminus\{0\}}$ such that

$$E[x_1, \dots ,x_r] \subset S_{R\smallsetminus\{0\}},$$

and $S_{R\smallsetminus\{0\}}$ is finitely generated as $E[x_1, \dots ,x_r]$–module. By clearing denominators, we can achieve that the x_i are elements of S, and $S_{R\smallsetminus\{0\}}$ is still a finitely generated $E[x_1, \dots ,x_r]$–module. Now consider

$$R[x_1, \dots ,x_r] \subset S.$$

This need not be a finitely generated extension. But we can do the following. Take generators $\alpha_1, \dots ,\alpha_s$ of $S_{R\smallsetminus\{0\}}$ as $E[x_1, \dots ,x_r]$–algebra. Again we may assume that in fact α_i in S. The elements α_i satisfy an integral equation with coefficients in $E[x_1, \dots ,x_r] = Q(R)[x_1, \dots ,x_r]$. Let f be the product of all the denominators that appear as coefficients in all such equations. Therefore, over $R_f[x_1, \dots ,x_r]$ every α_i is integral. We therefore forced that the extension

$$R_f[x_1, \dots ,x_r] \subset S_f$$

is integral. This was our goal. □

Theorem 2.3.13. *Consider an ideal* $I \subset K[x_1, \dots ,x_n,y_1, \dots ,y_m]$, *and let* $V = V(I) \subset K^{n+m}$. *Let* $p : K^{n+m} \longrightarrow K^n$ *be the projection on the first* n *factors. Then*

$$\overline{p(V)} = V(I \cap K[x_1, \dots ,x_n]).$$

The closure is taken here in the Zariski topology.

Proof. To show the inclusion $\overline{p(V)} \subset V(I \cap K[x_1, \dots ,x_n])$, it suffices to show $p(V) \subset V(I \cap K[x_1, \dots ,x_n])$, as the right hand side is closed. So let $a \in p(V)$. Hence there exists a $b \in K^m$ with $(a,b) \in V$. Let $f \in I \cap K[x_1, \dots ,x_n]$ then $f(a,b) = 0$. But f does not depend on the variables $y_1, \dots ,y_m$ at all, and therefore $f(a) = 0$. This holds for all $f \in I \cap K[x_1, \dots ,x_n]$ so that the inclusion $\subset$ follows.

For the converse inclusion consider

$$K[x_1,\dots,x_n]/(I\cap K[x_1,\dots,x_n])\subset K[x_1,\dots,x_n,y_1,\dots,y_m]/I,$$

so that $\overline{p(V)}$ is indeed given by the zero set of the ideal $I\cap K[x_1,\dots,x_n]$ by 2.3.12. □

Remark 2.3.14. Let $\varphi=(\varphi_1,\dots,\varphi_m):V\longrightarrow W\subset K^m$ be a regular map, $V=V(I)$. Consider the graph of φ:

$$\mathrm{Graph}(\varphi)=\{(x,\varphi(x)):x\in V\}\subset V\times W\subset V\times K^m.$$

Then the zero set of

$$(y_1-\varphi_1,\dots,y_m-\varphi_m)\subset K[V]\otimes K[W]$$

is the graph of φ. The graph of φ is canonically isomorphic to V: in the ideal $(y_1-\varphi_1,\dots,y_m-\varphi_m)$ one eliminates all y–variables, so that $\big(K[V]\otimes K[W]\big)/(y_1-\varphi_1,\dots,y_m-\varphi_m)$ is isomorphic to $K[V]$. Note that with this identification the map of K–algebras

$$K[W]\longrightarrow \big(K[V]\otimes K[W]\big)/(y_1-\varphi_1,\dots,y_m-\varphi_m)$$

gives the map $\mathrm{Graph}(\varphi)\longrightarrow W$, which maps $(x,\varphi(x))$ to $\varphi(x)$. We come to the conclusion that every map $\varphi:V\longrightarrow W$ can be viewed as the restriction of a projection mapping. This is a point of view which is sometimes useful. For example its local version, see 3.4.43 will be used in the proof of the Finite Mapping Theorem, see 6.3.5.

Example 2.3.15. Consider the regular map $\varphi:K\longrightarrow K^2$ given by $\varphi(t)=(t^2,t^3)$. Let the coordinates of K^2 be x,y. Then the graph of φ lies in K^3 and is the zero set of the ideal $I=(x-t^2,y-t^3)$ So the closure of the image of f is the zero set of the ideal $I'=(x-t^2,y-t^3)\cap K[x,y]$ by 2.3.14 and 2.3.13. One shows that $I'=(y^2-x^3)$. The proof of this is contained in Example 1.5.6.

Finally, we give a counterexample to the Going-Down Theorem for nonnormal rings.

Example 2.3.16. Let K be an algebraically closed field. Consider the map

$$\begin{array}{lcl}\varphi:K^2 & \longrightarrow & K^3\\ (t,s) & \longmapsto & (x,y,z)=(t^2-1,t^3-t,s).\end{array}$$

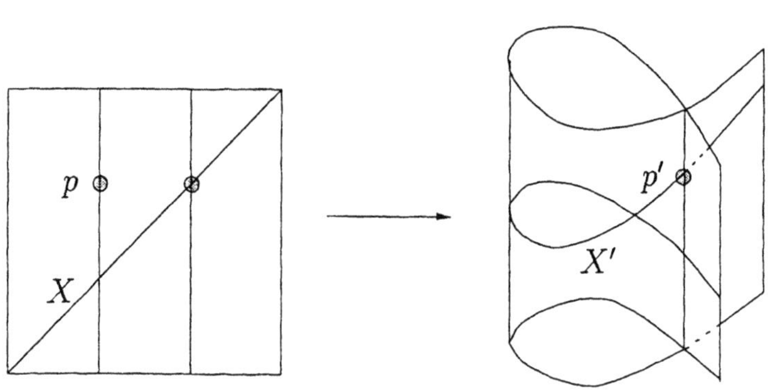

Then one shows that the image of φ is exactly given by $y^2 - x^2(x+1) = 0$. Moreover, via φ^*, the module $S := K[t,s]$ is a finitely generated $R := K[x,y,z]/(y^2 - x^2(x+1))$-module. Indeed, from the definition of the map it follows that $t^2 - 1 - x = 0$, hence t is integral. Notice, moreover, that $K[x,y,z]/(y^2 - x^2(x+1))$ is not a normal ring; the element $t = \frac{y}{x}$ is an integral element.

We now show that the Going-Down Theorem 1.5.26 does not hold for the ring extension $R \subset S$. Consider the prime ideal $\mathfrak{P} := (s-t)$ in S (in the picture $X = V(\mathfrak{P}) \subset K^2$), and the prime ideal $\mathfrak{p} := \mathfrak{P} \cap R$ in R (in the picture $X' = V(\mathfrak{p}) \subset K^3$). One easily calculates that $\mathfrak{p} = (x - z^2 + 1, y - xz)$. Now we claim that $\mathfrak{P}$ is the unique prime ideal in S which lies over $\mathfrak{p}$. This should be clear geometrically, but let us give a precise algebraic argument. Indeed, such a prime ideal $\mathfrak{P}$ contains at least $t^2 - 1 - s^2 + 1 = t^2 - s^2$, corresponding to $x - z^2 + 1$, the first generator of $\mathfrak{P}$. Hence, either $t - s \in \mathfrak{P}$, or $t + s \in \mathfrak{P}$. If $t - s \in \mathfrak{P}$, we are done. Therefore, suppose $t + s \in \mathfrak{P}$. As $y - xz \in \mathfrak{p}$, it follows that $t^3 - t - s(t^2-1) = 2t^3 - 2t + (1-t^2)(s+t) \in \mathfrak{P}$. Hence $\mathfrak{P}$ is either $(t-1, s+1)$ or (t,s) or $(t+1, s-1)$. This cannot be the case because these are all maximal and $\mathfrak{P}$ is not, see Exercise 2.2.37.

Now consider the maximal ideal $\mathfrak{q} = (x, y, z-1)$ of R, corresponding to the point $p' = (0,0,1)$ in the picture. There are two points in the preimage of $p' = (0,0,1)$ namely $(1,1)$ and $p = (-1,1)$. In particular the maximal ideal $\mathfrak{Q} : (t+1, s-1)$ corresponding to p is such that $\mathfrak{Q} \cap R = \mathfrak{q}$. However, $\mathfrak{P} \not\subset \mathfrak{Q}$, as otherwise $p \in X$. So we conclude that the statement of the Going-Down Theorem does not hold in this case.

Exercises

2.3.17.

Prove Theorem 2.3.9.
(Hint: Either use Theorem 2.3.5 to describe the fibers of y, and Exercise 2.2.37, or use the Graph Construction 2.3.14 and the Projection Theorem 2.2.8.)

2.3.18.

(1) Consider the map $f : K^2 \longrightarrow K^4$ given by $f(s,t) = (s, st, t^2, t^3) = (x,y,z,u)$. Show that the ideal of the image is equal to $I' = (y^2 - zx^2, yz - ux, z^2x - uy, u^2 - z^3)$.
(Hint: Show that any g is modulo I' equivalent to $ay + bu + c$ with $a,c \in K[x,z]$ and $b \in K[x]$.)

(2) Consider the map $f : K \longrightarrow K^3$, $f(t) = (t^3, t^4, t^5) = (x,y,z)$. Show that the image of f is given by the zero set of the ideal $(y^2 - xz, x^3 - yz, z^2 - x^2y)$.

(3) Use a computer algebra system, for example [Singular 2000] to check the results in (1) and (2).

3 Basics of Analytic Geometry

In this chapter we start studying local analytic geometry, that is, the zero sets of analytic functions in a (small) neighborhood of a point. To see why we want to do so, we look at the affine hypersurface in $\mathbb{C}^2$ defined by $f(x,y) = y^2 - x^2(x+1) = 0$. In a small neighborhood U of $(0,0)$ we see two "parts", or components of $f(x,y) = 0$.

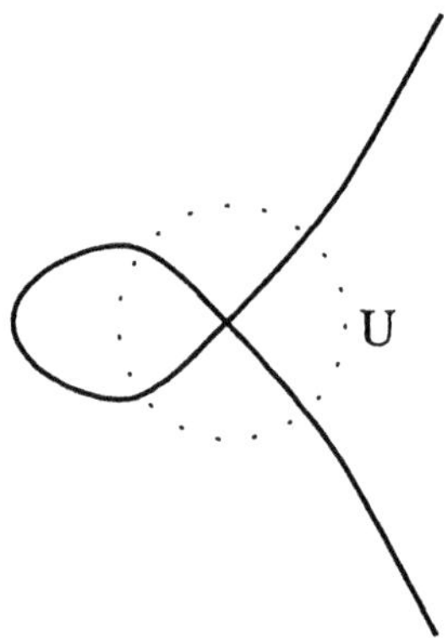

So we would like to say that the "singularity" consists of two components. But how small should we take U? The answer to this question is: as small as we need to. To make a precise notion, we define the notion of a germ of a space. A germ of a topological space X at a point p by definition is an equivalence class of open neighborhoods of p. But f, considered as element of $\mathbb{C}[x,y]$ is irreducible (Exercise 1.4.34), but reducible if considered as an element of $\mathbb{C}\{x,y\}$. Here $\mathbb{C}\{x,y\}$ is the ring of power series which converge in some neighborhood of 0.[1] Indeed, we have the following factorization in $\mathbb{C}\{x,y\}$

$$f(x,y) = \left(y - x\sqrt{x+1}\right)\left(y + x\sqrt{x+1}\right).$$

So we see that in a neighborhood on which $\sqrt{x+1}$ converges, the zero set $f = 0$ decomposes into two parts, namely $\{(x,y) : y = x\sqrt{x+1}\}$ and $\{(x,y) : y = -x\sqrt{x+1}\}$.

As we assume that the reader heard a course on holomorphic functions in one variable, we will, in Section 3.1, treat only the basic facts on holomorphic functions in several variables. Many of the statements and the proofs are similar or direct generalizations of the one variable case.

For polynomials $f \in \mathbb{C}[x_1, \ldots, x_n]$ one of the main invariants is the degree. It can be interpreted (in case f is reduced) as the number of intersection points of the set $\{p \in \mathbb{C}^n : f(p) = 0\}$ with a general line. By applying the Noether Normalization Theorem, one may assume that this line is given by $x_1 = a_1, \ldots, x_{n-1} = a_{n-1}$. In case of convergent power series f, its replacement is the *multiplicity* or *order* of the power series, and is defined as the *lowest* order term of f. Assuming that f converges on the (small enough) neighborhood U of 0, the multiplicity can be interpreted as the maximal number of intersection points of $f = 0$ with a general line L in U, in case f is reduced. This

[1] The neighborhood might depend on the power series.

statement follows from the so-called continuity of roots. So for example the element $f(x,y) = y^2 - x^2(x+1)$ has degree three considered as element of $\mathbb{C}[x,y]$, and has multiplicity two considered as element of $\mathbb{C}\{x,y\}$. There is a substitute for the Noether normalization for the case $f \in \mathbb{C}\{x_1,\dots,x_n\}$, so that we can always assume that the line L from above points in the x_n direction. In terms of the power series f of multiplicity b, this is equivalent to saying that the term x_n^b occurs with nonzero coefficient in the power series expansion of f. In this case one says that f is regular of order b in x_n.

The replacement for division with remainder, which played an important role in the proof of Study's Lemma, or more generally the Nullstellensatz, is the Weierstraß Division Theorem. The Weierstraß Division Theorem says the following: Let a convergent power series $f \in \mathbb{C}\{x_1,\dots,x_n\}$ be regular of order b in x_n, and a further power series $g \in \mathbb{C}\{x_1,\dots,x_n\}$ be given. Then there exist uniquely defined $q \in \mathbb{C}\{x_1,\dots,x_n\}$ and $r \in \mathbb{C}\{x_1,\dots,x_{n-1}\}[x_n]$, where r has, considered as polynomial in x_n, degree less than b, such that

$$g = q \cdot f + r.$$

The similarity with ordinary division with remainder is obvious. It follows from the Weierstraß Division Theorem, that the $\mathbb{C}\{x_1,\dots,x_{n-1}\}$–module $\mathbb{C}\{x_1,\dots,x_n\}/(f)$ is finitely generated.[2] As the ring $\mathbb{C}\{x_1,\dots,x_{n-1}\}$ is local, it even follows by an application of Nakayama's Lemma, that the Weierstraß Division Theorem follows from the fact that $\mathbb{C}\{x_1,\dots,x_n\}/(f)$ is a finitely generated $\mathbb{C}\{x_1,\dots,x_{n-1}\}$–module. In Section 3.2 we will prove a weak generalization of this seemingly weaker statement. If one looks closely at the proof, we will "only" show the existence of q and r where r has degree less than or equal to b. It turns out that in order to show the convergence of q and r, it is much easier to prove this last statement.

In Section 3.3, we give applications of the Weierstraß Theorems. To mention the most important ones, we show the Implicit and Inverse Function Theorems and Newton's Lemma (a "cooked-up" version of the Implicit Function Theorem). Furthermore, we show that the (formal) power series ring is Noetherian and factorial, prove the Noether Normalization Theorem, and Hensel's Lemma.

With this knowledge at hand we can start the study of germs of analytic spaces in Section 3.4. The main theorem, and the most difficult in this section is the Nullstellensatz. In the case of germs of analytic spaces, this theorem cannot be interpreted as a statement on maximal ideals. We will prove the analogous statement to Corollary 2.2.10. That is, let a germ of an analytic space (X,x) be given for which the ideal of (X,x) is prime, and let $\pi : (X,x) \longrightarrow (\mathbb{C}^k,y)$ be a Noether normalization. Then π is surjective. As in the affine case, the Nullstellensatz follows from this surjectivity. An essential ingredient in the proof of the surjectivity of π is the Theorem of Finiteness of Normalization 1.5.19. It will be used to describe the properties of a general projection of an irreducible germ (X,x) onto a hypersurface singularity. The precise properties of this projection of X onto a hypersurface are gathered in the *Local Parametrization Theorem.* This is a very important theorem, as it, in many cases, allows the reduction of general statements on germs of analytic spaces to the case of hypersurfaces, which are sometimes easier to handle. The Local Parametrization Theorem will be used several times in the next chapter.

[2] This module is even free.

3.1 Holomorphic Functions of Several Complex Variables

Before going on with geometry, we need some basic facts on holomorphic functions of several complex variables. The facts we mention here are either proved, or the proof can be easily generalized from the case of a function in one variable. In this case, no proof is given.

Definition 3.1.1. Let $n \in \mathbb{N}$, and ν_i, for $i = 1, \ldots, n$ be nonnegative integers, $x = (x_1, \ldots, x_n) \in \mathbb{C}^n$. We introduce the *multi index notation*

$$\nu := (\nu_1, \ldots, \nu_n); \; x^\nu := \prod_{i=1}^{n} x_i^{\nu_i}; \; |\nu| := \sum_{i=1}^{n} \nu_i.$$

The number $|\nu|$ is called the *degree* of ν. Let $r = (r_1, \ldots, r_n) \in \mathbb{R}_+^n$, $r_i > 0$ for $i = 1 \ldots, n$. The set

$$\overline{P}_r := \{x \in \mathbb{C}^n : |x_i| \leq r_i \text{ for } i = 1, \ldots, n\}$$

is called the *closed polydisc* (or *polycylinder*) with center $\{0\}$ and polyradius r. Similarly one defines open polydiscs P_r, and polydiscs with center $p \in \mathbb{C}^n$.

Definition 3.1.2. Let $U \subset \mathbb{C}^n$ be a domain[3], $f : U \longrightarrow \mathbb{C}$ a function, $p \in U$. Then f is called *complex differentiable* in p, if there exists an open neighborhood $V \subset U$ of p and functions $\Delta_1, \ldots, \Delta_n : V \longrightarrow \mathbb{C}$ which are continuous in p such that for all $x \in V$

$$f(x) = f(p) + \sum_{i=1}^{n} (x_i - p_i)\Delta_i(x).$$

f is called *holomorphic* in U if f is complex differentiable in p for all $p \in U$.

In this case it follows without much difficulty, by reducing it to the one-dimensional case, that the values $\Delta_i(p)$ are uniquely determined. Those values we denote by $\frac{\partial f}{\partial x_i}(p)$. Furthermore, sums, products, and quotients (whenever defined) of holomorphic functions are holomorphic. See Exercise 3.1.18.

It is known that one-variable functions are holomorphic if and only if they are locally represented by their Taylor series. To have the analogous statement for the multi-variable case, we need some definitions.

Definition 3.1.3.

(1) Let a_ν for $\nu \in \mathbb{N}^n$ be complex numbers. The expression[4]

$$\sum_{\nu=0}^{\infty} a_\nu (x - p_0)^\nu$$

is called a *formal power series* in the variables $x_1, \ldots, x_n$ around p_0. With the obvious addition and multiplication the set of formal power series is a ring, and even an $\mathbb{C}$-algebra. The ring of formal power series around $p_0 = 0$ is denoted by $\mathbb{C}[[x_1, \ldots, x_n]]$ or sometimes $\mathbb{C}[[x]]$ for short, and is called the *formal power series ring*.

[3] This means that U is an open and connected subset.

[4] This is an alternative notation for $\sum_{\nu \in \mathbb{N}^n} a_\nu (p - p_0)^\nu$.

(2) Let $p \in \mathbb{C}^n$, and $c \in \mathbb{C}$. One says that the formal power series $\sum_{\nu=0}^{\infty} a_\nu (x - p_0)^\nu$ *converges* in p with limit c if for all $\varepsilon > 0$ there exists a finite set $I_\varepsilon \subset \mathbb{N}^n$ such that for all finite sets I with $I_\varepsilon \subset I \subset \mathbb{N}^n$

$$\left| \sum_{\nu \in I} a_\nu (p - p_0)^\nu - c \right| < \varepsilon.$$

Note that $\sum_{\nu \in I} a_\nu (p - p_0)^\nu$ makes sense because the summation is over a finite set. In case that $\sum_{\nu=0}^{\infty} a_\nu (x - p_0)^\nu$ converges in p the number c is uniquely determined, and is called the value of the formal power series at p. We, therefore, simply write $\sum_{\nu=0}^{\infty} a_\nu (p - p_0)^\nu = c$.

The formal power series $\sum_{\nu=0}^{\infty} a_\nu (x - p_0)^\nu$ is called *absolutely convergent* in p if $\sum_{\nu=0}^{\infty} \mid a_\nu (x - p_0)^\nu \mid$ is convergent. An absolutely convergent power series is in fact convergent.

(3) Let $U \subset \mathbb{C}^n$, and $f : U \longrightarrow \mathbb{C}$ be a function. One says that the formal power series $\sum_{\nu=0}^{\infty} a_\nu (x - p_0)^\nu$ converges on U with *limit* f if for all p in U the formal power series $\sum_{\nu=0}^{\infty} a_\nu (x - p_0)^\nu$ converges in p with limit $f(p)$.

(4) Let again $U \subset \mathbb{C}^n$, and $f : U \longrightarrow \mathbb{C}$ be a function. One says that the formal power series $\sum_{\nu=0}^{\infty} a_\nu (x - p_0)^\nu$ converges *uniformly* to f on U if for all $\varepsilon > 0$ there exists a finite set $I_\varepsilon \subset \mathbb{N}^n$ such that for all finite sets I with $I_\varepsilon \subset I \subset \mathbb{N}^n$ and for all $p \in U$

$$\left| \sum_{\nu \in I} a_\nu (p - p_0)^\nu - f(p) \right| < \varepsilon.$$

The difference to ordinary convergence is that in case of ordinary convergence, the set I_ε might depend on p, whereas in case of uniform convergence, the set I_ε is not allowed to depend on p.

(5) Let $U \subset \mathbb{C}^n$ be a domain, $f : U \longrightarrow \mathbb{C}$ be a function. f is called *analytic* if for all $p_0 \in U$, there exists a neighborhood $V = V(p_0)$ of p_0 in U, and a power series $\sum_{\nu=0}^{\infty} a_\nu (x - p_0)^\nu$ which converges on V to f.

Let $p \in \mathbb{C}^n$ be fixed. It will turn out that the formal power series that converge on a neighborhood of p (the neighborhood might depend on the function) form a ring, called the *convergent power series ring* $\mathcal{O}_{n,p}$ at p. The ring of convergent power series at 0 is denoted by $\mathbb{C}\{x\}$ or $\mathbb{C}\{x_1, \ldots, x_n\}$ or also often by $\mathcal{O}_n$. instead of the "correct" $\mathcal{O}_{n,0}$.

(6) Let $0 \neq f \in \mathbb{C}[[x_1, \ldots, x_n]]$ be a formal power series. We can write

$$f = f_m + f_{m+1} + \ldots$$

where f_k are homogeneous polynomials of degree k in $x_1, \ldots, x_n$, and $f_m \neq 0$. Then we define $\operatorname{mult}(f) := \operatorname{ord}(f) := m$ and call it the *order* or *multiplicity* of f. One obviously has that $\operatorname{ord}(f \cdot g) = \operatorname{ord}(f) + \operatorname{ord}(g)$.

(7) More generally, let $\lambda = (\lambda_1, \ldots, \lambda_n) \in \mathbb{N}_+^n$ and let $0 \neq f \in \mathbb{C}[[x_1, \ldots, x_n]]$ be a formal power series. We can write

$$f = f_m + f_{m+1} + \cdots$$

where f_k are quasi-homogeneous polynomials (with respect to λ) of weighted degree k and $f_m \neq 0$. Then we define w-ord$(f) := m$ and call it the *weighted order* of f.

The first thing we want to show is that analytic functions are holomorphic. In order to show this, we need the following lemma.

Lemma 3.1.4. *Consider a formal power series $\sum_{\nu=0}^{\infty} a_\nu x^\nu$, which converges at $p \in \mathbb{C}^n$. Let $p = (p_1, \dots, p_n)$ and suppose that $p_i \neq 0$ for all i. Let the polyradius r be defined by $r_i = |p_i|$. Then the formal power series $\sum_{\nu=0}^{\infty} a_\nu x^\nu$ converges absolutely and uniformly on all compact subsets of the open poly-disc P_r.*

Proof. From the convergence of the series $\sum a_\nu p^\nu$ it follows that the set $\{a_\nu p^\nu\}$ is bounded, that is, there exists an $M \in \mathbb{R}$ with $|a_\nu p^\nu| = |a_\nu| r^\nu < M$. Consider a real number q with $0 < q < 1$. By definition of P_{qr} for all $x \in P_{qr}$ it follows that $|x^\nu| < q^{|\nu|} r^\nu$. Hence for all $x \in P_{qr}$

$$|a_\nu x^\nu| < |a_\nu| q^{|\nu|} r^\nu < M \cdot q^{|\nu|}.$$

But

$$M \sum_\nu q^{|\nu|} = M \left(\frac{1}{1-q}\right)^n,$$

see Exercise 3.1.19. The set $\mathbb{N}^n$ is countable, hence there exists a bijection $\alpha : \mathbb{N} \longrightarrow \mathbb{N}^n$. Put $b_k(x) := a_{\alpha(k)} x^{\alpha(k)}$. Hence, $\sum_{k=0}^{\infty} |b_k(x)| \leq M \left(\frac{1}{1-q}\right)^n$, so that this series is increasing and bounded. It follows that $\sum_{k=0}^{\infty} b_k(x)$ converges absolutely and uniformly on P_{qr} so that

$$\forall \varepsilon > 0 \text{ there exists } n_\varepsilon \in \mathbb{N} \text{ with } \sum_{k=n_\varepsilon}^{\infty} |b_k(x)| < \varepsilon \text{ for all } x \in P_{qr}.$$

Set $I_\varepsilon := \{\alpha(i) : i = 0, \dots, n_\varepsilon\}$ and choose I with $I \supset I_\varepsilon$. Then

$$\left|\sum_{\nu=0}^{\infty} |a_\nu x^\nu| - \sum_{\nu \in I} |a_\nu x^\nu|\right| = \left|\sum_{k=0}^{\infty} |b_k(x)| - \sum_{k \in \alpha^{-1}(I)} |b_k(x)|\right|$$
$$= \sum_{k \notin \alpha^{-1}(I)} |b_k(x)| \leq \sum_{k=n_\varepsilon+1}^{\infty} |b_k(x)| < \varepsilon.$$

This shows that the series converges absolutely and uniformly in P_{qr}. We still have to show that the series converges on all compact subsets K in P_r. But P_{qr} for $0 < q < 1$ is an open cover of K. As K is compact we have a finite subcover $P_{q_1 r}, \dots P_{q_s r}$. Let $q = \max\{q_1, \dots, q_s\}$. Hence $K \subset P_{qr}$, and thus the series converges uniformly on K. □

Remarks 3.1.5.

(1) Let f be analytic on a neighborhood of 0. As f is analytic, it follows from the proof of 3.1.4 that there exists a $\delta > 0$ such that

$$||f||_\delta := \sum_\alpha |f_\alpha| \delta^{|\alpha|} < \infty.$$

The fact that f is analytic is equivalent to this property. We could rephrase this as follows. For all $\delta > 0$ we put a seminorm[5] on $\mathbb{C}[[x_1,\dots,x_n]]$. Let $f = \sum f_\alpha x^\alpha \in \mathbb{C}[[x_1,\dots,x_n]]$ be a formal power series. Then we put for $\delta > 0$

$$||f||_\delta := \sum_\alpha |f_\alpha|\delta^{|\alpha|} \in \mathbb{R}_+ \cup \{\infty\}.$$

It follows from the proof of Lemma 3.1.4 that f is convergent exactly when there exists a $\delta > 0$ with $||f||_\delta < \infty$.

(2) Note that if $||f||_\delta < \infty$, and $0 < t < 1$, then $||f||_{t\delta} < \infty$. If, moreover, $f \in (x_1,\dots,x_n)$ an easy estimate shows that $||f||_{t\delta} \leq t \cdot ||f||_\delta$.

(3) Let $f = \sum_{k=0}^{\infty} f_k x_n^k$ be a convergent power series, with $f_k \in \mathbb{C}[[x_1,\dots,x_{n-1}]]$. From the absolute convergence of f in some domain in $\mathbb{C}^n$ it follows that f_i are absolute convergent in some domain of $\mathbb{C}^{n-1}$, see Exercise 3.1.23. It follows that $f_k \in \mathbb{C}\{x_1,\dots,x_{n-1}\}$.

Theorem 3.1.6. *Let $U \subset \mathbb{C}^n$ be open, $f : U \longrightarrow \mathbb{C}$ be analytic. Then f is holomorphic.*

Proof. Without loss of generality, it suffices to show that f is holomorphic in 0. Let $\sum_\nu a_\nu x^\nu$ be a power series that converges on an open neighborhood V of 0 to f. We write

$$\begin{aligned}\sum_\nu a_\nu x^\nu = a_{0\dots 0} &+ x_1 \sum_{\nu_1 \neq 0} a_\nu x_1^{\nu_1 - 1} x_2^{\nu_2} \cdots x_n^{\nu_n} \\ &+ x_2 \sum_{\substack{\nu, \nu_2 \neq 0 \\ \nu_1 = 0}} a_\nu x_2^{\nu_2 - 1} \cdots x_n^{\nu_n} \\ &\vdots \\ &+ x_n \sum_{\substack{\nu, \nu_n \neq 0 \\ \nu_1 = \cdots = \nu_{n-1} = 0}} a_\nu x_n^{\nu_n - 1} \\ = a_{0\dots 0} &+ x_1\Delta_1 + \dots + x_n\Delta_n.\end{aligned}$$

Take a point $p = (p_1,\dots,p_n)$ for which $\sum_\nu a_\nu p^\nu$ is absolutely convergent. Such a point exists by Lemma 3.1.4. We may assume that all the p_i are nonzero. It follows from the absolute convergence of $\sum_\nu a_\nu p^\nu$ that each of the sub series $p_i\Delta_i(p)$ are absolute convergent. As $p_i \neq 0$, the series $\Delta_i(p)$ are convergent. Applying Lemma 3.1.4, it follows that the Δ_i for $i = 1,\dots,n$ converge absolutely and uniformly on small neighborhoods of 0. From a standard theorem in analysis it then follows that the Δ_i are continuous. □

Theorem 3.1.7 (Osgood's Lemma). *Let U be an open set of $\mathbb{C}^n$, and $f : U \longrightarrow \mathbb{C}$ be a continuous function. The following conditions are equivalent.*

(1) f is analytic.

(2) f is holomorphic.

[5] A seminorm is a map $\mathbb{C}[[x_1,\dots,x_n]] \to R \cup \{\infty\}$ satisfying all conditions of a norm (with the obvious interpretation $a + \infty = \infty$).

(3) f is holomorphic in each variable.

Proof. We already showed that (1) implies (2) in Theorem 3.1.6. As (2) implies (3) is trivial, we only have to show (3) implies (1). Take a point $p \in U$. Without loss of generality, we may assume that $p = 0$. Let $\overline{P}_r$ be a closed poly-disc contained in U for $r = (r_1, \dots, r_n)$. Take a point $(\zeta_1, \dots, \zeta_{n-1})$ with $|\zeta_i| \leq r_i$. The function

$$f_n(x_n) := f(\zeta_1, \dots, \zeta_{n-1}, x_n)$$

by assumption is holomorphic in x_n. By the Cauchy integral formula in one variable we therefore have for all x_n with $|x_n| < r_n$

$$f(\zeta_1, \dots, \zeta_{n-1}, x_n) = \frac{1}{2\pi i} \int_{|\zeta_n| = r_n} \frac{f(\zeta_1, \dots, \zeta_{n-1}, \zeta_n)}{\zeta_n - x_n} \, d\zeta_n.$$

We can do the same trick with the penultimate variable,

$$f(\zeta_1, \dots, \zeta_{n-2}, x_{n-1}, x_n) = \frac{1}{2\pi i} \int_{|\zeta_{n-1}| = r_{n-1}} \frac{f(\zeta_1, \dots, \zeta_{n-1}, x_n)}{\zeta_{n-1} - x_{n-1}} \, d\zeta_{n-1}$$

$$= \left(\frac{1}{2\pi i}\right)^2 \int_{|\zeta_{n-1}| = r_{n-1}} \left[\int_{|\zeta_n| = r_n} \frac{f(\zeta_1, \dots, \zeta_{n-1}, \zeta_n)}{\zeta_n - x_n} \, d\zeta_n \right] \frac{d\zeta_{n-1}}{\zeta_{n-1} - x_{n-1}}$$

Continuing like this we obtain

$$f(x_1, \dots, x_n) =$$

$$\left(\frac{1}{2\pi i}\right)^n \int_{|\zeta_1| = r_1} \frac{1}{\zeta_1 - x_1} \left[\int_{|\zeta_2| = r_2} \frac{1}{\zeta_2 - x_2} \left[\cdots \left[\int_{|\zeta_n| = r_n} \frac{f(\zeta)}{\zeta_n - x_n} \, d\zeta_n \right] d\zeta_{n-1} \right] \cdots \right] d\zeta_1.$$

For fixed $(x_1, \dots, x_n)$ the integrand is continuous on $T := \{(\zeta_1, \dots, \zeta_n) : |\zeta_i| = r_i$ for $i = 1, \dots, n\}$, which is a compact set. By Fubini

$$f(x_1, \dots, x_n) = \left(\frac{1}{2\pi i}\right)^n \int_T \frac{f(\zeta)}{(\zeta_1 - x_1) \cdots (\zeta_n - x_n)} \, d\zeta_1 \cdots d\zeta_n.$$

It follows from Exercise 3.1.19 that for fixed $(x_1, \dots, x_n)$ the series

$$\frac{1}{(\zeta_1 - x_1) \cdots (\zeta_n - x_n)} = \sum_{\nu=0}^{\infty} \frac{x_1^{\nu_1} \cdots x_n^{\nu_n}}{\zeta_1^{\nu_1+1} \cdots \zeta_n^{\nu_n+1}}$$

converges absolutely and uniformly on T. As f is continuous, it is bounded on T. Therefore, for fixed $(x_1, \dots, x_n)$ the function

$$\frac{f(\zeta_1, \dots, \zeta_n)}{(\zeta_1 - x_1) \cdots (\zeta_n - x_n)} = \frac{f(\zeta_1, \dots, \zeta_n)}{\zeta_1 \cdots \zeta_n} \cdot \sum_{\nu=0}^{\infty} \left(\frac{x_1}{\zeta_1}\right)^{\nu_1} \cdots \left(\frac{x_n}{\zeta_n}\right)^{\nu_n}$$

converges absolutely and uniformly on T. It is a standard theorem from analysis, that one may in such a case interchange summation and integration. We get

$$
\begin{aligned}
f(x_1,\dots,x_n) &= \left(\frac{1}{2\pi i}\right)^n \int\limits_T \frac{f(\zeta)}{(\zeta_1-x_1)\cdots(\zeta_n-x_n)}\,d\zeta_1\cdots d\zeta_n \\
&= \sum_{\nu=0}^{\infty} x_1^{\nu_1}\cdots x_n^{\nu_n}\cdot\left(\frac{1}{2\pi i}\right)^n \int\limits_T \frac{f(\zeta)}{\zeta_1^{\nu_1+1}\cdots\zeta_n^{\nu_n+1}}\,d\zeta_1\cdots d\zeta_n \\
&= \sum_{\nu=0}^{\infty} a_\nu x^\nu, \text{ with} \\
a_\nu &= \left(\frac{1}{2\pi i}\right)^n \int\limits_T \frac{f(\zeta)}{\zeta_1^{\nu_1+1}\cdots\zeta_n^{\nu_n+1}}\,d\zeta_1\cdots d\zeta_n. \quad \square
\end{aligned}
$$

Remarks 3.1.8.

(1) It follows that the convergent power series is a ring. Indeed, if $f,g \in \mathbb{C}\{x_1,\dots,x_n\}$, then choose an open neighborhood U of 0 on which f converges, and an open neighborhood V of 0 on which g converges. Then f and g are holomorphic on $U \cap V$, so that $f \pm g$ and $f \cdot g$ are holomorphic on V, so define an analytic function. This fact could also have been proved directly, by using only Lemma 3.1.4. Note moreover that $\mathbb{C}\{x_1,\dots,x_n\}$ is a local ring with maximal ideal $(x_1,\dots,x_n)$. Indeed, for convergent power series f not in $(x_1,\dots,x_n)$ one has $f(0) \neq 0$. It follows that $\frac{1}{f}$ is also holomorphic in a neighborhood of 0, and therefore, by Osgood's Lemma, again is a convergent power series. So the elements of the complement of $(x_1,\dots,x_n)$ are units, and it follows from 1.3.3 that the power series ring is local.

(2) Let $U \subset \mathbb{C}^n$ be open. A map $F = (f_1,\dots,f_m) : U \longrightarrow \mathbb{C}^n$ is called holomorphic if each of the f_i is holomorphic. If $F(U) \subset W \subset \mathbb{C}^m$, and $G : W \longrightarrow \mathbb{C}^p$ is a holomorphic map, then the chain rule, see Exercise 3.1.22 gives that $G \circ F$ is holomorphic. This in particular gives the following. Let $g_1,\dots,g_n \in \mathbb{C}\{y_1,\dots,y_m\}$, with $g_i(0) = 0$. Let $f \in \mathbb{C}\{x_1,\dots,x_n\}$. Then $f(g_1,\dots,g_n) \in \mathbb{C}\{y_1,\dots,y_m\}$.

Theorem 3.1.9 (Identity Theorem). *Let $U \subset \mathbb{C}^n$ be a domain. Let $f : U \longrightarrow \mathbb{C}$ be a holomorphic function, such that the restriction of f to a nonempty open subset $V \subset U$ is the zero map. Then f is identically zero on U.*

Remark 3.1.10. In the case $n = 1$ we have a stronger version of the Identity Theorem, as follows.

Let $U \subset \mathbb{C}$ be a domain and $f : U \longrightarrow \mathbb{C}$ be a holomorphic function. Let $\{x_k\}$ be a convergent sequence, $x_k \in U$ and $\lim_{k\to\infty} x_k \in U$. If $f(x_k) = 0$ for all k, then f is identically zero on U.

Indeed, suppose $\lim x_k = 0$, and f not identically zero in a neighborhood of 0. Then $f = x^b \cdot u$, where $u(0) \neq 0$. Therefore, for p with $|p|$ small, it follows that $u(p) \neq 0$, and therefore, $f(p) \neq 0$. The zeros of f therefore lie isolated. This is a contradiction, and therefore, f vanishes in a neighborhood of 0.

One can reduce the proof of the Identity Theorem to the Identity Theorem in the one variable case. The idea is the same as in the proof of the *maximum principle*, which comes next. The proof of the Identity Theorem 3.1.9 is therefore left as Exercise 3.1.24.

Theorem 3.1.11 (Maximum Principle). *Consider a domain subset $U \subset \mathbb{C}^n$. Let $f : U \longrightarrow \mathbb{C}$ be a nonconstant holomorphic function. Then $|f|$ does not have a maximum on U.*

In other words, if $f : U \longrightarrow \mathbb{C}$ takes its maximum absolute value at a point p in U, then f is constant.

Proof. In case of a holomorphic function in one variable, this is well-known. We will prove the general statement by using the one-variable case and the Identity Theorem. Let $p \in U$ be a point on which $|f|$ attains its maximum. Consider a complex line L through p, and consider a small open ball B around p which lies inside U.

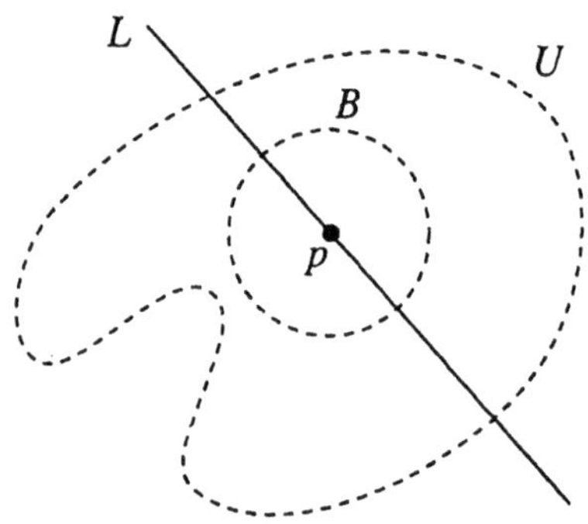

Then the restriction of f to $B \cap L$ is a holomorphic function of one variable. $B \cap L$ is a disc in L. By the maximum principle for the one variable case, the restriction of f to $B \cap L$ is constant. By varying the lines through p, we conclude that f is constant on B. By the Identity Theorem, f is constant on U. □

Definition 3.1.12.

(1) A set or space $X \subset \mathbb{C}^n$ is called *locally analytic*, if for any point $p \in X$, there exists an open subset V of p in $\mathbb{C}^n$, and *finitely* many holomorphic functions $f_1, \dots, f_s$ defined on V such that

$$X \cap V = \{x \in V : f_1(x) = \dots = f_s(x) = 0\}.$$

(2) Let U be an open subset of $\mathbb{C}^n$. A subset $X \subset U$ is called an *analytic subset* of U, if X is locally analytic, and closed in U.

(3) Let X be an analytic subset of U, $U \subset \mathbb{C}^n$ open. A function $f : X \longrightarrow \mathbb{C}$ is called *holomorphic* if for all $x \in X$ there exists an open neighborhood V of x in $\mathbb{C}^n$ such that $f_{|V \cap X}$ is the restriction of a holomorphic function on V.

Remarks 3.1.13.

(1) In this definition, we have to be more careful than in the affine case. Polynomials are always globally defined. But holomorphic functions are usually only defined in small open neighborhoods, so this is the reason why we have to choose open neighborhoods V.

(2) The set $X := \{(x,y) : y = 0, \ \mathrm{Im}(x) \geq 0\}$ in $\mathbb{C}^2$

is not locally analytic. Indeed, let V be a connected open neighborhood of $\{0\}$, and f be a function on V with

$$V \cap X \subset \{(x,y) : f(x,y) = 0.\}$$

Then we can restrict f to the set $V \cap \{y = 0\}$. By the Identity Theorem in one variable, f must vanish identically on $\{y = 0\} \cap V$. As this holds for all f, it follows that X is not locally analytic.

(3) In the definition of X to be an analytic subset of U, we need the condition X is closed in U to avoid phenomena as shown in the following picture.

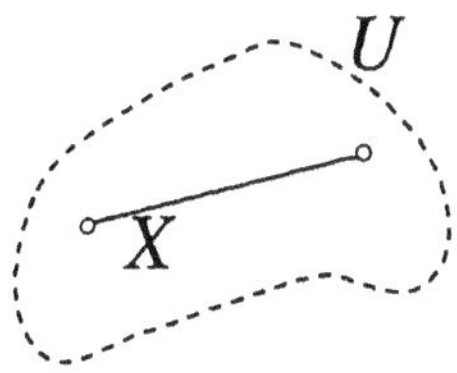

Morally speaking, one wants an analytic subset of U, except for discrete points, to "reach" the boundary. Even more severely, without the closedness condition, if X is the shaded part of

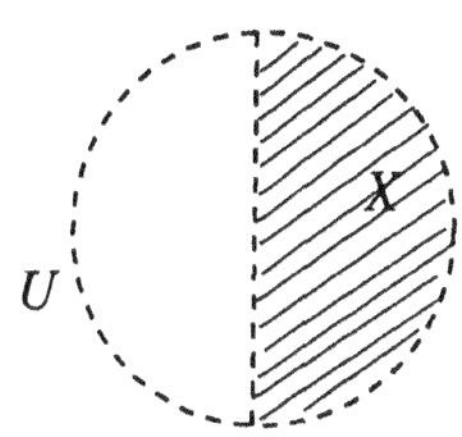

then X would be analytic.

(4) Consider a sequence $p_1, p_2, \ldots$ of nonzero points in $\mathbb{C}$ that converge to the origin. Then the union X of the points is not an analytic set. Indeed, any holomorphic function vanishing on $U \cap X$, for any open neighborhood U of 0 is the zero function. This follows from the Identity Theorem in one variable.

It is well-known (see Remark 3.1.10) that the zeros of a holomorphic function in one variable lie isolated. The following is a generalization of this fact.

Theorem 3.1.14. *Let $X \subset U$ be an analytic subset, with U connected. Suppose $X \neq U$. Then the closure*[6] *of $U \setminus X$ in U is U.*

Proof. The main idea of the proof is already contained in the proof of the fact that the sets in 3.1.13, (2) and (3) are not analytic. The claim is that X has no inner points of U, that is, we claim that the set $\mathring{X}$ of inner points of X is empty. Suppose the converse, that is, X contains a nonempty open subset $\mathring{X}$ of U. We claim that $\mathring{X}$ is closed. From

[6] In the topological sense.

this it would follow from the connectedness of U that $\mathring{X}$ is equal to U, in contradiction to the assumption. So let us prove $\mathring{X}$ is closed. Take a point p in the boundary of $\mathring{X}$ in U. As X is closed in U, the point p is in X. Hence, there exists an open, connected neighborhood V of p and holomorphic functions $f_1, \ldots, f_s$ defined on V such that

$$X \cap V = \{x \in V : f_1(x) = \ldots = f_s(x) = 0\}.$$

As p is in the boundary of $\mathring{X}$, and p is in the open set V, it follows that $V \cap \mathring{X} \neq \emptyset$.

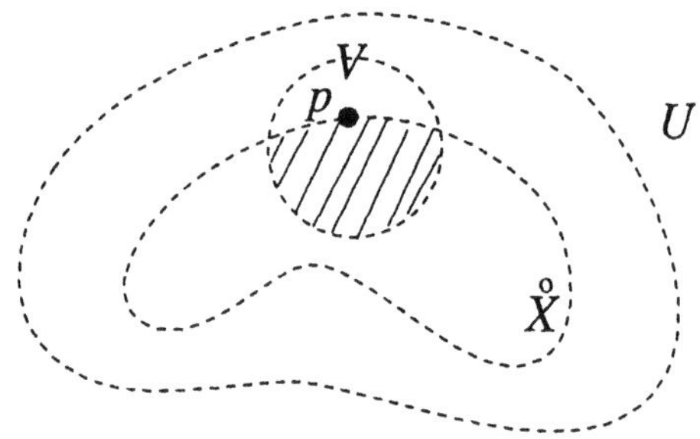

Therefore, for all $i = 1, \ldots, s$ the function f_i is identically zero on the (nonempty) open subset $V \cap \mathring{X}$, hence f_i is, by the Identity Theorem, identically zero on V. Hence $V \subset X$, and $p \in \mathring{X}$. So we proved that $\mathring{X}$ is closed. □

Theorem 3.1.15 (First Riemann Extension Theorem). *Let U be an open connected subset of $\mathbb{C}^n$, and $X \subset U$ an analytic subset. Consider*

$$f : U \setminus X \longrightarrow \mathbb{C}$$

holomorphic. Suppose f is locally bounded, that is, for all $p \in X$ there exists an open neighborhood V of p such that the restriction of f to $V \setminus (V \cap X)$ is bounded. *Then there exists a holomorphic extension of f to U. More precisely, there exists a holomorphic function $\tilde{f} : U \longrightarrow \mathbb{C}$ such that $\tilde{f}_{|U \setminus X} = f$.*

Proof. Let $p \in X$, and B_p an open ball around p. It suffices to extend f to a holomorphic function $\tilde{f}_p$ on B_p. Namely, if $B_p \cap B_q \neq \emptyset$, and if $\tilde{f}_p, \tilde{f}_q$ are extensions of f to B_p, respectively B_q, they must agree on $B_p \cap B_q$, as on $W := B_p \cap B_q \cap (U \setminus X)$ they coincide with f, and W is not empty because the closure of $U \setminus X$ in U is U, see Theorem 3.1.14. So we can use the Identity Theorem 3.1.9 to deduce that f_p and f_q coincide on $B_p \cap B_q$.

Consider a line L through p. Then $L \cap X$ is an analytic subset of X. As X contains no inner points, there exists a line not contained in X. As $L \cap X$ is an analytic subset of an open subset of $L \cong \mathbb{C}$, hence locally given by the zero-set of one analytic function, the points of $L \cap X$ lie isolated. By an affine change of coordinates, we may assume that the line L is the x_n-axis, given by $x_1 = \ldots = x_{n-1} = 0$, and p is equal to 0. Therefore, there exists an $r_n > 0$ such that

$$X \cap \{x = (x_1, \ldots, x_n) : x_1 = \ldots = x_{n-1} = 0, |x_n| \leq r_n\}$$

consists of just one point, namely 0. As the set $\{x = (x_1, \ldots, x_n) : x_1 = \ldots = x_{n-1} = 0, |x_n| = r_n\}$ is compact[7], and is contained in the open set $U \setminus X$, it follows without much difficulty that there exists an open neighborhood W of 0 in $\mathbb{C}^{n-1}$ such that $W \times \{|x_n| = r_n\} \subset U \setminus X$. So

[7] It is a circle.

$$X \cap (W \times \{|x_n| = r_n\}) = \emptyset. \tag{3.1}$$

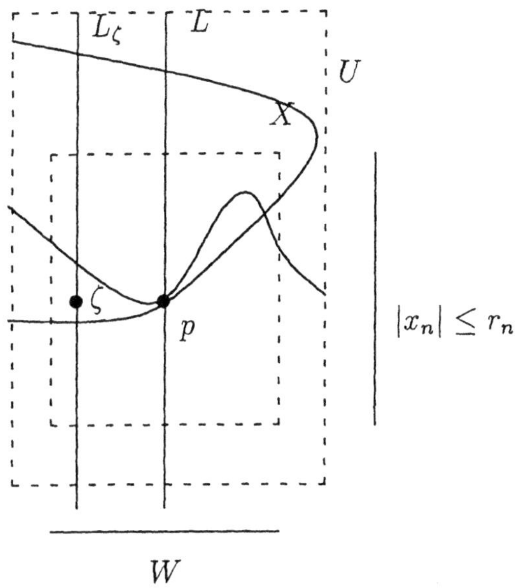

We consider a point $\zeta = (\zeta_1, \ldots, \zeta_{n-1}) \in W$ fixed. As X is given by the vanishing of holomorphic functions, and the restriction of such functions to the line L_ζ given by $x_1 = \zeta_1, \ldots, x_{n-1} = \zeta_{n-1}$ do not all vanish identically because of (3.1), it follows that the set $X \cap \{(\zeta_1, \ldots, \zeta_{n-1})\} \times \{|x_n| \leq r_n\}$ has only isolated points. The function $f(\zeta_1, \ldots, \zeta_{n-1}, x_n)$ is a holomorphic function in x_n, not defined in the finitely many points of X, and it is locally bounded. Therefore, by the Riemann Extension Theorem in one variable, it has a holomorphic extension to $\{(\zeta_1, \ldots, \zeta_{n-1})\} \times \{|x_n| \leq r_n\}$ and is, in fact, given by the integral

$$f(\zeta_1, \ldots, \zeta_{n-1}, x_n) = \int_{|\zeta_n| = r_n} \frac{f(\zeta)}{\zeta_n - x_n} \, d\zeta_n. \tag{3.2}$$

This holds for all $(\zeta_1, \ldots, \zeta_{n-1}) \in W$. Now we *define* $\tilde{f}$ for all $\zeta_1, \ldots, \zeta_{n-1}, x_n \in W \times \{|x_n| \leq r_n\}$ by the right hand side of equation (3.2). It follows that $\tilde{f}$ is holomorphic with respect to each variable. Therefore, by Osgood's Lemma 3.1.7, $\tilde{f}$ is holomorphic. □

Theorem 3.1.16 (Second Riemann Extension Theorem). *Let U be an open subset of $\mathbb{C}^n$, with $n \geq 2$. Let $p \in U$, and*

$$f : U \setminus \{p\} \longrightarrow \mathbb{C}$$

be a holomorphic function. Then f has a holomorphic extension to the whole of U.

Proof. Consider a small open ball B around p which is contained in U. We will show that f is bounded on $B \setminus \{p\}$, so that the First Riemann Extension Theorem gives that f can be extended to a holomorphic function on B. Consider $q \in B \setminus \{p\}$. As $n \geq 2$, we can find a complex line L through q which misses p.

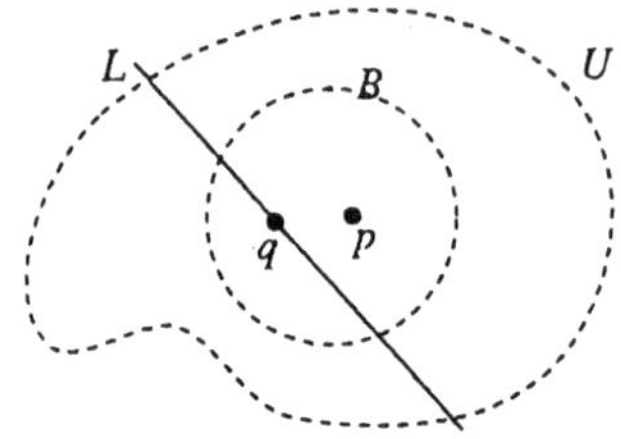

The restriction f to $B \cap L$ is holomorphic in one variable. By the maximum principle

$$|f(q)| \leq \max |f_{|\partial B \cap L}| \leq \max |f_{|\partial B}|.$$

As this holds for all $q \in B \setminus \{p\}$, it follows that f is bounded, as wanted. □

Remark 3.1.17. For $n = 1$ the theorem is not true: consider the function $\frac{1}{x}$ on $U = \mathbb{C} \setminus \{0\}$.

Exercises

3.1.18.

(1) Prove that the values $\Delta_i(p)$ in the definition of a holomorphic function are well-determined, by reducing it to a corresponding statement in the one variable case.

(2) Let $f = \sum_\nu a_\nu x^\nu$ be a convergent power series. Describe all partial derivatives of f in terms of the a_ν.

3.1.19. Let $q = (q_1, \ldots, q_n)$. Prove that

$$\sum_{\nu \in \mathbb{N}^n} q^\nu = \prod_{k=1}^{n} \frac{1}{1 - q_k}.$$

3.1.20. Consider the formal power series ring $\mathbb{C}[[x_1, \ldots, x_n]]$ and let $\mathfrak{m} = (x_1, \ldots, x_n)$.

(1) Show that $\mathfrak{m}$ is a maximal ideal.

(2) Let $f, g \in \mathbb{C}[[x_1, \ldots, x_n]]$. Show that $f = g$ if and only if $f \equiv g \mod \mathfrak{m}^k$ for all $k \in \mathbb{N}$.

(3) Show that in particular $\bigcap_{k=0}^{\infty} \mathfrak{m}^k = 0$.

(4) Let $g \in \mathfrak{m}$, and $a_k \in \mathbb{C}$. Show that $\sum_{k=0}^{\infty} a_k g^k$ is a well-defined element in $\mathbb{C}[[x_1, \ldots, x_n]]$.

(5) More generally, let elements $g_1, \ldots, g_s$ in $\mathfrak{m}$ be given. Show that there is a well-defined homomorphism of $\mathbb{C}$-algebras

$$\begin{array}{ccc} \mathbb{C}[[y_1, \ldots, y_s]] & \longrightarrow & \mathbb{C}[[x_1, \ldots, x_n]] \\ y_i & \mapsto & g_i. \end{array}$$

(6) Let $\varphi : \mathbb{C}[[y_1, \ldots, y_s]] \longrightarrow C[[x_1, \ldots, x_n]]$ be a $\mathbb{C}$-algebra homomorphism. Prove that for all f we have $\varphi(f) = f(\varphi(y_1), \ldots, \varphi(y_s))$.

(7) Show that $\mathbb{C}[[x_1, \ldots, x_n]]$ is a local ring.
(Hint: Let $g \in \mathfrak{m}$. What is $\frac{1}{1-g}$?)

3.1.21. Prove the following statements.

(1) Let f be holomorphic in p. Then f is continuous at p.

(2) Sums, products and quotients (when defined) of holomorphic functions are holomorphic.

3.1.22.

(1) Let $f \in \mathbb{C}\{x_1,\dots,x_n\}$, and $g_1,\dots,g_n \in \mathbb{C}\{y_1,\dots,y_m\}$ with $g_i(0) = 0$ for $i = 1,\dots,n$. Show that $f(g_1,\dots,g_m) \in \mathbb{C}\{y_1,\dots,y_m\}$ by using the seminorm $\|f\|_\delta$, see 3.1.5.

(2) Use Exercise 3.1.20 to prove that for a $\mathbb{C}$-algebra homomorphism $\varphi : \mathbb{C}\{y_1,\dots,y_s\} \longrightarrow C\{x_1,\dots,x_n\}$ for all f $\varphi(f) = f(\varphi(y_1),\dots,\varphi(y_s))$ holds.

3.1.23. Let $f = \sum f_k x_n^k$, $f_k \in \mathbb{C}[[x_1,\dots,x_{n-1}]]$, and $f \in \mathbb{C}\{x_1,\dots,x_n\}$. Prove that $f_k \in \mathbb{C}\{x_1,\dots,x_{n-1}\}$.
(Hint: Use 3.1.4 to show this for $k = 0$. Now use induction on k.)

3.1.24. Prove the Identity Theorem 3.1.9.

3.1.25. Let U be an open connected subset of $\mathbb{C}^n$, and $\mathcal{O}(U)$ the ring of holomorphic functions on U. Prove that $\mathcal{O}(U)$ is an integral domain.

3.1.26. Let f be a holomorphic function on an open subset in $\mathbb{C}^n$, with $n \geq 2$. Then f has no zeros which lie isolated.
(Hint: Look at $\frac{1}{f}$.)

3.1.27. Consider an ideal $I \subset \mathbb{C}[x_1,\dots,x_n]$, and $f \in \mathbb{C}[x_1,\dots,x_n]$. Suppose that $f \in I \cdot \mathbb{C}\{x_1,\dots,x_n\}$. Show that $f \in I$.

3.1.28. Let $I = (y^2 - xz, x^3 - yz, z^2 - x^2y)$. Show that $I \cap \mathbb{C}\{x\} = (0)$.
(Hint: Let $\alpha(y^2 - xz) + \beta(x^3 - yz) + \gamma(z^2 - x^2y) \in \mathbb{C}\{x\}$. Show that modulo (x) we can write $\alpha = Az, \beta = Ay + Cz, \gamma = Cy$. Take $\alpha' = Az + Cx^2, \beta' = Ay + Cz, \gamma' = Cx + Cy$, and subtract these from α,β and γ. Iterate this and show that even in the formal power series $\alpha(y^2 - xz) + \beta(x^3 - yz) + \gamma(z^2 - x^2y) = 0$.)

3.2 Weierstraß Division and Preparation Theorem

We now come to the famous Weierstraß Division Theorem, and its immediate corollary, the Weierstraß Preparation Theorem. Before stating it, we need some definitions.

Definition 3.2.1.

(1) An element $f \in \mathbb{C}\{x_1,\dots,x_n\}$ is called *regular of order* b in x_n if the power series in the variable x_n, defined by $f(0,\dots,0,x_n)$ has a zero of order b.

(2) An element

$$x_n^b + a_1 x_n^{b-1} + \dots + a_{b-1}x_n + a_b, \quad a_i \in \mathbb{C}\{x_1,\dots,x_n\}$$

is called a *Weierstraß polynomial* if $a_i(0) = 0$ for all $i = 1,\dots,b$. Equivalently, it is a monic polynomial of degree b in x_n which is, as convergent power series, regular of order b in x_n.

For general coordinate systems $(x_1,\dots,x_n)$ an $f \in \mathbb{C}\{x_1,\dots,x_n\}$ is regular in x_n. In fact, we have the following result, which is analogous to the Noether Normalization Theorem for Hypersurfaces, see 2.1.7. As the proof is similar it is left as Exercise 3.2.15.

Lemma 3.2.2. *Let $f \in \mathbb{C}\{x_1,\dots,x_n\}$ be of order b. Then after a general linear change of coordinates, f is regular of order b in x_n.*

We can now state the Weierstraß Theorems.

Theorem 3.2.3 (Weierstraß Division Theorem). *Let* $f,g \in \mathbb{C}\{x_1,\dots,x_n\}$, *and suppose that* f *is regular of order* b *in* x_n. *Then there exist uniquely determined* $q \in \mathbb{C}\{x_1,\dots,x_n\}$, *and* $r \in \mathbb{C}\{x_1,\dots,x_{n-1}\}[x_n]$, *with* r *of degree smaller than* b *in* x_n *such that*

$$g = qf + r.$$

Theorem 3.2.4 (Weierstraß Preparation Theorem). *Let* $f \in \mathbb{C}\{x_1,\dots,x_n\}$ *be regular of order* b *in* x_n. *Then there exists a unit* $u \in \mathbb{C}\{x_1,\dots,x_n\}$, *and a Weierstraß polynomial* p *of degree* b, *such that*

$$f = up.$$

These u *and* p *are uniquely determined by* f.

Proof. (Assuming the statement of the Weierstraß Division Theorem.) Apply the Weierstraß Division Theorem to $g = x_n^b$ to obtain a $q \in \mathbb{C}\{x_1,\dots,x_n\}$ and a polynomial r in x_n of degree smaller than b such that.

$$x_n^b = qf + r.$$

Now q is a unit, as otherwise in the expression $qf+r$ there does not appear the term x_n^b, which would be a contradiction to the equality $x_n^b = qf + r$. So now we can put $u = q^{-1}$ and $p = x_n^b - r$ and get the equality $f = up$. It remains to show that p is a Weierstraß polynomial. But if p were not a Weierstraß polynomial, one of the nonleading coefficients of r would be a unit, and f would be regular of order smaller than b in x_n. The fact that u and p are uniquely determined follows from the corresponding fact in the Weierstraß Division Theorem, see Exercise 3.2.16. □

Remark 3.2.5. The Weierstraß Division and Preparation Theorem also hold in the formal power series ring $\mathbb{C}[[x_1,\dots,x_n]]$. A proof for this case is much easier, as one does not have to cope with convergence questions. A proof will be given in Exercise 3.2.18.

Corollary 3.2.6. *Suppose that* $f \in \mathbb{C}\{x_1,\dots,x_n\}$ *is regular of order* b. *Then the* $\mathbb{C}\{x_1,\dots,x_{n-1}\}$*-module* $\mathbb{C}\{x_1,\dots,x_n\}/(f)$ *is finitely generated and free of rank* b. *A basis is given by* $1,x_n,\dots,x_n^{b-1}$.

The proof of this corollary is similar to the proof of 2.1.7 and therefore left as Exercise 3.2.17. It is in fact not so difficult to see that the Corollary is *equivalent* to the Weierstraß Division Theorem. We can even do something better by invoking Nakayama's Lemma.

Lemma 3.2.7. *Let* $f \in \mathbb{C}\{x_1,\dots,x_n\}$ *be regular of order* b. *If* $\mathbb{C}\{x_1,\dots,x_n\}/(f)$ *is a finitely generated* $\mathbb{C}\{x_1,\dots,x_{n-1}\}$*-module, then the Weierstraß Division Theorem holds for* f.

Proof. Put $M := \mathbb{C}\{x_1,\dots,x_n\}/(f)$. The maximal ideal of $\mathbb{C}\{x_1,\dots,x_{n-1}\}$ is the ideal $(x_1,\dots,x_{n-1})$. Now $M/(x_1,\dots,x_{n-1})M = \mathbb{C}\{x_n\}/(f(0,x_n))$, hence is a finite-dimensional $\mathbb{C}$-vector space of dimension b. Indeed, a basis of this vector space is given by the classes of $1,x_n,\dots,x_n^{b-1}$. By Corollary 1.3.6 (a corollary of Nakayama's Lemma), it follows that $1,x_n,\dots,x_n^{b-1}$ generate M as $\mathbb{C}\{x_1,\dots,x_{n-1}\}$-module. This exactly means that for all $g \in \mathbb{C}\{x_1,\dots,x_n\}$ there exist q and r such that

$$g = qf + r$$

where r is a polynomial in x_n of degree less that b with coefficients in $\mathbb{C}\{x_1, \dots, x_{n-1}\}$. To finish the proof, we need to show that q and r are uniquely determined. So assume that $g = qf + r = q'f + r'$, with the degree of both r and r' smaller than b. Then $(q - q')f + r - r' = 0$. It therefore suffices to show that from $qf + r = 0$ with degree of r less than b, it follows that $q = r = 0$. For this it obviously suffices to show that $q = 0$. We now write

$$f = \sum f_i x_n^i; \qquad q = \sum q_i x_n^i; \qquad r = \sum_{i=0}^{b-1} r_i x_n^i,$$

with f_i, q_i and r_i in $\mathbb{C}\{x_1, \dots, x_{n-1}\}$. As f is regular of order b in x_n, it follows that $\operatorname{ord}(f_b) = 0$. If $q \neq 0$, we can find a minimal k such that $\operatorname{ord}(q_k)$ is minimal. This means that for $j < k$ we have $\operatorname{ord}(q_j) > \operatorname{ord}(q_k)$, and for $j > k$ we have $\operatorname{ord}(q_j) \geq \operatorname{ord}(q_k)$. As f is regular of order b it follows that $\operatorname{ord}(f_i) > 0$ for $i = 0, \dots, b-1$. We now look at the coefficient of x_n^{b+k} of the power series $q \cdot f$. As the term x_n^{b+k} does not appear in r, and $qf + r = 0$, this coefficient is equal to

$$f_{b+k}q_0 + \cdots + f_{b+1}q_{k-1} + f_b q_k + f_{b-1}q_{k+1} + \cdots + f_0 q_{b+k} = 0. \tag{3.3}$$

We know the following

- $\operatorname{ord}(f_b q_k) = \operatorname{ord}(q_k)$,
- $\operatorname{ord}(f_{b+j}q_{k-j}) \geq \operatorname{ord}(q_{k-j}) > \operatorname{ord}(q_k)$ for $j > 0$,
- $\operatorname{ord}(f_{b-j}q_{k+j}) = \operatorname{ord}(f_{b-j}) + \operatorname{ord}(q_{k+j}) > \operatorname{ord}(q_k)$ for $j > 0$.

The only terms of degree $\operatorname{ord}(q_k)$ in formula (3.3) come from $f_b q_k$, so the expression (3.3) cannot be equal to zero. This is a contradiction. Hence $q = 0$ and the uniqueness is proved. □

So all we need to show now, in order to prove the Weierstraß Division Theorem, is that $\mathbb{C}\{x_1, \dots, x_n\}/(f)$ is a finitely generated $\mathbb{C}\{x_1, \dots, x_{n-1}\}$-module. As it is not more difficult, we will in fact prove a more general version. This more general version will play an important role in many arguments in this book. First we need a definition and a simple lemma.

Definition 3.2.8.

(1) An *analytic algebra* (also called analytic $\mathbb{C}$-algebra) is a $\mathbb{C}$-algebra of type $\mathbb{C}\{x_1, \dots, x_n\}/I$, where I is an ideal in $\mathbb{C}\{x_1, \dots, x_n\}$.

(2) A *formal* $\mathbb{C}$*-algebra* is a $\mathbb{C}$-algebra of type $\mathbb{C}[[x_1, \dots, x_n]]/I$, where I is an ideal in $\mathbb{C}[[x_1, \dots, x_n]]$.

Lemma 3.2.9. *Let* $(R, \mathfrak{m}_R)$, $(S, \mathfrak{m}_S)$ *be analytic (or formal)* $\mathbb{C}$*-algebras,* $\varphi : R \to S$ *a morphism of* $\mathbb{C}$*-algebras. Then* $\varphi(\mathfrak{m}_R) \subset \mathfrak{m}_S$.

Proof. Suppose the converse. Then there exists an $f \in \mathfrak{m}_R$ with $\varphi(f) \notin \mathfrak{m}_S$. Thus $\varphi(f)(0) = c \neq 0$. Hence $\varphi(f) - c \in \mathfrak{m}_S$ is not a unit. As φ is a ring homomorphism, it sends 1 to 1. As a ring homomorphism, it therefore sends units in R to units in S. However, as φ is a homomorphism of $\mathbb{C}$-vector spaces, $\varphi(f) - c = \varphi(f - c)$. Now $f - c$ is a unit in R, and, therefore, $\varphi(f) - c$ is a unit in S. This is a contradiction, proving the lemma. □

Theorem 3.2.10 (General Weierstraß Division Theorem). *Let R, S be analytic $\mathbb{C}$-algebras, $\varphi : R \to S$ a morphism of $\mathbb{C}$-algebras. One can, via φ, view S as an R-module. Then the following conditions are equivalent:*

(1) S is a finitely generated R-module.

(2) $\dim_{\mathbb{C}} S/\mathfrak{m}_R S < \infty$.

Analogous statements hold for formal $\mathbb{C}$-algebra's.

Before giving the proof of Theorem 3.2.10, we look at two examples.

Examples 3.2.11.

(1) We show that the Weierstraß Division Theorem follows from the General Weierstraß Division Theorem by using Lemma 3.2.7. So for $f \in \mathbb{C}\{x_1, \dots, x_n\}$ which is regular of order b in x_n we have to show that $\mathbb{C}\{x_1, \dots, x_n\}/(f)$ is a finitely generated $\mathbb{C}\{x_1, \dots, x_{n-1}\}$-module. We take $R = \mathbb{C}\{x_1, \dots, x_{n-1}\}$, and $S = \mathbb{C}\{x_1, \dots, x_n\}/(f)$. The map φ sends x_i to x_i for $i = 1, \dots, n-1$. Then $S/\mathfrak{m}_R S = \mathbb{C}\{x_1, \dots, x_n\}/(x_1, \dots, x_{n-1}, f) \cong \mathbb{C}\{x_n\}/(f(0, \dots, 0, x_n) \cong \mathbb{C}^b$. So the statement follows from the General Weierstraß Division Theorem.

(2) We consider $R = \mathbb{C}\{x\}$, $S = \mathbb{C}\{t\}$, and the map $\varphi : R \longrightarrow S$ be given by $\varphi(x) = t^2$. Then $S/\mathfrak{m}_R S = \mathbb{C}\{t\}/(t^2)$ is a two-dimensional vector space. We claim that $\mathbb{C}\{t\}$ is as $\mathbb{C}\{x\}$-module generated by 1 and t. Now let $f = \sum_{k=0}^{\infty} f_k t^k \in \mathbb{C}\{t\}$ be arbitrary, with $f_k \in \mathbb{C}$. Now define two power series

$$a_0 = \sum_{k=0}^{\infty} f_{2k} x^k; \; a_1 = \sum_{k=0}^{\infty} f_{2k+1} x^k.$$

Then $\varphi(a_0) = \sum_{k=0}^{\infty} f_{2k} t^{2k}$ and $\varphi(a_1) = \sum_{k=0}^{\infty} f_{2k+1} t^{2k}$, so that

$$f = \varphi(a_0) \cdot 1 + \varphi(a_1) \cdot t.$$

We still have to show that a_0 and a_1 are convergent power series. In this case this follows easily from the convergence of f, by using 3.1.4.

Proof of the General Weierstraß Division Theorem 3.2.10.
Step 1. We first show that (1) implies (2). This is quite obvious. Any set of generators of S as an R-module will, by taking classes, be a set of generators for $S/\mathfrak{m}_R S$ as $R/\mathfrak{m}_R = \mathbb{C}$-module. Hence, $S/\mathfrak{m}_R S$ is a finite-dimensional vector space.

Step 2. We first prove the implication (2) $\Longrightarrow$ (1) for the case that $R = \mathbb{C}\{x_1, \dots, x_n\}$, $S = \mathbb{C}\{y_1, \dots, y_m\}$. We put $g_i = \varphi(x_i)$ for $i = 1, \dots, n$. The assumption then is that

$$\mathbb{C}\{y_1, \dots, y_m\}/(g_1, \dots, g_n)$$

is a finite-dimensional vector space. So $S/(g_1, \dots, g_n)$ is an Artinian $\mathbb{C}$-algebra. Writing $\mathfrak{m} = \mathfrak{m}_S$ for short, it follows from Exercise 1.3.20 that there exists a $k \in \mathbb{N}$ such that $\mathfrak{m}^k \subset (g_1, \dots, g_n)$. Multiplying with $\mathfrak{m}$, and putting $r = k+1$, it follows that $\mathfrak{m}^r \subset \mathfrak{m}(g_1, \dots, g_n)$. In particular, for any monomial y^α of degree r there exist elements $p_{\alpha,i} \in \mathfrak{m}$ such that

(3.4) $$y^\alpha = \sum_{i=1}^n p_{\alpha,i} g_i.$$

The claim is that the monomials $\{y^\alpha : |\alpha| < r\}$ generate S as an R-module. This would prove the finite generatedness of S as R–module in this case.

Step 2a. We first prove this "formally". Let $f \in S$. We can write

(3.5) $$f = a(f) + \sum_{\alpha:|\alpha|=r} \widetilde{h}_\alpha(f) y^\alpha.$$

such that $\widetilde{h}_\alpha(f)y^\alpha$ and $\widetilde{h}_\beta(f)y^\beta$ have no common monomial if $\alpha \neq \beta$. Moreover $a(f)$ is a polynomial in the $y's$ of degree smaller than r. Plugging in (3.4) we get

(3.6) $$f = a(f) + \sum_{i=1}^n h_i(f) g_i,$$

where $h_i(f) = \sum_{\alpha:|\alpha|=r} p_{\alpha,i}\widetilde{h}_\alpha(f)$.We now iterate (3.6). For a multi-index $\beta = (\beta_1, \ldots, \beta_n)$ we define inductively

(1) $a_\beta(f) = a(f)$ for $\beta = (0, \ldots, 0)$,

(2) $h_\beta(f) = f$ for $\beta = (0, \ldots, 0)$,

(3) $h_\beta(f) = 0$ if some entry in β is negative,

(4) $a_\beta(f) = a(h_\beta(f))$,

(5) $h_\beta(f) = \sum_{i=1}^n h_i\big(h_{\beta-e_i}(f)\big)$. Here $e_1 = (1,0,\ldots,0)$ etc.

Note that the $a_\beta(f)$ are polynomials in the $y_1, \ldots, y_m$ of degree smaller than r. We claim that for all $k \in \mathbb{N}$

(3.7) $$f = \sum_{|\beta|<k} a_\beta(f) g^\beta + \sum_{|\beta|=k} h_\beta(f) g^\beta.$$

For $k = 1$ this is just formula (3.6). We now apply (3.6) to the coefficients $h_\beta(f)$. We get $h_\beta(f) = a(h_\beta(f)) + \sum_{i=1}^n h_i(h_\beta(f))g_i$. Plugging this in (3.7) and using the definition of $a_\beta(f)$ and $h_\beta(f)$ gives the induction step.

As $(g_1, \ldots, g_n) \in \mathfrak{m}$, it follows that for all $k \in \mathbb{N}$

$$f \equiv \sum_\beta a_\beta(f) g^\beta \mod \mathfrak{m}^k.$$

So in the formal power series ring we have the equality

$$f = \sum_\beta a_\beta(f) g^\beta.$$

The $a_\beta(f)$ are polynomials in the $y's$ of degree smaller than r. So we write

(3.8)
$$a_\beta(f) = \sum_{\alpha:|\alpha|<r} a_{\beta,\alpha}(f) y^\alpha, \quad a_{\beta,\alpha}(f) \in \mathbb{C}.$$

Now for all α with $|\alpha| < r$ look at the formal power series

(3.9)
$$A_\alpha(f) := \sum_\beta a_{\beta,\alpha}(f) x^\beta.$$

So if we look at the map $\varphi : \mathbb{C}[[x_1, \dots, x_n]] \longrightarrow \mathbb{C}[[y_1, \dots, y_m]]$ which sends x_i to g_i it follows that

$$f = \sum_{\alpha:|\alpha|<r} \varphi(A_\alpha(f)) \cdot y^\alpha,$$

showing the theorem in the formal case.

Step 2b. We have to show the convergence of $A_\alpha(f)$. We use the seminorm of 3.1.5. That is, for $f = \sum_\alpha f_\alpha x^\alpha$ we have $||f||_\delta = \sum_\alpha |f_\alpha| \delta^{|\alpha|}$. Now choose a $\delta > 0$ so small that

(1) $||f||_\delta$ is finite.

(2) Let C be the number of monomials of degree r. Then for the $p_{\alpha,i}$ of formula (3.4)

$$||p_{\alpha,i}||_\delta \le \tfrac{1}{C}.$$

We can find such a δ, because the elements $p_{\alpha,i}$ are in the maximal ideal, see 3.1.5 (2). We will show for all α with $|\alpha| < r$

(3.10)
$$\big\|A_\alpha(f)\big\|_\varepsilon < \infty \quad \text{for } \varepsilon = \frac{\delta^r}{2n^2}.$$

This will show the convergence of $A_\alpha(f)$. We have the following estimates for the terms in formula (3.5) and (3.6).

(1) $||a(f)||_\delta \le ||f||_\delta$,

(2) $||\widetilde{h}_\alpha(f)||_\delta \le \delta^{-r} ||f||_\delta$,

(3) $||h_i(f)||_\delta \le \delta^{-r} ||f||_\delta$.

The first two estimates follow from the definition of the seminorm, and the definition of $a(f)$ and $\widetilde{h}_\alpha(f)$. The final estimate follows because we have chosen δ so that $||p_{\alpha,i}||_\delta \le \frac{1}{C}$, C being the number of monomials of degree r. Inductively we get the estimates

(3.11)
$$||h_\beta(f)||_\delta \le n^{|\beta|-1} \cdot \delta^{-|\beta| r} \cdot ||f||_\delta,$$

(3.12)
$$||a_\beta(f)||_\delta \le n^{|\beta|-1} \cdot \delta^{-|\beta| r} \cdot ||f||_\delta.$$

Recall that $a_\beta(f) = \sum_{\alpha:|\alpha|<r} a_{\beta,\alpha}(f) y^\alpha$, for some $a_{\beta,\alpha}(f) \in \mathbb{C}$.
Obviously $|a_{\beta,\alpha}(f)| \le ||a_\beta(f)||_\delta \delta^{-|\alpha|}$. From (3.12) it follows with $C_\alpha = \frac{\delta^{-|\alpha|}}{n} ||f||_\delta$ that

$$|a_{\beta,\alpha}(f)| \le C_\alpha \big(n\delta^{-r}\big)^{|\beta|}.$$

With $\varepsilon = \frac{\delta^r}{2n^2}$ we get

$$\| \sum_\beta a_{\beta,\alpha}(f)x^\beta \|_\varepsilon = \sum_\beta |a_{\beta,\alpha}(f)|\varepsilon^{|\beta|} \leq C_\alpha \sum_\beta \left(\frac{n}{\delta^r}\right)^{|\beta|} \varepsilon^{|\beta|} = C_\alpha \sum_\beta \frac{1}{(2n)^{|\beta|}}.$$

As the number of β with $|\beta| = k$ is at most n^k this can be estimated as $C_\alpha \sum_{k=0}^\infty \frac{1}{2^k} < \infty$.

Step 3. The implication (2) $\Longrightarrow$ (1) will be proved for the case that $S = \mathbb{C}\{y_1, \ldots, y_m\}/J$, where $J = (f_1, \ldots, f_p)$ is a *finitely generated* ideal. This would be always the case if we knew that $\mathbb{C}\{y_1, \ldots, y_m\}$ is Noetherian. Now, as the Weierstraß Division Theorem is deduced from the General Weierstraß Division Theorem by using $J = (f)$, see Example 3.2.11 (1), and only the Weierstraß Division Theorem is used in the proof of the fact that $\mathbb{C}\{x_1, \ldots, x_n\}$ is Noetherian, see Lemma 3.2.7, this restriction is not essential.

Step 4. We now prove the general case, for J finitely generated.

Write $R = \mathbb{C}\{x_1, \ldots, x_n\}/I, S = \mathbb{C}\{y_1, \ldots, y_m\}/J$, where $J = (f_1, \ldots, f_p)$. Define new rings:

$$R' := \mathbb{C}\{x_1, \ldots, x_n, x_{n+1}, \ldots, x_{n+p}\}; \quad S' := \mathbb{C}\{y_1, \ldots, y_m\}.$$

Take representatives $g_1, \ldots, g_n \in S' = \mathbb{C}\{y_1, \ldots, y_m\}$ of $\varphi(x_1), \ldots, \varphi(x_n)$. The map φ' is then defined on generators by:

- $\varphi'(x_i) = g_i$ for $i = 1, \ldots, n$,
- $\varphi'(x_{n+j}) = f_j$ for $j = 1, \ldots, p$.

It is immediately clear that

$$S'/\mathfrak{m}_{R'}S' = \mathbb{C}\{y_1, \ldots, y_m\}/(g_1, \ldots, g_n, f_1, \ldots, f_p) = S/\mathfrak{m}_R S,$$

which by assumption is finite-dimensional. Therefore, by Step 2 we may conclude that S' is a finitely generated R'-module. Hence, there exist $v_1, \ldots, v_s \in S' = \mathbb{C}\{y_1, \ldots, y_m\}$, such that any $h \in S'$ can be written as:

$$h = \alpha_1 v_1 + \ldots + \alpha_s v_s \qquad \alpha_i \in \varphi'(R').$$

Reading this equation modulo $(f_1, \ldots, f_p)$, we see that S is an R-module generated by $v_1, \ldots, v_s$, so, in particular, finitely generated. $\square$

Now we know (under the assumptions of the theorem) that S is a finitely generated R-module, we can use Nakayama's Lemma to get the following somewhat stronger form of the theorem.

Corollary 3.2.12. *Keep the notations of the previous Theorem 3.2.10. Suppose we have elements $v_1, \ldots, v_s \in S$ whose classes generate $S/\mathfrak{m}_R S$ as a $\mathbb{C}$-vector space. Then $v_1, \ldots, v_s$ generate S as an R-module.*

Corollary 3.2.13. *Again keep the notation of Theorem 3.2.10.*

(1) Suppose $\mathfrak{m}_R S = \mathfrak{m}_S$. Then the map φ is surjective.

(2) Suppose S is an analytic algebra such that $\mathfrak{m}_S$ has s generators. Then S is a quotient of the power series ring $\mathbb{C}\{x_1, \ldots, x_s\}$.

Proof. The first statement follows from 3.2.12: indeed, it follows that 1 generates S as an R–module, and thus in particular, φ is surjective. For the second statement, suppose that $f_1,\dots,f_s$ generate $\mathfrak{m}_S$. Consider the map

$$\varphi : \mathbb{C}\{x_1,\dots,x_s\} \longrightarrow S,$$

which sends x_i to f_i. It follows from the first part that φ is surjective. □

There is a "direct" proof of the second part of this corollary, using only the statement of the Weierstraß Division Theorem. See Exercise 3.2.21.

Corollary 3.2.14. *Let R,S be analytic $\mathbb{C}$-algebras, $\varphi : R \longrightarrow S$ a morphism of $\mathbb{C}$-algebras. Let M be a finitely generated S–module such that* $\dim_{\mathbb{C}} M/\mathfrak{m}_R M < \infty$. *Then M is a finitely generated R–module.*

Proof. As M is an $S/\operatorname{Ann}_S(M)$–module, we may assume that $\operatorname{Ann}_S(M) = (0)$. Exercise 1.2.31 applied to $M/\mathfrak{m}_R M$ implies that $\dim_{\mathbb{C}} S/\mathfrak{m}_R S < \infty$. By the previous theorem S is a finitely generated R–module and, therefore, M is a finitely generated R–module. □

Exercises

3.2.15. Prove Lemma 3.2.2.

3.2.16. Prove that u and p in the Weierstraß Preparation Theorem are uniquely determined.

3.2.17. Prove that the Weierstraß Division Theorem is equivalent to the statement in Corollary 3.2.6.

3.2.18. Prove the Weierstraß Division Theorem for the formal power series ring.
(Hint: Consider a formal power series $f \in \mathbb{C}[[x_1,\dots,x_n]]$ which is regular of order b in x_n. Write

$$f = f_0 + x_n^b u_f. \tag{3.13}$$

As f is regular of order b in x_n it follows that

- f_0 is a polynomial of degree $\le b-1$ in x_n with coefficients in $(x_1,\dots,x_{n-1})$.
- u_f is a unit.

From (3.13) it follows that $x_n^b = \frac{f}{u_f} - \frac{f_0}{u_f}$ with $\frac{f_0}{u_f} \in (x_1,\dots,x_{n-1})$. Construct inductively r_i,q_i,g_i with the following properties.

(1) $g = q_i \cdot f + r_i + g_i$,

(2) $g_i, q_{i+1} - q_i, r_{i+1} - r_i \in (x_1,\dots,x_{n-1})^i$,

(3) r_i is a polynomial of degree $\le b-1$ in x_n.

Use as initialization $q_0 = r_0 = 0$, and $g_0 = g$, and divide g_i by x_n^b.)

3.2.19. Let f be a Weierstraß polynomial in x_n of degree b. Suppose that $\operatorname{ord}(f) = b$. Let g be a (formal) power series of order t. Apply the Weierstraß Division Theorem, so write $g = q \cdot f + r$, for r a polynomial of degree smaller than b in x_n. Prove that the order of r is at least t.
(Hint: Follow the steps in the proof of the formal Weierstraß Division Theorem, see Exercise 3.2.18.)

3.2.20.

(1) Let $f \in \mathbb{C}\{x_1,\dots,x_n\}$ be regular of order b in x_n, assume that $b = \operatorname{ord}(f)$ and let $g \in (x_1,\dots,x_n)^c\mathbb{C}\{x_1,\dots,x_n\}$. Let $g = qf + r$ as in 3.2.3. Prove that $q \in (x_1,\dots,x_n)^{c-b}$.

(2) Generalize the result to the weighted order (cf. 3.1.3): let $w_1, \ldots, w_n \in \mathbb{N}$ be weights for $x_1, \ldots, x_n$. Let $f, g \in \mathbb{C}\{x_1, \ldots, x_n\}$ be regular of order b in x_n, assume that w-ord$(f) = bw_n$ and w-ord$(g) = c$, let $g = qf + r$ as in 3.2.3. Prove that w-ord$(q) = c - bw_n$ if $q \neq 0$.

3.2.21.

(1) Suppose $I \subset \mathbb{C}\{x_1, \ldots, x_n\}$ contains an element which is regular of order one in x_n. Use the Weierstraß Division Theorem to show that $\mathbb{C}\{x_1, \ldots, x_n\}/I$ is isomorphic to a quotient of $\mathbb{C}\{x_1, \ldots, x_{n-1}\}$. So we can "eliminate" the variable x_n.

(2) Give a different proof of the second part of Corollary 3.2.13.

3.3 Applications

The first thing to note is that the Weierstraß Preparation Theorem implies the Implicit Function Theorem!

Theorem 3.3.1 (Implicit Function Theorem). *Let $f \in \mathbb{C}\{x_1, \ldots, x_n, y\}$ with $f(0) = 0$ and $\frac{\partial f}{\partial y}(0) \neq 0$. Then there exists a unique $\varphi \in (x_1, \ldots, x_n)\mathbb{C}\{x_1, \ldots, x_n\}$ with*

$$f(x_1, \ldots, x_n, y) = 0 \iff y = \varphi(x_1, \ldots, x_n).$$

Proof. Note that the condition $f(0) = 0$ and $\frac{\partial f}{\partial y}(0) \neq 0$ exactly means that f is regular of order one in y. By the Weierstraß Preparation Theorem

$$f(x_1, \ldots, x_n, y) = u \cdot (y - \varphi(x_1, \ldots, x_n))$$

for some unit $u \in \mathbb{C}\{x_1, \ldots, x_n, y\}$ and some $\varphi \in \mathbb{C}\{x_1, \ldots, x_n\}$. The Implicit Function Theorem follows from this equation. □

Example 3.3.2. Consider $f(x,y) = (y+1)^2 - x - 1 = y^2 + 2y - x$. Then f satisfies the conditions of the Implicit Function Theorem. Hence $y = -1 + \sqrt{x+1}$ exists as an analytic function.

From the uniqueness statements in the Weierstraß Preparation Theorem the following follows immediately.

Corollary 3.3.3. *Let $f \in \mathbb{C}\{x_1, \ldots, x_n\}$ be regular of order b in x_n, and suppose that $f = u \cdot p$, where $u, p \in \mathbb{C}[[x_1, \ldots, x_n]]$, u a unit and p a Weierstraß polynomial. Then u and p are in $\mathbb{C}\{x_1, \ldots, x_n\}$, that is, they are convergent.*

A similar statement holds for the Weierstraß Division Theorem.

Example 3.3.4. Consider $f(x,y) = 2y + y^2 - x$. Then f is regular of order one in y. Hence, by the Weierstraß Preparation Theorem $f(x,y) = u(y - \varphi(x))$ for some unit u. We can generate φ and u as indicated by a proof of the formal Weierstraß Division Theorem (Exercise 3.2.18).

$$\begin{aligned}
\varphi(x) &= \tfrac{1}{2}(2y + y^2 - x) + \tfrac{1}{2}x - \tfrac{1}{2}y^2 \\
&= (\tfrac{1}{2} - \tfrac{1}{4}y)(2y + y^2 - x) + \tfrac{1}{2}x - \tfrac{1}{4}xy + \tfrac{1}{4}y^3 \\
&= (\tfrac{1}{2} - \tfrac{1}{4}y - \tfrac{1}{8}x + \tfrac{1}{8}y^2)(2y + y^2 - x) + \tfrac{1}{2}x - \tfrac{1}{8}x^2 - \tfrac{1}{8}y^4 + \tfrac{1}{4}xy^2 \\
&= \ldots.
\end{aligned}$$

So we see the beginning of the power series

$$\varphi(x) = \tfrac{1}{2}x - \tfrac{1}{8}x^2 + \dots$$

which is the beginning of the power series of $\varphi(x) = -1 + \sqrt{x+1}$. Similarly, we obtain $u = y + 1 + \sqrt{x+1}$.

The Implicit Function Theorem holds just as well for the formal case. From the uniqueness statement in the Implicit Function Theorem the following fact follows.

Corollary 3.3.5. *Let $f \in \mathbb{C}\{x_1, \dots, x_n, y\}$ be regular of order 1 in y. Suppose $\varphi(x) \in \mathbb{C}[[x_1, \dots, x_n]]$ is such that $f(x_1, \dots, x_n, \varphi(x)) = 0$. Then $\varphi(x) \in \mathbb{C}\{x_1, \dots, x_n\}$.*

We might phrase this by saying that a formal solution to $f(x_1, \dots, x_n, y) = 0$ (solving for y), already is analytic.

By induction we get the Implicit Mapping Theorem from the Implicit Function Theorem.

Theorem 3.3.6 (Implicit Mapping Theorem).
Let $f_1, \dots, f_m \in \mathbb{C}\{x_1, \dots, x_n, y_1, \dots, y_m\}$ such that $f_i(0) = 0$ and $\det\left(\frac{\partial f_i}{\partial y_j}(0)\right) \neq 0$. Then there exist uniquely defined $\varphi_1, \dots, \varphi_m \in (x_1, \dots, x_n)\mathbb{C}\{x_1, \dots, x_n\}$ such that $f_i(x_1, \dots, x_n, \varphi_1, \dots, \varphi_m\} = 0$ for $i = 1, \dots, m$.

Proof. First suppose that the $m \times m$ matrix $\left(\frac{\partial f_i}{\partial y_j}(0)\right)$ is the identity matrix. We will prove the statement by induction on m. The case $m = 1$ is the Implicit Function Theorem. For the induction step, first apply the Implicit Function Theorem to the element f_m. We conclude that there exists a unique holomorphic function $\psi = \psi(x_1, \dots, x_n, y_1, \dots, y_{m-1})$ with $f_m(x_1, \dots, x_n, y_1, \dots, y_{m-1}, \psi) = 0$. It is left as an exercise to show that the functions

$$f_i(x_1, \dots, x_n, y_1, \dots, y_{m-1}, \psi), \quad i = 1, \dots m-1,$$

satisfy the induction hypothesis.

For the general case, let $A = \left(\frac{\partial f_i}{\partial y_j}(0)\right)^{-1}$. Then

$$A \cdot \left(\frac{\partial f_i}{\partial y_j}(0)\right)$$

is the identity matrix. Replacing $f_1, \dots, f_m$ by linear combinations of those given by A, we come back to the first case. □

The proof of the Inverse Function Theorem, which follows from the Implicit Mapping Theorem is standard, and left as Exercise 3.3.29.

Corollary 3.3.7 (Inverse Function Theorem). *Let $f_1, \dots, f_n \in \mathbb{C}\{x_1, \dots, x_n\}$ such that $f_1(0) = \dots = f_n(0) = 0$. Then $\det\left(\frac{\partial f_i}{\partial x_j}(0)\right) \neq 0$ if and only if the $\mathbb{C}$-algebra homomorphism*

$$\begin{array}{ccc} \mathbb{C}\{x_1, \dots, x_n\} & \longrightarrow & \mathbb{C}\{x_1, \dots, x_n\} \\ x_i & \longmapsto & f_i \end{array}$$

is an isomorphism. This again holds if and only if there exist open neighborhoods U,W of 0 such that

$$F = (f_1, \dots ,f_n)$$

defines a map $F : U \longrightarrow W$ and this map has a holomorphic inverse.

The Inverse Function Theorem is the basis to introduce coordinate functions and complex submanifolds of $\mathbb{C}^n$.

Definition 3.3.8.

(1) Let $f_1, \dots ,f_n$ be holomorphic on an open subset U. Let $p \in U$, and suppose $f_1(p) = \dots = f_n(p) = 0$. The set $\{f_1, \dots ,f_n\}$ is called a set of *coordinate functions* at p if $\det \left(\frac{\partial f}{\partial x_i}(p) \right) \neq 0$.

(2) A subset $X \subset \mathbb{C}^n$ is called a *complex submanifold* of $\mathbb{C}^n$ if for every $x \in X$ there exists an open subset U in $\mathbb{C}^n$ and coordinate functions $w_1, \dots ,w_n$ of x such that $X \cap U = \{y \in U : w_1(y) = \dots = w_m(y) = 0\}$ for some $m \leq n$.

Remark 3.3.9. A complex submanifold $X \subset \mathbb{C}^n$ locally looks like an open subset of $\mathbb{C}^k$ for some k.

Example 3.3.10. Let $f_1, \dots ,f_k$ be holomorphic functions on the open set $U \subset \mathbb{C}^n$. Then $X = \{(a_1, \dots ,a_n, f_1(a_1, \dots ,a_n), \dots ,f_k(a_1, \dots ,a_n)) : (a_1, \dots ,a_n) \in U\}$ is a complex submanifold of $\mathbb{C}^{n+k}$. To see this, let $x_1, \dots ,x_n,y_1, \dots y_k$ be coordinates on $\mathbb{C}^{n+k}$. Then X can be given by

$$y_1 - f_1(x_1, \dots ,x_n) = \dots = y_k - f_k(x_1, \dots ,x_n) = 0,$$

and one checks that these are part of a set of coordinate functions at every point of X.

We will see later in Chapter 4 that for an analytic subset of $\mathbb{C}^n$ almost all points have a neighborhood which is a zero set of coordinate functions. More precisely, an analytic subset contains as an open set a complex submanifold. This gives rise to the following definition.

Definition 3.3.11. Let $U \subset \mathbb{C}^n$ be an open set and $X \subset U$ an analytic subset. A point $x \in X$ is called *regular*, or X is called *smooth* at x, if there exists an open subset V in $\mathbb{C}^n$, $x \in V$, such that $X \cap V$ is a complex submanifold of $\mathbb{C}^n$. If $x \in X$ is not regular, then x is called *singular*, or a *singularity*. The set of singular points of X is denoted by $\mathrm{Sing}(X)$.

Examples 3.3.12.

(1) Every point in a complex submanifold of $\mathbb{C}^n$ is a regular point.

(2) Let $X = \{(x,y) \in \mathbb{C}^2 : x^2 - y^3 = 0\}$. Then X is an analytic subset of $\mathbb{C}^2$, $0 = (0,0) \in X$ is a singular point, see Exercise 3.3.37.

In chapter 4 we will obtain a characterization of the singular points in terms of the Jacobian matrix (cf. Theorems 4.3.6 and 4.3.15).

We now turn to Newton's Lemma. Newton's Lemma, as you learned it in a calculus course, is a method for approximating zeros of differentiable functions in one variable. Consider a differentiable function $y = f(x)$ in one real variable. The question is to find a zero of $f(x) = 0$. The idea is to take an approximate solution a_k, and to obtain a better approximation a_{k+1} by intersecting the tangent line of the graph of f at $(a_k, f(a_k))$ with the x-axis.

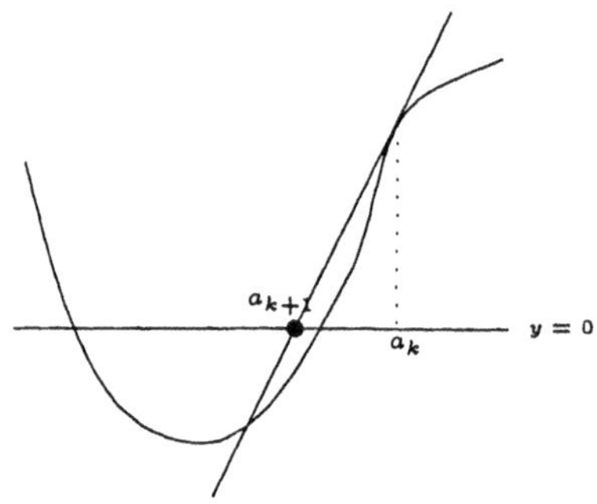

A simple calculation shows

$$a_{k+1} = a_k - \frac{f(a_k)}{f'(a_k)}. \tag{3.14}$$

Starting off with an initial approximation a_0 we get a sequence $a_1, a_2, \ldots$, which, under appropriate assumptions, converges to a number a with $f(a) = 0$.

We apply similar ideas to the equation $f(x,y) = y^2 - 1 - x$. We want to find a power series, $y = \varphi(x)$ with $\varphi(x)^2 = x + 1$, that is, we want to find a power series for $\sqrt{x+1}$. We start of with the initial approximation $\varphi_0(x) = 1$, and iterate with the obvious analog of (3.14):

$$\varphi_{k+1}(x) = \varphi_k(x) - \frac{f(x, \varphi_k(x))}{\frac{\partial f}{\partial y}(x, \varphi_k(x))}. \tag{3.15}$$

One calculates that

$$\varphi_1(x) = 1 + \frac{x}{2}, \quad \varphi_2(x) = 1 + \frac{x}{2} - \frac{\frac{x^2}{4}}{2(1 + x/2)} = 1 + \frac{x}{2} - \frac{x^2}{8} + \ldots,$$

so one sees the beginning of the power series of $\sqrt{1+x}$.

We now generalize.

Corollary 3.3.13 (Newton's Lemma: simple case).
Let $f \in \mathbb{C}\{x_1, \ldots, x_n, y\}$, $\overline{y}(x) \in \mathbb{C}\{x_1, \ldots, x_n\}$ and $c \geq 1$ with

$$f(x, \overline{y}(x)) \equiv 0 \mod \left(\frac{\partial f}{\partial y}(x, \overline{y}(x))\right)^2 (x)^c.$$

Then there exists a $y \in \mathbb{C}\{x_1, \ldots, x_n\}$ with

(1) $f(x,y(x)) = 0$.

(2) $y \equiv \overline{y} \mod \left(\frac{\partial f}{\partial y}(x,\overline{y}(x))\right)(x)^c$.

Proof. Write $\frac{\partial f}{\partial y}(x,\overline{y}(x)) = \alpha \in \mathbb{C}\{x_1,\dots,x_n\}$. We choose a new variable t and substitute

$$y = \overline{y}(x) + t \cdot \alpha$$

in $f(x,y)$. We therefore want to find a $t \in \mathbb{C}\{x_1,\dots,x_n\}$ with

$$g(x,t) := f\big(x,\overline{y}(x) + t\cdot\alpha\big) = 0.$$

The existence will be guaranteed by applying the Implicit Function Theorem. Indeed Taylor's formula gives

$$g(x,t) := f\big(x,\overline{y}(x) + t\alpha\big) = f\big(x,\overline{y}(x)\big) + t\alpha\frac{\partial f}{\partial y}\big(x,\overline{y}(x)\big) + \alpha^2 R = f\big(x,\overline{y}(x)\big) + t\alpha^2 + \alpha^2 R$$

with $R \in (t)^2$. By assumption $f(x,\overline{y}(x)) = \alpha^2 \cdot r$ with $r \in (x)^c$. Therefore $g(x,t) = \alpha^2(t + R + r)$. Hence it suffices to solve $t + R + r = 0$, with $R \in (t)^2$ and $r \in (x)^c$. Hence this equation satisfies the assumption of the Implicit Function Theorem. Even more, by Exercise 3.3.33 there exists a $t \in (x)^c$ with $g(x,t) = 0$. □

A more general version of Newton's Lemma, whose proof is only notationally more difficult, will be proved in Exercise 3.3.34. This generalization will, for example, be used in the proof of Artin's Approximation Theorem, see Chapter 8.

Corollary 3.3.14. *$\mathbb{C}\{x_1,\dots,x_n\}$ and $\mathbb{C}[[x_1,\dots,x_n]]$ are Noetherian rings.*

Proof. As the proofs are similar, we only give the proof for the analytic case. The proof is by induction on n. For the case $n = 0$ we have the field $\mathbb{C}$, which is obviously Noetherian. Now take an ideal $I \subset \mathbb{C}\{x_1,\dots,x_n\}$. We have to prove that I is finitely generated. If $I = (0)$, this is obviously the case. If $I \neq (0)$, take $0 \neq f \in I$. After a general linear change of coordinates we may assume that f is regular in x_n, say of order b, see 3.2.2. By 3.2.6, which is a corollary of the Weierstraß Division Theorem, it follows that $\mathbb{C}\{x_1,\dots,x_n\}/(f)$ is a finitely generated (even free) $\mathbb{C}\{x_1,\dots,x_{n-1}\}$–module. Now $\mathbb{C}\{x_1,\dots,x_{n-1}\}$ is Noetherian by induction, so that $\mathbb{C}\{x_1,\dots,x_n\}/(f)$ is a Noetherian $\mathbb{C}\{x_1,\dots,x_{n-1}\}$–module by 1.2.15. In particular I modulo (f) is finitely generated as $\mathbb{C}\{x_1,\dots,x_{n-1}\}$–module, and therefore certainly finitely generated as $\mathbb{C}\{x_1,\dots,x_n\}$–module, say by the classes of $f_1,\dots,f_t$. It follows that $f,f_1,\dots,f_t$ generate I. □

Remarks 3.3.15.

(1) Analogous ideas also give a new proof for the Hilbert Basis Theorem for the polynomial ring $\mathbb{C}[x_1,\dots,x_n]$.

(2) An alternative proof for the induction in the proof of Corollary 3.3.14 *uses* the Hilbert Basis Theorem. It runs as follows. Let $I \neq (0)$ be an ideal, and $0 \neq f \in I$. Again, we may assume that $\mathbb{C}\{x_1,\dots,x_n\}/(f)$ is a finitely generated $\mathbb{C}\{x_1,\dots,x_{n-1}\}$–module. We may also assume that f is a Weierstraß polynomial in x_n. Hence $\mathbb{C}\{x_1,\dots,x_n\}/(f) = \mathbb{C}\{x_1,\dots,x_{n-1}\}[x_n]/(f)$. It follows from the Weierstraß Division Theorem that

$$(I \cap \mathbb{C}\{x_1,\dots,x_{n-1}\}[x_n])\,\mathbb{C}\{x_1,\dots,x_n\} = I.$$

By induction and the Hilbert Basis Theorem, the ring $\mathbb{C}\{x_1,\dots,x_{n-1}\}[x_n]$ is Noetherian. Hence I is finitely generated as $\mathbb{C}\{x_1,\dots,x_{n-1}\}[x_n]$–ideal, and in particular finitely generated as $\mathbb{C}\{x_1,\dots,x_{n-1},x_n\}$–ideal.

The next thing we want to show is that the power series ring is a unique factorization domain. For the proof we need the following fact, which is interesting in its own right.

Lemma 3.3.16. *Let $h \in \mathbb{C}\{x_1,\dots,x_{n-1}\}[x_n]$ be a Weierstraß polynomial. Then*

$$h \text{ is reducible in } \mathbb{C}\{x_1,\dots,x_{n-1}\}[x_n] \iff h \text{ is reducible in } \mathbb{C}\{x_1,\dots,x_n\}.$$

In particular, every irreducible factor of h in $\mathbb{C}\{x_1,\dots,x_n\}$ is x_n–regular and, therefore, every decomposition of h in $\mathbb{C}\{x_1,\dots,x_n\}$ gives a decomposition of h into Weierstraß polynomials by multiplying with suitable units.

Similar statements hold for the formal power series ring.

Proof. We again only consider the analytic case, the formal case being similar.

$\Longleftarrow$: Let $h = g_1 \cdot g_2$ be a factorization in $\mathbb{C}\{x_1,\dots,x_n\}$ with g_1 and g_2 nonunits. As h is regular in x_n it follows that g_1 and g_2 are regular in x_n. As they are both nonunits, they are regular of order at least one. By the Weierstraß Preparation Theorem we can write $g_1 = u_1h_1$, and $g_2 = u_2h_2$, where u_1 and u_2 are units and h_1 and h_2 are Weierstraß polynomials in x_n. Hence $h = u_1u_2h_1h_2$. Now h_1h_2 is a Weierstraß polynomial. It follows from the uniqueness statement in the Weierstraß Preparation Theorem that $u_1u_2 = 1$. Therefore, $h = h_1h_2$, and h is reducible in $\mathbb{C}\{x_1,\dots,x_{n-1}\}[x_n]$.

$\Longrightarrow$: Suppose $h = g_1 \cdot g_2$, with g_1 and g_2 nonunits in $\mathbb{C}\{x_1,\dots,x_{n-1}\}[x_n]$. We have to show that g_1 and g_2 are also nonunits in $\mathbb{C}\{x_1,\dots,x_n\}$. But suppose, for example, that g_1 is a unit. As g_1 is not a unit in $\mathbb{C}\{x_1,\dots,x_{n-1}\}[x_n]$ it follows that $\deg_{x_n}(g_1) \geq 1$. So $\deg_{x_n}(g_2) < \deg_{x_n}(h)$. But $h = g_1 \cdot g_2$, so g_2 is regular in x_n. This implies that h is x_n–regular of order at most $\deg_{x_n}(g_2)$ which is a contradiction to the fact that h is a Weierstraß polynomial. □

Corollary 3.3.17. *$\mathbb{C}\{x_1,\dots,x_n\}$ and $\mathbb{C}[[x_1,\dots,x_n]]$ are factorial rings.*

Proof. Again we only treat the convergent case. The proof is by induction on n. The case $n = 0$ is trivial. Suppose therefore that $\mathbb{C}\{x_1,\dots,x_{n-1}\}$ is factorial. Then by the Lemma of Gauß (1.4.4) $\mathbb{C}\{x_1,\dots,x_{n-1}\}[x_n]$ is factorial. Let $f \in \mathbb{C}\{x_1,\dots,x_n\}$. After a general linear change of coordinates we may assume that f is regular in x_n. By the Weierstraß Preparation Theorem $f = uh$, where u is a unit and h is a Weierstraß polynomial. As by the previous lemma, the polynomial factors of h in both $\mathbb{C}\{x_1,\dots,x_{n-1}\}[x_n]$ and $\mathbb{C}\{x_1,\dots,x_n\}$ are equal, it follows that there is a unique factorization of h. □

For the case of two variables we obtain the following result which will be very useful in Chapter 5.

Corollary 3.3.18. *Let $f \in \mathbb{C}\{x,y\}$ and $\overline{y} \in (x)\mathbb{C}[[x]]$ such that $f(x,\overline{y}) = 0$. Then $\overline{y} \in \mathbb{C}\{x\}$.*[8]

[8] This corollary is also true for n variables $x_1,\dots,x_n$ and can be proved, for instance, using Artin's Approximation Theorem from Chapter 8. In fact, it follows from Exercise 8.1.7.

Proof. Suppose the converse, that is, assume that $\overline{y} \notin \mathbb{C}\{x\}$. First of all, we may assume that $f = y^n + a_{n-1}(x)y^{n-1} + \cdots + a_0(x)$ is an irreducible Weierstraß polynomial by (3.3.16). As $\overline{y} \notin \mathbb{C}\{x\}$ it follows that $n > 1$. From Lemma 1.4.5 it follows that f is irreducible in $Q(\mathbb{C}\{x\})[y]$. The ring $Q(\mathbb{C}\{x\})[y]$ is a principal ideal domain, hence $(f, \frac{\partial f}{\partial y}) = (1)$. By clearing denominators, we can find $p,q \in \mathbb{C}\{x\}[y]$ and $0 \neq h \in \mathbb{C}\{x\}$ with $qf + p\frac{\partial f}{\partial y} = h$. Plugging in $y = \overline{y}(x)$ it follows that $\frac{\partial f}{\partial y}(x,\overline{y}) \neq 0$. Let $\operatorname{ord}(\frac{\partial f}{\partial y}(x,\overline{y})) = m$ and choose $y_1 \in \mathbb{C}[x]$ such that $\operatorname{ord}(\overline{y} - y_1) \geq 2m + 1$. Then $f(x,y_1) \equiv 0 \mod (\frac{\partial f}{\partial y}(x,y_1))^2(x) = (x^{2m+1})$.

Using Newton's Lemma 3.3.13 we obtain an $y_2 \in \mathbb{C}\{x\}$ such that $f(x,y_2) = 0$ and $\overline{y} - y_2 \in (x)(\frac{\partial f}{\partial y}(x,y_1))$. Now $f(x,y) = g(x,y)(y - y_2)$, and $g(x,\overline{y}) = 0$. Hence g is not a unit, an we obtain a contradiction to the fact that f is irreducible. □

We now come to the Noether Normalization Theorem. Its proof is analogous to the proof in the affine case, see 2.2.9, but because it is such an important result, we spell it out.

Corollary 3.3.19 (Noether Normalization). *Let $I \subset \mathbb{C}\{x_1,\ldots,x_n\}$ be an ideal. Then after a general linear coordinate change*[9] *we have that there exists a k such that*

$$\mathbb{C}\{x_1,\ldots,x_k\} \subset \mathbb{C}\{x_1,\ldots,x_n\}/I$$

and such that $\mathbb{C}\{x_1,\ldots,x_n\}/I$ is a finitely generated $\mathbb{C}\{x_1,\ldots,x_k\}$-module. If we have such a situation we call

$$\mathbb{C}\{x_1,\ldots,x_k\} \subset \mathbb{C}\{x_1,\ldots,x_n\}/I$$

a Noether normalization. *If moreover $I \neq (0)$, then $k < n$.*

Proof. If $I = (0)$, we can take $k = n$. Otherwise, the theorem is proved by induction on n. The case $n = 0$ is trivial. Take $0 \neq g \in I$. By 3.2.2 we have, after a general linear change of coordinates, that g is regular of order b in x_n. By Corollary 3.2.6, we know that $\mathbb{C}\{x_1,\ldots,x_n\}/g$ is a finitely generated $\mathbb{C}\{x_1,\ldots,x_{n-1}\}$-module. Because $g \in I$, then certainly $\mathbb{C}\{x_1,\ldots,x_n\}/I$ is a finitely generated $\mathbb{C}\{x_1,\ldots,x_{n-1}\}/(I \cap \mathbb{C}\{x_1,\ldots,x_{n-1}\})$-module. We apply the induction hypothesis to the ideal $I \cap \mathbb{C}\{x_1,\ldots,x_{n-1}\}$. After a general linear coordinate change in $x_1,\ldots,x_{n-1}$ we have that

$$\mathbb{C}\{x_1,\ldots,x_k\} \subset \mathbb{C}\{x_1,\ldots,x_{n-1}\}/(I \cap \mathbb{C}\{x_1,\ldots,x_{n-1}\})$$

is finitely generated. We compose this with the inclusion

$$\mathbb{C}\{x_1,\ldots,x_{n-1}\}/(I \cap \mathbb{C}\{x_1,\ldots,x_{n-1}\}) \subset \mathbb{C}\{x_1,\ldots,x_n\}/I.$$

As a composition of finite rings extension is finite, see 1.5.2, we deduce that the ring extension $\mathbb{C}\{x_1,\ldots,x_k\} \subset \mathbb{C}\{x_1,\ldots,x_n\}/I$ is finitely generated. This proves the corollary. □

Due to the fact that for general coordinate changes we get a Noether normalization, we can get the following situation, which will play an important role in the proof of the Nullstellensatz.

[9] More precisely, there exists a Zariski open subset U of $\mathrm{Gl}_n(\mathbb{C})$ such that the statement holds for all coordinate transformations in U.

Corollary 3.3.20. *Suppose that* $\mathfrak{p} \subset \mathbb{C}\{x_1,\dots,x_n\}$ *is a prime ideal. Then we can get a Noether normalization*

$$\mathbb{C}\{x_1,\dots,x_k\} \subset \mathbb{C}\{x_1,\dots,x_n\}/\mathfrak{p},$$

with the property that the class of x_{k+1} *in* $Q(\mathbb{C}\{x_1,\dots,x_n\}/\mathfrak{p})$ *is a primitive element of the field extension*

$$Q(\mathbb{C}\{x_1,\dots,x_k\}) \subset Q(\mathbb{C}\{x_1,\dots,x_n\}/\mathfrak{p}).$$

We call a Noether normalization with the above property a primitive Noether normalization.

Proof. We know from the proof of the Theorem of the Primitive Element, see for example [Van der Waerden 1971], that the class of $c_{k+1}x_{k+1}+\dots+c_nx_n$ in $Q(\mathbb{C}\{x_1,\dots,x_n\}/\mathfrak{p})$ is a primitive element for the field extension $Q(\mathbb{C}\{x_1,\dots,x_n\}/\mathfrak{p}) \supset Q(\mathbb{C}\{x_1,\dots,x_k\})$ for general $c_{k+1},\dots,c_n \in \mathbb{C}$. Doing the coordinate change $x'_{k+1} = c_{k+1}x_{k+1}+\dots+c_nx_n$ the corollary follows. □

Corollary 3.3.21 (Hensel's Lemma). *Let* $F \in \mathbb{C}\{x_1,\dots,x_n\}[T]$ *and suppose that* $F \equiv (T-c_1)^{s_1}\cdots(T-c_e)^{s_e} \mod (x_1,\dots,x_n)$, *where the* $c_i \in \mathbb{C}$ *are pairwise different.*

Then $F = F_1\cdots F_e$, $F_i \in \mathbb{C}\{x_1,\dots,x_n\}[T]$, $F_i \equiv (T-c_i)^{s_i} \mod (x_1,\dots,x_n)$, *and the* F_i *have degree* s_i *in* T.

Proof. We use induction on e. The case $e=1$ is trivial. We may assume that $c_1 = 0$ (otherwise we perform the transformation $T \mapsto T + c_1$). Then $F(0,T) = T^{s_1}\cdot H(T)$ and $H(0) \neq 0$. Now we use the Weierstraß Preparation Theorem (3.2.4) and obtain $F = E\cdot F_1$, where F_1 is a Weierstraß polynomial of degree s_1 with respect to T, and E is a unit in $\mathbb{C}\{x_1,\dots,x_n,T\}$.

On the other hand, E also has to be a polynomial in T: divide F by F_1 with remainder in $\mathbb{C}\{x_1,\dots,x_n\}[T]$, $F = E_1F_1 + R_1$ with $E_1,R_1 \in \mathbb{C}\{x_1,\dots,x_n\}[T]$ and $\deg_T R_1 < s_1$. By the uniqueness statements in the Weierstraß Division Theorem, we obtain $R_1 = 0$ and $E = E_1$. Now $F_1(0,T) = T^{s_1}$ and, consequently, $E(0,T) = (T-c_2)^{s_2}\cdots(T-c_e)^{s_e}$. The corollary is proved by induction hypothesis.

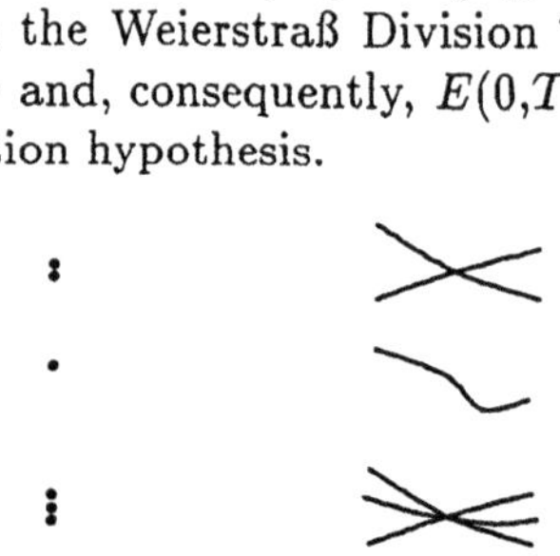

Zero set of $F(0,T)$ Zero set of F

□

Corollary 3.3.21 is equivalent to the following seemingly more general lemma:

Remark 3.3.22 (General Hensel's Lemma). *Let* $F,G,H \in \mathbb{C}\{x_1,\dots,x_n\}[T]$ *such that*

(1) $F \equiv G\cdot H \mod (x_1,\dots,x_n)$;

(2) G is a monic polynomial;

(3) $(G,H,x_1,\dots,x_n) = \mathbb{C}\{x_1,\dots,x_n\}[T]$.

Then there exist $G_1,H_1 \in \mathbb{C}\{x_1,\dots,x_n\}[T]$ *such that* $F = G_1 \cdot H_1$*, Moreover* $G_1 \equiv G \mod (x_1,\dots,x_n)$*, and* $H_1 \equiv H \mod (x_1,\dots,x_n)$.

The proof of the equivalence is left as part of Exercise 3.3.36.

Corollary 3.3.23. *Let f be a polynomial,*

$$f = x_n^k + a_1 x_n^{k-1} + \ldots + a_k, \quad a_i \in \mathbb{C}\{x_1,\dots,x_{n-1}\}.$$

Consider an open set $U \subset \mathbb{C}^{n-1}$ *on which the* a_i *converge, and a point* $p \in U$ *such that the polynomial*

$$x_n^k + a_1(p)x_n^{k-1} + \ldots + a_k(p)$$

has k different roots in $\mathbb{C}$*. Then there exists an open neighborhood* $V \subset U$ *of p, and k different holomorphic functions* $\alpha_1,\dots,\alpha_k$ *defined on V such that for all* $(x_1,\dots,x_n) \in V \times \mathbb{C}$

$$f(x_1,\dots,x_n) = \prod_{j=1}^{k} (x_n - \alpha_j(x_1,\dots,x_{n-1})).$$

Proof. This is in fact Hensel's Lemma for the case that all s_i are equal to one. □

The following corollary we need in the proof of the Finite Mapping Theorem 6.3.5,

Corollary 3.3.24. *Consider a monic polynomial* $F \in \mathbb{C}\{x_1,\dots,x_n\}[T]$*. Suppose that* $F(0,T) = (T-c_1)^{s_1}\cdots(T-c_e)^{s_e}$ *for pairwise different* $c_i \in \mathbb{C}$*. Let* $s = \sum_{i=1}^{e} s_i$ *be the degree of F. Let* $p_i := (0,\dots,0,c_i) \in \mathbb{C}^{n+1}$*. Then for all* $g_1,\dots,g_e$ *with* $g_i \in \mathcal{O}_{n+1,p_i}$ *there exist uniquely determined* $q_i \in \mathcal{O}_{n+1,p_i}$ *and a polynomial* $r \in \mathbb{C}\{x_1,\dots,x_n\}[T]$ *of degree less than s such that*

$$g_i = q_i F + r$$

holds in $\mathcal{O}_{n+1,p_i} = \mathbb{C}\{x_1,\dots,x_n,T-c_i\}$ *for* $i = 1,\dots,e$.

Proof. We first show existence. By Hensel's Lemma, we may write $F = F_1 \cdots F_e$, with $F_i(0,T) = (T-c_i)^{s_i}$, that is, the F_i are Weierstraß polynomials with respect to $T - c_i$. We write $E_i = \prod_{j\neq i} F_j$, so that $F = E_i \cdot F_i$. Note that E_i is a unit in $\mathcal{O}_{n+1,p_i}$. We apply the Weierstraß Division Theorem in $\mathcal{O}_{n+1,p_i}$ for $i = 1,\dots,e$:

$$E_i^{-1} g_i = q_i' F_i + r_i$$

where $r_i \in \mathbb{C}\{x_1,\dots,x_n\}[T]$ has degree smaller than s_i. So $g_i = q_i' F + r_i E_i$. We set $r = r_1 E_1 + \ldots + r_e E_e$. Thus $g_i = q_i' F + r - \sum_{j\neq i} r_j E_j$. In the ring $\mathcal{O}_{n+1,p_i}$ the elements F_j for $j \neq i$ are units, so we get

$$g_i = q_i' F + r - \sum_{j\neq i} r_j F_j^{-1} F.$$

Hence we may take $q_i = q_i' - \sum_{j\neq i} r_j F_j^{-1}$.

To show uniqueness, it suffices to show that from $q_i F - r = 0$ for $i = 1, \dots, e$ with degree of r in T smaller than s, it follows that $r = 0$. Thus we may write $r = F_1 \cdots F_e q_i$ with $q_i \in \mathcal{O}_{n+1,p_i}$. We define elements

$$P_1 := \frac{r}{F_1}, \quad P_2 := \frac{r}{F_1 F_2}, \quad \dots, \quad P_e := \frac{r}{F_1 F_2 \cdots F_e}$$

Note that as $r = F_1 \cdots F_e q_i$ the P_i are elements of $\mathcal{O}_{n+1,p_i}$. Note that

$$r = P_1 F_1; \quad P_{j-1} = P_j F_j \quad \text{for } j = 2, \dots, e.$$

As F_i and r are polynomials in $T - c_i$ it follows from the uniqueness in the Weierstraß Division Theorem in $\mathcal{O}_{n+1,p_1}$ and the fact we can do polynomial division that P_1 is a polynomial in T. Inductively, we get that $P_2, \dots, P_e$ are polynomials in T. So we get $r = P_e F_1 F_2 \cdots F_e = P_e F$. As the degree of F is equal to s, and r has smaller degree, it follows that $r = P_e = 0$. □

Corollary 3.3.25. *Let $R = \mathcal{O}_n/I$ be an analytic algebra, and S be an R-algebra, which is finitely generated as R-module. Suppose that S is a local ring. Then S is also an analytic algebra.*

Proof. First of all, take a Noether normalization $\mathbb{C}\{x_1, \dots, x_k\} = \mathcal{O}_k \hookrightarrow R$. Therefore, we may assume that $R = \mathcal{O}_k$ from the beginning. Let $\mathfrak{m}_R$ be the maximal ideal of R, and $\mathfrak{m}_S$ be the maximal ideal of S. We have to prove that $S \cong \mathbb{C}\{y_1, \dots, y_m\}/J$ for some m and J. Now $S/\mathfrak{m}_R S$ is a finitely generated $R/\mathfrak{m}_R = \mathbb{C}$-module, hence a finitely generated vector space. As $\mathfrak{m}_R S \subset \mathfrak{m}_S$, it follows that $S/\mathfrak{m}_S$ is a finite extension of the field $\mathbb{C}$. As $\mathbb{C}$ is algebraically closed it follows that $S/\mathfrak{m}_S = \mathbb{C}$. Therefore for all $u \in S$, there exists a $c \in \mathbb{C}$ such that $u - c \in \mathfrak{m}_S$. It follows that we can take generators $1 = u_0, u_1, \dots, u_t$ of S as R-module with $u_i \in \mathfrak{m}_S$ for $i = 1, \dots, t$. We look at the surjective $\mathbb{C}$-algebra homomorphism

$$\begin{array}{rcl} \Psi : \mathcal{O}_k[y_1, \dots, y_t] & \longrightarrow & S \\ y_i & \mapsto & u_i \end{array}$$

and conclude that $S \cong \mathcal{O}_k[y_1, \dots, y_t]/J$ for $J = \operatorname{Ker}(\Psi)$. But we want to show that S is a quotient of a *power series* ring. Therefore, we look at the canonical inclusion $\mathcal{O}_k[y_1, \dots, y_t] \subset \mathcal{O}_{k+t}$, and put $\widetilde{J} := J \cdot \mathcal{O}_{k+t}$. We obtain a canonical map:

$$\theta : \mathcal{O}_k[y_1, \dots, y_t]/J \to \mathcal{O}_{k+t}/\widetilde{J}.$$

Therefore, it suffices to show that θ is an isomorphism. So we have to prove two things.

Step 1: θ is surjective. As S is an integral extension of $\mathcal{O}_k$, there exists for all i a monic polynomial $p_i \in \mathcal{O}_k[X]$ such that $p_i(u_i) = 0$. This shows, by definition of the ideal J, that $p_i(y_i) \in J$. By the Weierstraß Division Theorem, any element in $\mathcal{O}_{k+t}$ is modulo $\widetilde{J}$ equivalent to a *polynomial* in $y_1, \dots, y_t$. This shows that the map θ is surjective.

Step 2: θ is injective. Take generators $p_1, \dots, p_s$ for the ideal J. Take any element $p \in \operatorname{Ker}(\theta) = \mathcal{O}_k[y_1, \dots, y_t] \cap \widetilde{J}$. We have to show that $p \in J$. For this it suffices to show by definition of J that $\Psi(p) = 0$. Now p is a polynomial in $y_1, \dots, y_t$, and we can write:

$$p = g_1 p_1 + \dots + g_s p_s, \quad g_i \in \mathcal{O}_{k+t}.$$

The g_i are power series in general, so it does not make sense to talk about $\psi(g_i)$. We therefore calculate modulo high powers of the y_i, that is, for every natural number k we write $g_i = g_{ik} + h_{ik}$, where $h_{ik} \in (y_1, \dots, y_t)^{k+1}$ and the g_{ik} are polynomials. Because $\Psi(y_i) = u_i \in \mathfrak{m}_S$ it follows that modulo $\mathfrak{m}_S^{k+1}$:

$$\Psi(p) = \Psi(g_{ik}p_1 + \dots + g_{sk}p_s).$$

But $\Psi(p_i) = 0$, because the p_i were supposed to be generators of J. It follows $\Psi(p) \in \mathfrak{m}_S^{k+1}$ for all k. It follows from Krull's Intersection Theorem 1.3.5 that $\Psi(p) = 0$, or, what is the same, $p \in J$. □

At the end of the section we look at generalization of this result, namely at the case S is not necessarily local.

Corollary 3.3.26. *Let R be an analytic algebra, and S be an R–algebra, which is finitely generated as R–module. Then*

$$S \cong \bigoplus_{i=1}^{r} S_i$$

and the S_i are analytic algebras. So S is a finite direct sum of analytic algebras.

The proof uses the case that S is a local ring, see 3.3.25 and the following lemma.

Lemma 3.3.27. *Let $R = \mathcal{O}_n$, and $S \supset R$ be an R–algebra, finitely generated as R–module. Let $\mathfrak{m}$ be the maximal ideal of R. Let $y \in S$ be an idempotent modulo $\mathfrak{m}S$, that is, $y^2 - y \in \mathfrak{m}S$. Then y can be lifted to an idempotent in S, that is, there exists a $z \in S$ with $z^2 = z$ and $z \equiv y \mod \mathfrak{m}S$.*

Proof. As S is a finitely generated R–module, and $y^2 - y \in \mathfrak{m}S \subset \sqrt{\mathfrak{m}S}$ it follows from Lemma 1.5.11 that $y^2 - y$ is integral over $\mathfrak{m}$. From this integral equation it follows that

$$(y^2 - y)^r = \sum_{k=0}^{m} c_k y^k, \text{ for some } c_k \in \mathfrak{m}.$$

Define

$$F(T) := (T^2 - T)^r - \sum_{k=0}^{m} c_k T^k \in R[T].$$

So $F(T) \equiv T^r(T-1)^r \mod \mathfrak{m}$. We can now apply Hensel's Lemma 3.3.21 and get $F = G \cdot H$, $G(T) \equiv T^r \mod \mathfrak{m}$ and $H(T) \equiv (T-1)^r \mod \mathfrak{m}$. In particular, from $y^2 \equiv y \mod \mathfrak{m}S$, we obtain $G(y) \equiv y^r \equiv y \mod \mathfrak{m}S$ and $H(y) \equiv (y-1)^r \equiv (-1)^r(1-y) \mod \mathfrak{m}S$. Hence

$$1 = y + (1 - y) \equiv G(y) + (-1)^r H(y) \mod \mathfrak{m}S.$$

Let $\mathfrak{M}$ be a maximal ideal of S. Then $\mathfrak{M} \cap R = \mathfrak{m}$ is the maximal ideal of R. This can be proved directly, or follows from the Going-Up Theorem. It follows that $\mathfrak{m}S$ is contained in all maximal ideals. Hence all elements of $1 + \mathfrak{m}S$ are units in S, and the inverse is also congruent to 1 modulo $\mathfrak{m}S$, see 1.3.25. Thus $\big(G(y) + (-1)^r H(y)\big)\big(1 + \alpha\big) = 1$ for some $\alpha \in \mathfrak{m}S$. We get

$$1 = c \cdot G(y) + dH(y)$$

for suitable $c,d \in S$, and $c \equiv 1 \mod \mathfrak{m}S$. We now define $z := c \cdot G(y)$. Then $z \equiv y \mod \mathfrak{m}S$, and

$$z = \big(c \cdot G(y) + dH(y)\big)z = z^2 + dH(y)cG(y) = z^2$$

because $0 = F(y) = G(y)H(y)$. This is what we had to show. □

Proof of Corollary 3.3.26. Let $\mathfrak{m}$ be the maximal ideal of R. The proof is by induction on $\dim_{\mathbb{C}}(S/\mathfrak{m}S)$. The case that S is local has already been considered. So assume that S is not a local ring. As S is a finitely generated R–module we get that $S/\mathfrak{m}S$ is an Artinian $\mathbb{C}$–algebra (1.1.6). Because of Theorem 1.4.26 we have[10]

$$S/\mathfrak{m}S \cong \bigoplus_{i=1}^{r} C_i, \quad C_i \text{ nonzero local Artinian } \mathbb{C}\text{–algebras.}$$

The number r is bigger than one, as $S/\mathfrak{m}S$ is not a local ring. In particular $S/\mathfrak{m}S$ contains an idempotent different from 1. (For example $(1,0,\dots,0)$ in $\oplus_{i=1}^{r} C_i$).

This idempotent can be lifted by Lemma 3.3.27 to an idempotent $y \in S$, $y \neq 1$. We apply the Chinese Remainder Theorem 1.1.12 and get

$$S \cong S/(y) \times S/(1-y).$$

To apply the Chinese Remainder Theorem, we need that $(y) + (1-y) = R$ (this is obvious), and $(y) \cap (1-y) = (0)$. To show this, let $a \in (y) \cap (1-y)$. Then we can write $a = \alpha y = \beta(1-y)$. Multiply with y and get $\alpha y^2 = 0$. As $y^2 = y$ it follows $a = \alpha y = 0$. Note that both y and $(1-y)$ are not in $\mathfrak{m}S$. Indeed, they were chosen such that modulo $\mathfrak{m}S$ the element y is equivalent to $(1,0,\dots,0)$, and $1-y$ is equivalent to $(0,1,\dots,1)$. So it follows that $\dim_{\mathbb{C}}(S/\mathfrak{m}S) > \dim_{\mathbb{C}}\Big(S/\big(\mathfrak{m}S + (y)\big)\Big)$. Similarly $\dim_{\mathbb{C}}(S/\mathfrak{m}S) > \dim_{\mathbb{C}}\Big(S/\big(\mathfrak{m}S + (1-y)\big)\Big)$. Thus we can apply the induction hypothesis, and the corollary follows. □

Exercises

3.3.28. Compute the power series of $y = -1 + \sqrt{x+1}$ using the ideas of Example 3.3.4.

3.3.29. Prove the Inverse Function Theorem 3.3.7.

3.3.30 (General Noether Normalization). Let $I \subset \mathbb{C}\{x_1,\dots,x_n\}$ be an ideal. Then, after a general linear coordinate change, we find that there exists a k such that

$$\mathbb{C}\{x_1,\dots,x_k\} \subset \mathbb{C}\{x_1,\dots,x_n\}/I$$

and for all $j > k$ there are monic polynomials $f_j \in I \cap \mathbb{C}\{x_1,\dots,x_{j-1}\}[x_j]$ (with respect to x_j) with $\deg_{x_j}(f_j) = \operatorname{ord}(f_j)$ (considered as a power series in $\mathbb{C}\{x_1,\dots,x_j\}$).

If, moreover, I is a prime ideal then $Q(\mathbb{C}\{x_1,\dots,x_n\}/I) = Q(\mathbb{C}\{x_1,\dots,x_k\})[x_{k+1}]/(f_{k+1})$. (Hint: Assume in the proof of 3.3.19 that the order of $g \in I$ is b and apply 3.2.2 and the Weierstraß Preparation Theorem 3.2.4. In case of I being a prime ideal, combine this idea with the proof of 3.3.20.)

[10] Recall that in the direct sum of algebras the multiplication is component wise.

3.3.31. Consider a general Noether normalization $\mathcal{O}_k \subset \mathcal{O}_{X,x} = \mathcal{O}_n/\mathfrak{p}$ in the sense of 3.3.30. Let f be a power series of order t. Show that modulo $\mathfrak{p}$ the element f is equivalent to a polynomial in $x_{k+1},\dots,x_n$ which has order at least t.
(Hint: Reduce to the case that $\mathfrak{p} = (P_{k+1},\dots,P_n)$, where the P_j are Weierstraß polynomials in x_j with coefficients in $\mathbb{C}\{x_1,\dots,x_k\}$. For the case $\mathfrak{p} = (P_n)$, use Exercise 3.2.19, and apply induction to the coefficients.)

3.3.32. Consider a prime ideal $\mathfrak{p} \in \mathbb{C}\{x_1,\dots,x_n\}$, and a Noether normalization $\mathbb{C}\{x_1,\dots,x_k\} \subset \mathbb{C}\{x_1,\dots,x_n\}/\mathfrak{p}$. Let $P_j(X)$ be an integral equation of x_j for $j = k+1,\dots,n$ of minimal degree. Prove that the P_j are Weierstraß polynomials.
(Hint: Use the Weierstraß Preparation Theorem.)

3.3.33. Let $f \in \mathbb{C}\{x_1,\dots,x_n,y\}$ with $f(0) = 0$ and $\frac{\partial f}{\partial y}(0) \neq 0$. Suppose moreover that $f \in (y) + (x_1,\dots,x_n)^c$. Then there exists a $\varphi \in (x_1,\dots,x_n)^c$ with

$$f(x_1,\dots,x_n,\varphi) = 0.$$

Generalize this to the case of m variables $y_1,\dots,y_m$. (Hint: Follow the steps in a proof of the formal Implicit Function Theorem.)

3.3.34 (Newton's Lemma). Let $f_1,\dots,f_m \in \mathbb{C}\{x_1,\dots,x_n, y_1,\dots,y_N\}$, $m \leq N$ and $J = J(f_1,\dots,f_m)$ the ideal generated by the m–minors of the Jacobian matrix $\left(\frac{\partial f_i}{\partial y_j}\right)$. Suppose $f_i(x,\overline{y}(x)) \equiv 0 \mod J^2(\overline{y}(x))(x_1,\dots,x_n)^c$ for $i = 1,\dots,m$, $\overline{y}(x) \in \mathbb{C}\{x_1,\dots,x_n\}^N$ and $c \geq 1$.
Then there exists a $y \in \mathbb{C}\{x_1,\dots,x_n\}^N$ such that

$$\begin{aligned} &f_i(x,y(x)) = 0, \quad i = 1,\dots,m \\ &y \equiv \overline{y} \mod J(\overline{y}(x))(x_1,\dots,x_n)^c. \end{aligned}$$

Prove this lemma along the following lines.

(1) Let $M_1,\dots,M_s$, with $s = \binom{N}{m}$ be the m–minors of $\left(\frac{\partial f_i}{\partial y_j}(\overline{y}(x))\right)$. Take new variables t_{ij} for $1 \leq i \leq N$ and $1 \leq j \leq s$. Substitute $y_i = \overline{y}_i + \sum_{j=1}^{s} t_{ij}M_j$. Let $f = \begin{pmatrix} f_1 \\ \vdots \\ f_m \end{pmatrix}$. Show that it suffices to solve for t_{ij} an equation of type

$$0 = f(y) = \sum M_iM_jr_{ij} + \sum M_iM_jh_{ij} + \left(\frac{\partial f_i}{\partial y_j}(\overline{y})\right) \begin{pmatrix} \sum_j t_{1j}M_j \\ \vdots \\ \sum_j t_{Nj}M_j \end{pmatrix}. \tag{3.16}$$

where the r_{ij} is a column vector of size m with entries in $(x_1,\dots,x_n)^c$ and h_{ij} is a column vector of size m with entries in $(t_{11},\dots,t_{Ns})^2$.

(2) Consider an $m \times N$ matrix A with entries in a ring R. Let M be an m–minor of A. Show that there exists a $N \times m$ matrix B such that $AB = MId_m$, where Id_m is the $m \times m$ identity matrix.
(Hint: If $m = N$, one can take for B the adjoint of A, see 1.2.9.)

(3) Use this to replace M_iM_j in (3.16) by $M_j\left(\frac{\partial f_i}{\partial y_j}(\overline{y})\right)B_i$ for certain matrices B_i.

(4) Show that for all j with $1 \leq j \leq s$ it suffices to solve for t_{ij} the equation

$$\sum_{i=1}^{s} B_i(r_{ij} + h_{ij}) + \begin{pmatrix} t_{1j} \\ \vdots \\ t_{Nj} \end{pmatrix} = 0.$$

Apply the Implicit Mapping Theorem to solve this system of equations.

3.3.35. Newton's Lemma cannot be strengthened in the following sense, that is, the following statement is not true in general.

Let $f \in \mathbb{C}\{x_1,\dots,x_n,y\}$, and $\overline{y}(x) \in \mathbb{C}\{x_1,\dots,x_n,y\}$ with

$$f(x,\overline{y}(x)) \equiv 0 \mod \left(\frac{\partial f}{\partial y}(x,\overline{y}(x))\right)(x)^c.$$

Then there exists a $y \in \mathbb{C}\{x_1,\dots,x_n,y\}$ with

(1) $f(x,y(x)) = 0$.

(2) $y \equiv \overline{y} \mod (x)^c$.

Indeed, show that $f(x,y) = y^2 - x^3$ is a counterexample.

3.3.36. A local $\mathbb{C}$-algebra R with maximal ideal $\mathfrak{m}$ is called Henselian if the statement of Hensel's Lemma (see 3.3.21), where $\mathbb{C}\{x_1,\dots,x_n\}$ is replaced by R, and $(x_1,\dots,x_n)$ is replaced by $\mathfrak{m}$, holds. Prove the following statements.

(1) Let R be an analytic algebra. Then R is Henselian.

(2) Show that $K[x]_{(x)}$ is not Henselian, by looking at the polynomial $F = T^2 - x - 1$.

(3) Let $\mathfrak{p}$ be a prime ideal. Prove that more generally $K[x_1,\dots,x_n]_{\mathfrak{p}}$ is not Henselian.

(4) Prove Remark 3.3.22.

3.3.37. Prove that the point $0 \in X$ in Example 3.3.12 (2) is singular.

3.4 Germs of Analytic Spaces

In this section we study germs of analytic spaces. We already know what an analytic space is, see 3.1.12, but we still have to define the notion of a germ.

Definition 3.4.1.

(1) Let X be a topological space and $p \in X$. Subsets A and B of X are called *equivalent* at p if there exists a neighborhood U of p such that $A \cap U = B \cap U$. It is easy to check that this is indeed an equivalence relation, see Exercise 3.4.33. The equivalence class of A at p we call the *germ of A at p*. A is called a representative of the germ. We write (A,p) for the germ of A at p.

(2) Let (X,x) and (Y,x) be germs. We define $(X,x) \subset (Y,x)$ if there are representatives X of (X,x) and Y of (Y,x) such that $X \subset Y$. It is not so difficult to see that $(X,x) = (Y,x)$ if and only if $(X,x) \subset (Y,x)$ and $(Y,x) \subset (X,x)$.

(3) Let $(X,x) \subset (Y,x)$ and $(Z,x) \subset (Y,x)$ be germs. Then we define $(X,x) \cap (Z,x)$ to be the germ of $X \cap Z$ at x for any representative X of (X,x), and Z of (Z,x) at x.

We define $(X,x) \cup (Z,x)$ to be the germ of $X \cup Z$ at x for any representative X of (X,x), and Z of (Z,x) at x.

We now define germs of analytic spaces.

Definition 3.4.2.

(1) A germ of an analytic space (X,x) is a germ at x of a locally analytic subset of $\mathbb{C}^n$.

(2) Let $f \in \mathcal{O}_{n,x}$. Then we define a *germ of an analytic hypersurface* $(V(f),x)$ as follows. Consider an open neighborhood U of x on which f converges. Consider the analytic subset $V(f) := \{p \in U : f(p) = 0\}$ of U. Then $(V(f),x)$ is the germ of $V(f)$ at zero, and is called the *zero set* of f.

(3) Let $I = (f_1, \dots ,f_s) \subset \mathcal{O}_{n,x}$ be an ideal. Then we define the germ of the analytic space $(V(I),x)$ by

$$(V(I),x) = \bigcap_{i=1}^{s} (V(f_i),x).$$

This definition is independent of the choice of generators f_i of I, as is easily checked. Any *germ of an analytic space* is a germ of type $(V(I),o)$ for some ideal $I \subset \mathbb{C}\{x_1, \dots ,x_n\}$.

(4) Let (X,o) be a germ of an analytic space. Then define

$$\mathscr{I}(X,x) = \{f \in \mathcal{O}_{n,x} : (X,x) \subset (V(f),o)\}.$$

The inclusion here is an inclusion of *germs*. So $f \in \mathscr{I}(X,x)$ if there exists a representative X of (X,o) and an open neighborhood U of x such that X is an analytic subset of U, f converges on U, and its restriction to X is the zero function.

Remark 3.4.3. We had to be a little careful with the definition of $(V(I),x)$. The definition of $(V(I),x)$ as $\cap_{f\in I}(V(f),x)$ would not make sense. In fact, in general it does not make sense to talk about infinite intersections of germs. Namely, if we have germs (X_i,x) for $i \in I$, and if representatives are defined on open sets U_i, then it is very well possible that $\cap_{i\in I}U_i = \{0\}$, if the set I has infinitely many elements. If I is a finite set however then $\cap_{i\in I}U_i$ is still an open neighborhood of x. Note that in the definition of $(V(I),x)$ we already used the basic fact that I is finitely generated, which follows from the fact that $\mathbb{C}\{x_1, \dots ,x_n\}$ is Noetherian.

We have properties similar to the statements in 2.2.2, except that, keeping in mind the remark above, *we are only allowed to take finite unions and intersections!* The proofs are all similar, and left to the reader. We also have:

Theorem 3.4.4 (Nullstellensatz). $\mathscr{I}(V(I),x) = \sqrt{I}$.

It obviously suffices to prove the Nullstellensatz for $x = o$. For germs of analytic spaces the Nullstellensatz cannot be interpreted as a statement on maximal ideals. This is because the power series ring is a local ring, that is, it just has one maximal ideal! Moreover, it seems that also the idea of the proof of the Projection Theorem 2.2.8 does not work in the local case. The point is (in the notation of the Projection Theorem) that we first have to define representatives, and in particular generators $h_1, \dots ,h_t$ for $I' = I \cap \mathbb{C}\{x_1, \dots ,x_{n-1}\}$, which converge on some open neighborhood U of 0. The proof of the Projection Theorem provides us with a holomorphic function g on U, which vanishes on X and such that $g(a) \neq 0$. So it follows that the germ of g is in I', but what is not clear is that g is in the ideal generated by the $h_1, \dots ,h_t$ all over U.

It is therefore no surprise that proofs of the Nullstellensatz for the power series ring tend to be somewhat more difficult.

As in the proof of the Nullstellensatz in the algebraic case, it suffices to prove the Nullstellensatz for *prime ideals* $\mathfrak{p}$. As the proof is completely similar, see 2.2.12, we leave the proof to the reader, and only formulate the result.

Lemma 3.4.5. *Suppose the Nullstellensatz holds for prime ideals. Then the Nullstellensatz holds for all ideals.*

So from now on we will look at prime ideals. As in the proof of the Nullstellensatz in the affine case, we will look at a Noether normalization,

$$\mathbb{C}\{x_1,\dots,x_k\} \subset \mathbb{C}\{x_1,\dots,x_n\}/\mathfrak{p}.$$

Put $(X,o) = (V(\mathfrak{p}),o)$. Then we would like to show that the linear projection $(X,o) \longrightarrow (\mathbb{C}^k,o)$ is surjective. This should be analogous to 2.2.10, where we showed that for $\pi : X \longrightarrow K^r$ a Noether normalization, the map π is surjective. This was the essential step in the proof of the Nullstellensatz. Note that we did, until now, not even define what a map between germs of spaces is!

Definition 3.4.6. Let (X,p) and (Y,q) be two germs of topological spaces. A *germ of a continuous map* $f : (X,p) \longrightarrow (Y,q)$ is defined as an equivalence class of maps $f : U \longrightarrow W$, with $f(p) = q$, and where U and W are representatives of (X,p) and (Y,p) respectively. Two such maps $f_1 : U_1 \longrightarrow W$ and $f_2 : U_1 \longrightarrow W$ are called *equivalent* if they agree on an open neighborhood of p contained in $U_1 \cap U_2$.

The problem with germs of continuous maps is, that it makes in general no sense to talk about the *image*. By this we mean that in general it might happen that for every subgerm $(Z,q) \subset (Y,q)$, there does not exist an open neighborhood U of p in X such that (Z,q) is represented by $f(U)$. For an example, see Exercise 3.4.34. In particular, it does not make sense to talk about surjectivity. For special kind of maps, so-called finite maps, this problem does not occur. Let us define this notion.

Definition 3.4.7. Let $f : X \to Y$ be a continuous map between topological spaces.

(1) f is called *closed* if the image $f(A) \subset Y$ is closed for all closed subspaces $A \subset X$.

(2) f is called *quasi-finite* if for all $p \in Y$ the fiber $f^{-1}(p)$ consists of a finite number of points.

(3) f is called *finite* if it is both closed and quasi-finite.

The simplest example of a mapping which is quasi-finite but not finite probably is the inclusion mapping $\mathbb{C} \setminus \{0\} \to \mathbb{C}$. (Here $\mathbb{C}$ has the standard Euclidean topology.)

Lemma 3.4.8. *Let X and Y be topological spaces, and $f : X \longrightarrow Y$ be closed. Let $p \in X$ and let $q = f(p)$. Assume that $f^{-1}(f(p)) = \{p\}$. Let $A,B \subset X$ with $p \in A$, $p \in B$, defining the same germ at p. Then $(f(A),q) = (f(B),q)$, that is, $f(A)$ and $f(B)$ define the same germ in q.*

Proof. By the definition of a germ, we have to find an open subset V of q in Y with

$$V \cap f(A) = V \cap f(B).$$

Step 1. We first show that there exists a neighborhood basis of p in X such that for each element U of this basis we have the equality $U = f^{-1}(f(U))$. So let W be an open neighborhood of p. Then $X \setminus W$ is closed, so that $f(X \setminus W)$ is closed by assumption. Then $T := Y \setminus f(X \setminus W)$ is open. Define $U = f^{-1}(T)$. Then U is open and we have the following properties:

(1) As $U = f^{-1}(T)$, it follows $f(U) = f \circ f^{-1}(T) \subset T$. Hence $f^{-1}(f(U)) \subset f^{-1}(T) = U$. As the converse inclusion is trivial, it follows that $f^{-1}(f(U)) = U$.

(2) $U \subset W$, as, otherwise, there exists an $x \in U$, $x \notin W$. By definition of T, it immediately follows from $x \notin W$ that $f(x) \in f(X \setminus W)$, so that $f(x) \notin T$. But we have $f(U) \subset T$, so this is a contradiction.

(3) We have $p \in U$. To show this, note that as $p \in W$ and $f^{-1}(f(p)) = \{p\}$ it follows $f(p) \notin f(X \setminus W)$ and then $f(p) \in T = Y \setminus f(X \setminus W)$. Thus $p \in f^{-1}(T) = U$.

Step 2. Let W be an open neighborhood of p such that $A \cap W = B \cap W$. Take $U \subset W$ and $T \subset Y$ as in Step 1. In particular $U \cap A = U \cap B$. We claim that

$$f(U \cap A) = T \cap f(A).$$

As $f(U) \subset T$, the inclusion $\subset$ is a tautology. To show the converse inclusion, take a point $a \in T \cap f(A)$. Then $a = f(b)$ for some $b \in A$. Moreover, $a \in T$, so that $b \in f^{-1}(T) = U$. This shows that $b \in U \cap A$, showing the converse inclusion.

As similarly $f(U \cap B) = T \cap f(B)$, and $U \cap A = U \cap B$, it follows that $T \cap f(A) = T \cap f(B)$. This is what we had to show. □

The lemma is the basis for the following definition:

Definition 3.4.9. Let (X,p) and (Y,q) be two germs of topological spaces. Let $f : (X,p) \longrightarrow (Y,q)$ be a germ of a continuous map.

(1) f is called *finite*, if there exists a representative $f : X \longrightarrow Y$ which is finite.

(2) Assume that f is finite. Then the *image* of f is given by $\mathrm{Im}(f) := (f(X),q)$, where $f : X \longrightarrow Y$ is a representative such that $f^{-1}(f(p)) = \{p\}$ and f is finite.

Note that because of Lemma 3.4.8 $\mathrm{Im}(f)$ is well-defined.

(3) Assume that f is finite. Then f is called *surjective* if $\mathrm{Im}(f) = (Y,q)$. Note that f is surjective, if and only if $f : X \longrightarrow Y$ is surjective for suitable representatives such that f is finite and $f^{-1}(f(p)) = \{p\}$.

The first example we give of a finite map is the following.

Lemma 3.4.10. *Let $P = y^r + a_1 y^{r-1} + \ldots + a_r$ be a polynomial, with coefficients $a_i \in \mathcal{O}_k$. Consider an open neighborhood $U \subset \mathbb{C}^k$ of 0 such that the power series $a_1, \ldots, a_r$ converge. Define:*

$$X := \{P = 0\} \cap (U \times \mathbb{C}).$$

Then the canonical projection $\pi : X \longrightarrow U$ is finite.

Proof. As the map obviously is quasi-finite, we only have to show that π is closed. Therefore, let $A \subset X$ be a closed subset, and take a point $p \in \overline{\pi(A)}$. We have to show that $p \in \pi(A)$. As $p \in \overline{\pi(A)}$ we have a sequence

$$(p_i,y_i) \qquad i = 1,2,\ldots$$

with $p_i \in \pi(A)$, $y_i \in \mathbb{C}$, $(p_i,y_i) \in A$, and $\lim_{i\to\infty} p_i = p$. Because $(p_i,y_i) \in A \subset X$, we have

$$y_i^r + a_1(p_i)y_i^{r-1} + \ldots + a_r(p_i) = 0 \quad \text{for } i \geq 1.$$

As the a_j are holomorphic, in particular continuous, we have that $\lim_{i\to\infty} a_j(p_i) = a_j(p)$. Therefore, the sequence $a_j(p_i)$, $i = 1,2,\ldots$ is bounded. We claim that the sequence y_i is bounded, too. Indeed, $|y_i| < 1$ or $|y_i| \geq 1$. But in the second case, we obviously have $\left|\frac{a_j(p_i)}{y_i^{j-1}}\right| \leq |a_j(p_i)|$, so that:

$$|y_i| = \left|\frac{a_r(p_i)}{y_i^{r-1}} + \ldots + a_1(p_i)\right| \leq |a_r(p_i)| + \ldots + |a_1(p_i)|,$$

hence the sequence y_i is bounded. There exists therefore a convergent subsequence $y_{i_1}, y_{i_2}, \ldots$, converging to say $\overline{y}$. As A is closed, the point $(p,\overline{y}) = \lim_{j\to\infty}(p_j, y_{i_j})$ belongs to A. Therefore, $p \in \pi(A)$, which is what we had to prove. □

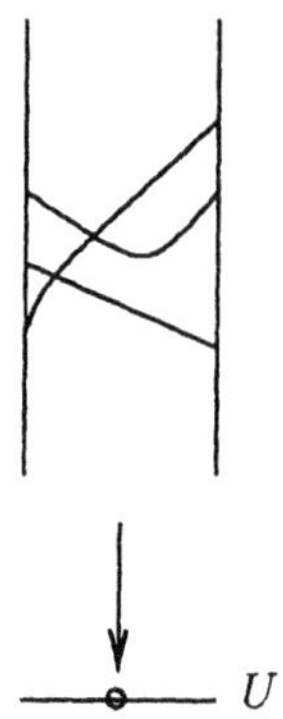

Typical picture of the zero set of a monic polynomial.

Lemma 3.4.11 (Continuity of Roots). *Consider a Weierstraß polynomial:*

$$P(x,y) = y^r + a_1 y^{r-1} + \ldots + a_r; \qquad a_i \in \mathfrak{m} \subset \mathbb{C}\{x_1, \ldots, x_k\}.$$

(1) Then for every arbitrary small neighborhood V of 0 in $\mathbb{C}$ there exists a small neighborhood U of 0 in $\mathbb{C}^k$ such that the power series $a_1, \ldots, a_r$ converge on U and such that for every p in U the r zeros (counted with multiplicity) of

$$y^r + a_1(p)y^{r-1} + \ldots + a_r(p) = 0$$

lie in V.

(2) Let $\pi : \mathbb{C}^{k+1} \longrightarrow \mathbb{C}^k$ be the projection on the first k coordinates. Then $\pi : (V(P), 0) \longrightarrow (\mathbb{C}^k, 0)$ is surjective.

(3) Suppose that P is irreducible. Let Δ be the discriminant of P and let $U \subset \mathbb{C}^k$ be an open set such that $a_1, \ldots, a_r$ converge on U. Let $X = \{(x,y) \in U \times \mathbb{C} : P(x,y) = 0\}$, $D = \{x \in U : \Delta(x) = 0\}$ and let $\pi : X \longrightarrow U$ be the projection on the first k coordinates. Then $\overline{X \setminus \pi^{-1}(D)} = X$.

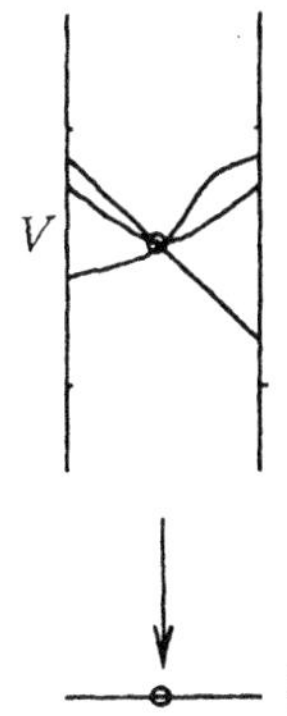

Typical picture of the zero set of of a Weierstraß polynomial.

Proof.
(1) The proof runs as the proof of the previous lemma. Suppose the converse. Then there exists a small neighborhood V of 0 in $\mathbb{C}$ such that for all neighborhoods U of 0 in $\mathbb{C}^k$ on which the power series $a_1,\ldots,a_r$ converge, there exists a point $p \in U$ and a root of the polynomial $P(p,y) = 0$, which is not in V. Take a sequence U_i of open neighborhoods of 0 whose intersection is $\{0\}$, and points $p_i \in U_i$ such that there is a root y_i of $P(p_i,y_i) = 0$ which does not lie in V. So there exist a $C > 0$, infinitely many y_i with $|y_i| \geq C$ and infinitely many $p_i \in \mathbb{C}^k$ such that $\{p_i\}$ converges to 0 and $P(p_i,y_i) = 0$. But then it follows as in the proof of 3.4.10 that

$$|y_i| \leq \tfrac{1}{C^{r-1}}\,|a_r(p_i)| + \tfrac{1}{C^{r-2}}\,|a_{r-1}(p_i)| + \ldots + |a_1(p_i)|\,.$$

By assumption $\lim_{i\to\infty} a_j(p_i) = 0$ for all j. So we can make y_i as small as we want, in contradiction to $|y_i| \geq C$.

(2) Let $U \subset \mathbb{C}^k$ be an open neighborhood of 0 such that the power series $a_1,\ldots,a_r$ converge. Define:

$$X := \{(x,y) \in U \times \mathbb{C} : P(x,y) = 0\}.$$

The canonical map $\pi : X \longrightarrow U$ is finite (Lemma 3.4.10) and, obviously, surjective. Moreover, because P is a Weierstraß polynomial, we have $\pi^{-1}(0) = 0$. This implies that $(V(P),\mathrm{o}) \longrightarrow (\mathbb{C}^k,\mathrm{o})$ is surjective, see 3.4.9(3).

(3) We have to prove that for any point $(a,b) \in \pi^{-1}(D)$ there exists a sequence $(a_i,b_i) \in X \setminus \pi^{-1}(D)$ with $\lim_{i\to\infty}(a_i,b_i) = (a,b)$. By Theorem 3.1.14 there exists a sequence a_i in $U \setminus D$ with $\lim_{i\to\infty} a_i = a$. We consider P at the point (a,b). Here we can write $P = U\cdot Q$, $U \in \mathbb{C}\{\overline{x},\overline{y}\}$ a unit, Q a Weierstraß polynomial with respect to $\overline{x} = x - a$ and $\overline{y} = y - b$. Now we use (1) and obtain for all i that there exists an open neighborhood V_i of a such that for $a_i \in V_i$ all roots of $Q(a_i,\overline{y}) = 0$ satisfy $|\overline{y}| < \frac{1}{i}$. Therefore, we may choose a sequence b_i such that $(a_i,b_i) \in X$ converging to (a,b) and $a_i \notin D$. □

Remarks 3.4.12.

(1) Alternatively, Lemma 3.4.11 could have been proved by using Rouché's Theorem of function theory. Even simpler, one could use the integral $\int \frac{f'}{f}$, which counts the number of zeroes of a holomorphic function in one variable and a continuity argument. The lemma in fact is the starting point for an alternative proof of the Weierstraß Division and Preparation Theorem, see the book of Gunning and Rossi, [Gunning-Rossi 1965].

(2) We will use the Nullstellensatz to generalize 3.4.11 (3). In fact we will prove that, if X is a small representative of $(V(\mathfrak{p}),0)$ for a prime ideal $\mathfrak{p} \subset \mathbb{C}\{x_1,\dots,x_n\}$, and $h \in \mathbb{C}\{x_1,\dots,x_n\}$, $h \notin \mathscr{I}(X,o)$, then $\overline{X \setminus \{h=0\}} = X$.

Lemma 3.4.13. *Let $I \subset \mathbb{C}\{x_1,\dots,x_n\}$ be an ideal.*

(1) Suppose that there exists an $f \in I$ which is regular in x_n. Let

$$p : \mathbb{C}^n \longrightarrow \mathbb{C}^{n-1}; \quad (x_1,\dots,x_n) \mapsto (x_1,\dots,x_{n-1}).$$

Then there exists a representative X of $(V(I),o)$ and an open neighborhood U of $0 \in \mathbb{C}^{n-1}$ such that the map $p : X \longrightarrow U$ is a finite map.

(2) Suppose more generally, that $\mathbb{C}\{x_1,\dots x_n\}/I$ is a finitely generated $\mathbb{C}\{x_1,\dots,x_k\}$-module. Then there exists a representative X of $(V(I),o)$ and an open neighborhood U of $\mathbb{C}^k$ such that the projection $p : X \longrightarrow U$ is a finite map.

Proof. By the Weierstraß Preparation Theorem, we may assume that $f = up$, where p is a Weierstraß polynomial, and u is a unit. So we may assume that f is a Weierstraß polynomial. By 3.4.10 there exists an open neighborhood U of 0 in $\mathbb{C}^{n-1}$ and a representative W of $(V(f),o)$ such that $p : W \longrightarrow U$ is finite. Consider a representative $X \subset W$ of $(V(I),o)$. As X is a closed subset of W (being the zero set of finitely many holomorphic functions), it follows that $p : X \longrightarrow U$ is finite too.

For the second statement, it follows that each of the classes of x_j for $j = k+1,\dots,n$ satisfy an integral equation with coefficients in $\mathbb{C}\{x_1,\dots,x_k\}$. Hence there exist Weierstraß polynomials $f_j(x_j)$ for $j \geq k+1$, with coefficients in $\mathbb{C}\{x_1,\dots,x_k\}$ which are contained in I. By Exercise 3.4.35 there exists an open neighborhood $U \subset \mathbb{C}^k$ such that the projection map $p : X \longrightarrow U$ is finite for a suitable representative X of $(V(f_{k+1},\dots,f_n),o)$. So we can argue as above to deduce the lemma. □

We will generalize the second part of Lemma 3.4.11 in the Local Parametrization Theorem 3.4.14. In particular we will prove

Let $\mathfrak{p} \subset \mathbb{C}\{x_1,\dots,x_n\}$ be a prime ideal and $\pi : (V(\mathfrak{p}),o) \longrightarrow (\mathbb{C}^k,o)$ be a primitive Noether normalization. Then π is finite and surjective.

This is the basis to prove the Nullstellensatz.

Proof of the Nullstellensatz 3.4.4. (Using the Local Parametrization Theorem 3.4.14.) It suffices to show the Nullstellensatz for prime ideals $\mathfrak{p}$. We choose a primitive Noether normalization $\pi : (V(\mathfrak{p}),o) \longrightarrow (\mathbb{C}^k,o)$ and use the fact that π is surjective. Then the proof is completely analogous to the proof of the Nullstellensatz in the affine case, see 2.2.11. □

Recall that we want to prove that for some Noether normalization $(V(\mathfrak{p}),o) = (X,o) \longrightarrow (\mathbb{C}^k,o)$ is surjective. This will be proved in two steps, namely we will first show that (X,o) projects surjectively onto a hypersurface $(V(P),o)$ in $(\mathbb{C}^{k+1},o)$, and then we project further onto $(\mathbb{C}^k,o)$. The first projection is "generically one to one". This projection is an important tool in several proofs in this book, and is called the Local Parametrization Theorem.

Theorem 3.4.14 (Local Parametrization Theorem). *Let* $(X,o) = (V(\mathfrak{p}),o)$ *for* $\mathfrak{p} \subset \mathbb{C}\{x_1,\ldots,x_n\}$ *a prime ideal and* $\mathbb{C}\{x_1,\ldots,x_k\} \subset \mathbb{C}\{x_1,\ldots,x_n\}/\mathfrak{p}$ *be a primitive Noether normalization (cf. 3.3.20). Let* $\pi : \mathbb{C}^n \longrightarrow \mathbb{C}^{k+1}$ *be the projection on the first* $k+1$ *coordinates,* $p : \mathbb{C}^{k+1} \longrightarrow \mathbb{C}^k$ *be the projection on the first* k *coordinates. Then we have the following statements.*

(1) $\mathfrak{p} \cap \mathbb{C}\{x_1,\ldots,x_{k+1}\} = (P)$, *for* $P \in \mathbb{C}\{x_1,\ldots,x_k\}[x_{k+1}]$ *a Weierstraß polynomial.*

(2) Let Δ *be the discriminant of* P *with respect to* x_{k+1}. *Then*

(a) there exist $q_{k+2},\ldots,q_n \in \mathbb{C}\{x_1,\ldots,x_k\}[x_{k+1}]$ *such that* $Q_j := \Delta x_j - q_j \in \mathfrak{p}$ *for* $j = k+2,\ldots,n$;

(b) there exist $f_1,\ldots,f_s \in \mathbb{C}\{x_1,\ldots,x_k\}[x_{k+1},\ldots,x_n]$ *such that*

i. $(f_1,\ldots,f_s) = \mathfrak{p}$ *in the ring* $\mathbb{C}\{x_1,\ldots,x_n\}$.

ii. $f_j \in \mathbb{C}\{x_1,\ldots,x_k\}[x_{k+j}]$ *are Weierstraß polynomials for* $j = 1,\ldots,n-k$ *and* $P,Q_{k+2},\ldots,Q_n \in \{f_1,\ldots,f_s\}$;

iii. $(P,Q_{k+2},\ldots,Q_n)\mathbb{C}\{x_1,\ldots,x_k\}_\Delta[x_{k+1},\ldots,x_n] = (f_1,\ldots,f_s)\mathbb{C}\{x_1,\ldots,x_k\}_\Delta[x_{k+1},\ldots,x_n]$.

(3) There is an open neighborhood U *of* 0 *in* $\mathbb{C}^k$ *such that* $P,\Delta,Q_{k+2},\ldots,Q_n,f_1,\ldots,f_s$ *converge on* $U \times \mathbb{C}^{n-k}$ *and for*

$$\begin{aligned} X &= \{x \in U \times \mathbb{C}^{n-k} : f_1(x) = \cdots = f_s(x) = 0\} \\ Y &= \{y \in U \times \mathbb{C}^{n-k} : P(y) = Q_{k+2}(y) = \cdots = Q_n(y) = 0\} \\ X' &= \{x' \in U \times \mathbb{C} : P(x') = 0\} \\ D &= \{x \in U : \Delta(x) = 0\} \end{aligned}$$

the following statements hold:

(a) $Y \setminus (p\circ\pi)^{-1}(D) = X \setminus (p\circ\pi)^{-1}(D)$;[11]

(b) $\pi : X \setminus (p\circ\pi)^{-1}(D) \longrightarrow X' \setminus p^{-1}(D)$ *is biholomorphic;*

(c) $X \setminus (p\circ\pi)^{-1}(D)$ *is a complex submanifold of* $U \times \mathbb{C}^{n-k}$;

(d) the map $p\circ\pi : X \longrightarrow U$ *is surjective and finite and* $(p\circ\pi)^{-1}(0) = \{0\}$.

(4) The map $p\circ\pi : (X,o) \longrightarrow (\mathbb{C}^k,o)$ *is finite and surjective.*

(5) The map $\pi : (X,o) \longrightarrow (V(P),o)$ *is finite and surjective.*

Before giving the proof of the Local Parametrization Theorem we give an example. In this example all the power series are in fact polynomials. The neighborhood U can be taken to be the whole of the complex vector space.

[11] The equation implies that $X \setminus (p\circ\pi)^{-1}(D) = \left\{\left(p,\frac{q_{k+2}(p)}{\Delta(p)},\ldots,\frac{q_n(p)}{\Delta(p)}\right) : p \in U \times \mathbb{C},\, P(p) = 0\right\}$. This (together with 3.3.23) is the reason to call 3.4.14 Local Parametrization Theorem.

Example 3.4.15. Consider the map $\mathbb{C} \longrightarrow \mathbb{C}^3$ given by $t \mapsto (t^3,t^4,t^5) = (x,y,z)$. Its image can be calculated, see 2.3.18, to be the zero set of the ideal $I = (y^2 - xz, x^3 - yz, z^2 - x^2y)$. A Noether normalization map is given by the projection on the x–axis. Indeed modulo I one has $y^3 = xyz = x^4$, and $z^3 = x^2yz = x^5$. This gives the integral equation $z^3 - x^5 = 0$, and $y^3 - x^4 = 0$.[12] To show that this is a Noether normalization one has to show that $I \cap \mathbb{C}\{x\}$ is zero. For this, see Exercise 3.1.28. We in fact have a primitive Noether normalization. Here y plays the role of x_{k+1}. This follows because in the quotient field $z = \frac{y^2}{x}$. The polynomial P is equal to $y^3 - x^4$. The discriminant is $\Delta = 27x^8$. So the Local Parametrization Theorem says that x^8z is a polynomial in y with coefficients in $\mathbb{C}\{x\}$. Indeed, even $xz = y^2$. So in this case the polynomial q is equal to $27x^7y^2$. The space X' is given by $y^3 - x^4 = 0$. For each $a \neq 0$, and each $y = b$ with $b^3 - a^4$, there is a unique point on X, with z–coordinate $z = \frac{b^2}{a}$. Thus the map π in the Local Parametrization Theorem is indeed surjective. In fact, it is easily seen that for $x \neq 0$, the analytic set is a complex submanifold.

Proof of the Local Parametrization Theorem 3.4.14.

Step 1. We first show (1) and (2a). We apply the Finiteness of Normalization Theorem 1.5.19. We obtain that $\mathfrak{p} \cap \mathbb{C}\{x_1,\dots,x_k\}[x_{k+1}] = (P)$, where P is the minimal polynomial of the element $x_{k+1} \mod \mathfrak{p}$ in the field extension $Q(\mathbb{C}\{x_1,\dots,x_n\}/\mathfrak{p}) \supset Q(\mathbb{C}\{x_1,\dots,x_k\})$. By Exercise 3.3.32, P is a Weierstraß polynomial. Using the Weierstraß Division Theorem 3.2.3, we obtain $\mathfrak{p} \cap \mathbb{C}\{x_1,\dots,x_{k+1}\} = (P)$, as, otherwise, we would find a nonzero element in $\mathfrak{p}$ which is a polynomial in x_{k+1} of degree smaller than the degree of P. This is in contradiction to the fact that P is the minimal polynomial of the field extension. This proves (1). To prove (2a) we apply again 1.5.19. So for all $f \in \mathcal{O}_n$ there exists $q_f \in \mathbb{C}\{x_1,\dots,x_k\}[x_{k+1}]$ with $\Delta \cdot f - q_f \in \mathfrak{p}$. In particular, this holds for $x_{k+2},\dots,x_n$. Therefore, there exist polynomials $q_{k+2},\dots,q_n \in \mathbb{C}\{x_1,\dots,x_k\}[x_{k+1}]$ such that $Q_j := \Delta x_j - q_j \in \mathfrak{p}$ for $j = k+2,\dots,n$. This shows (2a).

Step 2. We now prove (2b). Consider the minimal polynomials $P_j \in \mathbb{C}\{x_1,\dots,x_k\}[x_j]$ of x_j for $j = k+2,\dots,n$. Thus $P_{k+1} := P$.

Step 2a. We first show that $\mathfrak{p}$ has a system of generators in $\mathbb{C}\{x_1,\dots,x_k\}[x_{k+1},\dots,x_n]$. Indeed, it follows that the extension $\mathbb{C}\{x_1,\dots,x_k\} \subset \mathbb{C}\{x_1,\dots,x_n\}/(P_{k+1},\dots,P_n)$ is finite. We put $\overline{P}_i := P_i(0,\dots,0,x_{k+1},\dots,x_n)$. It then easily follows that the vector space $\mathbb{C}\{x_{k+1},\dots,x_n\}/(\overline{P}_{k+1},\dots,\overline{P}_n)$ is generated by monomials in $x_{k+1},\dots,x_n$. It follows from Corollary 3.2.12 that every element in $\mathbb{C}\{x_1,\dots,x_n\}$ is modulo $(P_{k+1},\dots,P_n)$ equivalent to an element in $\mathbb{C}\{x_1,\dots,x_k\}[x_{k+1},\dots,x_n]$. This then certainly holds modulo $\mathfrak{p}$. Hence we can choose generators $f_1,\dots,f_s$ of $\mathfrak{p}$ such that

- $f_1 = P_{k+1} = P,\dots,f_{n-k} = P_n$;
- $f_{n-k+1},\dots,f_s \in \mathbb{C}\{x_1,\dots,x_k\}[x_{k+1},\dots,x_n]$;
- $Q_{k+2},\dots,Q_n \in \{f_1,\dots,f_s\}$.

Step 2b. Put

[12] A different way to see this is by applying the general Weierstraß Division Theorem 3.2.10. Consider the $\mathbb{C}\{x\}$–module $\mathbb{C}\{x,y,z\}/I$. Divide out the maximal ideal (x) of $\mathbb{C}\{x\}$. We get the $\mathbb{C}$–vector space $\mathbb{C}\{y,z\}/(y^2,yz,z^2)$, which is finite-dimensional. A basis is given by the classes of $1,y,z$.

$$J_\Delta := (P, Q_{k+2}, \dots, Q_n)\mathbb{C}\{x_1, \dots, x_k\}_\Delta[x_{k+1}, \dots, x_n]$$
$$I_\Delta := (f_1, \dots, f_s)\mathbb{C}\{x_1, \dots, x_k\}_\Delta[x_{k+1}, \dots, x_n].$$

To prove *iii* of 2b, We have to show $J_\Delta = I_\Delta$. The inclusion $J_\Delta \subset I_\Delta$ is obvious and, therefore, we have a surjection

$$\text{(3.17)} \qquad \mathbb{C}\{x_1, \dots, x_k\}_\Delta[x_{k+1}, \dots, x_n]/J_\Delta \twoheadrightarrow \mathbb{C}\{x_1, \dots, x_k\}_\Delta[x_{k+1}, \dots, x_n]/I_\Delta.$$

The equation $Q_j = 0$ says that $x_j = \frac{q_j(x_1, \dots, x_{k+1})}{\Delta}$, so we can eliminate the variables $x_{k+2}, \dots, x_n$ in $\mathbb{C}\{x_1, \dots, x_k\}_\Delta[x_{k+1}, \dots, x_n]/J_\Delta$. After having done that, formula (3.17) gives a surjection

$$\mathbb{C}\{x_1, \dots, x_k\}_\Delta[x_{k+1}]/(P) \twoheadrightarrow \mathbb{C}\{x_1, \dots, x_k\}_\Delta[x_{k+1}, \dots, x_n]/I_\Delta.$$

We claim that is in fact a bijection, that is, $I_\Delta \cap \mathbb{C}\{x_1, \dots, x_k\}_\Delta[x_{k+1}] = (P)$. As I generates $\mathfrak{p}$, it follows from the first part of the Theorem that $I \cap \mathbb{C}\{x_1, \dots, x_k\}[x_{k+1}] = (P)$. As (P) is a prime ideal, and $\Delta \notin (P)$, the claim follows.

Step 3. The proof of (3a) now is easy. Let $f_1, \dots, f_s \in \mathbb{C}\{x_1, \dots, x_k\}[x_{k+1}, \dots, x_n]$ be the generators of $\mathfrak{p}$ we constructed in Step 2. By (2b) we have for some $b = b(i)$:

$$\text{(3.18)} \qquad \Delta^b f_i = \sum_{j=k+2}^{n} \alpha_j(f_i)Q_j + \alpha(f_i)P$$

for certain $\alpha_j(f_i), \alpha(f_i) \in \mathbb{C}\{x_1, \dots, x_k\}[x_{k+1}, \dots, x_n]$. Choose $U \subset \mathbb{C}^k$ an open neighborhood of 0 such that $P, \Delta, Q_{k+2}, \dots, Q_n$, $\alpha(f_i), \alpha_j(f_i)$ converge on $U \times \mathbb{C}^{n-k}$ for all i,j. Let $p \in Y$ with $\Delta(p) \neq 0$ be given. Then from (3.18) and $P(p) = Q_j(p) = 0$ for $j = k+2, \dots, n$, it follows that $f_i(p) = 0$ for $i = 1, \dots, s$. This exactly means that

$$Y \setminus (p \circ \pi)^{-1}(D) = X \setminus (p \circ \pi)^{-1}(D).$$

Step 4. We now prove (3b) and (3c). Consider a point $a \in X' \setminus p^{-1}(D)$. From 3.3.23 X' is in a small neighborhood U' of a a complex submanifold of $U' \times W$, given by the graph of a function $x_{k+1} = \alpha(x_1, \dots, x_n)$. By definition of $Q_j = \Delta x_j - q_j$. Therefore, there is a unique point of $X \setminus (p \circ \pi)^{-1}(D)$ mapping to a. If the coordinates of a are $(a_1, \dots, a_{k+1})$ the unique point in $\pi^{-1}(p)$ has coordinates

$$\left(a_1, \dots, a_k, a_{k+1}, \frac{q_{k+2}(a_1, \dots, a_{k+1})}{\Delta(a_1, \dots, a_k)}, \dots, \frac{q_n(a_1, \dots, a_{k+1})}{\Delta(a_1, \dots, a_k)}\right).$$

So we have an inverse to the map $\pi : X \setminus (p \circ \pi)^{-1}(D) \longrightarrow X' \setminus p^{-1}(D)$, which we see to be biholomorphic. Indeed, locally in $U' \times W \times \mathbb{C}^{n-k-1}$, the set $Y \setminus (p \circ \pi)^{-1}(D)$ is given by the set of equations

$$x_{k+1} - \alpha(x_1, \dots, x_k) = x_{k+2} - \frac{q_{k+2}(x_1, \dots, x_{k+1})}{\Delta(x_1, \dots, x_k)} = \dots = x_n - \frac{q_n(x_1, \dots, x_{k+1})}{\Delta(x_1, \dots, x_k)} = 0,$$

and is therefore a complex submanifold.

Step 5. We prove (3d), (4) and (5). By Lemma 3.4.13 the map $p \circ \pi : X \longrightarrow U$ is finite. The map $p : X' \setminus p^{-1}(D) \longrightarrow U \setminus D$ is, obviously, surjective and because of (3b) the map $p \circ \pi : X \setminus (p \circ \pi)^{-1}(D) \longrightarrow U \setminus D$ is surjective. As $p \circ \pi$ is closed, the closure of $U \setminus D$ is in $p \circ \pi(X)$. Using Theorem 3.1.14 we obtain that $\overline{U \setminus D} = U$ and, therefore, $\pi(X) = U$.

Now (4) and (5) are immediate consequences (cf. 3.4.9 and 3.4.13). □

Remarks 3.4.16. Let $\mathfrak{p}$ be a prime ideal, and $(X,o) = \big(V(\mathfrak{p}),o\big)$.

(1) Note that in the proof of the Local Parametrization Theorem, we showed that the number of points in the general fiber of a Noether normalization $\pi : X \longrightarrow U$ is equal to the degree of P, that is, the degree of the field extension

$$Q(\mathcal{O}_k) \subset Q(\mathcal{O}_{X,o}).$$

In fact, this holds generally for "finite maps between irreducible germs of analytic spaces", as will be shown in the exercises.

(2) We showed that locally, at a point $a \in X \setminus (p \circ \pi)^{-1}(D)$, the ideal $\mathcal{I}(X,a) \subset \mathcal{O}_{n,a}$ is generated by $Q_{k+2},\dots,Q_n,P$. Moreover, $\mathcal{O}_{n,a}/\mathcal{I}(X,a)$ is isomorphic to $\mathcal{O}_k$. It follows that if $(f_1,\dots,f_s) = \mathcal{I}(X,o)$, the ideal of (X,o), then there exists a neighborhood U of 0 in X such that the f_i converge on U and for all $a \in U \cap (X \setminus D)$ the ideal $\mathcal{I}(X,a)$ is generated by $f_1,\dots,f_s$.[13]

(3) In the notation of the Local Parametrization Theorem, write $\mathcal{O}_{X',o} = \mathcal{O}_{k+1}/(P)$. The cokernel $K := \mathcal{O}_{X,o}/\mathcal{O}_{X',o}$ is an $\mathcal{O}_k$–module. It was shown in the proof of the Local Parametrization Theorem, that there exists an $a \in \mathbb{N}$ such that $\Delta^a \cdot K = (0)$. In particular, we can view K as an $\mathcal{O}_k/(\Delta^a)$–module.

Definition 3.4.17. Let (X,x) be a germ of an analytic space. Then (X,x) is called *irreducible* if from $(X,x) = (X_1,x) \cup (X_2,x)$, with (X_1,x) and (X_2,x) germs of analytic spaces it follows that either $(X,x) = (X_1,x)$ or $(X,x) = (X_2,x)$.

Corollary 3.4.18. *Let (X,x) be a germ of an analytic space.*

(1) (X,x) is irreducible if and only if $\mathcal{I}(X,x)$ is a prime ideal.

(2) Let $(X,x) \subset (\mathbb{C}^n,x)$ be irreducible, and $h \in \mathcal{O}_n$, $h \notin \mathcal{I}(X,x)$. Then for small enough representatives $\overline{X \setminus \{h = 0\}} = X$.

(3) There is a, up to permutation, unique decomposition $(X,x) = (X_1,x) \cup \dots \cup (X_r,x)$, with (X_i,x) irreducible and $(X_j,x) \not\subset (X_i,x)$ for $i \neq j$. This is called the irreducible decomposition *of (X,x). The (X_i,x) are called* irreducible components *of (X,x).*

Proof. The proof runs as in the algebraic case, see 2.2.15, 2.2.16 and 2.2.17. But note that in the proof the Nullstellensatz is used! □

Definition 3.4.19. Let $(X,x) \subset (\mathbb{C}^n,x)$ be a germ of an analytic space. A *germ of an analytic function* $f : (X,x) \longrightarrow (\mathbb{C},y)$ is a germ of a map $f : (X,x) \longrightarrow (\mathbb{C},y)$ such that some representative is the restriction to X of an analytic function on an open neighborhood of x in $\mathbb{C}^n$. As is easily shown, the germs of analytic functions on (X,x) form a $\mathbb{C}$–algebra, which we denote by $\mathcal{O}_{X,x}$. This is called the *ring of analytic functions* on (X,x). Sometimes $\mathcal{O}_{\mathbb{C}^n,0}$, which is equal to $\mathbb{C}\{x_1,\dots,x_n\}$ is denoted by $\mathcal{O}_n$. A *germ of an analytic map* or simply map, $\varphi = (f_1,\dots,f_m) : (X,x) \longrightarrow (Y,y \subset (\mathbb{C}^m,y)$ is similarly defined. The map $\varphi : (X,x) \longrightarrow (Y,y)$ is called an *isomorphism* if φ has a two-sided inverse $\psi : (Y,y) \longrightarrow (X,x)$ which is an analytic map. (X,x) and (Y,y) are called *isomorphic* if there exists an isomorphism $\varphi : (X,x) \longrightarrow (Y,y)$.

[13] In fact, this holds for all $a \in U$. This is the Coherence Theorem of Oka-Cartan to be proved in Chapter 6.

The following lemma is easy, and left to the reader.

Lemma 3.4.20.

(1) Let $(X,x) \subset (\mathbb{C}^n,x)$ be a germ of an analytic space, and $\mathscr{I}(X,x)$ be the ideal of X. Then $\mathcal{O}_{X,x} = \mathcal{O}_{n,x}/\mathscr{I}(X,x)$.

(2) Let $(X,x) \subset (\mathbb{C}^n,x)$ be a germ of a submanifold. Then $\mathcal{O}_{X,x} \cong \mathbb{C}\{x_1,\dots,x_k\}$ for some k.

(3) Let $\varphi : (X,x) \longrightarrow (Y,y)$ be a germ of an analytic mapping. Then by composition φ induces a map of $\mathbb{C}$-algebras

$$\varphi^* : \mathcal{O}_{Y,y} \longrightarrow \mathcal{O}_{X,x} : f \mapsto f \circ \varphi.$$

Note that $(\varphi \circ \psi)^* = \psi^* \circ \varphi^*$ for two germs of analytic mappings φ and ψ.

The converse of the third part of the lemma also holds:

Theorem 3.4.21. *Let $(X,x) \subset (\mathbb{C}^n,x)$ and $(Y,y) \subset (\mathbb{C}^m,y)$ be germs of analytic spaces. Let $\alpha : \mathcal{O}_{Y,y} \longrightarrow \mathcal{O}_{X,x}$ be a $\mathbb{C}$-algebra homomorphism. Then there exists a unique germ of an analytic mapping $\varphi : (X,x) \longrightarrow (Y,y)$ with $\varphi^* = \alpha$.*

Proof. . Without loss of generality, we may assume that $x = 0$ and $y = 0$. In 3.2.9 it was proved that $\alpha(\mathfrak{m}_{Y,0}) \subset \mathfrak{m}_{X,0}$. As α is a ring homomorphism, we get

$$\alpha(\mathfrak{m}_{Y,0}^k) \subset \mathfrak{m}_{X,0}^k$$

for all $k > 0$. Take generators $w_1,\dots,w_m$ of the maximal ideal of $\mathcal{O}_m$. We have a surjective map $\mathcal{O}_m \longrightarrow \mathcal{O}_{Y,0} = \mathcal{O}_m/\mathscr{I}(Y,0)$, sending w_i to $\overline{w}_i$. Let

$$f_1 = \alpha(\overline{w}_1),\dots,f_m = \alpha(\overline{w}_m).$$

So the f_i are elements of $\mathcal{O}_{X,0}$. Define

$$\varphi = (f_1,\dots,f_m) : (X,0) \longrightarrow (\mathbb{C}^m,0).$$

Hence we get a map $\varphi^* : \mathcal{O}_m \longrightarrow \mathcal{O}_{X,0}$. Moreover consider the composition $\widetilde{\alpha} : \mathcal{O}_m \longrightarrow \mathcal{O}_{Y,0} \xrightarrow{\alpha} \mathcal{O}_{X,0}$. We want to prove the following:

(1) $\varphi^* = \widetilde{\alpha}$.

(2) $\varphi((X,0)) \subset (Y,0)$.

(1) By construction $\varphi^*(w_i) = \widetilde{\alpha}(w_i)$. As they are both $\mathbb{C}$-algebra homomorphisms, it follows that $\varphi^*(g) = \widetilde{\alpha}(g)$ for all *polynomials* $g \in \mathbb{C}[w_1,\dots,w_m]$. But, we want to show that $\varphi^*(g) = \widetilde{\alpha}(g)$ for all $g \in \mathcal{O}_m$. Given such a g, we can, for all $k \in \mathbb{N}$ write

$$g = g_k + g_k',$$

where g_k is a polynomial of degree smaller than k and $g_k' \in \mathfrak{m}^k$. (Here $\mathfrak{m}$ is the maximal ideal of $\mathcal{O}_m$.) Now

$$\varphi^*(g_k) = \widetilde{\alpha}(g_k).$$

As g_k is a polynomial, and both φ^* and $\widetilde{\alpha}$ map $\mathfrak{m}^k$ into $\mathfrak{m}^k_{X,o}$, it follows that

$$\varphi^*(g) - \widetilde{\alpha}(g) = \varphi^*(g'_k) - \widetilde{\alpha}(g'_k) \in \mathfrak{m}^k_{X,o}.$$

As this holds for all k it follows that $\varphi^*(g) - \widetilde{\alpha}(g) \in \cap_k \mathfrak{m}^k_{X,o}$. But as $\mathcal{O}_{X,o}$ is a Noetherian local ring, it follows that $\cap_k \mathfrak{m}^k_{X,o} = 0$, by Krull's Intersection Theorem 1.3.5. Therefore, $\varphi^*(g) = \widetilde{\alpha}(g)$.

(2) Because $(Y, o) = \big(V\big(\mathscr{I}(Y, o)\big), o\big)$, it suffices to show that for all $g \in \mathscr{I}(Y, o)$, the map

$$g \circ \varphi : (X, o) \longrightarrow (\mathbb{C}, o),$$

is the zero map. As $\widetilde{\alpha}$ is induced by $\alpha : \mathcal{O}_{Y,o} \longrightarrow \mathcal{O}_{X,o}$, the ideal $\mathscr{I}(Y, o)$ is in the kernel of $\varphi^* : \mathcal{O}_n \longrightarrow \mathcal{O}_{X,o}$. So the claim follows.

The uniqueness of φ is clear. □

Corollary 3.4.22. *Two germs of analytic spaces (X,x) and (Y,y) are isomorphic if and only if $\mathcal{O}_{X,x}$ and $\mathcal{O}_{Y,y}$ are isomorphic. So the study of germs of analytic spaces is equivalent to the study of reduced analytic algebras. In particular, a germ of a complex submanifold is isomorphic to $(\mathbb{C}^k, o)$ for some k.*

Proof. The Nullstellensatz says that there is a $1-1$ correspondence between reduced analytic algebras and germs of analytic spaces. The previous theorem implies that the isomorphism classes of reduced analytic algebras correspond to isomorphism classes of germs of analytic spaces. □

Remark 3.4.23. More generally, we will later on consider germs of complex spaces (X,x). By this, we mean nothing else than analytic algebras $\mathcal{O}_{X,x}$ which are, in general, nonreduced. They pop up naturally, when one considers intersections, as we will see later. Associated to such a germ of a complex space (X,x) is its reduction (X_{red},x), with local ring $\mathcal{O}_{X_{red},x}$ the reduction of $\mathcal{O}_{X,x}$, obtained by dividing out the nilpotent elements.

We now characterize finite maps algebraically.

Theorem 3.4.24. *Let $f : (X,x) \longrightarrow (Y,y)$ be a map between germs of analytic spaces. Then the following conditions are equivalent:*

(1) f is finite.

(2) $\mathcal{O}_{X,x}$ a finitely generated $\mathcal{O}_{Y,y}$-module.

(3) $\mathcal{O}_{X,x}/\mathfrak{m}_{Y,y}\mathcal{O}_{X,x}$ is a finite-dimensional $\mathbb{C}$-vector space.

(4) $f^{-1}(y) = \{x\}$.

Proof.
(1) $\Longrightarrow$ (4) is easy. As $f^{-1}(y)$ consist of a finite number of points, we can make X so small that it just contains x.

(4) $\Longrightarrow$ (3). As $f^{-1}(y) = \{x\}$, the zero set of $\mathfrak{m}_{Y,y}\mathcal{O}_{X,x}$ is $\{x\}$, and it follows from the Nullstellensatz, see Exercise 3.4.32, that $\mathcal{O}_{X,x}/\mathfrak{m}_{Y,y}\mathcal{O}_{X,x}$ is a finite-dimensional vector space.

(3) $\Longleftrightarrow$ (2). This is the general Weierstraß Division Theorem, see 3.2.10.

(2) $\Longrightarrow$ (1). As $(Y,y) \subset (\mathbb{C}^n, o)$ for some n, that is, we have a surjection $\mathcal{O}_{\mathbb{C}^n,o} \twoheadrightarrow \mathcal{O}_{Y,y}$, we may reduce to the case $(Y,y) = (\mathbb{C}^n, o)$. This case was treated in 3.4.13(2). $\square$

Example 3.4.25. A *hypersurface singularity* by definition is a germ of an analytic space $(V(f), o)$ defined by a single square free $f \in \mathfrak{m} \subset \mathbb{C}\{x_1, \dots, x_n\}$. The isomorphism class of special such f have special names.

- $A_k : f(x_1, \dots, x_n) = x_1^{k+1} + x_2^2 + \dots + x_n^2, \quad k \geq 1,$
- $D_k : f(x_1, \dots, x_n) = x_1 x_2^2 + x_1^{k-1} + x_3^2 + \dots + x_n^2, \quad k \geq 4,$
- $E_6 : f(x_1, \dots, x_n) = x_1^3 + x_2^4 + x_3^2 + \dots + x_n^2,$
- $E_7 : f(x_1, \dots, x_n) = x_1^3 + x_1 x_2^3 + x_3^2 + \dots + x_n^2,$
- $E_8 : f(x_1, \dots, x_n) = x_1^3 + x_2^5 + x_3^2 + \dots + x_n^2.$

Also for $k = 0$ the A_0 is a germ of an analytic space, for which the local ring is isomorphic to $\mathbb{C}\{x_2, \dots, x_n\}$. So the A_0 "singularity" defines a germ of a smooth space. The *A*-*D*-*E*-singularities are the "simplest" singularities, and will be studied further in chapter nine. We draw pictures of the A_k and D_k for $n = 2$ and small k.

A_1

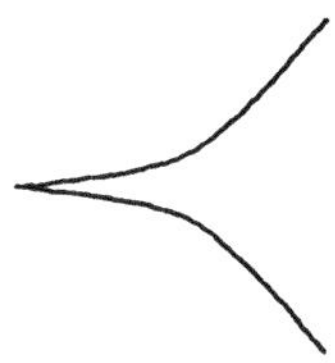

A_2

A_3

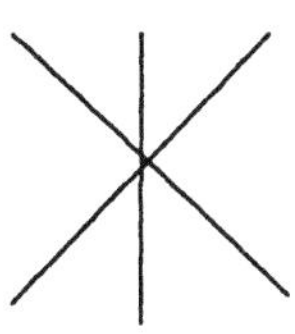

D_4

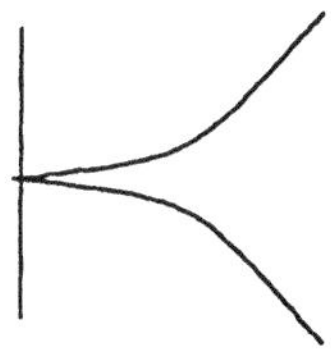

D_5

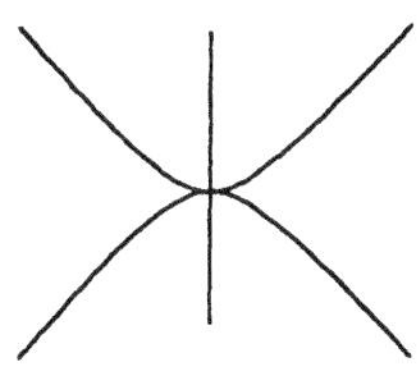

D_6

Definition 3.4.26. Let $f \in \mathbb{C}\{x_1, \dots, x_n\}$.

(1) We define the *Jacobian ideal* by $J(f) := (\frac{\partial f}{\partial x_1}, \dots, \frac{\partial f}{\partial x_n})$.

(2) The *Milnor algebra* of f by definition is the $\mathbb{C}$-algebra $\mathbb{C}\{x_1, \dots, x_n\}/J(f)$. The *Milnor number* of f, $\mu(f)$, by definition is the $\mathbb{C}$-vector space dimension of the Milnor algebra:

$$\mu(f) = \dim_{\mathbb{C}} \left(\mathbb{C}\{x_1, \dots, x_n\}/J(f) \right).$$

(3) The *Tjurina algebra* of f is by definition the $\mathbb{C}$-algebra $\mathbb{C}\{x_1, \dots, x_n\}/(f, J(f))$. The *Tjurina number* of f, $\tau(f)$ by definition is the $\mathbb{C}$-vector space dimension of the Tjurina algebra.

The Milnor number and Tjurina number might be infinite.

Theorem 3.4.27. *Let f and g be in $\mathbb{C}\{x_1,\dots,x_n\}$. Suppose that $(V(f),o)$ and $(V(g),o)$ are isomorphic. Then the Tjurina algebras of f and of g are isomorphic.*[14] *In fact, if φ is an automorphism of $\mathbb{C}\{x_1,\dots,x_n\}$ such that $\varphi((f)) = (g)$, then $\varphi((f)+J(f)) = (g+J(g))$. In particular the Tjurina numbers $\tau(f)$ and $\tau(g)$ are equal.*

Proof. Let $\varphi = (\varphi_1,\dots,\varphi_n) : \mathbb{C}\{x_1,\dots,x_n\} \longrightarrow \mathbb{C}\{x_1,\dots,x_n\}$ be an automorphism such that $(\varphi(f)) = (g)$. It is not difficult to see that $(g,J(g)) = (ug,J(ug))$ for a unit u. Therefore, we may assume that $\varphi(f) = g$. The chain rule says that for all i

$$\frac{\partial g}{\partial x_i} = \sum_j \Big(\frac{\partial f}{\partial x_j}\circ\varphi\Big)\cdot\frac{\partial \varphi_j}{\partial x_i} = \sum_j \varphi\Big(\frac{\partial f}{\partial x_j}\Big)\cdot\frac{\partial \varphi_j}{\partial x_i}.$$

So we see that $J(g) \subset \varphi(J(f))$. As φ^{-1} is also an isomorphism, it follows by symmetry $J(f) \subset \varphi^{-1}(J(g))$. Compose this with φ. We get $\varphi(J(f)) \subset J(g)$. As moreover $\varphi((f)) = (g)$, it follows that $\varphi((f)+J(f)) \subset (g+J(g))$. The converse inclusion follows because of symmetry. □

Example 3.4.28. The Tjurina numbers of the A_k,D_k and E_k singularities are equal to k. The singularities A_k and A_l are isomorphic if and only if $k = l$. Now let $(X,o) = (V(f),o)$ for $f \neq 0$ be a germ of a *regular* point. Then one calculates directly that $\tau(f) = 0$. In particular, all hypersurface singularities $(V(f),o)$ with $0 < \tau(f) < \infty$ are genuine singularities. So now we proved that the A_k for $k \geq 1$ are indeed singularities. Similarly D_k and D_l are isomorphic if and only if $k = l$, and these are indeed singularities.

It is also true that the Milnor number is an invariant of the hypersurface singularity (X,o). A proof however, is surprisingly much more difficult, and will be proved as Corollary 6.4.10.

A_1–singularities can conveniently be described by the Tjurina, or Milnor number.

Theorem 3.4.29. *Let $(X,o) = (V(f),o)$ be a hypersurface singularity. Then $\mu(f) = \tau(f) = 1$ if and only if (X,o) is isomorphic to an A_1–singularity.*

Proof. Suppose $f \in \mathfrak{m}\setminus\mathfrak{m}^2$. Then $\mu(f) = \tau(f) = 0$, as one easily checks. Hence we may assume that $f \in \mathfrak{m}^2$, so that $\tau(f) \geq 1$. A direct calculation shows that for an A_1–singularity $\mu(f) = \tau(f) = 1$.

Conversely, as $\tau(f) \leq \mu(f)$ it follows that either $\mu(f) = 0$ or 1. But $\mu(f) = 0$ cannot occur, as $f \notin \mathfrak{m}\setminus\mathfrak{m}^2$. So we may assume that $\mu(f) = 1$. We may write (nonuniquely) $f = \sum_{i,j} x_i x_j H_{ij}$, and we may assume that H_{ij} is symmetric. We put $h_{ij} = H_{ij}(0)$. Then

$$\frac{\partial f}{\partial x_i} = 2\sum_j x_j h_{ij} \text{ modulo } \mathfrak{m}^2.$$

As $\mu(f) = 1$, it follows that $J(f) = \mathfrak{m} = (x_1,\dots,x_n)$. So there exist α_{ki} with

$$x_k = \sum_i \alpha_{ki}\frac{\partial f}{\partial x_i}.$$

[14] It is a famous theorem of Mather and Yau that the converse holds in case the Tjurina number is finite, see 9.1.8.

Putting $a_{ij} = \alpha_{ij}(0)$ it follows that

$$x_k = 2\sum_{i,j} a_{ki}h_{ij}x_j.$$

So it follows that $\sum_i a_{ki}h_{ij} = \frac{1}{2}\delta_{kj}$ so that $h_{ij} = \left(\frac{\partial^2 f}{\partial x_i \partial x_j}(0)\right)$ is an invertible matrix. The theorem therefore follows from the famous Morse Lemma, to be proved next. □

Lemma 3.4.30 (Morse Lemma). *Let $f \in \mathbb{C}\{x_1,\dots,x_n\}$, $f \in \mathfrak{m}^2$ and suppose that $\left(\frac{\partial^2 f}{\partial x_i \partial x_j}(0)\right)$ has maximal rank n. Then there exists an automorphism φ of $\mathbb{C}\{x_1,\dots,x_n\}$ such that $\varphi(f) = x_1^2 + \dots + x_n^2$.*

Proof. We write

$$f = \sum_{i,j=1}^{n} x_i x_j H_{ij}(x_1,\dots,x_n).$$

We may suppose that $H_{ij} = H_{ji}$ for all i,j. The rank of $\left(\frac{\partial^2 f}{\partial x_i \partial x_j}(0)\right)$ is equal to the rank of $H_{ij}(0)$. It follows from linear algebra that this rank is independent of the coordinates. By induction on s we suppose we found coordinates $(y_1,\dots,y_n)$ such that

$$f(y_1,\dots,y_n) = y_1^2 + \dots + y_{s-1}^2 + \sum_{i,j\geq s} y_i y_j H_{ij}(y_1,\dots,y_n).$$

The case $s = 0$ is trivial. From the assumption that we have maximal rank, it follows that at least one of the $H_{ij}(0)$ is not equal to zero. After a linear coordinate change in the $y_s,\dots,y_n$ we may suppose that $H_{ss}(0) \neq 0$. One can, therefore, by the Implicit Function Theorem, consider the square root $g := \sqrt{H_{ss}}$ which is in fact a unit. We take new coordinates

$$y_i = w_i \text{ for } i \neq s; \quad y_s(w_1,\dots,w_n) = \frac{1}{g(w_1,\dots,w_n)} w_s - \frac{1}{2}\sum_{i>s} \frac{H_{is}(w_1,\dots,w_n)}{H_{ss}(w_1,\dots,w_n)} w_i.$$

It follows from the Inverse Function Theorem that this is an invertible map. Now one calculates that $f(w_1,\dots,w_n)$ is of the desired form. □

Exercises

3.4.31. Suppose $X \subset \mathbb{C}^n$, and $Y \subset \mathbb{C}^m$ are affine varieties. Let $\varphi : X \longrightarrow Y$ be a morphism and suppose that via φ^* the coordinate ring $K[X]$ is a finitely generated $K[Y]$-module. Let $x \in X$, and $y = f(x) \in Y$. Show that for the germs of the analytic spaces (X,x) and (Y,y) the ring $\mathcal{O}_{X,x}$ is a finitely generated $\mathcal{O}_{Y,y}$ module.
(Hint: Use the general Weierstraß Division Theorem 3.2.10.)

3.4.32. Let $I \subset \mathbb{C}\{x_1,\dots,x_n\}$ be an ideal. Use the Nullstellensatz to prove that $(V(I), 0) = \{0\}$ if and only if $\dim_{\mathbb{C}} \mathbb{C}\{x_1,\dots,x_n\}/I < \infty$. Prove that this is the case if and only if I is an $\mathfrak{m}$–primary ideal. Here $\mathfrak{m} = (x_1,\dots,x_n)$.

3.4.33. Prove that the relation in definition 3.4.1 is an equivalence relation.

3.4.34. Let $f : (X,p) \longrightarrow (Y,q)$ be a germ of a continuous map. In general, it does not make sense to talk about the *image* of the map f. Indeed, look at the map

$$f : \mathbb{R}^2 \longrightarrow \mathbb{R}^2, \quad (x,y) \mapsto (z,w) = (x,xy).$$

Show that there exist arbitrary small representatives X of (X,p) and Y of (Y,q) such that for all U open in $\mathbb{R}^2$

$$U \cap f(X) \neq U \cap f(Y).$$

So it is impossible to define the image as the class of the image of a small enough representative. Note that in this example, f is not finite.

3.4.35. Formulate and prove Lemma 3.4.10 and Lemma 3.4.11 for the case of finitely many Weierstraß polynomials. (For every Weierstraß polynomial one takes a new variable).

3.4.36. Let X,Y be topological spaces, $f : X \longrightarrow Y$ be a closed map. Then for each open neighborhood U in X of a fiber $f^{-1}(y)$ there exists an open neighborhood V of y in Y such that $f^{-1}(V) \subset U$. Prove this.

3.4.37. Let X be an irreducible affine variety. Formulate and prove a Local Parametrization Theorem for X.

3.4.38. Let (X,x) be a germ of an analytic space, $\mathcal{O}_{X,x}$ be its local ring. Define for $I \subset \mathcal{O}_{X,x}$ the germ of the analytic space $\big(V(I),x\big) \subset (X,x)$. For a subgerm $(A,x) \subset (X,x)$, define $\mathscr{I}(A,x)$. Prove the standard properties of these operations, including the Nullstellensatz.

3.4.39. Let $(X,0) = \big(V(f),0\big)$ be a hypersurface singularity. One says that $(X,0)$ has an *isolated singularity*, if there exists a representative of X such that $X \setminus \{0\}$ is a complex submanifold. Using the Nullstellensatz, show that $(X,0)$ has an isolated singularity if and only if $\tau(f) < \infty$.

3.4.40. Let $\varphi : (X,x) \longrightarrow (Y,y)$ be finite map between germs of analytic spaces. The map φ is called *generically s to* 1 if there exists a hypersurface (D,y) in (Y,y) such that for all small enough representatives Y of (Y,y) and D of (D,y), the set $\pi^{-1}(p)$ has cardinality s for all $p \in X \setminus D$.

(1) Show that if φ is generically s to 1 for some $s \geq 1$, then φ is surjective.

(2) Show that φ^* induces an isomorphism of total quotient rings $\varphi^* : Q(\mathcal{O}_{Y,y}) \cong Q(\mathcal{O}_{X,x})$ if and only φ is generically $1-1$.

(Hint: First reduce to the case (X,x) and (Y,y) irreducible. Furthermore, if $(Y,y) \longrightarrow (\mathbb{C}^k,0)$ is a primitive Noether normalization of (Y,y), then the induced map $(X,x) \longrightarrow (\mathbb{C}^k,0)$ is a primitive Noether normalization of (X,x). Apply the Local Parametrization Theorem.)

(3) Suppose that $\mathcal{O}_{Y,y}$ is a normal ring. Show that φ is generically s to 1, where s is equal to the degree of the field extension $Q(\mathcal{O}_{X,x}) \supset Q(\mathcal{O}_{Y,y})$.

(Hint: First recall from 1.5.7 that $\mathcal{O}_{Y,y}$ is an integral domain. Then check that in the proof of the Local Parametrization Theorem we can replace $\mathbb{C}\{x_1,\dots,x_k\}$ by $\mathcal{O}_{Y,y}$.)

(4) Show that this result can be generalized to general (Y,y) as soon as we know that a "normalization" exists, see Section 4.4. This remark also holds for the next exercise.

This final item also holds for the next exercise

3.4.41. Let $\varphi : (X,x) \longrightarrow (Y,y)$ be a finite analytic map.

(1) Suppose that $\varphi^* : \mathcal{O}_{Y,y} \longrightarrow \mathcal{O}_{X,x}$ is injective. Show that φ is surjective, in the following two cases.

(a) (X,x) and (Y,y) are irreducible, and φ^* induces an isomorphism on quotient fields.

(b) $\mathcal{O}_{Y,y}$ is a normal ring.

(2) Suppose φ is surjective. Show that φ^* is injective.

3.4.42.

(1) Let (X,x) and (Y,y) be isomorphic germs of analytic spaces. Show that for all $k \in \mathbb{N}$ the $\mathbb{C}$-algebras $\mathcal{O}_{X,x}/\mathfrak{m}_{X,x}^k$ and $\mathcal{O}_{Y,y}/\mathfrak{m}_{Y,y}^k$ are isomorphic. In particular they are isomorphic as $\mathbb{C}$-vector spaces.

(2) Reprove that A_k is isomorphic to A_l if and only if $k = l$. In fact, show that all of the A-D-E-singularities give different isomorphism classes.

3.4.43. Let $(X,0) \subset (\mathbb{C}^n, 0)$ and $(Y,0) \subset (\mathbb{C}^m, 0)$ be germs of analytic spaces, and let $\varphi : (X,0) \longrightarrow (Y,0)$ be a germ of an analytic map. Define the graph $\mathrm{Graph}(\varphi) \subset (\mathbb{C}^{n+m}, 0)$, which is isomorphic to $(X,0)$. Show that we can view φ as the projection on the second factor.

4 Further Development of Analytic Geometry

In this chapter we will continue with the basics of the theory of germs of analytic spaces (and of affine varieties). Probably the most important invariant of (irreducible) germs of analytic spaces is the *dimension*. There are various ways to introduce it. We decided to give, in fact, *four* definitions of dimension. We introduce this many of them, so that we have a flexible development of the theory.

Three of these definitions are given in Section 4.1, namely the Weierstraß dimension, the Krull dimension and the Chevalley dimension. They all have a geometric motivation. The Weierstraß dimension of a germ of an analytic space (X,x) is given by considering a Noether normalization $(X,x) \longrightarrow (\mathbb{C}^k, o)$. It turns out that k is independent of the Noether normalization of (X,x). This map is surjective and has finite fibers, so it should be geometrically clear to the reader that (X,x) has the same dimension as $(\mathbb{C}^k, o)$, which should be k. The Krull dimension is defined as the maximal length k of a chain of irreducible germs $(X_0,x) \subsetneqq (X_1,x) \subsetneqq \cdots \subsetneqq (X_k,x) \subset (X,x)$. The Chevalley dimension is defined by the minimal number k of hypersurfaces $(H_1,x), \ldots ,(H_k,x)$ one has to cut (X,x) with in order to get a point. The main part of the first section is devoted to the fact that these three definitions coincide. As suggested by the definition of Chevalley dimension, we will have to take intersections. Now it might happen that, even if one starts with a reduced space (X,x), the intersection with every hypersurface is nonreduced. Algebraically, this means it might happen that for a reduced local ring $(R,\mathfrak{m})$ and every $f \in \mathfrak{m}$ the ring $R/(f)$ might be nonreduced. This is why we have to consider germs of complex spaces, that is, nonreduced spaces. Algebraically, this just means that we study analytic algebras.

Our discussion on dimension applies just as well to algebraic sets. One has to keep in mind however, that dimension is a *local* property. For example, take $X = V(zx,zy)$ and the point p on the line as indicated in the following figure.

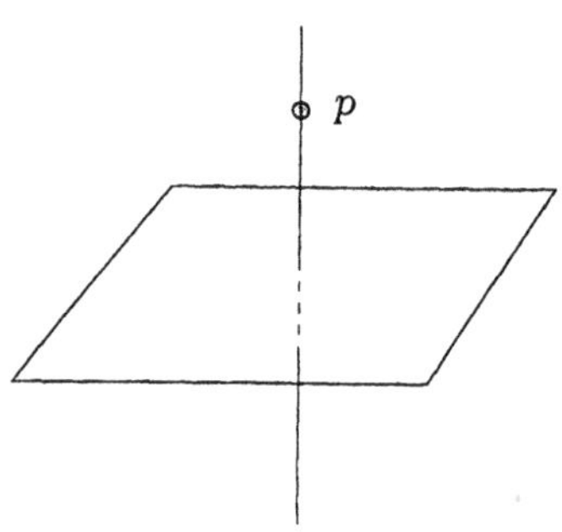

Then one would like to say that the dimension of X at the point p is one. If, however, p is as in one of the following two pictures,

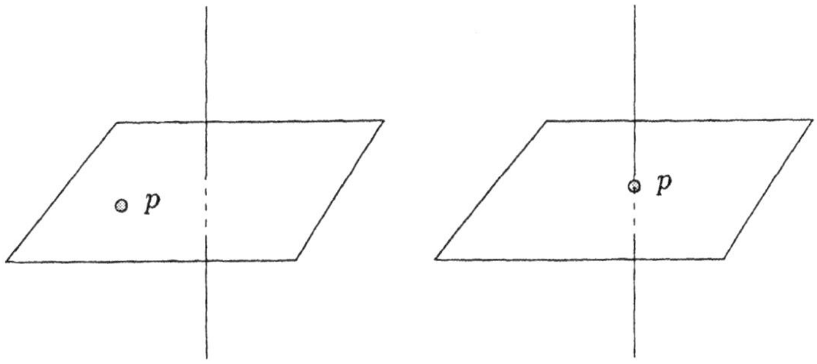

then the dimension of X at p is should be two. Therefore, when discussing dimension of algebraic sets, we will consider localizations $K[V]_{\mathfrak{m}}$. A Noether normalization for this case is discussed in Exercise 2.2.35.

In Section 4.2 we study a fourth characterization of dimension, by means of the Hilbert-Samuel polynomial. This definition is less geometrically motivated, but it has the advantage that in many interesting examples it can be calculated by means of a computer. The idea of the Hilbert-Samuel polynomial is to "count" the number of functions defined on the germ of the analytic space (X,x). More precisely, we will study the behavior of the function $d \mapsto \dim_{\mathbb{C}} (\mathcal{O}_{X,x}/\mathfrak{m}^d)$, the so-called Hilbert-Samuel function. The main result is that for $d \gg 0$, the Hilbert-Samuel function behaves like a polynomial, whose degree is equal to the dimension of (X,x). As it is not more difficult, we will study more generally the Hilbert-Samuel function of an arbitrary local ring, or even finitely generated modules over a local ring. In order to prove that the degree of the Hilbert-Samuel polynomial is smaller than the (Chevalley) dimension, we have to study even more generally Hilbert-Samuel functions with respect to an $\mathfrak{m}$–primary ideal $\mathfrak{q}$, that is, we have to study the function $k \mapsto \dim_{\mathbb{C}} (\mathcal{O}_{X,x}/\mathfrak{q}^d)$.

It turns out that also the leading coefficient of the Hilbert-Samuel polynomial has a nice geometrical interpretation. More precisely, if k is the dimension of (X,x), then $k!$ times the leading coefficient of the Hilbert-Samuel polynomial is called the multiplicity of (X,x). If our space (X,x) has a representative X in a (small) open subset U in $\mathbb{C}^n$, then the multiplicity is the number of intersections points of X with a general linear space of dimension $n-k$. We have already studied this situation for hypersurfaces, see 3.4.11. There we showed that the intersection of a general line near zero with the zero set of a reduced power series of order d has d intersection points. It is, in fact, not difficult to show that the multiplicity in this case is equal to d. The general case will be reduced to the hypersurface case by using the Local Parametrization Theorem. Like the dimension, the multiplicity can, in many cases, be calculated by computer.

In Section 4.3 we will consider tangent spaces, regular and singular points. We already defined in Chapter 3 what a smooth point of an analytic space is, see Definition 3.3.11. The first result is that the set of singular points is contained in a proper analytic subset. Essential use is made of the Local Parametrization Theorem. We have to postpone the proof that the set of singular points in fact is an analytic subset to Chapter 6. However, as a preliminary version, we will prove the Jacobian criterion. This says that the set of singular points is contained in the analytic subset given by the maximal minors of the Jacobian matrix. The fact that it is equal, uses the Coherence Theorem of Oka-Cartan, which is a deep theorem, to be proved in Chapter 6. We will introduce the Zariski tangent space and prove that singular points can be characterized by those points for which the embedding dimension, which is equal to the vector space dimension of the Zariski tangent space, is not equal to the dimension of the space. Finally we will prove the general Jacobian criterion, which gives a statement on when a general point of an analytic subset is smooth.

In the final section of this chapter we study the normalization of a germ of an analytic space. In case (X,x) is irreducible, this is a normal analytic space $(\widetilde{X},\widetilde{x})$ together with a finite degree one map $n : (\widetilde{X},\widetilde{x}) \longrightarrow (X,x)$. For reducible (X,x), we need the notion of a multi-germ. First of all, we will show that a normalization $(\widetilde{X},\widetilde{x})$ of a space (X,x) always exists. The proof uses the Finiteness of Normalization Theorem 1.5.19, Corollary 3.3.25, and Corollary 3.3.26. These are used to show that the normalization of $\mathcal{O}_{X,x}$ is a finite direct sum of analytic algebras. We continue to study properties of normal analytic spaces. It will be shown that the singular points of a normal analytic space are contained in a subset of codimension at least two. In particular, normal curve singularities are smooth. As the normalization always exists, this means that for irreducible curve singularities (X,x) there exists a germ of a finite, generically $1-1$ map $n : (\mathbb{C},\mathrm{o}) \longrightarrow (X,x)$. Hence, irreducible curve singularities can be parameterized! This property of curve singularities will be studied further in the next chapter. The final topic in this chapter is a function theoretic interpretation of normal analytic spaces, which can be phrased by saying that an analytic space (X,x) is normal, if and only if the "Riemann Extension Theorem" holds for (X,x). This means that if f is a holomorphic function on $X \setminus Y$, where Y contains the singular locus of X, and such that f is locally bounded, then there exists an extension of f to a holomorphic function on the whole of X. Again in the proof, Noether normalization is used to reduce the statement to the case of $\mathbb{C}^k$, a case we already know, see 3.1.15.

4.1 Dimension Theory

We start with a generalization of the notion of the germ of an analytic space. We need this, because in our development of dimension theory nonreduced analytic algebras occur naturally.

Definition 4.1.1. A *germ of a complex space* (X,x) consists of

(1) a germ of an analytic space (X_{red},x), called the reduction (X,x) and,

(2) an analytic algebra $\mathcal{O}_{X,x}$ such that $(\mathcal{O}_{X,x})_{red} = \mathcal{O}_{X_{red},x}$.

Example 4.1.2. Let $I = (x^2) \subset \mathbb{C}\{x\}$. Then $\big(V(I),\mathrm{o}\big) = (X_{red},\mathrm{o})$ is the point o, and $\mathcal{O}_{X_{red},x} = \mathbb{C}$. Now I defines a germ of a complex space (X,o) using (X_{red},o) and $\mathcal{O}_{X,x} = \mathbb{C}\{x\}/I$.

Next we give three definitions of dimension.

Definition 4.1.3. Let (X,x) be a germ of a complex space and $(\mathcal{O}_{X,x},\mathfrak{m})$ be its local ring.

- The *Weierstraß dimension* of (X,x) is the least number k, such that there exists a Noether normalization $\mathcal{O}_k \subset \mathcal{O}_{X,x}$ of (X,x). (It will turn out that the number k in a Noether normalization $\mathcal{O}_k \subset \mathcal{O}_{X,x}$ is independent of the Noether normalization.)

- The *Chevalley dimension* of (X,x) is the least number of generators for an $\mathfrak{m}$-primary ideal of $\mathcal{O}_{X,x}$. If k is the Chevalley dimension of (X,x), and $f_1,\ldots,f_k$ generate an $\mathfrak{m}$-primary ideal of $\mathcal{O}_{X,x}$, then $f_1,\ldots,f_k$ is called a *system of parameters* for (X,x). Note that the definition of Chevalley dimension works for all (Noetherian) local rings.

- The *Krull dimension* of (X,x) is the maximal length k of chains of prime ideals

$$\mathfrak{p}_0 \subsetneqq \mathfrak{p}_1 \subsetneqq \cdots \subsetneqq \mathfrak{p}_k$$

in $\mathcal{O}_{X,x}$. Note that this definition also works for all (Noetherian) rings.

Although we will prove in this section that for germs of analytic spaces the three definitions of dimension coincide, we call the Krull dimension simply the dimension of (X,x), notation $\dim(X,x)$.

To motivate the definition of Weierstraß dimension, consider a Noether normalization of (X,x). We have seen that we obtain a map $(X,x) \to (\mathbb{C}^k,o)$ with *finite* fibers. It would be natural to call k the *dimension* of (X,x). However, one has to show that the number k is independent of the Noether normalization. This needs proof, and the proof will follow from the fact that the Weierstraß dimension is equal to the Krull dimension. In order to be careful, we defined the Weierstraß dimension to be the *minimal* k such that there exists a Noether normalization $(X,x) \longrightarrow (\mathbb{C}^k, 0)$.

If k is the Chevalley dimension of (X,x), then there exists a system of parameters $f_1,\ldots,f_k$ and an $s \in \mathbb{Z}$ such that $\mathfrak{m}^s \subset (f_1,\ldots,f_k)$, as $(f_1,\ldots,f_k)$ is $\mathfrak{m}$-primary. In fact, the conditions are equivalent in this case, as obviously every element of $\mathcal{O}_{X,x}/(f_1,\ldots,f_k)$ which is not a unit is nilpotent. It follows from the Nullstellensatz that the intersection of (X,x) with the k hypersurfaces $\{f_1 = 0\} = \ldots = \{f_k = 0\}$ consists of just one point, which should have dimension zero. The idea is that if (X,x) has dimension k, then a general hypersurface section should have dimension $k-1$, so that one has to intersect (X,x) with at least k hypersurfaces to cut down to a single point.

By taking zero sets, it follows that the Krull dimension of (X,x) is the supremum of the length n of a chain of irreducible subvarieties of (X,x):

$$\{0\} \subsetneqq (X_1,x) \subsetneqq \cdots \subsetneqq (X_n,x) \subseteqq (X,x).$$

For the case $\mathcal{O}_{X,x} = \mathcal{O}_n$, consider the following chain of prime ideals:

$$(0) \subsetneqq (x_1) \subsetneqq (x_1,x_2) \subsetneqq \cdots \subsetneqq (x_1,\ldots,x_n)$$

from which it follows that the Krull dimension of $(\mathbb{C}^n, 0)$ is at least n. It needs proof, and is in fact nontrivial that the dimension of $(\mathbb{C}^n, 0)$ is indeed equal to n. Of course, it follows from the fact that the Krull dimension is equal to the Weierstraß dimension, or alternatively, it is equal to the Chevalley dimension. We now start off proving that the three definitions coincide.

Lemma 4.1.4. *Let $R \subset S$ be Noetherian local rings, $\mathfrak{m}_R \subset \mathfrak{m}_S$. Suppose that S is a finitely generated R-module. Then the Krull dimension of S is equal to the Krull dimension of R. In particular, if $\varphi : (X,x) \longrightarrow (Y,y)$ is a finite surjective map of germs of complex spaces, then* $\dim(X,x) = \dim(Y,y)$.

Proof.
Step 1. First let $\mathfrak{p}_0 \subsetneq \cdots \subsetneq \mathfrak{p}_d$ be a chain of prime ideals in R. By the Going-Up Theorem 1.5.25 there exists a chain of prime ideals $\mathfrak{P}_0 \subsetneq \cdots \subsetneq \mathfrak{P}_d$ with $\mathfrak{P}_i \cap R = \mathfrak{p}_i$. This shows that the Krull dimension of S is at least the Krull dimension of R.

Step 2. Now let $\mathfrak{P}_0 \subsetneq \cdots \subsetneq \mathfrak{P}_d$ be a chain of prime ideals in S. Let $\mathfrak{p}_i := \mathfrak{P}_i \cap R$. Then of course, the $\mathfrak{p}_i$ are prime ideals in R. It remains to show that for all $i = 0, \ldots, d-1$ we have a proper inclusion $\mathfrak{p}_i \subsetneq \mathfrak{p}_{i+1}$. By considering the quotient rings $R/\mathfrak{p}_i$ and $S/\mathfrak{P}_i$, which are integral domains, it obviously suffices to prove the following statement.

Let $R \subset S$ be an integral extension, and $(0) \neq J \subset S$ be an ideal containing a nonzerodivisor. Then $J \cap R \neq (0)$.

To prove this, let $x \in J$ be a nonzerodivisor. It satisfies an integral equation, which we take of minimal degree

$$x^n + \ldots + a_{n-1}x + a_n = 0, \quad a_i \in R.$$

Since x is a nonzerodivisor, it follows that $a_n \neq 0$, as otherwise we would get the integral equation $x^{n-1} + \ldots + a_{n-1} = 0$. Thus $a_n \in J \cap R$ is a nonzero element. □

Corollary 4.1.5. *Let (X,x) be a germ of a complex space, and $\mathcal{O}_k \subset \mathcal{O}_{X,x}$ be a Noether normalization. Then the Krull dimension of (X,x) is equal to k. In particular, the Weierstraß dimension is equal to the Krull dimension.*

Proof. From the previous lemma, it follows that the Krull dimension of (X,x) is equal to the Krull dimension of $(\mathbb{C}^k, o)$. It remains to show that the Krull dimension of $(\mathbb{C}^k, o)$ is equal to k. This will be shown by induction on k. If $k = 0$, this is obvious. We already showed that the Krull dimension of $(\mathbb{C}^k, o)$ is at least k. Consider a chain of prime ideals $\mathfrak{p}_0 \subsetneq \mathfrak{p}_1 \subsetneq \cdots \subsetneq \mathfrak{p}_d$ in $\mathcal{O}_k$ of maximal length. We have to show $d \leq k$. Consider the ring $\mathcal{O}_k/\mathfrak{p}_1$, which has Krull dimension $d-1$. Since $\mathfrak{p}_1 \neq (0)$, there exists a Noether normalization $\mathcal{O}_s \subset \mathcal{O}_k/\mathfrak{p}_1$ with $s < k$. By induction, the Krull dimension of $\mathcal{O}_s$ is equal to s which, by the previous lemma, is equal to the Krull dimension of $\mathcal{O}_k/\mathfrak{p}_1$. Therefore, $d - 1 = s < k$, so that $d \leq k$. □

Lemma 4.1.6. *Let $(R,\mathfrak{m})$ be a Noetherian local ring.*

(1) The Krull dimension of R is equal to the Krull dimension of R_{red}. A similar statement holds for the Chevalley dimension.

(2) The Krull dimension of R is zero, if and only if the Chevalley dimension of R is zero.

Proof.
Step 1. A chain of prime ideals in R_{red} lifts to a chain of prime ideals in R. On the other hand, the nilpotent elements of a ring are contained in all prime ideals of R. Therefore, a chain of prime ideals in R gives also rise to a chain of prime ideals in R_{red}. Thus we showed that the Krull dimensions of R and R_{red} are equal.

Step 2. Next we show that the Chevalley dimensions of R and R_{red} are equal. Let $(f_1, \ldots, f_d)$ be an $\mathfrak{m}$-primary ideal of R. Then, obviously, the classes of $f_1, \ldots, f_d$ also generate an $\mathfrak{m}$-primary ideal in R_{red}.

On the other hand, suppose that we have elements $f_1, \dots, f_d$ in R that generate an $\mathfrak{m}$–primary ideal $\mathfrak{q} = (f_1, \dots, f_d)$ in R_{red}. This means that there exists a $n \in \mathbb{N}$ with $\mathfrak{m}^n + N \subset \mathfrak{q} + N$, where N is the set of nilpotent elements of R. However, as R is Noetherian, there exists an $s \in \mathbb{N}$ with $(\mathfrak{q}+N)^s \subset \mathfrak{q}$. It follows that $\mathfrak{m}^{ns} \subset (\mathfrak{m}^n + N)^s \subset \mathfrak{q}$. Thus $\mathfrak{q}$ is an $\mathfrak{m}$–primary ideal.

Step 3. First suppose that the Chevalley dimension is zero. We may now suppose that the ring is reduced. We will show that (0) is a maximal ideal. Because the Chevalley dimension is zero, it follows that there exists an $\mathfrak{m}$–primary ideal with 0 generators. Thus the ideal (0) is $\mathfrak{m}$–primary. From the definition of primary, it follows that every zero divisor is nilpotent, and hence zero, since R is reduced. Hence there are no zerodivisors different from 0, which means that (0) is a prime ideal. As, moreover, (0) is $\mathfrak{m}$–primary, it is equal to $\mathfrak{m}$. Hence the Krull dimension is zero.

Step 4. Suppose the Krull dimension is zero. Then $\mathfrak{m}$ is the only prime ideal of R, as otherwise we can construct a chain of length at least one. As the nilpotent elements of R is the intersection of all minimal prime ideals, it follows that all elements of $\mathfrak{m}$ are nilpotent. Thus (0) is an $\mathfrak{m}$–primary ideal, which exactly says that the Chevalley dimension of R is zero. □

We now prove that the Chevalley dimension is smaller than or equal to the Krull dimension. The proof works for a general Noetherian local ring.

Lemma 4.1.7. *Let $(R,\mathfrak{m})$ be a Noetherian local ring. Then the Chevalley dimension is smaller than or equal to the Krull dimension.*

The main step in the proof is given by the following proposition.

Proposition 4.1.8. *Let R be a Noetherian ring of Krull dimension d. Suppose that f is an* active element *of R, that is, f is not contained in any of the minimal prime ideals of R* [1]*. Then the Krull dimension of $R/(f)$ is at most $d-1$.*

Proof. Consider the ring $R/(f)$. Any chain of prime ideals $\bar{\mathfrak{p}}_1 \subsetneq \cdots \subsetneq \bar{\mathfrak{p}}_s$ in $R/(f)$ gives a chain of prime ideals in R, with $(f) \subset \mathfrak{p}_1 \subsetneq \cdots \subsetneq \mathfrak{p}_s$. The ideal $\mathfrak{p}_1$ is not minimal, because otherwise f would be contained in a minimal prime ideal, contrary to the assumption. Therefore, we can find a minimal prime ideal $\mathfrak{p}_0$ contained in $\mathfrak{p}_1$. Hence $s \le d$, and it follows that the Krull dimension of $R/(f)$ is at most $d-1$. □

Proof of Lemma 4.1.7. Let d be the Krull dimension of R. The proof goes by induction on d. The case $d = 0$ was proved in Lemma 4.1.6.

Therefore, we may suppose $d > 0$. Then $\mathfrak{m}$ is not a minimal prime ideal. Take an element $f \in \mathfrak{m}$, which is not contained in $\cup \mathfrak{p}_i$, where the $\mathfrak{p}_i$ are the minimal prime ideals of R. Such an element exists because of Prime Avoidance, see 1.1.13. Applying the previous lemma, we get that the Krull dimension of $R/(f)$ is at most $d-1$. By induction we have elements $f_1, \dots, f_s$, with $s \le d-1$ generating an $\mathfrak{m}$–primary ideal in $R/(f)$. Then $f, f_1, \dots, f_s$ generate an $\mathfrak{m}$–primary ideal in R. Thus the Chevalley dimension is smaller than or equal to $s + 1 \le d$. □

Theorem 4.1.9. *Let (X,x) be a germ of a complex space. Then the Weierstraß dimension of (X,x) is equal to the Chevalley dimension of (X,x), and thus all three definitions of dimension coincide for germs of complex spaces.*

[1] This exactly means that f is not a zerodivisor of R_{red}, see 1.4.24.

Proof. By Corollary 4.1.5 and Lemma 4.1.7 it remains to show that the Weierstraß dimension is less than or equal to the Chevalley dimension of (X,x). This will be proved in two steps.

Step 1. Let $\varphi : \mathcal{O}_k \longrightarrow \mathcal{O}_{X,x}$ be a $\mathbb{C}$-algebra homomorphism, not necessarily injective. Suppose that $\mathcal{O}_{X,x}$ is, via φ, a finitely generated $\mathcal{O}_k$-module. Then we claim that the Weierstraß dimension of (X,x) is less than or equal to k.

Indeed, consider the injective map $\mathcal{O}_k/\operatorname{Ker}(\varphi) \subset \mathcal{O}_{X,x}$. This is an integral extension. Now $\mathcal{O}_k/\operatorname{Ker}(\varphi)$ is an analytic algebra. So there exists an $s \leq k$ and a Noether normalization $\mathcal{O}_s \subset \mathcal{O}_k/\operatorname{Ker}(\varphi)$. It follows that $\mathcal{O}_s \subset \mathcal{O}_{X,x}$ is a finitely generated extension, hence a Noether normalization. The claim follows.

Step 2. Now let $f_1, \dots f_s$ be generators of an $\mathfrak{m}$-primary ideal of $\mathcal{O}_{X,x}$. We consider the map

$$\begin{aligned} \varphi : \mathbb{C}\{y_1, \dots ,y_s\} &\longrightarrow \mathcal{O}_{X,x} \\ y_i &\mapsto f_i . \end{aligned}$$

By assumption $\mathcal{O}_{X,x}/(f_1, \dots ,f_s)$ is a finitely generated $\mathbb{C}$-vector space, so that, by the general Weierstraß Division Theorem 3.2.10, $\mathcal{O}_{X,x}$ is a finitely generated $\mathbb{C}\{y_1, \dots ,y_s\}$-module. By step 1 it follows that the Weierstraß dimension is less than or equal to s. □

Corollary 4.1.10 (Active Lemma). *Let $f \in \mathfrak{m} \subset \mathcal{O}_{X,x}$ be an active element, $(Y,x) := (V(f),x) \subset (X,x)$. Then $\dim(Y,x) = \dim(X,x) - 1$.*

Proof. We have shown in 4.1.8, $\dim(Y,x) \leq \dim(X,x) - 1$. It cannot be smaller, as otherwise we would be able to find s generators of a primary ideal with $s < \dim(X,x)$. □

Example 4.1.11. The condition that f is an active element is essential. For example, the germ consisting coordinate axes $(X,0) = (V(xy),0)$ has dimension one. Take $f = x$. Then $(Y,x) = (V(x),0)$ also has dimension one.

The following corollary is typical for the germ $(\mathbb{C}^n,0)$ and certainly does not hold for general germs of analytic spaces.

Corollary 4.1.12 (Characterization of Hypersurfaces). *Let $(X,0) \subset (\mathbb{C}^n,0)$ be irreducible, and suppose $\dim(X,0) = n-1$. Then there exists an f with $\mathcal{I}(X,0) = (f)$, that is $(X,0)$ is a hypersurface.*

More generally, if every irreducible component of $(X,0)$ has dimension $n-1$, it follows that $(X,0)$ is a hypersurface.

Proof. It follows that $\mathcal{I}(X,0)$ is a prime ideal, so that we can take an irreducible $f \in \mathcal{I}(X,0)$. We claim that $(f) = \mathcal{I}(X,0)$. Indeed, otherwise there would be a $g \in \mathcal{I}(X,0) \setminus (f)$. In particular g would be a nonzerodivisor modulo (f), and by the Active Lemma $\dim(X,0) \leq \dim(V(f,g),0) = n-2$, a contradiction.

To prove the second statement, let $(X,0) = (X_1,0) \cup \dots \cup (X_r,0)$ be an irreducible decomposition. Then we just proved that $(X_i,0) = (V(f_i),0)$ for some f_i. Thus $(V(f_1 \cdots f_r),0) = (V(f_1),0) \cup \dots \cup (V(f_r),0) = (X,0)$. □

Theorem 4.1.13. *Suppose (X,x) is a germ of a complex space. Let $(X,x) = (X_1,x) \cup \dots \cup (X_r,x)$ be an irreducible decomposition of (X,x). Then*

$$\dim(X,x) = \max\{\dim(X_i,x),\ i = 1, \dots ,r\}.$$

Proof. We may assume that $\mathcal{O}_{X,x}$ is reduced. We use the Krull dimension. The inequality $\geq$ is obvious, since a chain of irreducible germs in (X_i,x) gives one in (X,x). Let $\mathcal{I}(X_i,x) = \mathfrak{p}_i$. Then $\mathfrak{p}_i$ is a minimal prime ideal of $\mathcal{O}_{X,x}$. Let (Y,x) be an irreducible subgerm of (X,x). Then there exists an i with $(Y,x) \subset (X_i,x)$. This is because $\mathcal{I}(Y,x)$ is a prime ideal, and therefore there exists a minimal prime ideal $\mathfrak{p}_i \subset \mathcal{I}(Y,y)$. Taking zero sets gives $(Y,x) \subset (X_i,x)$. So a maximal chain of irreducible subgerms is contained in some irreducible component of (X,x). This shows the theorem. □

Now consider an irreducible germ of an analytic space (X,x), and let $0 \neq f \in \mathfrak{m} \subset \mathcal{O}_{X,x}$. Put $(Y,x) := (V(f),x) \subset (X,x)$. It may very well happen that (Y,x) is reducible. Note that we proved (Active Lemma) that the dimension of (Y,x) is equal to $\dim(X,x)-1$, but we did not prove that each irreducible component of (Y,x) has dimension $\dim(X,x)-1$. We only proved that there is at least one component of this dimension. However, the statement that for an active element f each irreducible component of $V(f)$ has codimension one is in fact true, and holds for general Noetherian rings. This is Krull's Principal Ideal Theorem, to be proved in the next section, see 4.2.16. In this section, we give a proof for germs of analytic spaces which uses the Going-Down Theorem 1.5.26. First we need a definition.

Definition 4.1.14. Let R be a ring.

(1) Let $\mathfrak{p}$ be prime ideal. We define the *height* $\operatorname{ht}(\mathfrak{p})$ of $\mathfrak{p}$, to be the maximal m such that there exists a chain of prime ideals

$$\mathfrak{p}_0 \subsetneq \mathfrak{p}_1 \subsetneq \cdots \subsetneq \mathfrak{p}_m = \mathfrak{p}.$$

(2) For a general ideal I in R we define the *height* of I by $\operatorname{ht}(I) := \min\{\operatorname{ht}(\mathfrak{p}) : \mathfrak{p} \in \operatorname{Ass}(I)\}$.

Remark 4.1.15. Note that from the interpretation of prime ideals in a localization, see 1.3.15, it follows that the height of a prime ideal is equal to the Krull dimension of $R_{\mathfrak{p}}$. Furthermore, we have the inequality

$$\operatorname{ht}(I) + \dim(R/I) \leq \dim(R), \tag{4.1}$$

as follows directly from the definition. Consider a germ of a complex space (X,x), and let $I \subset \mathcal{O}_{X,x}$ be an ideal. Equality in (4.1) would allow us to interpret the height of I as the *codimension* of the zero set of I in (X,o). The codimension of a germ of a complex (Y,x) in (X,x) is simply defined as $\dim(X,x) - \dim(Y,x)$. Of course, it suffices to show this for irreducible germs . This will be shown now.

There exist examples of Noetherian rings, for which we have strict inequality in (4.1), see Exercise 4.1.30.

Theorem 4.1.16. *Let $(X,x) \subset (\mathbb{C}^n,o)$ be an irreducible germ of a complex space. All maximal chains of prime ideals in $\mathcal{O}_{X,x}$ have the same length, namely $\dim(X,x)$. In particular, formula (4.1) holds for $R = \mathcal{O}_{X,x}$.*

Proof. Let $\mathcal{O}_k \subset \mathcal{O}_{X,x}$ be a Noether Normalization, and consider a chain of prime ideals

$$\mathfrak{P}_0 \subsetneq \mathfrak{P}_1 \subsetneq \cdots \subsetneq \mathfrak{P}_s$$

in $\mathcal{O}_{X,x}$ which cannot be extended to a larger chain. We will show that $s = k = \dim(X,x)$. Obviously $s \leq k$.

Step 1. Let $\mathfrak{p}$ be a prime ideal of height one in $\mathcal{O}_k$. We will first show that the dimension of $\mathcal{O}_k/\mathfrak{p}$ is $k-1$. Now $\mathfrak{p} \neq (0)$, and there exists an element $0 \neq f \in \mathfrak{p}$. As $\mathfrak{p}$ is prime, there exists an irreducible factor of f which lies in $\mathfrak{p}$. So we may assume that we have an irreducible f which lies in $\mathfrak{p}$. As $\mathcal{O}_k$ is a unique factorization domain, (f) is a prime ideal. Since $\mathfrak{p}$ has height one, it follows that $\mathfrak{p} = (f)$. By the Active Lemma 4.1.10, the dimension of $\mathcal{O}_k/\mathfrak{p}$ is $k-1$.

Step 2. Let $\mathfrak{p}_i = \mathfrak{P}_i \cap \mathcal{O}_k$. Then we have seen in the proof of Lemma 4.1.4 that

$$\mathfrak{p}_0 \subsetneqq \mathfrak{p}_1 \subsetneqq \cdots \subsetneqq \mathfrak{p}_s \tag{4.2}$$

is a chain of prime ideals in $\mathcal{O}_k$. We claim that this chain cannot be extended to a larger chain of prime ideals in $\mathcal{O}_k$. Suppose the converse, and suppose that there exists a prime ideal $\mathfrak{q}$ with say $\mathfrak{p}_i \subsetneqq \mathfrak{q} \subsetneqq \mathfrak{p}_{i+1}$. (Here we take the convention $\mathfrak{p}_{s+1} = \mathcal{O}_k$.) Take a Noether normalization

$$\mathcal{O}_t \subset \mathcal{O}_k/\mathfrak{p}_i.$$

We thus have $\mathcal{O}_t \subset \mathcal{O}_k/\mathfrak{p}_i \subset \mathcal{O}_{X,x}/\mathfrak{P}_i$. As $\mathcal{O}_k/\mathfrak{p}_i \subset \mathcal{O}_{X,x}/\mathfrak{P}_i$ is also a finite extension, we have a Noether normalization $\mathcal{O}_t \subset \mathcal{O}_{X,x}/\mathfrak{P}_i$. The prime ideals $\mathfrak{q}$ and $\mathfrak{p}_{i+1}$ induce prime ideals in $\mathcal{O}_k/\mathfrak{p}_i$ which we denote by the same name. Moreover, $(0) \subsetneqq \mathfrak{q} \subsetneqq \mathfrak{p}_{i+1}$. Intersecting this chain with $\mathcal{O}_t$, we get a chain of prime ideals in $\mathcal{O}_t$

$$(0) \subsetneqq \mathfrak{q}' \subsetneqq \mathfrak{p}'_{i+1}.$$

Moreover, by construction, $\mathcal{O}_t \cap \mathfrak{P}_{i+1} = \mathfrak{p}'_{i+1}$. Furthermore, $\mathcal{O}_t$ is a normal ring (it is even factorial), so that we can apply the Going-Down Theorem 1.5.26: there exists a prime ideal $\mathfrak{Q}$ in $\mathcal{O}_{X,x}/\mathfrak{P}_i$ with $(0) \subsetneqq \mathfrak{Q} \subsetneqq \mathfrak{P}_{i+1}$. We lift this prime ideal to a prime ideal in $\mathcal{O}_{X,x}$, which we also call $\mathfrak{Q}$. So $\mathfrak{P}_i \subsetneqq \mathfrak{Q} \subsetneqq \mathfrak{P}_{i+1}$ which is in contradiction to our assumption that the chain $\mathfrak{P}_o \subsetneqq \mathfrak{P}_1 \subsetneqq \cdots \subsetneqq \mathfrak{P}_s$ is maximal. Hence the chain (4.2) is maximal after all.

Step 3. We now show $s = k = \dim(X,x)$ by induction on k. The case $k=0$ is obvious. If $k>0$, then the maximal ideal of $\mathcal{O}_{X,x}$ is nonzero. So in a maximal chain $\mathfrak{P}_s$ is the maximal ideal and $s \geq 1$. By step 1, $\mathcal{O}_k/\mathfrak{p}_1$ has dimension $k-1$. Now $\mathfrak{p}_1 \subsetneqq \cdots \subsetneqq \mathfrak{p}_s$ induces a maximal chain in $\mathcal{O}_k/\mathfrak{p}_1$ by step 2, and by induction, $s-1 = k-1$. □

Corollary 4.1.17 (Krull's Principal Ideal Theorem for $\mathcal{O}_{X,x}$). *Let (X,x) be an irreducible germ of an analytic space, and $f \in \mathcal{O}_{X,x}$. Then all irreducible components of $(V(f),x) \subset (X,x)$ have dimension* $\dim(X,x)-1$.

Definition 4.1.18.

(1) Let $(X,o) \subset (\mathbb{C}^n,o)$ be a germ of an analytic space, and let $\mathscr{I}(X,o)$ be the ideal of (X,o), and k be the minimal number of generators of $\mathscr{I}(X,o)$. Then (X,o) is called a *complete intersection* if the dimension of (X,o) is $n-k$.

(2) A germ of an analytic space (X,o) is called a *set-theoretic complete intersection* if $(X,o) = (V(I),o)$, and I can be generated by $n-\dim(X,o)$ elements.

(3) Let R be a local ring. A sequence of nonunits $\{f_1,\dots,f_k\}$ is called a *regular sequence* if for all $i=1,\dots,k$ (the class of) f_i is a nonzero divisor of $R/(f_1,\dots,f_{i-1})$.

Example 4.1.19. We consider (X,o), the coordinate axes in $\mathbb{C}^3$, given by the zero set of the ideal $I=(xy,xz,yz)$. This is a radical ideal as it is the intersection of the prime ideals (x,y), (x,z) and (y,z). So $I=\mathscr{I}(X,o)$, and by Nakayama's Lemma it follows that the minimal number of generators of I is equal to three. It is easy to see (using for example the Chevalley dimension), that the dimension of (X,o) is one. Therefore, (X,o) is not a complete intersection. It is however a set-theoretic complete intersection. Indeed as one easily checks, $(X,o)=\big(V(J),o\big)$, where $J=(xy,zy+zx)$.

Corollary 4.1.20. *Let (X,o) be a germ of a complex space, and suppose $\mathscr{I}(X,o)$ is minimally generated by $f_1,\dots,f_k\in\mathcal{O}_n$. Then (X,o) is a complete intersection, if and only if $\{f_1,\dots,f_k\}$ form a regular sequence in $\mathcal{O}_n$. In this case,* $\dim(X,o)=n-k$.

The proof is left as Exercise 4.1.25.

Definition 4.1.21. Let $I\subset R$ be an ideal.

(1) I is called *unmixed* if all associated primes of I have the same height.

(2) I (or R/I) is called *equidimensional*, if for all minimal associated primes $\mathfrak{p}$ we have $\dim(R_\mathfrak{p}/IR_\mathfrak{p})=\dim(R/I)$.

In case $I\subset\mathcal{O}_n$ is unmixed, I does not have embedded components and is equidimensional. By abuse of language, we also say that (X,o) is unmixed, if the ideal I defining (X,o) is unmixed. Note that it very well might happen that unmixed ideals are not radical. As a corollary of dimension theory, we can give a nice characterization of unmixed ideals in $\mathcal{O}_n$ in terms of a Noether normalization.

Theorem 4.1.22. *Let $I\subset\mathcal{O}_n$ be an ideal and $\mathcal{O}_{X,o}=\mathcal{O}_n/I$. Suppose we have a Noether normalization $\mathcal{O}_k\subset\mathcal{O}_{X,o}$ of (X,o). Then:*

$$I \text{ is unmixed} \iff \text{Every } f\in\mathcal{O}_k \text{ is a nonzero divisor of } \mathcal{O}_{X,o}.$$

Proof. Consider a primary decomposition of I:

$$I=\mathfrak{q}_1\cap\ldots\cap\mathfrak{q}_s.$$

Let $\mathfrak{p}_i:=\sqrt{\mathfrak{q}_i}$ for $i=1,\dots,s$. By the Weierstraß characterization of dimension the ideal I is unmixed, if and only if for all $i=1,\dots,s$ the map

$$\mathcal{O}_k\longrightarrow\mathcal{O}_n/\mathfrak{p}_i,$$

defined by the composition $\mathcal{O}_k\subset\mathcal{O}_n/I\longrightarrow\mathcal{O}_n/\mathfrak{q}_i\longrightarrow\mathcal{O}_n/\mathfrak{p}_i$, is injective. Therefore, I is unmixed if and only if for all $f\in\mathcal{O}_k$ we have $f\notin\mathfrak{p}_i$ for all i, that is, f is a nonzerodivisor according to the characterization of nonzerodivisors, see Theorem 1.4.24. □

Example 4.1.23. Consider $I=(x^2+xy,xz)$. Then $(X,o)=\big(V(I),o\big)$ consists of two components. The first one is given by $x=0$, and the second by $z=0$, $x=-y$. So (X,o) is not equidimensional, and therefore not unmixed. A Noether normalization is given by $\mathbb{C}\{y,z\}\subset\mathbb{C}\{x,y,z\}/(x^2+xy,xz)$. So geometrically we are projecting on the y,z plane. It is indeed an inclusion. Any function which vanishes on the union of the plane and the line must also vanish on the plane $x=0$. The extension is integral. Indeed, x satisfies the integral equation $x^2+xy=0$. So we see from the Noether normalization that (X,o) is not unmixed, since the element z is a zerodivisor.

Theorem 4.1.24 (Second Riemann Extension Theorem). *Let $U \subset \mathbb{C}^n$ be open, and consider an analytic set X in U of codimension at least 2, that is, for all points $p \in X$, we have $\dim(X,p) \leq n-2$. Let*

$$f : U \setminus X \longrightarrow \mathbb{C}$$

be a holomorphic function. Then there exists a unique holomorphic extension of f to U.

Proof. The basic idea of the proof is the same as in the proof of the case that X consists of a single point, see Theorem 3.1.16. We will show that f is locally bounded on $U \setminus X$, so that we can apply the First Riemann Extension Theorem 3.1.15. Consider a Noether normalization at $p \in X$:

$$(X,p) \longrightarrow (\mathbb{C}^k, o).$$

The map is induced by the natural projection $\pi : U \longrightarrow \mathbb{C}^k$. By assumption $k \leq n-2$.

Step 1. We claim that for all small open balls $V \subset \mathbb{C}^{n-k}$ there exists a small closed neighborhood B of $0 \in \mathbb{C}^k$, such that $B \times \overline{V} \subset U$ and $\pi^{-1}(B) \cap X \subset B \times V$. In particular $(\pi^{-1}(B) \cap X) \cap (B \times \partial V) = \emptyset$.

To show this, note that, by the Noether Normalization Theorem, we may assume that $\mathcal{O}_{X,p} \cong \mathbb{C}\{x_1, \ldots, x_n\}/J \supset \mathbb{C}\{x_1, \ldots, x_k\}$ is finite. We consider Weierstraß polynomials $P_{k+1}, \ldots, P_n \in J$, with $P_i \in \mathbb{C}\{x_1, \ldots, x_k\}[x_i]$. We may assume that the P_i converge on U, and that $X \subset \{q \in U : P_{k+1}(q) = \cdots = P_n(q) = 0\}$. We now apply 3.4.11 to each of the P_i and obtain an open ball $B_0 \subset \mathbb{C}^k$, $0 \in B_0$, such that $\pi^{-1}(B_0) \cap X \subset B_0 \times V$. Let $B \subset B_0$ be a closed ball with $0 \in B$. Then $\pi^{-1}(B) \cap X \subset B \times V$.

Step 2. Now we choose a point $q \in (B \times V) \setminus X$. Then $\{\pi(q)\} \times V$ is also an open ball. As the restriction of π to X is finite, the intersection of $\{\pi(q)\} \times V$ with X consists of a finite number of points. We therefore can find a line L_q through q in $\{\pi(q)\} \times V$ which misses $(\{\pi(q)\} \times V) \cap X$, because of the fact that V has at least dimension 2.

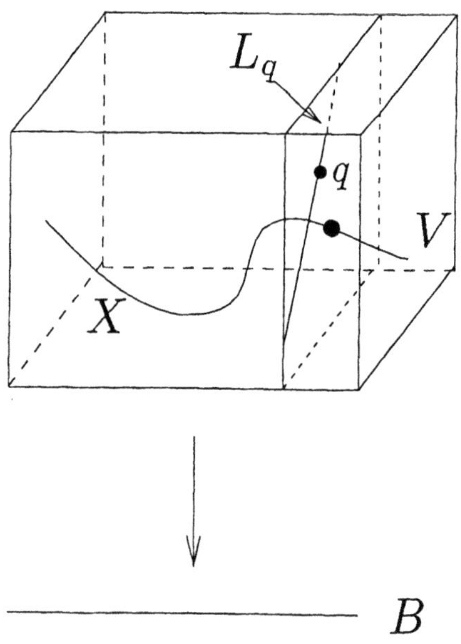

The line L_q intersects the boundary $\{\pi(q)\} \times \partial V$. The restriction of f to $L_q \cap (\{\pi(q)\} \times V)$ is a function of one complex variable, so attains maximum absolute value on the boundary $\{\pi(q)\} \times \partial V$. Hence

$$|f(q)| \leq \max | f_{|\{\pi(q)\} \times \partial V} | \leq \max | f_{|B \times \partial V} | .$$

As f is continuous, and $B \times \partial V$ is compact, it follows that f is locally bounded. So we can apply the First Riemann Extension Theorem 3.1.15. □

Exercises

4.1.25. Prove Corollary 4.1.20.

4.1.26. Let (X,x) and (Y,y) be germs of analytic spaces. Show that

$$\dim(X \times Y,(x,y)) = \dim(X,x) + \dim(Y,y).$$

4.1.27. Let $X \subset \mathbb{C}^n$ be an affine algebraic set, $\mathfrak{m} = (x_1, \ldots, x_n)$. Let $\mathfrak{m}_k = (x_1, \ldots, x_k) \subset \mathbb{C}[x_1, \ldots, x_k]$ and suppose

$$\mathbb{C}[x_1, \ldots, x_k]_{\mathfrak{m}_k} \subset \mathbb{C}[X]_{\mathfrak{m}}$$

is a Noether normalization. Consider the germ of the analytic space $(X, 0)$. Show that

$$\mathcal{O}_k \subset \mathcal{O}_{X,0}$$

is a Noether normalization, cf. 2.2.35. Deduce that the dimension of X at 0, considered as an algebraic set, is equal to the dimension of $(X, 0)$,

4.1.28. Let $(X,0) = (V(f_1, \ldots, f_r), 0), f_i \in \mathcal{O}_n$ be a germ of an analytic space. Consider the map:

$$\varphi : \mathcal{O}_n^r \longrightarrow \mathcal{O}_n,$$

given by the matrix $(f_1, \ldots, f_r)$. Show that for all $i < j$ the element $r_{ij} = f_j e_i - f_i e_j$ (here $e_1, \ldots, e_r$ is the canonical basis of $\mathcal{O}_n^r$) is in the kernel of φ. Show that $(X,0)$ has codimension r in $(\mathbb{C}^n, 0)$ if and only if the kernel of φ is generated by the r_{ij}.

4.1.29. Let $\mathfrak{p}$ be a prime ideal in $\mathbb{C}\{x_1, \ldots, x_n\}$, with $\operatorname{ht}(\mathfrak{p}) = k$. Show that there exists a regular sequence $f_1, \ldots, f_k$ in $\mathfrak{p}$. Geometrically, this means that an irreducible space of codimension k is contained as component in a complete intersection of codimension k.

4.1.30. Consider the ring $R = \mathbb{C}[[x]][y]$. Show that the dimension of R is two. Consider the principal ideal $I := (xy - 1)$. Show that $\operatorname{ht}(I) = 1$, and that I is a maximal ideal. Hence in (4.1) we have a strict inequality.

4.1.31. Prove that for an irreducible affine space X, $\dim(X,p)$ is independent of the chosen point $p \in X$.

4.1.32. Let $X \subset \mathbb{C}^n$ be an irreducible affine variety and let $\mathbb{C}(X)$ be the quotient field of the coordinate ring $\mathbb{C}[X]$. Prove that the dimension of X is equal to the *transcendence degree* of the field extension $\mathbb{C} \subset \mathbb{C}(X)$. (In some books, this is used as the definition of dimension.)

4.2 Hilbert-Samuel Function and Multiplicity

Definition 4.2.1. Let $(R,\mathfrak{m})$ be a Noetherian local ring. The function

$$\begin{aligned} \mathrm{HS}_R : \mathbb{N} &\longrightarrow \mathbb{N} \\ d &\mapsto \dim_{R/\mathfrak{m}}(R/\mathfrak{m}^d) \end{aligned}$$

is called the *Hilbert-Samuel function* of R. In case $R = \mathcal{O}_{X,x}$ is the local ring of a germ of a complex space (X,x) we also write $\mathrm{HS}_{X,x}$ and call it the Hilbert-Samuel function of (X,x).

Example 4.2.2. We take $(X,x) = (\mathbb{C}^n, 0)$. Then $R/\mathfrak{m}^d$ is equal to the vector space of all polynomials of degree less than d in n variables. This is a vector space of dimension $\binom{n+d-1}{n}$. This can be proved by induction for example, but the easiest way to see this is as follows. To a monomial $x_1^{\alpha_1} \cdots x_n^{\alpha_n}$ with $\sum_i \alpha_i < d$ we assign a monomial of degree $d-1$ in $n+1$ variables, namely $x_1^{\alpha_1} \cdots x_n^{\alpha_n} \cdot x_0^{d-1-\sum_i \alpha_i}$. So it suffices to count all monomials of degree $d-1$ in $n+1$ variables. A monomial $x_0^{\alpha_0} \cdots x_n^{\alpha_n}$ we represent by circles and crosses. We first put α_0 circles, then a cross, then α_1 circles, then a cross, etc.

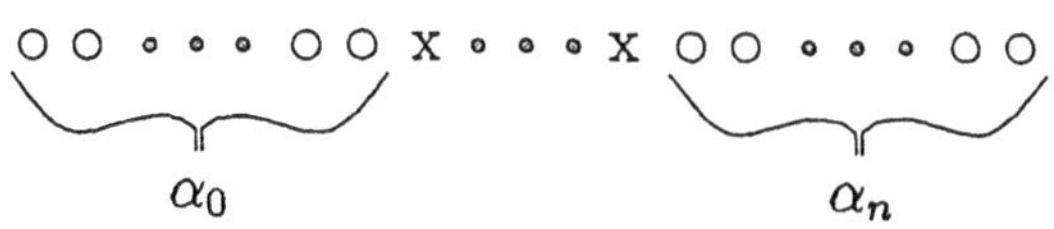

On the other hand, one can assign a polynomial to such a diagram. So we have to choose n crosses out of $n+d-1$ possibilities, and we get $\binom{n+d-1}{n}$ monomials. It follows that the Hilbert-Samuel function $\mathrm{HS}_{\mathbb{C}^n,0}$ of $(\mathbb{C}^n, 0)$ is given by $d \mapsto \binom{n+d-1}{n}$.

The aim of this section is to prove that for $d \gg 0$, the Hilbert-Samuel function behaves like a polynomial, and interpret its degree as the dimension of the space. Finally, in case R is the local ring of a singularity we will interpret the leading coefficient of the polynomial as its "multiplicity".

Theorem 4.2.3. *Let $(R,\mathfrak{m})$ be a Noetherian local ring.*

(1) There exists a polynomial $\mathrm{HSP}_R \in \mathbb{Q}[t]$, such that $\mathrm{HSP}_R(d) = \mathrm{HS}_R(d)$ for d sufficiently large. We call HSP_R the Hilbert-Samuel polynomial of R.

In case R is the local ring of (X,x), we also write $\mathrm{HSP}_{X,x}$ for the Hilbert-Samuel polynomial of R, and call it the Hilbert-Samuel polynomial of (X,x).

(2) $\deg(\mathrm{HSP}_R) = \dim(R)$.

The proof of 4.2.3 will be given on page 144. For the proof, we need to introduce a much more general notion of Hilbert-Samuel function and polynomial.

Definition 4.2.4. Let $(R,\mathfrak{m})$ be a Noetherian local ring, $\mathfrak{q}$ be an $\mathfrak{m}$-primary ideal and M be a finitely generated R-module. A *$\mathfrak{q}$-filtration* $\{M_i\} = \{M_i : i \in \mathbb{N}\}$ is a sequence of submodules $M = M_0 \supset M_1 \supset \ldots$ such that $\mathfrak{q}M_d \subset M_{d+1}$ for all d. It is called a *stable $\mathfrak{q}$-filtration* (or $\mathfrak{q}$-stable filtration) if moreover there exists an n_0 such that $\mathfrak{q}M_d = M_{d+1}$ for all $d \geq n_0$.

Let $\{M_i\}$ be a $\mathfrak{q}$–filtration. Then $\mathrm{HS}_{\{M_i\}} : \mathbb{N} \longrightarrow \mathbb{N}$ with

$$\mathrm{HS}_{\{M_i\}}(d) := \dim_{R/\mathfrak{m}} M/M_d$$

is called the *Hilbert-Samuel function of the filtration* $\{M_i\}$. (We will show in Theorem 4.2.6 that indeed $\dim_{R/\mathfrak{m}} M/M_d < \infty$, so that the definition makes sense.)

If we take the filtration $\{M_i\}$ with $M_i = \mathfrak{m}^i \cdot M$, we simply write HS_M for $\mathrm{HS}_{\{M_i\}}$, and this is called the Hilbert-Samuel function of M. The Hilbert-Samuel function of Definition 4.2.1 is the case $M = R$, $\mathfrak{q} = \mathfrak{m}$, and $M_i = \mathfrak{m}^i$.

Remarks 4.2.5.

(1) We need the general concept of the Hilbert-Samuel function with respect to a filtration to obtain additivity. Namely, let $N \subset M$ be a submodule, then the sequence

$$0 \longrightarrow N/N \cap \mathfrak{m}^n M \longrightarrow M/\mathfrak{m}^n M \longrightarrow (M/N)/\mathfrak{m}^n(M/N) \longrightarrow 0$$

is exact and, therefore,

$$\mathrm{HS}_M = \mathrm{HS}_{M/N} + \mathrm{HS}_{\{N \cap \mathfrak{m}^n M\}} \,.$$

But usually $\mathrm{HS}_N \neq \mathrm{HS}_{\{N \cap \mathfrak{m}^n M\}}$.

(2) We need the Hilbert-Samuel function with respect to an $\mathfrak{m}$–primary ideal $\mathfrak{q}$ in order to be able to use the Noether Normalization Theorem. Namely, let $\mathcal{O}_k \subset \mathcal{O}_{X,x}$ be a Noether normalization, and $\mathfrak{m}$ be the maximal ideal of $\mathcal{O}_{X,x}$. Then $\mathfrak{q} := (x_1, \ldots, x_k)\mathcal{O}_{X,x}$ is $\mathfrak{m}$–primary. $\{\mathfrak{q}^i\}$ is obviously a $(x_1, \ldots, x_k)$–stable filtration of $\mathcal{O}_{X,x}$. Thus $\mathrm{HSP}_{\{\mathfrak{q}^i\}}$ is the Hilbert-Samuel polynomial of the $\mathcal{O}_k$–module $\mathcal{O}_{X,x}$ with respect to $(x_1, \ldots, x_k)$, and at the same time it is the Hilbert-Samuel polynomial of $\mathcal{O}_{X,x}$ with respect to the $\mathfrak{m}$–primary ideal $\mathfrak{q}$.

Moreover, we will prove that $\{\mathfrak{m}^i\}$ is a stable $(x_1, \ldots, x_k)$–filtration of $\mathcal{O}_{X,x}$ if we have a general Noether normalization in the sense of 3.3.30. Therefore, the Hilbert-Samuel polynomial $\mathrm{HSP}_{X,x}$ can also be calculated as the Hilbert-Samuel polynomial $\mathrm{HSP}_{\{\mathfrak{m}^i\}}$ of the stable $(x_1, \ldots, x_k)$–filtration $\{\mathfrak{m}^i\}$ of the $\mathcal{O}_k$–module $\mathcal{O}_{X,x}$.

The general notion of Hilbert-Samuel function with respect to an $\mathfrak{m}$– primary ideal $\mathfrak{q}$ is also used in order to prove second part of 4.2.3, using the Chevalley dimension. In order to prove the first part of 4.2.3, we need to prove a more general version, as written down in the next theorem.

Theorem 4.2.6. *Let $(R,\mathfrak{m})$ be a local ring $\mathfrak{q}$ be a primary ideal, and M be a finitely generated R–module. Let r be the minimal number of generators of $\mathfrak{q}$, and let $\{M_i\}$ be a stable $\mathfrak{q}$-filtration of M. Then*

(1) $\mathrm{HS}_{\{M_i\}}(d) < \infty$ for all $d \geq 0$,

(2) there exists a polynomial $\mathrm{HSP}_{\{M_i\}}$ of degree $\leq r$, such that for all d sufficiently large $\mathrm{HS}_{\{M_i\}}(d) = \mathrm{HSP}_{\{M_i\}}(d)$,

(3) the degree and the leading coefficient of $\mathrm{HSP}_{\{M_i\}}$ only depends on M and $\mathfrak{q}$, and not on the chosen stable $\mathfrak{q}$–filtration.

The proof needs some more preparations, and will be given on page 142. First note that, obviously $\mathrm{HS}_{\{M_i\}}(d) = \sum_{k=0}^{d-1} \dim_{R/\mathfrak{m}}(M_k/M_{k+1})$. This leads to the following definition.

Definition 4.2.7. Let $(R,\mathfrak{m})$ be a local ring, $\mathfrak{q}$ an $\mathfrak{m}$–primary ideal and $\{M_i\}$ be a stable $\mathfrak{q}$–filtration. The *associated graded ring* $\mathrm{Gr}_\mathfrak{q} = \mathrm{Gr}_\mathfrak{q}(R)$ by definition is $\mathrm{Gr}_\mathfrak{q}(R) := \oplus_{d\geq 0}\mathfrak{q}^d/\mathfrak{q}^{d+1}$. The *associated graded module* $\mathrm{Gr}_{\{M_i\}} = \mathrm{Gr}_{\{M_i\}}(M)$ by definition is $\mathrm{Gr}_{\{M_i\}} := \oplus_{d\geq 0} M_d/M_{d+1}$. Note that $\mathrm{Gr}_\mathfrak{q}$ is a graded ring, and that $\mathrm{Gr}_{\{M_i\}}$ is a graded $\mathrm{Gr}_\mathfrak{q}$–module. This follows directly from the definition of a $\mathfrak{q}$–filtration.

The following lemma motivates the condition in the definition of a stable $\mathfrak{q}$–filtration.

Lemma 4.2.8. *Let M be a finitely generated module over the Noetherian local ring $(R,\mathfrak{m})$, $\mathfrak{q}$ be an $\mathfrak{m}$–primary ideal and $\{M_i\}$ be a stable $\mathfrak{q}$–filtration of M. Then $\mathrm{Gr}_\mathfrak{q}$ is a Noetherian ring, and $\mathrm{Gr}_{\{M_i\}}$ is a finitely generated $\mathrm{Gr}_\mathfrak{q}$–module.*

Proof.
Step 1. We first prove that $\mathrm{Gr}_\mathfrak{q}$ is Noetherian. By assumption R is Noetherian and, therefore, $\mathfrak{q}$ is finitely generated, say by $a_1, \ldots ,a_t$. The algebra map $(R/\mathfrak{q})[x_1, \ldots ,x_t] \longrightarrow \mathrm{Gr}_\mathfrak{q}$ which sends x_i to the class of a_i is surjective. The ring $(R/\mathfrak{q})[x_1, \ldots ,x_t]$ is Noetherian, and from the surjectivity it follows that $\mathrm{Gr}_\mathfrak{q}$ is Noetherian.

Step 2. Now we show that $\mathrm{Gr}_{\{M_i\}}$ is a finitely generated $\mathrm{Gr}_\mathfrak{q}$–module. Because $\{M_i\}$ is a stable $\mathfrak{q}$–filtration, there exists an $n_0 \in \mathbb{N}$ such that $M_{n_0+\ell} = \mathfrak{q}^\ell M_{n_0}$ for all $\ell \geq 0$. This says that $\mathrm{Gr}_{\{M_i\}}$ is generated by $\oplus_{k\leq n_0}(M_k/M_{k+1})$. Furthermore, because R is Noetherian, the module M_k/M_{k+1} is a finitely generated R–module. It follows that M_k/M_{k+1} is a finitely generated $R/\mathfrak{q}$–module. Therefore, $\oplus_{k\leq n_0}(M_k/M_{k+1})$ is a finitely generated $R/\mathfrak{q}$–module. So indeed $\mathrm{Gr}_{\{M_i\}}$ is a finitely generated $\mathrm{Gr}_\mathfrak{q}$–module. □

This motivates us to study the graded situation.

Definition 4.2.9. Let K be a field, $A = \oplus_{i\geq 0} A_i$ be a graded K–algebra. Suppose that $\dim_K A_0 < \infty$ and that $A_1 = (x_1, \ldots ,x_r)A_0$ generates A as an A_0–algebra. This means that A is the ring $A_0[x_1, \ldots ,x_r]$.[2] Let $M = \oplus_{i\geq 0} M_i$ be a finitely generated graded A–module. We define the *Hilbert function* $\mathrm{H}_M : \mathbb{N} \longrightarrow \mathbb{N}$ of M by

$$\mathrm{H}_M(n) = \dim_K M_n.$$

The formal power series with integer coefficients

$$\mathrm{HP}_M(t) = \sum_k \mathrm{H}_M(k)t^k \in \mathbb{Z}[[t]]$$

is called the *Hilbert-Poincaré* series of M.

Proposition 4.2.10. *With the notations of 4.2.9, there exists a polynomial $Q_M \in \mathbb{Z}[t]$ such that*

$$\mathrm{HP}_M(t) = \frac{Q_M(t)}{(1-t)^r},$$

that is, the Hilbert-Poincaré series is a rational function in t.

[2] More precisely, via the map $y_i \mapsto x_i$ the algebra A is a quotient of the polynomial ring $A_0[y_1, \ldots ,y_r]$.

Proof. The proof is by induction on r.

Step 1. We first consider the case $r = 0$. In this case $A_1 = 0$, so that $A = A_0$. Therefore, M is a finitely generated module over $A = A_0$. Hence, from $\dim_K(A_0) < \infty$, it follows that $\dim_K M < \infty$. Thus $M_k = 0$ for large k. Hence, by definition of the Hilbert-Poincaré series it follows that $\mathrm{HP}_M(t) \in \mathbb{Z}[t]$.

Step 2. The induction step. Let $r > 0$. Consider the following exact sequence of graded modules:

$$0 \longrightarrow \mathrm{Ker}(\varphi) \longrightarrow M(-1) \xrightarrow{\varphi} M \longrightarrow \mathrm{Coker}(\varphi) \longrightarrow 0.$$

Here $M(-1)$ is isomorphic to M, but has different grading: $M(-1)_k \doteq M_{k-1}$, see 1.2.17. The map φ is multiplication with x_r. If we have an exact sequence

$$0 \longrightarrow M_1 \longrightarrow M_2 \longrightarrow M_3 \longrightarrow M_4 \longrightarrow 0$$

of finite-dimensional K-vector spaces, then $\dim_K(M_1) + \dim_K(M_3) = \dim_K(M_2) + \dim_K(M_4)$, see Exercise 1.2.26. It follows from the definition of the Hilbert-Poincaré series that

$$\mathrm{HP}_{\mathrm{Ker}(\varphi)}(t) + \mathrm{HP}_M(t) = \mathrm{HP}_{M(-1)}(t) + \mathrm{HP}_{\mathrm{Coker}(\varphi)}(t).$$

From the definition of φ, it follows that $\mathrm{Ker}(\varphi)$ and $\mathrm{Coker}(\varphi)$ are finitely generated A/x_r-modules. By induction, we therefore have

$$\mathrm{HP}_{\mathrm{Ker}(\varphi)}(t) = \frac{Q_{\mathrm{Ker}(\varphi)}(t)}{(1-t)^{r-1}}, \quad \text{and} \quad \mathrm{HP}_{\mathrm{Coker}(\varphi)}(t) = \frac{Q_{\mathrm{Coker}(\varphi)}(t)}{(1-t)^{r-1}}.$$

From the definitions of $M(-1)$ and HP it follows immediately that $\mathrm{HP}_{M(-1)}(t) = t\,\mathrm{HP}_M(t)$, and we obtain

$$\begin{aligned}(1-t)\,\mathrm{HP}_M(t) &= \mathrm{HP}_M(t) - \mathrm{HP}_{M(-1)}(t)\\ &= \mathrm{HP}_{\mathrm{Coker}(\varphi)}(t) - \mathrm{HP}_{\mathrm{Ker}(\varphi)}(t)\\ &= \frac{Q_{\mathrm{Coker}(\varphi)}(t) - Q_{\mathrm{Ker}(\varphi)}(t)}{(1-t)^{r-1}}.\end{aligned}$$

This proves the proposition. □

Corollary 4.2.11. *With the conditions as in Definition 4.2.9 we have the following statements.*

(1) *There exists a polynomial* $\mathrm{P}_M \in \mathbb{Q}[t]$ *of degree* $\leq r-1$*, the so-called Hilbert polynomial of* M*, such that* $\mathrm{P}_M(d) = \mathrm{H}_M(d)$ *for* d *sufficiently large.*

(2) *Let* $N = \oplus_{i\geq 0} N_i$ *be a finitely generated graded* A*-module and* $\varphi : M \longrightarrow N$ *be a surjective homomorphism of graded* A*-modules. Then* $\deg \mathrm{P}_M \geq \deg \mathrm{P}_N$.

(3) $\deg \mathrm{P}_M \leq \deg \mathrm{P}_A$.

(4) *If* $\mathrm{Ann}(m) = 0$ *for some homogeneous* $m \in M$*, then* $\deg \mathrm{P}_M = \deg \mathrm{P}_A$.

Proof. (1) By definition $H_M(d)$ is the coefficient of t^d in the Hilbert-Poincaré series $\mathrm{HP}_M(t) = \frac{Q_M(t)}{(1-t)^r}$. We cancel powers of $1-t$ in this fraction and get

$$\mathrm{HP}_M(t) = \frac{G(t)}{(1-t)^s}, \quad G(1) \neq 0,$$

with $G(t) = \sum_{i=0}^k g_i t^i \in \mathbb{Z}[t]$ and $s \leq r$. We expand $(1-t)^{-s}$ in a power series. We leave it as an exercise to show that the power series expansion of $(1-t)^{-s}$ is

$$(1-t)^{-s} = \sum_{i=0}^{\infty} \binom{s-1+i}{s-1} t^i.$$

By multiplying G with $(1-t)^{-s}$ and taking the coefficient of t^d we obtain for $d \gg 0$

$$\mathrm{H}_M(d) = \sum_{i=0}^{k} g_i \binom{s-1+d-i}{s-1}.$$

This is a polynomial in d of degree $s-1$, which proves the first statement of the corollary.

(2) The surjectivity implies that $\mathrm{H}_M(d) = \dim_K M_d \geq \dim_K N_d = \mathrm{H}_N(d)$. Hence $\mathrm{P}_M(d) \geq \mathrm{P}_N(d) \geq 0$ for d sufficiently large. This is only possible if $\deg \mathrm{P}_M \geq \deg \mathrm{P}_N$.

(3) Let M be generated by $m_1, \ldots, m_s$. Let d_i be the degree of m_i. Then we have a surjective map of graded modules $\oplus_{i=1}^m A(d_i) \twoheadrightarrow M$. By the second part, $\deg \mathrm{P}_{\oplus A(d_i)} \geq \deg \mathrm{P}_M$. But $\mathrm{P}_{A(d_i)}(d) = \mathrm{P}_A(d+d_i)$ and $\mathrm{P}_{\oplus A(d_i)} = \sum_{i=1}^s \mathrm{P}_{A(d_i)}$. Therefore $\deg \mathrm{P}_{\oplus A(d_i)} = \deg \mathrm{P}_A$.

(4) If $\mathrm{Ann}(m) = 0$, for some homogeneous $m \in M_k$, then there exists an injection of graded A–modules

$$A(k) \longrightarrow M : \qquad 1 \mapsto m \in M_k.$$

This implies $\mathrm{H}_A(d+k) = \mathrm{H}_{A(k)}(d) \leq \mathrm{H}_M(d)$. This proves the corollary. □

Example 4.2.12. Let $A = \mathbb{C}[x_1, \ldots, x_n]$, $A_i = \{f \in A : f \text{ homogeneous of degree } i\}$. Then $\mathrm{H}_A(d) = \mathrm{P}_A(d) = \binom{n+d-1}{n-1}$. Hence, $\deg(\mathrm{P}_A) = n-1$ and

$$\mathrm{HP}_A(t) = \sum_{i=0}^{\infty} \binom{i+n-1}{n-1} t^i = \frac{1}{(1-t)^n}.$$

We are now able to prove Theorem 4.2.6.

Proof of Theorem 4.2.6. (1) We have seen that for all k the $R/\mathfrak{q}$ module M_k/M_{k+1} is finitely generated. So we have a surjection $(R/\mathfrak{q})^s \twoheadrightarrow M_k/M_{k+1}$. As $\mathfrak{q}$ is $\mathfrak{m}$–primary, $\dim_{R/\mathfrak{m}}(R/\mathfrak{q}) < \infty$. Hence, $\dim_{R/\mathfrak{m}}(M_k/M_{k+1}) < \infty$. It follows that $\mathrm{HS}_{\{M_i\}}(d) = \sum_{k=0}^{d-1} \dim_{R/\mathfrak{m}}(M_k/M_{k+1}) < \infty$.

(2) Define $A_i = \mathfrak{q}^i/\mathfrak{q}^{i+1}$ for all $i \geq 0$. Then $\mathrm{Gr}_\mathfrak{q} = \oplus_{i \geq 0} A_i$, and satisfies the condition of Definition 4.2.9, and $\mathrm{Gr}_{\{M_i\}}$ is a finitely generated $\mathrm{Gr}_\mathfrak{q}$–module by 4.2.8. Let r be the minimal number of generators of $\mathfrak{q}$. In the proof of 4.2.8, we saw that then the minimal number of generators of A_1 is at most r. It follows that the the Hilbert polynomial $\mathrm{HP}_{\mathrm{Gr}_{\{M_i\}}}$ has degree at most $r-1$ by Corollary 4.2.11. Note that $\mathrm{HP}_{\mathrm{Gr}_{\{M_i\}}}(d) = \mathrm{HS}_{\{M_i\}}(d+1) - \mathrm{HS}_{\{M_i\}}(d)$ for large d. It is a nice exercise, see 4.2.25, that this implies (2).

(3) Let $\{\widetilde{M_i}\}$ be another stable $\mathfrak{q}$–filtration of M. It is not difficult to show, see Exercise 4.2.31, that there exists an n_0 such that $M_{n+n_0} \subset \widetilde{M_n}$ and $\widetilde{M}_{n+n_0} \subset M_n$ for all n. This implies $\mathrm{HS}_{\{M_i\}}(d) \le \mathrm{HS}_{\{\widetilde{M_i}\}}(d+n_0)$ and $\mathrm{HS}_{\{\widetilde{M_i}\}}(d) \le \mathrm{HS}_{\{M_i\}}(d+n_0)$. As the degree of $d \mapsto \mathrm{HSP}_{\{M_i\}}(d+n_0)$ in d is equal to the degree of $\mathrm{HSP}_{\{M_i\}}$, and similar for $\{\widetilde{M_i}\}$, this shows that the degrees of $\mathrm{HSP}_{\{M_i\}}$ and $\mathrm{HSP}_{\{\widetilde{M_i}\}}$ are equal. On the other hand the inequalities give

$$\frac{\mathrm{HSP}_{\{M_i\}}(d)}{\mathrm{HSP}_{\{\widetilde{M_i}\}}(d+n_0)} \le 1, \quad \text{and} \quad \frac{\mathrm{HSP}_{\{\widetilde{M_i}\}}(d)}{\mathrm{HSP}_{\{M_i\}}(d+n_0)} \le 1,$$

for d large. This implies that the leading coefficients of $\mathrm{HSP}_{\{M_i\}}$ and $\mathrm{HSP}_{\{\widetilde{M_i}\}}$ are equal and proves the theorem. □

Before giving the proof of Theorem 4.2.3 we need two more results.

Lemma 4.2.13 (Artin-Rees Lemma). *Let $(R,\mathfrak{m})$ be a Noetherian local ring, $\mathfrak{q}$ be an $\mathfrak{m}$-primary ideal, and $\{M_i\}$ be a $\mathfrak{q}$-stable filtration of a finitely generated R-module M. Let $N \subset M$ be a submodule. Then the filtration $\{N_i\}$ defined by $N_i := M_i \cap N$ is a stable $\mathfrak{q}$-filtration of N.*

Proof.
Step 1. Consider the ring $R^* := \oplus_{k\ge 0}\mathfrak{q}^k$. This is a graded ring in the obvious way. If $\mathfrak{q}$ is generated by $a_1, \dots, a_r$, then we get a surjective algebra map $R[x_1, \dots, x_r] \longrightarrow R^*$ by sending x_i to a_i. Hence R^* is Noetherian.

Step 2. Now consider a filtration $\{M_i\}$ of M, with $\mathfrak{q}M_k \subset M_{k+1}$, but which is not necessarily $\mathfrak{q}$–stable. Then we can form the graded R^*–module $M^* := \oplus_{k\ge 0}M_k$. We claim:

$$\{M_i\} \text{ is a stable } \mathfrak{q}\text{–filtration} \iff M^* \text{ is a finitely generated } R^*\text{–module.}$$

To see this, consider for each n the submodule M'_n of M^* generated by $\oplus_{k=0}^n M_k$. It is equal to

$$M'_n = M_0 \oplus \cdots \oplus M_n \oplus \mathfrak{q}M_n \oplus \mathfrak{q}^2 M_n \oplus \cdots .$$

The M'_n form an ascending chain of submodules of M^*. As R^* is Noetherian by Step 1, this chain becomes stable, if and only if M^* is a finitely generated R^*–module. This is the case, if and only if $M'_{n_0} = M^*$ for some n_0. This again is the case if and only if $\mathfrak{q}^r M_{n_0} = M_{n_0+r}$ for some n_0 and all $r \ge 0$, which is equivalent to the stability of the filtration.

Step 3. We now can prove the Artin-Rees Lemma. We have $\mathfrak{q}N_k = \mathfrak{q}(N \cap M_k) \subset \mathfrak{q}N \cap \mathfrak{q}M_k \subset N \cap M_{k+1} = N_{k+1}$, for all $k \ge 0$ showing that $\{N_k\}$ is a $\mathfrak{q}$–filtration. Therefore $N^* := \oplus_{k\ge 0}N_k$ is an R^*–module, and in fact a submodule of M^*. As $\{M_i\}$ is $\mathfrak{q}$–stable, the module M^* is finitely generated, and therefore Noetherian by the previous step. Therefore, N^* is finitely generated, and the filtration $\{N_i\}$ is $\mathfrak{q}$–stable, again by step 2. □

Proposition 4.2.14. *Let $(R,\mathfrak{m})$ be a local ring, $\mathfrak{q}$ be an $\mathfrak{m}$-primary ideal, and $\{M_i\}$ be the stable $\mathfrak{q}$-filtration defined by $M_i = \mathfrak{q}^i M$. Let $x \in R$ be a nonzerodivisor of M, $M' := M/(x \cdot M)$, and $\{M'_i\}$ be the filtration on M' defined by $M'_i = \mathfrak{q}^i M'$. Then*

$$\deg(\mathrm{HSP}_{\{M_i'\}}) \le \deg(\mathrm{HSP}_{\{M_i\}}) - 1.$$

Proof. Consider $N = x \cdot M \subset M$. Because x is not a zerodivisor, N is as R-module isomorphic to M. Consider the filtration $\{N_i\}$ given by $N_i = N \cap \mathfrak{q}^i M$. By the Artin Rees Lemma 4.2.13, $\{N_i\}$ is a $\mathfrak{q}$-stable filtration. So by 4.2.6 the degree and the leading coefficient of $\mathrm{HSP}_{\{N_i\}}$ and of $\mathrm{HSP}_{\{M_i\}}$ are the same. We have for all k the short exact sequence

$$0 \longrightarrow N/N_k \longrightarrow M/M_k \longrightarrow M'/M_k' \longrightarrow 0.$$

Therefore, for $d \gg 0$ we have that $\mathrm{HSP}_{\{M_i'\}}(d) = \mathrm{HSP}_{\{M_i\}}(d) - \mathrm{HSP}_{\{N_i\}}(d)$, so we see that $\mathrm{HSP}_{\{M_i'\}}$ has smaller degree than $\mathrm{HSP}_{\{M_i\}}$. □

Proof of Theorem 4.2.3. We apply Theorem 4.2.6 to the case $M = R$, and get the first part. Now we show that the degree of the Hilbert-Samuel polynomial is equal to the dimension of R. This is done in several steps.

Step 1. We first prove that the degree of HSP_R is smaller than or equal to the Chevalley dimension of R. Let $\mathfrak{q}$ be any $\mathfrak{m}$-primary ideal, and consider the filtration $\mathfrak{q}^i R$. We claim that $\deg(\mathrm{HSP}_{\{\mathfrak{q}^i\}}) = \deg(\mathrm{HSP}_R)$ which was by definition equal to $\deg(\mathrm{HSP}_{\{\mathfrak{m}^i R\}})$. To see this, first note that for some $s \in \mathbb{N}$ we have $\mathfrak{m} \supset \mathfrak{q} \supset \mathfrak{m}^s$. Thus for all $d \ge 0$, we have $\mathfrak{m}^d \supset \mathfrak{q}^d \supset \mathfrak{m}^{sd}$. Thus for large d we get

$$\mathrm{HSP}_R(d) \le \mathrm{HSP}_{\{\mathfrak{q}^i\}}(d) \le \mathrm{HSP}_R(sd).$$

Now $\mathrm{HSP}_R(sd)$ is also a polynomial in d, and has the same degree as $\mathrm{HSP}_R(d)$. Thus the degrees of the polynomials are equal, proving the claim. The second part of Theorem 4.2.6, says that this degree is at most the minimal number of generators of $\mathfrak{q}$, which is the Chevalley dimension of R.

Step 2. We now prove that the Krull dimension of R is less than or equal to $\deg(\mathrm{HSP}_R)$. As we already showed that the Chevalley dimension is less than or equal to the Krull dimension, see 4.1.7, this suffices to prove the theorem. The inequality is proved by induction on $\deg(\mathrm{HSP}_R)$.

Step 2a. Suppose that $\deg(\mathrm{HSP}_R) = 0$. It follows that $\dim(R/\mathfrak{m}^d)$ is independent of d for $d \gg 0$. Hence $\mathfrak{m}^d = \mathfrak{m}^{d+1}$ for $d \gg 0$. It follows from Nakayama's Lemma that $\mathfrak{m}^d = 0$.Thus the zero ideal is $\mathfrak{m}$-primary. So the Chevalley dimension, and hence the Krull dimension is zero.

Step 2b. We now do the induction step. Let $s = \dim(R)$, and $\mathfrak{p}_0 \subsetneqq \cdots \subsetneqq \mathfrak{p}_s$ be a maximal chain of prime ideals in R. As the case $s = 0$ is trivial, we may suppose $s > 0$. In particular $\mathfrak{p}_1 \neq 0$. Take an element $x \in \mathfrak{p}_1 \setminus \mathfrak{p}_0$, and consider the ring $S := R/\mathfrak{p}_0$.[3] We claim that $\deg(\mathrm{HSP}_{S/(x)}) \le \deg(\mathrm{HSP}_R) - 1$. Indeed, x is a nonzerodivisor so that by 4.2.14 $\deg(\mathrm{HSP}_{S/(x)}) \le \deg(\mathrm{HSP}_S) - 1$. Moreover, we have a surjective map $R/\mathfrak{m}^d \twoheadrightarrow S/\mathfrak{m}^d$. Therefore, $\deg(\mathrm{HSP}_S) \le \deg(\mathrm{HSP}_R)$. Thus $\deg(\mathrm{HSP}_{S/(x)}) \le \deg(\mathrm{HSP}_R) - 1$. We apply the induction hypothesis. Any chain in $S/(x)$ has length at most $\deg(\mathrm{HSP}_R) - 1$. But the image of the chain $\mathfrak{p}_1 \subsetneqq \cdots \subsetneqq \mathfrak{p}_s$ in $S/(x)$ gives a chain of length $s - 1$. So $\dim(R) - 1 = s - 1 \le \deg(\mathrm{HSP}_R) - 1$. This proves the theorem. □

[3] Note that if R is an integral domain, then $S = R$, because $\mathfrak{p}_0 = (0)$ is a prime ideal.

Remark 4.2.15. Note that in this proof we showed that:

(1) for any Noetherian local ring the Krull dimension is equal to the Chevalley dimension;

(2) the degree of the Hilbert-Samuel polynomial of an R-module M does not depend on the choice of the filtration on M, and is smaller than or equal to the dimension of R. (Simply replace in the proof R by M.)

Theorem 4.2.16 (Krull's Principal Ideal Theorem). *Let R be a Noetherian ring, $f \in R$, and $\mathfrak{p} \supset (f)$ be a minimal associated prime ideal of (f). Then* $\operatorname{ht}(\mathfrak{p}) \leq 1$. *If moreover f is an active element of R, for example a nonzerodivisor, then* $\operatorname{ht}(\mathfrak{p}) = 1$. *In particular,* $\dim(R/(f) = \dim(R) - 1$.

Proof. Consider an irredundant primary decomposition

$$(f) = \mathfrak{q}_1 \cap \cdots \cap \mathfrak{q}_r .$$

So $\sqrt{\mathfrak{q}_i} = \mathfrak{p}$ for some i. Since $\mathfrak{p}$ is minimal it follows that for all $j \neq i$ we have $\sqrt{\mathfrak{q}_j} \not\subset \mathfrak{p}$. Consider the local ring $R_\mathfrak{p}$, with maximal ideal $\mathfrak{p}R_\mathfrak{p}$. So we get the primary decomposition

$$(f)R_\mathfrak{p} = \mathfrak{q}_i R_\mathfrak{p} .$$

Therefore $(f)R_\mathfrak{p}$ is a primary ideal in $R_\mathfrak{p}$, and is even $\mathfrak{p}R_\mathfrak{p}$-primary. It follows from the definition that the (Chevalley) dimension of $R_\mathfrak{p}$ is at most one. Thus $\operatorname{ht}(\mathfrak{p}) \leq 1$, by using the Krull dimension and the first part of Remark 4.2.15.

Suppose moreover that f is an active element. If $\operatorname{ht}(\mathfrak{p}) = 0$, then $\mathfrak{p}$ is a minimal prime ideal. As $f \in \mathfrak{p}$, this is in contradiction to the assumption that f is an active element. □

For a different proof of Krull's Principal Ideal Theorem, which does not use the Hilbert-Samuel function, see Exercise 4.2.32.

Definition 4.2.17. Let $(R,\mathfrak{m})$ be a Noetherian local ring of dimension e. We write $\operatorname{HSP}_R(t) = \sum_{k=0}^{e} a_k t^k$. Then $m(R) := e! \cdot a_e$ is called the *multiplicity of R* .

If $R = \mathcal{O}_{X,x}$ is the local ring of a singularity (X,x), then we also write $m(X,x)$ for $m(\mathcal{O}_{X,x})$ and call this the multiplicity of (X,x), or the multiplicity of X in x.

Remark 4.2.18. The multiplicity $m(X,x)$ is positive, because $\frac{m(X,x)}{e!}$ is the coefficient of the monomial of highest degree in $\operatorname{HSP}_{X,x}$ and $\operatorname{HSP}_{X,x}(d) > 0$ for d sufficiently large. We leave it is as Exercise 4.2.25, to show that $m(X,x)$ is indeed an integer.

Examples 4.2.19.

(1) Let $(X,x) = (\mathbb{C}^n, 0)$. Then $\operatorname{HSP}_{X,x}(d) = \binom{n+d-1}{n} = \frac{(n+d-1)\cdots d}{n!} = \frac{1}{n!}d^n + \ldots$, which is a polynomial in d of degree n. Therefore, $m(X,x) = 1$.

(2) Consider the A_2-singularity with local ring $R = \mathbb{C}\{x,y\}/(y^2 - x^3)$. Consider an element $f \in \mathbb{C}\{x,y\}$. Then modulo $(y^2 - x^3)$ f has a representative P_f which is a polynomial in y of degree at most one. It is even true that $\operatorname{ord}(P_f) \geq \operatorname{ord}(f)$, see Exercise 3.2.19. So we get that $R/\mathfrak{m}^d = \mathbb{C}\{x,y\}/(y^2 - x^3, x^{d-1}y, x^d)$, and it follows that $\operatorname{HSP}_R(d) = 2d - 1$. So the cusp singularity has dimension one, and multiplicity 2.

We now generalize Example 4.2.19(2).

Lemma 4.2.20. *Let* $f \in \mathbb{C}\{x_1,\dots,x_n\}$, $\operatorname{ord}(f) = m$ *and* $\mathcal{O}_{X,x} = \mathcal{O}_n/(f)$. *Then*

$$\mathrm{HSP}_{X,x}(d) = \sum_{j=1}^{m} \binom{n+d-j-1}{n-1}.$$

In particular the dimension of (X,x) *is* $n-1$, *and the multiplicity is* m. *Note that we have a Noether normalization* $\mathcal{O}_{n-1} \subset \mathcal{O}_n/(f) = \mathcal{O}_{X,x}$, *so that In particular, if* f *irreducible, the multiplicity of* (X,x) *is equal to the degree of the field extension* $Q(\mathcal{O}_{X,x}) \supset Q(\mathcal{O}_{n-1})$.

Proof. Without loss of generality we may suppose that f is a Weierstraß polynomial in x_n of degree equal to $m = \operatorname{ord}(f)$. As in the example, any element $g \in \mathcal{O}_{X,x}$ has a representative P_g which is a polynomial in x_n of degree at most $m-1$. Furthermore, $\operatorname{ord}(g) \leq \operatorname{ord}(P_g)$. Now $\mathrm{HS}_R(d) = \dim_{\mathbb{C}} \mathbb{C}\{x_1,\dots,x_n\}/((f) + \mathfrak{m}^d)$ is equal to

$$\dim_{\mathbb{C}} \mathbb{C}\{x_1,\dots,x_n\}/(f, \{x^\alpha : \alpha_1 + \dots + \alpha_n = d, \alpha_n < m\}).$$

For $d > m$ a basis of this vector space is $B = \cup_{j=1}^m B_j$ with $B_j = \{x^\alpha : \alpha_n = j-1, \alpha_1 + \dots + \alpha_{n-1} \leq d - j\}$. By Example 4.2.2 B_j has $\binom{n+d-j-1}{n-1}$ elements. □

We now turn our attention to a geometric interpretation of the multiplicity. First we need a lemma.

Lemma 4.2.21. *Let* (X,x) *be an irreducible germ of an analytic space defined by a prime ideal* $\mathfrak{p} \subset \mathbb{C}\{x_1,\dots,x_n\}$. *Let* $\mathcal{O}_k \subset \mathcal{O}_{X,x} := \mathbb{C}\{x_1,\dots,x_n\}/\mathfrak{p}$ *be a general Noether normalization, that is, there exist monic polynomials* $P_j \in \mathbb{C}\{x_1,\dots,x_{j-1}\}[x_j] \cap \mathfrak{p}$ *for* $j = k+1,\dots,n$, *such that*

- $\deg_{x_j}(P_j) = \operatorname{ord}(P_j) =: n_j$,
- $Q(\mathcal{O}_{X,x}) = Q(\mathcal{O}_k)[x_{k+1}]/(P_{k+1})$.

Such a general Noether normalization exists according to Exercise 3.3.30. Let $\mathfrak{m}$ *be the maximal ideal of* $\mathcal{O}_{X,x}$, *and* $\mathfrak{m}_k$ *be the maximal ideal of* $\mathcal{O}_k$. *Consider the filtration* $\{\mathfrak{m}^i\}$ *of* $\mathcal{O}_{X,x}$ *Then* $\{\mathfrak{m}^i\}$ *is a stable* $\mathfrak{m}_k$*-filtration.*

Proof. Obviously, $\{\mathfrak{m}^i\}$ is an $\mathfrak{m}_k$-filtration. Let $f \in \mathbb{C}\{x_1,\dots,x_n\}$ be a power series of order t. It follows from Exercise 3.3.31 that modulo $\mathfrak{p}$ f is equivalent to a polynomial

$$P_f = \sum_{\nu_j < n_j} f_{\nu_{k+1},\dots,\nu_n} x_{k+1}^{\nu_{k+1}} \cdots x_n^{\nu_n}, \quad f_{\nu_{k+1},\dots,\nu_n} \in \mathcal{O}_k.$$

Moreover, P_f has order at least t. So we see that, if $t \geq \sum_{j \geq k+1} n_j$, we can write every element of $\mathfrak{m}^{t+1}$ as an element of $\mathfrak{m}_k \cdot \mathfrak{m}^t$. This is what we needed to show. □

Example 4.2.22. $\mathbb{C}\{x\} \subset \mathbb{C}\{x,y\}/(y^2 - x) = \mathcal{O}_{X,x} \cong \mathbb{C}\{y\}$ is not a general Noether normalization (in the sense of 3.3.30) because $\deg_y(y^2 - x) = 2 > \operatorname{ord}(y^2 - x) = 1$. Now, $\mathfrak{m}_k = (x)$, $\mathfrak{m} = (x,y)$ and $\mathfrak{m}^{2i} = (y^{2i}) = (x^i)$ implies $\mathfrak{m}_k \cdot \mathfrak{m}^{2i} = (x^{i+1}) = \mathfrak{m}^{2i+2} \neq \mathfrak{m}^{2i+1}$ for all i. This shows that $\{\mathfrak{m}^i\}$ is not a stable $\mathfrak{m}_k$-filtration of $\mathcal{O}_{X,x}$.

Proposition 4.2.23. *Let (X,x) be an irreducible germ of an analytic space, and $\mathcal{O}_k \subset \mathcal{O}_{X,x}$ be a general Noether normalization. Then*

$$m(X,x) = [Q(\mathcal{O}_{X,x}) : Q(\mathcal{O}_k)],$$

that is, the multiplicity is equal to the degree of the field extension $Q(\mathcal{O}_{X,x}) \supset Q(\mathcal{O}_k)$.

Proof. We apply the Local Parametrization Theorem 3.4.14, in order to reduce the statement to the case of a hypersurface singularity. This case was already considered in Lemma 4.2.20. Let Δ be the discriminant of $P := P_{k+1}$. Let $\mathcal{O}_{X,x} := \mathcal{O}_n/\mathfrak{p}$, $\mathcal{O}_{X',x'} := \mathcal{O}_k[x_{k+1}]/(P)$, and $K = \mathcal{O}_{X,x}/\mathcal{O}_{X',x'}$. It was shown in the first part of the Local Parametrization Theorem that $\mathfrak{p} \cap \mathcal{O}_k[x_{k+1}] = (P)$ so that we have an exact sequence of $\mathcal{O}_k$-modules.

$$0 \longrightarrow \mathcal{O}_k[x_{k+1}]/(P) \longrightarrow \mathcal{O}_n/\mathfrak{p} \longrightarrow K \longrightarrow 0. \tag{4.3}$$

Let $\mathfrak{m}_k$ be the maximal ideal of $\mathcal{O}_k$. We consider the following filtrations on these three $\mathcal{O}_k$-modules.

- $\{\mathfrak{m}_k^i \mathcal{O}_{X,x}\}$ on $\mathcal{O}_{X,x}$,
- $\{\mathfrak{m}_k^i \mathcal{O}_{X,x} \cap \mathcal{O}_{X',x'}\}$ on $\mathcal{O}_{X',x'}$,
- $\{\mathfrak{m}_k^i K\}$ on K.

These are all stable $\mathfrak{m}_k$-filtrations, where the second one follows from the Artin-Rees Lemma 4.2.13. It follows from exact sequence (4.3) that

$$\mathrm{HSP}_{\{\mathfrak{m}_k^i \mathcal{O}_{X,x}\}} = \mathrm{HSP}_{\{\mathfrak{m}_k^i \mathcal{O}_{X,x} \cap \mathcal{O}_{X',x}\}} + \mathrm{HSP}_{\{\mathfrak{m}_k^i K\}}.$$

Let $\mathfrak{m}$ be the maximal ideal of $\mathcal{O}_{X,x}$, and $\mathfrak{m}'$ be the maximal ideal of $\mathcal{O}_{X',x'}$. Then the filtrations $\{\mathfrak{m}^i\}$ and $\{\mathfrak{m}'^i\}$ are $\mathfrak{m}_k$-stable by Lemma 4.2.21.

It was noted in 3.4.16, that as a consequence of the Local Parametrization Theorem, the module K is in fact an $\mathcal{O}_k/\Delta^a$-module for some $a \in \mathbb{N}$. The ring $\mathcal{O}_k/\Delta^a$ has dimension $k-1$. It follows from the second part of Theorem 4.2.6, and the second part of Remark 4.2.15 that the Hilbert-Samuel polynomial of K has degree at most $k-1$. By definition $m(X,x)$ is determined by the leading term of $\mathrm{HSP}_{\{\mathfrak{m}^i\}}$, and $m(X',x')$ is determined by the leading term of $\mathrm{HSP}_{\{\mathfrak{m}'^i\}}$, which both have degree k, and those are independent of the stable $\mathfrak{m}_k$-filtration by Theorem 4.2.6. From these two facts it follows that $m(X,x) = m(X',x')$. So we reduced the problem to the case of an irreducible hypersurface, as promised. □

We finally come to the geometric interpretation of multiplicity.

Theorem 4.2.24. *Let $(X,x) \subset (\mathbb{C}^n,0)$ be an irreducible a germ of an analytic space, and let $\dim(X,x) = k$. There exists a Zariski open subset $U \subset M_{n-k,n}(\mathbb{C})$[4] with the following property: given $M \in U$, there exists*

- *an $\varepsilon > 0$ and a representative $\widetilde{X}$ of (X,x) in the open ball B_ε with center 0 and radius ε,*

[4] $M_{n-k,n}(\mathbb{C})$ is the vector space of $(n-k) \times n$-matrices with entries in $\mathbb{C}$.

- *an open and dense subset V of B_ε,*

such that for $p \in V$ the plane defined by

$$M\begin{pmatrix} x_1 \\ \vdots \\ x_n \end{pmatrix} = p^t$$

intersects $\widetilde{X}$ in exactly $m(X,x)$ points.

To put it more down to earth, if $k = \dim(X,x)$, a general $n-k$–dimensional plane near the origin cuts a representative $\widetilde{X}$ of (X,x) in exactly $m(X,x)$ points.

Proof. Let $\widetilde{U} \subset Gl_n(\mathbb{C})$ be the Zariski open subset such that for $\widetilde{M} \in \widetilde{U}$ the corresponding linear coordinate change defines, via the projection $\pi : \mathbb{C}^n \longrightarrow \mathbb{C}^k$ onto the first k coordinates, a general Noether normalization $(X,x) \longrightarrow (\mathbb{C}^k, 0)$.

Let M be the part of $\widetilde{M}$ consisting of the first $n-k$ rows. This part of $\widetilde{M}$ corresponds to the direction of the projection π. Let $U = \{M \mid \widetilde{M} \in \widetilde{U}\}$.

Choosing $M \in U$ restricts the situation to the case that a general primitive Noether normalization $\mathcal{O}_{X,x} \supset \mathcal{O}_k$ is given and we may assume $\mathcal{O}_{X,x} = \mathbb{C}\{x_1,\ldots,x_n\}/\mathfrak{p} \supset \mathbb{C}\{x_1,\ldots,x_k\}$ and $M = \begin{pmatrix} 1 & & & 0\,0 \ldots 0 \\ & \ddots & & \vdots \quad \vdots \\ 0 & & & 1\,0 \ldots 0 \end{pmatrix}$.

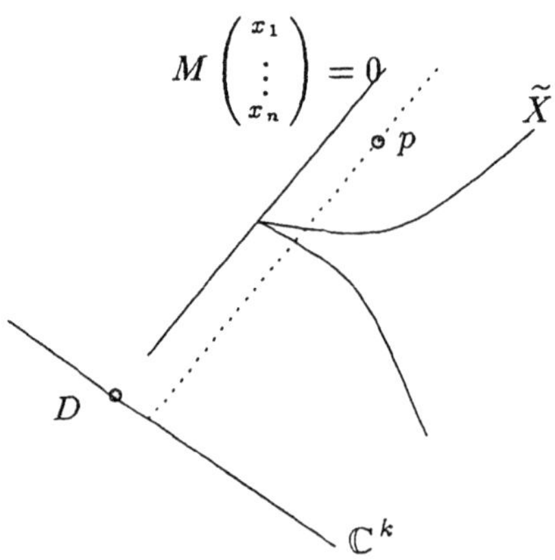

Let $V = \{p \mid \pi(p) \notin D\}$, where D is the discriminant, given by $V(\Delta)$, with Δ the discriminant of the minimal polynomial of the field extension $Q(\mathcal{O}_{X,x}) \supset Q(\mathcal{O}_k)$. We now apply 3.4.16 (1) to obtain that the $(n-k)$–dimensional plane E defined by $M\begin{pmatrix} x_1 \\ \vdots \\ x_n \end{pmatrix} = p$, that is $E = \left\{\begin{pmatrix} x_1 \\ \vdots \\ x_n \end{pmatrix} : x_1 = p_1, \ldots, x_k = p_k\right\} = \pi^{-1}(\pi(p))$, intersects a representative of (X,x) in $[Q(\mathcal{O}_{X,x}) : Q(\mathcal{O}_k)]$ points. Now $[Q(\mathcal{O}_{X,x}) : Q(\mathcal{O}_k)]$ is equal to $m(X,x)$ by Proposition 4.2.23. $\square$

Exercises

4.2.25. Define for all $k \in \mathbb{Z}$ the polynomial $\binom{t}{k} := \frac{1}{k!}t(t-1)\cdots(t-k+1) \in \mathbb{Q}[t]$. For $f \in \mathbb{Q}[t]$ define Δf by $\Delta f(t) := f(t+1) - f(t) \in \mathbb{Q}[t]$.

(1) Prove that $\Delta\binom{t}{k} = \binom{t}{k-1}$.

(2) For any $f \in \mathbb{Q}[t]$ we can write $f(t) = c_0\binom{t}{k} + c_1\binom{t}{k-1} + \ldots + c_k$ for suitable $c_i \in \mathbb{Q}$. Show this.

(3) Suppose moreover that $f(n) \in \mathbb{Z}$ for $n \gg 0$, and $k = \deg(f)$. Prove that $c_i \in \mathbb{Z}$ for $i = 0, \ldots, k$.
(Hint: Use induction on k, and consider Δf.)

(4) Let $f : \mathbb{N} \longrightarrow \mathbb{N}$ be any function. Suppose that $d \mapsto \Delta f(d) = f(d+1) - f(d)$ is a polynomial $Q(d)$ for $d \gg 0$ of degree $k-1$. Prove that $d \mapsto f(d)$ is, for $d \gg 0$, a polynomial of degree k.
(Hint: Write $\Delta f(t) = c_0\binom{t}{k-1} + c_1\binom{t}{k-2} + \ldots + c_{k-1}$, and consider $P = c_0\binom{t}{k} + c_1\binom{t}{k-1} + \ldots + c_{k-1}\binom{t}{1}$. Prove that $\Delta(f-P)(d)$ is a constant for $d \gg 0$.)

4.2.26. Compute the Hilbert-Samuel function and polynomial of $R = \mathbb{C}\{x,y,z\}/(y^2 - xz, z^3 - x^5, z^2 - x^2y)$. Check the result by using the computer algebra system SINGULAR.

4.2.27. Let $I = (xz - y^2,\ xw - yz,\ yw - z^2) \subset \mathbb{C}\{x,y,z,w\}$. Let (X,x) be the germ at 0 defined by I. Compute the Hilbert-Samuel polynomial $\mathrm{HSP}_{X,x}$.

4.2.28. Let $I \subset \mathbb{C}[x_1, \ldots, x_r] = \mathbb{C}[x]$ be an ideal, define $\mathbb{C}[x]_{\le s} = \{f \mid \deg(f) \le s\}$ and $I_{\le s} = I \cap \mathbb{C}[x]_{\le s}$. Define the affine Hilbert function by $\mathrm{AH}_I(n) = \dim_{\mathbb{C}} \mathbb{C}[x]_{\le n}/I_{\le n}$.
Prove that for sufficiently large n $\mathrm{AH}_I(n)$ is a polynomial $\mathrm{AP}_I(n) = \sum\limits_{i=0}^{d} b_i\binom{s}{i}$, $b_i \in \mathbb{Z}, b_d > 0$.
(Hint: Let $I^h \subset \mathbb{C}[t,x_1, \ldots, x_r]$ be the homogenization of I. Prove that (with the canonical grading of $\mathbb{C}[t,x_1,\ldots,x_r]$) $H_{\mathbb{C}[t,x_1,\ldots,x_r]/I^h}(n) = \mathrm{AH}_I(n)$.)

4.2.29. With the notations of Exercise 4.2.28 prove that $\deg \mathrm{AP}_I = \dim \mathbb{C}[x_1, \ldots, x_r]/I$.
(Hint: Let I^h be as in 4.2.28. Prove that

$$\dim(\mathbb{C}[t,x_1,\ldots,x_r]/I^h) = \dim(\mathbb{C}[x_1,\ldots,x_r]/I) + 1,$$

by using Noether normalization.)

4.2.30. Prove the following version of the Artin-Rees Lemma:
Let $(R,\mathfrak{m})$ be a local ring, $\mathfrak{q}$ be an $\mathfrak{m}$–primary ideal, and M be a finitely generated module. If $N \subset M$ a submodule then there exists an n_0 such that

$$(\mathfrak{q}^n M) \cap N = \mathfrak{q}^{n-n_0}((\mathfrak{q}^{n_0}M) \cap N).$$

4.2.31. Prove: if $\{M_n\}$, $\{\widetilde{M}_n\}$ are stable $\mathfrak{q}$–filtrations of M then there exists an n_0 such that $\widetilde{M}_{n+n_0} \subset M_n$ and $\widetilde{M}_{n+n_0} \subset M_n$ for all $n \in \mathbb{N}$.
(Hint: It suffices to prove this for the special case $\widetilde{M}_n = \mathfrak{q}^n M$ for all n.)

4.2.32. Give a different proof of Krull's Principal Ideal Theorem 4.2.16 along the following steps.

(1) Reduce the second statement to the first, by using the characterization of zerodivisors.

(2) By looking at the localization $R_{\mathfrak{p}}$, reduce to the case that R is local, with maximal ideal $\mathfrak{m} = \mathfrak{p}$. From now one we suppose this.

(3) From now suppose $\mathfrak{q} \neq \mathfrak{m}$ is a prime ideal. Show that it suffices to show that $\mathrm{height}(\mathfrak{q}) = \dim R_{\mathfrak{q}} = 0$.

(4) Show that it suffices to show that $\mathfrak{q}R_{\mathfrak{q}}$ is nilpotent.

(5) Show that in $R/(f)$ every descending chain of ideals becomes stationary using the fact that $\mathfrak{m}^\rho \subset (f)$ for some ρ.

(6) Consider the symbolic powers $\mathfrak{q}^{(n)} := \mathfrak{q}^n R_\mathfrak{q} \cap R$, see Exercise 1.3.23. Show that the chain of ideals

$$(f) + \mathfrak{q}^{(1)} \supset (f) + \mathfrak{q}^{(2)} \supset (f) + \mathfrak{q}^{(3)} \supset \dots$$

becomes stationary, that is, there exists an $n \in \mathbb{N}$ with $(f) + \mathfrak{q}^{(n)} = (f) + \mathfrak{q}^{(n+1)}$.

(7) Prove that for this n we have $\mathfrak{q}^{(n)} = (f)\mathfrak{q}^{(n)} + \mathfrak{q}^{(n+1)}$. (Use the second part of Exercise 1.3.23.)

(8) Apply Nakayama to see that $\mathfrak{q}^{(n)} = \mathfrak{q}^{(n+1)}$.

(9) Use the definition of symbolic power and Nakayama to see that $\mathfrak{q}^n R_\mathfrak{q} = 0$.

(10) Deduce the theorem.

4.2.33. Let (X,x) be a germ of an analytic space and $(X,x) = \bigcup_{i=1}^n (X_i,x)$ be an irreducible decomposition. We set $I = \{i : \dim(X_i,x) = \dim(X,x)\}$. Prove that $m(X,x) = \sum_{i \in I} m(X_i,x)$. (Hint: Study the situation $(X,x) = (X_1,x) \cup (X_2,x)$, with (X_i,x) reduced and defined by the ideals $I_i \subset \mathcal{O}_n$. Compare the Hilbert-Samuel functions by looking at the exact sequence

$$0 \to \mathcal{O}_n/I_1 \cap I_2 \to \mathcal{O}_n/I_1 \oplus \mathcal{O}_n/I_2 \to \mathcal{O}_n/I_1 + I_2 \to 0.$$

One may assume that $\dim(\mathcal{O}_n/I_1 + I_2) < \max\{\dim(\mathcal{O}_n/I_1), \dim(\mathcal{O}_n/I_2)\}$.)

4.3 Regular Local Rings and the Jacobian Criterion

We now study regular and singular points of analytic spaces. According to Definition 3.3.11 a point x of an analytic space X is called regular if a neighborhood of this point is a complex submanifold of $\mathbb{C}^n$. In particular, the local ring $\mathcal{O}_{X,x}$ of a smooth point x of X is isomorphic to $\mathbb{C}\{x_1, \dots, x_k\}$ for some k, which is an integral domain. Since for a reducible germ (X,x) we have that $\mathcal{O}_{X,x}$ is not an integral domain the following proposition follows.

Proposition 4.3.1. *Let X be an analytic space, $x \in X$, and (X,x) be a reducible germ. Then $x \in \mathrm{Sing}(X)$. In particular, if $X = X_1 \cup \dots \cup X_r$ with $X_i \not\subset X_j$ for $i \neq j$, then* $\mathrm{Sing}(X) = \bigcup_i \mathrm{Sing}(X_i) \bigcup \left(\cup_{i \neq j} (X_i \cap X_j) \right)$.

It might not be a big surprise for the reader that "almost all" points of an analytic space are smooth. The proof however is a little bit tricky. For an irreducible space this is proved by means of the Local Parametrization Theorem 3.4.14 (cf. Remark 3.4.16 (2)), which essentially allows us to reduce the statement to the hypersurface case.

Proposition 4.3.2. *Let U be an open subset of $\mathbb{C}^n$, and X be an analytic subset of U. Then the set of singular points* $\mathrm{Sing}(X)$ *is contained in a proper analytic subset of X.*

Proof. This is a local statement, that is, for all $x \in X$ we have to prove it for a neighborhood of $x \in X$. If (X,x) is an irreducible germ, it was proved in the Local Parametrization Theorem that there exists a hypersurface (D,x) in (X,x), and representatives X and D such that $X \setminus D$ is a complex submanifold of an open set in $\mathbb{C}^n$.

Now if (X,x) is reducible, we can take an irreducible decomposition $(X_1,x) \cup \cdots \cup (X_r,x)$ of (X,x). For suitable representatives, we have $X = X_1 \cup \ldots \cup X_r$ with $X_i \not\subset X_j$ for $i \neq j$. We therefore have that $X_i \cap X_j$ for $i \neq j$ is a proper analytic subset of X. We may assume that $\mathrm{Sing}(X_i) \subset V_i$, V_i a proper analytic subset of X_i. On the other hand, we have $\mathrm{Sing}(X) = \cup_i \mathrm{Sing}(X_i) \cup \left(\cup_{i\neq j}(X_i \cap X_j)\right)$ by the previous proposition. Now $\cup_{i\neq j}(X_i \cap X_j)$ is a proper analytic subset of X. Therefore, if x is not contained in the proper analytic subset $\cup_i V_i \cup \left(\cup_{i\neq j}(X_i \cap X_j)\right)$ then x is a smooth point of X. □

Our next goal would be to show that the set of singular points in fact forms an *analytic* subset of X. However, as we will explain later, we are not able to prove this in this section, since we will need the Oka-Cartan Theorem, see 6.3.2, concerning the coherence of the ideal sheaf. So the fact that the singular locus is an analytic subset is surprisingly deep. What we will show is, that we characterize the regular points by means of the Jacobian Criterion.

Definition 4.3.3. Let $(R,\mathfrak{m})$ be a Noetherian local ring. The *embedding dimension* $\mathrm{edim}(R)$ of R is defined by

$$\mathrm{edim}(R) := \dim_{R/\mathfrak{m}}(\mathfrak{m}/\mathfrak{m}^2).$$

If (X,x) is a germ of an analytic space, then $\mathrm{edim}(X,x) := \mathrm{edim}(\mathcal{O}_{X,x})$. The ring R is called a *regular local ring*, if the embedding dimension is equal to the dimension, that is, if

$$\mathrm{edim}(R) = \dim(R).$$

By Nakayama's Lemma, the embedding dimension is the minimal number of generators of the maximal ideal. As the dimension of R is equal to the Chevalley dimension of R, see Remark 4.2.15, it follows that the number of generators of the maximal ideal of a Noetherian local ring cannot be less than $\dim(R)$.

Example 4.3.4. Consider the power series ring $\mathbb{C}\{x_1,\ldots,x_n\}$. This is a ring of dimension n. The maximal ideal $(x_1,\ldots,x_n)$ is generated by $n = \dim(\mathbb{C}\{x_1,\ldots,x_n\})$ generators, so the power series ring is a regular local ring. Similarly, the formal power series ring $\mathbb{C}[[x_1,\ldots,x_n]]$, and the localization $\mathbb{C}[x_1,\ldots,x_n]_{\mathfrak{m}}$, $\mathfrak{m} = (x_1,\ldots,x_n)$ are regular local rings.

The following lemma motivates the name embedding dimension.

Lemma 4.3.5. *Let (X,x) be a germ of an analytic space, and let $n \in \mathbb{N}$ be the minimal number such that there exists an injective analytic map $(X,x) \hookrightarrow (\mathbb{C}^n,0)$. Then $\mathrm{edim}(X,x) = n$. In particular, (X,x) is smooth at x if and only if $\mathcal{O}_{X,x}$ is a regular local ring.*

Proof. Suppose $\varphi : (X,x) \hookrightarrow (\mathbb{C}^n,0)$. So we have a surjective $\mathbb{C}$-algebra map $\varphi^* : \mathcal{O}_{\mathbb{C}^n,0} \twoheadrightarrow \mathcal{O}_{X,x}$. In particular φ induces a surjective map of vector spaces $\mathfrak{m}_{\mathbb{C}^n,0}/\mathfrak{m}^2_{\mathbb{C}^n,0} \twoheadrightarrow \mathfrak{m}_{X,x}/\mathfrak{m}^2_{X,x}$. This shows that the minimal number of generators of $\mathfrak{m}_{X,x}$ is at most n, so that $\mathrm{edim}(X,x) \leq n$. On the other hand, let $s = \mathrm{edim}(X,x)$. Then s is the number of generators of $\mathfrak{m}_{X,x}$, and it was proved in Corollary 3.2.13 that $\mathcal{O}_{X,x}$ is a quotient of $\mathbb{C}\{x_1,\ldots,x_s\}$. By dualizing we get an injective analytic map $\varphi : (X,x) \hookrightarrow (\mathbb{C}^s,0)$, so that $s \geq n$. This proves the lemma. □

We now come to the Jacobian criterion.

Theorem 4.3.6. *Let $(X, o) \subset (\mathbb{C}^n, o)$ be a germ of an analytic space, and let the ideal of (X, o) be generated by $f_1, \dots, f_s \in \mathbb{C}\{x_1, \dots, x_n\}$. We denote by* $\operatorname{rank}_o(f_1, \dots, f_s)$ *the rank of the* Jacobian matrix

$$\left(\frac{\partial f_i}{\partial x_j}(0)\right) := \left(\frac{\partial f_i}{\partial x_j}(0)\right)_{1 \le i \le s, 1 \le j \le n}.$$

Then

$$\operatorname{edim}(X, o) + \operatorname{rank}_o(f_1, \dots, f_s) = n.$$

Proof. Put $e := \operatorname{edim}(X, o)$ and $r = \operatorname{rank}_o(f_1, \dots, f_s)$.

Step 1. Suppose we have functions $f_1, \dots, f_s$ that generate the maximal ideal $\mathfrak{m}$ of $\mathcal{O}_n$. Then we claim that the Jacobian matrix

$$\left(\frac{\partial f_i}{\partial x_j}(0)\right)_{1 \le i \le s, 1 \le j \le n}$$

has rank n. It follows from Nakayama's Lemma that $s \ge n$, and in fact already n of the f_i generate the maximal ideal $\mathfrak{m}$. Without loss of generality, $f_1, \dots, f_n$ are generators. Now write $f_i = \sum_j A_{ij} x_j$ modulo $\mathfrak{m}^2$, for $A_{ij} \in \mathbb{C}$. As the f_i generate $\mathfrak{m}$, we conversely have $x_i = \sum_j B_{ij} f_j \mod \mathfrak{m}^2$, for some $B_{ij} \in \mathbb{C}$. Therefore (A_{ij}) is an invertible matrix with inverse (B_{ij}). Now $\left((\frac{\partial f_i}{\partial x_j}(0)\right) = (A_{ij})$, so the Jacobian matrix has maximal rank n.

Step 2. We now prove $r \ge n - e$. Let $g_1, \dots, g_e \in \mathcal{O}_{\mathbb{C}^n, o}$ project to generators of the maximal ideal $\mathfrak{m}_{X,o}$. Now

$$\mathbb{C} \cong \mathcal{O}_{X,o}/\mathfrak{m}_{X,o} = \mathcal{O}_{\mathbb{C}^n,o}/(f_1, \dots, f_s, g_1, \dots, g_e).$$

So $f_1, \dots, f_s, g_1, \dots, g_e$ generate the maximal ideal of $\mathcal{O}_{\mathbb{C}^n, o}$. Hence the Jacobian matrix of $(f_1, \dots, f_s, g_1, \dots, g_e)$ has rank n by step 1. Deleting a column reduces the rank by at most one. So deleting the columns corresponding to $g_1, \dots, g_e$ shows that the rank r of the Jacobian matrix $\left((\frac{\partial f_i}{\partial x_j}(0)\right)$ is at least $n - e$.

Step 3. It remains to show that $e \le n - r$. This will be proved as an application of the Implicit Mapping Theorem. After renumbering the f_i and x_j we may suppose that

$$\det\left(\frac{\partial f_i}{\partial x_j}(0)\right)_{n-r+1 \le i,j \le n} \ne 0.$$

We can therefore apply the Implicit Mapping Theorem 3.3.6 and deduce that there exist functions $\varphi_{n-r+1}(x_1, \dots, x_{n-r}), \dots, \varphi_n(x_1, \dots, x_{n-r})$ such that

$$f_i\big(x_1, \dots, x_{n-r}, \varphi_{n-r+1}(x_1, \dots, x_{n-r}), \dots, \varphi_n(x_1, \dots, x_{n-r})\big) = 0,$$

for $i = n - r + 1, \dots, n$. This shows that (X, o) is a subgerm of the graph of the map $\varphi : \mathbb{C}^{n-r} \longrightarrow \mathbb{C}^r$, that is, a subgerm of

$$\Big(V\big(x_{n-r+1} - \varphi_{n-r+1}, \dots, x_n - \varphi_n\big), o\Big).$$

This is a smooth germ of (embedding) dimension $n - r$. Thus $e \le n - r$, as was to be proved. □

Remark 4.3.7. Let X be an analytic subset of an open subset U of $\mathbb{C}^n$, say locally defined by holomorphic functions $f_1,\ldots,f_s$ on U. Suppose that for all $x \in X$, the germ (X,x) has dimension $n-c$. It follows directly from the Jacobian Criterion and Lemma 4.3.5 that the singular locus of X is contained in the zero set of the c-minors of the Jacobian matrix, which is an analytic set. However, a priori these two sets need not be equal. Indeed, to complete the proof, one needs that for all $x \in X$, the classes of $f_1,\ldots,f_s$ generate the ideal of (X,x), so it has to be a radical ideal. Now even if for some $x \in X$, the ideal generated by $f_1,\ldots,f_s$ is radical in $\mathcal{O}_{\mathbb{C}^n,x}$, it is not clear that for all y in a small neighborhood of x, the ideal generated by the classes of $f_1,\ldots,f_s$ is radical in $\mathcal{O}_{\mathbb{C}^n,y}$. That this is in fact the case, is surprisingly difficult to prove. It is equivalent to the Coherence Theorem of Oka-Cartan, see 6.3.2. From this Coherence Theorem it therefore follows that the singular locus of an analytic set itself is analytic.

Definition 4.3.8. Let (X,p) be a germ of an analytic subset in $\mathbb{C}^n$, and let $f_1,\ldots,f_s$ generate the ideal of (X,p). The (Zariski) *tangent space* $T_{X,p}$ of X at the point $p = (p_1,\ldots,p_n)$ is the affine space in $\mathbb{C}^n$ defined by

$$T_{X,p} = p + \left\{ q = (q_1,\ldots,q_n) \in \mathbb{C}^n : \sum_{j=1}^{n} \frac{\partial f_i}{\partial x_j}(p) \cdot q_j = 0,\ i = 1,\ldots,s \right\}.$$

Lemma 4.3.9. *$T_{X,p}$ is independent of the choice of $f_1,\ldots,f_s$.*

Proof. Suppose that $g_1,\ldots,g_m$ also generate $\mathcal{I}(X,p)$. Then we have an equality

$$g_i = \sum_{k=1}^{s} \xi_{ik} \cdot f_k$$

for some $\xi_{ik} \in \mathbb{C}\{x_1,\ldots,x_n\}$. We differentiate. By the product rule, and because $f_i(p) = 0$ for all i, it follows

$$\frac{\partial g_i}{\partial x_j}(p) = \sum_{k=1}^{s} \xi_{ik}(p) \frac{\partial f_k}{\partial x_j}(p).$$

Now let $(q_1,\ldots,q_n) \in \mathbb{C}^n$ be such that $\sum_{j=1}^{n} \frac{\partial f_i}{\partial x_j}(p) q_j = 0$ for $i = 1,\ldots,s$. Then

$$\sum_{j=1}^{n} \frac{\partial g_i}{\partial x_j}(p) q_j = \sum_{k=1}^{s} \xi_{ik}(p) \sum_{j=1}^{n} \frac{\partial f_k}{\partial x_j}(p) q_j = 0.$$

This shows one inclusion. The other inclusion follows by symmetry. □

Note that the rank of $\left(\frac{\partial f_i}{\partial x_j}(p)\right)$ is $n - \mathrm{edim}(X,p)$ by 4.3.6, so that the tangent space has dimension $\mathrm{edim}(X,x)$. In particular, by 4.3.3, 4.3.5 and 3.4.20 we have the following.

Theorem 4.3.10. *Let (X,p) be a germ of an analytic space. Then the following conditions are equivalent.*

(1) p is a regular point of X.

(2) $\mathcal{O}_{X,p}$ is a regular local ring.

(3) $\mathcal{O}_{X,p} \cong \mathbb{C}\{x_1,\ldots,x_k\}$ *for some* k.

(4) $\dim(\mathcal{O}_{X,p}) = \dim_{\mathbb{C}}(T_{X,p})$.

Examples 4.3.11.

(1) Let $X = \{(x,y) \in \mathbb{C}^2 : x^2 - y = 0\}$. Then X is an analytic subset of $\mathbb{C}^2$, $0 = (0,0) \in X$ and $T_{X,0} = \{(q,0) : q \in \mathbb{C}\} \cong \mathbb{C}$.

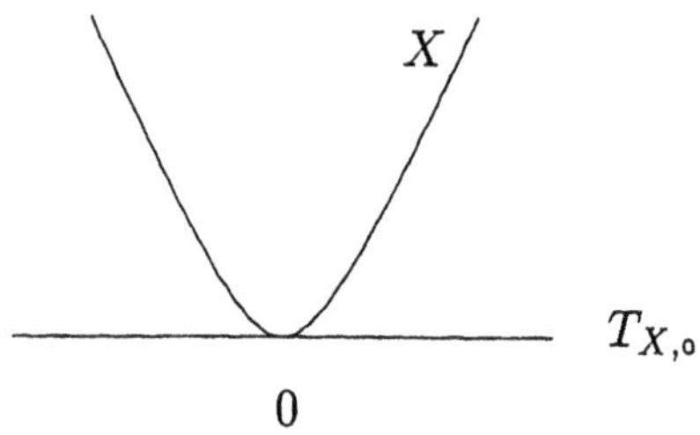

(2) Let $X = \{(x,y) \in \mathbb{C}^2 : x^3 - y^2 = 0\}$. Then again X is an analytic subset of $\mathbb{C}^2$ and $0 = (0,0) \in X$, but $T_{X,0} = \mathbb{C}^2$. On the other hand, $(1,1) \in X$ and $T_{X,(1,1)} = \{(q, \frac{3}{2}q - \frac{1}{2}), q \in \mathbb{C}\}$.

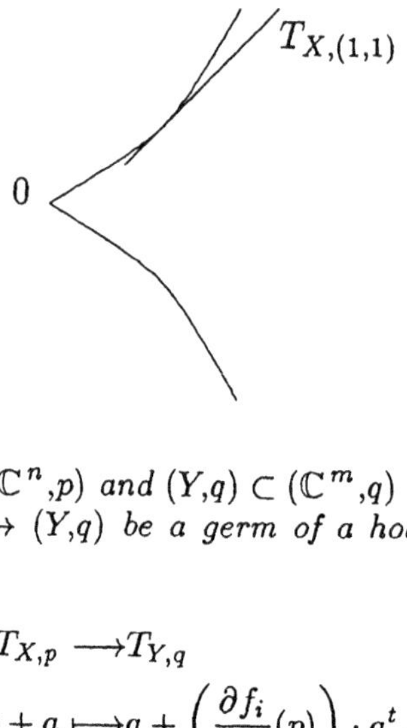

Theorem 4.3.12. *Let* $(X,p) \subset (\mathbb{C}^n,p)$ *and* $(Y,q) \subset (\mathbb{C}^m,q)$ *be germs of analytic subsets and* $f = (f_1,\ldots,f_m) : (X,p) \longrightarrow (Y,q)$ *be a germ of a holomorphic map. Then there exists a canonical linear map*

$$\begin{aligned} T_{f,p} : T_{X,p} &\longrightarrow T_{Y,q} \\ p + a &\longmapsto q + \left(\frac{\partial f_i}{\partial x_j}(p)\right) \cdot a^t. \end{aligned}$$

If $g : (Y,q) \longrightarrow (Z,r)$ *is a further germ of a holomorphic map, then*

$$T_{g \circ f,p} = T_{g,q} \circ T_{f,p}.$$

As $T_{Id,p} = Id_{T_{X,p}}$, *it follows that if* f *is an isomorphism, then* $T_{f,p}$ *is an isomorphism.*

Proof. Without loss of generality, $p = 0$ and $q = 0$. Let $\mathcal{I}(X,0) = (g_1,\ldots,g_s)$, and $\mathcal{I}(Y,0) = (h_1,\ldots,h_t)$. Then

$$T_{X,o} = \{a \in \mathbb{C}^n : \sum_{j=1}^{n} \frac{\partial g_i}{\partial x_j}(0) a_j = 0, \quad i = 1, \dots, s\}$$

$$T_{Y,o} = \{b \in \mathbb{C}^m : \sum_{j=1}^{m} \frac{\partial h_i}{\partial y_j}(0) b_j = 0, \quad i = 1, \dots, t\}.$$

Now $h_1 \circ f, \dots, h_t \circ f$ vanish on X, so they are in the ideal generated by $g_1, \dots, g_s$. Hence there exist $s_{ik} \in \mathcal{O}_{\mathbb{C}^n,o}$ such that

$$h_i \circ f = \sum_{k=1}^{s} s_{ik} g_k.$$

We differentiate and get, using the chain rule,

$$\sum_{k=1}^{m} \frac{\partial h_i}{\partial y_k}(0) \frac{\partial f_k}{\partial x_j}(0) = \frac{\partial}{\partial x_j}(h_i \circ f)(0) =$$

$$\sum_{k=1}^{s} \frac{\partial s_{ik}}{\partial x_j}(0) g_k(0) + \sum_{k=1}^{s} s_{ik}(0) \frac{\partial g_k}{\partial x_j}(0) = \sum_{k=1}^{s} s_{ik}(0) \frac{\partial g_k}{\partial x_j}(0).$$

Now consider a point $a = (a_1, \dots, a_n) \in T_{X,o}$. So $\sum_{j=1}^{n} \frac{\partial g_k}{\partial x_j}(0) a_j = 0$. It follows that

$$\sum_{k=1}^{m} \frac{\partial h_i}{\partial y_k}(0) \sum_{j=1}^{n} \frac{\partial f_k}{\partial x_j}(0) a_j = 0.$$

But this says exactly that $\left(\frac{\partial f_k}{\partial x_j}(0)\right) \cdot a^t \in T_{Y,o}$.

The second statement follows from the chain rule, and is left as an exercise. □

Examples 4.3.13.

(1) Let $\pi : \mathbb{C}^n \longrightarrow \mathbb{C}^k$ be the projection given by $\pi(x_1, \dots, x_n) = (x_1, \dots, x_k)$, and $p = (p_1, \dots, p_n) \in \mathbb{C}^n$. Then

$$T_{\pi,p} : T_{\mathbb{C}^n,p} = p + \mathbb{C}^n \longrightarrow T_{\mathbb{C}^k,\pi(p)} = \pi(p) + \mathbb{C}^k$$

is defined by $T_{\pi,p}(p + a) = \pi(p) + \pi(a)$ for $a \in \mathbb{C}^n$.

(2) Let $I = \{f_1, \dots, f_m\} \subset \mathbb{C}\{x_1, \dots, x_n\}$ be an ideal and $i : (V(I), o) \longrightarrow (\mathbb{C}^n, o)$ be the canonical inclusion. Then $T_{V(I),o} = \{a = (a_1, \dots, a_n) \in \mathbb{C}^n : \sum_{j=1}^{n} \frac{\partial f_i}{\partial x_j}(0) a_j = 0, i = 1, \dots, m\}$, and $T_{i,o} : T_{V(I),o} \longrightarrow \mathbb{C}^n = T_{\mathbb{C}^n,o}$ is the canonical inclusion.

(3) Let $f : \mathbb{C} \longrightarrow \mathbb{C}^2$ be defined by $f(t) = (t^2, t^3)$,

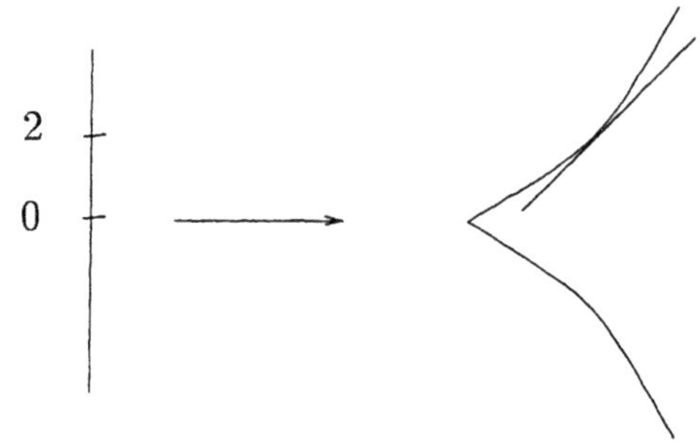

Then $T_{f,o} : T_{\mathbb{C},o} = \mathbb{C} \longrightarrow T_{\mathbb{C}^2,(o,o)} = \mathbb{C}^2$ is the zero map. On the other hand $T_{f,2} : T_{\mathbb{C},2} = 2 + \mathbb{C} \longrightarrow T_{\mathbb{C}^2,(4,8)} = (4,8) + \mathbb{C}^2$ is defined by

$$T_{f,2}(2+a) = (4,8) + (4a,12a).$$

The following theorem we only need in Chapter nine.

Theorem 4.3.14. *Let $X,Y \subset \mathbb{C}^n$ be irreducible affine algebraic sets, and $\varphi : X \longrightarrow Y$ be a dominant map, see 2.3.11. Then there exists a point $p \in X$ such that with $q = f(p)$*

$$T_{\varphi,p} : T_{X,p} \longrightarrow T_{Y,q}$$

is surjective.

Proof. In the proof of 2.3.12, it was shown that there exists an $f \in \mathbb{C}[Y] \subset \mathbb{C}[X]$, and an $r \in \mathbb{N}$ such that φ factorizes on $X \setminus V(f)$ as follows

$$X \setminus V(f) \xrightarrow{\tilde{\varphi}} (Y \setminus V(f)) \times \mathbb{C}^r \xrightarrow{\pi} Y \setminus V(f).$$

Moreover $\mathbb{C}[X]_f$ is a finitely generated $\mathbb{C}[Y]_f[x_1, \dots, x_r]$–module. Here π is the projection on the second factor, and thus for all $q' \in Y \setminus V(f)$ and $a \in \mathbb{C}^r$, the tangent map $T_{\pi,(q',a)}$ is surjective. So it suffices to show that there exists a point $p \in X \setminus V(f)$ with $T_{\tilde{\varphi},p}$ is surjective, as an application of 4.3.12. As both $X \setminus V(f)$ and $(Y \setminus V(f)) \times \mathbb{C}^r$ are affine, see 2.3.10, we reduced the problem to the case of a finite map $\varphi : X \longrightarrow Y$ of affine algebraic sets, such that $\mathbb{C}[X]$ is a finitely generated $\mathbb{C}[Y]$–module.

So take a smooth point $q' \in Y$. By Theorem 2.3.9 the map φ is surjective, so there exists a point p' in $\varphi^{-1}(q')$. It follows from 3.4.24 that the germ of the analytic map $\varphi : (X,p') \longrightarrow (Y,q')$ is finite. Thus this is a Noether normalization, as by assumption q' is a smooth point of Y. It obviously suffices to prove the statement for an irreducible germ (X,p'). It was proved in the Local Parametrization Theorem, see 3.4.14 that for "almost all" points p in an open neighborhood of p' the germ of the map $\varphi : (X,p) \longrightarrow (Y,q)$ is an isomorphism. In particular the tangent map $T_{\varphi,p}$ is an isomorphism. This concludes the proof of the theorem. □

For later applications, for example in the interpretation of normal rings, and in the proof of Artin's Approximation Theorem, we need a more general version of the Jacobian Criterion.

Theorem 4.3.15 (Jacobian Criterion for Smoothness). *Let* $I = (f_1,\dots,f_m) \subset \mathbb{C}\{x_1,\dots,x_n\}$*,* $\mathfrak{p}$ *be a prime ideal associated to* I*, and* $\mathfrak{q} \supset \mathfrak{p}$ *be a prime ideal containing* $\mathfrak{p}$*. Then*[5]

$$\mathbb{C}\{x_1,\dots,x_n\}_\mathfrak{q}/I_\mathfrak{q} \text{ is a regular local ring} \iff \operatorname{rank}\Big(\big(\tfrac{\partial f_i}{\partial x_j} \mod \mathfrak{q}\big)\Big) = \operatorname{ht}(\mathfrak{p}).$$

Furthermore, if $\operatorname{ht}(\mathfrak{p}) = k$ *and* $\operatorname{rank}\Big(\big(\tfrac{\partial f_i}{\partial x_j} \mod \mathfrak{q}\big)_{1\le i\le k}\Big) = k$ *then* $f_1,\dots,f_k$ *generate the ideal* $I_\mathfrak{q}$.

We first discuss how to interpret this theorem.

Examples 4.3.16. (1) For example, take $\mathfrak{q} = \mathfrak{m}$ the maximal ideal. Then $\left(\frac{\partial f_i}{\partial x_j}\right) \mod \mathfrak{m}$ is just the Jacobian matrix $\left(\frac{\partial f_i}{\partial x_j}(0)\right)$. As all elements in $\mathbb{C}\{x_1,\dots,x_n\} \setminus \mathfrak{m}$ are already invertible in $\mathbb{C}\{x_1,\dots,x_n\}$ we have the equality $\mathbb{C}\{x_1,\dots,x_n\}_\mathfrak{m}/I_\mathfrak{m} = \mathbb{C}\{x_1,\dots,x_n\}/I$. It follows that $\mathbb{C}\{x_1,\dots,x_n\}/I$ is a regular local ring if and only the Jacobian matrix has maximal rank. This we have already seen in 4.3.6.

(2) Suppose that $\mathbb{C}\{x_1,\dots,x_n\}_\mathfrak{q}/I_\mathfrak{q}$ is a regular local ring. Put $k = \operatorname{ht}(\mathfrak{p})$. Then k is the codimension of $V(\mathfrak{p})$. Suppose $\mathfrak{p}$ is a minimal associated prime of I. Consider the k–minors $\Delta_1,\dots,\Delta_s$ of the Jacobian matrix $\left(\frac{\partial f_i}{\partial x_j}\right)$. Then $\operatorname{rank}\Big(\big(\frac{\partial f_i}{\partial x_j} \mod \mathfrak{q}\big)\Big) = \operatorname{ht}(\mathfrak{p})$ exactly means that there exists a minor Δ_i which is not in $\mathfrak{q}$. It follows that for all a in an open subset of $V(\mathfrak{q})$ the variety $V(I)$ is smooth at the point a. Namely $\big(V(\Delta_1,\dots,\Delta_s),0\big) \cap \big(V(\mathfrak{q}),0\big) \subsetneq \big(V(\mathfrak{q}),0\big)$ is a proper analytic subset. Now take representatives of all spaces involved. It follows that for all $a \in V(\mathfrak{q})\setminus V(\Delta_1,\dots,\Delta_s)$ there is also a Δ_i with $\Delta_i(a) \neq 0$.

(3) Note that in the special case $I = \mathfrak{q} = \mathfrak{p}$, that $\mathbb{C}\{x_1,\dots,x_n\}_\mathfrak{p}/\mathfrak{p}_\mathfrak{p}$ is a field. As a field obviously is a regular local ring, it follows that almost all points of $V(\mathfrak{p})$ are nonsingular. This we have already seen in the Local Parametrization Theorem, see 3.4.14. In the proof of the Jacobian criterion, this special case will be treated first and, in fact, in the proof the Local Parametrization Theorem is used.

(4) We consider the explicit example of the Whitney umbrella defined by $x^2 - y^2z = 0$. So $\mathfrak{p} = I = (x^2 - y^2z)$. We take $\mathfrak{q} = (x,y)$. Then $\mathfrak{q}$ is the ideal of the singular locus. So we expect that $\mathbb{C}\{x,y,z\}_\mathfrak{q}/(x^2 - y^2z)_\mathfrak{q}$ is not a regular local ring. Indeed, the minimal number of generators of ideal $(x,y)_\mathfrak{q}$ is two, as one checks by using Nakayama's Lemma. However the dimension of $\mathbb{C}\{x,y,z\}_\mathfrak{q}/(x^2 - y^2z)_\mathfrak{q}$ is one.

On the other hand, looking at the y–axis, a "general" point of the y–axis is smooth. We take the ideal $\mathfrak{q} = (x,z)$, in which we localize. This is the same as localizing in the multiplicative set $S = \{1,y,y^2,\dots\}$. So y becomes a unit, and we can eliminate z. Hence $\mathbb{C}\{x,y,z\}_\mathfrak{q}/(x^2 - y^2z) \cong \mathbb{C}\{x,y\}_y$, which one directly checks to be a regular local ring of dimension one.

Before we can give the proof of the general Jacobian criterion, we need some basic facts on regular local rings.

Theorem 4.3.17. *Let* $(R,\mathfrak{m})$ *be a regular local ring. Then* R *is an integral domain.*

[5] Here $\operatorname{rank}\Big(\big(\frac{\partial f_i}{\partial x_j} \mod \mathfrak{q}\big)\Big)$ is the rank of the matrix over the quotient field $Q(\mathbb{C}\{x_1,\dots,x_n\}/\mathfrak{q})$.

Proof. The proof is by induction on the Krull-dimension of R.

Step 1. If $\dim(R) = 0$, then $\operatorname{edim}(R) = 0$, hence the maximal ideal has zero generators. Thus $\mathfrak{m} = 0$, hence $R = R/\mathfrak{m}$ is a field, which obviously has no zerodivisors.

Step 2. Now suppose $\dim(R) > 0$. Let $\mathfrak{p}_1, \dots, \mathfrak{p}_s$ be the minimal prime ideals of R. If we prove that $\mathfrak{p}_i = (0)$, for some i, then $i = s = 1$ (because of minimality), and therefore $R = R/\mathfrak{p}_1 = R/(0)$ is an integral domain.

As $\dim(R) > 0$, it follows by Nakayama that $\mathfrak{m} \neq \mathfrak{m}^2$. By prime avoidance 1.1.13, we can find an $x \in \mathfrak{m}\backslash\mathfrak{m}^2$ with $x \notin \cup_i \mathfrak{p}_i$. The last condition means that x is an active element of R. By Krull's Principal Ideal Theorem 4.2.16, $\dim(R/(x)) = \dim(R) - 1$. Moreover, $\mathfrak{m}/((x) + \mathfrak{m}^2)$ is a vector space of dimension $\dim(R) - 1$. Hence $R/(x)$ is a regular local ring, and by induction hypothesis it is an integral domain. Therefore, the ideal (x) is a prime ideal. It cannot be a minimal prime ideal, since we have chosen $x \notin \cup_i \mathfrak{p}_i$. Therefore, there must be a minimal prime ideal, say $\mathfrak{p}_i$, which is contained in (x). Let $y \in \mathfrak{p}_i$ be arbitrary. It follows that $y = ax$ for some $a \in R$. As $\mathfrak{p}_i$ is prime, it follows $a \in \mathfrak{p}_i$. Hence we proved the equality $x\mathfrak{p}_i = \mathfrak{p}_i$. From Nakayama's Lemma it follows that $\mathfrak{p}_i = 0$, which is what we had to prove. □

Corollary 4.3.18. *Let $(R,\mathfrak{m})$ be an r-dimensional regular local ring and $x_1, \dots, x_i \in \mathfrak{m}$. The following conditions are equivalent.*

(1) There exist $x_{i+1}, \dots, x_r \in R$ such that $\mathfrak{m} = (x_1, \dots, x_r)$.

(2) The images of $x_1, \dots, x_i$ in $\mathfrak{m}/\mathfrak{m}^2$ are linearly independent over $R/\mathfrak{m}$.

(3) $R/(x_1, \dots, x_i)$ is an $(r - i)$-dimensional regular local ring.

Proof.
Step 1. We first show (1) $\Longrightarrow$ (2). The images of $x_1, \dots, x_r$ generate $\mathfrak{m}/\mathfrak{m}^2$ as $R/\mathfrak{m}$–vector space. Since $\dim_{R/\mathfrak{m}}(\mathfrak{m}/\mathfrak{m}^2) = r$ they are linearly independent.

Step 2. (2) $\Longrightarrow$ (3). We choose $x_{i+1}, \dots, x_r \in \mathfrak{m}$ such that the images of $x_1, \dots, x_r$ form a basis of $\mathfrak{m}/\mathfrak{m}^2$. Nakayama's Lemma implies that $\mathfrak{m} = (x_1, \dots, x_r)$.

Now x_1 is not a zero divisor of R by Theorem 4.3.17 . Thus $\dim(R/(x_1)) = \dim(R) - 1$ by Krull's Principal Ideal Theorem. The images of $x_2, \dots, x_r$ generate the maximal ideal in $R/(x_1)$. This implies that $R/(x_1)$ is regular. By induction we get that $R/(x_1, \dots, x_i)$ is regular.

Step 3. We finally show (3) $\Longrightarrow$ (1). Let $\mathfrak{m}/(x_1, \dots, x_i)$ be generated by the images of $x_{i+1}, \dots, x_r \in \mathfrak{m}$. Then $\mathfrak{m} = (x_1, \dots, x_i, x_{i+1}, \dots, x_r)$. □

Theorem 4.3.19. *Let $\mathfrak{p}$ be a prime ideal in $\mathbb{C}\{x_1, \dots, x_n\}$. Then $\mathbb{C}\{x_1, \dots, x_n\}_\mathfrak{p}$ is a regular local ring.*

Proof. The ring $\mathbb{C}\{x_1, \dots, x_n\}_\mathfrak{p}$ has dimension $\operatorname{ht}(\mathfrak{p})$. Consider a primitive Noether normalization

$$\mathbb{C}\{x_1, \dots, x_k\} \longrightarrow \mathbb{C}\{x_1, \dots, x_n\}/\mathfrak{p}.$$

Then $\mathrm{ht}(\mathfrak{p})$, which is the codimension of $V(\mathfrak{p})$, is equal to $n-k$. Hence we have to find $n-k$ generators of the maximal ideal $\mathfrak{p}\mathbb{C}\{x_1,\dots,x_n\}_{\mathfrak{p}}$. Let $P \in \mathbb{C}\{x_1,\dots,x_k\}[x_{k+1}]$ be the minimal polynomial of $\overline{x}_{k+1} \in \mathbb{C}\{x_1,\dots,x_n\}/\mathfrak{p}$, and $\Delta \in \mathbb{C}\{x_1,\dots,x_k\}$ be the discriminant. In particular $\Delta \notin \mathfrak{p}$. Moreover, let $Q_j = x_j\Delta - q_j(x_{k+1}) \in \mathfrak{p}$ for $j = k+2,\dots,n$. Here the q_j are polynomials with coefficients in $\mathbb{C}\{x_1,\dots,x_k\}$ which exist according to the Finiteness of Normalization Theorem 1.5.19. Then $Q_n,\dots,Q_{k+2},P \in \mathfrak{p}$, and we proved already in 3.4.14 that they generate the ideal $\mathfrak{p}\mathbb{C}\{x_1,\dots,x_n\}_{\Delta}$. As $\Delta \notin \mathfrak{p}$, they in particular generate the ideal $\mathfrak{p}\mathbb{C}\{x_1,\dots,x_n\}_{\mathfrak{p}}$. □

Remark 4.3.20. It is quite generally true that localizations of regular local rings in prime ideals are regular local rings. A proof of this more general result uses homological methods, see for example [Eisenbud 1995], Corollary 19.14. The proof which we gave works for the polynomial case too, and is a simple consequence of the Local Parametrization Theorem.

Proof of Theorem 4.3.15.
Step 1. We first consider the case $(f_1,\dots,f_m) = I = \mathfrak{p} = \mathfrak{q}$.

The ring $\mathbb{C}\{x_1,\dots,x_n\}_{\mathfrak{p}}/\mathfrak{p}_{\mathfrak{p}}$ is a field and hence a regular local ring. Therefore, in this case we have to prove $\mathrm{rank}((\frac{\partial f_i}{\partial x_j} \mod \mathfrak{p})_{i\le m, j\le n}) = \mathrm{ht}(\mathfrak{p})$. Take a primitive Noether normalization

$$\mathbb{C}\{x_1,\dots,x_k\} \longrightarrow \mathbb{C}\{x_1,\dots,x_n\}/\mathfrak{p}.$$

So we have to prove that the Jacobian matrix has rank $n-k$. We use the Local Parametrization Theorem, and keep the notations of the proof of the previous Theorem 4.3.19. We noted in the previous proof that the ideal $\mathfrak{p}\mathbb{C}\{x_1,\dots,x_n\}_{\mathfrak{p}}$ is generated by $(P,Q_{k+2},\dots,Q_n)$. By Exercise 4.3.24 the rank of the Jacobian matrix is independent of the generators of $\mathfrak{p}\mathbb{C}\{x_1,\dots,x_n\}_{\mathfrak{p}}$. As this ideal is generated by $n-k$ elements, it follows that the rank of the Jacobian matrix modulo $\mathfrak{p}$ is at most $n-k$. Now the final $n-k$ columns of the Jacobian matrix for the generators $(P,Q_{k+2},\dots,Q_n)$ look like

$$J := \begin{pmatrix} \frac{\partial P}{\partial x_{k+1}} & \cdots & \frac{\partial P}{\partial x_n} \\ \frac{\partial Q_{k+2}}{\partial x_{k+1}} & \cdots & \frac{\partial Q_{k+2}}{\partial x_n} \\ \vdots & & \vdots \\ \frac{\partial Q_n}{\partial x_{k+1}} & \cdots & \frac{\partial Q_n}{\partial x_n} \end{pmatrix} = \begin{pmatrix} \frac{\partial P}{\partial x_{k+1}} & 0 & & \cdots & 0 \\ -\frac{\partial q_{k+2}}{\partial x_{k+1}} & \Delta & 0 & \cdots & 0 \\ \vdots & & \ddots & & \\ \vdots & & & \ddots & \\ -\frac{\partial q_n}{\partial x_{k+1}} & & & & \Delta \end{pmatrix}.$$

So $\det(J) = \Delta^{n-k-1}\cdot\frac{\partial P}{\partial x_{k+1}}$. Because P is the minimal polynomial of the field extension $\mathbb{C}\{x_1,\dots,x_n\}_{\mathfrak{p}}/\mathfrak{p}_{\mathfrak{p}} \supset Q(\mathbb{C}\{x_1,\dots,x_k\})$ we have $\frac{\partial P}{\partial x_{k+1}} \notin \mathfrak{p}$. So $\det(J) \notin \mathfrak{p}$. This implies that

$$\mathrm{rank}((\frac{\partial f_i}{\partial x_j} \mod \mathfrak{p})_{i\le m, j\le n}) = \mathrm{rank}(J \mod \mathfrak{p}) = n-k = \mathrm{ht}(\mathfrak{p}).$$

This shows the theorem in the special case $I = \mathfrak{p} = \mathfrak{q}$.

Step 2. Now suppose that $\mathbb{C}\{x_1,\dots,x_n\}_\mathfrak{q}/I_\mathfrak{q}$ is a regular local ring. Then we claim that

$$I_\mathfrak{q} = \mathfrak{p}_\mathfrak{q}. \tag{4.4}$$

Indeed, $\mathfrak{p} \supset I$. By assumption, $\mathbb{C}\{x_1,\dots,x_n\}_\mathfrak{q}/I_\mathfrak{q}$ is a regular local ring, so in particular an integral domain by 4.3.17. So in the localization, $\mathbb{C}\{x_1,\dots,x_n\}_\mathfrak{q}$ the ideal $I_\mathfrak{q}$ is a prime ideal. Formula (4.4) now follows from 1.4.50 (5), which says how a primary decomposition behaves under localization.

So if we suppose that $\mathbb{C}\{x_1,\dots,x_n\}_\mathfrak{q}/I_\mathfrak{q}$ is a regular local ring, we may suppose that $I = \mathfrak{p}$ is a prime ideal. Put $a := \mathrm{ht}(\mathfrak{p})$, and $b := \mathrm{ht}(\mathfrak{q})$. The dimension of $\mathbb{C}\{x_1,\dots,x_n\}_\mathfrak{q}/I_\mathfrak{q}$ is equal to $b - a$. It is not difficult to see, see 4.3.23, that the ideal $\mathfrak{p}_\mathfrak{q}$ in the ring $\mathbb{C}\{x_1,\dots,x_n\}_\mathfrak{q}$ can be generated by elements $u_1,\dots,u_a \in \mathfrak{q}\setminus\mathfrak{q}^2$. By 4.3.18 this sequence can be extended to a regular sequence $u_1,\dots,u_b \in \mathfrak{q}\setminus\mathfrak{q}^2$ which generates the ideal $\mathfrak{q}_\mathfrak{q}$ in $\mathbb{C}\{x_1,\dots,x_n\}_\mathfrak{q}$. We apply step 1 and get

$$\mathrm{rank}\Big(\big(\frac{\partial u_i}{\partial x_j} \mod \mathfrak{q}\big)_{i\le b,\, j\le n}\Big) = \mathrm{ht}(\mathfrak{q}) = b,$$

which is the maximal possible rank. Deleting the columns corresponding to $u_{a+1},\dots,u_b$ we obtain

$$\mathrm{rank}\Big(\big(\frac{\partial u_i}{\partial x_j} \mod \mathfrak{q}\big)_{i\le a,\, j\le n}\Big) = a = \mathrm{ht}(\mathfrak{p}),$$

which is again the maximal possible rank. We deduce that the Jacobian matrix for the f's also has maximal rank by applying again Exercise 4.3.24.

Step 3. We now prove the converse. Therefore, assume $\mathrm{rank}\Big(\big(\frac{\partial f_i}{\partial x_j} \mod \mathfrak{q}\big)_{i\le m,\, j\le n}\Big) = \mathrm{ht}(\mathfrak{p}) = a$. By renumbering, we may assume

$$\mathrm{rank}\Big(\big(\frac{\partial f_i}{\partial x_j} \mod \mathfrak{q}\big)_{i\le a,\, j\le a}\Big) = a.$$

We first show that $f_1,\dots,f_a$ is part of a generating set for $\mathfrak{q}_\mathfrak{q}$. Assume

$$\sum_{i=1}^{a} c_i f_i \in \mathfrak{q}^2$$

for some $c_1,\dots,c_a \in \mathbb{C}$. Then

$$\frac{\partial}{\partial x_j}\Big(\sum_{i=1}^{a} c_i f_i\Big) = \sum_{i=1}^{a} c_i \frac{\partial f_i}{\partial x_j} \in \mathfrak{q}.$$

As $\mathrm{rank}\Big(\big(\frac{\partial f_i}{\partial x_j} \mod \mathfrak{q}\big)_{i\le a,\, j\le a}\Big) = a$ it follows that $c_i = 0$ for $i = 1,\dots,a$. By Theorem 4.3.19 and Corollary 4.3.18, $\mathbb{C}\{x_1,\dots,x_n\}_\mathfrak{q}/(f_1,\dots,f_a)_\mathfrak{q}$ is a regular local ring. In particular $(f_1,\dots,f_a)$ is a prime ideal in $\mathbb{C}\{x_1,\dots,x_n\}_\mathfrak{q}$. As $f_1,\dots f_a \in I \subset \mathfrak{p}$ we have inclusions

$$(f_1,\dots,f_a)_\mathfrak{q} \subset I_\mathfrak{q} \subset \mathfrak{p}_\mathfrak{q}.$$

But as the outer ideals have the same height, and are both prime in $\mathbb{C}\{x_1,\dots,x_n\}_\mathfrak{q}$ all the ideals must be equal. In particular $(f_1,\dots,f_a)_\mathfrak{q} = I_\mathfrak{q}$. Therefore, $\mathbb{C}\{x_1,\dots,x_n\}_q/I_\mathfrak{q}$ is a regular local ring. This proves the theorem. □

Exercises

4.3.21. Prove that $T_{g\circ f,p} = T_{g,q} \circ T_{f,p}$, see 4.3.12.

4.3.22. Let $X \subset \mathbb{C}^n$, and $Y \subset \mathbb{C}^m$ be algebraic set. Suppose $f : X \longrightarrow Y$ is defined by $(f_1, \ldots, f_m)$. Consider the tangent map $T_{f,p} : T_{X,p} \longrightarrow T_{Y,f(p)}$. Let ε be a new variable with $\varepsilon^2 = 0$. Suppose $p + a \in T_{X,p}$, and $T_{f,p}(p+a) = f(p) + b$. Prove that $f(p+\varepsilon a) = f(p) + \varepsilon b$. Make this to a precise statement!

4.3.23. Let $(R,\mathfrak{m})$ be a regular local ring, $\mathfrak{p} \subset R$ a nonzero prime ideal.

(1) Suppose $\mathfrak{p}$ is contained in $\mathfrak{m}^2$. Show that $R/\mathfrak{p}$ is not a regular local ring. (Hint: Use Krull's Principal Ideal Theorem to prove that $\dim(R/\mathfrak{p}) < \dim(R)$, and look at the generators of the maximal ideal in $R/\mathfrak{p}$.)

(2) Suppose $R/\mathfrak{p}$ is a regular local ring. Show that $\mathfrak{p}$ can be generated by a regular sequence $u_1, \ldots u_s$, with $u_i \in \mathfrak{m} \setminus \mathfrak{m}^2$.

4.3.24. Let $\mathfrak{q} \subset \mathbb{C}\{x_1, \ldots, x_n\}$ be a prime ideal, and $I = (f_1, \ldots, f_m)$ be a further ideal. Let $g_1, \ldots, g_s \in I$ such that $I_\mathfrak{q} = (g_1, \ldots, g_s)_\mathfrak{q}$. Prove that

$$\operatorname{rank}\left(\left(\frac{\partial f_i}{\partial x_j} \mod \mathfrak{q}\right)\right) = \operatorname{rank}\left(\left(\frac{\partial g_i}{\partial x_j} \mod \mathfrak{q}\right)\right).$$

4.3.25. Let $X \subset \mathbb{C}^n$ be an affine set. Prove that $\operatorname{Sing}(X)$ is an affine algebraic subset of X. (Hint: Use the Primary Decomposition Theorem.)

4.3.26. Let X be an analytic space. Prove the Identity Theorem, cf. 3.1.9, for X. (Hint: Use the fact that the singular points of X are contained in a proper analytic subset.)

4.3.27. Prove the following statement. Let $(X,0) \subset (\mathbb{C}^n,0)$ be a germ of an analytic subset. Let $\mathfrak{m} \subset \mathcal{O}_{X,0}$ be the maximal ideal. There is a canonical isomorphism of affine spaces $T_{X,0} \cong (\mathfrak{m}/\mathfrak{m}^2)^*$.
(Hint: Choose a representative $Y \subset U$ of $(X,0)$, $U \subset \mathbb{C}^n$ open, and $f_1, \ldots, f_s$ holomorphic functions on U such that $Y = \{y \in U : f_1(y) = \cdots = f_s(y) = 0\}$ and $\mathcal{I}(X,0) = (f_1, \ldots, f_s)$. Then $T_{X,0} = \{a \in \mathbb{C}^n : \left(\frac{\partial f_i}{\partial x_j}(0)\right) \cdot a^t = 0\}$.

On the other hand, $\mathfrak{m}$ is the maximal ideal in $\mathbb{C}\{x_1, \ldots, x_n\}/(f_1, \ldots, f_s)$ generated by $\overline{x}_1, \ldots, \overline{x}_n$, with $\overline{x}_i = x_i \mod (f_1, \ldots, f_s)$. Define $\phi : T_{X,0} \longrightarrow (\mathfrak{m}/\mathfrak{m}^2)^*$ by $\phi(a)(\overline{x}_i) = a_i$, for $a = (a_1, \ldots, a_n)$.

Prove that ϕ is well-defined, that is $\sum_{i=1}^n c_i\overline{x}_i = 0$ implies $\sum_{i=1}^n c_i a_i = 0$. Use that $\sum_{i=1}^n c_i\overline{x}_i = 0$ implies $\sum_{i=1}^n c_i x_i + \sum_{i=1}^s \xi_i f_i \in (x_1, \ldots, x_n)^2$ for suitable $\xi_i \in \mathbb{C}\{x_1, \ldots, x_n\}$.

Prove that ϕ is an isomorphism using the fact that $\sum_{j=1}^n \frac{\partial f_i}{\partial x_j}(0)\overline{x}_j = 0$ in $\mathfrak{m}/\mathfrak{m}^2$.)

4.4 Normalization of Germs of Analytic Spaces

The first aim of this section is to show that every germ of an analytic space has a *normalization.* We start with the definition of a normal germ.

Definition 4.4.1. A germ of an analytic space (X,x) is called *normal* if the local ring $\mathcal{O}_{X,x}$ is normal.

The first property of normal germs is that they are irreducible.

Proposition 4.4.2. *Let (X,x) a normal germ of an analytic space. Then (X,x) is irreducible.*

Proof. This is an immediate consequence of Theorem 1.5.7. □

So we see that normality is a nontrivial condition. The first goal in this section is to prove that for all germs there exists a normalization. But in order to reach this goal, we have to allow more general spaces then germs of analytic spaces: we have to consider multi-germs.

Definition 4.4.3. A *multi-germ* of analytic spaces (X,x) is a finite disjoint union $(X,x) = (X_1,x_1) \cup \ldots \cup (X_r,x_r)$ of germs of analytic spaces. The ring $\mathcal{O}_{X,x}$ by definition is equal to $\oplus_{i=1}^r \mathcal{O}_{X_i,x_i}$.

Let $(Y,y) = (Y_1,y_1) \cup \ldots \cup (Y_s,y_s)$ be a further multi-germ. A map $\varphi : (X,x) \longrightarrow (Y,y)$ is given by a system of maps of $\varphi_i : (X_i,x_i) \longrightarrow (Y_{\alpha(i)},y_{\alpha(i)})$ for $i = 1,\ldots,r$, and some $\alpha(i) \in \{1,\ldots,s\}$. Such a map φ induces, and is induced by a $\mathbb{C}$-algebra map $\varphi^* : \mathcal{O}_{Y,y} \longrightarrow \mathcal{O}_{X,x}$. In the obvious way, one defines such a map to be finite, generic s to 1, etc.

A multi-germ (X,x) is called *normal* if $\mathcal{O}_{X,x}$ is a normal ring.

Remarks 4.4.4.

(1) It is not so difficult to see that $\mathcal{O}_{X,x}$ is a semi-local ring, see Exercise 4.4.16. This, by definition, means that $\mathcal{O}_{X,x}$ has only finitely many maximal ideals.

(2) It is an exercise, see 4.4.16, to show that a multi-germ $(X,x) = (X_1,x_1) \cup \ldots \cup (X_r,x_r)$ is normal if and only if (X_i,x_i) is normal for $i = 1,\ldots,r$.

Now we can define what we mean by a normalization of a germ of an analytic space.

Definition 4.4.5. Let (X,x) be a germ of an analytic space. A *normalization* of (X,x) is a normal multi-germ $(\widetilde{X},\widetilde{x})$ together together with a finite, generically $1-1$ map $n : (\widetilde{X},\widetilde{x}) \longrightarrow (X,x)$.

Remark 4.4.6. It was proved in Exercise 3.4.40 that a generically $1-1$ map φ between irreducible germs of analytic spaces induces an isomorphism of quotient fields. This was a simple consequence of the Local Parametrization Theorem. This holds more generally for multi-germs, see Exercise 4.4.16. As a normalization should be generically $1-1$, it follows that the ring $\mathcal{O}_{\widetilde{X},\widetilde{x}}$ is the normalization of the ring $\mathcal{O}_{X,x}$. So a normalization, assuming it exists, is uniquely determined. Note that the normalization $(\widetilde{X},\widetilde{x})$ has the same dimension as (X,x). Furthermore, note that if (X,x) is irreducible, then the normalization $(\widetilde{X},\widetilde{x})$ must be a germ of an analytic space. This is because the total quotient ring of $\mathcal{O}_{X,x}$ is equal to the total quotient ring of $\mathcal{O}_{\widetilde{X},\widetilde{x}}$, which in case (X,x) is irreducible, is a field.

Examples 4.4.7.

(1) A smooth germ $(\mathbb{C}^n, o)$ is normal. This is because $\mathbb{C}\{x_1,\ldots,x_n\}$ is a unique factorization domain by Corollary 3.3.17, and unique factorization domains are normal, see Theorem 1.5.5,[6]

(2) Consider the A_1-singularity, defined by the equation $x^2 - y^2 = 0$. So $\mathcal{O}_{X,x} = \mathbb{C}\{x,y\}/(x^2 - y^2)$. The normalization $(\widetilde{X},\widetilde{x})$ of (X,x) consists of two copies of smooth spaces (X_1,x_1) with parameter t_1 and (X_2,x_2) with parameter t_2. The normalization map n is given by two maps $n_1 : (X_1,x_1) \longrightarrow (X,x)$ which sends t_1 to (t_1,t_1), and $n_2 : (X_2,x_2) \longrightarrow (X,x)$ which sends t_2 to $(t_2, -t_2)$.

[6] More generally, regular local rings are normal because they are unique factorization domains. But we do not prove this.

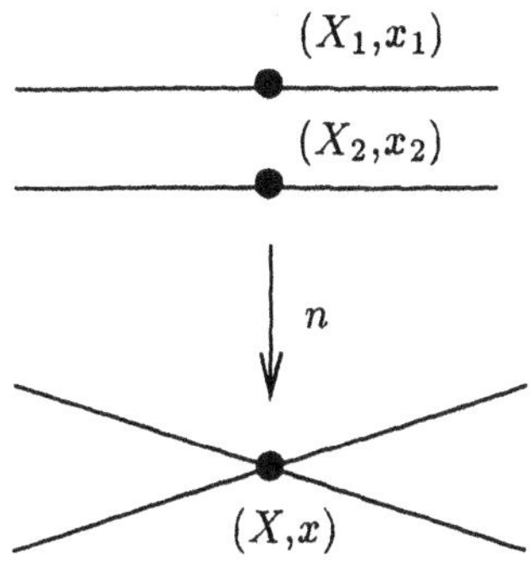

(3) Let (X,x) be the A_2–singularity, with local ring $\mathcal{O}_{X,x} = \mathbb{C}\{x,y\}/(y^2-x^3)$. Then the normalization of $\mathcal{O}_X$ is equal to $\mathbb{C}\{t\}$, where $t = \frac{y}{x}$. Compare example 1.5.6. The normalization $(\widetilde{X},\widetilde{x})$ is a smooth one-dimensional space with parameter t, and the normalization map is given by $t \mapsto (x(t),y(t)) = (t^2,t^3)$. Note that in this example the normalization map is "bijective", but not an isomorphism.

(4) Let (X,x) be the A_1–singularity in three space, defined by the equation $z^2 - xy$. Then X is already normal, cf. Exercise 1.5.33. The normalization of (X,x) is (X,x) itself, and the normalization map is the identity.

(5) Let (X,x) be the "Whitney umbrella" defined by $x^2 - zy^2 = 0$.

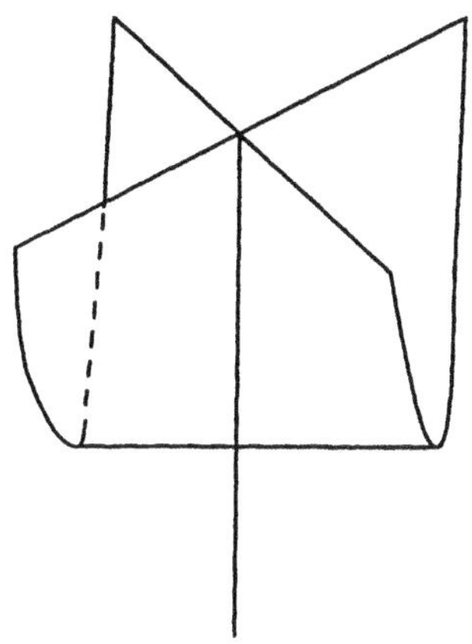

The element $s := \frac{x}{y}$ is integral. Indeed, $s^2 = \frac{x^2}{y^2} = \frac{zy^2}{y^2} = z$. We consider the ring $\mathcal{O}_{X,x}[s]$ generated by $\mathcal{O}_{X,x}$ and s in the quotient ring $Q(\mathcal{O}_{X,x})$ and claim that it is equal to the normalization. Indeed, by applying the Weierstraß Division Theorem, it is not so difficult to see that

$$\mathcal{O}_{X,x}[s] = \mathbb{C}\{x,y,z,s\}/(x^2 - zy^2, ys - x, s^2 - z).$$

We can eliminate x and z and see that $\mathcal{O}_{X,x}[s]$ is isomorphic to $\mathbb{C}\{y,s\}$, which is normal. So the normalization is smooth, with parameters y,s and the normalization map is given by $(y,s) \mapsto (ys,y,s^2)$.

Note that in this example the normalization map is indeed generically 1–1: for all points (x,y,z) on the Whitney umbrella not on the z–axis there is only one point in the preimage, but for all points on the z–axis except the origin there are *two* points in the preimage of the normalization map.

In the third example we saw that the normalization of a cusp singularity is smooth. This turns out to be the case for all curve singularities. So normality is a very strong condition for curve singularities! It follows that for an irreducible curve singularity (X,x) there always exists a finite generically 1–1 map from $(\mathbb{C},o)$ to (X,x). Such a map can even be given in a particularly simple form, which is then called *Puiseux expansion*. This will be studied in chapter five.

More generally the singular locus of a normal space must have at least codimension two, compare the fifth example. For hypersurface singularities, this condition is also sufficient as suggested by the fourth example, but we are not able to prove this at the moment, but only in the section on Cohen-Macaulay spaces, see in particular Corollary 6.5.9. So in higher dimensions the normality condition is not as strong as in the case of curve singularities.

As noted before, if a normalization $(\widetilde{X},\widetilde{x})$ of (X,x) exists, the ring $\mathcal{O}_{\widetilde{X},\widetilde{x}}$ must be equal to the normalization of the ring $\mathcal{O}_{X,x}$. The normalization of a ring always exists, and it is our aim to show that this ring is a finite direct sum of analytic $\mathbb{C}$-algebras.

Theorem 4.4.8 (Existence of the Normalization). *Let (X,x) be a germ of an analytic space. Then there exists a normalization $n : (\widetilde{X},\widetilde{x}) \longrightarrow (X,x)$ of (X,x).*

If moreover (X,x) is irreducible, then $(\widetilde{X},\widetilde{x})$ is also an irreducible germ.

Proof. Consider an irreducible decomposition $(X,x) = (X_1,x)\cup\ldots\cup(X_r,x)$ of (X,x). By Splitting of Normalization, see Theorem 1.5.20, we have that $\widetilde{\mathcal{O}}_{X,x} = \widetilde{\mathcal{O}}_{X_1,x}\oplus\ldots\oplus\widetilde{\mathcal{O}}_{X_r,x}$. Thus we reduce to the case that (X,x) is irreducible. By the Finiteness of Normalization Theorem 1.5.19 we deduce that $\widetilde{\mathcal{O}}_{X,x}$ is a finitely generated $\mathcal{O}_{X,x}$-module. It was proved in Corollary 3.3.26 that $\widetilde{\mathcal{O}}_{X,x}$ is isomorphic to a direct sum $\oplus_{i=1}^{s} S_i$ of nonzero analytic algebras. As the total quotient ring of $\widetilde{\mathcal{O}}_{X,x}$ is a field, we have that $\widetilde{\mathcal{O}}_{X,x}$ is an integral domain. Therefore $s = 1$, and it follows that $\widetilde{\mathcal{O}}_{X,x}$ is an analytic algebra. Thus it is equal to $\mathcal{O}_{\widetilde{X},\widetilde{x}}$ for some germ of an irreducible analytic space $(\widetilde{X},\widetilde{x})$.

This normalization map is finite by 3.4.24, and it is generically 1-1 because of Exercise 3.4.40. □

To prove that curve singularities are not normal, it suffices to prove the following purely algebraic theorem.

Theorem 4.4.9. *Let R be a Noetherian local ring of dimension one. Then*

$$R \text{ is regular} \iff R \text{ is normal.}$$

Proof. Step 1. We first prove "$\Longrightarrow$". As principal ideal domains are unique factorization domains, see Example 1.4.3, and those are normal by 1.5.5, it suffices to show that R is a principal ideal domain. The definition of regularity gives that $\dim_{\mathbb{C}} \mathfrak{m}/\mathfrak{m}^2 = 1$, so $\mathfrak{m}$ is a principal ideal. To show that an arbitrary ideal $I \subset R$ is principal, first write $\mathfrak{m} = (t)$ for some t. Take an element $x \in I$, $x \neq 0$. We claim that there exists a unit $u \in R$ and a natural number $k(x)$ such that $x = ut^{k(x)}$. This is obvious if x is a unit. Otherwise put $x = x_1 t$ for some $x_1 \in R$, and do the same with x_1. Either x_1 is a unit, and we are done, or $x_1 = x_2 t$, etc. This process must stop because, otherwise, $x = x_k t^k \in \mathfrak{m}^k$ for all k. But $\cap\mathfrak{m}^k = 0$ by Krull's Intersection Theorem, which is in contradiction to the choice of x. This shows the claim.

Now take k to be the minimum of all $k(x)$ where x runs over the nonzero elements of I. It follows then $I = (t^k)$, so, in particular, principal. Thus R is a principal ideal domain, as was to be proved.

Step 2. "$\Longleftarrow$". As a normal local ring, R is an integral domain by Theorem 1.5.7. Assume that R is not regular. We consider the R–module $\mathfrak{m}^* = (R : \mathfrak{m}) = \{x \in Q(R) : x\mathfrak{m} \subset R\}$. Obviously, $R \subset \mathfrak{m}^*$. We claim that this inclusion is strict.

To prove the claim, take an arbitrary $a \in \mathfrak{m}$, $a \neq 0$. Then $\sqrt{(a)} = \mathfrak{m}$. Indeed, by the Primary Decomposition Theorem, $\sqrt{(a)}$ is an intersection of prime ideals. But, by assumption on $\dim(R) = 1$, and because R is an integral domain, there is just one nontrivial prime ideal, namely $\mathfrak{m}$. Because R is Noetherian, it follows that there exists an $s \in \mathbb{N}$ with $\mathfrak{m}^s \subset (a)$. Let s be minimal with this property. If $s = 1$, then $\mathfrak{m} = (a)$, hence R would be a regular local ring. Take a nonzero element $b \in \mathfrak{m}^{s-1} \setminus (a)$. Then $\frac{b}{a} \notin R$. However, $\frac{b}{a}\mathfrak{m} \subset R$, because $b\mathfrak{m} \subset \mathfrak{m}^s \subset (a)$. Thus $\frac{b}{a} \in \mathfrak{m}^* \setminus R$. This proves the claim.

From the definition of $\mathfrak{m}^*$ it follows that $\mathfrak{m} \subset \mathfrak{m}\mathfrak{m}^* \subset R$. So either $\mathfrak{m}\mathfrak{m}^* = \mathfrak{m}$ or $\mathfrak{m}\mathfrak{m}^* = R$. We claim that the first equality cannot occur. Indeed, it would follow that $\frac{b}{a}\mathfrak{m} \subset \mathfrak{m}$. Therefore, $\mathfrak{m}$ would be an $R[\frac{b}{a}]$–module, finitely generated as an R–module. By the Cayley-Hamilton Theorem 1.5.8, $\frac{b}{a}$ would be an integral element, which, by assumption, would have to be in R. But $\frac{b}{a}$ was constructed in such a way that it does not belong to R.

Therefore $\mathfrak{m}\mathfrak{m}^* = R$. Now take any element $t \in \mathfrak{m} \setminus \mathfrak{m}^2$, which exists according to Nakayama's Lemma. Then $t\mathfrak{m}^* \subset R$. It is impossible to have $t\mathfrak{m}^* \subset \mathfrak{m}$, for, otherwise, $tR = t\mathfrak{m}^*\mathfrak{m} \subset \mathfrak{m}^2$, in contradiction to the choice of t. Therefore, $t\mathfrak{m}^* = R$, and $(t) = tR = t\mathfrak{m}^*\mathfrak{m} = \mathfrak{m}$. This shows that $\mathfrak{m}$ is principal. □

Corollary 4.4.10. *Let $R = \mathbb{C}\{x_1, \ldots, x_n\}/I$ be a one-dimensional integral domain. Then its normalization $\widetilde{R}$ is isomorphic to $\mathbb{C}\{t\}$.*

Proof. Because of Theorem 4.4.8 $\widetilde{R}$ is a one-dimensional analytic algebra, which is normal. By Theorem 4.4.9, $\widetilde{R}$ is regular of dimension one, hence isomorphic to $\mathbb{C}\{t\}$. □

With a little bit more effort, one can give the following beautiful characterization of a normal ring, due to Serre.

Theorem 4.4.11 (Serre's R1 and S2 Criterion). *Let R be a reduced ring. Then R is normal, if and only if the following two conditions are satisfied:*

(R1) *for each prime ideal $\mathfrak{p}$ of height one, $R_\mathfrak{p}$ is a regular local ring;*

(S2) *let $f \in R$ be a nonzerodivisor, then the ideal (f) in R has no embedded primes.*

Proof.
Step 1. Suppose first that R is normal. Then $R_\mathfrak{p}$, for a prime ideal $\mathfrak{p}$, is normal too, by Exercise 1.5.32. Moreover, $R_\mathfrak{p}$ has dimension one, so is regular by 4.4.9.

To check the second condition, suppose, on the contrary, that we have an embedded prime $\mathfrak{p}$ for some (f). In particular, $\mathfrak{p}$ is not a minimal prime ideal with $\mathfrak{p} \supset (f)$. By Lemma 1.4.17 we can find an element $b \in R$ such that $\mathfrak{p} = (f) : (b)$. Let $\mathfrak{m} = \mathfrak{p}R_\mathfrak{p}$. Considering $\mathfrak{m}^* := (R_\mathfrak{p} : \mathfrak{m}) = \{x \in Q(R) : x \cdot \mathfrak{m} \subset R_\mathfrak{p}\}$, one sees that $\frac{b}{f} \in \mathfrak{m}^*$. It follows, as in the proof of 4.4.9, that if $\mathfrak{m}\mathfrak{m}^* = R_\mathfrak{p}$, then $\mathfrak{m}$ is principal. But then it follows that the dimension of $R_\mathfrak{p}$ is less than or equal to one. This is in contradiction to the fact that $\mathfrak{p}$ is not minimal over (f), as the dimension of $R_\mathfrak{p}$ is equal to the height of $\mathfrak{p}$.

So we may assume that $\mathfrak{m}\mathfrak{m}^* = \mathfrak{m}$. We can now apply the Cayley-Hamilton Theorem 1.5.8: all elements of $\mathfrak{m}^*$ are integral over $R_\mathfrak{p}$. As we assumed R, and hence $R_\mathfrak{p}$, to be normal, we deduce that $\mathfrak{m}^* = R_\mathfrak{p}$. In particular, as $\frac{b}{f} \in \mathfrak{m}^*$. we have $\frac{b}{f} \in R_\mathfrak{p}$. Therefore, we can write $\frac{b}{f} = \frac{c}{a}$ for some $a \notin \mathfrak{p}$. Now

$$\mathfrak{p} = (f) : (b) = (fa) : (ba) = (fa) : (cf) = (a) : (c).$$

In particular, $a \in \mathfrak{p}$, which is a contradiction. So our assumption that we had an embedded prime $\mathfrak{p}$ for (f) is wrong. This proves one direction.

Step 2. Suppose on the other hand, that (R1) and (S2) hold, and suppose $\frac{b}{f}$ is integral over R. Consider an irredundant primary decomposition of (f):

$$(f) = \mathfrak{q}_1 \cap \ldots \cap \mathfrak{q}_t.$$

As (f) does not have embedded primes by (S2), all the $\mathfrak{p}_i := \sqrt{\mathfrak{q}_i}$ are minimal associated primes. By Krull's Principal Ideal Theorem, see 4.2.16, all the $\mathfrak{p}_i$ are prime ideals of height one. By (R1) it follows that the $R_{\mathfrak{p}_i}$ are regular. Moreover they have dimension one, so the $R_{\mathfrak{p}_i}$ are normal by Theorem 4.4.9. It follows that $\frac{b}{f} \in R_{\mathfrak{p}_i}$ for all i. To put it in another way: $b \in fR_{\mathfrak{p}_i}$ for all i. By the Second Uniqueness Theorem of Primary Decomposition 1.4.23 we have $\mathfrak{q}_i = R \cap fR_{\mathfrak{p}_i}$ for all i. Therefore, $b \in \mathfrak{q}_i$ for all i. Hence, $b \in \bigcap \mathfrak{q}_i = (f)$, and thus $\frac{b}{f} \in R$, which had to be proved. □

We will now consider a function-theoretic interpretation of normal spaces. For this, we will show that a space is normal if and only if "Riemann's Extension Theorem" holds for X. First we need a definition.

Definition 4.4.12. Let X be an analytic space, and $\mathrm{Sing}(X)$ be the singular locus of X. A function $f : X \setminus \mathrm{Sing}(X) \longrightarrow \mathbb{C}$ is called *weakly holomorphic* on X if

(1) f is holomorphic on $X \setminus \mathrm{Sing}(X)$;

(2) f is locally bounded.

Let (X,x) be a germ of an analytic space. A *weakly holomorphic function germ* f on (X,x) is an equivalence class of weakly holomorphic functions $f : X \setminus \mathrm{Sing}(X) \longrightarrow \mathbb{C}$. Here X is a representative of (X,x). Two such functions are called equivalent if they agree on the intersection of the corresponding representatives.

Remarks 4.4.13.

(1) The weakly holomorphic functions on X form a ring. In fact, as $\mathrm{Sing}(X)$ is contained in a proper analytic subset of X, it follows by continuity that the ring of holomorphic functions on X embeds in the ring of weakly holomorphic functions on X (an analytic function on X is already determined by its values on $X \setminus \mathrm{Sing}(X)$).

(2) The germs of weakly holomorphic functions form a ring which we denote by $\mathcal{O}'_{X,x}$. As above one sees that $\mathcal{O}_{X,x} \subset \mathcal{O}'_{X,x}$.

Examples 4.4.14. (1) Consider the coordinate axes, defined by $xy = 0$. Look at the function f which is identically 1 on the the x–axis, except at the origin, and identically 0 on the y–axis except at the origin. Then f cannot be extended to the origin (not even as a continuous function), but is bounded. Note that we can write $f(x,y) = \frac{x}{x+y}$.

(2) Consider the analytic subset $X = \{y^2 - x^3 = 0\} \subset \mathbb{C}^2$. The function $f = \frac{y}{x}$ is holomorphic, whenever $x \neq 0$. So f is a holomorphic function on $X \setminus \{(0,0)\}$. We have the estimate $\left|\frac{y}{x}\right| \leq |\sqrt{x}|$, hence f defines a germ of a weakly holomorphic function on (X,x). In this case, f extends to a continuous function on X, by $f(0,0) = 0$. However f is not holomorphic.

Theorem 4.4.15.

(1) Let (X,x) be a germ of an analytic space. Then there is a canonical isomorphism $\mathcal{O}'_{X,x} \cong \widetilde{\mathcal{O}}_{X,x}$. To put it in another way, f is a weakly holomorphic function germ if and only if f is in the integral closure of $\mathcal{O}_{X,x}$.

(2) A germ of an analytic space (X,x) is normal if and only if every weakly holomorphic function germ on (X,x) can be extended to a holomorphic function on (X,x). To put it in another way, a space (X,x) is normal, if and only if the First Riemann Extension Theorem holds for (X,x).

Proof. The second statement follows immediately from the first one. The proof of the first statement is quite long, and will be divided into several steps. Suppose the germ (X,x) is in $(\mathbb{C}^n, o)$.

Step 1. First we prove $\widetilde{\mathcal{O}}_{X,x} \subset \mathcal{O}'_{X,x}$. Take $f \in \widetilde{\mathcal{O}}_{X,x}$. Then f satisfies an equation

$$f^s + a_1 f^{s-1} + \ldots + a_s = 0, \quad a_i \in \mathcal{O}_{X,x}. \tag{4.5}$$

Consider a representative X of (X,x). Since in particular $f \in Q(\mathcal{O}_{X,x})$ we can write $f = \frac{g}{h}$, for some $g \in \mathcal{O}_{X,x}$ and $h \in \mathcal{O}_{X,x}$ a nonzerodivisor. As h is a nonzerodivisor it does not vanish identically on any of the components of (X,x).

By definition, $X \setminus \operatorname{Sing}(X)$ is a locally complex submanifold of an open subset in $\mathbb{C}^n$, so that Riemann's Extension Theorem holds for $X \setminus \operatorname{Sing}(X)$ (cf. Theorem 3.1.15). Therefore, it suffices to prove that f is *bounded* on $X \setminus \operatorname{Sing}(X)$. This will be proved as an application of the Continuity of Roots. We take X so small that the a_i of Equation (4.5) are holomorphic on X, and that X is an analytic subset of U. So we may lift the a_i to holomorphic functions on U. Now look at the equation

$$F := T^s + a_1 T^{s-1} + \ldots + a_s = 0. \tag{4.6}$$

We plug in 0 in the $a_1, \ldots, a_s$ and factorize

$$F(0,T) = (T - c_1)^{s_1} \cdots (T - c_e)^{s_e}.$$

By Hensel's Lemma 3.3.21 F factorizes as $F = F_1 \cdots F_e$, with $F_i(0,T) = (T - c_i)^{s_i}$, and F_i a polynomial in T of degree s_i. In particular $F_i(x_1, \ldots, x_n, T + c_i)$ is a Weierstraß polynomial in T. Take a small open neighborhood V_i of c_i in $\mathbb{C}$, for $i = 1, \ldots, e$. By the Continuity of Roots, there exists an open neighborhood U_i of 0, such for all $p \in U_i$, all s_i zeros (counted with multiplicity) of $F_i(p,T)$ lie in V_i. So if we take $V = V_1 \cup \ldots \cup V_e$, and $U = U_1 \cap \ldots \cap U_e$ it follows that for all $p \in U$ the $s = \sum s_i$ zeros of $F(p,T)$ lie in V. Now formula (4.5) says that $f(p)$ satisfies the equation $F(p,T)$. Therefore, if $p \in U$, and $f(p)$ is defined then $f(p) \in V$, in particular it is bounded. So we can extend f to a holomorphic function on $X \setminus \operatorname{Sing}(X)$. By continuity, this extension is also bounded.

Step 2. We now turn our attention to the inclusion $\mathcal{O}'_{X,x} \subset \widetilde{\mathcal{O}}_{X,x}$. We claim that it suffices to show this for irreducible (X,x). To see this let $(X,x) = (X_1,x) \cup \ldots \cup (X_r,x)$ be the irreducible decomposition of (X,x). Let f be a weakly holomorphic function on a representative $X = X_1 \cup \ldots \cup X_r$ of (X,x). As f is a genuine function outside the singular locus of X, it follows that f is a holomorphic function on each $X_i \setminus (\mathrm{Sing}(X_i) \cup \cup_{j \neq i}(X_i \cap X_j))$. By the Riemann Extension Theorem in $\mathbb{C}^n$, the function f can be extended to a weakly holomorphic function on X_i. So, once we have shown the inclusion for an irreducible (X,x), f defines an element of $\widetilde{\mathcal{O}}_{X_i,x}$ for each each i. But by the Splitting of Normalization, see 1.5.20, we get $\widetilde{\mathcal{O}}_{X,x} = \oplus_{i=1}^r \widetilde{\mathcal{O}}_{X_i,x}$. This implies that $f \in \widetilde{\mathcal{O}}_{X,x}$.

Step 3. We now reduce to the case that (X,x) is an irreducible *hypersurface*. By the Local Parametrization Theorem we have a surjective map π from (X,x) onto a hypersurface (X',x) which has the following properties:

- $\mathcal{O}_{X',x} \subset \mathcal{O}_{X,x}$.
- $Q(\mathcal{O}_{X,x}) = Q(\mathcal{O}_{X',x})$.
- There exists a representative $\pi : X \longrightarrow X'$ and a proper analytic subset D of X' such that $\pi : X \setminus \pi^{-1}(D) \longrightarrow X' \setminus D$ is biholomorphic.
- $\mathrm{Sing}(X') \subset D$.

Let f be a weakly holomorphic function on X. Then f gives a bounded holomorphic function on $X' \setminus D$. This is a smooth space, so by Riemann's Extension Theorem 3.1.15 f can extend to a bounded holomorphic function on $X' \setminus \mathrm{Sing}(X')$. We obtain a weakly holomorphic function on X'. So if the theorem is proved for a hypersurface we get $f \in Q(\mathcal{O}_{X',x'})$ and an integral equation for f with coefficients in $\mathcal{O}_{X',x}$. As $\mathcal{O}_{X',x} \subset \mathcal{O}_{X,x}$ this shows that f is in $\widetilde{\mathcal{O}}_{X,x}$, since $f \in Q(\mathcal{O}_{X',x}) = Q(\mathcal{O}_{X,x})$.

Step 4. Now let (X,x) be a germ of an analytic hypersurface, defined by a Weierstraß polynomial

$$P = x_n^s + a_1 x_n^{s-1} + \ldots + a_s = 0.$$

So a_i is in the maximal ideal of $\mathcal{O}_{n-1}$. We will first show that $\mathcal{O}'_{X,x} \subset Q(\mathcal{O}_{X,x})$. Let Δ be the discriminant of P, $(D,x) = \big(V(\Delta),x\big)$ and f be a weakly holomorphic function germ on (X,x). We will show that on some representatives X and D we can write f as $\frac{g}{\Delta}$, for a holomorphic function g on X.[7] So we have to show that $\Delta \cdot f$ has a holomorphic extension to X. Consider the projection

$$\pi : X \longrightarrow U \subset \mathbb{C}^{n-1}.$$

We choose U so small, so that there exists a constant $C > 0$ such that the absolute value of the x_n-th coordinate of each point is smaller than C. This can be done by Continuity of Roots, see 3.4.11. Furthermore, we may assume that $|f(a)| < C$ for all $a \in X \setminus \mathrm{Sing}(X)$, because by assumption f is locally bounded. Take a point $p \in U \setminus D$. Locally around p, there exists an open neighborhood V of p, and s holomorphic functions

[7] In view of the fact that we try to prove that f is in the integral closure of $\mathcal{O}_{X,x}$, and the fact that any element in the integral closure has as universal denominator Δ, see 1.5.19, this is the obvious thing to do.

$$\alpha_1, \ldots ,\alpha_s : V \longrightarrow \mathbb{C},$$

such that $\pi^{-1}(V)$ is the disjoint union of s complex submanifolds, given by the equations $x_n = \alpha_i(x_1, \ldots ,x_{n-1})$, see 3.3.23. U was chosen to be so small that the α_i are bounded. Let $q \in V$, and consider the Lagrange interpolation polynomial in x_n

$$\sum_{i=1}^{s} f(q,\alpha_i(q)) \prod_{j \neq i} \frac{x_n - \alpha_j(q)}{\alpha_i(q) - \alpha_j(q)}.$$

For $q \in V$, this is indeed a well-defined polynomial in x_n. By construction, the Lagrange interpolation polynomial has the same values as f on the s points $(q,\alpha_1(q)), \ldots , (q,\alpha_s(q))$. We now vary q, and we thus may assume that on $\pi^{-1}(V)$ f is equal to the above polynomial in x_n whose coefficients are holomorphic functions on V. Now on V the discriminant is equal to

$$\Delta = \prod_{j \neq i} (\alpha_i - \alpha_j).$$

Thus we see that $\Delta \cdot f$ has as coefficients holomorphic functions on V. As one easily checks, these functions are bounded by a number $A(C)$, which only depends on C. Doing this for all $p \in U \setminus D$, we see that $\Delta \cdot f$ is a polynomial in x_n whose coefficients are holomorphic functions on $U \setminus D$ which are bounded. By the Riemann Extension Theorem for $\mathbb{C}^{n-1}$, see 3.1.15, it follows that we can extend these to holomorphic functions on U. Hence $\Delta \cdot f$ can be extended to a holomorphic function on all of X.

Step 5. We keep the notations of Step 4, and we finally show that any weakly holomorphic function is integral over $\mathcal{O}_{X,x}$. For $p \in U \setminus D$ and V an open neighborhood of p, look at the holomorphic functions $\alpha_1, \ldots ,\alpha_s : V \longrightarrow \mathbb{C}$, as above. The function $q \mapsto f(q,\alpha_i(q))$ defines a holomorphic function on V. Moreover on $\pi^{-1}(V)$ the holomorphic function

$$\Big(f - f(x_1, \ldots ,x_{n-1},\alpha_1)\Big) \cdots \Big(f - f(x_1, \ldots ,x_{n-1},\alpha_s)\Big)$$

is the zero function. Let b_i be the i-th symmetric polynomial in the $f(x_1, \ldots ,x_{n-1},\alpha_j)$. Then the b_i define holomorphic functions on V, and we have an equation

$$f^s + b_1 f^{s-1} + \ldots + b_s = 0, \tag{4.7}$$

as a holomorphic function on $\pi^{-1}(V)$. Moreover, the b_i are bounded on V, because the $f(x_1, \ldots ,x_{n-1},\alpha_i)$ are. Doing this for all p, we get bounded functions $b_1, \ldots ,b_s$ on $U \setminus D$. By the Riemann Extension Theorem 3.1.15, they extend to a holomorphic function on all of U. Now write $f = \frac{g}{\Delta}$. This we can do by Step 4. As Δ does not vanish on $U \setminus D$, we can multiply equation (4.7) with Δ^s and get

$$g^s + b_1 \Delta g^{s-1} + \ldots + b_s \Delta^s = 0, \tag{4.8}$$

as holomorphic functions on $X \setminus \pi^{-1}(D)$. By continuity, it follows that (4.8) holds on X. We can now read (4.8) as an equation in $\mathcal{O}_{X,x}$. Upon dividing (4.8) by Δ^s we get an integral equation for $f = \frac{g}{\Delta}$. □

Exercises

4.4.16. Let $(X,x) = (X_1,x_1) \cup \cdots \cup (X_r,x_r)$ be a multi-germ of analytic spaces.

(1) Show that $\mathcal{O}_{X,x}$ has exactly r maximal ideals, so in particular $\mathcal{O}_{X,x}$ is a semi-local ring. Describe those maximal ideals.

(2) Show that (X,x) is normal if and only if (X_i,x_i) is normal for $i = 1,\ldots,r$.

(3) Let $\varphi : (X,x) \longrightarrow (Y,y)$ be a map between multi-germs of analytic spaces. Show that φ is generically $1-1$ if and only if φ induces an isomorphism of total quotient rings $\varphi^* : Q(\mathcal{O}_{Y,y}) \longrightarrow Q(\mathcal{O}_{X,x})$.

(4) Let $\varphi : (X,x) \longrightarrow (Y,y)$ be a finite map between irreducible germs of analytic spaces. Suppose that $\varphi^* : \mathcal{O}_{Y,y} \longrightarrow \mathcal{O}_{X,x}$ is injective and let s be the degree of the field extension $Q(\mathcal{O}_{X,x}) \supset Q(\mathcal{O}_{Y,y})$. Show that φ is generically s to 1 and that φ is surjective.
(Hint: Use 3.4.41 and 3.4.40)

4.4.17. Let $\varphi : (X,x) \longrightarrow (Y,y)$ be a finite map of germs of analytic spaces. Prove that the image of φ is a germ of an analytic space.
(Hint: Let $I = \mathrm{Ker}(\varphi^*)$. Show that the image of φ consists of all points $a \in Y$ with $f(a) = 0$ for all $f \in I$.)

4.4.18. The normalization of a local domain is not necessarily local. The local ring R, obtained from $\mathbb{C}[x,y]/(y^2 - x^3 - x^2)$ by localizing in the maximal ideal (x,y) is an integral domain. Show that the normalization of R is not local.

4.4.19. Let (X,x) be a germ of an analytic space, and $n : (\widetilde{X},\widetilde{x}) \longrightarrow (X,x)$ be a normalization. Let (Y,y) be a normal multi-germ, and $\varphi : (Y,y) \longrightarrow (X,x)$ be a map of multi-germs. Show that there exists a uniquely determined map of multi-germs $\widetilde{\varphi} : (Y,y) \longrightarrow (\widetilde{X},\widetilde{x})$ making the following diagram commutative.

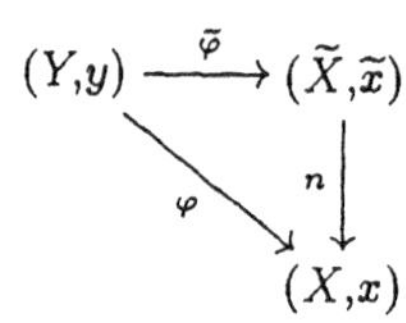

(Hint: Prove a corresponding statement for rings.)

4.4.20. Let (X,x) be a germ of an analytic space, and $\mathrm{Sing}(X,x)$ be the singular locus of (X,x). Let $I \subset \mathcal{O}_{X,x}$ be the ideal of germs of analytic functions vanishing on the singular locus. Show that (X,x) is normal if and only if $\mathrm{Hom}_{\mathcal{O}_{X,x}}(I,I) = \mathcal{O}_{X,x}$.
(Hint: Use the fact that regular analytic rings are normal. Follow the proof of 1.5.13, but now use the Nullstellensatz 3.4.4 to prove $I \subset \sqrt{J}$ (notations of 1.5.13).)

4.4.21. Let $f \in \mathbb{C}\{x_1,\ldots,x_n\}$. Prove that $f^k \in \left(\frac{\partial f}{\partial x_1},\ldots,\frac{\partial f}{\partial x_n}\right)$ for some $k \geq 0$.
(Hint: Let $(X,0) = (V(f,\frac{\partial f}{\partial x_1},\ldots,\frac{\partial f}{\partial x_n}),0)$ and $(Y,0) = (V(\frac{\partial f}{\partial x_1},\ldots,\frac{\partial f}{\partial x_n}),0)$. Show that $(X,0) = (Y,0)$ in the following way. Suppose $(X,0) \subsetneq (Y,0)$. Find a map $n : (\mathbb{C},0) \longrightarrow (Y,0)$ with $n((\mathbb{C},0)) \not\subset (X,0)$ by using the normalization of a curve $(C,0) \subset (Y,0)$ which is not contained in $(X,0)$. Compute $\frac{d}{dt}(f \circ n)$.)

4.4.22. Let $(C,0)$ be an irreducible curve singularity. Let $n : (\widetilde{C},\widetilde{0}) \longrightarrow (C,0)$ be given such that $(\widetilde{C},\widetilde{0})$ is smooth. Suppose that $\dim_{\mathbb{C}}(\mathcal{O}_{\widetilde{C},\widetilde{0}}/\mathcal{O}_{C,0}) < \infty$. Show that n is the normalization mapping.

Generalize this result to reducible germs.
(Hint: Show that the degree of the corresponding field extension is one.)

5 Plane Curve Singularities

This chapter is devoted to the study of plane curve singularities. First of all, from the existence of the normalization, and the fact that one-dimensional normal singularities are smooth, which are results proved in the previous chapter, it follows that one can parameterize curve singularities. Thus, if R is the local ring of an irreducible plane curve singularity, we have an injection $R \subset \mathbb{C}\{t\}$. From the Finiteness of Normalization Theorem 1.5.19 it follows without much difficulty that the codimension of R in $\mathbb{C}\{t\}$ (as $\mathbb{C}$-vector space) is finite. This codimension is called the δ-invariant. Here we prove the converse of this statement, that is, given a subring R of $\mathbb{C}\{t\}$ of finite codimension, generated by two elements, say $x(t)$ and $y(t)$, then R the local ring of an irreducible plane curve singularity, say given by $f(x,y) = 0$. In order to obtain information on the order of f, we introduce the intersection multiplicity of two (irreducible) plane curve singularities (C, o) and (D, o). This number can be calculated as follows. Consider the normalization $\mathbb{C}\{t\}$ of $\mathcal{O}_{C,o}$, and let (D, o) be given by $g = 0$. Via the inclusion $\mathcal{O}_{C,o} \subset \mathbb{C}\{t\}$, we can consider at g as an element of $\mathbb{C}\{t\}$. Then the intersection multiplicity is given by the vanishing order of g. From this it easily follows that if $R = \mathbb{C}\{x(t),y(t)\}$, and $f \in \mathbb{C}\{x,y\}$ is irreducible with $f(x(t),y(t)) = 0$, then the order of f in y is equal to $\mathrm{ord}_t(x(t))$, and similar for x. As an application we prove that if we have a convergent power series $f \in \mathbb{C}\{x,y\}$, which is irreducible when considered as a formal power series, then it is already irreducible as a convergent power series. This in fact holds for power series in any number of variables, and is an easy consequence of the Artin Approximation Theorem, which will be proved in Chapter 8. Next we discuss Newton's method. Here we start with an $f \in \mathbb{C}\{x,y\}$, and try to find $x(t), y(t)$ with $f(x(t),y(t)) = 0$. We give two proofs of the convergence. The first one is a consequence of Corollary 3.3.18, and is, to our knowledge, new. The second one is standard. Finally, we discuss the irreducibility of f in terms of its Newton polygon. It is shown that for an irreducible f the Newton polygon has only one face. Our proof is algorithmic, that is, if the Newton polygon has more than one face, we will give an algorithm to find a factorization of f.

In Section 5.2 we will study several invariants of an irreducible plane curve singularity R. As above, we may assume that $R \subset \mathbb{C}\{t\}$, which is the normalization. We already mentioned the δ-invariant. Moreover, we introduce the semigroup $\Gamma(R) \subset \mathbb{N}$, which consists of the vanishing orders of the elements in R. Furthermore, the conductor $c(R)$ is introduced. It is the smallest natural number α with $\alpha + \mathbb{N} \subset \Gamma(R)$. Our first aim is to show the Theorem of Gorenstein, which says that $c(R) = 2 \cdot \delta(R)$. The proof uses duality theory for fractional ideals. Fractional ideals are R-modules in the quotient field $Q(R)$, which have a universal denominator. Given a fractional ideal I, we can look at its dual $I^* := \mathrm{Hom}_R(I,R)$. For example, the normalization $\widetilde{R}$ and the conductor ideal $\mathrm{Ann}_R(\widetilde{R}/R)$ are fractional ideals, which are dual to each other. We have a canonical map $I \longrightarrow I^{**}$, which one easily sees to be injective. One may ask oneself whether this map is an isomorphism for all fractional ideals. In general, this turns out not to be the case. Those curve singularities for which it is true are called Gorenstein, and we will prove that irreducible plane curve singularities are indeed Gorenstein. We then discuss how to compute the minimal generators of the semigroup $\Gamma(R)$ from the parametrization. In general, this would be much too difficult, so we first suppose that the parametrization is of type $x(t) = t^n$, $y(t) = \sum_{i\geq m} a_i t^i$ with $m \geq n$. After a coordinate change in t, and interchanging x and y we may suppose that our parametrization is of this type. Characteristic exponents $k_0, \ldots, k_g$ are introduced, in terms of n, and the $i \in \mathbb{N}$ with $a_i \neq 0$. A priori these are only invariants of the parametrization, but we will show that they determine, and are determined by, the minimal generators $\beta_0, \ldots, \beta_g$ of the semigroup $\Gamma(R)$. We then discuss the Inversion Theorem, which is of importance in the next section. This gives a method for computing the characteristic exponents in case $n > m$.

In Section 5.3 we consider resolutions of irreducible plane curve singularities. First of all, the blowing-up $\pi : Bl_0\mathbb{C}^2 \longrightarrow \mathbb{C}^2$ of $\mathbb{C}^2$ in a point, say 0, is introduced. The space $Bl_0\mathbb{C}^2$ is a complex manifold of dimension one, in which the point 0 has been replaced by all directions through 0. The set $\pi^{-1}(0)$ is a projective line, called the exceptional curve or exceptional divisor. If one now has a plane curve singularity, which has representative C in some small open neighborhood U of 0, we can look at the strict transform $C^{(1)} := \overline{\pi^{-1}(C \setminus \{0\})}$, and iterate this process. That is, we blow up in the singular point of $C^{(1)}$, if present. The Theorem of Resolutions of Plane Curve Singularities says that this process stops after finitely many steps. The main ingredient of the proof is the existence of a parametrization of the irreducible plane curve singularity, as studied in the first section. After some more blowing-ups, one can arrange that the strict transform intersects just one exceptional curve, and intersects this curve transversally. This is then called the standard resolution. Having such a resolution process, one can introduce two invariants. First of all, one has the multiplicity sequence. If the standard resolution is obtained after k blowing-ups, one has a sequence of numbers $(m_0, \ldots, m_{k-1})$, where m_0 is the multiplicity of C, and m_i is the multiplicity of the strict transform $C^{(i)}$ after i blowing-ups. Furthermore, the resolution graph is introduced. It is a weighted graph with one vertex for each exceptional divisor. The weight of a vertex is i if the curve is created under the i-th blowing-up. Furthermore, for the strict transform we add a star $*$. Two vertices are connected by an edge, if the corresponding divisors intersect. Furthermore, the star is connected to the vertex corresponding to the unique exceptional divisor which it intersects. The main result of the section is that the multiplicity sequence, the resolution graph, the characteristic exponents, and the semigroup determine each other. In the proof, essential use of the Inversion Theorem is made.

In Section 5.4 we study reducible plane curve singularities. Again we can study a standard resolution. Its existence easily follows from the irreducible case. As a new invariant the contact number between two irreducible components, also called branches of the plane curve singularity, is introduced. The contact number of two irreducible plane curve singularities is the number of blowing-ups one needs to separate the strict transforms. It turns out that this number is finite. As in the irreducible case, we introduce a resolution graph, but of course we need a star $*$ for each branch. We then show that the resolution graph determines the resolution graphs of the irreducible components and the contact numbers between the branches, and conversely. Finally, as an application of the standard resolution we give two results. First of all, given irreducible plane curve singularities, we show how to compute the intersection number from the multiplicity sequences and the contact numbers. Secondly, for irreducible plane curve singularities, we will prove a theorem of Max Noether, which gives a formula for the δ-invariant of an irreducible plane curve singularity in terms of its multiplicity sequence. This result will be generalized to the reducible case in the exercises.

5.1 Puiseux Expansion

From the results of the previous chapter it is easy to see that one can parameterize irreducible (plane) curve singularities:

Theorem 5.1.1 (Puiseux Expansion). *Let* $0 \neq f \in \mathbb{C}\{x,y\}$ *be irreducible. Then there exist* $x(t)$, $y(t) \in \mathbb{C}\{t\}$ *such that*

(1) $f(x(t),y(t)) = 0$,

(2) $\dim_{\mathbb{C}}(\mathbb{C}\{t\}/\mathbb{C}\{x(t),y(t)\}) < \infty$.

Proof. Let R be the normalization of $\mathbb{C}\{x,y\}/(f)$. Because of Corollary 4.4.10 R is isomorphic to $\mathbb{C}\{t\}$. Therefore, we have an injective map

$$\mathbb{C}\{x,y\}/(f) \longrightarrow \mathbb{C}\{t\},$$

and $\mathbb{C}\{t\}$ is a finitely generated $\mathbb{C}\{x,y\}/(f)$-module via this map. Let $x(t)$ respectively $y(t)$ be the images of $x \mod (f)$ respectively $y \mod (f)$ in $\mathbb{C}\{t\}$. Then obviously (1) holds.

To prove (2) we may assume that f is a Weierstraß polynomial in y. Let Δ be its discriminant. As f is irreducible, it does not have multiple factors. In particular $\Delta \notin (f)$. It follows from the Theorem of Finiteness of Normalization, see 1.5.19, that $\Delta\mathbb{C}\{t\} \subset \mathbb{C}\{x(t),y(t)\}$. We can view Δ as an element of $\mathbb{C}\{t\}$. Let k be the vanishing order of Δ, that is, $\Delta = t^k u$ for some unit u. Then it follows that $t^k\mathbb{C}\{t\} \subset \mathbb{C}\{x(t),y(t)\}$, so that $\{1,\dots,t^{k-1}\}$ generate $\mathbb{C}\{t\}/\mathbb{C}\{x(t),y(t)\}$ as $\mathbb{C}$-vector space. □

Example 5.1.2. For the A_2-singularity $y^2 - x^3 = 0$ we can take $x(t) = t^2$, and $y(t) = t^3$.

We have the following converse to 5.1.1.

Theorem 5.1.3. *Let two power series* $x(t)$, $y(t) \in t\mathbb{C}\{t\}$ *be given. Define* $R = \mathbb{C}\{x(t),y(t)\}$ *and suppose that* $\dim_{\mathbb{C}}(\mathbb{C}\{t\}/R) < \infty$. *Then there exists an irreducible* $f \in \mathbb{C}\{x,y\}$ *with* $R \cong \mathbb{C}\{x,y\}/(f)$. *Moreover,* $R \subset \mathbb{C}\{t\}$ *is the normalization of* R.

Proof. Consider the map $\varphi : \mathbb{C}\{x,y\} \longrightarrow \mathbb{C}\{t\}$ defined by $\varphi(x) = x(t)$, $\varphi(y) = y(t)$. Then $\mathbb{C}\{x,y\}/\operatorname{Ker}(\varphi) \cong R$. From $\dim_{\mathbb{C}}(\mathbb{C}\{t\}/R) < \infty$ it follows that $R \subset \mathbb{C}\{t\}$ is a finite ring extension, and as $\mathbb{C}\{t\}$ has dimension one, it follows that R has dimension one, see 4.1.4. Because $R \subset \mathbb{C}\{t\}$, it is an integral domain. Thus R is the local ring of an irreducible plane curve singularity. Therefore $\operatorname{Ker}(\varphi)$ is a principal ideal by the Characterization of Hypersurfaces 4.1.12. So $\operatorname{Ker}(\varphi) = (f)$ for some $f \in \mathbb{C}\{x,y\}$. As R is an integral domain, we get that $\operatorname{Ker}(\varphi)$ is a prime ideal, and it follows that f is irreducible.

It remains to prove that $\mathbb{C}\{t\}$ is the normalization of R. As $\dim_{\mathbb{C}}(\mathbb{C}\{t\}/R) < \infty$, it follows that $I := \operatorname{Ann}_R(\mathbb{C}\{t\}/R) \neq 0$ (Exercise 5.1.30). Now I is an ideal in $R = \mathbb{C}\{x(t),y(t)\}$, but from the definition it follows immediately that I is also an ideal in $\mathbb{C}\{t\}$. As $\mathbb{C}\{t\}$ is a principal ideal domain it follows that $I = t^c \cdot \mathbb{C}\{t\}$ for some $c > 0$. This implies that $t^c\mathbb{C}\{t\} \subset \mathbb{C}\{x(t),y(t)\}$. Let $\psi \in t\mathbb{C}\{t\}$, then for all $k \geq c$ we have $\psi^k \in R$. In particular $\psi = \frac{\psi^{k+1}}{\psi^k}$ is in the quotient field of $\mathbb{C}\{x(t),y(t)\}$. As certainly the constants are in the quotient field of $\mathbb{C}\{x(t),y(t)\}$ too, it follows that the quotient fields of $\mathbb{C}\{t\}$ and $\mathbb{C}\{x(t),y(t)\}$ are equal. As $\mathbb{C}\{t\}$ is a finitely generated $\mathbb{C}\{x(t),y(t)\}$-module, every element in $\mathbb{C}\{t\}$ is integral over $\mathbb{C}\{x(t),y(t)\}$. Therefore $\mathbb{C}\{t\}$ is in the normalization of $\mathbb{C}\{x(t),y(t)\}$. As $\mathbb{C}\{t\}$ is normal itself, it follows that $\mathbb{C}\{t\}$ is the normalization of $\mathbb{C}\{x(t),y(t)\}$. □

In this theorem we showed that we can find such an f, but we still have no information on its order. In order to answer this question, we introduce the intersection multiplicity.

Definition 5.1.4. Let (X,o) and (Y,o) be plane curve singularities, not necessarily reduced, not having a component in common. Let (X,o) be defined by $f=0$, and let (Y,o) be defined by $g=0$. The *intersection multiplicity* or *intersection number* of (X,o) and (Y,o) at 0 is defined by

$$(X,o)\cdot(Y,o) := \dim_{\mathbb{C}}\left(\mathbb{C}\{x,y\}/(f,g)\right).$$

In case (X,o) and (Y,o) are reduced, this intersection multiplicity has a geometric interpretation as the number of intersection points in a small neighborhood of 0 after a small perturbation of f and g. We are not able to prove this at the moment, but it will be done in Chapter 6. The following important lemma tells us how to compute the intersection number with the help of the normalization in case one of the curves, say (X,o), is irreducible.

Lemma 5.1.5. *Suppose that (X,o) is an irreducible plane curve singularity. Let $\mathcal{O}_{X,o} \subset \mathcal{O}_{\overline{X},\overline{o}} \cong \mathbb{C}\{t\}$ be its normalization. Let (Y,o) be a plane curve singularity, defined by $g=0$. We can consider the class of g in $\mathcal{O}_{X,o}$ and, therefore, we can view g as an element of $\mathbb{C}\{t\}$. Then*

$$(X,o)\cdot(Y,o) = \operatorname{ord}_t(g).$$

Proof. Consider the diagram

$$\begin{array}{ccc} \mathcal{O}_{\overline{X},\overline{o}} & \xrightarrow{\cdot g} & \mathcal{O}_{\overline{X},\overline{o}} \\ \uparrow & & \uparrow \\ \mathcal{O}_{X,o} & \xrightarrow{\cdot g} & \mathcal{O}_{X,o}. \end{array}$$

Then $\operatorname{ord}_t(g) = \dim_{\mathbb{C}}\left(\operatorname{Coker}(\mathcal{O}_{\overline{X},\overline{o}} \xrightarrow{\cdot g} \mathcal{O}_{\overline{X},\overline{o}})\right)$, and $(X,o)\cdot(Y,o) = \dim_{\mathbb{C}}\left(\operatorname{Coker}(\mathcal{O}_{X,o} \xrightarrow{\cdot g} \mathcal{O}_{X,o})\right)$. The problem is now reduced to a problem in linear algebra, and its solution is left as Exercise 5.1.31. □

Corollary 5.1.6. *Under the conditions of Theorem 5.1.3, suppose that* $\operatorname{ord}(x(t)) = n$, *and that* $\operatorname{ord}(y(t)) = m$. *Let $f \in \mathbb{C}\{x,y\}$ be irreducible with $\mathbb{C}\{x(t),y(t)\} \cong \mathbb{C}\{x,y\}/(f)$. Then*

(1) f is regular in y of order n, and regular in x of order m.

(2) The multiplicity $\operatorname{ord}(f)$ *of f is equal to* $\min\{n,m\}$.

Proof. Let (X,o) is the singularity defined by $f=0$, and (L,o) the germ of the line defined by $x=0$. The first statement follows because, by Lemma 5.1.5,

$$\operatorname{ord}_y(f) = \dim_{\mathbb{C}}\left(\mathbb{C}\{x,y\}/(f,x)\right) = (X,o)\cdot(L,o) = \operatorname{ord}_t(x(t)).$$

Similarly for x.

To prove the second statement we may without loss of generality assume that $n \leq m$, and that f is regular in y of order $\operatorname{ord}(f)$. This is always possible after a linear coordinate change. It follows from the first part that

$$\operatorname{ord}(f) = \operatorname{ord}_y(f) = (X,0) \cdot (L,0) = \operatorname{ord}_t(x(t)) = n = \min\{n,m\},$$

as claimed. □

Theorem 5.1.7. *Let ε be a primitive n-th root of unity. Consider the ring extension $\mathbb{C}\{x,y\} \subset \mathbb{C}\{t,y\}$, where $x = t^n$.*

(1) Let $f = y^n + a_{n-1}(x)y^{n-1} + \cdots + a_0(x)$ be an irreducible Weierstraß polynomial. Then there exists a $y(t) \in t\mathbb{C}\{t\}$ such that

$$f = \prod_{i=1}^{n} (y - y(\varepsilon^i t)).$$

One also writes this as $f = \prod_{i=1}^{n} (y - y(\varepsilon^i x^{1/n}))$.

(2) Let $y(t) \in t\mathbb{C}\{t\}$ such that $\dim_{\mathbb{C}} (\mathbb{C}\{t\}/\mathbb{C}\{t^n,y(t)\}) < \infty$ be given. Then $f = \prod_{i=1}^{n} (y - y(\varepsilon^i t))$ is an irreducible Weierstraß polynomial in $\mathbb{C}\{x,y\}$.

Similar statements hold in the formal case.

Proof. (1) We first prove the first part. By Theorem 5.1.1 there exist $x(t)$ and $y(t)$ in $\mathbb{C}\{t\}$ such that $f(x(t),y(t)) = 0$. By Corollary 5.1.5 it follows that $\operatorname{ord}(x(t)) = n$, and we may after a coordinate change in t, assume that $x(t) = t^n$. Consider $f(t^n,y) \in \mathbb{C}\{t,y\}$. As $f(t^n,y(t)) = 0$, it follows from the Weierstraß Division Theorem that $y - y(t)$ is a factor of $f(t^n,y)$. For all i with $1 \leq i \leq n$ it follows that $f((\varepsilon^i t)^n, y(\varepsilon^i t)) = f(t^n, y(\varepsilon^i t)) = 0$. So $y - y(\varepsilon^i t)$ is also a factor of $f(t^n,y)$. We claim that all these factors are different. That is, we claim

$$y(\varepsilon^i t) = y(\varepsilon^j t) \iff n \mid (i-j).$$

The direction "$\Longleftarrow$" is trivial. For the converse, write $y(t) = \sum_{i \geq m} a_i t^i$. It is an elementary fact that for $h \in \mathbb{C}\{t^n,y(t)\}$ the order $\operatorname{ord}_t(h)$ is in the semigroup generated by n and all i with $a_i \neq 0$. As $\mathbb{C}\{t\}/\mathbb{C}\{t^n,y(t)\}$ is a finite-dimensional vector space we can find $k_1 \ldots, k_s \in \mathbb{N}$ with $\gcd(n,k_1,\ldots,k_s) = 1$, and $a_{k_\ell} \neq 0$ for $\ell = 1,\ldots,s$. Now $y(\varepsilon^i t) = y(\varepsilon^j t)$ implies that $\varepsilon^{k_\ell i} = \varepsilon^{k_\ell j}$ for $\ell = 1,\ldots,s$. Thus $n \mid k_\ell(i-j)$ for $\ell = 1,\ldots,s$. It follows that $n \mid (i-j)$. (This is elementary number theory.) So the first part follows, because $\mathbb{C}\{t,y\}$ is a unique factorization domain.

(2) Consider the kernel of the map $\varphi : \mathbb{C}\{x,y\} \longrightarrow \mathbb{C}\{t\}$ which sends x to $x(t)$ and y to $y(t)$. Now in Theorem 5.1.3 it was shown that $\operatorname{Ker}(\varphi)$ is principal, and in fact can be generated by an irreducible Weierstraß polynomial in y of degree n by Corollary 5.1.6. This polynomial we call f. From the first part it follows that f has in $\mathbb{C}\{t,y\}$ the desired factorization. □

The following corollary is most easily proved as an application of Artin's Approximation Theorem. The proof we give now is more elementary, however.

Corollary 5.1.8. *Consider an $f \in \mathbb{C}\{x,y\}$. Then we have the following statements.*

(1) *Suppose f is irreducible in $\mathbb{C}\{x,y\}$. Then f is also irreducible when considered as element of $\mathbb{C}[[x,y]]$.*

(2) *Suppose $f = f_1 \cdot f_2$ with $f_1, f_2 \in \mathbb{C}[[x,y]]$, and suppose f_1 is a Weierstraß polynomial in y. Then f_1 and f_2 are convergent.*

(3) *Suppose $y = y(x)$ is a formal solution of $f(x,y) = 0$. Then $y(x)$ is convergent.*[1]

Proof. (1) We may assume that f is an irreducible Weierstraß polynomial. As f is irreducible we can, by Corollary 5.1.7, write $f = \prod_{i=1}^{n}(y - y(\varepsilon^i t))$, where $x = t^n$. Moreover $\dim_{\mathbb{C}}(\mathbb{C}\{t\}/\mathbb{C}\{t^n,y(t)\})$ is finite. So for all $k \gg 0$ the element t^k can be written as a power series in t^n and $y(t)$. Thus it follows that $\dim_{\mathbb{C}}(\mathbb{C}[[t]]/\mathbb{C}[[t^n,y(t)]])$ is finite. Applying the second part of Theorem 5.1.7 in the formal case, shows that f is irreducible in $\mathbb{C}[[x,y]]$.

(2) Take a factorization $f = u \cdot g_1 \cdots g_n$ of f, where the g_i are Weierstraß polynomials and u is a unit. Then u and $g_1 \ldots, g_n$ are uniquely determined. By the first part, the g_i are also irreducible in $\mathbb{C}[[x,y]]$. Taking similar factorizations of f_1 and f_2. It follows that f_1 is a product of the $g's$, hence convergent. But then $f_2 = \frac{f}{f_1}$ is also convergent.

(3) If $y = y(x)$ is a solution of $f(x,y) = 0$, we have that $y - y(x)$ is a factor of f. Thus (3) is a special case of the second part. □

Now we wish to explain an explicit construction of the parametrization obtained in Theorem 5.1.1. As a tool we need to introduce the Newton polygon.

Definition 5.1.9. Let $f = \sum a_{ij}x^iy^j \in \mathbb{C}\{x,y\}$. We call

$$\operatorname{supp}(f) = \{(i,j) : a_{ij} \neq 0\}$$

the *support* of f. Sometimes we will identify the pair (i,j) with the monomial x^iy^j. Let

(1) $\Gamma_+(f) = \bigcup_{(i,j)\in \operatorname{supp}(f)} ((i,j) + \mathbb{R}^2_+)$.

(2) $\Gamma(f)$ the boundary of the convex hull of $\Gamma_+(f)$ in $\mathbb{R}^2$.

Then

(1) $\Gamma(f)$ is called the *Newton polygon* of f.

(2) The compact segments of $\Gamma(f)$ we call *faces*. The slope of the line through a face is called the *slope* of the face.

(3) Let Δ be a face of $\Gamma(f)$. Then the *restriction* of f to Δ by definition is $f_\Delta = \sum_{(i,j)\in\Delta} a_{ij}x^iy^j$.

(4) The face Δ induces a grading on $\mathbb{C}\{x,y\}$ in the following way: if Δ has slope $-\mu = -\frac{w_1}{w_2}$, in lowest terms, and w_1, $w_2 > 0$. Then we set w-deg$(x) = w_1$, and w-deg$(y) = w_2$. Note that f_Δ is quasi-homogeneous with respect to these weights.

Example 5.1.10. Let $f = x^5 + x^3y^3 + y^3 + xy$. Then $\operatorname{Supp}(f) = \{(5,0),(3,3),(0,3),(1,1)\}$.

[1] This we already proved as 3.3.18.

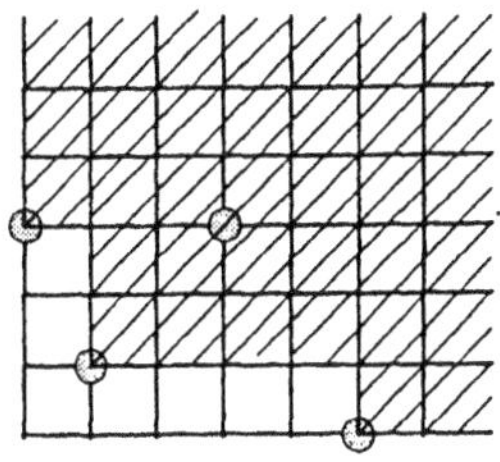

$\Gamma_+(f)$

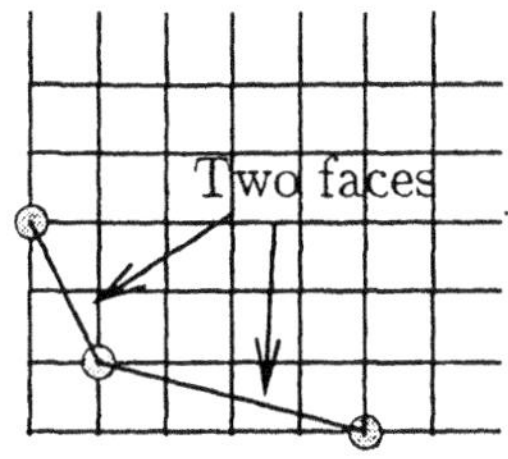

$\Gamma(f)$

Let Δ be the face defined by (1,1) and (5,0). Then $f_\Delta = x^5 + xy$. The weights corresponding to the grading defined by Δ are w-deg$(x) = 1$ and w-deg$(y) = 4$.

Obviously, different f can have the same Newton polygon. For example, $x^5 + x^3y^3 + y^3 + xy$ and $x^5 + y^3 + xy$ have the same Newton polygon. The element (3,3) does not enlarge the Newton polygon.

Example 5.1.11. $f = x^3y^3 + x^2y^4$.

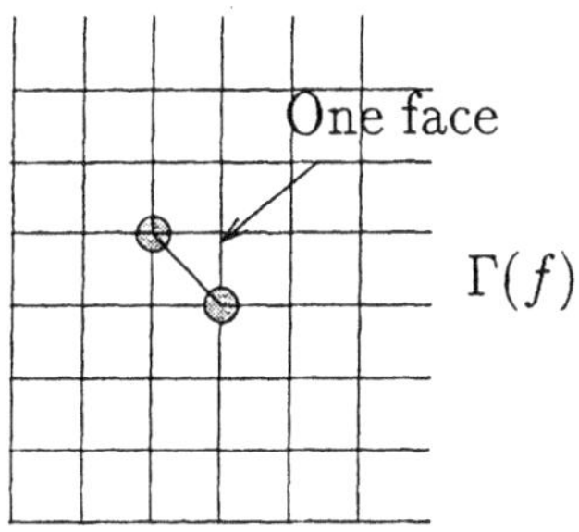

Definition 5.1.12. f is called *convenient* if there exist positive integers n,m such that $(n,0),(0,m) \in \operatorname{supp}(f)$, that is, the noncompact parts of $\Gamma(f)$ are on the coordinate axis.

Remark 5.1.13. f in Example 5.1.10 is convenient, f in Example 5.1.11 is not convenient. For any $f \in \mathbb{C}\{x,y\}$, we can always write $f = x^a \cdot y^b \cdot f_1$, where f_1 is convenient. Therefore, from the point of view of the zeros of f the convenient case is the important one.

We now come to Newton's method to find a parametrization of a branch of a plane curve singularity.

Theorem 5.1.14. *Let $f \in \mathbb{C}\{x,y\}$, and suppose that f is convenient. We have the following algorithm for finding a parametrization of an irreducible component (also called branch) of $f = 0$.*

(1) Let Δ be the face of $\Gamma(f)$ with smallest slope $-\frac{w_1}{w_2}$, which is in lowest terms, and with $w_1, w_2 > 0$. Then f_Δ is quasi-homogeneous of say, degree d_0. Consider a zero of $f_\Delta(1,y)$ which we call a_0. Take new variables x_1 and y_1 and substitute

$$x = x_1^{w_1}, \quad y = x_1^{w_2}(a_0 + y_1).$$

Then

$$f^{(1)}(x_1,y_1) := \frac{1}{x_1^{d_0}}\left(f(x_1^{w_1},x_1^{w_2}(a_0+y_1))\right) \in \mathbb{C}\{x_1,y_1\}.$$

(2) Inductively find a parametrization $x_1 = t^a$, $y_1 = \varphi(t)$ of a branch of $f^{(1)}(x_1,y_1)$ in case $f^{(1)} \neq 0$. Then

$$x = t^{aw_1},\ y = a_0 t^{aw_2} + t^{aw_2}\varphi(t),$$

gives a parametrization of a branch of $f = 0$.

Before giving the proof we give an example.

Example 5.1.15. Let $f(x,y) = y^4 - 2x^3y^2 + x^6 - x^7$. The Newton polygon consists of one face.

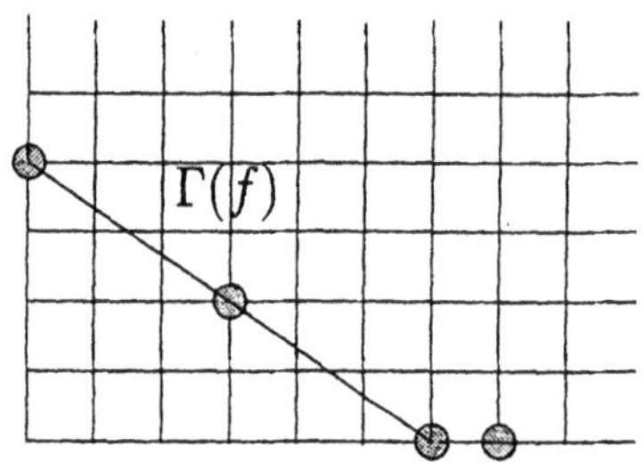

We have $w_1 = 2$, $w_2 = 3$. Then $d_0 = 12$ and $f_\Delta(x,y) = (y^2 - x^3)^2$. Thus $f_\Delta(1,y) = (y^2-1)^2$. We have the solution $a_0 = 1$. We set

$$x = x_1^2, \quad y = x_1^3(1+y_1).$$

We plug this into $f(x,y)$ to get $x_1^{12}f^{(1)}(x_1,y_1)$. We obtain as result

$$f^{(1)}(x_1,y_1) = y_1^4 + 4y_1^3 + 4y_1^2 - x_1^2.$$

Of course, we can go on as in the first step, but we can also use some common sense. Indeed,

$$f^{(1)}(x_1,y_1) = (2y_1 + y_1^2)^2 - x_1^2,$$

so that $x_1 = \pm(2y_1 + y_1^2)$. We, therefore, see two solutions for y_1:

$$y_1 = -1 + \sqrt{1 - x_1}; \quad y_1 = -1 + \sqrt{1 + x_1}.$$

Hence, we obtain the solutions (setting $x_1 = t$)

(1) $x = t^2, \quad y = t^3\sqrt{1-t}$,

(2) $x = t^2, \quad y = t^3\sqrt{1+t}$.

Hence, we can deduce easily, that f is reducible, as we can find $g_1 \in \mathbb{C}\{x,y\}$ with $g_1(t^2,t^3\sqrt{1-t}) = 0$, and g_1 is a Weierstraß polynomial in y. Similarly we can find a g_2 with $g_2(t^2,t^3\sqrt{1+t}) = 0$. Note that the plane curve singularities $(V(g_1),0)$ and $(V(g_2),0)$ are such that their intersection is zero. Indeed $(t^2,t^3\sqrt{1-t}) = (t^2,t^3\sqrt{1-t})$ for t small if and only if $t = 0$, as a simple calculation shows. Thus $(g_1) \neq (g_2)$, and we get the factorization $f = g_1 \cdot g_2$, since $\deg_y(g_i) = 2$ for $i = 1,2$.

Proof of Theorem 5.1.14. As f is convenient, it is in particular regular in y, of order say b.

Step 1. First of all, we show that the definition of $f^{(1)}$ makes sense. We can write

$$f = \sum_{d \geq d_0} f_d,$$

where f_d is quasi-homogeneous of degree d with weights w_1 and w_2. From the definition of quasi-homogeneity it follows that

$$f(x_1^{w_1}, x_1^{w_2}(a_0 + y_1)) = \sum_{d \geq d_0} x_1^d f_d(1, a_0 + y_1),$$

so that indeed the definition of $f^{(1)}$ makes sense. Obviously, if we can find a parametrization of a branch of $f^{(1)} = 0$, we get a parametrization of a branch of $f = 0$.

Step 2. We now look at $f^{(1)}$. Its definition was

$$f^{(1)}(x_1, y_1) = \sum_{d \geq d_0} x_1^{d-d_0} f_d(1, a_0 + y_1). \tag{5.1}$$

We claim that $f^{(1)}$ is again regular in y_1. In view of formula (5.1) we have to show that $f_{d_0}(1, a_0 + y_1)$ is regular in y_1. We can factorize $f_{d_0}(1, y)$: (Recall that f is regular of order b.)

$$f_{d_0}(1, y) = c \cdot (y - a_0) \cdots (y - a_{b-1}),$$

for some constants $c, a_0, \ldots, a_{b-1} \in \mathbb{C}$. From this we see that

$$f_{d_0}(1, a_0 + y_1) = c \cdot y_1 (y_1 + a_0 - a_1) \cdots (y_1 + a_0 - a_{b-1}),$$

so $f_{d_0}(1, a_0 + y_1)$ is regular in y_1. Thus $f^{(1)}$ is also regular in y_1, of order at most b. There are two cases to consider.

(1) $f^{(1)}$ is also regular of order b. Then it follows that $a_0 = a_1 = \cdots = a_{b-1}$, so that $f_{d_0}(1, y) = c \cdot (y - a_0)^b$. As f_{d_0} is quasi-homogeneous and convenient it follows that $f_{d_0} = c \cdot (y - a_0 x^m)^b$. So $w_1 = 1$, and $m \cdot b = d_0$. Furthermore $w_2 = m$, and $x_1 = x$, $y = x_1^m(a_0 + y_1) = x^m(a_0 + y_1)$.

(2) $f^{(1)}$ is regular of order $b_1 < b$.

So if we iterate this procedure, that is, we define

$$x_{i+1} := (x_i)_1,\ y_{i+1} := (y_i)_1,\ f^{(i+1)} := (f^{(i)})^{(1)},$$

then the second case above can occur only finitely many times in the iteration. For the rest of the proof, we may therefore assume that we are always in the first case, that is, we may assume that $x = x_1 = x_2 = \cdots$. So in this case one tries to find a parametrization of type $x = t$, $y = \varphi(t)$, that is a smooth branch $y - y(x) = 0$ of $f = 0$.

Step 3. So assuming we are always in the first case, we want to show that the algorithm gives a solution $y = \overline{y}(x)$ of $f(x, y) = 0$. By Corollary 3.3.18, or alternatively, the third part of Corollary 5.1.8 it suffices to show that $y = \overline{y}(x)$ is a formal solution of $f(x, y) = 0$. So if a_i is a solution of $f^{(i)}(0, y_i) = 0$, the iteration gives

$$y = x^{m_0}(a_0 + y_1), \quad y_1 = x^{m_1}(a_1 + y_2), \text{ etc},$$

for certain $m_0, m_1, \ldots$. Plugging in we get

$$y = x^{m_0}\left(a_0 + x^{m_1}(a_1 + y_2)\right) = a_0 x^{m_0} + a_1 x^{m_0+m_1} + y_2 x^{m_0+m_1} = \cdots .$$

We claim that the formal power series

$$y(x) = \sum_{k=0}^{\infty} a_k x^{m_0+\ldots+m_k}$$

is a formal solution of $f(x,y) = 0$. Indeed, $f(x,y) = x^{d_0} f^{(1)}(x,y_1) = \cdots = x^{kd_0} f^{(k)}(x,y_k)$. Plugging in $y = y(x)$ we get $f(x,y(x)) = x^{kd_0} f^{(k)}(x,y_k(x))$ for some power series $y_k(x)$. Thus $f(x,y(x))$ is divisible by all powers of x, and hence is zero. □

The next thing we study in this section is the irreducibility of a power series f, in terms of the Newton polygon. Let us look at the simplest case. Suppose $h = f \cdot g$, f and g convenient, and suppose that the Newton polygon of both f and g has just one face. Let the slope of the face of $\Gamma(f)$ be given by the endpoints $(p_1,0)$ and $(0,q_1)$, and similarly the slope of the face of $\Gamma(g)$ by the endpoints $(p_2,0)$ and $(0,q_2)$. We assume that the slope of the face of $\Gamma(f)$ is less than or equal to the slope of the face of $\Gamma(g)$. Then $h = f \cdot g$ is convenient. The Newton polygon of h has two faces, resp. one face if the slopes are equal. The first face has endpoints $(0,q_1+q_2)$, and (p_1,q_2), the second has endpoints (p_1,q_2) and $(p_1+p_2,0)$.

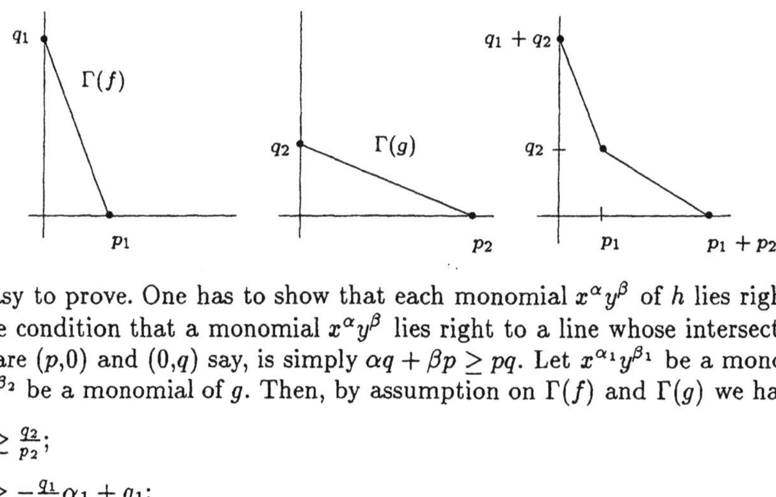

This is easy to prove. One has to show that each monomial $x^\alpha y^\beta$ of h lies right to both faces. The condition that a monomial $x^\alpha y^\beta$ lies right to a line whose intersections with the axes are $(p,0)$ and $(0,q)$ say, is simply $\alpha q + \beta p \geq pq$. Let $x^{\alpha_1} y^{\beta_1}$ be a monomial of f and $x^{\alpha_2} y^{\beta_2}$ be a monomial of g. Then, by assumption on $\Gamma(f)$ and $\Gamma(g)$ we have

(1) $\frac{q_1}{p_1} \geq \frac{q_2}{p_2}$;

(2) $\beta_1 \geq -\frac{q_1}{p_1}\alpha_1 + q_1$;

(3) $\beta_2 \geq -\frac{q_2}{p_2}\alpha_2 + q_2$.

This implies that $\beta_1+\beta_2 \geq -\frac{q_1}{p_1}(\alpha_1+\alpha_2)+q_1+q_2$ and $\beta_1+\beta_2 \geq -\frac{q_2}{p_2}(\alpha_1+\alpha_2)+\frac{q_2}{p_2}(p_1+p_2)$. The first inequality follows by adding (2) and (3) and using (1). The second inequality follows by adding $\frac{p_1}{p_2}q_2$ to (3) and using the inequality $\alpha_1 q_2 + \beta_1 p_2 \geq p_1 q_2$. Therefore, $x^{\alpha_1+\alpha_2} y^{\beta_1+\beta_2}$ lies right to both faces. Moreover if $x^{\alpha_1+\alpha_2} y^{\beta_1+\beta_2}$ lies on both faces, it follows that $\beta_1 = 0$, from which $\alpha_1 = p_1$, $\beta_2 = q_2$ and $\alpha_2 = 0$ easily follow. So the monomial $x^{p_1} y^{q_2}$ indeed occurs in the product $f \cdot g$.

More generally, we need the following lemma. The proof is just more technical, and therefore left to the reader.

Lemma 5.1.16. *Let $f_1,f_2 \in \mathbb{C}\{x,y\}$ be convenient. Assume $\Gamma(f_2)$ has one face $\overline{\Delta}_1$ defined by $(p_2,0)$ and $(0,q_1-q_2)$. Denote the faces of f_1 by $\overline{\Delta}_2,\dots,\overline{\Delta}_r$ and let $\overline{\Delta}_i$ be defined by $(\overline{p}_i,\overline{q}_i)$ and $(\overline{p}_{i+1},\overline{q}_{i+1})$ with $\overline{p}_2=0$, $\overline{q}_2=q_2$ and $\overline{q}_{r+1}=0$. Assume furthermore that the slope of $\overline{\Delta}_1$ is smaller than the slopes of the $\overline{\Delta}_i$, $i\geq 2$.*

Then the Newton polygon $\Gamma(f)$ of $f=f_1\cdot f_2$ has faces $\Delta_1,\dots,\Delta_r$, which are given by:

- *Δ_1 defined by $(0,q_1)$ and (p_2,q_2);*
- *Δ_i defined by $(\overline{p}_i+p_2,\overline{q}_i)$ and $(\overline{p}_{i+1}+p_2,\overline{q}_{i+1})$ for $i=2,\dots,r$.*

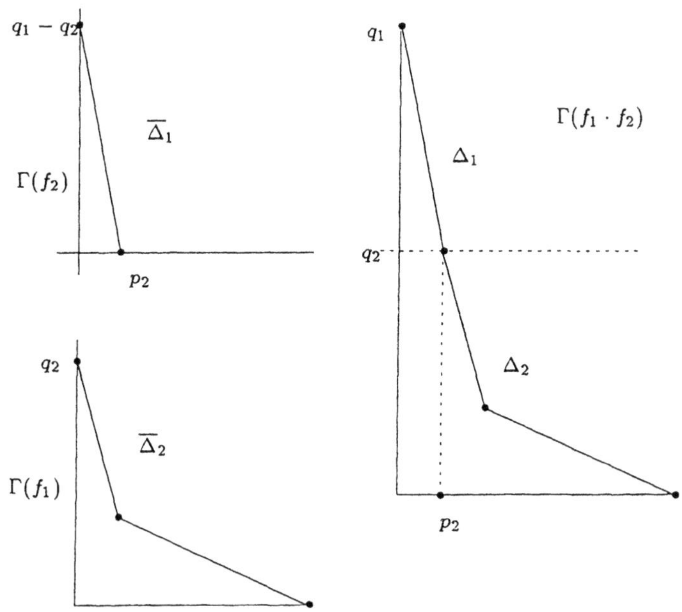

We now come to a converse of this lemma. It follows without much difficulty from the explicit construction of the Puiseux expansion, that the converse must hold. We want an explicit construction of the factorization, however.

Theorem 5.1.17. *Let $f\in\mathbb{C}\{x,y\}$ be convenient, $\Delta_1,\dots,\Delta_r$ be the faces of $\Gamma(f)$ and d_i the slope of Δ_i. Then $f=f_1\cdots f_r$, where f_i is convenient such that $\Gamma(f_i)$ has only one face of slope d_i, $i=1,\dots,r$.*

In the proof of this theorem, we use the following lemma.

Lemma 5.1.18. *Consider the polynomial ring $\mathbb{C}[x,y]$, with grading defined by* w-deg$(x)=w_1>0$ *and* w-deg$(y)=w_2>0$. *Let $g=x^b+\dots$ be quasi-homogeneous of weighted degree bw_1. Let $h\in\mathbb{C}[x,y]$ be quasi-homogeneous with weighted degree at least bw_1+aw_2 for some $a\in\mathbb{N}$. Then $h\in(y^a,g)$.*

Proof. We apply Division with Remainder, see Theorem 2.1.5, and get

$$h=q\cdot g+r,$$

where r is a polynomial in x of degree at most $b-1$. From the algorithm of division with remainder, it follows that also r is quasi-homogeneous, in fact w-deg(r) = w-deg(h) if $r \neq 0$. Take a monomial $x^i y^j$ in r. Then $i < b$. Hence

$$bw_1 + aw_2 \leq \text{w-deg}(r) = \text{w-deg}(x^i y^j) = iw_1 + jw_2 < bw_1 + jw_2.$$

It follows that $j > a$, and every monomial of r is divisible by y^a. Thus indeed $h \in (y^a, g)$. □

Proof of Theorem 5.1.17. Let $f = \sum a_{ij} x^i y^j$, assume $r > 1$ and let Δ_1 be the face with minimal slope. Because f is convenient, Δ_1 is the line segment from $(0,q_1)$ to (p_2,q_2):

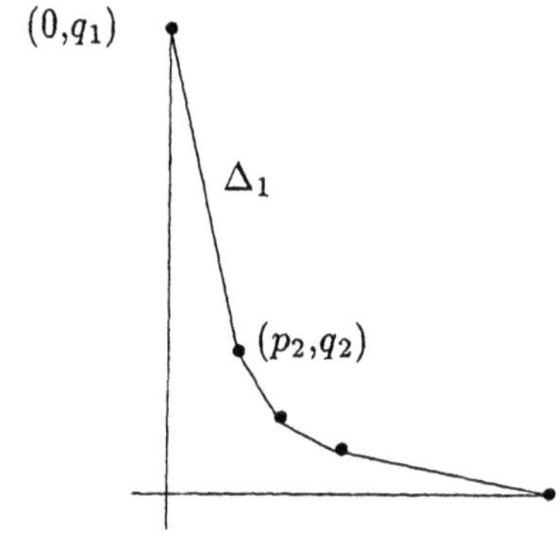

We define

$$g := \frac{1}{y^{q_2}} \cdot f_{\Delta_1}, \quad h := f - y^{q_2} g.$$

We obviously have $f = y^{q_2} g + h$.

As in Definition 5.1.9 we introduce the weighted degree on the monomials of $\mathbb{C}\{x,y\}$ corresponding to Δ_1, that is, we put w-deg$(x) = w_1 = (q_1 - q_2)/c$, and w-deg$(y) = w_2 = p_2/c$, where $c = \gcd(q_1 - q_2, p_2)$. With these weights, g is weighted homogeneous of degree $(q_1 - q_2)p_2/c$. We consider the elements of $\mathbb{C}\{x,y\}$ to be ordered by the corresponding weighted order function, w-ord$(f) = \min\{\text{w-deg}(x^i y^j) : (i,j) \in \text{supp}(f)\}$.

We will now inductively construct $f_1^{(n)}$, $f_2^{(n)}$ and $h^{(n)}$ such that

- $f = f_1^{(n)} \cdot f_2^{(n)} + h^{(n)}$,
- $f_1^{(n)} = y^{q_2} +$ terms of higher w-order not divisible by y^{q_2},
- $f_2^{(n)} = g\ +$ terms of higher w-order,
- $p_2 q_1/c <$ w-ord$(h) <$ w-ord$(h^{(2)}) < \cdots <$ w-ord$(h^{(n)})$.

The initialization of course is $f_1^{(1)} = y^{q_2}$, $f_2^{(1)} = g$ and $h^{(1)} = h$. Note that the weighted order of h is indeed bigger than $p_2 q_1/c$, as w-ord$(h) >$ w-ord$(y^{q_2} g) = p_2 q_1/c$. This is because all monomials of h lie right to the line through the face Δ_1. Now let k be the quasi-homogeneous part of $h^{(n)}$ of w–degree equal to w-ord$(h^{(n)})$. Because of w-deg$(k) >$ w-ord$(y^{q_2} g)$ we can find by Lemma 5.1.18 weighted homogeneous ξ, η such that $k = \xi y^{q_2} + \eta g$. Moreover one can arrange that no monomial of η is divisible by y^{q_2}. Now set

- $f_1^{(n+1)} = f_1^{(n)} + \eta$,
- $f_2^{(n+1)} = f_2^{(n)} + \xi$,
- $h^{(n+1)} = f - f_1^{(n+1)} \cdot f_2^{(n+1)}$.

It is not difficult to check, using $\text{w-ord}(f_1^{(n)} - y^{q_2}) > p_2 q_2/c$ and $\text{w-ord}(f_2^{(n)} - g) > p_2(q_1 - q_2)/c$, that

$$\text{w-ord}(h^{(n+1)}) > \text{w-ord}(h^{(n)}).$$

Continuing this way, we get formal power series f_1 and f_2 with $f = f_1 \cdot f_2$. In particular, by construction, f_1 is a Weierstraß polynomial with respect to y. Therefore, the convergence of f_1 and f_2 follows from Corollary 5.1.8.

The Newton polygon $\Gamma(f_2) = \Gamma(g)$ looks like,

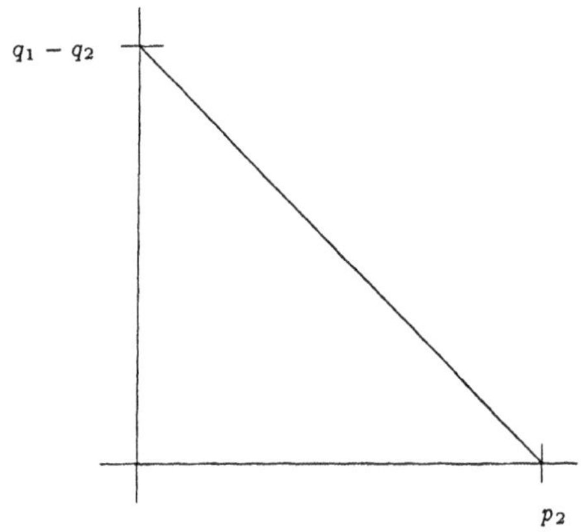

Using Lemma 5.1.16, the lemma follows by induction on r. □

Example 5.1.19. We consider $f = y^7 + x^3y^5 + x^2y^2 + x^5$.

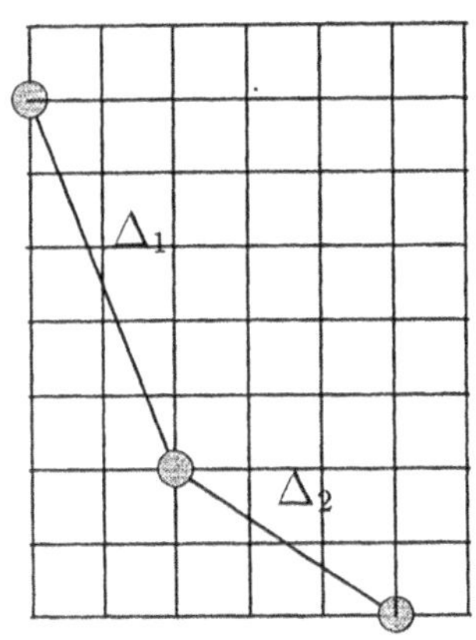

We have $w_1 = 5$, $w_2 = 2$, $g = f_2^{(1)} = \frac{1}{y^2} \cdot f_{\Delta_1} = y^5 + x^2$. So with $f_1^{(1)} = y^2$ and $h^{(1)} = x^5 + x^3y^5$ we get

$$f = f_1^{(1)} f_2^{(1)} + h^{(1)}.$$

Now $h^{(1)}$ is weighted homogeneous of degree 25. The monomials of $h^{(1)}$ all are divisible by x^3. In fact $h^{(1)} = (x^2+y^5)\cdot x^3 = g\cdot x^3$, so that

$$f = (y^2+x^3)(y^5+x^2).$$

We have the following generalization.

Theorem 5.1.20. *Let $f \in \mathbb{C}\{x,y\}$ be convenient, and suppose $\Gamma(f)$ has just one face Δ. Suppose that there exist $g = \sum_{\nu\geq d} g_\nu$, $h = \sum_{\nu\geq d} h_\nu$ and R such that*

(1) $f = g\cdot h + R$;

(2) the g_ν and h_ν are homogeneous of degree ν with respect to the grading introduced by Δ;

(3) $g_\nu = h_\nu$ for $\nu = d,\dots,p-1$, and $g_p \neq h_p$;

(4) (g_d,h_p) is (x,y)-primary;

(5) $\text{w-ord}(R) > d+p$.

Then f is reducible.

The proof runs as in the previous theorem, and is therefore left to the reader as an exercise. One needs the following lemma, the proof of which will be given in Chapter seven, see Lemma 7.3.3.

Lemma 5.1.21. *Let $h_1^{(0)}$ and $h_2^{(0)}$ be homogeneous and assume that $(h_1^{(0)},h_2^{(0)})$ is (x,y)-primary. Let $h_1 = h_1^{(0)} + h_1^{(1)}$, $h_2 = h_2^{(0)} + h_2^{(1)}$, with $\text{w-ord}(h_i^{(1)}) > \text{w-ord}(h_i^{(0)})$. Suppose that $iw_1 + jw_2 > \text{w-ord}(h_1\cdot h_2)$ for some $i,j\in\mathbb{N}$. Then $x^iy^j \in (h_1,h_2)$.*

Examples 5.1.22. (1) In the previous example we were "lucky" to get factors of f which are polynomials. In general of course, one cannot expect this. We now look at the standard example $f = y^2 - x^2(x+1)$. We get $f = (y-x)(y+x) + h^{(1)}$. Then $k = h^{(1)} = -x^3 \in (y-x,y+x)$. In fact $-x^3 = \frac{1}{2}x^2(y-x) - \frac{1}{2}x^2(y+x)$. Hence,

$$f = (y - x - \tfrac{1}{2}x^2)(y + x + \tfrac{1}{2}x^2) + \tfrac{1}{4}x^4.$$

Now $\frac{1}{4}x^4 = -\frac{1}{8}x^3(y-x) + \frac{1}{8}x^3(y+x)$. Hence

$$f = (y - x - \tfrac{1}{2}x^2 + \tfrac{1}{8}x^3)\cdot(y + x + \tfrac{1}{2}x^2 - \tfrac{1}{8}x^3) - \tfrac{1}{8}x^5 + \tfrac{1}{64}x^6,$$

etc. This process will not stop, because we get as the factorization $f(x,y) = (y - x\sqrt{x+1})(y + x\sqrt{x+1})$.

(2) We consider $f = y^4 - 2x^3y^2 + x^6 - x^7$. We have already seen in Example 5.1.15 that f is reducible. To apply the theorem, we note that

$$f = (y^2 - x^3 - x^2y + \tfrac{1}{2}x^4)(y^2 - x^3 + x^2y + \tfrac{1}{2}x^4) - \tfrac{1}{4}x^8.$$

so that the conditions of Theorem 5.1.20 are satisfied, and we get a second proof that f is reducible.

Exercises

5.1.23. Use the algorithm for the Puiseux expansion to find a solution of $y^2+2y-x=0$, that is, find the power series expansion of $y=-1+\sqrt{1+x}$.

5.1.24. Find a factorization of $f(x,y)=y^4-2x^3y^2+x^6-x^7$ in $\mathbb{C}\{x,y\}$, see Example 5.1.22.

5.1.25. Find the Puiseux expansions of the following (irreducible) power series.

(1) $f=y^4-2x^3y^2-4x^5y+x^6-x^7$.

(2) $f=x^{33}-3x^{22}y^2+6x^{21}y-x^{20}+3x^{11}y^4+2x^{10}y^3-y^6$.

5.1.26. Prove Lemma 5.1.16.

5.1.27. Let $\mathbb{C}((x))$ be the quotient field of $\mathbb{C}\{x\}$ and $\overline{\mathbb{C}((x))}$ its algebraic closure. Prove that $\overline{\mathbb{C}((x))}=\bigcup_{n\in\mathbb{N}}\mathbb{C}((x^{1/n}))$.

5.1.28. This is an exercise for the reader who knows about resultants. Let (X,o) and (Y,o) be plane curve singularities, which are given by Weierstraß polynomials f and g in y. Thus in particular $f,g\in\mathbb{C}\{x\}[y]$. Consider the resultant $R_{f,g}\in\mathbb{C}\{x\}$. Prove that

$$(X,o)\cdot(Y,o)=\operatorname{ord}_x(R_{f,g})$$

(Hint: Reduce to the case that f and g are irreducible. Let $x=t^n$, $y=\varphi(t)$ be a Puiseux expansion of (X,o), and factorize f in $\mathbb{C}\{t\}[y]$. Compute the resultant in $\mathbb{C}\{t\}$.)

5.1.29. This exercise uses the previous exercises. Let (X,o) be a plane curve singularity given by a Weierstraß polynomial f in y. Let $\Delta\in\mathbb{C}\{x\}$ be the discriminant of f. Prove

$$\operatorname{ord}_x(\Delta)=\mu(f)+(X,o)\cdot(L,o),$$

where (L,o) is the line given by $y=0$, and $\mu(f)$ is the Milnor number.
(Hint: Consider (Y,o) given by $\frac{\partial f}{\partial x}=0$. Compute the intersection number $(X,o)\cdot(Y,o)$ in two ways. One way is by using the previous exercise, the other way is by using Puiseux expansion for the branches of (Y,o): if $(\varphi_1(t),\varphi_2(t))$ is a parametrization of a branch then look at the vanishing order of $t\mapsto\frac{d}{dt}(f(\varphi_1(t),\varphi_2(t)))$, which differs by one from the intersection multiplicity.)

5.1.30. Let $(R,\mathfrak{m})$ be an analytic $\mathbb{C}$-algebra and M a finitely generated R-module. Assume that $\dim_{\mathbb{C}} M<\infty$. Prove that $\operatorname{Ann}_R(M)\neq\{0\}$.
(Hint: Choose $x\in\mathfrak{m}$, $x\neq 0$ and consider $M\supset xM\supset\ldots$.)

5.1.31. Consider a vector space V over k and a subspace $W\subset V$. Let $T:V\longrightarrow V$ be a linear map, such that $T(W)\subset W$ and $\operatorname{Ker}(T)\subset W$. Suppose that both V/W and $V/T(V)$ are finite-dimensional. Prove that $\dim_k(V/T(V))=\dim_k(W/T(W))$.

5.1.32. Prove Theorem 5.1.20.

5.2 Invariants

In this Section we introduce and study several invariants of irreducible (plane) curve singularities. An invariant is a map from the set of isomorphism classes of singularities to $\mathbb{Z}^k$ for some k, or any other kind of mathematical object, for example, the dimension of some vector space associated to the singularity, as we will soon see. Invariants play an important role in studying singularities.

Let $R=\mathcal{O}_{X,x}$ be the local ring of an irreducible curve singularity and $\mathfrak{m}$ its maximal ideal. By the results of the previous chapter, we may assume that $R\subset\widetilde{R}=\mathbb{C}\{t\}$, where $\widetilde{R}$ is the normalization of R. We introduce our first invariants.

Definition 5.2.1.

(1) We call $\delta(R) := \dim_{\mathbb{C}}(\widetilde{R}/R)$ the *delta-invariant*, or δ-invariant of R.

(2) We call $\Gamma(R) := \{\mathrm{ord}_t(a) \mid a \in R, a \neq 0\}$ the *semigroup of values* of R.

(3) We define the *conductor* $c(R)$ by $c(R) := \min\{\alpha \in \Gamma(R) : \alpha + \mathbb{N} \subset \Gamma(R)\}$. The conductor ideal is the ideal $(t^{c(R)})$ in $\mathbb{C}\{t\}$.

Lemma 5.2.2.

(1) $\mathrm{Ann}_R(\widetilde{R}/R) = t^{c(R)}\widetilde{R}$.

(2) $\delta(R) = \#(\mathbb{N} \setminus \Gamma(R))$.

Proof. (1) In the proof of 5.1.3 we saw already that for plane curve singularities the ideal $\mathrm{Ann}_R(\widetilde{R}/R)$ is an ideal in both R and $\widetilde{R}$. This in fact holds, with a similar proof, for general curve singularities. Hence $\mathrm{Ann}_R(\widetilde{R}/R) = t^c\widetilde{R}$ for some c. We have to prove that $c-1 \notin \Gamma(R)$. But $c-1 \in \Gamma(R)$ and $t^c\widetilde{R} \subset R$ implies that $t^{c-1}\widetilde{R} \subset R$ and, consequently, $t^{c-1} \in \mathrm{Ann}_R(\widetilde{R}/R)$, which contradicts $\mathrm{Ann}_R(\widetilde{R}/R) = t^c\widetilde{R}$.

(2) Let $\gamma = \#(\mathbb{N}\setminus\Gamma(R))$. Suppose that $a_1, \ldots, a_\delta$ is a basis of $\widetilde{R}/R$. If $\mathrm{ord}(a_i) = \mathrm{ord}(a_j) = k$ for some $i \neq j$, we can write $a_i = c_i t^k + \ldots$, $a_j = c_j t^k + \ldots$ with $c_i c_j \neq 0$. We may then replace a_j by $a_j - \frac{c_j}{c_i} a_i$, which has higher vanishing order. Using this operation several times we can get a basis $a_1, \ldots, a_\delta$ of $\widetilde{R}/R$ with $\mathrm{ord}(a_i) \neq \mathrm{ord}(a_j)$ for $i \neq j$. Similarly, one can show that one can take the a_i such that that $\mathrm{ord}(a_i) \notin \Gamma(R)$, so $\mathrm{ord}(a_1), \ldots, \mathrm{ord}(a_\delta)$ are different elements of $\mathbb{N} \setminus \Gamma(R)$. Hence $\delta \leq \gamma$.

On the other hand, if we have different elements $c_1 < \ldots < c_\gamma$ of $\mathbb{N} \setminus \Gamma(R)$, then $b_1 t^{c_1} + \ldots + b_\gamma t^{c_\gamma}$ with $b_i \in \mathbb{C}$ has vanishing order $\min\{c_i : b_i \neq 0\}$. So $b_1 t^{c_1} + \ldots + b_\gamma t^{c_\gamma} \in R$ only when $b_1 = \ldots = b_\gamma = 0$. This means that $t^{c_1}, \ldots, t^{c_\gamma}$ are linearly independent in $\widetilde{R}/R$, so that $\delta \geq \gamma$. □

Examples 5.2.3. (1) Consider the A_2-singularity given by $R = \mathbb{C}\{x,y\}/(y^2 - x^3)$. Then $\widetilde{R} = \mathbb{C}\{t\}$, and $R \subset \widetilde{R}$ is given by the map $x \mapsto t^2$ and $y \mapsto t^3$. Thus the semigroup $\Gamma(R)$ is generated by 2 and 3. Therefore $4 = 2+2$ and $5 = 3+2$ are in $\Gamma(R)$. It follows that $c(R) = 2$, and the unique element of $\mathbb{N} \setminus \Gamma(R)$ is one. Hence, $\delta(R) = 1$.

(2) Consider the E_8 singularity given by $R = \mathbb{C}\{x,y\}/(y^3 - x^5) \subset \mathbb{C}\{t\}$ via the map $x \mapsto t^5$ and $y \mapsto t^3$. Thus the elements 3,5,6,8,9,10,... are in $\Gamma(R)$. Thus we see that $c(R) = 8$. Moreover 1,2,4 and 7 are the elements of $\mathbb{N} \setminus \Gamma(R)$, so that $\delta(R) = 4$. Note that in this example, and also the previous one, there is a symmetry. One has $\alpha \in \Gamma(R) \iff c(R) - 1 - \alpha \notin \Gamma(R)$. For a discussion on this matter, see Exercise 5.2.25.

In these examples the conductor was always two times the δ-invariant. This is, indeed, a general fact for plane curve singularities.

Theorem 5.2.4 (Gorenstein).
$c(R) = 2\delta(R)$.

The theorem does not in general hold for non-plane curve singularities.

Example 5.2.5. Consider $\mathbb{C}\{x,y,z\} \xrightarrow{\varphi} \mathbb{C}\{t\}$ with maps $x \mapsto t^3$, $y \mapsto t^4$, and $z \mapsto t^5$. Then $R = \mathbb{C}\{x,y,z\}/(\operatorname{Ker}(\varphi))$ is the local ring of a curve singularity in three-space. Here $c(R) = 3$, and $\delta(R) = 2$. Thus $c(R) \neq 2\delta(R)$. This curve singularity will be further studied in Exercise 5.2.32.

For the proof of Theorem 5.2.4, we need some preparations.

Definition 5.2.6.

(1) Let R be as above, K the quotient field of R. A set $I \subset K$ is called a *fractional ideal* if

   (a) I is via the multiplication in K an R-module, that is $r \cdot a \in I$ for all $r \in R$ and $a \in I$ and, of course, $a + a' \in I$ for all $a,a' \in I$.

   (b) There exists a "universal denominator", that is, there exists a $d \in R$, $d \neq 0$, with $d \cdot I \subset R$.

(2) Let $I \subset K$ be a fractional ideal. Then $I^* := \operatorname{Hom}_R(I,R)$ is called the *dual* of I.

(3) The ring R, or the curve singularity, is called *Gorenstein* if for all fractional ideals I the canonical map

$$I \longrightarrow I^{**} = \operatorname{Hom}_R(\operatorname{Hom}_R(I,R),R)$$
$$a \mapsto \hat{a},$$

where $\hat{a}$ is defined by $\hat{a}(\varphi) := \varphi(a)$ for all $\varphi \in \operatorname{Hom}_R(I,R)$, is an isomorphism.

Remarks 5.2.7.

(1) For finite-dimensional vector spaces V over a field k, one knows that the canonical map $V \longrightarrow V^{**}$ is an isomorphism. The corresponding statement for modules over rings is false in general, even if one takes only fractional ideals. In Exercise 5.2.32 you will be asked to find a counterexample. It is easy to prove that the canonical map is injective, see Exercise 5.2.24, but the map need not be surjective.

(2) As a fractional ideal is as R-module *isomorphic* to an ideal in R, it follows that if one wants to show that R is Gorenstein one has to prove the duality only for ideals.

(3) Any ideal in R is a fractional ideal. The normalization $\widetilde{R}$ of R is a fractional ideal, which, in general, is not an ideal. Note that it is indeed a fractional ideal, because $t^{c(R)}\widetilde{R} \subset R$, so the second condition is satisfied.

(4) For fractional ideals I,J different from (0) one has the equality

$$\operatorname{Hom}_R(I,J) = (J : I)_K := \{x \in K : x \cdot I \subset J\}.$$

In the remaining part of this Chapter we will use $(J : I)$ for $(J : I)_K$ to simplify the notation.

Indeed, the map $(J : I) \longrightarrow \operatorname{Hom}_R(I,J)$, sending an element $a \in (J : I)$ to the multiplication a in the field K, is injective.

Conversely, a homomorphism $\alpha \in \mathrm{Hom}_R(I,J)$ can be extended to an R–linear map $\overline{\alpha}$ from K to K. Indeed, any element of K can be written as $\frac{r}{s}$ with $r \in I$ (multiply numerator and denominator with an element in I), so that we can define $\overline{\alpha}(\frac{r}{s}) := \frac{1}{s}\alpha(r)$. Then one shows that the element $\overline{\alpha}(1) \in (J : I)$, giving the inverse map.

In particular, for $(0) \neq J \subset I$ we always have $I^* \subset J^*$.

(5) Note that for any fractional ideal I and all $0 \neq x \in R$ we have $(xI)^* = \frac{1}{x}I^*$, or more generally $(I : xJ) = \frac{1}{x}(I : J)$ for a further fractional ideal J. Similarly $(xI : J) = x(I : J)$ holds.

If I is an ideal in R and $x \in I$ we also have $(x) : I$ is an ideal in R. Indeed, if $a \in (x) : I$, then in particular $a \cdot x \in (x)$, hence $a \in R$.

(6) $(\widetilde{R})^* = (R : \widetilde{R}) = \{c \in K : c\widetilde{R} \subset R\} = \{c \in R : c\widetilde{R} \subset R\} = \mathrm{Ann}_R(\widetilde{R}/R) = t^{c(R)}\widetilde{R}$.

Similarly, $\left(t^{c(R)}\widetilde{R}\right)^* = \widetilde{R}$. So the normalization and the conductor ideal are dual to each other.

(7) If I is a fractional ideal, then $I^*$is a fractional ideal.

Lemma 5.2.8. *Let R be as above. Then R is Gorenstein if and only if* $\dim_{\mathbb{C}}(\mathfrak{m}^*/R) = 1$.

Proof. First suppose that R is Gorenstein, but $\dim_{\mathbb{C}}(\mathfrak{m}^*/R) \neq 1$. The first case is $\dim_{\mathbb{C}}(\mathfrak{m}^*/R) = 0$, that is $\mathfrak{m}^* = R$. As R is Gorenstein, it follows that $\mathfrak{m} = \mathfrak{m}^{**} = R^* = R$, which is a contradiction. Thus $\dim_{\mathbb{C}}(\mathfrak{m}^*/R) \geq 2$, and hence we can find a $\mathbb{C}$–vector space I with $R \subsetneqq I \subsetneqq \mathfrak{m}^*$. We claim that I is a fractional ideal. The fact that I has a universal denominator follows immediately from the fact that $\mathfrak{m}^*$ has a universal denominator. So we only have to show that I is an R–module. Obviously $a + a' \in I$ if $a,a' \in I$. Let $f \in R$, and $a \in I$. Then we can write $f = c + r$, with $c \in \mathbb{C}$ and $r \in \mathfrak{m}$. Then $r \cdot a \in R$, as $a \in \mathfrak{m}^*$. Furthermore $ca \in I$ as I is a vector space. Hence $fa \in I$. By assumption R is Gorenstein, so it follows that $\mathfrak{m} \subsetneqq I^* \subsetneqq R$. This is a contradiction, as $\dim_{\mathbb{C}}(R/\mathfrak{m}) = 1$.

The converse is more difficult, and is proved in three steps. So for the rest of the proof assume $\dim_{\mathbb{C}}(\mathfrak{m}^*/R) = 1$.

Step 1. Let $(0) \neq J \subset I$ be nonzero fractional ideals. Then we claim that $\dim_{\mathbb{C}}(I/J) \geq \dim_{\mathbb{C}}(J^*/I^*)$.[2] If $J \subset L \subset I$ then we have

$$\begin{aligned}\dim_{\mathbb{C}}(I/J) &= \dim_{\mathbb{C}}(I/L) + \dim_{\mathbb{C}}(L/J), \quad \text{and}\\ \dim_{\mathbb{C}}(J^*/I^*) &= \dim_{\mathbb{C}}(J^*/L^*) + \dim_{\mathbb{C}}(L^*/I^*).\end{aligned}$$

If $\dim_{\mathbb{C}}(I/J) \geq 2$, we can always find a fractional ideal L with $J \subsetneqq L \subsetneqq I$, and $\mathfrak{m}L \subset J$. Indeed, take any vector space L such that J has codimension one in L. By Nakayama's Lemma, it follows that $\mathfrak{m}L \subset J$. It follows as above that L is a fractional ideal.

Therefore, we can use induction and it suffices to prove the claim for the case $\dim_{\mathbb{C}}(I/J) = 1$. Assuming this, we will construct an injective $\mathbb{C}$–linear map

$$\varphi : J^*/I^* \longrightarrow \mathfrak{m}^*/R,$$

[2] Because of R being one-dimensional we have $\dim_{\mathbb{C}}(I/J) < \infty$.

which, using the assumption, proves the claim. Take an element $a \in I$ with $(J,a) = I$. Then $\mathfrak{m}(I/J) \subsetneqq (I/J)$ by Nakayama's Lemma. This implies $\mathfrak{m}(I/J) = 0$ because $\dim_{\mathbb{C}}(I/J) = 1$. In particular $a \cdot \mathfrak{m} \subset J$. We now construct the map φ. Let $b \in J^*$. Then $ba \cdot \mathfrak{m} = b \cdot a\mathfrak{m} \subset b \cdot J \subset R$. Thus $ba \in \mathfrak{m}^*$. Thus we may define the map φ by sending b to ba. This gives a well-defined map, as $b \in I^*$ sends $a \in I$ to an element in R.

We now prove φ is injective. So suppose $ba \in R$, for some $b \in J^*$. But as $I = (J,a)$, the element b then sends all elements of I to R. Thus $b \in I^*$.

Step 2. Let I be a nonzero fractional ideal. We may assume that I is an ideal in R. Take an element $0 \neq x \in I$. We claim that we have the following equalities:

$$\dim_{\mathbb{C}}(I/(x)) = \dim_{\mathbb{C}}\Big(R/((x):I)\Big), \tag{5.2}$$

$$\dim_{\mathbb{C}}(R/I) = \dim_{\mathbb{C}}(I^*/R). \tag{5.3}$$

Indeed, applying Step 1 and using 5.2.7 (5) gives

$$\begin{aligned}\dim_{\mathbb{C}}(R/(x)) &= \dim_{\mathbb{C}}(R/I) + \dim_{\mathbb{C}}\big(I/(x)\big)\\ &\geq \dim_{\mathbb{C}}(I^*/R) + \dim_{\mathbb{C}}\big((x)^*/I^*\big)\\ &= \dim_{\mathbb{C}}\big((x)^*/R\big)\\ &= \dim_{\mathbb{C}}(\frac{1}{x}R/R)\\ &= \dim_{\mathbb{C}}\big(R/(x)\big).\end{aligned}$$

So all inequalities have to be equalities, in particular (5.3) follows. As we just proved, $\dim_{\mathbb{C}}\big(I/(x)\big) = \dim_{\mathbb{C}}\big((x)^*/I^*\big)$ implies

$$\dim_{\mathbb{C}}\big(I/(x)\big) = \dim_{\mathbb{C}}\big((x)^*/I^*\big) = \dim_{\mathbb{C}}\big(x(x)^*/xI^*\big) = \dim_{\mathbb{C}}\big(R/((x):I)\big),$$

which is formula (5.2).

Step 3. We now apply (5.2) and (5.3) to the ideal $\big((x):I\big)$ and obtain

$$\dim_{\mathbb{C}}(R/I) = \dim_{\mathbb{C}}(I^*/R) = \dim_{\mathbb{C}}\Big(((x):I)/(x)\Big) = \dim_{\mathbb{C}}\Big(R/((x):((x):I))\Big). \tag{5.4}$$

Now $\big((x):I\big) = xI^*$ and $\big((x):((x):I)\big) = \big((x):(xI^*)\big) = x(xI^*)^*$. Furthermore $(xI^*)^* = \frac{1}{x}I^{**}$, so that (5.4) now reads

$$\dim_{\mathbb{C}}(R/I) = \dim_{\mathbb{C}}(R/I^{**}).$$

As $I \subset I^{**}$ is obvious, the equality $I = I^{**}$ follows. This is what we had to prove. □

Corollary 5.2.9. *Let R be the local ring of a plane curve singularity. Then R is Gorenstein.*

Proof. By the previous theorem, it suffices to show that that $\dim_{\mathbb{C}}(\mathfrak{m}^*/R) = 1$. This will be done in two steps.

Step 1. Let $0 \neq \varphi \in \mathfrak{m} \subset R$. We consider the so-called *socle* of $R/(\varphi)$:

$$\operatorname{Soc}\big(R/(\varphi)\big) := \{g \in R/(\varphi) : \ \mathfrak{m} \cdot g = 0 \in R/(\varphi)\}.$$

We claim

$$\operatorname{Soc}(R/(\varphi)) \cong \mathfrak{m}^*/R.$$

Indeed, given $g \in \operatorname{Soc}(R/(\varphi))$ we can assign to this the element $\frac{g}{\varphi}$. This is an element of $\mathfrak{m}^*$, because for all $f \in \mathfrak{m}$ it follows that $g \cdot f \in (\varphi)$ and, hence, $\frac{g}{\varphi} \cdot f \in R$.

On the other hand, suppose we have an element $\frac{a}{b} \in \mathfrak{m}^*/R$. Then we claim that $\frac{a}{b}\varphi \in \operatorname{Soc}(R/(\varphi))$. Let $f \in \mathfrak{m}$ be arbitrary. Then $\frac{a}{b}\varphi \cdot f = \frac{a}{b} \cdot f \cdot \varphi \in (\varphi)$. As the maps are obviously inverse to each other, the claim follows.

Step 2. It remains to show that for a plane curve singularity R, the socle of $R/(\varphi)$ has dimension one for some $0 \neq \varphi \in \mathfrak{m}$. Now assume that R is defined by a Weierstraß polynomial f in y of degree m. It suffices to show that the socle of $\mathbb{C}\{x,y\}/(f,x)$ has vector-space dimension 1. But as $\mathbb{C}\{x,y\}/(f,x) \cong \mathbb{C}\{y\}/(y^m)$, we can easily calculate the socle. It is equal to $\mathbb{C} \cdot y^{m-1}$, hence has dimension 1. □

Theorem 5.2.4 follows from this corollary and the next theorem.

Theorem 5.2.10. *Let R be a Gorenstein curve singularity. Then $c(R) = 2\delta(R)$.*

Proof. By duality $R/I \cong I^*/R$ as $\mathbb{C}$-vector spaces for every nonzero ideal I. We apply this to the conductor ideal, which by Remark 5.2.7 (6) is $I = (\widetilde{R})^*$. Then $I^* = \widetilde{R}$, and we get $R/I \cong \widetilde{R}/R$. The conductor by definition is equal to $\dim_{\mathbb{C}}(\widetilde{R}/I)$, $\delta(R)$ by definition is equal to $\dim_{\mathbb{C}}(\widetilde{R}/R)$. Thus

$$c(R) = \dim_{\mathbb{C}}(\widetilde{R}/I) = \dim_{\mathbb{C}}(\widetilde{R}/R) + \dim_{\mathbb{C}}(R/I) = \delta(R) + \delta(R).$$

□

The next thing we do is to give a method for computing the semigroup of plane curve singularities, given the parametrization.

Definition 5.2.11 (Generators of the Semigroup). Let

$$\beta_0 = \min\{a \in \Gamma(R) : a \neq 0\},$$

$$\beta_\nu = \min\{a \in \Gamma(R) : \gcd(a,\beta_0,\ldots,\beta_{\nu-1}) < \gcd(\beta_0,\ldots,\beta_{\nu-1})\}, \text{ for } \nu \geq 1.$$

By this construction we obviously obtain only finitely many β_ν, say $\beta_0,\ldots,\beta_{g'}$. Let $G(R) := \{\beta_0,\ldots,\beta_{g'}\}$.

The following lemma is easy, and left to the reader as an exercise.

Lemma 5.2.12.

(1) The set $G(R)$ is the minimal set of generators of $\Gamma(R)$.

(2) $\gcd(\beta_0,\ldots,\beta_{g'}) = 1$.

(3) Let $\varphi : R \longrightarrow R'$ be an isomorphism. Then $G(R) = G(R')$.

Examples 5.2.13.

(1) Let the cusp singularity be given by the parametrization $x = t^2$, $y = t^3$. So $R = \mathbb{C}\{t^2,t^3\}$, and the semigroup $\Gamma(R)$ is generated by $\beta_0 = 2$ and $\beta_1 = 3$.

(2) We consider the parametrization $x = t^4$, $y = t^6 + t^7$. Then obviously $\beta_0 = 4$ and $\beta_1 = 6$ are in $\Gamma(R)$. But also $\beta_2 = 13 \in \Gamma(R)$, as it is the vanishing order of $y^2 - x^3 = 2t^{13} + t^{14}$. We will soon see that $\Gamma(R)$ is generated by 4, 6 and 13.

To answer the question how to compute generators of the semigroup of a plane curve singularity, we introduce another set of invariants, the so-called characteristic exponents. Note that by coordinate changes in t and interchanging x and y we we can always arrange that our curve has a parametrization

$$x = t^n; \quad y = \sum_{i \geq n} a_i t^i.$$

Definition 5.2.14 (Characteristic Exponents).
Let $R = \mathbb{C}\{t^n, y(t)\}$, be a plane curve singularity. Write $y(t) = \sum_{i \geq n} a_i t^i$. We define

$$k_0 = n,$$
$$k_\nu = \min\{i : \ a_i \neq 0, \ \gcd(i, k_0, \ldots, k_{\nu-1}) < \gcd(k_0, \ldots, k_{\nu-1})\}, \quad \text{for } \nu \geq 1.$$

Again we obviously obtain only finitely many k_ν, say $k_0, \ldots, k_g$. Note that $k_0 < k_1 < \ldots < k_g$. Moreover $\gcd(k_0, \ldots, k_g) = 1$ follows from $\dim_{\mathbb{C}}(\mathbb{C}\{t\}/R) < \infty$, see Exercise 5.2.27. The set $E(R) := \{k_0, k_1, \ldots, k_g\}$ is called the set of *characteristic exponents* of R.

Examples 5.2.15.

(1) A curve with just one characteristic exponent $k_0 = 1$ is smooth.

(2) Consider the curve with parametrization $x = t^4$, $y = t^6 + t^7$. The characteristic exponents are $k_0 = 4$, $k_1 = 6$ and $k_2 = 7$.

Note that we defined the characteristic exponents by using a parametrization of R, so a priori they might depend on the parametrization. But indeed, they are invariants of R, because the next theorem says that they can be calculated from the minimal set of generators $\beta_0, \ldots, \beta_{g'}$ of $\Gamma(R)$.

Theorem 5.2.16. *Let $x(t) = t^n$ and $y(t) = \sum_{i \geq n} a_i t^i$ such that $R = \mathbb{C}\{x(t), y(t)\}$ is the local ring of a plane curve singularity. Then the minimal set of generators of $\Gamma(R)$, $G(R) = \{\beta_0, \ldots, \beta_{g'}\}$, and the set of characteristic exponents, $E(R) = \{k_0, \ldots, k_g\}$, satisfy the relations:*

(1) $$g = g',$$

(2) $$k_i = \beta_i - \sum_{j=1}^{i-1} \left(\frac{\gcd(\beta_0, \ldots, \beta_{j-1})}{\gcd(\beta_0, \ldots, \beta_j)} - 1 \right) \beta_j, \quad i = 1, \ldots, g,$$

(3) $$\beta_i = k_i + \sum_{j=1}^{i-1} \left(\frac{\gcd(k_0, \ldots, k_{j-1})}{\gcd(k_0, \ldots, k_{i-1})} - \frac{\gcd(k_0, \ldots, k_j)}{\gcd(k_0, \ldots, k_{i-1})} \right) k_j, \quad i = 1, \ldots, g.$$

In particular we always have $k_0 = \beta_0$ and $k_1 = \beta_1$.

It is left as Exercise 5.2.27 to show that the formulas in (2) and (3) are equivalent.

Example 5.2.17.

(1) Suppose $\beta_0 = 4$, $\beta_1 = 6$, and $\beta_2 = 13$. Then $k_0 = 4$, $k_1 = 6$ and

$$\begin{aligned} k_2 &= \beta_2 - \left(\frac{\gcd(\beta_0)}{\gcd(\beta_0,\beta_1)} - 1\right)\beta_1 \\ &= 13 - \left(\frac{4}{2} - 1\right)\cdot 6 = 7. \end{aligned}$$

(2) Suppose on the other hand $k_0 = 4$, $k_1 = 6$, and $k_2 = 7$. Then $\beta_0 = 4$, $\beta_1 = 6$, and

$$\begin{aligned} \beta_2 &= k_2 + \left(\frac{\gcd(k_0)}{\gcd(k_0,k_1)} - \frac{\gcd(k_0,k_1)}{\gcd(k_0,k_1)}\right) k_1 \\ &= 7 + \left(\frac{4}{2} - 1\right)\cdot 6 = 13. \end{aligned}$$

In order to prove Theorem 5.2.16 the following lemma is of crucial importance.

Lemma 5.2.18. *Let $R = \mathbb{C}\{x,y\}/(f) = \mathbb{C}\{t^n,y(t)\}$, with $y(t) = \sum_{i\geq m} a_i t^i$, $a_m \neq 0$, but not necessarily $n \leq m$. In case $n > m$ the k_i are also defined by the formulae in Definition 5.2.14. Suppose f is a Weierstraß polynomial in y. Then*

(1) $\operatorname{ord} \frac{\partial f}{\partial y}(t^n,y(t)) = \sum_{j=1}^{g} (\gcd(k_0,\ldots,k_{j-1}) - \gcd(k_0,\ldots,k_j))k_j.$

(2) $\operatorname{ord} \frac{\partial^j f}{\partial y^j}(t^n,y(t)) \geq \operatorname{ord} \frac{\partial f}{\partial y}(t^n,y(t)) - k_1 - \ldots - k_{j-1}.$

Proof. We consider $\mathbb{C}\{x,y\} \longrightarrow \mathbb{C}\{t,y\}$ which sends x to t^n, and y to y. The image of f in $\mathbb{C}\{t,y\}$ has the following factorization,

$$f = \prod_{i=1}^{n} (y - y(\varepsilon^i t)),$$

where ε is a primitive n–th root of unity, see Theorem 5.1.7. We get

$$\frac{\partial f}{\partial y} = \sum_{j=1}^{n} \prod_{i\neq j} (y - y(\varepsilon^i t)).$$

Its image in R is obtained by replacing y by $y(t)$, so

$$\frac{\partial f}{\partial y}(t^n,y(t)) = \prod_{i=1}^{n-1} (y(t) - y(\varepsilon^i t)).$$

It thus suffices to calculate the vanishing order for the different $y(t) - y(\varepsilon^i t)$, their sum being the vanishing order of $\frac{\partial f}{\partial y}(t^n,y(t))$.

Now $k_0 = n$, and we may expand $y(t) - y(\varepsilon^i t)$ in a power series:

$$y(t) - y(\varepsilon^i t) = \sum_{\nu\geq m} a_\nu (1 - \varepsilon^{i\nu}) t^\nu. \tag{5.5}$$

Define for $j = 1, \dots, g$

$$N_j = \left\{ i : \tfrac{k_0}{\gcd(k_0,\dots,k_{j-1})} \mid i, \tfrac{k_0}{\gcd(k_0,\dots,k_j)} \nmid i \quad \text{for } 1 \le i \le k_0 - 1 \right\}.$$

As $\gcd(k_0, \dots, k_g) = 1$, it follows that

$$\bigcup_{j=1}^{g} N_j = \{1, \dots, k_0 - 1\}.$$

We claim that for all $i \in N_j$ we have

$$\operatorname{ord}\big(y(t) - y(\varepsilon^i t)\big) = k_j. \tag{5.6}$$

The claim is proved in two steps.

Step 1. Suppose $\operatorname{ord}\big(y(t) - y(\varepsilon^i t)\big) = \nu$. We want to show that $\nu \ge k_j$. Suppose the converse. From $\nu < k_j$ and the definition of k_j it follows that $\gcd(k_0, \dots, k_{j-1}) \mid \nu$. As $i \in N_j$ this would imply $n = k_0 \mid i\nu$. This is not the case, as otherwise $\operatorname{ord}\big(y(t) - y(\varepsilon^i t)\big) = \nu$ would imply $1 - \varepsilon^{i\nu} \ne 0$, a contradiction. Thus $\nu \ge k_j$.

Step 2. By definition $a_{k_j} \ne 0$. So in order to prove formula (5.6) it suffices to show that $\varepsilon^{ik_j} \ne 1$, that is $n = k_0 \nmid i \cdot k_j$. We write $q = \frac{\gcd(k_0,\dots,k_{j-1})}{\gcd(k_0,\dots,k_j)}$ for short. Then

$$\gcd\left(q, \tfrac{k_j}{\gcd(k_0,\dots,k_j)}\right) = \gcd\left(\tfrac{k_0}{\gcd(k_0,\dots,k_j)}, \dots, \tfrac{k_j}{\gcd(k_0,\dots,k_j)}\right) = 1. \tag{5.7}$$

By assumption $i \in N_j$, so that $\frac{qk_0}{\gcd(k_0,\dots,k_{j-1})} = \frac{k_0}{\gcd(k_0,\dots,k_j)} \nmid i$. Therefore

$$q \nmid \tfrac{i \gcd(k_0,\dots,k_{j-1})}{k_0}.$$

As $i \in N_j$ we have that indeed the right hand side is an integer. From (5.7) we deduce that

$$q \nmid \tfrac{i \gcd(k_0,\dots,k_{j-1}) k_j}{k_0 \gcd(k_0,\dots,k_j)} = \frac{iqk_j}{k_0}.$$

So $k_0 \nmid ik_j$ follows. This proves formula (5.6).

It remains to show that $\#N_j = \gcd(k_0, \dots, k_{j-1}) - \gcd(k_0, \dots, k_j)$. This is not difficult. Define

$$M_j = \left\{ i : \quad \tfrac{k_0}{\gcd(k_0,\dots,k_j)} \mid i, \quad 1 \le i \le k_0 - 1 \right\}.$$

It is an easy fact from number theory that $\#M_j = \gcd(k_0, \dots, k_j) - 1$. Furthermore, $N_j = M_{j-1} \setminus M_j$. Thus the desired formula for $\#N_j$ follows.

The proof of the second part is easier and is left to the reader as Exercise 5.2.28. □

Proof of Theorem 5.2.16. We define $\beta_0, \dots, \beta_g$ by the formulas in Theorem 5.2.16(3). We have to show that $\{\beta_0, \dots, \beta_g\}$ is the minimal set of generators of $\Gamma(R)$. This we will do in several steps.

Step 1. By Corollary 5.1.6 there exists a Weierstraß polynomial $f \in \mathbb{C}\{x\}[y]$ such that $f(t^n, y(t)) = 0$. Moreover $\deg_y(f) = n = k_0$. For an arbitrary $h \in \mathbb{C}\{x,y\}$ we can by the Weierstraß Division Theorem write

$$h = qf + r,$$

where $r \in \mathbb{C}\{x\}[y]$ a polynomial of degree $\deg_y(r) < n = k_0$. Now $f = 0$ in R, so if we want to compute the vanishing order of h, we may assume it is a polynomial in y of degree less than n.

Step 2. For $d = 1, \dots, g$ we will construct a function $\overline{f} \in \mathbb{C}\{x\}[y]$ with $\deg_y(\overline{f}) = \frac{n}{K_{d-1}}$ and $\operatorname{ord}(\overline{f}) = \beta_d$. Here we write $K_j := \gcd(k_0, \dots, k_j)$. Note that $K_0 = n$ and $K_g = 1$. We define a new curve $\overline{R}$ by the following parametrization:

$$\overline{x}(t) = t^{\frac{n}{K_{d-1}}}; \quad \overline{y}(t) = \sum_{i<k_d} a_i t^{\frac{i}{K_{d-1}}}.$$

This definition makes sense, as by definition of k_d we have $K_{d-1} \mid i$ for all $i < k_d$ with $a_i \neq 0$. By 5.1.6 there exists an $\overline{f} \in \mathbb{C}\{x\}[y]$ with $\deg_y(\overline{f}) = \frac{n}{K_{d-1}}$ such that $\overline{f}(\overline{x}(t), \overline{y}(t)) = 0$. It remains to show that $\operatorname{ord}(\overline{f}) = \beta_d$. Now, by Taylor's formula

$$\begin{aligned} \overline{f}(x(t), y(t)) =& \overline{f}(t^n, \overline{y}(t^{K_{d-1}}) + \sum_{i\geq k_d} a_i t^i) \\ =& \overline{f}(t^n, \overline{y}(t^{K_{d-1}})) \\ &+ \frac{\partial \overline{f}}{\partial y}(t^n, \overline{y}(t^{K_{d-1}})) \cdot \sum_{i\geq k_d} a_i t^i \\ &+ \sum_{j\geq 2} \frac{1}{j!}\frac{\partial^j \overline{f}}{\partial y^j}(t^n, \overline{y}(t^{K_{d-1}})) \cdot \left(\sum_{i\geq k_d} a_i t^i\right)^j . \end{aligned}$$

The first term is zero by definition of $\overline{f}$. We calculate the vanishing order of the second term. The curve $\overline{R}$ by definition has characteristic exponents

$$\frac{k_0}{K_{d-1}}, \dots, \frac{k_{d-1}}{K_{d-1}}$$

so that by the first part of Lemma 5.2.18 the second term has vanishing order

$$k_d + K_{d-1} \cdot \sum_{j=1}^{d-1} \left(\frac{K_{j-1}}{K_{d-1}^2} - \frac{K_j}{K_{d-1}^2} \right) \cdot k_j,$$

which is equal to β_d.

It remains to show that the third term has vanishing order bigger than β_d. But this follows from the second part of Lemma 5.2.18, as we have $k_0 < k_1 < \cdots < k_g$, and we have $d \geq 2$.

Step 3. We will show by induction on $d \leq g$ that

$$\left\{ \operatorname{ord}_t h : \ h \in \mathbb{C}\{x\}[y], \ \deg_y(h) < \tfrac{n}{K_d} \right\} = \sum_{\nu=0}^{d} \beta_\nu \mathbb{N}. \tag{5.8}$$

By using Step 1, we see that the case $d = g$ gives that $\beta_0, \dots, \beta_g$ generate $\Gamma(R)$. The case $d = 0$ is easy, as then h with $\deg_y(h) < \frac{n}{K_d} = 1$ means that $h \in \mathbb{C}\{x\} = \mathbb{C}\{t^n\}$.

By induction we now may assume that for all $h \in \mathbb{C}\{x\}[y]$ with $\deg_y(h) < \frac{n}{K_{d-1}}$

$$\operatorname{ord}(h(t^n, y(t))) \in \sum_{j=0}^{d-1} \beta_j \mathbb{N}.$$

Now suppose $h \in \mathbb{C}\{x\}[y]$, $\deg_y(h) < \frac{n}{K_d}$. By successive application of the Division with Remainder Theorem for polynomials we can write

$$h = \sum_{i=0}^{s} h_i(x,y)\overline{f}^i$$

with $h_i \in \mathbb{C}\{x\}[y]$, $\deg_y(h_i) = i < \frac{n}{K_{d-1}}$ and $s = \dfrac{K_{d-1}}{K_d} - 1$. We claim that

(5.9) $\quad \operatorname{ord}(h) = \min\{\operatorname{ord}(h_i) + i\beta_d : i = 1, \dots s\} = \min\{\operatorname{ord}(h_i \overline{f}^i) : i = 1, \dots, s\}.$

(Here we take the convention that the order of the zero function is equal to infinity.) This suffices to show the induction step. To show (5.9) we claim that no canceling occurs, that is, for $i \neq j$

$$\operatorname{ord}(h_i \overline{f}^i) \neq \operatorname{ord}(h_j \overline{f}^j).$$

Indeed, suppose the converse. Without loss of generality we suppose $i > j$. Then we have $(i-j)\beta_d = \operatorname{ord}(h_i) - \operatorname{ord}(h_j)$. By induction it follows that $(i-j)\beta_d \in \sum_{j=0}^{d-1} \beta_j \mathbb{N}$. By Exercise 5.2.27, $\gcd(\beta_0, \dots, \beta_{d-1}) = K_{d-1}$. It follows that $K_{d-1} \mid (i-j)\beta_d$. From the definition of β_d it then follows that $K_{d-1} \mid (i-j)k_d$. Therefore $\frac{K_{d-1}}{K_d} \mid i-j$, because $K_d = \gcd(K_{d-1}, k_d)$. But as $i - j < \frac{K_{d-1}}{K_d} - 1$, this implies $i = j$, a contradiction.

Step 4. It remains to prove that $\{\beta_0, \dots, \beta_g\}$ is a *minimal* set of generators. This is elementary number theory, and is left as Exercise 5.2.30. □

We get the following corollary.

Corollary 5.2.19. *Let R be a plane curve singularity. We may assume that R is given by $R = \mathbb{C}\{t^n, y(t)\}$, with $y(t) = \sum_{i>n} a_i t^i$. Then the characteristic exponents introduced in 5.2.14 are independent of the parametrization, and therefore invariants of R.*

In the next section, parametrizations of plane curve singularities

$$x = t^n; \quad y = \sum_{i \geq m} a_i t^i, \quad a_m \neq 0,$$

with not necessarily $n \leq m$ will appear naturally. It will of importance to be able to calculate the characteristic exponents in this case.

Proposition 5.2.20. *Let the plane curve singularity R be given by the parametrization*

$$x = t^n; \quad y = \sum_{i \geq m} a_i t^i, \quad a_m \neq 0,$$

and suppose $n > m$. Let

$$k_0 = m,$$

$$k_1 = \begin{cases} \min\{i :\ a_i \neq 0,\ \gcd(i,k_0) < k_0\} + n - m & \text{if } m \mid n, \\ n & \text{if } m \nmid n, \end{cases}$$

$$k_\nu = \min\{i :\ a_i \neq 0,\ \gcd(i,k_0,\dots,k_{\nu-1}) < \gcd(k_0,\dots,k_{\nu-1})\} + n - m \quad \text{for } \nu \geq 2.$$

Then $k_0, \dots ,k_g$ are the characteristic exponents of R.

Proof. As the formula's (2) and (3) in Theorem 5.2.16 are equivalent it suffices to prove

$$\beta_i = k_i + \sum_{j=1}^{i-1} \left(\frac{\gcd(k_0,\dots,k_{j-1})}{\gcd(k_0,\dots,k_{i-1})} - \frac{\gcd(k_0,\dots,k_j)}{\gcd(k_0,\dots,k_{i-1})} \right) k_j, \tag{5.10}$$

as then we can calculate the k_i from the β_i in the same way as the characteristic exponents are calculated from the β_i. So we again have to show that if we *define* β_i by formula (5.10), that then $\{\beta_1, \dots ,\beta_g\}$ is the minimal set of generators of $\Gamma(R)$. The proof of this statement runs largely as the proof of Theorem 5.2.16, but there are some subtle changes.

Step 1. First of all, it is clear that $\beta_0 = k_0 = m$.

Step 2. We consider the case that $g = 1$, that is we have $\gcd(k_0,k_1) = 1$. We claim that we can arrange that $R = \mathbb{C}\{t^{k_0},t^{k_1} + \dots\}$ using suitable coordinate changes in x, y and t.

The claim is obvious for the case $m \nmid n$, as this situation can be arranged by interchanging x and y and replacing t by ut for some unit u. Now consider the case $m \mid n$, say $n = r \cdot m$. We first show that for some unit u in $\mathbb{C}\{y\}$ we have that $u \cdot y = t^{k_0} + a_{\overline{k}_1} t^{\overline{k}_1} + \dots$ where $\overline{k}_1 = \min\{i :\ a_i \neq 0,\ \gcd(i,k_0) < k_0\}$. Note that as $k_0 = m \mid n$, we have that $\gcd(\overline{k}_1,k_0) = \gcd(k_1,k_0) = 1$. If $u = 1$ does not work, then one has

$$y = t^{k_0} + a_b t^b + \dots + a_{\overline{k}_1} t^{\overline{k}_1} + \dots ,$$

and $a_b \neq 0$. It follows from the definition that $m = k_0 \mid b$, say $s \cdot k_0 = b$ for some $s > 1$. Then $a_b y^s = a_b t^b + \sum_{j>b} \widetilde{a}_j t^j$ for some $\widetilde{a}_j$. Note that if $\widetilde{a}_j \neq 0$, then $k_0 \mid j$. So

$$y(1 - a_b y^{s-1}) = t^{k_0} + \overline{a}_{b'} t^{b'} + \dots ,$$

and if $\overline{a}_{b'} \neq 0$ and $b' < \overline{k}_1$ we must have $k_0 \mid b'$. Going on like this we get the desired unit in $\mathbb{C}\{y\}$.

We now may assume that $u = 1$. As $m \mid n$, we can write $m = r \cdot n$ for some $r \in \mathbb{N}$. Then

$$\begin{aligned} x - y^r &= t^n - \left(t^{k_0} + a_{\overline{k}_1} t^{\overline{k}_1} + \dots\right)^r \\ &= -t^{k_0(r-1)} \cdot a_{\overline{k}_1} t^{\overline{k}_1} + \dots . \end{aligned}$$

So we do the coordinate change which replaces x by $x - y^r$. Now interchange x and y. If we moreover replace t by $v \cdot t$, for some unit $v \in \mathbb{C}\{t\}$ we get $x = t^{k_0}$, $y = \sum_{j \geq k_1} a_j t^j$ with $a_{k_1} \neq 0$. Moreover, $\gcd(k_0,k_1) = 1$. We now know from m 5.2.16 that there are two characteristic exponents, namely k_0 and k_1. This concludes the proof for the case $g = 1$.

Step 3. Now consider the case $g \geq 2$. Define $K_j = \gcd(k_0, \dots, k_j)$. Note that $K_g = 1$. We first construct a function $\overline{f}$ of degree $\frac{n}{K_{d-1}}$ which has vanishing order β_d. This is done by considering the curve $\overline{R}$ with parametrization

$$\overline{x} = t^{\frac{n}{K_{d-1}}}; \quad \overline{y} = \sum_{m \leq i < \overline{k}_d} a_i t^{\frac{i}{K_{d-1}}}.$$

Here $\overline{k}_d = k_d - n + m$. Then there exists an $\overline{f} \in \mathbb{C}\{x\}[y]$ with $\overline{f}(\overline{x},\overline{y}) = 0$. Moreover, the degree of $\overline{f}$ in y is $\frac{n}{K_{d-1}}$. It remains to show that $\operatorname{ord}(\overline{f}) = \beta_d$. This is done as in the proof of Theorem, 5.2.16, but now one has to use Exercise 5.2.31 instead of Lemma 5.2.18.

Step 4. Similarly, as in the proof of Theorem 5.2.16 we prove by induction on $1 \leq d \leq g$ that

$$\left\{\operatorname{ord}_t h : \ h \in \mathbb{C}\{x\}[y], \quad \deg_y(h) < \tfrac{n}{K_d}\right\} = \sum_{\nu=0}^{d} \beta_\nu \mathbb{N}. \tag{5.11}$$

The case $d = 1$ is reduced to the case $g = 1$ by considering the curve with parametrization

$$x = t^{\frac{n}{K_1}}; \quad y = \sum_{m \leq i < \overline{k}_2} a_i t^{\frac{i}{K_1}}.$$

For this curve, we have $g = 1$, and the case $d = 1$ can be reduced to it. Details are left to the reader.

The induction Step now runs exactly as in the proof of Theorem 5.2.16. Details are again left to the reader. □

Theorem 5.2.21 (Inversion Theorem). *Let $x = t^n$ and $y(t) = \sum_{i \geq m} a_i t^i$ with $a_m \neq 0$ be a parametrization of a plane curve singularity, with $n > m$. Let $k_0, \dots, k_g$ be the characteristic exponents of R, and consider the following invariants of this parametrization*

$$k'_0 = n,$$
$$k'_\nu = \min\{i : \ a_i \neq 0, \ \gcd(i,k'_0,\dots,k'_{\nu-1}) < \gcd(k'_0,\dots,k'_{\nu-1})\}, \ \text{for } 1 \leq \nu \leq g'.$$

Then we have the following statements:

(1) Suppose $m \nmid n$. Then $g = g'$, $k_0 = m$, $k_1 = n$, and $k_\nu = k'_\nu + n - m$ for $\nu = 2, \dots, g$.

(2) Suppose $m \mid n$. Then $g' = g + 1$, $k_0 = m$, and $k_\nu = k'_{\nu+1} + n - m$ for $\nu = 1, \dots, g$.

Proof. This is just a reformulation of Proposition 5.2.20. Indeed, if $m \nmid n$, this proposition says that $k_0 = m$ and $k_1 = n$. Moreover, because $\gcd(k_0,k_1) = \gcd(m,n) \mid m - n$, it easily follows that $\gcd(k_0, \dots, k_{\nu-1}) = \gcd(k'_0, \dots k'_{\nu-1})$. So $k_\nu = k'_\nu + n - m$ follows from 5.2.20.

If $m \mid n$, then again $k_0 = m$. Note that $k'_1 = m$. So we see that $k'_2 = \min\{i : a_i \neq 0, \ \gcd(i,n,m) = \gcd(i,m) = \gcd(i,k_0) < k_0\}$. Hence $k_1 = k'_2 + n - m$ follows from Proposition 5.2.20. The rest of the proof goes similar to the case $m \nmid n$. □

Very closely related to the characteristic exponents are the Puiseux pairs.

Definition 5.2.22 (Puiseux pairs). Let R be a plane curve singularity, and $E(R) = \{k_0,\dots,k_g\}$ be the characteristic exponents of R. Define $P(R) = \{(n_1,m_1),\dots,(n_g,m_g)\}$, the set of *Puiseux pairs* as follows: Let (n_1,m_1) be defined by $\frac{k_1}{k_0} = \frac{n_1}{m_1}$ in lowest terms, and define (n_ν,m_ν) for $\nu = 2,\dots,g$ by $\frac{n_\nu}{m_\nu} = \frac{k_\nu \cdot m_1 \cdots m_{\nu-1}}{k_0}$ in lowest terms.

To digest this definition a little bit, let us write the Puiseux expansion as

$$y = \sum_{\overline{k}\in\mathbb{Q}} a_{\overline{k}} x^{\overline{k}}.$$

If the curve is not smooth, there is a smallest $\overline{k}_1$ with $a_{\overline{k}_1} \neq 0$ and $\overline{k}_1$ not an integer. We write $\overline{k}_1 = \frac{n_1}{m_1}$ in lowest terms, and then (m_1,n_1) is the first Puiseux pair of the curve. To find the second Puiseux pair, we look for the smallest $\overline{k}_2 > \overline{k}_1$ which cannot be written in the form $\frac{a}{m_1}$. We write $\overline{k}_2 = \frac{n_2}{m_1 \cdot m_2}$ with $\gcd(m_2,n_2) = 1$ and call (m_2,n_2) the second Puiseux pair. To find the third, we look for the smallest $\overline{k}_3 > \overline{k}_2$ which cannot be written in the form $\frac{a}{m_1 \cdot m_2}$. Write $\overline{k}_3 = \frac{n_3}{m_1 \cdot m_2 \cdot m_3}$ with $\gcd(m_3,n_3) = 1$. Then (m_3,n_3) the third Puiseux pair, etc.

Remark 5.2.23. $k_0 = m_1 \cdots m_g$ and $k_i = k_0 \overline{k}_i$, $i = 1,\dots,g$. Therefore, $P(R)$ determines also $E(R)$.

Exercises

5.2.24. Prove that for a fractional ideal I the canonical map $I \longrightarrow I^{**}$ is injective.

5.2.25. Let R be the local ring of an irreducible curve singularity, and $\Gamma(R)$ be its semigroup.

(1) Let $\alpha \in \Gamma(R)$. Prove that $c(R) - 1 - \alpha \notin \Gamma(R)$.

(2) Prove that $2\delta(R) \leq c(R)$.

(3) Prove that $2\delta(R) = c(R)$ if and only if $\Gamma(R)$ is symmetric, that is,

$$\alpha \in \Gamma(R) \iff c(R) - 1 - \alpha \notin \Gamma(R).$$

(4) Check by an explicit computation, that the semigroup of the E_8-singularity is symmetric.

5.2.26. Let $\varphi : R \longrightarrow R'$ be an isomorphism, then $\Gamma(R) = \Gamma(R')$.
(Hint: Use the fact that φ extends to an isomorphism $\overline{\varphi} : \mathbb{C}\{t\} \longrightarrow \mathbb{C}\{t\}$ of the normalizations of R and R', and $\overline{\varphi}(t) = t \cdot$ unit.)

5.2.27. We use the notations of 5.2.11 and 5.2.14.

(1) Show that $\gcd(k_0,\dots,k_g) = 1$.

(2) Show that (2) $\iff$ (3) in 5.2.16.

(3) Show that for $d = 1,\dots,g$ we have $\gcd(k_0,\dots,k_d) = \gcd(\beta_0,\dots,\beta_d)$.

5.2.28. Prove the second part of Lemma 5.2.18.

5.2.29. Fill in the details of the proof of Theorem 5.2.21.

5.2.30. Prove that $G(R) = \{\beta_0,\dots,\beta_g\}$ for $\beta_0,\dots,\beta_g$ defined by (3) of Theorem 5.2.16.
(Hint: We know that $\beta_0,\dots,\beta_g$ generate $\Gamma(R)$. Use (3) to prove that

$$\gcd(\beta_0,\dots,\beta_i) < \gcd(\beta_0,\dots,\beta_{i-1}) \text{ for } i = 1,\dots,g, \text{ and } \gcd(\beta_0,\dots,\beta_g) = 1.)$$

5.2.31. Let $R = \mathbb{C}\{x,y\}/(f) = \mathbb{C}\{t^n,y(t)\}$, with $y(t) = \sum_{i\geq m} a_i t^i$ and $a_m \neq 0$. Suppose $n > m$. Suppose f is a Weierstraß polynomial in y. Let $k_0,\dots,k_g$ be the characteristic exponents of R. Prove that

$$\operatorname{ord} \frac{\partial f}{\partial y}(t^n,y(t)) = \sum_{j=1}^{g} (\gcd(k_0,\dots,k_{j-1}) - \gcd(k_0,\dots,k_j))k_j + n - m.$$

(Hint:
Let $\overline{k}_i$ be defined by $\overline{k}_0 = n$ and $\overline{k}_\nu = \min\{i : a_i \neq 0,\ \gcd(i,\overline{k}_0,\dots,\overline{k}_{\nu-1}) < \gcd(\overline{k}_0,\dots,\overline{k}_{\nu-1})\}$. Then in 5.2.18 we proved how to express the order of $\frac{\partial f}{\partial y}(t^n,y(t))$ in the $\overline{k}_i$. Now rewrite this formula in the k_i.)

5.2.32. Consider the curve singularity given by $R = \mathbb{C}\{x,y,z\}/(y^2 - xz, yz - x^3, z^2 - x^2 y)$.

(1) Show that $R \cong \mathbb{C}\{t^3,t^4,t^5\}$.

(2) Show that R is not Gorenstein by using Lemma 5.2.8.

(3) Compute the semigroup $\Gamma(R)$. Show that $\Gamma(R)$ is not symmetric.

(4) Find an ideal $I \subset R$ such that $I \subset I^{**}$ is a proper inclusion.

5.3 Resolutions of Irreducible Plane Curve Singularities

This section is devoted to the study of embedded resolutions of plane curve singularities, which will be done by successive "blowing-ups" of the curve singularity. In order to define the notion of blowing-up we need some other definitions first.

Definition 5.3.1. We define the 1-*dimensional projective space* $\mathbb{P}^1 = \mathbb{P}^1_{\mathbb{C}}$ to be the set of lines in $\mathbb{C}^2$ which pass through the origin.

One can define $\mathbb{P}^n = \mathbb{P}^n_{\mathbb{C}}$ for $n \geq 2$ in an analogous way. Alternatively, but very similarly, one could define $\mathbb{P}^1$ as

$$\mathbb{P}^1 = (\mathbb{C}^2 \setminus \{0,0\})/\sim .$$

Here $(x,y) \sim (x',y')$ if and only if there exists a $\lambda \in \mathbb{C} \setminus \{0\}$ with $(x',y') = \lambda(x,y)$. The equivalence class of $(x,y) \neq (0,0)$ is denoted by $(x : y)$. An equivalence class is called a *point* of $\mathbb{P}^1$. The $(x : y)$ are called *homogeneous coordinates* of $\mathbb{P}^1$.

As any line through $(0,0)$ has a nonempty intersection with the unit sphere in $\mathbb{C}^2 = \mathbb{R}^4$ we also could have defined $\mathbb{P}^1$ as

$$\{(x,y) \in \mathbb{C}^2 : |x|^2 + |y|^2 = 1\}/\sim\ = S^3/\sim ,$$

where $(x,y) \sim (x',y')$ if and only if $\exists\ \lambda \in \mathbb{C}$ with $|\lambda| = 1$ so that $(x,y) = \lambda(x',y')$. We give $\mathbb{P}^1$ the quotient topology. As a quotient of a compact topological space, $\mathbb{P}^1$ is compact. Moreover, it is left to the reader to show that $\mathbb{P}^1$ is Hausdorff. In fact, $\mathbb{P}^1$ is homeomorphic to S^2. This can be seen as follows. Consider a line with homogeneous coordinates $(x : y)$ and suppose that $y \neq 0$. Then there is a representative with $y = 1$. We see that the set of lines $(x : y)$ with $y \neq 0$ can be identified with the elements of $\mathbb{C}$. So this subset of $\mathbb{P}^1$ is homeomorphic to $\mathbb{C}$. There is only one point of $\mathbb{P}^1$ with $y = 0$, as all $(x,0)$ are equivalent to $(1,0)$. Thus, it can be seen that $\mathbb{P}^1_{\mathbb{C}}$ is homeomorphic to the one point compactification of $\mathbb{C} = \mathbb{R}^2$, hence to S^2. Another way to see this is by considering an open covering

$$\mathbb{P}^1 = U_1 \cup U_2,$$

where $U_1 = \{(x : y) \in \mathbb{P}^1 : x \neq 0\}$, and $U_2 = \{(x : y) \in \mathbb{P}^1 : y \neq 0\}$. Now U_1 is homeomorphic to $\mathbb{C}$ via the map $f_1 : U_1 \longrightarrow \mathbb{C}$ which sends $(x : y) \mapsto \frac{y}{x}$, and similarly U_2 is homeomorphic to $\mathbb{C}$ via the map $f_2 : U_2 \longrightarrow \mathbb{C}$ which sends $(x : y) \mapsto \frac{x}{y}$. Note that

$$f_1 \circ f_2^{-1} : \mathbb{C} \setminus \{0\} \longrightarrow \mathbb{C} \setminus \{0\}$$

is the map which sends z to $\frac{1}{z}$, so is holomorphic. Therefore, $\mathbb{P}^1$ carries the structure of a complex manifold. A *complex manifold* in general is a topological Hausdorff space X which has an open covering $X = \cup_{i \in I} U_i$, where the U_i are homeomorphic to an open subset V_i in $\mathbb{C}^n$ via, say f_i. That is, $f_i : U_i \longrightarrow V_i$ is a homeomorphism. Moreover, for all $i,j \in I$, the map

$$f_i \circ f_j^{-1} : f_j(U_i \cap U_j) \longrightarrow f_i(U_i \cap U_j)$$

should be holomorphic.[3] This makes sense, as these sets are subsets of $\mathbb{C}^n$.

We now can define the blowing-up of $\mathbb{C}^2$ in the origin.

Definition 5.3.2. The *blowing-up* $\mathrm{Bl}_0\,\mathbb{C}^2$ of $\mathbb{C}^2$ in 0 is the subset of $\mathbb{C}^2 \times \mathbb{P}^1$ defined by

$$\mathrm{Bl}_0\,\mathbb{C}^2 = \{(p,\ell) : \ p \in \ell\}.$$

So a pair of a point p and a line ℓ are in $\mathrm{Bl}_0\,\mathbb{C}^2$ exactly when p lies on ℓ. Similarly one can define $\mathrm{Bl}_q\,\mathbb{C}^2$ for an arbitrary point $q \in \mathbb{C}^2$.

There is a natural map

$$\begin{aligned} \pi : \mathrm{Bl}_0\,\mathbb{C}^2 &\longrightarrow \mathbb{C}^2 \\ (p,\ell) &\mapsto p, \end{aligned}$$

which is called the *blowing-up map.* Note that for $p \in \mathbb{C}^2$ with $p \neq 0$ there is exactly one element in $\pi^{-1}(p)$. Indeed there is just one line going through 0 and p. On the other hand, $\pi^{-1}(0)$ is a $\mathbb{P}^1$, as 0 lies on all the lines through the origin. This $\mathbb{P}^1$ is called the *exceptional divisor*, and is usually denoted by E. We can describe $\mathrm{Bl}_0\,\mathbb{C}^2$ in coordinates:

$$\mathrm{Bl}_0\,\mathbb{C}^2 = \{(x,y,\, u : v) : \ xv = yu\} \subset \mathbb{C}^2 \times \mathbb{P}^1_{\mathbb{C}}.$$

Here x,y are coordinates on $\mathbb{C}^2$, whereas $(u : v)$ are homogeneous coordinates on $\mathbb{P}^1$. This is quite obvious, as $xv = yu$ exactly says that (x,y) is a multiple of (u,v).

Remark 5.3.3. Note that we can cover $\mathrm{Bl}_0\,\mathbb{C}^2$ by two charts, just as we could for $\mathbb{P}^1$. Indeed, let $\mathbb{P}^1_{\mathbb{C}} = U_1 \cup U_2$, $U_1 = \{(1 : v) \in \mathbb{P}^1_{\mathbb{C}}\}$, $U_2 = \{(u : 1) \in \mathbb{P}^1_{\mathbb{C}}\}$. Then

$$\mathrm{Bl}_0\,\mathbb{C}^2 \cap (\mathbb{C}^2 \times U_1) = \{(x,xv,1 : v) : x,v \in \mathbb{C}\} \cong \mathbb{C}^2 \quad \text{with coordinates } (x,v),$$
$$\mathrm{Bl}_0\,\mathbb{C}^2 \cap (\mathbb{C}^2 \times U_2) = \{(yu,y,u : 1) : y,u \in \mathbb{C}\} \cong \mathbb{C}^2 \quad \text{with coordinates } (y,u).$$

Let $U \subset \mathbb{C}^2$ be an open neighborhood of 0. We then define $\mathrm{Bl}_0\, U$ by $\pi^{-1}(U)$.

[3] For a definition for two complex manifolds to be isomorphic we refer to Section 6.2.

Definition 5.3.4. Let $(C,0)$ be a curve singularity and $C \subset U$ be a representative. The closure $C^{(1)}$ of $\pi^{-1}(C \setminus 0)$ is called the *strict transform* of C. By abuse of notation, we denote by $C^{(1)}$ the germ $(C^{(1)},p)$, where p is the intersection point of $C^{(1)}$ with the exceptional divisor.

The inverse image $\pi^{-1}(C)$ is called the *total transform* of C.

Remark 5.3.5. In the sequel we will consider strict transforms of curve singularities. Whenever we do so, we implicitly choose a chart which contains all points of intersection with the exceptional divisor.

Examples 5.3.6. (1) Let us consider a line $y = a \cdot x$. We look in the first chart

$$\mathrm{Bl}_0\, \mathbb{C}^2 \cap (\mathbb{C}^2 \times U_1) = \{(x,xv,1:v) : x,v \in \mathbb{C}\}.$$

This is a $\mathbb{C}^2$ with coordinates (x,v). So the total transform has equation

$$xv = ax.$$

Note that $\pi(x,v) = 0$ exactly when $x = 0$. So the exceptional divisor is given by $x = 0$, and we can decompose

$$\pi^{-1}(C) = E \cup \{(x,v) :\ v = a\}.$$

So we see that $C^{(1)} = \{(x,v) :\ v = a\}$ is a line, which intersects E in the point with homogeneous coordinates $(1:a)$. This is exactly the point which corresponds to the line C. Similarly on $\mathrm{Bl}_0\, \mathbb{C}^2 \cap (\mathbb{C}^2 \times U_2)$ one calculates that $C^{(1)} = \{(y,u) :\ u \cdot a = 1\}$.

(2) Let C be the union of two different lines through the origin. Then $C^{(1)}$ is the union of two different lines, which do not intersect. For example, if one line is given by $y = ax$, and the other by $y = bx$, and $a \neq b$, we get the following picture.

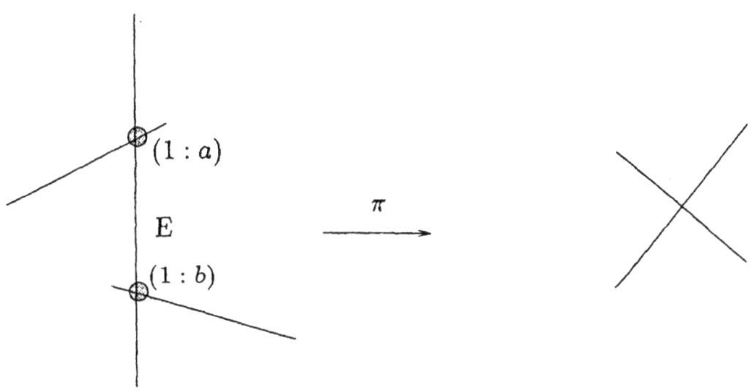

So we see that the blowing-up tears apart the different directions through the origin.

(3) Consider two smooth curves C and D given by $y = 0$ and $y = x^k$.

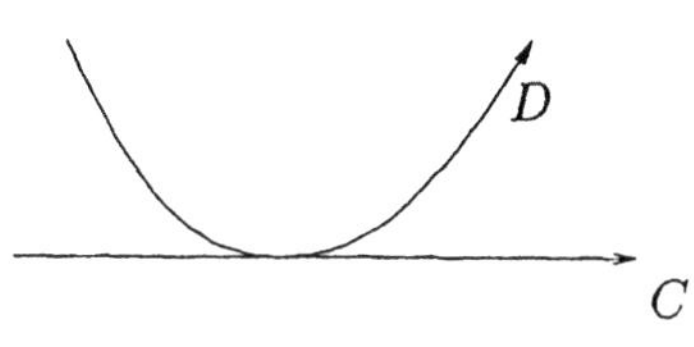

Those curves have the same tangents, so we do not expect that the strict transforms separate after one blowing-up. We blow up, and look at the chart $u = 1$, so $y = xv$. The strict transform of C is given by $v = 0$. The total transform of D has equation $xv - x^k = 0$. The exceptional divisor E_1 has equation $x = 0$, so that $D^{(1)}$ has equation $v = x^{k-1}$. We get the following picture.

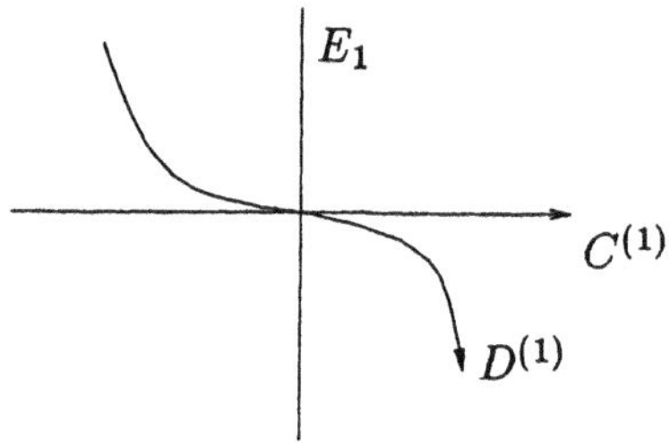

We can continue blowing-up. After k blowing-ups we get the following picture.

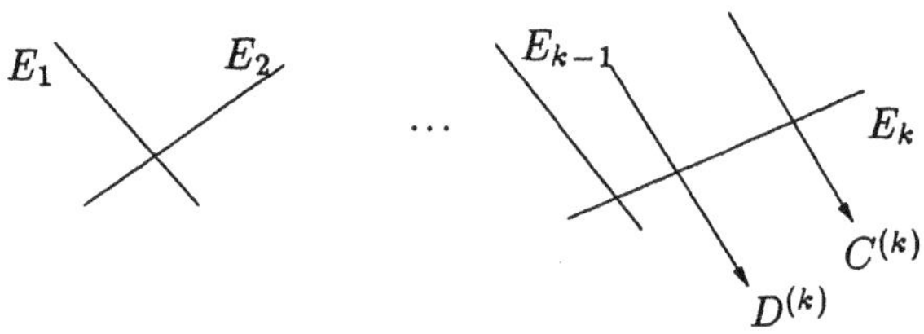

Now if we have two smooth curve (C, o) and (D, o) we always can choose local coordinates such that (C, o) is given by $y = 0$, and (D, o) is given by $y = x^k$. So we see from this example, that after one blowing up the intersection multiplicities of the strict transforms has decreased by one. In particular, we see that after finitely many blowing-ups, the strict transforms of (C, o) and (D, o) become separated.

(4) Consider the curve C given by the equation $y^2 - x^3 = 0$.

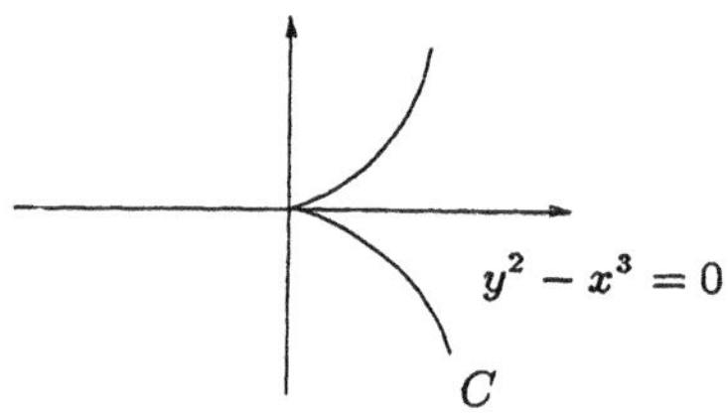

We consider the blowing-up, which we call X_1, and look at the chart $u = 1$, so $y = xv$. We have coordinates x,v. The total transform has equation

$$(xv)^2 - x^3 = x^2(v^2 - x) = 0.$$

The exceptional divisor E_1 is given by $x = 0$, so that the strict transform has equation $v^2 - x = 0$.

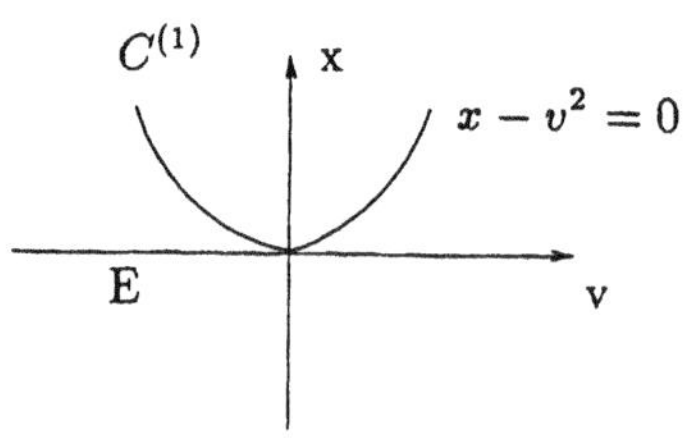

So we see that $C^{(1)}$ is smooth. This is quite satisfactory, but is not good enough for our purposes. Namely, the curves $C^{(1)}$ and E do not have transverse intersection. So we blow up X_1. As we do not want to introduce to much notation we rename v into y. So we see that we got the total transform by replacing (x,y) by (x,xy). In the other chart we have to replace (x,y) by (xy,y). Now $C^{(1)}$ is given by $y^2 - x = 0$, and we replace (x,y) by (xy,x) to get X_2 in a chart. The total transform is given by

$$y^2 - xy = y(y - x) = 0.$$

The exceptional divisor now is given by $y = 0$. The strict transform of E_1 is given by $x = 0$. This strict transform we also call E_1. We get the following picture.

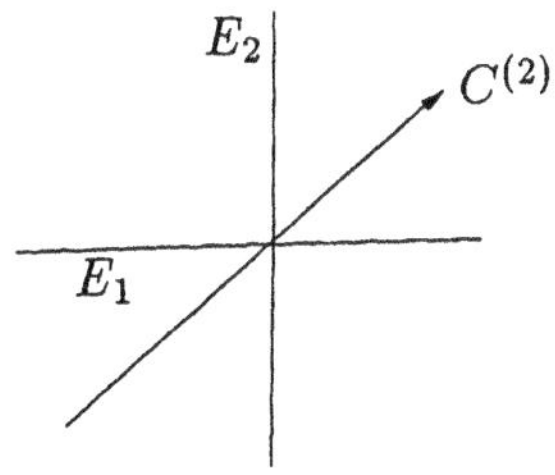

We are still not satisfied, since $C^{(2)}$ intersects two different exceptional divisors. One more blowing up gives:

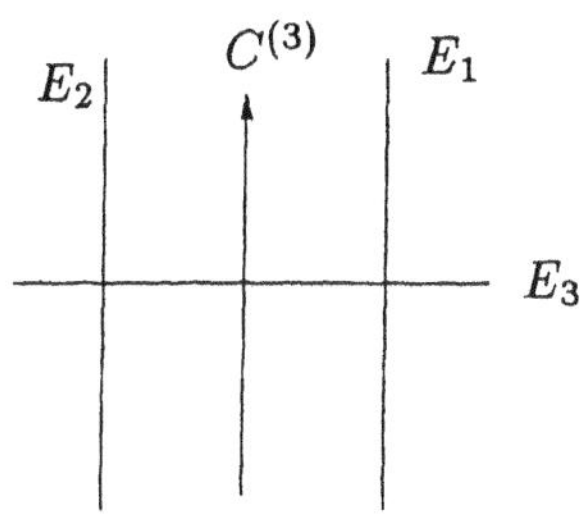

This is the situation we were seeking for.

Definition 5.3.7. Let (C, o) be an irreducible curve singularity, and $C \subset U$ be a representative. Consider a sequence of blowing-ups.

$$\begin{array}{ccccccccc} X_k & \xrightarrow{\pi_k} & X_{k-1} & \longrightarrow & \dots & \longrightarrow & X_1 & \xrightarrow{\pi_1} & U \\ \cup & & \cup & & & & \cup & & \cup \\ C^{(k)} & \longrightarrow & C^{(k-1)} & \longrightarrow & \dots & \longrightarrow & C^{(1)} & \longrightarrow & C = C^{(0)} \end{array}$$

Then we define $E^{(i)} := (\pi_1 \circ \dots \circ \pi_i)^{-1}(0)$, and the i-th strict transform $C^{(i)}$ by the closure of $(\pi_1 \circ \dots \pi_i)^{-1}(C \setminus \{0\})$. By abuse of notation the germ $(C^{(j)}, p_j)$, where p_j is the intersection point of $C^{(j)}$ with $E^{(j)}$, is simply denoted by $C^{(j)}$.

The above sequence of blowing-ups is called a *standard resolution* if either (C, o) is smooth and $k = 0$ or for $i = 2, \dots, k$ either

(1) $C^{(i-1)} \subset X_{i-1}$ has one singular point, and π_i is the blowing up of X_{i-1} in this point, or

(2) $C^{(i-1)}$ is smooth, but the intersection with $E^{(i-1)}$ in X_{i-1} is not transversal, or

(3) $C^{(i-1)}$ is smooth, intersects $E^{(i-1)}$ transverse, but does intersect more than one component of $E^{(i-1)}$,

and, furthermore, $C^{(k)}$ is smooth, intersects $E^{(k)}$ transversally, and intersects only one component of $E^{(k)}$.

For $j = 1, \ldots, k-1$, let m_j be the multiplicity of $C^{(j)}$ at the intersection point with $E^{(j)}$, an put $m_0 = m$ the multiplicity of (C, o). Then $(m_0, \ldots, m_{k-1})$ is called the *multiplicity sequence* of (C, o).[4]

Our first goal in this section is to prove that a standard resolution always exists.

Theorem 5.3.8. *Let (C, o) be an irreducible plane curve singularity. Then a standard resolution of (C, o) exists.*

Proof. If (C, o) is smooth then there is nothing to do.

We may choose coordinates x, y and t such that (C, o) is given by a parametrization

$$x = t^p, \quad y = t^q + \ldots$$

with $p < q$ and $p \nmid q$. In particular, the multiplicity of (C, o) is equal to p. We blow up, and look at the chart $u = 1$, so $y = vx$. We have coordinates (x,v). Let C be a sufficiently small representative. Then $\pi^{-1}(C \setminus \{0\})$ is given by the parametrization

$$x = t^p, \quad v = t^{q-p} + \ldots$$

for $t \neq 0$. Namely, we have in this chart $y = vx = t^q + \ldots$, and $x = t^p$ so that $v = t^{q-p} + \ldots$ The parametrization extends for $t = 0$ to give a parametrization for $C^{(1)}$. Let μ be minimal with $q - \mu p < p$. Then after μ blowing-ups we have as parametrization of $C^{(\mu)}$

$$x = t^p, \quad v = t^{q-\mu p} + \ldots.$$

The multiplicity of $C^{(\mu)}$ dropped: it is equal to $q - \mu p$. So we proved that after a finite number of blowing-ups the multiplicity drops. By doing a change of coordinates we have that $C^{(\mu)}$ has parametrization

$$x = t^{p-\mu q}, \quad y = t^{q_1} + \ldots$$

with $p - \mu q < q_1$ and $p - \mu q \nmid q_1$. Going on like this, we arrive at the case that for some j the multiplicity is one, so $C^{(j)}$ is smooth. Now it might happen that $C^{(j)}$ does not intersect an exceptional curve transversally. We then can do further blowing-ups, so that $C^{(\ell)}$ for some $\ell > j$ intersects transversally, cf. the third part of Example 5.3.6. It might still happen that $C^{(\ell)}$ intersects more than one exceptional component transversally.[5] One more blowing up gives the desired standard resolution, cf. the fourth part of Example 5.3.6. □

Example 5.3.9. Let $(C, o) \subset (\mathbb{C}^2, o)$ be defined by

$$(y^2 - x^3)^2 - 4x^5y - x^7 = 0.$$

[4] For the case of a *smooth* germ (C, o) the multiplicity sequence is empty.

[5] One can easily convince oneself that $C^{(\ell)}$ can intersect at most two exceptional curves.

This is a curve singularity of multiplicity 4.

(1) We blow up, and look in the chart where we replace[6] (x,y) by (x,xy). We get as total transform

$$((xy)^2 - x^3)^2 - 4x^5xy - x^7 = 0.$$

The exceptional curve E_1 is given by $x = 0$. We pull out the factor x^4, and so the strict transform $C^{(1)}$ is given by

$$(y^2 - x)^2 - 4x^2y - x^3 = 0. \tag{5.12}$$

Thus $m(C^{(1)}) = 2$.

(2) We look at the second blowing-up, replacing (x,y) by (xy,y) in (5.12). We get as total transform

$$(y^2 - xy)^2 - 4(xy)^2y - (xy)^3 = 0.$$

The exceptional curve E_2 is given by $y = 0$. Pulling out y^2 gives $C^{(2)}$

$$(y - x)^2 - 4x^2y - x^3y = 0. \tag{5.13}$$

We see that $m(C^{(2)}) = 2$.

(3) Now we change coordinates $y \mapsto y + x$. In these new coordinates (5.13) is given by

$$y^2 - 4x^3 - 4x^2y - x^4 - x^3y = 0. \tag{5.14}$$

In fact, by doing further coordinate changes in (5.14) one sees that $(C^{(2)}, o)$ is isomorphic to an A_2-singularity. In Example 5.3.6 we saw that we have to blow up three more times. Thus we get as multiplicity sequence (4,2,2,1,1).

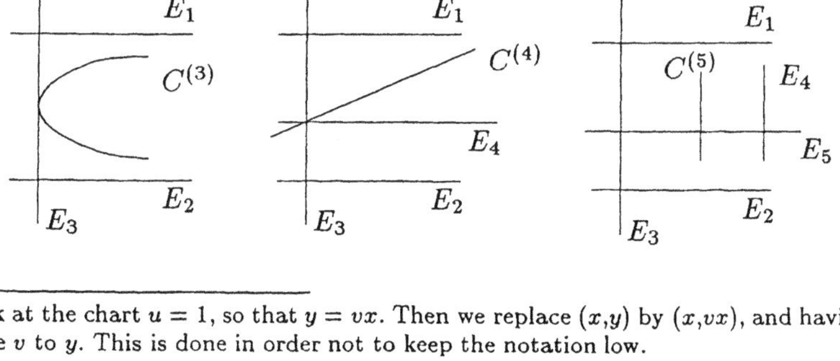

[6] So we look at the chart $u = 1$, so that $y = vx$. Then we replace (x,y) by (x,vx), and having done this, we rename v to y. This is done in order not to keep the notation low.

Definition 5.3.10. Let $\pi : X_k \xrightarrow{\pi_k} X_{k-1} \longrightarrow \cdots \xrightarrow{\pi_1} U$ be a standard resolution of the singular point $0 \in C \subset U$. Let $\pi^{-1}(0) = E = \cup_{i=1}^k E_i$ be the exceptional divisor and $C^{(k)}$ the strict transform of C. To this resolution we can associate a weighted graph, the *resolution graph.* The vertices of the graph are the exceptional lines E_i and the curve $C^{(k)}$. We connect two vertices E_i and E_j if $E_i \cap E_j \neq \emptyset$ and, similarly, we connect a vertex E_i to $C^{(k)}$ if $E_i \cap C^{(k)} \neq \emptyset$, that is, we connect $C^{(k)}$ to E_k. The weight of E_i in the graph is i, if $\pi_i \circ \cdots \circ \pi_k(E_i)$ is a point, but $\pi_{i+1} \circ \cdots \circ \pi_k(E_i)$ is isomorphic to $\mathbb{P}^1$. We will draw "•" for an E_i and "*" for $C^{(k)}$.

Example 5.3.11. In Example 5.3.9 we obtain as resolution graph:

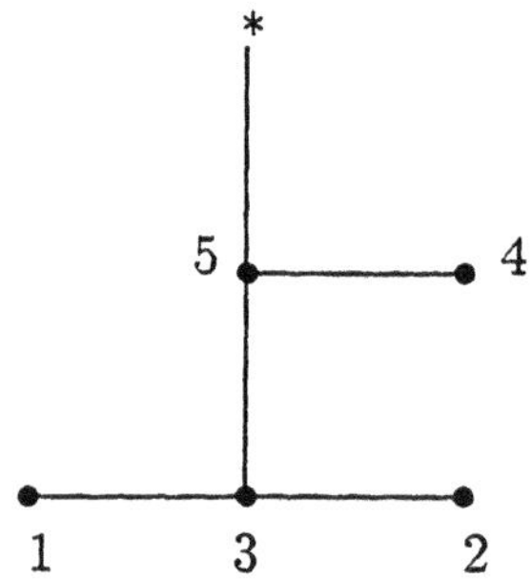

For an irreducible curve singularity, we introduced several invariants. The question arises, how they are related to each other. We first show that the multiplicity sequence and the characteristic exponents determine each other.

Theorem 5.3.12. *Let $(C, 0)$ be an irreducible curve singularity with characteristic exponents $k_0, \ldots, k_g$. Then the multiplicity sequence can be computed by the following algorithm.*

If $g = 0$ then the multiplicity sequence is empty. If $g > 0$ we perform the Euclidean algorithm for $\mathcal{X}_1 := k_1$ and $r_{1,1} := k_0$.

$$\begin{aligned} \mathcal{X}_1 &= \mu_{1,1} r_{1,1} + r_{1,2}, \\ r_{1,1} &= \mu_{1,2} r_{1,2} + r_{1,3}, \\ &\vdots \\ r_{1,w(1)-1} &= \mu_{1,w(1)} r_{1,w(1)}. \end{aligned}$$

For $i = 2, \ldots, g$, let $\mathcal{X}_i = k_i - k_{i-1}$ and define inductively $r_{i,1} := r_{i-1,w(i-1)}$.

We perform the Euclidian algorithm with $\mathcal{X}_i$ and $r_{i,1}$ and obtain:

$$\begin{aligned} \mathcal{X}_i &= \mu_{i,1} r_{i,1} + r_{i,2}, \\ r_{i,1} &= \mu_{i,2} r_{i,2} + r_{i,3}, \\ &\vdots \\ r_{i,w(i)-1} &= \mu_{i,w(i)} r_{i,w(i)}. \end{aligned}$$

Then the multiplicity sequence of $(C, 0)$ contains $\mu_{i,j}$ times the multiplicity $r_{i,j}$.[7] *In particular, the characteristic exponents determine the multiplicity sequence. On the other hand, reversing the procedure allows to compute the $k_0, k_1, \ldots, k_g$ inductively from the multiplicity sequence.*

[7] Here it is understood that the $r_{i-1,w(i-1)} = r_{i,1}$ which occurs in two Euclidean algorithms is counted both times.

Example 5.3.13. Let (C, o) have characteristic exponents $k_0 = 28$, $k_1 = 64$, $k_2 = 66$, $k_3 = 77$. We perform the first Euclidean algorithm.

$$\begin{aligned} 64 &= 2 \cdot 28 + 8, \\ 28 &= 3 \cdot 8 + 4, \\ 8 &= 2 \cdot 4. \end{aligned}$$

We perform the second Euclidean algorithm with $k_2 - k_1 = 2$ and $4 = r_{1,3} = r_{2,1}$.

$$\begin{aligned} 2 &= 0 \cdot 4 + 2, \\ 4 &= 2 \cdot 2. \end{aligned}$$

Finally the third Euclidean algorithm with $k_3 - k_2 = 11$ and $2 = r_{2,2} = r_{3,1}$.

$$\begin{aligned} 11 &= 5 \cdot 2 + 1, \\ 2 &= 2 \cdot 1. \end{aligned}$$

So we get the multiplicity sequence (28,28,8,8,8,4,4,2,2,2,2,2,2,2,1,1).

On the other hand, starting from this multiplicity sequence, we show how to compute the characteristic exponents. First of all, $k_0 = 28$, and as by definition $k_0 \nmid k_1$, the right hand side of first Euclidean algorithm has to start with

$$k_1 = 2 \cdot 28 + 8.$$

So we calculated $k_1 = 64$. The Euclidean algorithm has to continue as follows.

$$\begin{aligned} 28 &= 3 \cdot 8 + 4, \\ 8 &= 2 \cdot 4. \end{aligned}$$

The right hand side of the next Euclidean algorithm has to be $0 \cdot 4 + 2$, as all the fours of the multiplicity sequence have been used by the previous Euclidean algorithm. So we get

$$k_2 - k_1 = 0 \cdot 4 + 2,$$

so $k_2 = 66$. Then the Euclidean algorithm continues as follows.

$$4 = 2 \cdot 2.$$

The third Euclidean algorithms starts with

$$k_3 - k_2 = 5 \cdot 2 + 1,$$

as of the seven 2's in the multiplicity sequence, two have been used in the previous Euclidean algorithm. Thus $k_3 = 77$, and so we calculated the characteristic exponents from the multiplicity sequence.

Proof of Theorem 5.3.12. The proof is by induction on the number of blowing-ups needed to get the standard resolution. The case of a smooth curve is trivial, as we have no Euclidean algorithm at all: the multiplicity sequence is empty. Now we may assume that (C, o) has parametrization

$$x = t^{k_0}, \quad y = t^{k_1} + \dots.$$

The strict transform under the first blowing-up has parametrization

$$x = t^{k_0}, \quad y = t^{k_1 - k_0} + \dots.$$

There are three cases to consider.

(1) If $k_0 \leq k_1 - k_0$, then $k_0 < k_1 - k_0$ and we obtain as characteristic exponents for $C^{(1)}$:

$$\overline{k}_0, \dots, \overline{k}_g = k_0, k_1 - k_0, \dots, k_g - k_0.$$

(2) If $k_0 > k_1 - k_0$ and $k_1 - k_0 \nmid k_0$, then we have as characteristic exponents for $C^{(1)}$:

$$\overline{k}_0, \dots, \overline{k}_g = k_1 - k_0, k_0, k_2 + k_0 - k_1, \dots, k_g + k_0 - k_1.$$

(3) If $k_0 > k_1 - k_0$ and $k_1 - k_0 \mid k_0$, then $k_1 - k_0 = \gcd(k_0, k_1)$ and we obtain as characteristic exponents for $C^{(1)}$:

$$\overline{k}_0, \dots, \overline{k}_{g-1} = k_1 - k_0,\ k_2 + k_0 - k_1, \dots, k_g + k_0 - k_1.$$

So in this case the number of characteristic exponents drops.

The first case follows from the definition of the characteristic exponents, whereas the second and third case follow from the Inversion Theorem 5.2.21. Of course the multiplicity sequence of $(C^{(1)}, o)$ is obtained from the multiplicity sequence of (C, o) by deleting the first entry. Let us consider the three different cases.

(1) In this case we have that $\overline{k}_1 = k_1 - k_0$, whereas $\overline{k}_0 = k_0$. So the first row in the first Euclidean algorithm changes as follows

$$\overline{k}_1 = (\mu_{1,1} - 1) r_{1,1} + r_{1,2},$$

but the remainder of the first Euclidean algorithm does not change. As moreover $\overline{k}_i - \overline{k}_{i-1} = k_i - k_{i-1}$ for $i \geq 2$, the other $g - 1$ Euclidean algorithms do not change. So the multiplicity sequence of $C^{(1)}$ is, by induction, as claimed.

(2) In this case, the multiplicity k_0 occurs just once in the multiplicity sequence of (C, o), as by assumption the multiplicity of $C^{(1)}$ is smaller. So we have by definition of $r_{1,2}$ that $k_1 = 1 \cdot k_0 + r_{1,2}$. Now $\overline{k}_1 = k_0 = r_{1,1}$, and $\overline{k}_0 = k_1 - k_0 = r_{1,2}$. So the first line of the first Euclidean algorithm of $C^{(1)}$ starts of with the second line of the first Euclidean algorithm of (C, o):

$$r_{1,1} = \mu_{1,2} r_{1,2} + r_{1,3}.$$

As again $\overline{k}_i - \overline{k}_{i-1} = k_i - k_{i-1}$ for $i \geq 2$, the other $g - 1$ Euclidean algorithms do not change. Induction proves the second case.

(3) In the final case, by assumption on the characteristic exponents, we have $r_{1,2} = k_1 - k_0 \mid k_0 = r_{1,1}$. Hence the first Euclidean algorithm stops after two divisions! Recall that $r_{1,1} = k_0$.

$$k_1 = 1 \cdot r_{1,1} + r_{1,2}$$
$$k_0 = r_{1,1} = \mu_{1,2} r_{1,2}.$$

As $r_{2,1} = r_{1,2}$ the first line of the second Euclidean algorithm starts as follows

$$k_2 - k_1 = \mu_{2,1} r_{1,2} + r_{2,2}.$$

So we see that the claimed multiplicity sequence of C starts with k_0 and then $\mu_{1,2} + \mu_{2,1}$ times the multiplicity $r_{1,2} = k_1 - k_0$. Now $\overline{k}_1 = k_2 + k_0 - k_1 = k_2 - r_{1,2}$ and $\overline{k}_0 = k_1 - k_0 = r_{1,2}$. So the first line of the first Euclidean algorithm for $C^{(1)}$ reads

$$k_2 - k_1 + k_0 = (\mu_{1,2} + \mu_{2,1}) r_{1,2} + r_{2,2},$$

and the rest of this Euclidean algorithm continues as the second Euclidean algorithm of (C, o). As moreover $\overline{k}_i - \overline{k}_{i-1} = k_{i+1} - k_i$, we see that the rest of the calculations remain unchanged, and we are again done by induction. □

Now we come to the description of the resolution graph.

Theorem 5.3.14. *Let (C, o) be an irreducible plane curve singularity, and let $k_0, \dots, k_g$ be the characteristic exponents of (C, o). Let $\mu_{i,j}$ be determined by the g Euclidean algorithms of the previous theorem. Here $1 \le i \le g$, and $1 \le j \le w(i)$. The resolution graph consists of g* Puiseux chains *P_i for $i \dots g$. These chains are constructed from $w(i)$ building blocks S_i^j:*

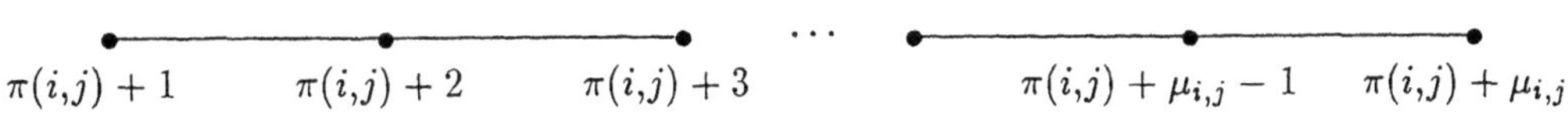

with

$$\pi(1,1) = 0,$$
$$\pi(i,1) = \pi(i-1, w(i-1)) + \mu_{i-1, w(i-1)}, \qquad \text{for } i = 2, \dots, g,$$
$$\pi(i,j) = \pi(i, j-1) + \mu_{i,j-1}, \qquad \text{for } 2 \le j \le w(i) \text{ and } i = 1, \dots, g.$$

The point with the smallest weight (respectively largest weight) in a building block S_1^j is called the start point (respectively end point) of the graph S_i^j.

The Puiseux chain P_i is constructed by connecting the end point of S_i^k with the start point of S_i^{k+2} as long as it is possible, that is, for $k \le w(i) - 2$. Furthermore the end point of $S_i^{w(i)-1}$ is connected to the end point of $S_i^{w(i)}$. This ends the construction of the Puiseux chain P_i.

The Puiseux chains are put into one graph by connecting the end point of $S_i^{w(i)}$ with

(1) the start point of S_{i+1}^1 if $\mu_{i+1,1} \neq 0$, or

(2) with the start point of S_{i+1}^3 if $\mu_{i+1,1} = 0$ and $w(i+1) \ge 3$,

(3) with the end point of S_{i+1}^2 if $\mu_{i+1,1} = 0$ and $w(i+1) = 2$.

(4) Finally, we connect the end point of $S_g^{w(g)}$ with a $*$.

Because $w(i) \geq 2$, *the end point of* $S_i^{w(i)}$ *is, therefore, connected with three other points, and it is called a* contact point *of the graph.*

The procedure leads to a graph which looks like a mobile:

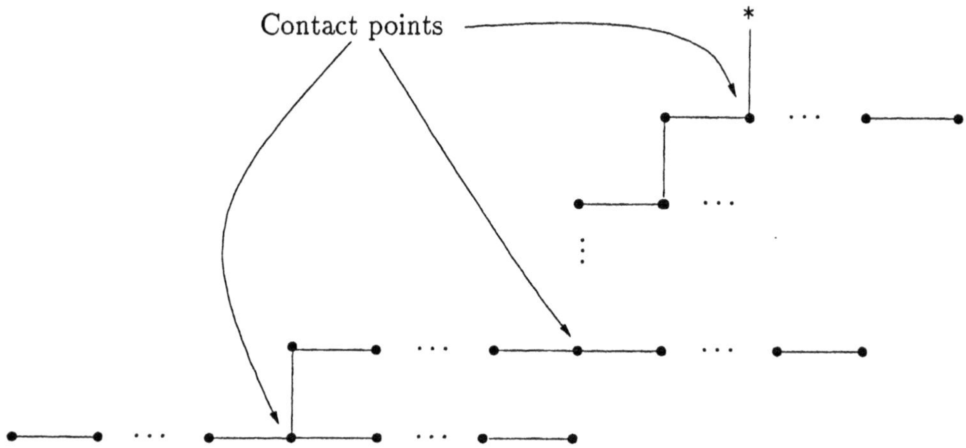

We have g contact points, and we draw the graph in such a way that each contact point creates a new level.

Remark 5.3.15. It is quite obvious how to obtain the characteristic exponents from the resolution graph: g is the number of contact points, points which are connected to three other points. One can identify the S_i^j as the maximal connected subsets of the graph such that the weights are strictly increasing (respectively decreasing). Starting with multiplicity 1 which occurs $\mu_{g,w(g)}$–times in $S_g^{w(g)}$ we obtain $r_{g,w(g)-1} = \mu_{g,w(g)} \cdot 1$. This way, going from bottom to top in the Euclidean algorithm we can reconstruct the k_i.

Let us now look at two examples.

Example 5.3.16.

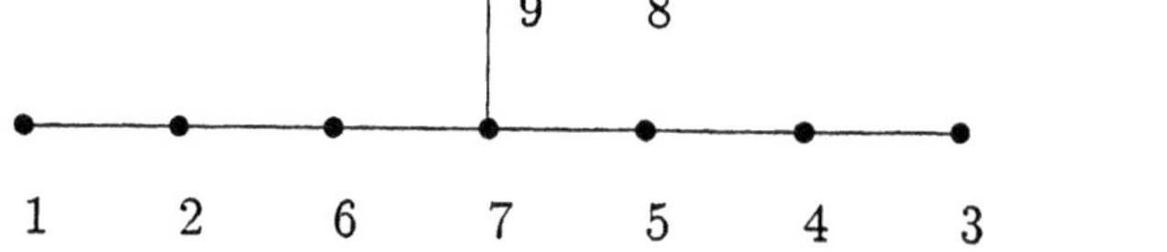

From the diagram we can extract the number g of levels and the number $w(i)$ of blocks in each level i, as well the blocks S_i^j themselves. In particular we get the $\mu_{i,j} = \#S_i^j$. Note that it might very well happen that $S_{i+1}^1 = \emptyset$. This happens exactly if $\mu_{i+1,1} = 0$. In the particular example we see that $S_2^1 = \emptyset$, as the contact point is connected to a point not of minimal weight in the Puiseux chain P_2. The remaining data in the following table have to be calculated:

$g=3$				
$w(1)=3$	$S_1^1=\{1,2\}$ $\mu_{1,1}=2$ $r_{1,1}=28$	$S_1^2=\{3,4,5\}$ $\mu_{1,2}=3$ $r_{1,2}=8$	$S_1^3=\{6,7\}$ $\mu_{1,3}=2$ $r_{1,3}=4$	$\chi_1=64$
$w(2)=2$	$S_2^1=\emptyset$ $\mu_{2,1}=0$ $r_{2,1}=4$	$S_2^2=\{8,9\}$ $\mu_{2,2}=2$ $r_{2,2}=2$		$\chi_2=2$
$w(3)=2$	$S_3^1=\{10,11,12,13,14\}$ $\mu_{3,1}=5$ $r_{3,1}=2$	$S_3^2=\{15,16\}$ $\mu_{3,2}=2$ $r_{3,2}=1$		$\chi_3=11$

In order to do the calculations we proceed as follows: The initialization step in level three is done by setting

$$r_{3,2} = r_{g,w(g)} = 1.$$

We then go on with the following formulas:

$$\begin{aligned} r_{3,1} &= \mu_{3,2}\cdot r_{3,2} = 2\cdot 1 = 2,\\ k_3-k_2 = \chi_3 &= \mu_{3,1}\cdot r_{3,1}+r_{3,2} = 5\cdot 2+1 = 11. \end{aligned}$$

This finishes level three, and we may continue with level two:

$$\begin{aligned} r_{2,2} &= r_{3,1} = 2,\\ r_{2,1} &= \mu_{2,2}\cdot r_{2,2} = 2\cdot 2 = 4,\\ k_2-k_1 = \chi_2 &= \mu_{2,1}\cdot r_{2,1}+r_{2,2} = 0\cdot 4+2 = 2. \end{aligned}$$

Finally, in level one we obtain:

$$\begin{aligned} r_{1,3} &= r_{2,1} = 4,\\ r_{1,2} &= \mu_{1,3}\cdot r_{1,3} = 2\cdot 4 = 8,\\ k_0 = r_{1,1} &= \mu_{1,2}\cdot r_{1,2}+r_{1,3} = 3\cdot 8+4 = 28,\\ k_1 = \chi_1 &= \mu_{1,1}\cdot r_{1,1}+r_{1,2} = 2\cdot 28+8 = 64. \end{aligned}$$

We, therefore, find as characteristic exponents:

$$k_0=28,\ k_1=64,\ k_2=66,\ k_3=77.$$

Example 5.3.17. In Example 5.3.9 we had the multiplicity sequence (4,2,2,1,1) and

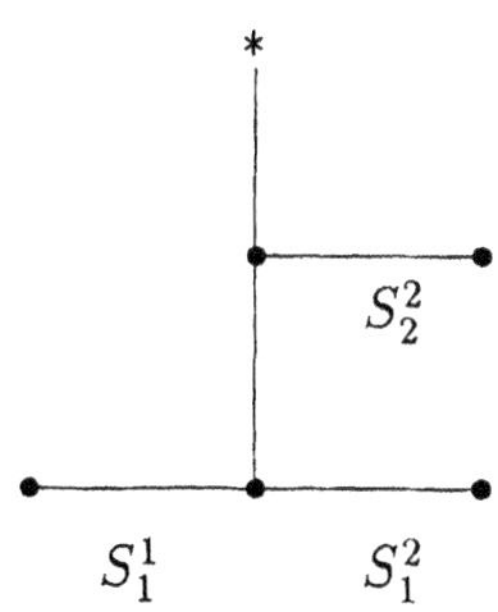

as resolution graph. The curve singularity (C,o) was defined by $f = y^4 - 2x^3y^2 - 4x^5y + x^6 - x^7$. As we have already seen in Exercise 5.1.25 and Example 5.2.15, (C,o) has the parametrization $x = t^4$, $y = t^6 + t^7$ with the characteristic exponents $k_0 = 4$, $k_1 = 6$, $k_2 = 7$. We want now to calculate the $r_{i,j}$ and the $\mu_{i,j}$ using their definition. The initial values in the first level are given by

$$\chi_1 = k_1 = 6 \text{ and } r_{1,1} = k_0 = 4.$$

Proceeding along the lines of the algorithm, i. e. doing division with remainder, we find:

$$\begin{array}{lllll} \chi_1 & = & 6 = 1 \cdot 4 + 2 = \mu_{1,1} \cdot r_{1,1} + r_{1,2} & \Rightarrow & \mu_{1,1} = 1,\ r_{1,2} = 2, \\ r_{1,1} & = & 4 = 2 \cdot 2 = \mu_{1,2} \cdot r_{1,2} & \Rightarrow & \mu_{1,2} = 2,\ w(1) = 2. \end{array}$$

The second level thus starts with

$$\chi_2 = k_2 - k_1 = 1 \text{ and } r_{2,1} = r_{1,w(1)} = r_{1,2} = 2.$$

Doing division with remainder we then get:

$$\begin{array}{lllll} \chi_2 & = & 1 = 0 \cdot 2 + 1 = \mu_{2,1} \cdot r_{2,1} + r_{2,2} & \Rightarrow & \mu_{2,1} = 0,\ r_{2,2} = 1, \\ r_{2,1} & = & 2 = 2 \cdot 1 = \mu_{2,2} \cdot r_{2,2} & \Rightarrow & \mu_{2,2} = 2,\ w(2) = 2. \end{array}$$

And this finishes the algorithm. One easily sees that starting from the diagram we would get the same values for the $\mu_{i,j}$ and for the $r_{i,j}$.

Proof of Theorem 5.3.14. By using induction on the number of blowing-ups needed to get the standard resolution, we may assume that the singularity after one blowing up has the resolution graph as described in the theorem. Denote the corresponding building blocks by $\overline{S}_i^j$. We have the invariants $\mu_{i,j}$ of (C,o). The corresponding invariants of the strict transform we denote by $\overline{\mu}_{i,j}$. We again suppose that our curve has parametrization

$$x = t^{k_0}, \quad y = t^{k_1} + \dots .$$

The strict transform has parametrization

$$x = t^{k_0}, \quad y = t^{k_1 - k_0} + \dots .$$

The exceptional curve is in these coordinates given by $x = 0$. We consider the same three cases as in the proof of Theorem 5.3.12.

Case 1. Suppose $k_0 < k_1 - k_0$, then the exceptional curve $x = 0$ intersects the strict transform transversally.

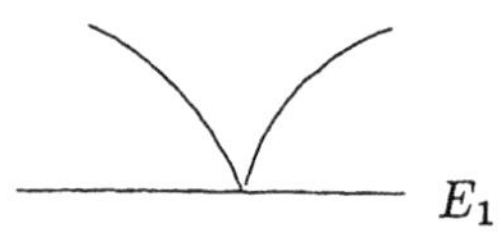

We thus have that the next blowing up gives

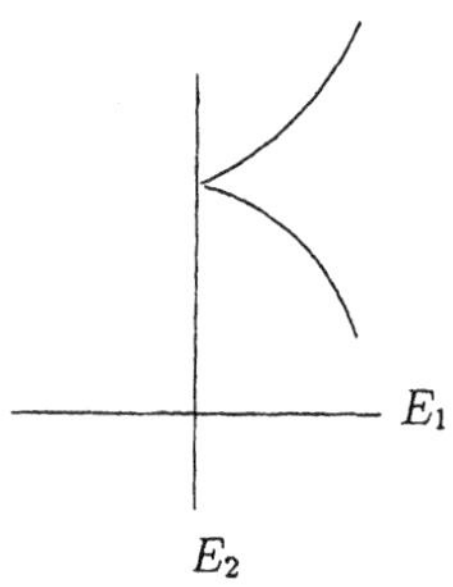

or

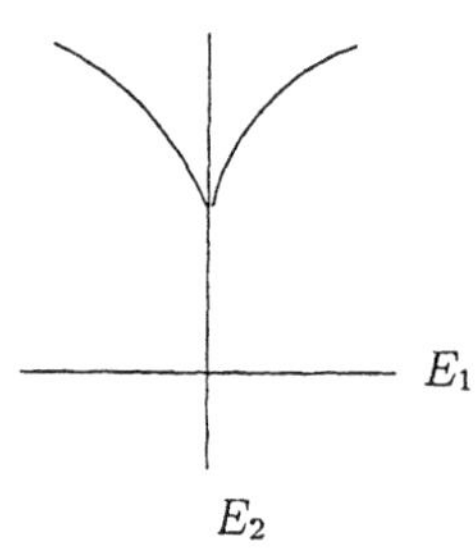

which means that E_1 and the strict transform separate. This implies $\overline{S}_i^j = S_i^j$ (with the weights shifted by one) for $(i,j) \neq (1,1)$ and if $\overline{S}_1^1$ is

then S_1^1 is:

Case 2. Suppose $k_0 > k_1 - k_0$, but $k_0 \nmid k_1 - k_0$. In this case the multiplicity of the strict transform drops, and the intersection of the strict transform with the exceptional divisor E_1 is not transverse. By a coordinate change in t we may assume that the strict transform has parametrization

$$x = t^{k_0} + \dots, \quad y = t^{k_1 - k_0}.$$

The next blowing-up gives:

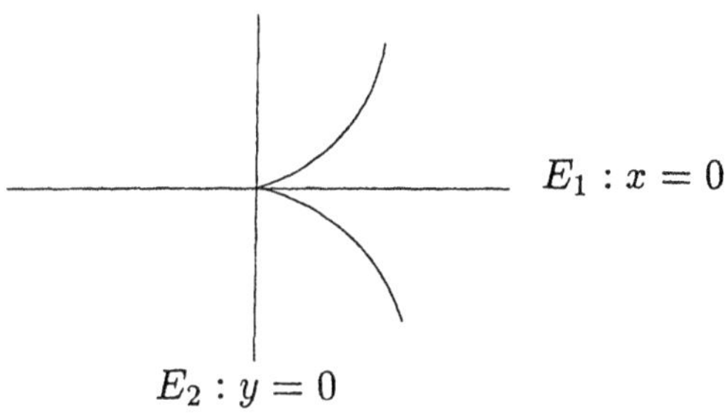

Indeed, in the (xy,y)–chart the strict transform will be given by the parametrization

$$x = t^{k_0 - k_0 + k_1} = t^{k_1}, \quad y = t^{k_1 - k_0} + \dots.$$

The sequence of blowing up continues like

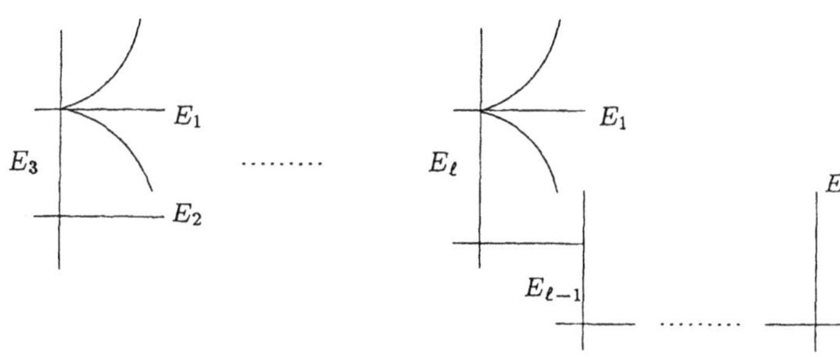

up to the next multiplicity change which occurs after $\ell := \mu_{1,2}$ times. Indeed after $\ell = \mu_{1,2}$ blowing ups we have as parametrization of the strict transform of (C, o)

$$x = t^{r_{1,3}} + \dots, \quad y = t^{k_1 - k_0}.$$

As now $r_{1,3} < k_1 - k_0$, the curve E_1, given by $x = 0$, intersects the strict transform transversally. On the other hand, the curve $E_{\ell+1}$, given by $y = 0$, does not intersect the curve transversally. So, after the $\ell + 1$–st this blowing-up we have the following picture:

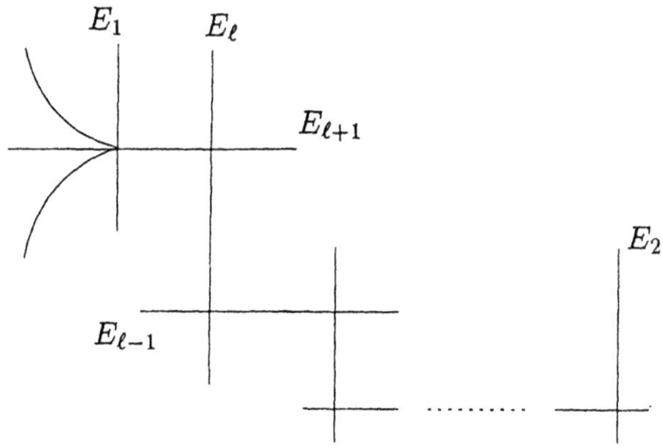

It follows that after the next blowing-up, the strict transform of (C, o) and E_1 are separated. So E_1 intersects $E_{\ell+2}$, and we can conclude this case by induction.

Case 3. Suppose $k_0 > k_1 - k_0$, but $k_1 - k_0 \mid k_0$. It was shown in the proof of 5.3.12 that the number of characteristic exponents of the strict transform $C^{(1)}$ is equal to $g - 1$. In fact we have for the $\mu_{i,j}$ of (C, o) and the corresponding $\overline{\mu}_{i,j}$ for $C^{(1)}$ the following relations:

(1) $\overline{\mu}_{i,j} = \mu_{i+1,j}$ for $(i,j) \neq (1,1)$,

(2) $\overline{\mu}_{1,1} = \mu_{1,2} + \mu_{2,1}$.

It follows from the definition that the graph constructed in the theorem for $C^{(1)}$ is obtained from that of (C, o) by deleting S_1^1, and by lowering the weights by one. This is a simple case by case check. It remains to show that S_1^1 intersects on the resolution the end curve of S_1^2, and that S_1^2 is a chain consisting of $\mu_{1,2}$ curves. We change coordinates again as in case 2, and see that after $\ell = \mu_{1,2} - 1 = \frac{k_0}{k_1 - k_0} - 1$ further blowing-ups, the strict transform has parametrization

$$x = t^{k_1 - k_0} + \dots, \quad y = t^{k_1 - k_0},$$

so that E_1, given by $x = 0$, does intersect the strict transform transversally, as does $E_{\ell+1}$. Thus we have the following picture:

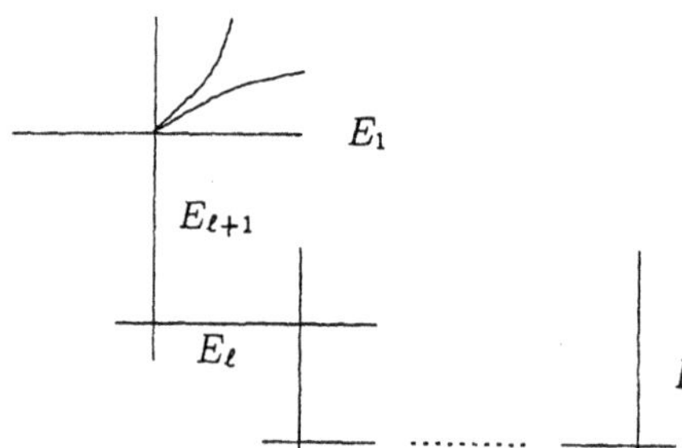

So indeed, during the next blowing-up E_1 and the strict transform separate, while E_1 and $E_{\ell+2}$ intersect transversally. So E_1 intersects the end curve of S_1^2.

□

Putting all results together we obtain the following theorem.

Theorem 5.3.18. *For a given irreducible plane curve singularity the following data determine each other:*

(1) the Puiseux pairs $P(R)$;

(2) the characteristic exponents $E(R)$;

(3) the minimal set of generators of the semigroup $G(R)$;

(4) the multiplicity sequence $(m_0, \ldots, m_{k-1})$;

(5) the resolution graph.

Exercises

5.3.19. Compute the resolution graph, the characteristic exponents, the semigroup and the multiplicity sequence for the following singularities:

(1) E_8 given by $y^3 - x^5 = 0$, and

(2) the singularity given by $x^{33} - 3x^{22}y^2 + 6x^{21}y - x^{20} + 3x^{11}y^4 + 2x^{10}y^3 - y^6 = 0$.

5.4 Reducible Plane Curve Singularities

As in the irreducible case, we can define the standard resolution of a reducible plane curve singularity, as follows:

Definition 5.4.1. Let $C = \cup_{i=1}^r C_i \subset U$ be a small representative of a reducible plane curve singularity, with branches $C_1, \ldots, C_r$, $r \geq 2$. Assume that $X_i \xrightarrow{\pi_i} \ldots \xrightarrow{\pi_2} X_1 \xrightarrow{\pi_1} U$ is already defined. Denote by $E^{(i)} = (\pi_1 \circ \cdots \circ \pi_i)^{-1}(0)$ the exceptional divisor and by $C^{(i)} = \overline{(\pi_1 \circ \cdots \circ \pi_i)^{-1}(C \setminus \{0\})}$ the strict transform of C.

Let $X_{i+1} \xrightarrow{\pi_{i+1}} X_i$ be the blowing-up of X_i in all points of $C^{(i)} \cap E^{(i)}$ which are still singular on $C^{(i)}$ or nontransversal intersection of $C^{(i)}$ with $E^{(i)}$, that is, points with intersection multiplicity of $C^{(i)}$ and $E^{(i)}$ greater than one, or where two exceptional divisors and $C^{(i)}$ meet.

Then $X_k \xrightarrow{\pi_k} \cdots \xrightarrow{\pi_1} X_1 \longrightarrow U$ is called a *standard resolution* of (C, o) if all branches of $C^{(k)}$ are smooth, do not intersect each-other, do intersect just one component of $E^{(k)}$, and do intersect this component transversally.

Theorem 5.4.2. *Every germ of a plane curve singularity has a standard resolution.*

Proof. We have to prove that the process of constructing $X_{i+1} \xleftarrow{\pi_{i+1}} X_i$ as described above stops after a while, that is, there exists an n such that $C^{(n)}$ has no singularities on $C^{(n)} \cap E^{(n)}$ and intersects $E^{(n)}$ transversally. From the irreducible case, we know that we can resolve the singularities of all branches such that after a while we have at most nontransversal intersections of $C^{(n)}$ and $E^{(n)}$. Now it might still happen that two (smooth) branches of $C^{(n)}$ intersect each other. But we studied the intersections of smooth curves under blowing-up, and we know that the intersection multiplicity drops with each blowing-up, see 5.3.6. This proves the theorem. □

Definition 5.4.3. Consider the standard resolution of a reducible plane curve singularity $(C, 0) = \cup(C_i, 0)$. Write $C_i^{(0)} := C_i$. Furthermore, for some i and k assume that $C_i^{(j)} \cap C_k^{(j)} \neq \emptyset$, but $C_i^{(j+1)} \cap C_k^{(j+1)} = \emptyset$. Then $\gamma(C_i, C_k) := j + 1$ is called the *contact number* of the branches C_i and C_k.

Definition 5.4.4. With the notations of Definition 5.4.1 we consider the following weighted graph, the *resolution graph* of C:

(1) To each component of $E^{(k)}$ a point $\bullet$ is associated.

(2) To each component of $C^{(k)}$ a point $*$ is associated.

(3) Two points are connected by an edge if the corresponding components intersect.

(4) The points of type (1) are weighted. Let E be a component of $E^{(k)}$. We give to the corresponding point the weight i, if E is created in the i–th level of the blowing ups, that is, i is minimal such that $\pi_{i+1} \circ \cdots \circ \pi_n(E)$ is not a point.

Theorem 5.4.5. *Let* $(C, 0) = \cup_{i=1}^r (C_i, 0)$ *be a plane curve singularity. The following data determine each other:*

(1) The resolution graph of C.

(2) The resolution graph of the branches $(C_i, 0)$, *together with the contact numbers* $\gamma(C_i, C_j)$ *for* $i \neq j$.

Proof.
Step 1. First suppose we know the resolution graph of C. Take a branch $(C_i, 0)$, and delete all stars from the resolution graph which belong to the branches $(C_j, 0)$ for $j \neq i$. Call a point of the remaining graph contractible, if it satisfies one of the following conditions:

(1) It is not connected to the star, and connected only to points of lower weight.

(2) It is connected to a star, and to just one other point of lower weight.

This exactly means that the corresponding curve was created in a blowing-up, which would not have been necessary for getting the standard resolution of the branch $(C_i, 0)$. From the resolution process, it follows that such a point is connected to at most two points in the graph. If it is connected to just one other point, then delete the point of highest weight, and the edge connected to it. If it is connected to two points A and B say, then delete the point of highest weight, the connecting edges, and connect A and B by an edge. This corresponds to "blowing-down" the curve corresponding to this highest weight. Now iterate this process, until it is no longer possible. We then arrive at the resolution graph of $(C_i, 0)$. For example, if we have the following resolution graph

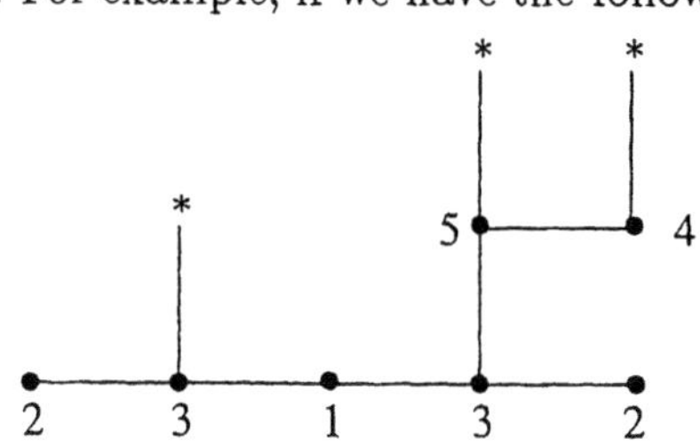

we can get the resolution graph of the curve corresponding to the star on the left by the following process

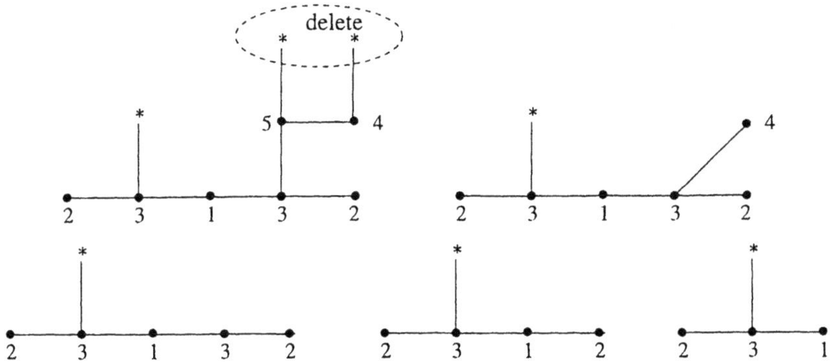

To get the contact number $\gamma(C_i,C_j)$, we remove all the stars corresponding to the branches C_k with $k \neq i,j$, and then successively remove points with property (1) and (2). If this is no longer possible, then we are left with the resolution graph of the union of $(C_i,0)$ and $(C_j,0)$. Consider the following numbers.

(1) The maximal natural number k such that the numbers $1,\ldots,k-1,k$ occur as weight just once in the resolution graph of the union of $(C_i,0)$ and $(C_j,0)\}$.

(2) The weights of the vertices connected to the stars in the resolution graph of the union of $(C_i,0)$ and $(C_j,0)$.

The contact number of $(C_i,0)$ and $(C_j,0)$ is the minimum of these numbers. So for example in the resolution graph above, the contact number between the curves belonging to the right two stars is equal to $\min\{5,5,4\} = 4$, between the curve belonging to the star on the left and the other two is one.

Step 2. Conversely, suppose we know the resolution graphs of the $(C_i,0)$ and the contact numbers. We give an algorithm for finding the resolution graph of $(C,0)$. The proof goes by induction on r. The case $r = 1$ is trivial. So suppose that the resolution graph Γ of $\cup_{i=1}^{r-1}(C_i,0)$ has been constructed out off the resolution graphs of the $(C_i,0)$ and the contact numbers. For $i = 1,\ldots,r-1$, define $\ell(i) = \gamma(C_r,C_i)$, and let $l(r)$ be the maximum of $\gamma(C_i,C_r)$ for $i \neq r$. Let $w(i)$ be the weight of the vertex in Γ belonging to the exceptional curve which intersects the strict transform of C_i, for $i = 1,\ldots r-1$. Let $w(r)$ be the vertex of highest weight in the resolution graph Γ_r of $(C_r,0)$. It is clear that the point in the resolution graph connected to the star, which corresponds to $(C_i,0)$ has weight $\max\{\ell(i),w(i)\}$. So if $\ell(i) > w(i)$, we need, starting from the standard resolution, $\ell(i) - w(i)$ extra blowing-ups. Therefore, if $\ell(i) > w(i)$, we modify the graph Γ as suggested by the following picture.

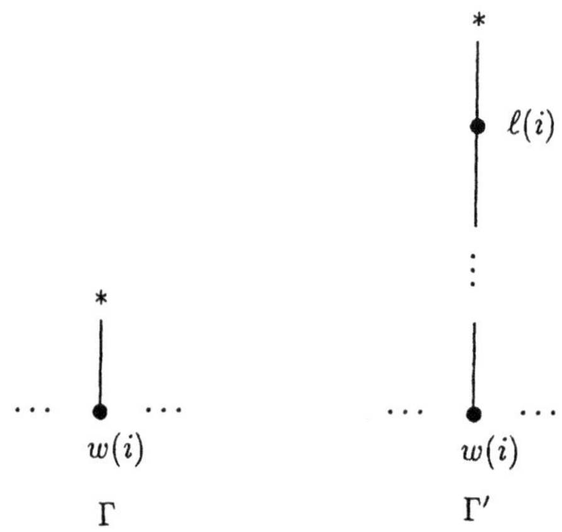

If $\ell(i) \leq w(i)$ we leave Γ unchanged at the star corresponding to $(C_i\, o)$. We get a modified graph Γ'. Similarly, we get a modified graph Γ'_r. Let $k = \max\{\ell(j) : j < r\}$ be the maximum of the contact numbers of (C_r, o) with (C_j, o) for $j < r$. Without loss of generality, we may suppose that this maximum is achieved for $j = 1$. Let $\pi_1 \circ \cdots \circ \pi_k : X_k \longrightarrow U$ be the map which occurs in the standard resolution process of $\cup_{i=1}^r (C_i, o)$ after k steps. So by definition of contact number, the strict transforms $C_1^{(k)}$ and $C_r^{(k)}$ intersect a common component $E_1^{(k)}$ of $E^{(k)}$, but do not intersect each other. In fact, by choice of k, $C_r^{(k)}$ does not intersect any of the $C_j^{(k)}$. Let p be the point of Γ' which corresponds to $E_1^{(k)}$. If $k > 1$, it follows from the resolution process that there is a unique component E' of $E^{(k)}$ which intersects $E_1^{(k)}$ and was created under π_{k-1}. The point of Γ' which corresponds to E' we call q. It has weight $k - 1$. There are two possible cases.

(1) $k > 1$, and the curve $C_r^{(k)}$ intersects E'. Note that the points p and q in Γ' are connected by an edge. The following is a possible picture of a part of the resolution graph after k blowing-ups.

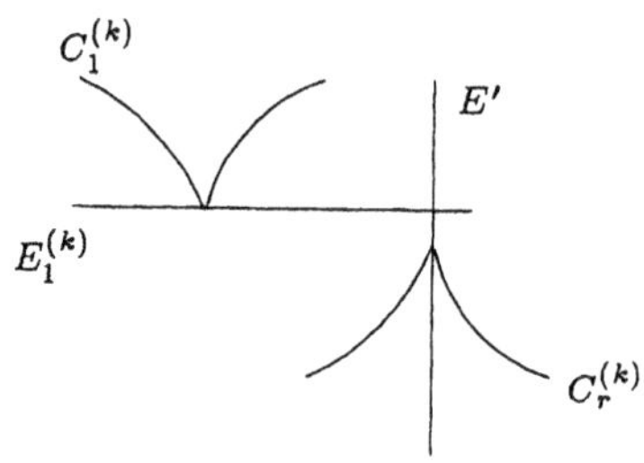

(2) $k = 1$, or the curve $C_r^{(k)}$ does not intersect E'.

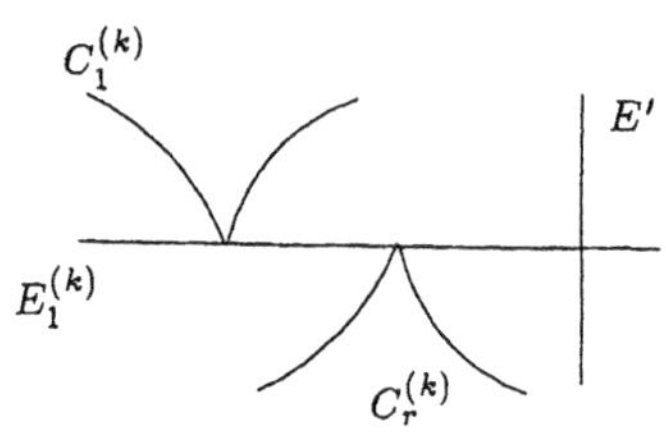

In the second case, consider the subgraph Γ_r'' of Γ_r' which one gets by deleting all points of weight less than k. So this is the resolution graph of $C_r^{(k-1)}$ with all weights shifted by $k-1$. As we already noted that none of the $C_j^{(k)}$ intersects $C_r^{(k)}$, it follows that the resolution graph of (C, o) is obtained by identifying the unique point of Γ_r'' of weight k with p, which also has weight k.

In the first case, consider the subgraph Γ_r'' of Γ_r' which one gets by deleting all points of weight less than $k-1$. The graph Γ_r'' looks like

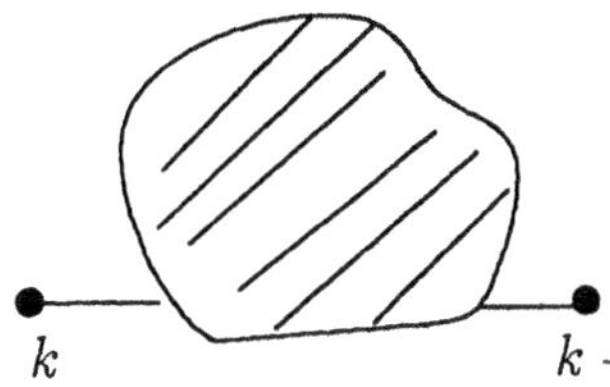

and we get the resolution graph (C, o) by doing surgery: In Γ' one replaces the subgraph

by Γ_r''. So we see that we can construct the resolution graph (C, o) from the resolution graphs of $(C_i\, o)$ and the contact numbers. □

Example 5.4.6. We consider the reducible plane curve singularity given by $f = 0$, where $f = f_1 f_2 f_3$ with

$$f_1 = (y^2 - x^3)^2 - 4x^5y - x^7,$$
$$f_2 = y^2 - x^3,$$
$$f_3 = y^3 - x^2$$

The second and third branch are cusp singularities, and we know how the resolution for those looks like, see the fourth part of Example 5.3.6. The resolution of the first branch has been studied in Example 5.3.9. So we have to put the following three resolution graphs into one graph.

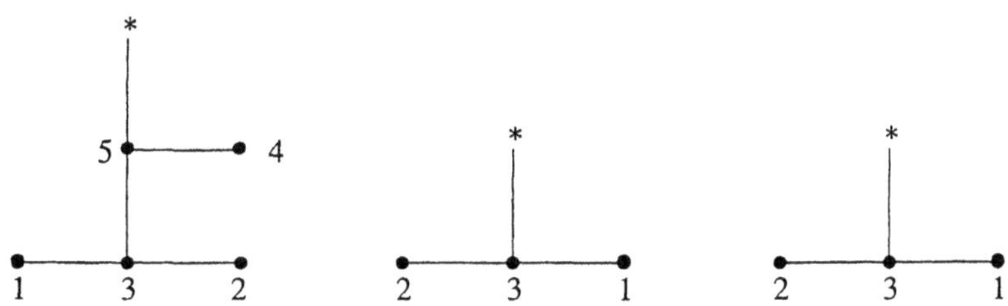

For this we have to determine the contact numbers.

The interesting parts of the strict transforms $C_1^{(1)}$ and $C_2^{(1)}$ of the blowing-up of the branches (C_1, o) and (C_2, o) are in the first chart:

$C_1^{(1)}$: defined by $(y^2 - x)^2 - 4x^2y - x^3 = 0$,
$C_2^{(1)}$: defined by $y^2 - x = 0$,
E_1 : defined by $x = 0$.

In the second chart we obtain:
$C_3^{(1)}$: defined by $y - x^2$,
E_1 : defined by $y = 0$.
So we have the following picture

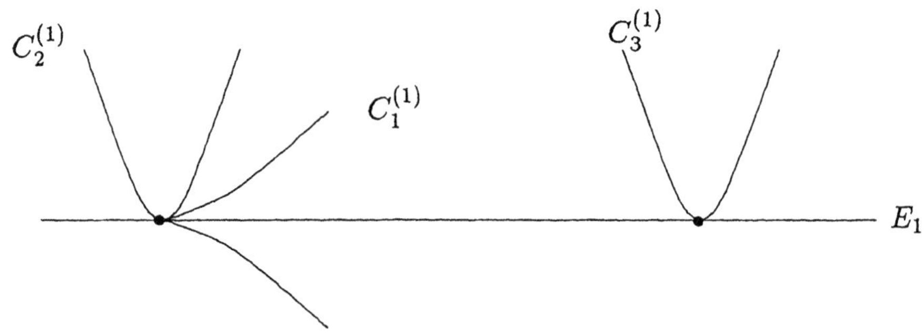

and deduce that the contact number of (C_1, o) and (C_3, o) is equal to one. Similarly, the contact number of (C_2, o) and (C_3, o) is equal to one. We still have to determine the contact number of (C_1, o) and (C_2, o). So we look at the resolution of the union of (C_1, o) and (C_2, o). One can calculate that the sequence of blowing-ups continues as suggested in the following picture.

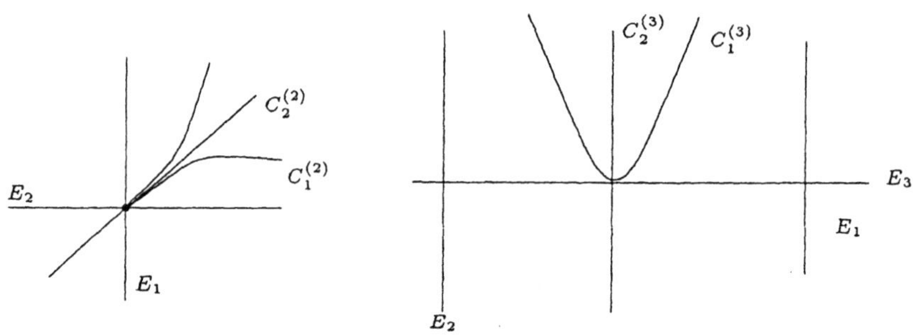

We see that the on the fourth blowing-up the branches $C_1^{(4)}$ and $C_2^{(4)}$ are separated. Thus the contact number is four. It follows that the resolution graph of the curve (C, o) is equal to

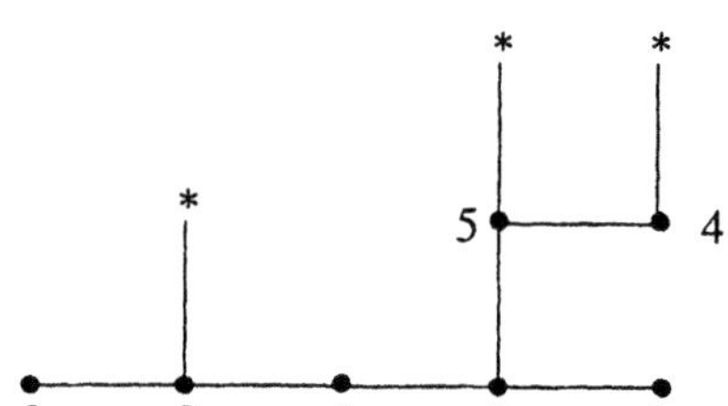

by using the algorithm described in the proof. Of course, we could also have gotten this resolution graph by simply doing the whole resolution process.

Finally, summarizing all results, we obtain the following theorem:

Theorem 5.4.7. *Let (C,o) be a germ of a plane curve singularity. The following data determine each other:*

(1) the multiplicity sequences of the branches of C and the contact numbers between the branches;

(2) the resolution graphs of the branches of C and the contact numbers between the branches;

(3) the resolution graph of C.

Next we want to characterize the contact numbers of the branches by their intersection multiplicity.

Theorem 5.4.8. *Let (C,o) respectively (D,o) be irreducible plane curve singularities with multiplicity sequences $(m_0,\dots,m_g)$ respectively $(n_0,\dots,n_h)$. For $k \geq g+1$ define $m_k = 1$, and similarly, for $k \geq h+1$ define $n_k = 1$. Then*

$$(C,o)\cdot(D,o) = \sum_{i=0}^{\gamma(C,D)} m_i \cdot n_i.$$

Proof. Of course we use Lemma 5.1.5 and induction on $\gamma(C,D)$. After a general linear coordinate change we may assume that (D,o) is given by a Weierstraß polynomial $f = y^n + a_1y^{n-1} + \ldots + a_n$ of degree $n = n_0$ in y, and that (C,o) has as parametrization

$$x(t) = t^m,\ y(t) = t^q + \ldots, \quad \text{where } m = m_0.$$

By Lemma 5.1.5 we have $(C,o)\cdot(D,o) = \mathrm{ord}_t(f)$. We look at what happens after blowing-up. Then

$$x(t) = t^m,\ v(t) = t^{q-m} + \ldots.$$

is a parametrization of the strict transform $C^{(1)}$. An equation for the strict transform $D^{(1)}$ of (D,o) is given by $\overline{f}(x,v) = 0$, where $\overline{f}(x,v)$ is defined by

$$\overline{f}(x,v) = x^{-n} f(x,xv).$$

Therefore,

$$\begin{aligned}\mathrm{ord}_t\Big(f\big(x(t),y(t)\big)\Big) &= \mathrm{ord}_t\Big(f\big(x(t),x(t)v(t)\big)\Big)\\ &= \mathrm{ord}_t\big(x(t)^n\big)\cdot \mathrm{ord}_t(\overline{f}(x(t),v(t)))\\ &= m_0\cdot n_0 + C^{(1)}\cdot D^{(1)}\end{aligned}$$

by induction. □

Example 5.4.9. Consider the curve (C,o) given by $(y^2-x^3)^2 - 4x^5y - x^7 = 0$, and (D,o) given by $g = y^2 - x^3 = 0$. The multiplicity sequence of (C,o) is (4,2,2,1,1), and the multiplicity sequence of (D,o) is (2,1,1). We have seen that the contact number is four, see Example 5.4.6. It follows that the intersection multiplicity is equal to $4\cdot2+2\cdot1+2\cdot1+1\cdot1 = 13$. The curve (C,o) is given by the parametrization $x = t^4$, $y = t^6 + t^7$. So g is in $\mathbb{C}\{t\}$ equal to $(t^6+t^7)^2 - (t^4)^3 = 2t^{13} + t^{14}$, so its order is thirteen. Therefore, conversely, one can now deduce that the contact number is four.

As in the irreducible case, we define the δ–invariant.

Definition 5.4.10. Let (C,o) be a curve singularity and $(\widetilde{C},\widetilde{o})$ its normalization. Then we define the *δ–invariant* by $\delta(C,o) = \dim_{\mathbb{C}}(\mathcal{O}_{\widetilde{C},\widetilde{o}}/\mathcal{O}_{C,o})$.

The following easy lemma is useful if one wants to calculate the δ–invariant for reducible singularities.

Lemma 5.4.11 (Lemma of Hironaka). *Let (C,o), and (D,o) in $(\mathbb{C}^n,o)$ be reduced curve singularities which do not have a component in common. Then*

$$\delta((C,o)\cup(D,o)) = \delta(C,o)+\delta(D,o)+(C,o)\cdot(D,o).$$

Proof. Set $I = \mathscr{I}(C,o) \subset \mathcal{O}_n$ and $J = \mathscr{I}(D,o) \subset \mathcal{O}_n$. Consider the following exact sequence

$$0 \longrightarrow \mathcal{O}_n/(I\cap J) \longrightarrow \mathcal{O}_n/I \oplus \mathcal{O}_n/J \longrightarrow \mathcal{O}_n/(I+J) \longrightarrow 0.$$

We conclude from this exact sequence that the vector space dimension of $(\mathcal{O}_{C,o}\oplus\mathcal{O}_{D,o})/\mathcal{O}_{C\cup D,o}$ and the intersection multiplicity $(C,o)\cdot(D,o)$ are equal. By splitting of the normalization we know that $\mathcal{O}_{\widetilde{C},\widetilde{o}}\oplus\mathcal{O}_{\widetilde{D},\widetilde{o}} = \mathcal{O}_{\widetilde{C\cup D},o}$. Therefore $\delta(C\cup D,o)$ is equal to $\dim_{\mathbb{C}}(\mathcal{O}_{\widetilde{C},\widetilde{o}}\oplus\mathcal{O}_{\widetilde{D},\widetilde{o}}/\mathcal{O}_{C\cup D,o})$, which is equal to

$$\begin{aligned}\dim_{\mathbb{C}}\big((\mathcal{O}_{\widetilde{C},\widetilde{o}}\oplus\mathcal{O}_{\widetilde{D},\widetilde{o}})/(\mathcal{O}_{C,o}\oplus\mathcal{O}_{D,o})\big) &+ \dim_{\mathbb{C}}\big((\mathcal{O}_{C,o}\oplus\mathcal{O}_{D,o})/\mathcal{O}_{C\cup D,o}\big)\\ &= \dim_{\mathbb{C}}(\mathcal{O}_{\widetilde{C},\widetilde{o}}/\mathcal{O}_{C,o}) + \dim_{\mathbb{C}}(\mathcal{O}_{\widetilde{D},\widetilde{o}}/\mathcal{O}_{D,o}) + (C,o)\cdot(D,o)\\ &= \delta(C,o)+\delta(D,o)+(C,o)\cdot(D,o). \quad\square\end{aligned}$$

We wish to finish this chapter with a proof of Max Noether's Theorem. This theorem tells us how to compute the δ–invariant from the multiplicity sequence of an irreducible plane curve singularity. In the proof of Noether's Theorem, we need the following lemma.

Lemma 5.4.12. *Let $(C,o) = (V(f),o)$ be an irreducible plane curve singularity. Let the multiplicity be $m = mult(C,o)$, and $c = c(C,o)$ be the conductor. Suppose $g \in \mathcal{O}_{C,o} \subset \mathcal{O}_{\widetilde{C},o} = \mathbb{C}\{t\}$ with $\mathrm{ord}_t(g) \geq c$. Then the multiplicity of g is at least $m-1$.*

Proof. Suppose the converse. Then after a general linear coordinate change, we may assume that g is regular in y of order $k < m-1$ in y. We may also assume that f is regular of order m in y, so that (C,o) has parametrization $x = t^m$, $y = t^s + \ldots$ We now look at $h := \frac{y}{x}\cdot g \in \mathcal{O}_{\widetilde{C},o}$. This is indeed in $\mathcal{O}_{\widetilde{C},o}$, since $\frac{y}{x} = t^{s-m}\cdot$ unit $\in \mathbb{C}\{t\} = \mathcal{O}_{\widetilde{C},o}$. Obviously $\mathrm{ord}_t(h) \geq c(C,o)$, hence $h \in \mathcal{O}_{C,o}$. Thus in $\mathcal{O}_{C,o}$ we have the equality $yg = xh$. Therefore,

$$qf = yg - xh$$

for some $q \in \mathbb{C}\{x,y\}$. Now on the right-hand side a term y^{k+1} occurs, as it occurs in yg but not in xh. But it cannot occur on the left-hand side, as the multiplicity of f is m, and $k+1 < m$. This is a contradiction. $\square$

Theorem 5.4.13 (M. Noether). *Let (C,o) be an irreducible plane curve singularity, $(m_0,\ldots,m_r)$ be the multiplicity sequence of (C,o). Then*

$$\delta(C,o) = \sum_{i=0}^{r}\frac{m_i(m_i-1)}{2}.$$

Proof. As $c(C,o) = 2\delta(C,o)$ by 5.2.4, it suffices to show $c(C,o) = \sum_{i=0}^{r} m_i(m_i - 1)$. The formula, obviously, holds for the case of a smooth curve, since $c(C,o) = 0$ and $m_0 = 1$. Let $C^{(1)}$ be the strict transform of (C,o) under blowing-up U in the origin for some sufficiently small neighborhood U. We claim that if the formula holds for $C^{(1)}$, then it holds for (C,o). By induction on r, this suffices to prove the theorem. Suppose (C,o) is given by the parametrization

$$x = t^m,\ y = t^s + \ldots, \quad s > m.$$

Step 1. We first show $\alpha \in \Gamma(C^{(1)}) \implies \alpha + m(m-1) \in \Gamma(C,o)$.

The strict transform $C^{(1)}$ has parametrization ($x = x, y = xv$):

$$x = t^m,\ v = t^{s-m} + \ldots.$$

Let $\overline{g} = \overline{g}(x,v) \in \mathcal{O}_{C^{(1)}}$. As we are looking at the vanishing order of $\overline{g}$ in $\mathcal{O}_{\widetilde{C},o} = \mathbb{C}\{t\}$, and $C^{(1)}$ is the zero set of a function which is regular of order $\leq m$, we may, by applying the Weierstraß Division Theorem, assume that $\deg_v(\overline{g}) \leq m-1$. We write

$$\overline{g} = a_{m-1}v^{m-1} + a_{m-2}v^{m-2} + \ldots + a_0, \quad a_i \in \mathbb{C}\{x\}.$$

Multiplying $\overline{g}$ with x^{m-1} we get a function that we call $g(x,y)$:

$$g(x,y) := x^{m-1}\overline{g}(x,v) = a_{m-1}y^{m-1} + xa_{m-2}y^{m-2} + \ldots + a_0x^{m-1}.$$

As (C,o) and $C^{(1)}$ have the same normalization, it follows that $g(x,y) \in \mathcal{O}_{C,o}$ and has vanishing order $\operatorname{ord}_t(\overline{g}) + m(m-1)$. Hence it follows that

$$\alpha \in \Gamma(C^{(1)}) \implies \alpha + m(m-1) \in \Gamma(C,o),$$

as claimed.

Step 2. To finish the proof we show that

$$\nexists\, g \in \mathcal{O}_{C,o} \text{ with } \operatorname{ord}_t(g) = c(C^{(1)}) + m(m-1) - 1.$$

Suppose the converse, that is, let $g \in \mathcal{O}_{C,o}$ with $\operatorname{ord}_t(g) = c(C^{(1)}) + m(m-1) - 1$. It is left as Exercise 5.4.15 to show that we may take such g which is moreover reduced. By the first step we know that $c(C,o) \leq c(C^{(1)}) + m(m-1) - 1$. By the previous lemma, we know that the multiplicity of g, say q, is at least $m-1$. Let $\overline{g}(x,v) = x^{-q}g(x,xv)$. Then 5.4.8 gives that $(C,o) \cdot (V(g),o) = C^{(1)} \cdot (V(\overline{g}),o) + qm$. Hence

$$C^{(1)} \cdot (V(\overline{g}),o) = c(C^{(1)}) + m(m-1) - 1 - qm,$$

because $(C,o) \cdot (V(g),o) = \operatorname{ord}_t(g)$. Thus, we have found a function $\overline{g} \in \mathcal{O}_{C^{(1)}}$ which has vanishing order $\operatorname{ord}_t(\overline{g}) = c(C^{(1)}) + m(m-1) - qm - 1$. Now $q \geq m-1$, so that $x^{q-m+1} \cdot \overline{g} \in \mathcal{O}_{C^{(1)}}$. This function has vanishing order $c(C^{(1)}) - 1$. This is a contradiction to the definition of the conductor $c(C^{(1)})$. □

Example 5.4.14. Consider the plane curve singularity given by $(y^2 - x^3)^2 - 4x^5y - x^7 = 0$. The multiplicity sequence is $(4,2,2,1,1)$, so the delta–invariant is equal to $6+1+1 = 8$. This can also be calculated immediately, with the aid of the characteristic exponents, which are 4,6 and 7. Thus one can calculate the minimal set of generators of the semigroup to be 4,6 and 13. The fact that $\delta = 8$ is then easily seen by counting.

Exercises

5.4.15. Prove that one may take g in the proof of 5.4.13 to be reduced.
(Hint: Add elements of high order to some factors which are multiple.)

5.4.16. Let $(C, o) = \bigcup(C_i, o)$ be a curve singularity. Consider the standard resolution of (C, o), for some small U:

$$X_k \longrightarrow \ldots \longrightarrow X_1 \longrightarrow U.$$

Let $C^{(j)}$ be the strict transform of C in X_j. An infinitely near point of C is, by definition, a singular point of $C^{(j)}$ for some j, $C^{(0)} = C$. Prove that

$$\delta(C, o) = \sum_{\substack{p \text{ infinitely} \\ \text{near point}}} \frac{m_p(m_p - 1)}{2}$$

where m_p is the multiplicity of $C^{(j)}$ at p.
(Hint: Use the Lemma of Hironaka 5.4.11, Lemma 5.4.8 and Lemma 5.4.13.)

5.4.17. Let (C, o) be the plane curve singularity defined by $f = (x^3 - y^4)(x^3 - y^7)(y^{11} - x^5) = 0$. Compute the multiplicity sequences of the branches of (C, o), the contact numbers, and the resolution graph of (C, o). Also construct the resolution graph of (C, o) out off the resolution graphs of the branches and their contact numbers.

5.4.18. Compute the δ–invariant of

(1) the D_4-singularity: $y(y^2 - x^2) = 0$,

(2) the E_7-singularity: $y(y^2 - x^3) = 0$, and

(3) m different lines through the origin.

5.4.19. Let (C, o) be a reduced plane curve singularity given by a Weierstraß polynomial f in y of degree b and multiplicity m. Let $C^{(1)}$ be the strict transform under blowing-up, E the exceptional divisor. Prove:

(1) $\sum_{p \in C^{(1)} \cap E} (E, p) \cdot (C^{(1)}, p) = b$.

(2) $\mu(X, o) - 1 = \sum_{p \in C^{(1)} \cap E} \mu(C^{(1)}, p) + m(m - 1)$.
(Hint: Use Exercise 5.1.29, and study the behavior of the terms in that formula under blowing-up; use part 1 of this exercise and use Lemma 5.4.8.)

(3) Deduce Milnor's formula

$$\mu(C, o) = 2 \cdot \delta(C, o) - r(C, o) + 1,$$

where $r(C, o)$ is the number of branches of (C, o).

6 The Principle of Conservation of Number

The aim of this chapter is to study the behavior of certain invariants of singularities in a family. The idea is that the invariants should be constant in the following sense: if in the family a singular point splits into several singular points by varying the parameter, then the sum of the invariants of those singular points should be equal to the invariant of the point we started with.

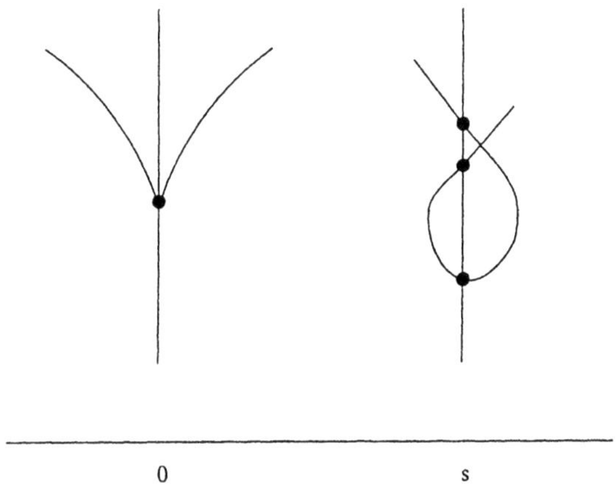

The picture describes a typical situation. In the fiber above the point 0, we have an intersection of a line with a cusp and the intersection multiplicity at that point is three. In a small neighborhood we have in the fiber three different intersection points and a transversal intersection at each point of multiplicity 1. Altogether we see a conservation of the intersection number. We will see that other invariants, such as the Milnor number and the δ-invariant have the same property.

The basis to prove such this property is the theory of sheaves. Therefore, we start the chapter with an overview of the basic facts of sheaf theory. In short, a presheaf $\mathscr{S}$ of rings (or modules, or ...) on a topological space gives for all open subsets U a ring $\mathscr{S}(U)$ (or module, or ...), with some natural conditions. An element of $\mathscr{S}(U)$ is called a section over U. One can talk about the germ of a section at any point of U. The set of germs at x is called the stalk at x, notation $\mathscr{S}_x$. A sheaf is a presheaf, which is, loosely speaking, determined by its stalks. The application we have in mind is that these stalks are modules whose vector space dimension gives an invariant of a germ of an analytic space. Thus, we want to compare the various stalks for a given sheaf. Now general sheaves are like "sand" but *coherent* sheaves are not. This is reflected in the statement that a map of sheaves which induces an isomorphism at the stalk at x, is an isomorphism in a small neighborhood of x. Loosely speaking, the stalk determines the sheaf in an open neighborhood. The problem with the theory of coherent sheaves is that it is difficult to give nontrivial examples. The only examples one gets from the definition are the zero sheaf, and finitely generated sheaves over a point. However, it is true that most sheaves we consider are coherent. For example $\mathcal{O}_X$, that is, the sheaf of holomorphic functions, the ideal sheaf of an analytic set and the normalization sheaf of an analytic space are coherent. Moreover, for finite maps coherence is preserved (Finite Mapping Theorem). These theorems are all proved in Section 6.3. In Section 6.2 we develop the technical tools. Here we prove the Meta-Theorem, which says that coherence of sheaves is preserved under all "reasonable" algebraic operations. In Section 6.2 we discuss these coherent sheaves, but give only two examples.

On this basis, in Section 6.4, the principle of conversion of number is studied. In particular we study the behavior of the invariants mentioned above. Finally, in Section 6.5, we study Cohen-Macaulay spaces. This notion is closely related to the principle of conservation of number. Hypersurfaces, more generally, complete intersections are Cohen-Macaulay.

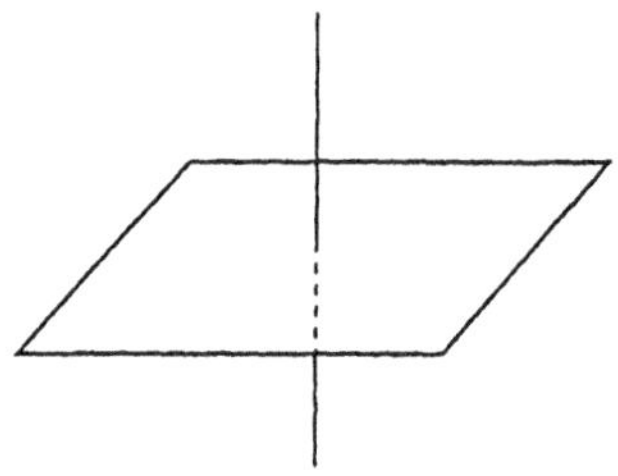

The space in the picture is not Cohen-Macaulay. There is an algebraic method of deciding whether spaces are Cohen-Macaulay. This method is by using resolutions and the Auslander-Buchsbaum formula. From this it follows that a space is Cohen-Macaulay if and only if the resolution has the right length.

For lack of a better site, we will finally discuss the Hilbert-Burch Theorem. It says that the ideal of Cohen-Macaulay codimension two spaces can be described by the t minors of a certain $t \times (t+1)$ matrix. It is used in Chapter 10 to give examples of singularities for which there exists a semi-universal deformation over a smooth space.

6.1 Sheaves

Definition 6.1.1. Let X be a topological space. A *presheaf* $\mathscr{S}$ of rings on X consists of the following data

(1) for each open set U of X a ring $\mathscr{S}(U)$; elements of $\mathscr{S}(U)$ we call *sections* of $\mathscr{S}$ over U, elements of $\mathscr{S}(X)$ are called *global sections* of $\mathscr{S}$;

(2) for each inclusion $V \subset U$ of open sets in X a ring homomorphism (called the *restriction map*):

$$\rho_{UV} : \mathscr{S}(U) \longrightarrow \mathscr{S}(V).$$

These data have to satisfy the following conditions:

1. $\mathscr{S}(\emptyset) = 0$,
2. $\rho_{UU} : \mathscr{S}(U) \longrightarrow \mathscr{S}(U)$ is the identity map,
3. if $W \subset V \subset U$ are three open subsets of X then $\rho_{UW} = \rho_{VW} \circ \rho_{UV}$.

For $s \in \mathcal{S}(U)$ one usually writes $s_{|V}$ instead of $\rho_{UV}(s)$. For all open subsets V we get in a natural way a presheaf $\mathcal{S}_{|V}$ which we call the restriction of $\mathcal{S}$ to V.

Similarly one defines presheaves of abelian groups, of vector spaces, $\mathbb{C}$-algebras, etc.

Examples 6.1.2.

(1) Let X be the topological space consisting of just one point. Then a presheaf of rings on X can be identified with a ring.

(2) Let X be a topological space. For all open $U \subset X$ put

$$\mathscr{C}(U) := \{f : U \longrightarrow \mathbb{C},\ f \text{ is continuous}\}.$$

For $V \subset U$, let ρ_{UV} be the obvious restriction of functions from U to V. Then one shows without difficulty that this defines a presheaf of rings, or even of $\mathbb{C}$-algebras on X.

(3) Consider an open set $X \subset \mathbb{C}^n$. Put for $U \subset X$ open:

$$\mathcal{O}(U) := \{f\colon U \longrightarrow \mathbb{C},\ f \text{ is holomorphic}\},$$

and again for an open subset V of U, the map ρ_{UV} is given by restriction. This is a presheaf of $\mathbb{C}$-algebras.

(4) Consider an open set $X \subset \mathbb{C}^n$. For $U \subset X$ open we put

$$\mathcal{O}'(U) := \{f\colon U \longrightarrow \mathbb{C},\ f \text{ is holomorphic}\},$$

but we define different restriction maps. For $V \subsetneqq U$ we put $\rho_{UV} = 0$, and $\rho_{UU} = \text{id}$. Then $\mathcal{O}'$ is a presheaf.

(5) Let A be a ring, X a topological space. The constant presheaf $\mathscr{C}$ is given by $\mathscr{C}(U) = A$ if $U \neq \emptyset$, and $\mathscr{C}(\emptyset) = 0$. The maps ρ_{UV} is the identity map if $V \neq \emptyset$. Moreover, $\rho_{U\emptyset}$ is the zero map for all open U.

Definition 6.1.3 (Sheaf Conditions). Let X be a topological space, $\mathcal{S}$ a presheaf on X. Then $\mathcal{S}$ is called a *sheaf* if the following two sheaf axioms are satisfied.

(1) Let U be an open set in X, and $U = \cup_{i \in I} V_i$ be an open covering of U. Let $s \in \mathcal{S}(U)$ be a section of $\mathcal{S}$ over U, such that $s_{|V_i} = 0$ for all i. Then $s = 0$.

(2) Let U be an open set in X, and $U = \cup_{i \in I} V_i$ be an open covering of U. Let $s_i \in \mathcal{S}(V_i)$ be given such that $s_{i|V_i \cap V_j} = s_{j|V_i \cap V_j}$ for all $i,j \in I$. Then there exists an $s \in \mathcal{S}(U)$ such that $s_{|V_i} = s_i$.

A presheaf which satisfies the sheaf axioms, so is in fact a sheaf, is also called a *canonical presheaf.*

Remark 6.1.4. It follows from the first condition that such an s, existing according to the second condition, is uniquely determined.

Example 6.1.5. In 6.1.2, the first three examples, in fact, are sheaves. The fourth example, the presheaf $\mathcal{O}'$ is not a sheaf, as it does not satisfy the first condition. Moreover, the *constant presheaf* A is, in general, not a sheaf. In fact, suppose U,V are two open subsets of X with $U \cap V = \emptyset$. Let a,$b \in A$, with $a \neq b$. Then the restrictions of a and b to $U \cap V = \emptyset$ are equal, but there does not exist a section of $U \cup V$, restricting to a and b, respectively.

Definition 6.1.6. Let $\mathscr{S}$ be a presheaf (of rings) on a topological space X and $p \in X$ be a point. Consider the collection of rings:

$$\{\mathscr{S}(U) : U \ni p \text{ an open neighborhood of } p\}.$$

We define an equivalence relation on the disjoint union of this collection:

$$s \in \mathscr{S}(U) \sim_p t \in \mathscr{S}(V) \iff \text{there exists an open neighborhood } W \subset U \cap V \text{ of } p \text{ such that } s_{|W} = t_{|W}.$$

The equivalence class of s in p we call the germ of s in p, notation s_p. The set of equivalence classes we call the *stalk* of $\mathscr{S}$ at p, and will be denoted by $\mathscr{S}_p$. It is not difficult to show that, in a natural way, $\mathscr{S}_p$ is a ring again, see Exercise 6.1.28. The *support* of $\mathscr{S}$ is the set of all points $p \in X$ such that $\mathscr{S}_p \neq 0$. The support of $\mathscr{S}$ is denoted by $\operatorname{supp}(\mathscr{S})$.

Definition 6.1.7. Let two presheaves $\mathscr{S}$ and $\mathscr{T}$ (of rings) on a topological space, with restriction mappings ρ_{UV} and ρ'_{UV} respectively, be given. A *morphism of presheaves* $f : \mathscr{S} \longrightarrow \mathscr{T}$ is a collection of ring homomorphisms

$$f_U : \mathscr{S}(U) \longrightarrow \mathscr{T}(U),$$

for all open subsets U of X, such that for all $V \subset U \subset X$ open, the diagram

$$\begin{array}{ccc} \mathscr{S}(U) & \xrightarrow{f_U} & \mathscr{T}(U) \\ {\scriptstyle \rho_{UV}}\downarrow & & \downarrow{\scriptstyle \rho'_{UV}} \\ \mathscr{S}(V) & \xrightarrow{f_V} & \mathscr{T}(V) \end{array}$$

commutes. A morphism of sheaves is given by a morphism of the underlying presheaves.

A subsheaf of a sheaf $\mathscr{F}$ is a sheaf $\mathscr{G}$ such that for all open subsets U of X $\mathscr{G}(U)$ is a subring of $\mathscr{F}(U)$, and the restriction homomorphism are those induced by $\mathscr{F}$, that is, the inclusion mapping is a morphism of presheaves.

Examples 6.1.8.

(1) Consider the presheaves $\mathscr{C}$ and $\mathcal{O}$ on $\mathbb{C}^n$ defined in Example 6.1.2 and $f : \mathcal{O} \longrightarrow \mathscr{C}$ defined by the canonical injection $\mathcal{O}(U) \subset \mathscr{C}(U)$. Then f is a morphism of sheaves.

(2) The map defined by the canonical injection $\mathcal{O}'(U) \subset \mathscr{C}(U)$ is not a morphism of presheaves.

Lemma 6.1.9. *Let $f : \mathscr{S} \longrightarrow \mathscr{T}$ be a morphism of presheaves on X. Then there is, for all $p \in X$, a naturally defined ring homomorphism*

$$f_p : \mathscr{S}_p \longrightarrow \mathscr{T}_p.$$

This map is defined as follows. Take an element $a_p \in \mathscr{S}_p$, and a representative $a \in \mathscr{S}(U)$ of a in some open neighborhood U of a. Then define $f_p(a_p)$ as the germ in p of $f(a)$.

The fact that this map is well defined is left as Exercise 6.1.29.

The second part of the following theorem says, loosely speaking, that a *sheaf* is determined by its stalks.

Theorem 6.1.10. *Let $f : \mathscr{S} \longrightarrow \mathscr{T}$ be a morphism of* sheaves *on a topological space X. Let, for all $p \in X$, $f_p : \mathscr{S}_p \longrightarrow \mathscr{T}_p$ be the induced map on stalks. Then*

(1) f_U is injective for all U open in X if and only if f_p is injective for all p;

(2) f is an isomorphism if and only if f_p is an isomorphism for all p.

Proof. In both cases, one direction is quite obvious. If f_U is injective (resp. an isomorphism) for all U, then f_p is injective (resp. an isomorphism) for all p.

(1) We prove that f_U is injective. Let $s \in \mathscr{S}(U)$, with $f_U(s) = 0$. In particular, for all $p \in U$, the germs $(f_U(s))_p = 0$. Thus $f_p(s_p) = 0$. By assumption $s_p = 0$ for all $p \in U$. This means that for all p there exists a neighborhood U_p of p with $s_{|U_p} = 0$. By the first sheaf axiom, it follows that $s = 0$. Thus f_U is injective.

(2) We need to show that for all U open in X, the map $f_U : \mathscr{S}(U) \longrightarrow \mathscr{T}(U)$ is an isomorphism. Namely, if this is the case, we can define the inverse map f^{-1} by $(f^{-1})_U := (f_U)^{-1}$. It is a direct check that this gives indeed an inverse map. The first part gives that f_U is injective. So it remains to prove the surjectivity of f_U. Take a section $t \in \mathscr{T}(U)$. Let $p \in U$, and consider t_p. As f_p is surjective by assumption, there exists an element s_p with $f_p(s_p) = t_p$. Take a representative $\widetilde{s}_p \in \mathscr{S}(U_p)$ of s_p for some small open neighborhood U_p of p. We may, by making U_p smaller if necessary, assume that $f_{U_p}(\widetilde{s}_p) = t_{|U_p}$. So we have

(1) an open covering $\{U_p\}_{p\in U}$ of U,

(2) sections $\widetilde{s}_p \in \mathscr{S}(U_p)$,

(3) such that $(\widetilde{s}_p)_{|U_p\cap U_q} = (\widetilde{s}_q)_{|U_p\cap U_q}$ for all $p,q \in U$.

The third condition follows as $(\widetilde{s}_p)_{|U_p\cap U_q}$ and $(\widetilde{s}_q)_{|U_p\cap U_q}$ are both sections of $\mathscr{S}(U_p \cap U_q)$ which are mapped under $f_{|U_p\cap U_q}$ to $t_{|U_p\cap U_q}$. They are equal, because of the injectivity proved in (1)!

It follows from the second sheaf property, that there exists a section $s \in \mathscr{S}(U)$ with $s_{|U_p} = \widetilde{s}_p$. By construction $f_U(s)_{|U_p} = t_{|U_p}$ for all p, so that we may conclude $f_U(s) = t$ by using the first sheaf axiom. □

This results motivates the following definition.

Definition 6.1.11. A sequence of morphisms of sheaves on X

$$\mathscr{R} \xrightarrow{f} \mathscr{S} \xrightarrow{g} \mathscr{T}$$

is called *exact*, if for all $p \in X$, the induced map on stalks

$$\mathscr{R}_p \xrightarrow{f_p} \mathscr{S}_p \xrightarrow{g_p} \mathscr{T}_p$$

is exact. In particular, a morphism $\mathscr{S} \longrightarrow \mathscr{T}$ is called *injective*, if for all $p \in X$, the induced map on stalks $f_p : \mathscr{S}_p \longrightarrow \mathscr{T}_p$ is injective. Similarly surjectivity is defined.

Remark 6.1.12. It is important to note that from the surjectivity of f_p for all p it does not follow that f_U is surjective for all U in general. To give an example, consider the following map of sheaves on $X = \mathbb{C} \setminus \{0\}$:

$$f : \mathcal{O}_X \longrightarrow \mathcal{O}_X^*, \quad f_U : g \mapsto e^{2\pi i g}.$$

Let U be an open subset of X, and $\mathcal{O}_X^*(U) = \{h \in \mathcal{O}(U)\colon h(p) \neq 0 \text{ for all } p \in U\}$. Then f_p is surjective for all p. Indeed, take $h_p \in \mathcal{O}_{X,p}^*$, and a small simply connected open neighborhood U of p on which h is a representative of h_p. By continuity, we may assume that U is so small that h is nonzero throughout U. From function theory, we know that $\log(h)$ exists, and $e^{2\pi i \log(h)} = h$. Thus $\log(h)_p$ maps to h_p, showing that the map of sheaves is surjective. However f_p is not injective. With $\log(h)$ also $\log(h) + k$ for all $k \in \mathbb{Z}$ maps to h. Let z be a coordinate on $\mathbb{C}$. Then $z \in \mathcal{O}^*(X)$ but z is not in the image of f, as $\log(z)$ is not a well defined holomorphic function on $X \setminus \{0\}$.

This example also shows that given a map $f : \mathscr{G} \longrightarrow \mathcal{H}$ of sheaves the presheaf defined by $U \mapsto \mathcal{H}(U)/f_U(\mathscr{G}(U))$ is not necessarily a sheaf. The first sheaf axiom is not satisfied: locally defined logarithms do not necessarily glue to a globally defined logarithm. Thus we have problems in defining the cokernel of a map as sheaf. These kinds of problems occur for several other constructions with sheaves as, for example, the tensor product. This is the reason for the following construction, called sheafifying.

Theorem 6.1.13. *Let X be a topological space and $\mathscr{F}$ be a presheaf on X. There exists a sheaf $\mathscr{F}^+$ and a morphism $j : \mathscr{F} \longrightarrow \mathscr{F}^+$ of presheaves with the following properties:*

(1) $j_p : \mathscr{F}_p \longrightarrow \mathscr{F}_p^+$ is an isomorphism for all p.

(2) $(\mathscr{F}^+, j)$ is universal, that is, for any morphism $\varphi : \mathscr{F} \longrightarrow \mathscr{G}$ of $\mathscr{F}$ to a sheaf $\mathscr{G}$ there exists a unique morphism $\psi : \mathscr{F}^+ \longrightarrow \mathscr{G}$ such that $\varphi = \psi \circ j$.

Definition 6.1.14. The (uniquely) determined sheaf $\mathscr{F}^+$ is called the *sheaf associated to the presheaf* $\mathscr{F}$. One says that $\mathscr{F}^+$ is obtained from $\mathscr{F}$ by *sheafifying*.

Proof of Theorem 6.1.13. Let U be an open set and define $\mathscr{F}^+(U)$ to be the set of functions $f : U \longrightarrow \bigcup_{x \in U} \mathscr{F}_x$ having the following two properties:

(1) for each $x \in U$, $f(x) \in \mathscr{F}_x$,

(2) for each $x \in U$ there is a neighborhood $W \subset U$ of x and an element $g \in \mathscr{F}(W)$ such that the germ g_y of g in $y \in W$ is equal to $f(y)$ for all $y \in W$.

Now it is not difficult to verify, see Exercise 6.1.36, that $\mathscr{F}^+$ together with the canonical restriction maps is a sheaf. It is also not difficult to check that the canonical map $j : \mathscr{F} \longrightarrow \mathscr{F}^+$ defined by $j_U(g)(y) = g_y$ has the universal property mentioned, see Exercise 6.1.36. □

Note that not all nonzero section of $\mathscr{F}(U)$ give nonzero sections of $\mathscr{F}^+(U)$. Namely, it might happen that all stalks are zero, but $\mathscr{F}(X) \neq 0$. On the other hand, in the example $f : \mathcal{O}_X \longrightarrow \mathcal{O}_X^*$, $f_U(g) = e^{2\pi i g}$ the function z is in $\mathrm{Im}(f)(U)$ as sheaf, but not as presheaf. So $\mathscr{F}^+(U)$ might have more sections. This is because by definition sections of $\mathscr{F}^+(U)$ are not sections of $\mathscr{F}(U)$, but functions which are *locally* equal to sections of $\mathscr{F}$.

We are now prepared to give a lot of examples for sheaves.

Definition 6.1.15. Let X be a topological space, $\mathscr{F}, \mathscr{G}$ be sheaves on X and $f : \mathscr{F} \longrightarrow \mathscr{G}$ a morphism.

(1) The *kernel* of f, $\mathrm{Ker}(f)$ is the sheaf defined by $\mathrm{Ker}(f)(U) = \mathrm{Ker}(f_U)$.

(2) The *cokernel* of f, $\mathrm{Coker}(f)$ is the sheaf associated to the presheaf defined by $U \mapsto \mathscr{G}(U)/f_U(\mathscr{F}(U))$. A cokernel of a map of sheaves is also called a *quotient sheaf.*

(3) The *image* of f, $\mathrm{Im}(f)$ is the sheaf associated to the presheaf defined by $U \mapsto f_U(\mathscr{F}(U))$.
(Note that $\mathrm{Im}(f)$ is a subsheaf of $\mathscr{G}$, Exercise 6.1.37.)

(4) Let A be a ring, the *constant sheaf* $\mathscr{C}_A$ is the sheaf associated to the constant presheaf defined by $U \mapsto A$ if $U \neq \emptyset$. In this case if U is the disjoint union of two connected subsets, we have $\mathscr{C}(U) = A \oplus A$. Sections of $\mathscr{C}_A$ over U are now allowed to differ on the two connected subsets.

The proof that the presheaf defined in (1) is already a sheaf will be given in Exercise 6.1.37.

We now relate (pre)sheaves which are defined on different topological spaces.

Definition 6.1.16. Let a continuous map $f : X \longrightarrow Y$ between topological spaces be given.

(1) Let $\mathscr{S}$ be a sheaf on X. We define a presheaf $f_*\mathscr{S}$ on Y by putting for an open set $U \subset Y$

$$f_*\mathscr{S}(U) := \mathscr{S}(f^{-1}(U)),$$

with the obvious restriction maps. (It can be proven without difficulty, see Exercise 6.1.30, that $f_*\mathscr{S}$ is in fact a sheaf.)

(2) Let $\mathscr{F}$ be a sheaf on Y. $f^{-1}\mathscr{F}$ is the sheaf on X associated to the presheaf defined by $U \mapsto \lim\limits_{V \supset f(U)} \mathscr{F}(V)$.[1]

(3) If f is an injection and $\mathscr{F}$ a sheaf on Y, we define the *restriction* $\mathscr{F}_{|X}$ of $\mathscr{F}$ to X by $f^{-1}\mathscr{F}$.

As a simple example, if f is the map which sends X to a point, then $f_*\mathscr{S}$ can be identified with a sheaf on a point, hence a ring. In this case we can identify $f_*\mathscr{S}$ with the global sections $\mathscr{S}(X)$ of X. On the other hand, if $f : p \longrightarrow X$ is the injection of the point p in the space X, then $f^{-1}\mathscr{F}$ is the sheaf on the point p, consisting of the stalk $\mathscr{F}_p$. More generally, if $f : X \longrightarrow Y$ is the injection of a closed subspace, then $f^{-1}\mathscr{G}$ is a sheaf whose stalks at $p \in X$ are equal to $\mathscr{G}_p$.

It is a direct and easy consequence of the definitions, that if one has a morphism $\alpha : \mathscr{S} \longrightarrow \mathscr{T}$ of sheaves on X, and a continuous map $f : X \longrightarrow Y$, then one has an induced map $\alpha_* : f_*\mathscr{S} \longrightarrow f_*\mathscr{T}$. One cannot hope in general, that for a surjective morphism of sheaves $\alpha : \mathscr{S} \longrightarrow \mathscr{T}$, the induced map $\alpha_* : f_*\mathscr{S} \longrightarrow f_*\mathscr{T}$ is surjective (look at the constant map f, that is, look at global sections). But, on the other hand, we have the following theorem.

[1] By definition $\lim\limits_{V \supset f(U)} \mathscr{F}(V) := \bigcup\limits_{V \supset f(U)} \mathscr{F}(V)/\sim$ where $s \sim s'$ for $s \in \mathscr{F}(V)$, $s' \in \mathscr{F}(W)$ if and only if $s_{|V \cap W} = s'_{|V \cap W}$. Note that this is a generalization of the notion of germ, which corresponds to the case that f is the injection of a point.

Theorem 6.1.17. *Let X and Y be Hausdorff topological spaces and $f : X \longrightarrow Y$ be a finite continuous map. Consider an exact sequence of sheaves on X:*

$$0 \longrightarrow \mathscr{S}_1 \longrightarrow \mathscr{S}_2 \longrightarrow \mathscr{S}_3 \longrightarrow 0.$$

Then the induced sequence of direct image sheaves

$$0 \longrightarrow f_*\mathscr{S}_1 \longrightarrow f_*\mathscr{S}_2 \longrightarrow f_*\mathscr{S}_3 \longrightarrow 0$$

is also exact.

One phrases this by saying that for finite f, the functor f_* is *exact.* This theorem will play an important role in the proof of the Finite Mapping Theorem, see 6.3.5. The theorem follows immediately from the fact that exactness means exactness of the stalks and the following theorem.

Theorem 6.1.18. *Let X,Y be Hausdorff spaces, and $f : X \longrightarrow Y$ be a finite continuous map. Let $\mathscr{S}$ be a sheaf on X. Let $y \in Y$, and $x_1, \dots ,x_t$ be the points in the fiber $f^{-1}(y)$. Then we have the equality of stalks:*

$$(f_*\mathscr{S})_y = \bigoplus_{i=1}^{t} \mathscr{S}_{x_i}.$$

Proof. As X is Hausdorff, we can choose pairwise disjoint open neighborhoods $U_1', \dots ,U_t'$ of $x_1, \dots ,x_t$ in X. As f is closed, there exists an open neighborhood V of y in Y, such that $f^{-1}(V) \subset \cup_{i=1}^t U_i'$, see Exercise 3.4.36.

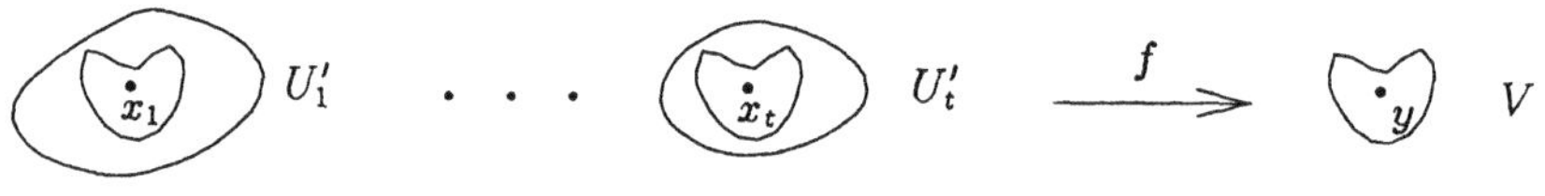

We define $U_i := f^{-1}(V) \cap U_i'$. Then

(1) The $U_1, \dots ,U_t$ are pairwise disjoint.

(2) $f^{-1}(V) = \cup_{i=1}^t U_i$.

Therefore,

$$f_*\mathscr{S}(V) := \mathscr{S}(f^{-1}(V)) = \mathscr{S}(\cup_{i=1}^t U_i) = \oplus_{i=1}^t \mathscr{S}(U_i).$$

The last equality is due to the fact that the U_i are pairwise disjoint, and because of the sheaf properties. Now use the definition of stalk to deduce the theorem. □

There is a method to produce a lot of examples of sheaves, if we know them locally.

Proposition 6.1.19 (Gluing Sheaves). *Let X be a topological space, $X = \bigcup_{i \in I} U_i$, U_i open subsets in X. For every i let $\mathscr{F}_i$ be a sheaf on U_i and $\varphi_{ij} : \mathscr{F}_{i|U_i \cap U_j} \xrightarrow{\sim} \mathscr{F}_{j|U_i \cap U_j}$ be isomorphisms such that*

(1) φ_{ii} is the identity,

(2) $\varphi_{ik} = \varphi_{jk} \circ \varphi_{ij}$ on $U_i \cap U_j \cap U_k$.

Then there exists a unique sheaf $\mathscr{F}$ *on* X *and isomorphisms* $\lambda_i : \mathscr{F}_{|U_i} \xrightarrow{\sim} \mathscr{F}_i$ *such that for all* i,j, $\varphi_{ij} \circ \lambda_i = \lambda_j$ *on* $U_i \cap U_j$.

In fact, for an open subset V of X we define

$$\mathscr{F}(V) := \left\{ \{s_i\}_{i\in I} \in \prod_{i\in I} \mathscr{F}_i(U_i \cap V) : \varphi_{ij}(s_{i|U_i\cap U_j\cap V}) = s_{j|U_i\cap U_j \cap V} \right\}.$$

The proof that this defines a sheaf, and the uniqueness is left as Exercise 6.1.40.

Definition 6.1.20. A ringed space $(X,\mathcal{O}_X)$ is a topological space X together with a *sheaf* of rings $\mathcal{O}_X$. Usually, if it is clear from the context what the sheaf of rings is, we simply write X for the pair $(X,\mathcal{O}_X)$.

A sheaf $\mathscr{F}$ on X is called an $\mathcal{O}_X$*-module* if $\mathscr{F}(U)$ is an $\mathcal{O}_X(U)$-module for all $U \subset X$ open, such that the restriction mapping $\mathscr{F}(U) \longrightarrow \mathscr{F}(V)$ is $\mathcal{O}_X(U)$-linear for $V \subset U$. (Note that $\mathscr{F}(V)$ for $V \subset U$ is also an $\mathcal{O}_X(U)$-module.)

An $\mathcal{O}_X$-module sheaf $\mathscr{F}$ which is a subsheaf of $\mathcal{O}_X$ is called an *ideal sheaf.*

Examples 6.1.21 (Ringed Spaces).

(1) Let U be an open set of $\mathbb{C}^n$ and $\mathcal{O}_U$ be the sheaf of holomorphic functions on U.

(2) Let U be an open set in $\mathbb{C}^n$, $f_1,\dots,f_t$ be holomorphic functions defined on U. Consider $X := V(f_1,\dots,f_t)$. Let $V \subset U$ be open. Consider the presheaf defined by $\mathcal{O}_X(V \cap X) := \mathcal{O}(V)/(f_1,\dots,f_t)\mathcal{O}(V)$ on X. This is easily checked to be a sheaf. The ringed space constructed in this way is called a complex model space.

Consider $X = \{x \in U : f_1(x) = \dots = f_t(x) = 0\}$. Consider the presheaf $\mathscr{I}(X)$ on U defined by $\mathscr{I}(X)(W) = \{f \in \mathcal{O}_{\mathbb{C}^n}(W) : f(x) = 0 \text{ for all } x \in W \cap X\}$. This is easily checked to be a sheaf and is called the ideal sheaf of X.

(3) Let $U \subset \mathbb{R}^n$ be an open subset, and $\mathcal{O}_U$ the sheaf of differentiable functions on U. Similarly let $\mathscr{C}_U$ be the sheaf of continuous functions on U.

(4) An affine variety X (with the Zariski topology) can also be seen as a ringed space by taking as canonical presheaf the presheaf which assigns to each $U \subset X$ the ring

$$\mathcal{O}_X(U) = \{f \in K(X) : \forall p \in U\ \exists g,h \in K[X] \text{ with } f = \frac{g}{h} \text{ and } h(p) \neq 0\}.$$

An element f in $\mathcal{O}_X(U)$ therefore can be viewed as a *rational function* $f : U \longrightarrow K$ which have a well-defined value for all $p \in U$.

(5) Spectrum of a ring: Let R be a ring (commutative with 1 as always). Then the spectrum $\operatorname{Spec}(R)$ is defined as follows:

- As a set $\operatorname{Spec}(R)$ is the set of prime ideals $\mathfrak{p} \subset R$.
- One defines a topology on $\operatorname{Spec}(R)$ by defining the closed sets to be the sets of type $V(I) := \{\mathfrak{p} \in \operatorname{Spec}(R) : I \subset \mathfrak{p}\}$. This topology is called the *Zariski topology* on $\operatorname{Spec}(R)$.
- We define a sheaf $\mathcal{O}$ of rings on $\operatorname{Spec}(R)$. Let $U = D(f) := \operatorname{Spec}(R) \setminus V((f))$ be an open set of $\operatorname{Spec}(R)$. Then $\mathcal{O}(D(f)) := R_f$. We have to show that this is independent of the chosen f. For this we refer to Exercise 6.1.33. As the $D(f)$ form a basis for the topology by Exercise 6.1.34 this induces a presheaf on Spec R. This presheaf is in fact a sheaf, see Exercise 6.1.34.

Definition 6.1.22. Let X be a ringed space and $\mathscr{F},\mathscr{G}$ be $\mathcal{O}_X$–module sheaves on X.

(1) $\mathscr{H}om_X(\mathscr{F},\mathscr{G})$ is the sheaf defined by $\mathscr{H}om_X(\mathscr{F},\mathscr{G})(U) = \mathrm{Hom}_{\mathcal{O}_{X|U}}(\mathscr{F}_{|U},\mathscr{G}_{|U})$.

(2) $\mathscr{F}\otimes_{\mathcal{O}_X}\mathscr{G}$ is the sheaf associated to the presheaf defined by $U \mapsto \mathscr{F}(U)\otimes_{\mathcal{O}_X(U)}\mathscr{G}(U)$.

(3) $\mathscr{F}\oplus\mathscr{G}$ is the sheaf defined by $(\mathscr{F}\oplus\mathscr{G})(U) = \mathscr{F}(U)\oplus\mathscr{G}(U)$.

(4) Let $\mathscr{F},\mathscr{G} \subset \mathcal{H}$, $\mathcal{H}$ an $\mathcal{O}_X$–module sheaf. Then $\mathscr{F}\cap\mathscr{G}$ is the sheaf defined by $(\mathscr{F}\cap\mathscr{G})(U) = \mathscr{F}(U)\cap\mathscr{G}(U)$. $\mathscr{F}+\mathscr{G}$ is the sheaf associated to the presheaf defined by $U \mapsto \mathscr{F}(U)+\mathscr{G}(U)$. $\mathscr{F}:\mathscr{G}$ is the sheaf associated to the presheaf $U \mapsto \mathscr{F}(U) : \mathscr{G}(U)$.

(5) $\mathscr{A}nn(\mathscr{F})$, the annihilator sheaf of $\mathscr{F}$, is the sheaf associated to the presheaf $U \mapsto \mathrm{Ann}(\mathscr{F}(U)) = \{f \in \mathcal{O}_X(U) : f\cdot\mathscr{F}(U) = 0\}$.

The proof that the presheaves defined in (1), (3) and (4) are already sheaves is left as Exercise 6.1.38.

Note that in all the examples considered above the stalk at each point is a *local ring.* We put this phenomenon in a definition.

Definition 6.1.23. A *locally ringed space* $(X,\mathcal{O}_X)$ is a ringed space, such that for all points $p \in X$ the stalk $\mathcal{O}_{X,p}$ is a local ring. A *morphism between locally ringed spaces* is a pair: $(f,f^*) : (X,\mathcal{O}_X) \to (Y,\mathcal{O}_Y)$, where $f : X \to Y$ is a continuous map, and f^* is a morphism of sheaves of $\mathcal{O}_Y$–modules:

$$f^* : \mathcal{O}_Y \to f_*\mathcal{O}_X$$

such that the induced map on stalks:

$$f_p^* : \mathcal{O}_{Y,f(p)} \to \mathcal{O}_{X,p}$$

sends the maximal ideal of $\mathcal{O}_{Y,f(p)}$ into the maximal ideal of $\mathcal{O}_{X,p}$. Usually, if there is no danger of confusion, we just write f for the pair (f,f^*).

Examples 6.1.24.

(1) The examples of Examples 6.1.21.

(2) The projective space $\mathbb{P}^n_{\mathbb{C}}$, the set of lines through the origin in $\mathbb{C}^{n+1}$. As a set we have $\mathbb{P}^n_{\mathbb{C}} := \{c \in \mathbb{C}^{n+1}, c \neq 0\}/\sim$ and $c \sim c'$ if there exists a $\lambda \in \mathbb{C}$ such that $\lambda c = c'$. The equivalence class of $c = (c_0,\ldots,c_n)$ we write as $(c_0 : c_1 : \cdots : c_n)$.

We consider $\mathbb{P}^n_{\mathbb{C}}$ with the quotient topology of the ordinary topology of $\mathbb{C}^{n+1}$. Let $U_i = \{c = (c_0 : \cdots : c_n) \in \mathbb{P}^n_{\mathbb{C}}, c_i \neq 0\}$. Then U_i is an open subset of $\mathbb{P}^n_{\mathbb{C}}$ and $\bigcup\limits_{i=0}^{n} U_i = \mathbb{P}^n_{\mathbb{C}}$. We have homeomorphisms $\varphi_i : U_i \longrightarrow \mathbb{C}^n$ defined by $\varphi_i((c_0 : \cdots : c_n)) = \left(\frac{c_0}{c_i},\frac{c_1}{c_i},\ldots,\frac{c_{i-1}}{c_i},\frac{c_{i+1}}{c_i},\ldots,\frac{c_n}{c_i}\right)$.

The map $\varphi_j \circ \varphi_i^{-1} : \varphi_i(U_i\cap U_j) \longrightarrow \varphi_j(U_i\cap U_j)$ maps $(z_1,\ldots,z_n)$ to $\left(\frac{z_1}{z_j},\ldots,\frac{z_{j-1}}{z_j},\frac{z_{j+1}}{z_j},\ldots,\frac{1}{z_j},\ldots,\frac{z_n}{z_j}\right)$ and is obviously holomorphic.

$\mathcal{O}_{\mathbb{P}^n}$ is the sheaf obtained by gluing the sheaves $\mathcal{O}_{\mathbb{C}^n}$ on U_i.

The second example can be generalized as follows:

Proposition 6.1.25. *Let X be a Hausdorff space and $X = \cup_{i\in I} U_i$ for open sets U_i. Let $\varphi_i : U_i \longrightarrow W_i$ be homeomorphisms to domains W_i in $\mathbb{C}^n$ such that for all i,j and $U_i \cap U_j \neq \emptyset$*

$$\varphi_j \circ \varphi_i^{-1} : W_i \cap \varphi_i(U_j) \longrightarrow W_j \cap \varphi_j(U_i)$$

is holomorphic. Then X is a locally ringed space.

Proof. the structure sheaf $\mathcal{O}_X$ is defined to be the gluing of the sheaves $\mathcal{O}_{U_i} = \varphi_i^* \mathcal{O}_{W_i}$ (Proposition 6.1.19). □

Definition 6.1.26. Let X be a Hausdorff space.

(1) Let $(X, \mathcal{O}_X)$ be a locally ringed space. $(X, \mathcal{O}_X)$ is called a *complex manifold* if every $x \in X$ has a neighborhood U such that $(U, \mathcal{O}_{X|U})$ is isomorphic to $(V, \mathcal{O}_V)$, $V \subset \mathbb{C}^n$ a domain.

(2) $(X, \mathcal{O}_X)$ is called an *analytic space* if every $x \in X$ has a neighborhood U such that $(U, \mathcal{O}_{X|U})$ is isomorphic to $(V, \mathcal{O}_V)$, V an analytic subset of an open set W in some $\mathbb{C}^n$, and $\mathcal{O}_V = (\mathcal{O}_W / \mathcal{I}(V))_{|V}$.

(3) $(X, \mathcal{O}_X)$ is called a *complex space* if if every $x \in X$ has a neighborhood U such that $(U, \mathcal{O}_{X|U})$ is isomorphic to $(V, \mathcal{O}_V)$, where $(V, \mathcal{O}_V)$ is a complex model space. This means that V is an analytic subset of an open set W in some $\mathbb{C}^n$ and $\mathcal{O}_V = (\mathcal{O}_W / J)_{|V}$, $J \subset \mathcal{O}_W$ an ideal sheaf such that for all $x \in V$, $\sqrt{J_x} = \mathcal{I}(V)_x$. One says that $(Y, \mathcal{O}_Y)$ is a *closed complex subspace of* $(X, \mathcal{O}_X)$, if there exists a map of locally ringed spaces $(i, i^*) : (Y, \mathcal{O}_Y) \longrightarrow (X, \mathcal{O}_X)$ such that i is injective, and i^* is surjective.

For the case $(X, \mathcal{O}_X) = (U, \mathcal{O}_U)$ for $U \subset \mathbb{C}^n$ open, we obtain that a closed complex subspace is a complex model space.

Exercises

6.1.27. Consider the following presheaf on $\mathbb{R}$. Let $U \subset X$ be open and

$$\mathcal{B}(U) := \{f : U \longrightarrow \mathbb{R} : f \text{ bounded}\},$$

and ρ_{UV} the ordinary restrictions. Show that $\mathcal{B}$ is not a sheaf.

6.1.28. Let $\mathcal{S}$ be a sheaf of rings on a topological space X. Show that for each point $x \in X$ the stalk $\mathcal{S}_x$ is in a natural way a ring again.

6.1.29. Prove Lemma 6.1.9.

6.1.30. Prove that for $f : X \longrightarrow Y$ continuous, $\mathcal{S}$ a sheaf on X, the direct image $f_* \mathcal{S}$ is also a sheaf.

6.1.31. Let R be a Noetherian ring. Prove that the topology on Spec(R) is indeed a topology by proving the following statements:

(1) $V((0)) = \text{Spec}(R)$; $V((1)) = \emptyset$,

(2) $V(I \cdot J) = V(I) \cup V(J)$,

(3) $V(\sum_j I_j) = \cap_j V(I_j)$.

6.1.32. For a subset A of Spec(R) let $\mathscr{I}(A) = \bigcap_{\mathfrak{p}\in A} \mathfrak{p}$. Prove the analog of Hilbert's Nullstellensatz: $\sqrt{I} = \mathscr{I}(V(I))$.
(Hint: Use primary decomposition of an ideal).

6.1.33. Let R be a ring and $f,g \in R$ such that $D(f) = D(g)$ (equivalently $V(f) = V(g)$). Prove that $R_f = R_g$.

6.1.34. Let R be a ring. Show that the open sets $D(f)$ form a basis for the topology of Spec(R). Show that the presheaf $\mathcal{O}$ on Spec(R), defined in 6.1.21, is in fact a sheaf.

6.1.35. Let $f : X \longrightarrow Y$ be a continuous map, $A \subset Y$ a closed subset, $\mathscr{S}$ a sheaf on X. Prove that

$$(f_*\mathscr{S})_{|A} = \left(f_{|f^{-1}(A)}\right)_* \mathscr{S}_{|f^{-1}(A)}.$$

6.1.36. Prove that $\mathscr{F}^+$ defined in the proof of Theorem 6.1.13 is a sheaf and the canonical map $j : \mathscr{F} \longrightarrow \mathscr{F}^+$ has the universal property.

6.1.37. Let $f : \mathscr{F} \longrightarrow \mathscr{G}$ be a morphism of sheaves.

(1) Prove that there is a canonical injection $i : \mathrm{Im}(f) \longrightarrow \mathscr{G}$.

(2) Prove that the presheaf defined by $U \mapsto \mathrm{Ker}(f_U)$ is already a sheaf.

6.1.38. Let X be a ringed space and $\mathscr{F},\mathscr{G}$ be $\mathcal{O}_X$-module sheaves on X.

(1) Suppose that both $\mathscr{F}$ and $\mathscr{G}$ are subsheaves of $\mathscr{H}$. Prove that the presheaf defined by $U \mapsto \mathscr{F}(U) \cap \mathscr{G}(U)$ is a sheaf.

(2) Prove that the presheaf defined by $U \mapsto \mathscr{F}(U) \oplus \mathscr{G}(U)$ is a sheaf.

(3) Prove that the presheaf defined by $U \mapsto \mathrm{Hom}_{\mathcal{O}_{X|U}}(\mathscr{F}_{|U},\mathscr{G}_{|U})$ is a sheaf.

6.1.39. Let $f : X \longrightarrow Y$ be a continuous map between the topological spaces X,Y, $\mathscr{F}$ be a sheaf on X and $\mathscr{G}$ be a sheaf on Y.

(1) Prove that $\mathscr{G}_{f(x)} = (f^{-1}\mathscr{G})_x$.

(2) Show that there are canonical maps $\lambda : f^{-1}f_*\mathscr{F} \longrightarrow \mathscr{F}$ and $\rho : \mathscr{G} \longrightarrow f_*f^{-1}\mathscr{G}$.

(3) Prove that $\mathrm{Hom}(f^{-1}\mathscr{G},\mathscr{F}) = \mathrm{Hom}(\mathscr{G},f_*\mathscr{F})$.

6.1.40. Prove Proposition 6.1.19.

6.1.41. One can view $X = \mathbb{C}^2 \setminus \{0\}$ as a locally ringed space $(X,\mathcal{O}_X)$ just as one can see an affine variety as a locally ringed space.

(1) Show that $\mathcal{O}_X(X) = \mathbb{C}[x,y]$.

(2) Show that X, as locally ringed space, is not isomorphic to an affine variety.

6.2 Fundamental Properties of Coherent Sheaves

In this section, unless otherwise stated explicitly, X is a locally ringed space, with structure sheaf $\mathcal{O}_X$. Our most important example is the case of X being a complex space, see Definition.6.1.26. $\mathscr{S},\mathscr{T}$, etc. will denote sheaves of $\mathcal{O}_X$-modules. If $(X,\mathcal{O}_X)$ is fixed we will often use $\mathcal{O}$ instead of $\mathcal{O}_X$, if there is no danger of confusion. For an open subset $U \subset X$, we sometimes write $\mathcal{O}_U$ instead of $\mathcal{O}_{X|U}$.

We first give the definition of a coherent sheaf.

Definition 6.2.1.

(1) The sheaf $\mathscr{S}$ is said to be of *finite type* if for every point $x \in X$ there is an open neighborhood U of X and a surjective morphism of sheaves:

$$\mathcal{O}^q_{X|U} \longrightarrow \mathscr{S}_{|U} \longrightarrow 0.$$

(2) The sheaf $\mathscr{S}$ is said to be of *relation finite type* if for *any* open set U and any morphism of sheaves: $\alpha : \mathcal{O}^q_{X|U} \longrightarrow \mathscr{S}_{|U}$ the kernel $\mathrm{Ker}(\alpha)$ is a sheaf of finite type.

(3) A sheaf $\mathscr{S}$ of $\mathcal{O}_X$-modules is called *coherent*, if $\mathscr{S}$ is both of finite and of relation finite type.

Examples 6.2.2.
(1) The zero sheaf is coherent. Indeed, it is generated by zero elements, and the kernel $\mathcal{O}^q_U \longrightarrow 0$ is equal to $\mathcal{O}^q_U$, so certainly finitely generated.

(2) Let X be a point, and $\mathcal{O}_X(X) = \mathbb{C}$. This is a locally ringed space. A sheaf of finite type on X can be identified with a finite-dimensional $\mathbb{C}$-vector space, say $\mathbb{C}^m$. As the kernel of a map $\mathbb{C}^p \longrightarrow \mathbb{C}^m$ certainly is a finitely generated $\mathbb{C}$-vector space, it follows that such a sheaf is of relation finite type. Thus a coherent sheaf on a point can be identified with a finite-dimensional $\mathbb{C}$-vector space.

(3) Consider a sequence of points $p_1, p_2, \ldots$ in the complex space $\mathbb{C}$ converging to 0. Let $\mathscr{T}$ be the subsheaf of $\mathcal{O}_{\mathbb{C}}$ consisting of all holomorphic functions *vanishing* at each point p_i. We will see that the sheaf $\mathscr{T}$ is not finitely generated. Let U be an open neighborhood of $\{0\}$. We claim that there does not exist a nontrivial map $\mathcal{O}^p_U \longrightarrow \mathscr{T}$ at all! It suffices to show this for $p = 1$. The image of $1 \in \mathcal{O}(U)$ is a holomorphic function defined on U vanishing on the p_i. By the Identity Theorem, this holomorphic function is zero, proving the claim. We see that the sheaf $\mathscr{T}$ is not "generated" at all, and in particular $\mathscr{T}$ is not coherent.

(4) To give an example of a sheaf of finite type, which is not of relation finite type, consider the sheaf $\mathscr{T}$ of the previous example, and let $\mathscr{S}$ be the quotient sheaf $\mathcal{O}/\mathscr{T}$. This sheaf certainly is of finite type. It is not of relation finite type. Indeed, consider the canonical projection $\mathcal{O} \twoheadrightarrow \mathscr{S} = \mathcal{O}/\mathscr{T}$. The kernel is equal to $\mathscr{T}$, and we have seen it is not a finitely generated sheaf.

The problem with the theory of coherent sheaves (in complex analysis, not in algebra, see Exercise 6.2.21) is that one has to work hard to produce nontrivial examples. As soon as we have proved that for a complex space X the sheaf $\mathcal{O}_X$ is coherent, see 6.3.1, the so-called *Meta-Theorem* for coherent sheaves allows us to construct many more coherent sheaves. The Meta-Theorem says that coherent sheaves behave well under most algebraic operations on sheaves:

Theorem 6.2.3 (Meta-Theorem for Coherent Sheaves). *Let $\mathscr{S}$ and $\mathscr{T}$ be coherent sheaves of $\mathcal{O}_X$-modules. Then every reasonable operation with $\mathscr{S}, \mathscr{T}$ (as taking kernel, cokernel, $\mathscr{H}om$, $\otimes$, finitely generated subsheaves ...) produces again a coherent sheaf.*

The Meta-Theorem will be proved in Lemma 6.2.14, Corollary 6.2.15, Lemma 6.2.16 and Lemma 6.2.17.

The following theorem describes a property of coherent sheaves which is the justification for the name *coherence*. It says that the stalk $\mathscr{S}_x$ of a coherent sheaf *determines* the sheaf in a sufficiently small open neighborhood U of x.

Theorem 6.2.4. *Let $\mathscr{S}, \mathscr{T}$ be* coherent *sheaves of $\mathcal{O}_X$-modules. Let $\alpha : \mathscr{S} \longrightarrow \mathscr{T}$ be a map of $\mathcal{O}_X$-modules. Suppose that at some point $x \in X$ the map of stalks*

$$\alpha_x : \mathscr{S}_x \longrightarrow \mathscr{T}_x$$

is an isomorphism. Then there exists an open neighborhood U of x such that for any point $p \in U$ the map of stalks

$$\alpha_p : \mathscr{S}_p \longrightarrow \mathscr{T}_p$$

is an isomorphism, that is, the sheaves $\mathscr{S}_{|U}$ and $\mathscr{T}_{|U}$ are isomorphic.

For the proof we need the following lemma.

Lemma 6.2.5. *Let $s_1, \dots, s_q$ be in $\mathcal{O}_X(U)$. Let $\mathscr{S}$ be a sheaf of finite type, and suppose that for some point $x \in U$ we have a surjection of stalks:*

$$\mathcal{O}^q_{X,x} \xrightarrow{(s_{1x}, \dots, s_{qx})} \mathscr{S}_x \longrightarrow 0.$$

Then on some open neighborhood $V \subset U$ of x, we have a surjection:

$$\mathcal{O}^q_{X|V} \xrightarrow{(s_1, \dots, s_q)} \mathscr{S}_{|V} \longrightarrow 0.$$

In particular, if the stalk at x of a sheaf of finite type is zero, then the sheaf is identically zero in some open neighborhood V of x.

Proof. Because $\mathscr{S}$ is of finite type there exists an open neighborhood W of x and sections $t_1, \dots, t_s$ generating $\mathscr{S}_{|V}$. But, because the s_{ix} generate the stalk at x, we can find germs f_{ijx} such that $t_{ix} = \sum_j f_{ijx} s_{jx}$. Take representatives f_{ij} on a sufficiently small open subset V. Because the t_i generate $\mathscr{S}$ on V, it follows that the s_j also generate $\mathscr{S}$ on V. □

Proof of 6.2.4. Consider the sheaf $\mathrm{Ker}(\alpha)$. By definition, $\mathrm{Ker}(\alpha)_x = \mathrm{Ker}(\alpha_x)$. By assumption its stalk is equal to (0), hence it is generated by zero elements. As $\mathrm{Ker}(\alpha)$ is coherent by the Meta-Theorem, it is in particular finitely generated. It follows from the previous lemma that $\mathrm{Ker}(\alpha)$ is generated by zero elements in some neighborhood U of 0. Thus $\mathrm{Ker}(\alpha)$ is the zero sheaf on some open neighborhood of 0. The argument for the cokernel is similar. □

So, in order to show that two coherent sheaves $\mathscr{S}$ and $\mathscr{T}$ are isomorphic *on some open set* containing a point x, one finds a map of sheaves $\alpha : \mathscr{S} \longrightarrow \mathscr{T}$, and then shows that α_x is an isomorphism. The last condition usually is pure algebra!

Example 6.2.6. Consider the subsheaf $\mathscr{T}$ of $\mathcal{O}_{\mathbb{C}}$ of functions vanishing on a sequence of points $p_1, p_2, \dots$, converging to 0 (cf. Example 6.2.2(3)). Then we have an inclusion $0 \longrightarrow \mathscr{T}$, and an isomorphism of stalks $0 \cong \mathscr{T}_0$ by the Identity Theorem. However, $\mathscr{T}$ is not the zero sheaf, and indeed, $\mathscr{T}$ is not coherent.

Definition 6.2.7. Let X be a complex space, and $\mathscr{J} \subset \mathcal{O}_X$ be an ideal sheaf. We define $V(\mathscr{J}) := \{x \in X : \mathscr{J}_x \subset \mathfrak{m}_{X,x}\}$.

Remark 6.2.8. Let X be a complex space, and $\mathscr{J} \subset \mathcal{O}_X$ be an ideal sheaf which is finitely generated. Then $V(\mathscr{J})$ is an analytic space. Indeed, if $\mathscr{J}(U) = (f_1, \dots, f_m)$ on some open neighborhood U of x, then $V(\mathscr{J}) \cap U = \{x \in U : f_1(x) = \dots = f_m(x) = 0\}$.

Corollary 6.2.9. *Let X be a complex space and $\mathscr{M}$ be a coherent $\mathcal{O}_X$-module. Then the support of $\mathscr{M}$:*

$$\operatorname{supp}(\mathscr{M}) := \{x \in X : \mathscr{M}_x \neq 0\}$$

is an analytic space.

Proof. Along with $\mathscr{M}$, the annihilator is coherent (cf. Theorem 6.2.3 or Lemma 6.2.17), so in particular it is finitely generated. The annihilator sheaf is an ideal sheaf. Now the support of $\mathscr{M}$ is given by the zero set of the annihilator of $\mathscr{M}$: $\operatorname{supp}(\mathscr{M}) = V(\mathscr{A}nn(\mathscr{M}))$.

Indeed, let $x \notin \operatorname{Supp}(\mathscr{M})$. Then $\mathscr{M}_x = 0$, so that $\operatorname{Ann}(\mathscr{M}_x) = \mathscr{A}nn(\mathscr{M})_x = \mathcal{O}_{X,x}$. Therefore, x does not lie in the zero set of $\mathscr{A}nn(\mathscr{M})$. Suppose on the other hand that $x \notin V(\mathscr{A}nn(\mathscr{M}))$. Then there exists an $f_x \in \operatorname{Ann}(\mathscr{M}_x)$ with $f_x(x) \neq 0$ and $f_x m = 0$ for all $m \in \mathscr{M}_x$. As $f_x(x) \neq 0$ it follows that the germ f_x is a unit, so that $m = 0$ for all $m \in \mathscr{M}_x$. Hence $\mathscr{M}_x = 0$, and thus $x \notin \operatorname{Supp}(\mathscr{M})$. Hence the support of $\mathscr{M}$ is equal to $V(\mathscr{A}nn(\mathscr{M}))$ and is therefore an analytic subset by Remark 6.2.8. □

So a typical application of coherence is to prove that certain sets are analytic sets. By using the Finite Mapping Theorem (this is not a trivial theorem, to be proved in the next Section), we get the following corollary, which we already proved in Exercise 4.4.17.

Corollary 6.2.10. *Let $f : X \longrightarrow Y$ be a finite map between analytic spaces. Then the image of f is an analytic subset of Y.*

Proof. The sheaf $f_*\mathcal{O}_X$ is a sheaf of $\mathcal{O}_Y$-modules. By the Finite Mapping Theorem 6.3.5 $f_*\mathcal{O}_X$ is *coherent*, and thus in particular its support is an analytic subset, which is closed. It remains to show that the image of f is equal to the support of $f_*\mathcal{O}_X$. This follows as we showed in 6.1.18 that $(f_*\mathcal{O}_X)_y = \oplus_{i=1}^t \mathcal{O}_{X,x_i}$, for $x_1, \dots, x_t$ being the elements in $f^{-1}(y)$. □

As another example, suppose that for an analytic space X there exists a normalization $n : \widetilde{X} \longrightarrow X$. Here n is a finite map, and $(\widetilde{X}, p)$ is a normal germ for all $p \in \widetilde{X}$. Then one sees that the support of $n_*\mathcal{O}_{\widetilde{X}}/\mathcal{O}_X$ is the set of all points x in X in which (X,x) is not normal. Thus one deduces that the set of nonnormal points in X is an analytic subset. This is really the world up-side down. In the next section we will prove the coherence of the normalization sheaf, and the main step is to show that the set of nonnormal points is an analytic subset.

Theorem 6.2.11. *Let X be a complex space and suppose that $\mathscr{S}$ is a coherent $\mathcal{O}_X$-module. Then the set $A = \{x \in X : \mathscr{S}_x$ is not a free $\mathcal{O}_{X,x}$-module$\}$ is a proper analytic subset of X.*

Proof. Let $x \in X$. As $\mathscr{S}$ is coherent, there exists an open neighborhood U of x such that $\mathscr{S}_{|U}$ has a representation

$$\mathcal{O}^q_{X|U} \xrightarrow{\varphi} \mathcal{O}^p_{X|U} \longrightarrow \mathscr{S}_{|U} \longrightarrow 0.$$

For $y \in U$, $\mathscr{S}_y$ is a free $\mathcal{O}_{X,y}$-module of rank $p-r$ if and only if for the Fitting ideals we have $I_r(\varphi_y) = \mathcal{O}_{X,y}$ and $I_{r+1}(\varphi_y) = (0)$. This has been proved in Proposition 1.3.8. Certainly those Fitting ideals are finitely generated. We may choose r so that on U, $I_{r+1}(\varphi) = (0)$ and $I_r(\varphi_y) \neq (0)$ for some $y \in U$. Then $A \cap U = V(I_r(\varphi)) \subsetneqq U$. □

Theorem 6.2.12. *Let X be a complex space, and $\mathscr{F}$ be a coherent $\mathcal{O}_X$-module. Suppose* $\text{supp}(\mathscr{F}) = \{y : \mathscr{F}_y \neq 0\} = \{x\}$. *Then*

(1) $\mathscr{F}(X) = \mathscr{F}_x$,

(2) $\dim_{\mathbb{C}} \mathscr{F}_x < \infty$.

Proof.
(1) The first property holds without using the coherence. Consider the map $\mathscr{F}(X) \longrightarrow \mathscr{F}_x$. We first show that it is injective. Take $s \in \mathscr{F}(X)$ and suppose $s_x = 0$. By definition of germ, there exists a neighborhood U of x such that $s_{|U} = 0$. Let $W := X \setminus \{x\}$. Because $\mathscr{F}_y = 0$ for all $y \in W$, we certainly have that $s_{|W} = 0$. By the first sheaf axiom, it follows that $s = 0$, as $X = U \cup W$. To prove the surjectivity let $f_x \in \mathscr{F}_x$ and choose a representative $f \in \mathscr{F}(U)$. With $W := X \setminus \{x\}$ we have $f_{|U \cap W} = 0$. So the section 0 on W and f on U glue to a section on X, by the second sheaf axiom.

(2) Let $I = \mathscr{A}nn(\mathscr{F})$ and assume that, in a neighborhood U of x, I is generated by $f_1, \ldots, f_r$. Here we used the Meta-Theorem to deduce the coherence of $\mathscr{A}nn(\mathscr{F})$ from the coherence of $\mathscr{F}$. We have already seen that $V(I)$ is the support of $\mathscr{F}$, which by assumption is x. In $\mathcal{O}_{X,x}$ we consider the primary decomposition of $I_x = \mathfrak{q}_1 \cap \cdots \cap \mathfrak{q}_s$. Now assume that $I_x \subset \mathfrak{q}$ is primary but $\sqrt{\mathfrak{q}} \neq \mathfrak{m}_{X,x}$. In this case $V(I) \supset V(\mathfrak{q}) \supsetneqq \{x\}$. This is a contradiction to the fact that $I_y = \mathcal{O}_{X,y}$ for all $y \neq x$. We conclude that $s = 1$ and $\mathfrak{q}_1$ is $\mathfrak{m}_{X,x}$-primary. Hence $\mathfrak{m}_{X,x}^c$ annihilates $\mathscr{F}_x$ for a suitable c and, therefore, $\mathscr{F}_x$ is a finitely generated $\mathcal{O}_{X,x}/\mathfrak{m}_{X,x}^c$-module. But $\dim_{\mathbb{C}} \mathcal{O}_{X,x}/\mathfrak{m}_{X,x}^c < \infty$ implies $\dim_{\mathbb{C}} \mathscr{F}_x < \infty$. □

We now turn our attention to the proof of the Meta-Theorem. We first note the following trivial lemma, whose proof we leave as Exercise 6.2.20.

Lemma 6.2.13.

(1) A quotient sheaf of a sheaf of finite type is a sheaf of finite type.

(2) A subsheaf of a sheaf of relation finite type is a sheaf of relation finite type.

(3) A finitely generated subsheaf of a coherent sheaf is coherent.

The following lemma is basic for all what follows.

Lemma 6.2.14. *Consider a short exact sequence of sheaves of $\mathcal{O}_X$-modules*

$$0 \longrightarrow \mathscr{S}_1 \xrightarrow{\phi} \mathscr{S}_2 \xrightarrow{\psi} \mathscr{S}_3 \longrightarrow 0.$$

Then all sheaves $\mathscr{S}_1, \mathscr{S}_2, \mathscr{S}_3$ are coherent if two of them are coherent.

Proof. In this proof, we will take a point $x \in X$. The notation $\mathcal{O}$ will be reserved for the sheaf $\mathcal{O}_U$ for some not specified open neighborhood U of x. Every time we make a statement like that there exists a map $\mathcal{O}^p \longrightarrow \mathscr{S}$, we in fact mean that there exists a small neighborhood U of x, smaller than the one considered until that moment, and a map of sheaves $\mathcal{O}_U^p \longrightarrow \mathscr{S}_U$.

There are three cases to consider.

(1) Suppose that $\mathscr{S}_1$ and $\mathscr{S}_2$ are coherent. Because $\mathscr{S}_2$ is a sheaf of finite type, it follows from Lemma 6.2.13 that $\mathscr{S}_3$ is of finite type. So we only have to show that $\mathscr{S}_3$ is of relation finite type. Therefore, consider a map $\chi : \mathcal{O}^p \longrightarrow \mathscr{S}_3$. We can lift this map over ψ:

$$\begin{array}{ccc} & & \mathcal{O}^p \\ & \overset{\sigma}{\swarrow} & \downarrow \chi \\ \mathscr{S}_2 & \xrightarrow{\psi} & \mathscr{S}_3. \end{array}$$

Indeed, for all basic vectors e_i, $i = 1, \dots p$, consider an element $s_i \in \mathscr{S}_2$ with $\psi(s_i) = \chi(e_i)$. These s_i exist by surjectivity of ψ. Now define $\sigma(e_i) = s_i$. This defines the map σ. Because $\mathscr{S}_1$ is finitely generated, we can take a surjective map $\tau : \mathcal{O}^q \longrightarrow \mathscr{S}_1$. Hence we get an exact sequence

$$\mathcal{O}^q \xrightarrow{\tau} \mathscr{S}_2 \xrightarrow{\psi} \mathscr{S}_3 \longrightarrow 0.$$

Because of coherence $\mathscr{S}_2$ the kernel of the map $\sigma \oplus \tau : \mathcal{O}^{p+q} \longrightarrow \mathscr{S}_2$ is finitely generated. Hence, also the projection on the first factor $\pi : \mathrm{Ker}(\sigma \oplus \tau) \longrightarrow \mathcal{O}^p$ is finitely generated. We are finished if we show that $\pi(\mathrm{Ker}(\sigma \oplus \tau)) = \mathrm{Ker}(\chi)$. So we have to show two things.

(i) Let $a \in \mathrm{Ker}(\chi)$ and consider $\sigma(a)$. As $\psi\sigma(a) = \chi(a) = 0$, it follows that $\sigma(a) \in \mathrm{Ker}(\psi)$, which by exactness is $\mathrm{Im}(\tau)$. Thus there exists an element $b \in \mathcal{O}^q$ with $\tau(b) = \sigma(a)$. Thus $(a, -b) \in \mathrm{Ker}(\sigma \oplus \tau)$, and $a = \pi(a, -b) \in \pi(\mathrm{Ker}(\sigma \oplus \tau))$.

(ii) Let $a \in \pi(\mathrm{Ker}(\sigma \oplus \tau))$. Hence there exists an element $b \in \mathcal{O}^q$ with $\sigma(a) + \tau(b) = 0$. Compose this with ψ. Then $\chi(a) = \psi\sigma(a) = \psi\sigma(a) + \psi\tau(b) = 0$ by exactness. Hence $a \in \mathrm{Ker}(\chi)$.

(2) Suppose that $\mathscr{S}_2$ and $\mathscr{S}_3$ are coherent. As $\mathscr{S}_1$ is a subsheaf of $\mathscr{S}_2$, which is of relation finite type, it follows from Lemma 6.2.13 that $\mathscr{S}_1$ is of relation finite type. So it remains to show that $\mathscr{S}_1$ is of finite type. As $\mathscr{S}_2$ is finitely generated we can take a surjection $\chi : \mathcal{O}^q \longrightarrow \mathscr{S}_2$, so $\mathscr{S}_2 = \mathrm{Im}(\chi)$. Because $\mathscr{S}_3$ is of relation finite type, we get that $\mathrm{Ker}(\psi\chi : \mathcal{O}^q \longrightarrow \mathscr{S}_3)$ is of finite type, so we get a surjection $\mathcal{O}^p \longrightarrow \mathrm{Ker}(\psi\chi)$ for some p. As χ is surjective, $\chi(\mathrm{Ker}(\psi\chi)) = \mathrm{Ker}(\psi)$. Thus we have a surjection $\mathcal{O}^p \longrightarrow \mathrm{Ker}(\psi)$ which is $\mathrm{Im}(\phi)$ by exactness. As ϕ is injective, $\mathrm{Im}(\phi) \cong \mathscr{S}_1$. Thus $\mathscr{S}_1$ is of finite type.

(3) Suppose that $\mathscr{S}_1$ and $\mathscr{S}_3$ are coherent. We first show that $\mathscr{S}_2$ is of relation finite type. We take a map $\chi : \mathcal{O}^p \longrightarrow \mathscr{S}_2$. Because $\mathscr{S}_3$ is of relation finite type, we get a surjective map $\sigma : \mathcal{O}^q \longrightarrow \mathrm{Ker}(\psi\chi : \mathcal{O}^p \longrightarrow \mathscr{S}_3)$. Consider $\chi\sigma : \mathcal{O}^q \longrightarrow \mathscr{S}_2$. The image of this map by construction lies in $\mathrm{Ker}(\psi)$, and this is by exactness equal to $\mathrm{Im}(\phi)$, which is isomorphic to $\mathscr{S}_1$. The sheaf $\mathscr{S}_1$ is of relation finite type, hence $\mathrm{Ker}(\chi\sigma)$ is finitely generated. As σ is surjective we have that σ maps $\mathrm{Ker}(\chi\sigma)$ surjectively onto $\mathrm{Ker}(\chi)$. Thus $\mathrm{Ker}(\chi)$ is finitely generated.

The proof that $\mathscr{S}_2$ is of finite type is easy, and left to the reader, cf. 1.2.14. □

Corollary 6.2.15.

(1) The direct sum of finitely many coherent sheaves is coherent.

(2) Let $\alpha : \mathscr{S} \longrightarrow \mathscr{T}$ *be a morphism of coherent sheaves. Then the sheaves* $\mathrm{Coker}(\alpha)$, $\mathrm{Ker}(\alpha)$ *and* $\mathrm{Im}(\alpha)$ *are coherent.*

(3) Let $\mathscr{S}$ *and* $\mathscr{T}$ *be coherent subsheaves of a coherent sheaf* $\mathscr{U}$*. Then the sum* $\mathscr{S} + \mathscr{T}$ *and the intersection* $\mathscr{S} \cap \mathscr{T}$ *are coherent sheaves.*

(4) Suppose that the structure sheaf $\mathcal{O}_X$ *is coherent. Then* $\mathscr{S}$ *is coherent, if and only if for every* $x \in X$ *there exists a small neighborhood* U *of* x *and an exact sequence:*

$$\mathcal{O}^q_{X|U} \longrightarrow \mathcal{O}^p_{X|U} \longrightarrow \mathscr{S}_{|U} \longrightarrow 0.$$

(5) If $\mathscr{S}$ *and* $\mathscr{T}$ *are coherent, then* $\mathscr{S} \otimes \mathscr{T}$ *is coherent.*

Proof. Exercise 6.2.19. □

It is somewhat more tricky to show that the sheaf $\mathscr{H}om$ of coherent sheaves is coherent:

Lemma 6.2.16. *Suppose that* $\mathscr{S}$ *and* $\mathscr{T}$ *are coherent* $\mathcal{O}$*-modules.*

(1) The natural map $\mathscr{H}om_{\mathcal{O}}(\mathscr{S},\mathscr{T})_x \longrightarrow \mathrm{Hom}_{\mathcal{O}_x}(\mathscr{S}_x,\mathscr{T}_x)$ *is bijective.*

(2) $\mathscr{H}om_{\mathcal{O}}(\mathscr{S},\mathscr{T})$ *is coherent.*

Proof.
(1) The injectivity is easy: let $\varphi : \mathscr{S}_{|U} \longrightarrow \mathscr{T}_{|U}$ be a map and assume that $\mathrm{Ker}(\varphi_x) = \mathscr{S}_x$. Theorem 6.2.4 implies that $\mathrm{Ker}(\varphi)_{|V} = \mathscr{S}_{|V}$ for an open neighborhood of x. This says that φ is the zero map in $\mathscr{H}om_{\mathcal{O}}(\mathscr{S},\mathscr{T})_x$. To show that the map is surjective, we have to lift any map $\varphi_x : \mathscr{S}_x \longrightarrow \mathscr{T}_x$ between germs to some open neighborhood U of x. This is easy for the case that $\mathscr{S} = \mathcal{O}^p_X$, because we then just have to take a lift in $\mathscr{T}(U)$ of the images of the basis elements. Because $\mathscr{S}$ is finitely generated, we can find a map $\psi : \mathcal{O}^p_{|U} \longrightarrow \mathscr{S}_{|U}$. As just noted, we can find a lift of the map $\chi_x := \varphi_x\psi_x$ to $\chi_U : \mathcal{O}^p_{|U} \longrightarrow \mathscr{T}_{|U}$, after possibly shrinking U. We have $\mathrm{Ker}\,\psi_x \subset \mathrm{Ker}\,\chi_x$. Because $\mathscr{S}$ is of relation finite type, $\mathrm{Ker}\,\psi$ is finitely generated. Thus $\mathrm{Ker}\,\psi_U \subset \mathrm{Ker}\,\chi_U$ after possibly shrinking U, cf. the proof of 6.2.5. Therefore, χ_U induces the desired map $\varphi_U : \mathscr{S}_{|U} = \mathcal{O}^p_{|U}/\mathrm{Ker}(\psi_U) \longrightarrow \mathscr{T}_{|U}$. Note that in this proof we only made use of the coherence of $\mathscr{S}$.

(2) It follows directly from the definition of coherence, that if $\mathscr{S}$ is coherent we (locally) have an exact sequence $\mathcal{O}^q \longrightarrow \mathcal{O}^p \longrightarrow \mathscr{S} \longrightarrow 0$. We want to show that

$$0 \longrightarrow \mathscr{H}om_{\mathcal{O}}(\mathscr{S},\mathscr{T}) \longrightarrow \mathscr{H}om_{\mathcal{O}}(\mathcal{O}^p,\mathscr{T}) \longrightarrow \mathscr{H}om_{\mathcal{O}}(\mathcal{O}^q,\mathscr{T}) \tag{6.1}$$

is exact. This suffices, because one easily sees that $\mathscr{H}om_{\mathcal{O}}(\mathcal{O}^s,\mathscr{T})$ is isomorphic to $\mathscr{T}^s$ for all s, which is coherent by assumption. As a kernel of a morphism between coherent sheaves is coherent again, it then follows that $\mathscr{H}om_{\mathcal{O}}(\mathscr{S},\mathscr{T})$ is coherent. To prove the exactness of (6.1), look at the following commutative diagram:

$$\begin{array}{ccccccc}
0 \longrightarrow & \mathscr{H}om_{\mathcal{O}}(\mathscr{S},\mathscr{T})_x & \longrightarrow & \mathscr{H}om_{\mathcal{O}}(\mathcal{O}^p,\mathscr{T})_x & \longrightarrow & \mathscr{H}om_{\mathcal{O}}(\mathcal{O}^q,\mathscr{T})_x \\
& \downarrow & & \downarrow & & \downarrow \\
0 \longrightarrow & \mathrm{Hom}_{\mathcal{O}_x}(\mathscr{S}_x,\mathscr{T}_x) & \longrightarrow & \mathrm{Hom}_{\mathcal{O}_x}(\mathcal{O}^p_x,\mathscr{T}_x) & \longrightarrow & \mathrm{Hom}_{\mathcal{O}_x}(\mathcal{O}^q_x,\mathscr{T}_x).
\end{array}$$

The sequence at the bottom is exact, because that is just a statement about rings and modules, and has been proved in Exercise 1.2.37. The left vertical arrow is an isomorphism by the first part. Because, as already remarked, $\mathcal{H}om_{\mathcal{O}}(\mathcal{O}^p,\mathcal{T})$ is isomorphic to $\mathcal{T}^p$, the fact that the other vertical arrows are isomorphisms is trivial. Therefore, the upper row is exact. As this is true for all $x \in X$, the lemma follows. □

The following now follows easily:

Lemma 6.2.17. *Suppose that the structure sheaf is coherent. Let $\mathcal{S}$ be coherent, and suppose that $\mathcal{P}$ and $\mathcal{Q}$ are coherent submodules of $\mathcal{S}$. Then the sheaves $\mathcal{A}nn(\mathcal{S})$ and the transporter sheaf $(\mathcal{Q} : \mathcal{P})$ are coherent.*

Proof. The annihilator sheaf is the kernel of the morphism

$$\mathcal{O} \longrightarrow \mathcal{H}om_{\mathcal{O}}(\mathcal{S},\mathcal{S})$$

by sending an element to multiplication with this element. Therefore, by the previous result, $\mathcal{H}om_{\mathcal{O}}(\mathcal{S},\mathcal{S})$ is coherent, and the therefore the annihilator sheaf, as kernel of a map between coherent sheaves, is coherent. The other result is Exercise 6.2.23. □

The final basic property of coherent sheaves is the following extension principle:

Theorem 6.2.18 (Extension Principle). *Let $(Y,\mathcal{O}_Y)$ be a* closed *subspace of the complex space $(X,\mathcal{O}_X)$, $i : Y \longrightarrow X$ the inclusion. Suppose that the sheaf $\mathcal{O}_X$ is coherent.*[2] *Let $\mathcal{S}$ be a sheaf of $\mathcal{O}_Y$-modules. Then*

(1) $\mathcal{S}$ is coherent if and only if $i_*\mathcal{S}$ is coherent.

(2) The sheaf $\mathcal{O}_Y$ is coherent.

Proof. The second statement is a special case of the first statement. Indeed, as $i_*\mathcal{O}_Y$ is coherent as the quotient of a coherent sheaf by a finitely generated ideal sheaf, it follows that $\mathcal{O}_Y$ itself is coherent.

Step 1. We first prove that $\mathcal{S}$ is of finite type if and only if $i_*\mathcal{S}$ is of finite type. Let $i_*\mathcal{O}_Y = \mathcal{O}_X/\mathcal{I}$, where $\mathcal{I}$ is a finitely generated (and hence coherent) $\mathcal{O}_X$-ideal sheaf. Suppose that $\mathcal{S}$ is a finitely generated sheaf of $\mathcal{O}_Y$-modules. Then we have, for each $y \in Y$, a surjective map $\mathcal{O}_V^p \longrightarrow \mathcal{S}_{|V}$ for some $p \in \mathbb{N}$, and an open set $V \ni y$ in Y. Then $V = Y \cap U$ for some open set U in X. As $\mathcal{O}_U \twoheadrightarrow i_*\mathcal{O}_V = \mathcal{O}_U/\mathcal{I}_{|U}$ is surjective, it follows that the induced map $\mathcal{O}_U^p \longrightarrow (i_*\mathcal{S})_{|U}$ is surjective.

On the other hand, if $\mathcal{O}_U^p \longrightarrow (i_*\mathcal{S})_{|U}$ is surjective, the fact that $(i_*\mathcal{S})_{|U}$ is an $i_*\mathcal{O}_V = \mathcal{O}_U/\mathcal{I}_{|U}$-module implies that the map induces a surjective map $\mathcal{O}_V^p \longrightarrow \mathcal{S}_{|V}$.

Step 2. Suppose that $i_*\mathcal{S}$ is $\mathcal{O}_X$-coherent. To show that $\mathcal{S}$ is coherent, it remains, by step 1, to show that $\mathcal{S}$ is of relation finite type. Therefore, consider an open set V in Y, and a map of $\mathcal{O}_V$-modules

$$\varphi : \mathcal{O}_V^p \longrightarrow \mathcal{S}_{|V}.$$

Now $V = U \cap Y$ for some open set U in X. Let $\pi : \mathcal{O}_U \twoheadrightarrow i_*\mathcal{O}_V$ be the canonical surjection. The composed map $i_*\varphi \circ \pi : \mathcal{O}_U^p \longrightarrow (i_*\mathcal{S})_{|U}$ of $\mathcal{O}_U$-modules has a finitely generated kernel, because we assumed that $i_*\mathcal{S}$ is $\mathcal{O}_X$-coherent. As π is surjective, it maps $\operatorname{Ker}(\varphi \circ \pi)$ surjectively onto $i_* \operatorname{Ker}(\varphi)$. Thus $i_* \operatorname{Ker}(\varphi)$ is finitely generated. By step 1, it follows that $\operatorname{Ker}(\varphi)$ is finitely generated.

[2] In the next section it will turn out that this is a superfluous condition.

Step 3. Now suppose that $\mathscr{S}$ is $\mathcal{O}_Y$–coherent. It follows that we can write $\mathscr{S}$ locally as a cokernel. Thus for all points $y \in Y$ there exists an open neighborhood V of y in Y and an exact sequence

$$\mathcal{O}_V^q \longrightarrow \mathcal{O}_V^p \longrightarrow \mathscr{S} \longrightarrow 0.$$

As $i : Y \hookrightarrow X$ is a finite map we get by 6.1.17 an exact sequence

$$i_*(\mathcal{O}_V^q) \longrightarrow i_*(\mathcal{O}_V^p) \longrightarrow i_*\mathscr{S} \longrightarrow 0.$$

Again we write $V = U \cap Y$ for an open set U in X. Since $i_*\mathcal{O}_V$ is quotient of $\mathcal{O}_U$ by a finitely generated ideal sheaf, it is $\mathcal{O}_U$–coherent, and therefore $i_*\mathcal{O}_V^q$ and $i_*\mathcal{O}_V^p$ are $\mathcal{O}_U$–coherent. It follows that $i_*\mathscr{S}$, as cokernel of two coherent $\mathcal{O}_U$–modules, itself is $\mathcal{O}_U$–coherent. □

Exercises

6.2.19. Prove Corollary 6.2.15.
(Hint: For (2) first prove that $\operatorname{Im}(\alpha)$ is coherent without using Lemma 6.2.14. For (5), consider an exact sequence $\mathcal{O}^q \longrightarrow \mathcal{O}^p \longrightarrow \mathscr{S} \longrightarrow 0$ and use "right exactness" of $\otimes$, see Exercise 1.2.43.)

6.2.20. Prove Lemma 6.2.13.

6.2.21. Let X be an affine variety, and M be a finitely generated $K[X]$-module. Show that the associated sheaf $\widetilde{M}$, defined by $\widetilde{M}(D(f)) = M_f$, is a coherent $\mathcal{O}_X$–module.

6.2.22. Assume that $\mathcal{O}_X$ is coherent. Prove that the support of a coherent sheaf $\mathscr{S}$ is an analytic set, by using the exact sequence

$$\mathcal{O}_{|U}^q \xrightarrow{M} \mathcal{O}_{|U}^p \longrightarrow \mathscr{S}_{|U} \longrightarrow 0$$

and looking at minors of M.

6.2.23. Prove the second assertion of 6.2.17.

6.3 The Four Basic Coherence Theorems

In this section we prove the four basic Coherence Theorems in complex analysis.

Theorem 6.3.1 (Oka). *The sheaf $\mathcal{O}_{\mathbb{C}^n}$ is coherent. More generally, for any complex space X, the sheaf $\mathcal{O}_X$ is coherent.*

Proof.
Step 1. A sheaf $\mathscr{S}$ is coherent, if for every point there exists an open neighborhood of this point such that $\mathscr{S}_{|U}$ is coherent. Thus to show that $\mathcal{O}_X$ is coherent, it suffices to show this for a complex model space $(D,\mathcal{O}_D)$. Such a complex model space injects, by definition, into an open subset of $\mathbb{C}^n$. By the extension principle 6.2.18 it suffices therefore to prove the coherence of $\mathcal{O}_{\mathbb{C}^n}$. This will be done by induction on n. The case $n = 0$ is trivial, see the second example of 6.2.2.

Step 2. For a short notation we write $\mathcal{O}_n$ instead of $\mathcal{O}_{\mathbb{C}^n}$ and, since there is no possible confusion, $\mathcal{O}_U$ instead of $\mathcal{O}_{\mathbb{C}^n|U}$ for an open subset $U \subset \mathbb{C}^n$. So assume that $\mathcal{O}_{n-1}$ is coherent. The statement is local, so we look at a neighborhood of a point $p \in \mathbb{C}^n$. Without loss of generality we may assume $p = 0$. As the sheaf $\mathcal{O}_n$ is obviously of finite type, we only have to show that it is of relation finite type. Therefore, we have to show that for any map of sheaves:

$$\mathcal{O}_n^s \xrightarrow{F=(f_1,\dots,f_s)} \mathcal{O}_n$$

the kernel Ker(F) is of finite type. Finding elements in Ker(F) means finding *relations*, that is, $r_1, \dots, r_s$ with $\sum_{i=1}^{s} r_i f_i = 0$. After a linear change of coordinates (if necessary), we may assume that $f_1, \dots, f_s$ are regular in x_n. As it is easy to relate the relations between $f_1, \dots, f_s$ and $u_1 f_1, \dots, u_s f_s$ for *units* u_i, we may, by applying the Weierstraß Preparation Theorem, assume that the f_i are Weierstraß polynomials in x_n. Let d_i be the degree of f_i in x_n, and put $d = \max\{d_i\}$. We may assume, without loss of generality, that the degree of f_s is d.

For any k we denote by $\mathcal{O}_{n-1}[x_n]_{\leq k}$ the free $\mathcal{O}_{n-1}$-module of rank $k+1$ of polynomials in x_n of degree less than or equal to k. By induction, $\mathcal{O}_{n-1}$ is coherent and, therefore, the free $\mathcal{O}_{n-1}$-module $\mathcal{O}_{n-1}[x_n]_{\leq k}$ is also coherent for all k by 6.2.15. By definition of coherence, the kernel of the following map of $\mathcal{O}_{n-1}$-modules:

$$\begin{aligned} (\mathcal{O}_{n-1}[x_n]_{\leq d})^s &\longrightarrow \mathcal{O}_{n-1}[x_n]_{\leq 2d} \\ (a_1, \dots, a_s) &\longmapsto a_1 f_1 + \dots + a_s f_s. \end{aligned}$$

is finitely generated. This exactly means that $\text{Ker}(F) \cap (\mathcal{O}_{n-1}[x_n]_{\leq d})^s$ is a finitely generated $\mathcal{O}_{n-1}$-module.

Step 3. It remains to show that in some small open neighborhood of 0, the sheaf of relations Ker(F) is as an $\mathcal{O}_n$-module generated by $\text{Ker}(F) \cap (\mathcal{O}_{n-1}[x_n]_{\leq d})^s$. This means the following: Take generators $h_1, \dots, h_l$ of $\text{Ker}(F) \cap \mathcal{O}_{n-1}[x_n]^s_{\leq d}$. Take a small open neighborhood U of 0 on which the $h_1, \dots, h_l$ and the $f_1, \dots, f_s$ converge. Then the germs $h_{1,p}, \dots h_{l,p}$ of the h_i in p generate the kernel of the map F_p for all $p \in U$. To prove this, first note that elements of the following type (so-called Koszul relations):

$$(-f_s, 0, \dots, 0, f_1), \quad \dots, \quad (0, \dots, 0, -f_s, f_{s-1})$$

are in $\text{Ker}(F) \cap \mathcal{O}_{n-1}[x_n]^s_{\leq d}$. Indeed, they are in Ker(F), and also of degree less than or equal to d in x_n. We take an arbitrary point $p = (p_1, \dots, p_n) \in U$. At the point p the function $f_{s,p}$ is regular of order $d' \leq d$ in $x_n - p_n$. Consider any relation between the f_i at the point p:

$$a_1 f_{1,p} + \dots + a_s f_{s,p} = 0.$$

For $i = 1, \dots, s-1$ we perform Weierstraß division with respect to f_s at p:

$$a_i = q_i f_{s,p} + c_i, \qquad \deg(c_i) < d' \leq d.$$

Moreover, if we define $c_s := a_s + \sum_{i=1}^{s-1} f_{i,p} q_i$ we see that

$$(a_1, \dots, a_s) = (c_1, \dots, c_s) - (-f_{s,p}, 0, \dots, 0, f_{1,p}) q_1 - \dots - (0, \dots, 0, -f_{s,p}, f_{s-1,p}) q_{s-1}.$$

From this it follows that $c_1 f_{1,p} + \dots + c_s f_{s,p} = 0$ and, therefore, we expressed our relation as an $\mathcal{O}_n$-linear combination of elements of $\text{Ker}(F) \cap \mathcal{O}_{n-1}[x_n]^s_{\leq d}$, as soon as we have proved that c_s is a polynomial of degree less than or equal to d. This is easy, because we have just proved that $c_s f_{s,p} = -\sum_{i=1}^{s-1} c_i f_{i,p}$. The right-hand side is a polynomial in x_n of degree less than $d + d'$. As we have assumed that the degree of f_s is d, it follows that c_s has degree at most $d' \leq d$. This concludes the proof of Oka's Theorem. □

Theorem 6.3.2 (Oka-Cartan). *Let $X \subset U \subset \mathbb{C}^n$ be an analytic set. Then the ideal sheaf $\mathscr{I} := \mathscr{I}(X)$ of X is coherent.*

Remark 6.3.3. Let the ideal $I = (f_1, \dots, f_s) \subset \mathbb{C}\{x_1, \dots, x_n\}$ be a radical ideal , and let $(X, o) = \big(V(I), o\big)$. Take small enough representatives. Then it is a tautology that for all $a \in X$

$$V(f_{1,a}, \dots f_{s,a}) = (X,a)$$

but it is not clear that the ideal of (X,a) is generated by the $f_{i,a}$. To put it in another way, it is not clear a priori that the ideal $(f_{1,a}, \dots f_{s,a})$ is a radical ideal in $\mathcal{O}_{X,a}$. That this is indeed the case is the content of the Coherence Theorem of Cartan and Oka.

Proof of Theorem 6.3.2.
Step 1. Consider a point $x \in X$. Without loss of generality, we may assume that $x = 0$. Consider the ideal $\mathscr{I}_0 \subset \mathcal{O}_{n,o} = \mathbb{C}\{x_1, \dots, x_n\}$ of the germ (X, o). As this ring is Noetherian, this ideal is finitely generated. We take finitely many generators:

$$\mathscr{I}_0 = (f_1, \dots, f_s)$$

and choose a small open neighborhood U of 0 such that all f_i converge on U. As a finitely generated subsheaf of the coherent sheaf $\mathcal{O}_n$ the sheaf generated by $(f_1, \dots, f_s)$ is coherent by Lemma 6.2.13. We have a map of sheaves $(f_1, \dots, f_s) \longrightarrow \mathscr{I}$. This is an isomorphism, if it induces an isomorphism on all stalks. So we only have to show

$$\mathscr{I}_p = (f_{1,p}, \dots, f_{s,p}) \tag{6.2}$$

for all points p in a small open neighborhood V of 0.

Step 2. We reduce to the case that the germ (X, o) is irreducible. Indeed, suppose $(X, o) = (X_1, o) \cup \dots \cup (X_r, o)$. Take representatives of (X_i, o), so that $X = X_1 \cup \dots \cup X_r$ in a small open set U. For the ideal sheaves on U we have

$$\mathscr{I}(X) = \mathscr{I}(X_1) \cap \dots \cap \mathscr{I}(X_r).$$

Therefore, if we show that the $\mathscr{I}(X_j)$ are coherent it follows that $\mathscr{I}(X)$ is coherent by by 6.2.15.

Step 3. Thus we may suppose that (X, o) is irreducible. By Remark 3.4.16, as a consequence of the Local Parametrization Theorem, there exists a representative X of (X, o) and a hypersurface $D = V(\Delta) \subset X$ such that for $p \in X \setminus D$, $\mathscr{I}_p = (f_1, \dots, f_s)_p$. Hence (6.2) holds for all points p outside a proper analytic subset. We now consider the following ideal sheaf on U:

$$\mathscr{J} := (f_1, \dots, f_s) : (\Delta) \subset \mathcal{O}_n.$$

This is a coherent sheaf. Indeed, since $(f_1, \dots, f_s)$ is a finitely generated subsheaf of the coherent sheaf $\mathcal{O}_n$, it is coherent by 6.2.13. Hence $\mathscr{J}$ is coherent by 6.2.17. Therefore, we can take finitely many generators $g_1, \dots, g_t$ for $\mathscr{J}$. Looking at the stalk at zero, these elements have the property that $\Delta_0 g_{i,0} \in (f_1, \dots, f_s) = \mathscr{I}_0$. But this ideal was supposed to be prime, hence $g_{i,0} \in (f_1, \dots, f_s) \subset \mathcal{O}_{n,0}$. By taking U smaller, we may assume that $g_{i,p} \in (f_{1,p}, \dots, f_{s,p})$ for all $p \in U$. Hence we reduced to the case that we have the following equality of ideal sheaves:

$$(f_1,\dots,f_s):(\Delta)=(f_1,\dots,f_s). \tag{6.3}$$

Step 4. Take a point p in U and an element $g\in\mathscr{I}_p$, the stalk of the ideal sheaf of X at p. Take a representative g vanishing on $V\cap X$, for a (very small) neighborhood $V\subset U$ of p. We have to show $g\in(f_{1,p},\dots,f_{s,p})$.

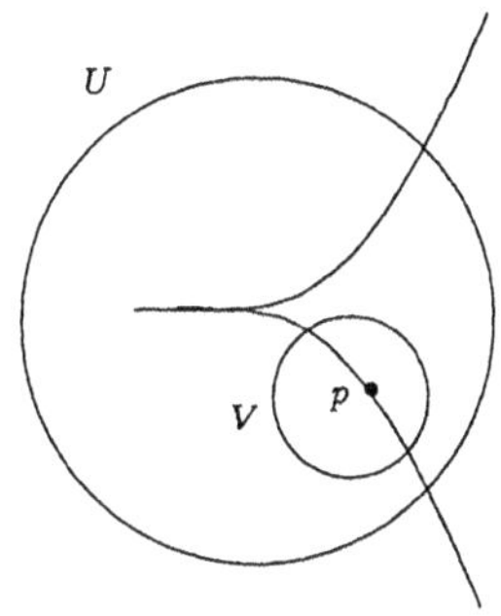

On V, the sheaf

$$\mathscr{S}:=(f_1,\dots,f_s):(g)$$

is coherent by 6.3.1, 6.2.13 and 6.2.17. Hence, $\mathscr{S}$ has finitely many generators, say $h_1,\dots,h_u$, for holomorphic functions h_i defined on V. Since (6.2) was true for all points $p\in X\setminus D$, it follows that $g_p\in(f_{1,p},\dots,f_{s,p})$ for all $p\in V\setminus D$, so that we deduce $\mathscr{S}_p=\mathcal{O}_{n,p}$ for all $p\in V\setminus D$. Thus, the zero set $V(h_1,\dots,h_u)\subset D\cap V$. As Δ also vanishes on D, it follows from the Nullstellensatz that there exists an $r\geq 0$ such that $\Delta^r\in(h_{1,p},\dots,h_{u,p})$ for all $p\in D\cap V$. From the defining properties of the h_i it follows that $\Delta^r g\in(f_{1,p},\dots,f_{s,p})$. Let $r\geq 0$ be minimal with this property. To prove the theorem, we have to show that $r=0$. Suppose the converse, that is, $r>0$. But then we have $\Delta^{r-1}g\in(f_{1,p},\dots,f_{s,p}):(\Delta)=(f_{1,p},\dots,f_{s,p})$, by (6.3), giving a contradiction. □

Corollary 6.3.4. *Let $X\subset\mathbb{C}^n$ be an analytic space. The set of singular points* Sing(X) *of X is an analytic subspace of X.*

Proof. Let $x\in X$. By applying Proposition 4.3.1, we may assume that (X,x) is irreducible. Now we may apply the Jacobian Criterion, 4.3.6, and the coherence of ideal sheaves, 6.3.2. □

Theorem 6.3.5 (Finite Mapping Theorem). *Consider a finite mapping $\pi:X\longrightarrow Y$ of complex spaces, and let $\mathscr{S}$ be a coherent $\mathcal{O}_X$-sheaf. Then $\pi_*\mathscr{S}$ is a coherent $\mathcal{O}_Y$-sheaf.*

Proof. Using the Extension principle (6.2.18) it is enough to prove the theorem for analytic spaces, that is, both X and Y are reduced. Let $y\in Y$ be a point. Since coherence is a local property it is enough to prove that $(\pi_*\mathscr{S})_{|U}$ is coherent for a sufficiently small neighborhood U of y. We may assume that $y=0$. The theorem is proved in several steps.

Step 1. We first consider the case that X is a small representative of $(X,0)=(V(f),0)$, where $f\in\mathbb{C}\{x_1,\dots x_n\}$ is a Weierstraß polynomial of degree b in x_n, Y an open subset of $\mathbb{C}^{n-1}$, $\pi:X\longrightarrow Y$ the projection on the first $n-1$ coordinates, and $\mathscr{S}=\mathcal{O}_X$.

Let $V \subset \mathbb{C}$ be a small open neighborhood of 0. We choose an open neighborhood $U \subset \mathbb{C}^{n-1}$ of 0 such that for all $p \in U$ the b zeros (counted with multiplicity) of $f(p,x_n)$ lie in V (continuity of roots, see 3.4.11). We may assume that

$$X = \{(p,x) : p \in U,\, f(p,x) = 0\} \subset U \times V$$

and $\pi : X \longrightarrow U$ is induced by the canonical projection. For an open subset $W \subset U$ we have a canonical map:

$$\mathcal{O}_U^b(W) \longrightarrow \mathcal{O}_U(W)[x_n] \subset \mathcal{O}_{U\times V}(W \times V) \longrightarrow \mathcal{O}_X(W \times V) = \pi_*\mathcal{O}_X(W)$$

$$(s_0, \dots s_{b-1}) \mapsto \sum_{j=0}^{b-1} s_j x_n^j.$$

We will prove that for all $p \in U$ the induced map of the stalks

$$\mathcal{O}_{U,p}^b \longrightarrow (\pi_*\mathcal{O}_X)_p \tag{6.4}$$

is an isomorphism. This will prove this step. Namely, as $\mathcal{O}_{n-1|U}^b$ is coherent by Oka's Theorem 6.3.1 , it follows that $\pi_*\mathcal{O}_{X|(U\times V)}$ is a coherent $\mathcal{O}_{n-1|U}$-module.

Now let $p \in U$, $x_1,\dots,x_t$ be the zeros of $f(p,x_n)$ and $b_1,\dots,b_t$ their multiplicities. Put $p_i := (p,x_i)$ for $i = 1,\dots,t$. In particular, $\sum_{i=1}^t b_i = b$. Using 6.1.18, we obtain $(\pi_*\mathcal{O}_X)_p = \oplus_{i=1}^t \mathcal{O}_{X,p_i}$. To prove that (6.4) is surjective, consider an element $(g_1,\dots,g_t) \in (\pi_*\mathcal{O}_X)_p = \oplus_{i=1}^t \mathcal{O}_{X,p_i}$. By Corollary 3.3.24 there exists an $r = \sum_{j=0}^{b-1} r_i x_n^j$, such that $r = g_i$ in $\mathcal{O}_{X,p_i}$ for $i = 1,\dots,t$. This shows the surjectivity. The injectivity of (6.4) follows from the uniqueness of the decomposition of 3.3.24.

Step 2. We consider now the situation as in Step 1 but now $\mathcal{S}$ is an arbitrary coherent sheaf. Because $\mathcal{S}$ is coherent, we have locally an exact sequence:

$$\mathcal{O}_X^q \longrightarrow \mathcal{O}_X^p \longrightarrow \mathcal{S} \to 0.$$

We use Theorem 6.1.17 and obtain an exact sequence:

$$\pi_*\mathcal{O}_X^q \longrightarrow \pi_*\mathcal{O}_X^p \longrightarrow \pi_*\mathcal{S} \to 0.$$

We proved in Step 1 that $\pi_*\mathcal{O}_X$ is coherent. Hence $\pi_*\mathcal{O}_X^q = (\pi_*\mathcal{O}_X)^q$ and $\pi_*\mathcal{O}_X^p$ are coherent. Therefore, $\pi_*\mathcal{S}$ as a cokernel is coherent.

Step 3. We now consider the case that $i : X \subset V(f)$ is a closed subspace, f regular in x_n, Y an open subset of $\mathbb{C}^{n-1}$, $\pi' : V(f) \longrightarrow Y$ the projection on the first $n-1$ coordinates and $\pi = \pi' \circ i$.

Then $i_*\mathcal{S}$ is coherent by the extension principle, and $\pi_*\mathcal{S} = (\pi' \circ i)_*\mathcal{S} = \pi'_*(i_*\mathcal{S})$ is coherent by the previous step.

Step 4. We now consider the case that X is a small representative of $(V(I), 0) \subset (\mathbb{C}^n, 0)$, where $\mathbb{C}\{x_1,\dots,x_n\}/I$ is a finitely generated $\mathbb{C}\{x_1,\dots,x_k\}$-module, Y is a small open neighborhood of 0 in $\mathbb{C}^k$, and $\pi : X \longrightarrow \mathbb{C}^k$ the obvious projection on the first k coordinates. This case follows from Step 3 and induction on $n-k$.

Step 5. We now consider the general case. We can (locally) view any finite mapping as a projection:

$$\begin{array}{ccc} X & \overset{i}{\hookrightarrow} & \mathbb{C}^n \\ \downarrow{\scriptstyle \pi} & & \downarrow{\scriptstyle \pi'} \\ Y & \overset{j}{\hookrightarrow} & \mathbb{C}^k. \end{array}$$

This is the graph construction, see 2.3.14 and Exercise 3.4.43. Here π' is the projection on the first k coordinates. By the extension principle $\pi_* \mathcal{S}$ is coherent if and only if $j_*(\pi_* \mathcal{S}) = \pi'_*(i_* \mathcal{S})$ is coherent. But the last sheaf is coherent by Step 4. □

For the proof of the final coherence theorem in this chapter, the following theorem is of crucial importance.

Theorem 6.3.6 (Oka). *Let X be an analytic space. Then the set of nonnormal points of X forms an analytic subspace of X. In particular, the set of normal points is open in X.*

Proof. Consider a point $0 \in X$, and let $I \subset R := \mathcal{O}_{X,0}$ be the ideal of the singular locus of X at the point 0. Then Exercise 4.4.20, which is the local version of Theorem 1.5.13, gives:

$$R \overset{=}{\hookrightarrow} Hom_R(I,I) \iff R \text{ is normal.}$$

We now sheafify. We have proved in 6.3.2 that the ideal sheaf $\mathcal{I}$ of the singular locus (it is an analytic space) is coherent. It follows from 6.2.16 that the sheaf $\mathcal{H}om_{\mathcal{O}_X}(\mathcal{I},\mathcal{I})$ is coherent, and that

$$\mathcal{H}om_{\mathcal{O}_X}(\mathcal{I},\mathcal{I})_p = \mathrm{Hom}_{\mathcal{O}_{X,p}}(\mathcal{I}_p,\mathcal{I}_p)$$

for all $p \in X$. It follows that the set of nonnormal points is given by the support of the sheaf $\mathcal{H}om_{\mathcal{O}_X}(\mathcal{I},\mathcal{I})/\mathcal{O}_X$. As this is the quotient of two coherent sheaves, it is a coherent sheaf and, therefore, its support is an analytic set, see 6.2.9. This proves the theorem. □

Theorem 6.3.7 (Oka). *Let $(X,\mathcal{O}_X)$ be an analytic space. The normalization sheaf $\widetilde{\mathcal{O}}_X$ is a sheaf whose stalk $\widetilde{\mathcal{O}}_{X,x}$ at x by definition is the normalization of $\mathcal{O}_{X,x}$. Then the normalization sheaf $\widetilde{\mathcal{O}}_X$ is $\mathcal{O}_X$-coherent.*

Proof. For simplicity of notation, we will assume that X is irreducible. We showed in 4.4.8 that there is a germ of a normal space $(\widetilde{X},\widetilde{x})$ and a finite map $n : (\widetilde{X},\widetilde{x}) \longrightarrow (X,x)$. In particular, there exist functions $f_1,\ldots,f_s$ in $Q(\mathcal{O}_{X,x})$ which generate $\mathcal{O}_{\widetilde{X},\widetilde{x}} = \widetilde{\mathcal{O}}_{X,x}$ as $\mathcal{O}_{X,x}$-module, that is, $\mathcal{O}_{\widetilde{X},\widetilde{x}} = \sum_{i=1}^s f_i \mathcal{O}_{X,x}$.

Let $\widetilde{X}$ be a representative of $(\widetilde{X},\widetilde{x})$ and X be a representative of (X,x) such that $\widetilde{X}$ is a normal analytic space, $f_1,\ldots,f_s$ are defined on $\widetilde{X}$, and the induced map $n : \widetilde{X} \longrightarrow X$ is finite. This is possible because the set of normal points is open (6.3.6). Then $n_* \mathcal{O}_{\widetilde{X}}$ is coherent by the Finite Mapping Theorem 6.3.5. As the f_i have a universal denominator Δ, multiplication with Δ gives an isomorphism $\sum_{i=1}^s f_i \mathcal{O}_X \cong \mathcal{S}$, where $\mathcal{S}$ is a finitely generated ideal sheaf. Thus $\sum_{i=1}^s f_i \mathcal{O}_X$ is coherent. To prove the theorem, it therefore suffices to show that $n_* \mathcal{O}_{\widetilde{X}} = \sum_{i=1}^s f_i \mathcal{O}_X$ and is equal to the normalization sheaf $\widetilde{\mathcal{O}}_X$. Using 6.2.16 the equality $n_* \mathcal{O}_{\widetilde{X}} = \sum_{i=1}^s f_i \mathcal{O}_X$ holds as both sheaves are coherent, and we can lift the isomorphism on stalks $\mathcal{O}_{\widetilde{X},\widetilde{x}} = \sum_{i=1}^s f_i \mathcal{O}_{X,x}$ to an isomorphism on a small open neighborhood.

To show that it is equal to the normalization sheaf, note that the f_i satisfy an integral equation

$$f_i^{k_i} + a_1^{(i)} f_i^{k_i - 1} + \ldots + a_k^{(i)} = 0. \tag{6.5}$$

Consider an even smaller open neighborhood U of x, on which the $a_j^{(i)}$ converge. Then the equation (6.5) holds on U. In particular, for all $a \in U$, we have $\sum_{i=1}^{s} f_i \mathcal{O}_{X,a} \subset \widetilde{\mathcal{O}}_{X,a}$. This shows that $\sum_{i=1}^{s} f_i \mathcal{O}_{X|U}$ is contained in the normalization sheaf. Now let $n^{-1}(a) = \{p_1, \ldots, p_t\}$. By construction of the germ of the analytic space $\widetilde{X}$ and Theorem 6.1.18 we have an equality

$$\sum_{i=1}^{s} f_i \mathcal{O}_{X,a} = n_*(\mathcal{O}_{\widetilde{X}})_a = \mathcal{O}_{\widetilde{X},p_1} \oplus \ldots \oplus \mathcal{O}_{\widetilde{X},p_t}.$$

All of the rings $\mathcal{O}_{\widetilde{X},p_i}$ for $i = 1, \ldots, t$ are normal by the choice of $\widetilde{X}$. Therefore, the direct sum, which is $\sum_{i=1}^{s} f_i \mathcal{O}_{X,a}$, is normal and, therefore, equal to the normalization $\widetilde{\mathcal{O}}_{X,a}$. This proves the theorem. □

Exercises

6.3.8. Prove that any compact complex subspace of a domain in $\mathbb{C}^n$ consists of a finite number of points. (Hint: Use the Finite Mapping Theorem, and induction on n.)

6.4 The Principle of Conservation of Number

To illustrate the principle of conservation of number, let us formulate three theorems, whose proofs will be given later in this Section.

Theorem 6.4.1. *Let (C, o) and (D, o) be two plane curve singularities, given by $f(x,y) = 0$ and $g(x,y) = 0$, respectively. Suppose that (C, o) and (D, o) do not have components in common. Consider deformations of f and g, that is, $F(x,y,s)$ and $G(x,y,s)$ in $\mathbb{C}\{x,y,s\}$ with $F(x,y,0) = f$ and $G(x,y,0) = g$.*

We can choose for all sufficiently small open neighborhoods U of 0 in $\mathbb{C}^2$, an open neighborhood V of 0 in $\mathbb{C}^2$ such that

(1) both F and G converge on $U \times V$,

(2) if C resp. D are representatives of (C, o) respectively (D, o), then C and D intersect only in 0 and $C \setminus \{0\}$ and $D \setminus \{0\}$ are smooth.

For s fixed, denote by C_s, respectively D_s the curve $\{F(x,y,s) = 0\} \subset U$, respectively $\{G(x,y,s) = 0\} \subset U$. Then, for all U sufficiently small there exists a V such that for all $s \in V$

$$(C, o) \cdot (D, o) = \sum_{p \in C_s} (C_s, p) \cdot (D_s, p).$$

Here $(C_s, p) \cdot (D_s, p)$ denotes the intersection number of C_s and D_s in the point p, see 5.1.4.

This theorem gives a geometric interpretation of the intersection number. Suppose that C_s and D_s intersect transversally, that is, the intersection number $(C_s \cdot D_s)_p = 1$ for all $p \in C_s \cap D_s$. The theorem then says, that the intersection number of C and D is equal to the number of intersection points appearing after a small perturbation of C and D.

Example 6.4.2. Let C be given by $f(x,y) = y^2 - x^3 = 0$, and D be given by $g(x,y) = y = 0$. Then

$$(C, 0) \cdot (D, 0) = \dim_{\mathbb{C}} \mathbb{C}\{x,y\}/(y^2 - x^3, y) = \dim_{\mathbb{C}} \mathbb{C}\{x\}/(x^3) = 3.$$

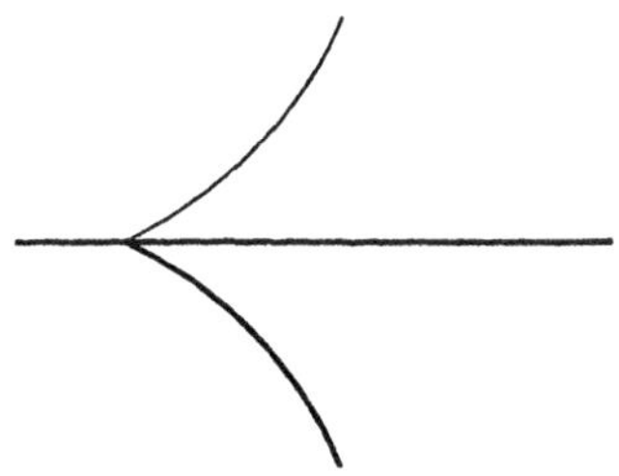

We consider the small perturbations:

$$F(x,y,s) = y^2 - x^2(x - s), \quad G(x,y,s) = y - s.$$

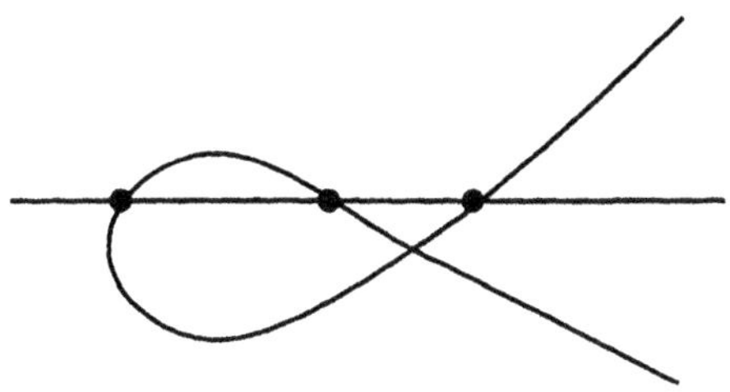

We see indeed three intersection points appearing.

Theorem 6.4.3. *Let $(C, 0) = \cup_{i=1}^r (C_i, 0)$ be a germ of a plane curve singularity, and let $f_i = 0$ be the equation of C_i. Consider the normalization $(\widetilde{C}, 0) = \cup_{i=1}^r (\widetilde{C}_i, 0) \longrightarrow (C, 0)$ of $(C, 0)$. Let $W \subset \mathbb{C}$ be an open neighborhood of 0 and*

$$(x(t_i), y(t_i)) : W \longrightarrow \mathbb{C}^2 \text{ for } i = 1, \dots, r$$

be a parametrization of the branch C_i, and consider a one-parameter deformation of the parametrization with base space $S \subset \mathbb{C}$

$$(x(t_i,s), y(t_i,s)) : W \times S \longrightarrow \mathbb{C}^2, \quad x(t_i,0) = x(t_i), \quad y(t_i,0) = y(t_i) \quad \text{for } i = 1, \dots, r.$$

Then we can take W and S so small that the image of $(x(t_i,s), y(t_i,s))$ is an analytic space in $W \times S$ given by $F_i(x,y,s) = 0$ with $F_i(x,y,0) = f_i$. Then $F(x,y,s) = \prod_{i=1}^r F_i(x,y,s)$ is the induced one-parameter deformation of $f = \prod_{i=1}^r f_i$.

Moreover for all sufficiently small neighborhoods U of 0 in W, there exists an open neighborhood V of 0 in S such that:

(1) F converges on $U \times V$.

(2) The only singular point of f on U is 0.

Denote for fixed s by C_s the curve $\{F(x,y,s)=0\}\cap U$, then for all $s\in V$ we have

$$\delta(C)=\sum_{p\in C_s}\delta(C_s,p),$$

where $\delta(C_s,p)$ denotes the δ-invariant of C_s in the point p.

Example 6.4.4. We consider the previous example. We consider the curve singularity $(C\cup D,o)$ given by the equation $y(y^2-x^3)=0$. The deformation in fact is given by the following deformation of the parametrization:

$$x(t_1,s)=t_1^2+s,\quad y(t_1,s)=t_1^3+st_1;\quad x(t_2)=t_2,\quad y(t_2)=s,$$

(check this). As the δ-invariant of an A_1-singularity is one, we see from the picture

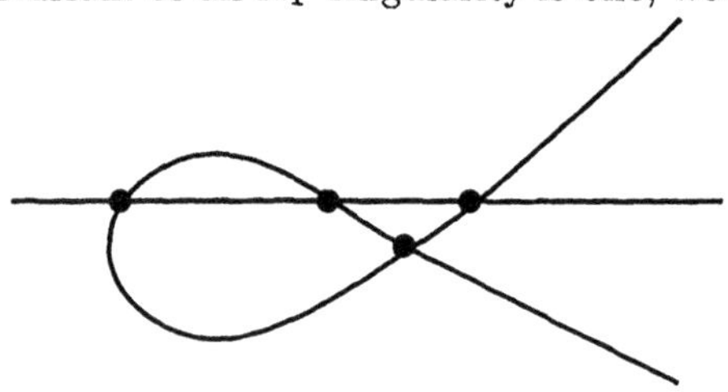

that $\delta(C\cup D,o)=4$.

Theorem 6.4.5. *Let $f=f(x_1,\dots,x_n):(\mathbb{C}^n,o)\longrightarrow(\mathbb{C},o)$ be a germ of a holomorphic function with isolated singularity, $\mu(f)$ its Milnor number. Consider a deformation of f:*

$$F(x_1,\dots,x_n,s):(\mathbb{C}^n\times S,o)\longrightarrow(\mathbb{C}\times S,o),\quad F(x_1,\dots,x_n,0)=f(x_1,\dots,x_n).$$

Then for all sufficiently small open neighborhoods U of 0 there exists an open neighborhood V of 0 such that for all $s\in V$:

$$\mu(f)=\sum_{p\in U}\mu(f_s,p)$$

where $\mu(f_s,p)$ is the Milnor number of the germ of the holomorphic function:

$$F(x_1,\dots,x_n,s)=f_s:(\mathbb{C}^n,p)\longrightarrow(\mathbb{C},f_s(p))$$

Example 6.4.6. We consider the E_7-singularity given by $f(x,y)=y(y^2-x^3)$, and the deformation $F(x,y,s)=(y-s)(y^2-x^2(x-s))$. The zero sets of f and f_s are given in the previous figures. In the picture of the zero set we see four A_1-singularities. But since a function on a compact set must have a maximum and a minimum, we have that in each compact region in the real plane bounded by the zero set of f_s there must be a singularity, too. One can in fact calculate that f_s has seven A_1-singularities, so that $\mu(f)=7$.

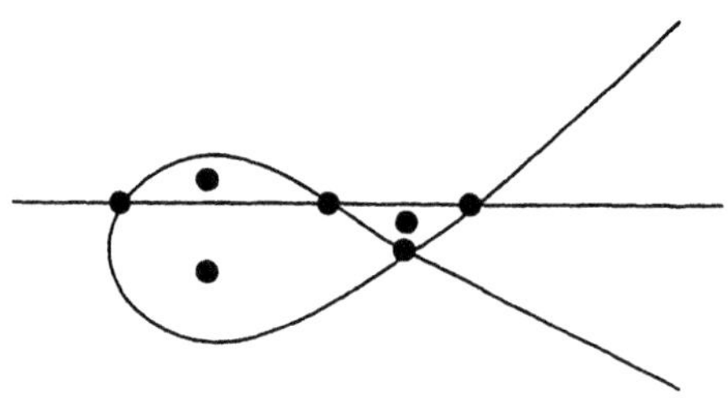

How to prove that "an invariant is constant under deformation", or, to put it in another way, when do we have conservation of number? Usually, a numerical invariant is defined by the vector space dimension of a certain module M over the local ring of a singularity $\mathcal{O}_{X,o}$. In the above theorems, we have respectively as local ring and as module:

(1) The local ring is $\mathcal{O}_{X,o} = \mathbb{C}\{x,y\}/(f,g)$, the module M is $\mathcal{O}_{X,o}$.

(2) The local ring is $\mathcal{O}_{C,o}$, the module M is $\mathcal{O}_{\tilde{C},o}/\mathcal{O}_{C,o}$.

(3) The local ring is $\mathcal{O}_n$, and the module M is $\mathcal{O}_n/J(f)$, where $J(f) = (\frac{\partial f}{\partial x_1}, \dots, \frac{\partial f}{\partial x_n})$ is the Jacobian ideal.

When deforming, the local ring $\mathcal{O}_{X,o}$ will be a quotient of the local ring $\mathcal{O}_{X_S,o}$, and similarly M is a factor module of M_S. Both the local ring and the module are in fact also $\mathbb{C}\{s\}$–modules. In the above examples:

(1) $\mathcal{O}_{X_S,o} = \mathbb{C}\{x,y,s\}/(F,G)$, $M_S = \mathcal{O}_{X_S,o}$.

(2) The local ring is $\mathcal{O}_{C_S,o} = \mathbb{C}\{x,y,s\}/F$, the module M_S is $\tilde{\mathcal{O}}_{C_S,o}/\mathcal{O}_{C_S,o}$.

(3) The local ring is $\mathbb{C}\{x_1, \dots, x_n, s\}$, and the module is $M_S = \mathbb{C}\{x_1, \dots, x_n, s\}/J(F)$, where $J(F) = (\frac{\partial F}{\partial x_1}, \dots, \frac{\partial F}{\partial x_n})$ is the "relative" Jacobian ideal.

In order to be able to compare the invariants at different points in the parameter space S, we need to sheafify.

Let the singularity (X, o) be embedded in $(\mathbb{C}^n, o)$, and let U be a sufficiently small open neighborhood of 0. Then we consider a representative X of (X, o) in U. Furthermore, we choose a small open neighborhood V of 0 in the parameter space $S = \mathbb{C}$. Let $\pi : U \times V \longrightarrow V$ be the canonical projection and $X_S \subset U \times V$ be a complex space such that $X_S \cap \pi^{-1}(0) = X$. For $s \in V$ we put $X_s := X_S \cap \pi^{-1}(s)$.

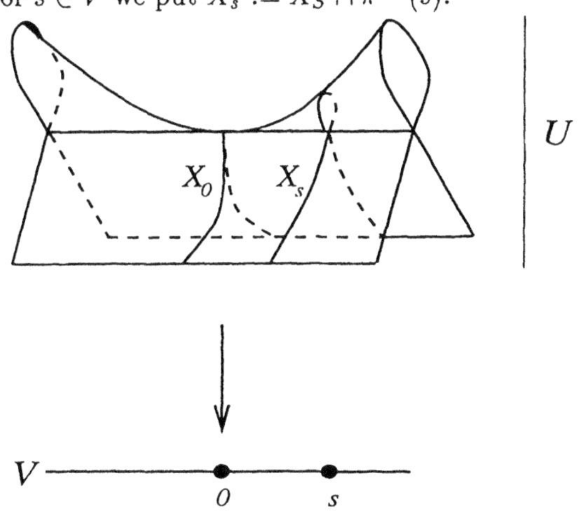

Theorem 6.4.7 (Principle of Conservation of Number). *Let $\mathcal{M}$ be a sheaf on X_S with the following properties:*

(1) $\mathcal{M}$ is a coherent $\mathcal{O}_{X_S}$-module.[3]

[3] The invariants we are interested in are the dimensions of the stalks of the restriction of $\mathcal{M}$ to the fibers.

(2) $\mathcal{M}_0$, the stalk of $\mathcal{M}$ at 0, is a free $\mathbb{C}\{s\}$-module of finite rank.

Then for all sufficiently small neighborhoods $U' \subset U$ of 0 in $\mathbb{C}^n$ there exists an open neighborhood $V' \subset V$ of 0 in $\mathbb{C}$ such that for all $s \in V'$:

$$\dim_{\mathbb{C}}\left((\mathcal{M}_{|X_0})_0\right) = \sum_{p \in X_s'} \dim_{\mathbb{C}}\left((\mathcal{M}_{|X_s})_p\right),$$

where $X_s' = X_s \cap (U' \times \{s\})$.

Proof. As support of a coherent analytic sheaf, $Z := \operatorname{Supp}(\mathcal{M})$ is an analytic space, see 6.2.9. We claim that the natural projection map $f : (Z, o) \longrightarrow (S, o)$ is finite. This we prove in three steps.

Step 1. We give Z the structure of a complex space by means of the ideal sheaf $\mathcal{A}nn(\mathcal{M})$. This means that $\mathcal{O}_Z = \mathcal{O}_{X_S}/\mathcal{A}nn(\mathcal{M})$. By Theorem 6.2.18 $\mathcal{M}$ is a coherent $\mathcal{O}_Z$-module. From the second assumption it follows that $\mathcal{M}_o/\mathfrak{m}_{S,o}\mathcal{M}_o$ is a finite-dimensional vector space.

Step 2. By construction $\operatorname{Ann}_{\mathcal{O}_{Z,o}}(\mathcal{M}_0) = (0)$. We claim

$$I := \operatorname{Ann}_{\mathcal{O}_{Z,o}/\mathfrak{m}_{S,o}\mathcal{O}_{Z,o}}(\mathcal{M}_o/\mathfrak{m}_{S,o}\mathcal{M}_o) \subset \sqrt{(0)},$$

that is, I is a nilpotent ideal. To show this, let $\mathcal{M}_0$ as $\mathbb{C}\{s\}$-module be generated by $m_1, \ldots, m_r$. Let $f \in \mathcal{O}_{Z,o}$ be a representative of an element in I, so that $f\mathcal{M}_0 \subset \mathfrak{m}_{S,o}\mathcal{M}_0$, that is, $fm_i = \sum_{j=1}^{r} \xi_{ij} m_j$, $i = 1, \ldots, r$ for suitable $\xi_{ij} \in \mathfrak{m}_{S,o}$. By Cramer's rule 1.2.8, we get that $\det\left((\xi_{ij}) - f\,\mathrm{Id}_r\right)m_k = 0$ for $k = 1, \ldots, r$. As $\operatorname{Ann}_{\mathcal{O}_{Z,o}}(\mathcal{M}_0) = (0)$, it follows that $\det\left((\xi_{ij}) - f\,\mathrm{Id}_r\right) = 0$. This says that $f^r \in \mathfrak{m}_{S,o}\mathcal{O}_{Z,o}$.

Step 3. Put $R = \mathcal{O}_{Z,o}/\mathfrak{m}_{S,o}\mathcal{O}_{Z,o}$ and $M = \mathcal{M}_0/\mathfrak{m}_{S,o}\mathcal{M}_0$. Thus M is a finite-dimensional $\mathbb{C}$-vector space, and an R/I algebra. As R/I-algebra, the annihilator of M is zero. It follows from Exercise 1.2.31 that R/I is a finite-dimensional vector space, hence R/I has dimension zero. On the other hand, because $I \subset \sqrt{(0)}$ the dimensions $\dim(R)$ and $\dim(R/I)$ are equal, hence $\dim(R)$ is zero. This implies that $\dim_{\mathbb{C}}(R) < \infty$ because R is a factor ring of a convergent power series ring. By Theorem 3.4.24 it follows that $f : (Z, o) \longrightarrow (S, o)$ is a finite map, as claimed.

We now finish the proof. We take a representative $f : Z \longrightarrow S$ such that $f^{-1}(0) = \{0\}$, and such that f is finite. From the Finite Mapping Theorem 6.3.5 it follows that $f_*\mathcal{M}$ is coherent, whose stalk at 0 by assumption is $\mathbb{C}\{s\}^t$ for a suitable t. From the coherence of $f_*\mathcal{M}$ it follows that there exists an open neighborhood V' of 0 in S such that $f_*\mathcal{M}_{|V'} \cong \mathcal{O}^t_{\mathbb{C}|V'}$ (consider the map $\mathcal{O}^t_{\mathbb{C}|V'} \longrightarrow f_*\mathcal{M}_{|V'}$ induced by $(f_*\mathcal{M})_0 \cong \mathbb{C}\{s\}^t$ on a neighborhood $V' \subset S$ and use Lemma 6.2.16).

Take a point $s_0 \in V'$, and let $q_1, \ldots, q_\ell \in Z$ be the points in the fiber of f. Because f is finite, one has by Theorem 6.1.18:

$$\mathbb{C}\{s\}^t \cong (f_*\mathcal{M})_{s_0} = \oplus_{i=1}^{\ell} \mathcal{M}_{q_i}.$$

Restricting this to the point $s_0 \in V'$ and applying Exercise 6.1.35 implies that

$$\mathbb{C}^t \cong \oplus_{i=1}^{\ell} (\mathcal{M}_{|X_{s_0}})_{q_i}.$$

This proves the theorem. □

Proofs of Theorems 6.4.1, 6.4.3 and 6.4.5.

Proof of 6.4.1. We have to show that we can apply the previous theorem to the sheaf $\mathcal{M} := \mathcal{O}_{U\times V}/(F,G)$. By the results of the previous section we know that $\mathcal{M}$ is coherent and, therefore, the first assumption is satisfied. So we have to check the second condition. Therefore, consider the stalk of $\mathcal{M}$ at 0 which is the ring $\mathbb{C}\{x,y,s\}/(F,G)$. This is a $\mathbb{C}\{s\}$–module. By assumption (C,o) and (D,o) have no component in common, so that by the Nullstellensatz 3.4.4 the vector space $\mathbb{C}\{x,y\}/(f,g)$ is finite-dimensional. By the general Weierstraß Division Theorem 3.2.10 it follows that $\mathbb{C}\{x,y,s\}/(F,G)$ is a finitely generated $\mathbb{C}\{s\}$–module. To prove that it is free, we have to show, by Theorem 1.3.9, that it is torsion free, hence it remains to show that s is a nonzerodivisor. As $(C,o)\cap(D,o)$ consists of one point by assumption, it follows from 4.1.20 that f, g is a regular sequence in $\mathbb{C}\{x,y\}$, that is, g is a nonzerodivisor of $\mathbb{C}\{x,y\}/(f)$. From Exercise 4.1.28 it follows that if

$$a\cdot f+b\cdot g=0,$$

then $a=Lg$ and $b=-Lf$ for some element L in $\mathbb{C}\{x,y\}$. Now suppose that $s\cdot Q\in(F,G)$. We have to show that $Q\in(F,G)$. We can write

$$s\cdot Q=A\cdot F+B\cdot G$$

for some $A,B\in\mathbb{C}\{x,y,s\}$. We put $a=A(x,y,0)$ and $b=B(x,y,0)$. By plugging in it follows that $af+bg=0$, so that $a=Lg$, $b=-Lf$. We now add $0=(-LG)F+(LF)G$ to the expression for sQ and get

$$s\cdot Q=(A-LG)F+(B+LF)G.$$

The power series $A-LG$ and $B+LG$ are divisible by s so that

$$Q=\frac{A-LG}{s}F+\frac{B+LF}{s}G.$$

Hence $Q\in(F,G)$ as we wanted to show. □

Proof of 6.4.3. As above, we have to show that the conditions of Theorem 6.4.7 are satisfied. For notational convenience we will only consider the case that the curve C is irreducible. So we have a map

$$W\times S\longrightarrow\mathbb{C}^2\times S\quad(t,s)\mapsto(x(t,s),y(t,s),s).$$

Let $C_S=\{F=0\}$ be the image. F is a generator of the kernel of the map $\mathbb{C}\{x,y,s\}\longrightarrow\mathbb{C}\{t,s\}$, $x\mapsto x(t,s)$, $y\mapsto y(t,s)$, $s\mapsto s$. In particular F is irreducible. Let $\mathfrak{m}=(x,y,s)$ be the maximal ideal in $\mathbb{C}\{x,y,s\}/(F)$. Then $\mathbb{C}\{t,s\}/\mathfrak{m}\mathbb{C}\{t,s\}=\mathbb{C}\{t\}/(x(t,0),y(t,0))\mathbb{C}\{t\}$ is a finite-dimensional $\mathbb{C}$–vector space. This implies, using the General Weierstraß Division Theorem 3.2.10, that $\mathbb{C}\{t,s\}$ is a finitely generated $\mathbb{C}\{x,y,s\}/(F)$–module. Using Theorem 3.4.24 we may assume (after perhaps shrinking W and S) that the map $n: W\times S\longrightarrow C_S$ is finite. Using Theorem 6.3.5 we obtain that $n_*\mathcal{O}_{W\times S}$ is a coherent $\mathcal{O}_{C_S}$–module and, therefore, $\mathcal{M}:=n_*\mathcal{O}_{W\times S}/\mathcal{O}_{C_S}$ is a coherent $\mathcal{O}_{C_S}$–module, too. The first condition of Theorem 6.4.7 is therefore fulfilled. The stalk at 0 is $\mathcal{M}_0=\mathcal{O}_{W\times S,o}/\mathcal{O}_{C_S,o}=\mathbb{C}\{t,s\}/\mathbb{C}\{x(t,s),y(t,s),s\}$. Because $\mathcal{M}_0/(s)\mathcal{M}_0=\mathbb{C}\{t\}/\mathbb{C}\{x(t,0),y(t,0)\}$ is a finite-dimensional $\mathbb{C}$–vector space, we can apply Corollary 3.2.14, and conclude that $\mathcal{M}_0$ is a finitely generated $\mathbb{C}\{s\}$–module.

We now show that s is a nonzerodivisor of $\mathcal{O}_{W\times S,o}/\mathcal{O}_{C_{S,o}}$. To show this, we claim that $F(x,y,0)$ is a defining equation for the curve C, that is, up to multiplication with a unit it is equal to f. Now we have the following isomorphisms:

$$\begin{aligned}\mathbb{C}\{x,y\}/(F(x,y,0),x) &\cong \mathbb{C}\{x,y,s\}/(F,x,s) \cong \mathbb{C}\{x(t,s),y(t,s),s\}/(x(t,s),s)\\ &\cong \mathbb{C}\{x(t),y(t)\}/(x(t)) \cong \mathbb{C}\{x,y\}/(f,x).\end{aligned}$$

This implies that both f and $F(x,y,0)$ are y–regular of the same order. So the fact that f is up to a unit equal to $F(x,y,0)$ now follows from 3.3.16. Now suppose $s\cdot Q \in \mathcal{O}_{C_{S,o}}$ for some $Q \in \mathcal{O}_{\mathbb{C}\times S,o}$. We have to show that $Q \in \mathcal{O}_{C_{S,o}}$. We can write

$$s\cdot Q = G(x(t,s),y(t,s),s)$$

for some $G \in \mathbb{C}\{x,y,s\}$. Plugging in $s=0$ yields that $G(x(t),y(t),0)=0$, so that $G(x,y,0)$ is divisible by f, say $G(x,y,0) = Lf$. We may now, without loss of generality, assume that $F(x,y,0)$ is equal to f. Now add $0=-LF$ to the expression for $s\cdot Q$. Then

$$s\cdot Q = (G-LF)(x(t,s),y(t,s),s).$$

In the ring $\mathbb{C}\{x,y,s\}$ we can divide $G-LF$ by s, so that $Q = \frac{G-LF}{s}(x(t,s),y(t,s),s)$. Hence Q is already an element of $\mathcal{O}_{C_{S,o}}$, so that s is a nonzerodivisor, as was to be shown.

It remains to show that $(\mathcal{M}_{|C_s})_p = \widetilde{\mathcal{O}}_{C_s,\tilde{p}}/\mathcal{O}_{C_s,p}$ for $s \in S$ and $p \in C_s$. Looking at the integral equations for $x(t,s)$ and $y(t,s)$ for *fixed* s it follows that the map $n : W\times\{s\} \longrightarrow C_s$ is finite. As we just proved that in particular $\mathcal{O}_{W\times\{s\}}/\mathcal{O}_{C_s}$ is a finite-dimensional vector space it follows from Exercise 4.4.22 that $n : W\times\{s\} \longrightarrow C_s$ is the normalization for all points in C_s. This is what we needed to show. □

Proof of 6.4.5. This is analogous to the proof of 6.4.1, and left as Exercise 6.4.11. □

We may ask ourselves whether, for example, in the case of the Milnor number, a deformation with only A_1–singularities always exists. This question is answered in the following theorem.

Theorem 6.4.8. *Consider a function $f \in \mathbb{C}\{x_1,\dots,x_n\}$ with an isolated singularity. Let $F : (\mathbb{C}^n\times\mathbb{C}^n,o) \longrightarrow (\mathbb{C},o)$ be defined by*

$$F(x_1,\dots,x_n,t_1,\dots,t_n) := f + t_1x_1 + \ldots + t_nx_n.$$

Then there exists a hypersurface $(D,o) \subset (\mathbb{C}^n,o)$, such that for all small enough representatives D and T of (D,o) and $(\mathbb{C}^n,o)$, there exists a small neighborhood U of 0 in $\mathbb{C}^n$ such that for all $t \in T\setminus D$

$$F_t(x_1,\dots,x_n,t_1,\dots,t_n) : U \longrightarrow \mathbb{C}$$

has only A_1–singularities. The number of these A_1–singularities is equal to the Milnor number $\mu(f)$ of f.

Proof. Let U be a neighborhood of 0 in $\mathbb{C}^n$ such that f converges on U and let $t \in \mathbb{C}^n$. A point $p \in U$ is a singular point of F_t exactly if all partial derivatives with respect to the x_i vanish. In short, if p is in the zero set of $(\frac{\partial F_t}{\partial x_1},\dots,\frac{\partial F_t}{\partial x_n})$. We therefore consider the space (Σ,o) in $(\mathbb{C}^n\times\mathbb{C}^n,o)$ defined by the ideal

$$\left(\frac{\partial F}{\partial x_1},\dots,\frac{\partial F}{\partial x_n}\right)=\left(\frac{\partial f}{\partial x_1}+t_1,\dots,\frac{\partial f}{\partial x_n}+t_n\right).$$

Let Σ be a representative of (Σ, o), We want a condition for a point (p,t) in Σ to have an A_1-singularity. For this, we look at the projection to the second factor

$$\pi:\Sigma\longrightarrow\mathbb{C}^n.$$

Suppose that given $(p,t)\in\Sigma$, there exists a neighborhood of (p,t) which is mapped biholomorphically under π onto an open subset T of $\mathbb{C}^n$. Then we claim that F_t has an A_1-singularity at p. Indeed, by assumption we have that the inclusion of stalks

$$\mathcal{O}_{T,t}\subset\mathcal{O}_{\Sigma,(p,t)}$$

is a isomorphism. Dividing out the local parameters at t, we have

$$\mathbb{C}\cong\mathcal{O}_{n,p}/\left(\frac{\partial F_t}{\partial x_1}(p),\dots,\frac{\partial F_t}{\partial x_n}(p)\right).$$

Hence F_t has Milnor number one at p. This implies that it is an A_1-singularity at p.

We therefore have to investigate the map $\pi:\Sigma\longrightarrow\mathbb{C}^n$. We first look at the space (Σ, o). In the local ring $\mathcal{O}_{\Sigma,o}$ of this space, we can eliminate the functions $t_1,\dots,t_n$. Hence $\mathcal{O}_{\Sigma,o}\cong\mathbb{C}\{x_1,\dots,x_n\}$, so we see that (Σ, o) is smooth. The local ring of the fiber of π above $0\in\mathbb{C}^n$ is given by

$$\mathcal{O}_{\mathbb{C}^n\times\mathbb{C}^n,o}/\left(\frac{\partial f}{\partial x_1}+t_1,\dots,\frac{\partial f}{\partial x_n}+t_n,t_1,\dots,t_n\right)=\mathcal{O}_{\mathbb{C}^n,o}/\left(\frac{\partial f}{\partial x_1},\dots,\frac{\partial f}{\partial x_n}\right),$$

which is a finite-dimensional vector space by assumption. Therefore, the map $\pi:(\Sigma,o)\longrightarrow(\mathbb{C}^n,o)$ is finite by 3.2.10 and 3.4.24. As $\dim(\Sigma,o)=\dim(\mathbb{C}^n,o)$ and π is closed, it follows that π is surjective. Therefore, $\pi:(\Sigma,o)\longrightarrow(\mathbb{C}^n,o)$ is a Noether normalization. We proved in 3.4.14 that there exists a hypersurface $D\subset T$, such that for all $t\in T\setminus D$, there exists a small neighborhood V of t, such that $\pi^{-1}(V)$ is the union of s open subsets in Σ, each of which is mapped biholomorphically on V. Therefore, by the discussion in the beginning of the proof, for all $t\in T\setminus D$, the function F_t has only (finitely) many A_1-singularities. The fact that this number is equal to $\mu(f)$ is proved by means of conservation of number, see 6.4.5. □

Example 6.4.9. Let $f=x^3$. Consider $F=x^3+tx$. Then Σ is given by

$$\frac{\partial F}{\partial x}=3x^2+t=0.$$

The map $\pi:\Sigma\longrightarrow T$ is given by the following picture.

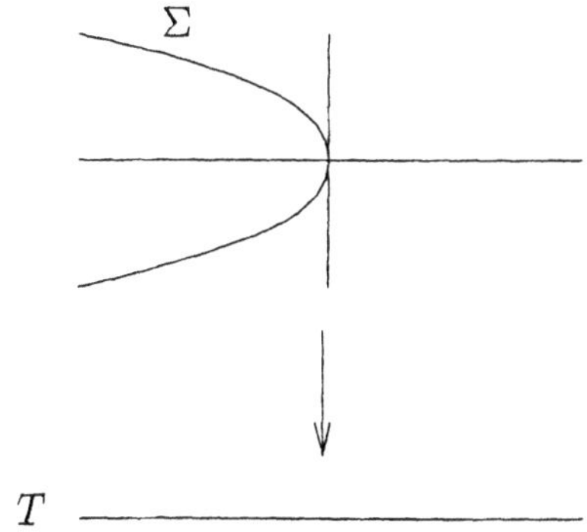

So we see that for $t \neq 0$ we get two A_1–singularities.

Corollary 6.4.10. *Consider $f \in \mathcal{O}_n$ with $\mu(f) < \infty$, and let $u \in \mathcal{O}_n$ be a unit. Then $\mu(uf) = \mu(f)$. In particular, for a germ of an analytic hypersurface (X, o) with an isolated singularity the Milnor number $\mu(X, o)$ is well defined.*

Proof. By symmetry, it suffices to show $\mu(uf) \geq \mu(f)$. Take a small open neighborhood V of 0 in $\mathbb{C}^n$. Consider a family $F_t \in \mathbb{C}\{x_1, \dots, x_n, t\}$ with $F(x_1, \dots x_n, 0) = f$. Let U be a small open neighborhood of 0, and suppose that the function F_t for t fixed has exactly $\mu(f)$ A_1–singularities in V. Then uF_t is a family of functions with $uF(x_1, \dots x_n, 0) = uf$. If F_t has an A_1–singularity at p, then uF_t also has an A_1–singularity at p. From the principle of conservation of number, it follows that $\mu(uf) \geq \mu(f)$. This is what we wanted to show. □

Exercises

6.4.11. Prove 6.4.5.

6.4.12. For plane curve singularities (X, o) and (Y, o), prove Hironaka's formula, see 5.4.11,using the principle of conservation of number, in particular Theorems 6.4.1 and 6.4.3.

6.4.13. In the Theorem of the Principle of Conservation of Number, replace the second condition by the condition that $\mathcal{M}_0$ is a finitely generated $\mathbb{C}\{s\}$–module, or, what is the same by Nakayama, that $(\mathcal{M}_{|X_0})_0$ is a finitely generated $\mathbb{C}$–vector space. Prove that

$$\dim_{\mathbb{C}}\left((\mathcal{M}_{|X_0})_0\right) \geq \sum_{p \in X_s'} \dim_{\mathbb{C}}\left((\mathcal{M}_{|X_s})_p\right).$$

6.4.14. An ordinary m–tuple point is by definition a plane curve singularity $(X, o) = \cup_{i=1}^m (X_i, o)$ such that all (X_i, o) are smooth, and the intersection number $(X_i, o) \cdot (X_j, o) = 1$ for all $i \neq j$. Prove that the δ–invariant of an ordinary m–tuple point is $\binom{m}{2}$ using

(1) Hironaka's formula 5.4.11 or

(2) the principle of conservation of number.

Find a $\mathbb{C}$–basis of $\mathcal{O}_{\bar{X},o}/\mathcal{O}_{X,o}$.

6.4.15. Consider an irreducible plane curve singularity $(X, o) = (V(f), o)$ in $(\mathbb{C}^2, o)$ of multiplicity m. Without loss of generality we may suppose that (X, o) is given by the parametrization

$$x(t) = t^m, \quad y(t) = t^m \varphi(t).$$

Consider a deformation of the parametrization as follows

$$x(t,s) = t^m + s, \quad y(t,s) = (t^m + s)\varphi(t).$$

Let $F \in \mathbb{C}\{x, y.x\}$ be irreducible with $F(x(t,s), y(t,s)) = 0$ and $F(x, y, 0) = f$.

(1) Show that one can take f as follows:

$$f(x,y) = y^m + a_1(x)y^{m-1} + \ldots + a_m(x) = 0,$$

where the order of a_i is at least i.

(2) Show that F looks like

$$F = y^m + \frac{a_1(x-s)}{x-s} y^{m-1} x + \ldots + \frac{a_m(x-s)}{(x-s)^m} x^m = 0.$$

(3) Show that for small $s \neq 0$ the zero set of $f_s(x,y) := F(x, y, s)$ has two singular points:

1. an ordinary m-tuple point.
2. the strict transform of the blowing-up of the origin in $\mathbb{C}^2$.
   For which values of t do these occur?

(4) Use the principle of conservation of number, in particular 6.4.3, to reprove Noether's formula 5.4.13.

(5) Generalize these results to *reducible* curves.

6.5 Cohen-Macaulay Spaces

The Cohen-Macaulay property (CM for short) of a space is a local property, related to the principle of conservation of number. In this section we deal with the first properties of Cohen-Macaulay spaces, hopefully showing that it is a good (and, therefore, quite strong) property a space can have. Nevertheless, we will prove that complete intersections (in particular hypersurface singularities) are Cohen-Macaulay. First we introduce the notion of depth, needed in order to define Cohen-Macaulay. We introduce these matters directly for *modules* over a local ring, in order to get a more flexible development of the theory.

Definition 6.5.1. Let $(R,\mathfrak{m})$ be a local ring and M be an R-module.

(1) A sequence $f_1, \dots ,f_r$ of elements in $\mathfrak{m}$ is called a *regular sequence of* M if f_1 is not a zerodivisor of M, and f_i is not a zerodivisor of $M/(f_1, \dots ,f_{i-1})M$ for $i = 2, \dots ,r$.

(2) Let $I \subset R$ be an ideal with $IM \neq M$. Then the I-*depth* of M, $\operatorname{depth}(I,M)$ is the maximal length of a regular sequence of M in I. If $IM = M$ we define $\operatorname{depth}(I,M) = \infty$.

(3) The depth of M, $\operatorname{depth}(M)$ is the maximal length of a regular sequence of M, that is, $\operatorname{depth}(M) = \operatorname{depth}(\mathfrak{m},M)$. If we want to emphasize the ring R, we will write $\operatorname{depth}_R(M)$.

(4) The module M is called *Cohen-Macaulay*, if the depth of M is equal to the dimension of M. The dimension of the module M is defined to be the dimension of $R/\operatorname{Ann}(M)$.

(5) R is called *Cohen-Macaulay*, if R is a *Cohen-Macaulay* R-module.

(6) A germ (X,x) is called *Cohen-Macaulay*, if $\mathcal{O}_{X,x}$ is Cohen-Macaulay.

As usual, the depth of a space at a point is defined as the depth of the corresponding local ring. Note that because of the Active Lemma 4.1.10 we always have the inequality $\operatorname{depth}(M) \le \dim(M)$. If $IM \neq M$ then $\operatorname{depth}(I,M) \le \operatorname{depth}(M)$. Notice that $\operatorname{depth}(I,M) > \operatorname{depth}(M)$ implies $IM = M$ and, especially for $M = R$, it implies $I = R$. Furthermore, it is immediate from the definition that the depth of a module is zero exactly if every element of $\mathfrak{m}$ is a zerodivisor for M. From 1.4.52 it follows that $\mathfrak{m}$ is an associated prime ideal of M, so that $\mathfrak{m} = M : f$ for some $f \in \mathfrak{m}$. Hence the following lemma follows.

Lemma 6.5.2. *Suppose that M is an R-module, and that* $\operatorname{depth}(M) = 0$. *Then $\mathfrak{m}$ is an associated prime of M, that is, there exists an $f \in M \setminus \{0\}$ such that $\mathfrak{m} \cdot f = 0$.*

In case that $M = R$ is the local ring of a germ of a complex space, the depth is zero exactly if R is a zero-dimensional ring, or the special point is an embedded point. More generally, the depth of a germ of a complex space at a point is the number of hypersurface sections one has to take before one gets an embedded point. Note that if a nonzero-dimensional germ of a complex space is reduced, it must have positive depth. The Cohen-Macaulay property, therefore, states that an embedded point will not occur when taking $\dim(X)$ successive hypersurface sections. It will turn out that the length of a maximal regular sequence is independent of the maximal regular sequence. Let us, before giving examples, prove this statement, for which we need the following lemma.

Lemma 6.5.3. *Let R be a Noetherian local ring, M a finitely generated R-module. Let $\{f,g\}$ be a regular sequence of M. Then $\{g,f\}$ is also a regular sequence.*

Proof. The proof will be given in two steps.

Step 1. We first show that g is a nonzerodivisor. Suppose $ga_0 = 0$, $a_0 \in M$. In particular, $ga_0 \in f \cdot M$. The assumption that $\{f,g\}$ is a regular sequence, allows us to deduce that $a_0 \in f \cdot M$. Write $a_0 = fa_1$, $a_1 \in M$. Then $ga_0 = fga_1 = 0$, and f is not a zero divisor, so $ga_1 = 0$, etc. Therefore, for all k there exists an $a_k \in M$ such that $a_0 = a_k f^k$. Therefore, $a_0 \in \bigcap \mathfrak{m}^k M$ which is zero according to Krull's Intersection Theorem (Exercise 1.3.24).

Step 2. We have to show that f is a nonzerodivisor of $M/g \cdot M$. Suppose $fa \in g \cdot M$, that is, $fa = gb$ for some $a,b \in M$. Since by assumption, g is a nonzerodivisor of $M/f \cdot M$, it follows that $b \in f \cdot M$, that is, $b = fc$ for some $c \in M$. Therefore, $f(a - gc) = 0$. Because f is not a zero divisor, $a - cg = 0$, that is, $a \in g \cdot M$. This is what we had to show. □

Lemma 6.5.4. *Let R be a Noetherian local ring, $I \subset R$ an ideal and M an R-module. Any two maximal regular sequences in I for M have the same number of elements.*

Proof. Let $f_1, \ldots, f_n$ and $g_1, \ldots, g_p$ be regular sequences in I, which cannot be extended. We may assume that $n \geq p$. We have to show that $p = n$. The lemma will be proved by induction on n. The case $n = 1$ is trivial. We now claim that there exists an $f \in I$ such that the sequences $f_1, \ldots, f_{n-1}$ and $g_1, \ldots, g_{p-1}$ can be extended to regular sequences $f_1, \ldots, f_{n-1}, f$ and $g_1, \ldots, g_{p-1}, f$. Let S be the union of the associated primes of $(f_1, \ldots, f_{n-1})M$ and of $(g_1, \ldots, g_{p-1})M$, then, by the characterization of nonzerodivisors, see 1.4.52, the ideal $(f_n, g_p) \not\subset S$, because $(f_n, g_p) \subset S$ would imply $(f_n, g_p) \subset \mathfrak{p}$, $\mathfrak{p}$ one of the associated primes (cf. 1.1.13, Prime Avoidance). But by assumption f_n or g_p are nonzerodivisors in $M/(f_1, \ldots, f_{n-1})M$, resp. $M/(g_1, \ldots, g_{p-1})M$. This implies that $af_n + bg_p =: f \notin S$ for suitable $a,b \in R$ and has, therefore, the required property. Applying the case $n = 1$ to the ring $R/(f_1, \ldots, f_{n-1})$ gives that the sequence $f_1, \ldots, f_{n-1}, f$ is maximal. Similarly $g_1, \ldots, g_{p-1}, f$ is a maximal regular sequence.
By repeatedly applying the previous lemma, we find that $f, f_1, \ldots, f_{n-1}$ and $f, g_1, \ldots, g_{p-1}$ are maximal regular sequences. Applying the induction hypothesis to the ring $R/(f)$ proves the lemma. □

Corollary 6.5.5. *Let R be a Noetherian local ring, M be an R-module, $I \subset R$ an ideal and $f \in I$ a nonzerodivisor of M. Then:*

(1) $\operatorname{depth}(I, M/fM) = \operatorname{depth}(I, M) - 1$,

(2) *M is Cohen-Macaulay $\Longleftrightarrow$ M/fM is Cohen-Macaulay.*

Proof. The first part follows from Lemma 6.5.4. For the second part, the implication "$\Longleftarrow$" is trivial. For the other implication, we use the first part. It remains to show $\dim(R/(\operatorname{Ann}(M),f) = \dim(R/\operatorname{Ann}(M))-1$. This holds, as from the assumption it follows easily that f is a nonzerodivisor of $\operatorname{Ann}(M)$, so that we can apply Krull's Principal Ideal Theorem 4.2.16. □

Examples 6.5.6.
(1) Smooth spaces are Cohen-Macaulay.

(2) By applying the second part of Corollary 6.5.5 we obtain that complete intersection singularities (in particular hypersurface singularities) are Cohen-Macaulay.

(3) Reduced curve singularities are Cohen-Macaulay. This is also obvious, as the depth is positive and, therefore, only can be one, because the depth cannot be bigger than the dimension.

(4) The singularity in three–space given by the ideal (xz,yz)

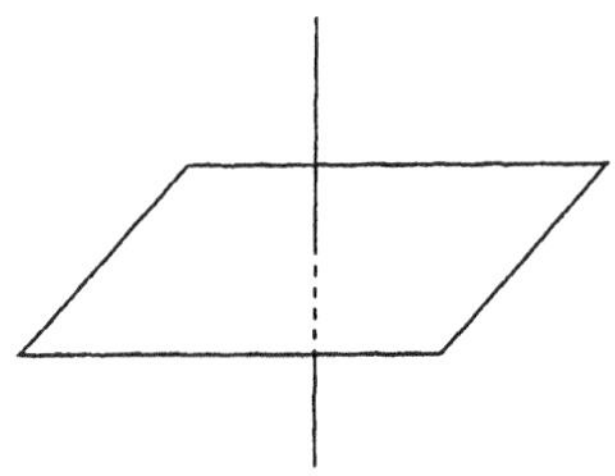

is reduced, but not Cohen-Macaulay. An example of a nonzerodivisor is $z-x$, which cuts down the space to the following space with embedded point.

The local ring of the intersection is $\mathbb{C}\{x,y,z\}/(xz,yz,x-z) \cong \mathbb{C}\{x,y\}/(x^2,xy)$, hence we see the embedded point appear. The depth of the germ the analytic space $(V(xz,yz),0)$ is one. This space is therefore not Cohen-Macaulay.

(5) Consider $\mathbb{C}^4$ with coordinates x,y,z,w, and consider the union X of the two planes $x=y=0$ and $z=w=0$, intersecting in one point. The ideal of this union is $(x,y) \cap (z,w) = (xz,xw,yz,yw)$. It is geometrically clear that if we intersect with a generic three-dimensional linear space, we have the union of two lines. But we obtain an embedded point. If we intersect with $y-w=0$, the local ring becomes:

$$\mathbb{C}\{x,y,z,w\}/(xz,xw,yz,yw,y-w) \cong \mathbb{C}\{x,y,z\}/(xz,xy,yz,y^2).$$

A primary decomposition of (xz,xy,yz,y^2) is given by:

$$(xz,xy,yz,y^2) = (y,x) \cap (y,z) \cap (y^2,x,z).$$

So we see an embedded point appearing on the A_1–singularity. The depth is one, and, therefore, $(X,0)$ is not Cohen-Macaulay.

(6) To give an example of an *irreducible* space which is not Cohen-Macaulay, we consider the map:

$$(\mathbb{C}^2,0) \longrightarrow (\mathbb{C}^4,0); \quad (s,t) \mapsto (x,y,z,u) = (s,st,t^2,t^3).$$

By Exercise 2.3.18 the ideal of the image is:

$$(y^2 - zx^2, yz - ux, z^2x - uy, u^2 - z^3).$$

A general hyperplane section is given by $x - z = 0$. This section gives a space given by the ideal

$$(y^2 - x^3, x(y-u), x^3 - uy, u^2 - x^3).$$

From $x(y-u) = 0$ it follows that either $x = 0$ or $y = u$. From the first possibility, it follows that $y = u = 0$. The second possibility gives $y = u$, $y^2 = x^3$, a cusp singularity. In total we get a cusp singularity with an embedded point. In fact, a primary decomposition of the ideal is:

$$(y - u, y^2 - x^3) \cap (x, y^2, uy, u^2).$$

Indeed, we need embedding dimension four to find an example of a reduced irreducible space which is not Cohen-Macaulay, due to the Principal Ideal Theorem and the second example.

The examples four and five are quite typical. In fact, a Cohen-Macaulay space is equidimensional, even unmixed. For the definition of unmixed, see 4.1.22. In the fifth example, the intersection of the two components has dimension zero. A necessary condition for the union to be Cohen-Macaulay is the condition that the intersection has codimension one, as we will see in 6.5.19. At least from these examples we see that the Cohen-Macaulay property is indeed a nontrivial one.

Proposition 6.5.7. *Let $(R, \mathfrak{m}_R)$ and $(S, \mathfrak{m}_S)$ be local rings, and $R \subset S$ be a finite ring extension of local rings. Let M be an S-module. Suppose* $\operatorname{depth}_R(M) = 0$. *Then* $\operatorname{depth}_S(M) = 0$.

Proof. As $\operatorname{depth}_R(M) = 0$, we have that $\mathfrak{m}$ is an associated prime of M. Thus there exists an $x \in M$ with $x \cdot \mathfrak{m} = 0$. Take an element $g \in \mathfrak{m}_S$. We have to show that g is a zerodivisor of M. By 1.5.11 g is integral over $\mathfrak{m}_R$, so we have an integral equation:

$$g^n + a_1 g^{n-1} + \ldots + a_n = 0, \quad a_i \in \mathfrak{m}_R.$$

As $\operatorname{depth}_R(M) = 0$, we have $a_i x = 0$ for $i = 1, \ldots, n$. Hence $g^n \cdot x = 0$. Thus g is a zerodivisor of M. □

This proposition will be generalized in Exercise 6.5.29.

Corollary 6.5.8.

(1) Let $\mathcal{O}_{X,0}$ be a singularity, $\mathcal{O}_k \hookrightarrow \mathcal{O}_{X,0}$ be a Noether normalization. Then $(X, 0)$ is Cohen-Macaulay if and only if $\mathcal{O}_{X,0}$ is a free $\mathcal{O}_k$-module.

(2) A Cohen-Macaulay space is unmixed.

Proof. (1) Recall from 1.3.9 that if M is a finitely generated R-module, $x \in \mathfrak{m}$ a nonzerodivisor of M, then M is a free R-module if and only if M/xM is a free $R/(x)$-module. By 6.5.7 we can find an $x \in \mathcal{O}_k$ which is a nonzerodivisor of $\mathcal{O}_{X,0}$. The first statement follows by induction, since Corollary 6.5.5 gives that $\mathcal{O}_{X,0}/(x)$ is also Cohen-Macaulay.

(2) We know by the first part, that $\mathcal{O}_{X,0}$ is a free $\mathcal{O}_k$-module, so, in particular, every element of $\mathcal{O}_k$ is a nonzerodivisor of $\mathcal{O}_{X,0}$. Therefore, $(X, 0)$ is unmixed, by Theorem 4.1.22. □

The second part of this corollary, applied to the case of a complete intersection is usually called Macaulay's Unmixedness Theorem.

Corollary 6.5.9. *Let (X,o) be an irreducible complete intersection singularity, and suppose the singular locus of (X,o) has codimension at least two. Then (X,o) is normal.*

Proof. By assumption, $\mathcal{O}_{X,o}$ satisfies (R1). We only need to show that it satisfies (S2) by 4.4.11. The condition (S2) means that every hypersurface in (X,o) has no embedded primes. Because of Corollary 6.5.5 this hypersurface is also Cohen-Macaulay, and therefore unmixed. Thus (S2) holds for Cohen-Macaulay singularities. □

Theorem 6.5.10. *Let (X,o) be Cohen-Macaulay germ of an analytic space and $\pi : (X,o) \longrightarrow (\mathbb{C}^k,o)$ be a Noether normalization. Then there exist representatives X of (X,o) and U of $(\mathbb{C}^k,o)$ and a hypersurface $D \subset U$ such that for all $t \in U \setminus D$ the fiber $\pi^{-1}(t)$ consists of $\dim_{\mathbb{C}} \mathcal{O}_{X,o}/\mathfrak{m}_{\mathbb{C}^k,o}\mathcal{O}_{X,o}$ many points.*

Proof. As a Cohen-Macaulay space is unmixed, in particular equidimensional, we can reduce to the case that (X,o) is irreducible. As in the proof of 6.4.7 we choose X and U such that $\pi_*\mathcal{O}_X \cong \mathcal{O}_U^s$. This is possible because $\pi_*\mathcal{O}_X$ is coherent and $(\pi_*\mathcal{O}_X)_o = \mathcal{O}_{X,o}$ is a free $\mathcal{O}_{\mathbb{C}^k,o}$-module of finite rank s (Corollary 6.5.8). Let $p \in U$ and $\pi^{-1}(p) = \{q_1, \dots, q_r\}$ then

$$\mathcal{O}_{U,p}^s \cong (\pi_*\mathcal{O}_X)_p = \oplus_{i=1}^r \mathcal{O}_{X,q_i}$$

and, therefore,

$$s = \sum_{i=1}^r \dim_{\mathbb{C}} \mathcal{O}_{X,q_i}/\mathfrak{m}_{\mathbb{C}^k,p}\mathcal{O}_{X,q_i}.$$

As in the proof of Theorem 6.4.8, we use the Local Parametrization Theorem 3.4.14 to find a hypersurface $D \subset U$ with the following property: every point $p \in U \setminus D$ has a neighborhood V such that $\pi^{-1}(V)$ is a union of open subsets of X each of which is mapped biholomorphically to V. This implies $\mathcal{O}_{X,q_i} \cong \mathcal{O}_{U,p}$ for all i and, therefore, $r = s$. □

Remark 6.5.11. The previous theorem does not hold if (X,o) is not Cohen-Macaulay, see the following examples. We can apply the theorem to the theory of intersection number. We consider f,g such that $\dim_{\mathbb{C}} \mathbb{C}\{x,y\}/(f,g) < \infty$, and deformations F_S, G_S of f and g, with $\mathcal{O}_{S,o} = \mathcal{O}_k$. Let $(X,o) = (V(F_S,G_S),o)$. Then $(X,o) \longrightarrow (\mathbb{C}^k,o)$ is a Noether normalization. (X,o) is a complete intersection, hence Cohen-Macaulay. The number of points in $X_s = V(F_s,G_s)$ for general s, therefore, is equal to $\dim_{\mathbb{C}} \mathbb{C}\{x,y\}/(f,g)$.

Examples 6.5.12.
(1) Let $I = (xy, x(x-z))$. Then $(X,o) = (V(I),o)$ is the union of a plane and a line. The projection $(X,o) \longrightarrow (\mathbb{C}^2,o)$ which sends (a,b,c) to (b,c) is a Noether normalization. On the level of rings we have a map

$$\mathbb{C}\{y,z\} \subset \mathbb{C}\{x,y,z\}/(xy, x^2 - xz).$$

The number of points in a general fiber obviously is one. But

$$\dim_{\mathbb{C}} (\mathbb{C}\{x,y,z\}/(I + (y,z))) = \dim_{\mathbb{C}} (\mathbb{C}\{x\}/(x^2)) = 2.$$

Therefore, (X,o) is not Cohen-Macaulay.

(2) We consider example 6.5.6 (6). So let $I = (y^2 - zx^2, yz - ux, z^2x - uy, u^2 - z^3)$. We consider the projection $\pi : (X,o) = (V(I),o) \longrightarrow (\mathbb{C}^2, o)$, which sends (a,b,c,d) to (a,c). Recall that (X,o) was given by the image of the map $(\mathbb{C}^2,o) \longrightarrow (\mathbb{C}^4,o)$ which sends (s,t) to (s,st,t^2,t^3). The composition is the map sending (s,t) to (s,t^2). So we see that $(X,o) \longrightarrow (\mathbb{C}^2,o)$ is surjective, and in fact in a general fiber of π we have two elements. Since π is surjective we have an inclusion of rings

$$\mathbb{C}\{x,z\} \hookrightarrow \mathbb{C}\{x,y,z,u\}/I.$$

Now $\mathbb{C}\{x,y,z,u\}/(I + (x,z)) = \mathbb{C}\{y,u\}/(y^2,uy,u^2)$ which is a three-dimensional vector space. By the General Division Theorem, it follows that $\mathbb{C}\{x,y,z,u\}/I$ is a finitely generated $\mathbb{C}\{x,z\}$–module. Hence $\pi : (X,o) \longrightarrow (\mathbb{C}^2,o)$ is a Noether normalization, has two points in a general fiber, but the vector space $\mathbb{C}\{x,y,z,u\}/(I + (x,z))$ has dimension three. Thus we see that (X,o) is not Cohen-Macaulay.

There is also a completely algebraic method by which one can decide whether a ring (or module) is Cohen-Macaulay. It is based on resolutions of modules. This method is often used in the literature for proving that for a certain invariant the principle of conservation of number holds.

Definition 6.5.13. Let $(R,\mathfrak{m})$ be a Noetherian local ring and M be a finitely generated R–module. A *free resolution* of M is an exact sequence (infinite of finite)

$$\cdots \longrightarrow F_{k+1} \xrightarrow{\alpha_{k+1}} F_k \longrightarrow \cdots \xrightarrow{\alpha_1} F_0 \xrightarrow{\alpha_0} M \longrightarrow 0, \tag{6.6}$$

with finitely generated free R–modules F_i for $i \geq 0$. One says that M has *finite projective dimension* if there exists an exact sequence

$$0 \longrightarrow F_n \xrightarrow{\alpha_n} F_{n-1} \longrightarrow \cdots \xrightarrow{\alpha_1} F_0 \xrightarrow{\alpha_0} M \longrightarrow 0, \tag{6.7}$$

with finitely generated free R–modules F_i. The minimal n with $F_n \neq 0$ is called the projective dimension of M, denoted by $\mathrm{pd}(M) = \mathrm{pd}_R(M)$. If M does not have finite projective dimension, we write $\mathrm{pd}_R(M) = \infty$. The resolution (6.6) is called *minimal*, if for all $k \geq 1$, the map α_k maps F_k into $\mathfrak{m}F_{k-1}$.

Theorem 6.5.14. *(Minimal) free resolutions of finitely generated modules M over a Noetherian local ring $(R,\mathfrak{m})$ exist. If a module M has a finite minimal resolution*

$$0 \longrightarrow F_n \xrightarrow{\alpha_n} F_{n-1} \longrightarrow \cdots \xrightarrow{\alpha_1} F_0 \xrightarrow{\alpha_0} M \longrightarrow 0,$$

with $F_n \neq (0)$, then $n = \mathrm{pd}_R(M)$. To put it in other words, two minimal resolutions have the same length. Moreover, the rank of F_k is independent of the minimal resolution, and is called the k-th Betti number of M.

Proof. Consider a minimal set of generators $b_1, \ldots, b_{s_0}$ of M. We get a surjective map

$$F_0 := R^{s_0} \xrightarrow{\alpha_0} M$$

with $\alpha_0(a_1, \ldots, a_{s_0}) = a_1 b_1 + \ldots + a_{s_0} b_{s_0}$. The map α_0 is surjective, because $b_1, \ldots, b_{s_0}$ generate M. By Nakayama, α_0 induces an isomorphism

$$\overline{\alpha}_0 : F_0/\mathfrak{m}F_0 \cong M/\mathfrak{m}M.$$

Let K_1 be the kernel of α_0. Then $K_1 \subset \mathfrak{m}F_0$, as $\overline{\alpha}_0$ is an isomorphism. As K_1 is a submodule of the Noetherian module F_0, it is finitely generated. Hence, there exists a surjective map

$$F_1 \cong R^{s_1} \longrightarrow K_1 \longrightarrow 0,$$

where s_1 is the minimal number of generators of K_1. We denote the composition $F_1 \longrightarrow K_1 \longrightarrow F_0$ by α_1. As $K_1 \subset \mathfrak{m}F_0$, it follows that $\alpha_1(F_1) \subset \mathfrak{m}F_0$. We get an exact sequence

$$F_1 \xrightarrow{\alpha_1} F_0 \xrightarrow{\alpha_0} M \longrightarrow 0.$$

Now put $K_2 = \text{Ker}(\alpha_1)$, take minimal number of generators of K_2 etc. Continuing this way, we get a minimal free resolution of M.

To show the invariance of the Betti-numbers, consider two minimal resolutions of M:

$$\cdots \longrightarrow F_{k+1} \xrightarrow{\alpha_{k+1}} F_k \longrightarrow \cdots \xrightarrow{\alpha_1} F_0 \xrightarrow{\alpha_0} M \longrightarrow 0,$$

$$\cdots \longrightarrow F'_{k+1} \xrightarrow{\alpha'_{k+1}} F'_k \longrightarrow \cdots \xrightarrow{\alpha'_1} F'_0 \xrightarrow{\alpha'_0} M \longrightarrow 0.$$

As both α_1 and α'_1 both have entries in $\mathfrak{m}$, we get isomorphisms $F_0/\mathfrak{m}F_0 \cong M/\mathfrak{m}M \cong F'_0/\mathfrak{m}F'_0$. Therefore $\text{rank}(F_0) = \text{rank}(F'_0)$.

Let $e_1,\dots,e_{s_0}$ be a basis of F_0, and $f_1,\dots,f_{s_0}$ be a basis of F'_0. As the $\alpha'_0(f_j)$ generate M, there exist $\beta_{ij} \in R$ with $\alpha_0(e_i) = \sum \beta_{ij}\alpha'_0(f_j)$. Consider the map $h_1 = (\beta_{ij}) : F_0 \longrightarrow F'_0$. So we get a commutative diagram

$$\begin{array}{ccc} F_0 & \xrightarrow{\alpha_0} & M \\ \downarrow{\scriptstyle h_1} & & \| \\ F'_0 & \xrightarrow{\alpha'_0} & M. \end{array}$$

Modulo $\mathfrak{m}$ the map $\overline{h}_1 : F_0/\mathfrak{m}F_0 \longrightarrow F'_0/\mathfrak{m}F'_0$ is an isomorphism, that is, $\det(\beta_{ij}) \bmod \mathfrak{m} \neq 0$. Then $\det(\beta_{ij})$ is a unit in R which implies that h_1 is an isomorphism. Hence h_1 induces an isomorphism

$$h_1 : \text{Ker}(\alpha_0) = K_1 \longrightarrow K'_1 = \text{Ker}(\alpha'_0).$$

As α_2 and α'_2 have entries in $\mathfrak{m}$, and we have surjections $F_1 \longrightarrow K_1$ and $F'_1 \longrightarrow K'_1$, it follows that $\text{rank}(F_1) = \text{rank}(F'_1)$, etc. □

Examples 6.5.15.

(1) An R–module M has projective dimension zero if and only if M is a free R–module.

(2) Let $R = \mathbb{C}\{x,y\}$, $I = (x,y)$, and $M = R/I$. We will show that

$$0 \longrightarrow R \xrightarrow{\binom{-y}{x}} R^2 \xrightarrow{(x\ y)} R \xrightarrow{res} R/I \longrightarrow 0$$

is a free resolution of R/I as R–module. Here res is just taking residue classes, and is obviously surjective. The kernel of res is generated by x and y. We have to compute the kernel of $R^2 \xrightarrow{(x\ y)} R$. So let (α,β) be in the kernel. Then $x \cdot \alpha + y \cdot \beta = 0$. As (x,y) is a regular sequence in $\mathbb{C}\{x,y\}$ it follows that $\alpha = y \cdot m$, for a certain m. Similarly $\beta = x \cdot \ell$ for a certain ℓ. Then $xy \cdot m + xy \cdot \ell = 0$, hence $\ell = -m$. Therefore, $(\alpha,\beta) = \ell(-y,x)$. So

$$R \xrightarrow{\binom{-y}{x}} R^2 \xrightarrow{(x\ y)} R$$

is exact. Moreover, it is easy to see that $R \xrightarrow{\binom{-y}{x}} R^2$ is injective.

(3) We consider the ring $R = \mathbb{C}\{x,y\}/(xy)$, and $M = \mathbb{C}\{y\} \oplus \mathbb{C}\{x\} = R/(x) \oplus R/(y)$. Then $M = \widetilde{R}$, the normalization of R. M can be described as the cokernel of the map

$$R^2 \xrightarrow{\alpha_1 = \left(\begin{smallmatrix} x & 0 \\ 0 & y \end{smallmatrix}\right)} R^2 \longrightarrow M \longrightarrow 0.$$

The kernel of α_1 is easily seen to be generated by $(y,0)$ and $(0,x)$. Hence we get an exact sequence

$$R^2 \xrightarrow{\alpha_2 = \left(\begin{smallmatrix} y & 0 \\ 0 & x \end{smallmatrix}\right)} R^2 \xrightarrow{\alpha_1 = \left(\begin{smallmatrix} x & 0 \\ 0 & y \end{smallmatrix}\right)} R^2 \longrightarrow M \longrightarrow 0.$$

Going on like this, we see that M has a minimal free resolution which is 2-periodic. In particular we see that M does not have finite projective dimension.

So one sees that a finitely generated module over a local ring does not, in general, have finite projective dimension. But it is true in the case that R is a regular local ring. The analogous statement for graded modules over a polynomial ring is a famous theorem of Hilbert.

Theorem 6.5.16 (Hilbert's Syzygy Theorem). *Let $(R,\mathfrak{m})$ be a regular local ring of dimension n, and M be a finitely generated R-module. Then $\mathrm{pd}_R(M) \leq n$. In particular, the projective dimension of M is finite.*

For the proof we need the following lemma.

Lemma 6.5.17. *Let $(R,\mathfrak{m})$ be a Noetherian local ring, let M be a finitely generated R-module, and $x \in \mathfrak{m}$ be a nonzerodivisor of R and M. Consider a minimal free resolution of M*

$$0 \longrightarrow F_n \longrightarrow F_{n-1} \longrightarrow \cdots \longrightarrow F_0 \longrightarrow M \longrightarrow 0.$$

Then the induced sequence

$$0 \longrightarrow F_n/xF_n \longrightarrow F_{n-1}/xF_{n-1} \longrightarrow \cdots \longrightarrow F_0/xF_0 \longrightarrow M/xM \longrightarrow 0 \tag{6.8}$$

is a minimal free resolution of M/xM as $R/(x)$-module.
Moreover, $\mathrm{pd}_R(M) = \mathrm{pd}_{R/(x)}(M/xM)$.

Proof. By 1.2.35 the sequence (6.8) is indeed exact. Moreover, from $F_n/xF_n = 0$ it would follow that $F_n = xF_n$, hence $F_n = 0$ by Nakayama. Hence $\mathrm{pd}_R(M) = \mathrm{pd}_{R/(x)}(M/xM)$. □

Proof of Hilbert's Syzygy Theorem 6.5.16. The proof is by induction on n. If $n = 0$, then R is a field and M is a finitely generated vector space over R. Hence M is free, and the projective dimension $\mathrm{pd}_R(M) = 0$. Otherwise, consider a minimal free resolution

$$\cdots \longrightarrow F_k \longrightarrow F_{k-1} \longrightarrow \cdots \longrightarrow F_1 \longrightarrow F_0 \longrightarrow M \longrightarrow 0$$

of M. We split this exact sequence into two exact sequences

$$\cdots \longrightarrow F_k \longrightarrow F_{k-1} \longrightarrow \cdots \longrightarrow F_1 \longrightarrow K \longrightarrow 0,$$

$$0 \longrightarrow K \longrightarrow F_0 \longrightarrow M \longrightarrow 0,$$

with $K = \mathrm{Ker}(F_0 \to M)$. Take any element $x \in \mathfrak{m} \setminus \mathfrak{m}^2$. As R is an integral domain, see 4.3.17, x is a nonzerodivisor of F_0. As K is a submodule of F_0, x is a nonzerodivisor of K. Therefore x is a nonzerodivisor of F_i for all $i \geq 1$ and of K, so that we get an exact sequence

$$\cdots \longrightarrow F_k/xF_k \longrightarrow F_{k-1}/xF_{k-1} \longrightarrow \cdots \longrightarrow F_1/xF_1 \longrightarrow K/xK \longrightarrow 0$$

The ring R is Noetherian, so that K/xK is a finitely generated $R/(x)$-module. As $R/(x)$ is a regular local ring of dimension $n-1$, it follows by induction that $\mathrm{pd}_{R/(x)}(K/xK) \leq n-1$. Hence $F_k/xF_k = 0$ for $k \geq n+1$, because the resolution is minimal. By Nakayama, it follows that $F_k = 0$ for $k \geq n+1$. This proves the theorem. □

Lemma 6.5.18 (Depth Lemma). *Let R be a Noetherian local ring, and consider a short exact sequence of R-modules*

$$0 \longrightarrow M_1 \longrightarrow M_2 \longrightarrow M_3 \longrightarrow 0.$$

Then

$$\mathrm{depth}(M_2) \geq \min\bigl(\mathrm{depth}(M_1), \mathrm{depth}(M_3)\bigr).$$

In case this inequality is strict we have, furthermore,

$$\mathrm{depth}(M_1) = \mathrm{depth}(M_3) + 1.$$

Proof. Suppose that all three modules have positive depth. From Prime Avoidance, see Exercise 6.5.30, we can find an $f \in R$ which is a nonzerodivisor of M_1,M_2 and M_3. By the Snake Lemma 1.2.12, it follows that

$$0 \longrightarrow M_1/fM_1 \longrightarrow M_2/fM_2 \longrightarrow M_3/fM_3 \longrightarrow 0$$

is also exact. Since the depth drops by one if one divides out a nonzerodivisor (Corollary 6.5.5), we can immediately reduce to the case that the depth of one of the three modules M_1,M_2 or M_3 is zero. So we distinguish three cases.

Case 1. Suppose $\mathrm{depth}(M_1) = 0$. We will prove that $\mathrm{depth}(M_2) = 0$. But this is obvious, as any nonzerodivisor of M_2 can be viewed as a nonzerodivisor of M_1, as M_1 may be viewed as a submodule of M_2.

Case 2. Suppose $\mathrm{depth}(M_2) = 0$. Then we have to show that either $\mathrm{depth}(M_1) = 0$ or $\mathrm{depth}(M_3) = 0$. Suppose not, and let (by prime avoidance) $f \in R$ be a nonzerodivisor of M_1 and M_3. Therefore, the multiplication by f on M_1 and M_3 has a trivial kernel. From the Snake Lemma it follows that multiplication with f on M_2 has trivial kernel, so f is a nonzerodivisor of M_2. But as $\mathrm{depth}(M_2) = 0$ this cannot be the case, contradiction.

Case 3. Suppose $\mathrm{depth}(M_3) = 0$. We have to show that if, moreover, $\mathrm{depth}(M_2) > 0$, then $\mathrm{depth}(M_1) = 1$. Take a nonzerodivisor f of M_2. As M_1 is a submodule of M_2 it is also a nonzerodivisor of M_1. Hence, $\mathrm{depth}(M_1) \geq 1$. By the Snake Lemma we obtain an inclusion

(6.9) $$0 \longrightarrow M' \longrightarrow M_1/fM_1,$$

where $M' := \mathrm{Ker}(M_3 \xrightarrow{\cdot f} M_3)$. As $\mathrm{depth}(M_3) = 0$, f is a zerodivisor. Thus $M' \neq 0$. One even has that there exists an element $g \in M_3 \setminus \{0\}$ with $\mathfrak{m}g = 0$. Hence, in particular, $g \in M'$. Any nonzerodivisor of M_1/fM_1 would give a nonzerodivisor of M' by the inclusion (6.9). But M' cannot have one by definition of $g \in M_3$. Therefore, $\mathrm{depth}(M_1/fM_1) = 0$, and $\mathrm{depth}(M_1) = 1$. □

Corollary 6.5.19. *Let $I,J \subset \mathcal{O}_n$ be radical ideals. Suppose that $(X, o) = \big(V(I), o\big)$ and $(Y, o) = \big(V(J), o\big)$ both have dimension k. Suppose (X, o) intersects (Y, o) properly, that is, $\dim\big((X \cap Y), o\big) \leq k - 1$. Suppose (X, o) and (Y, o) are Cohen-Macaulay. Then*

$(X \cup Y, o)$ is Cohen-Macaulay $\Longleftrightarrow$ $(X \cap Y, o)$ is Cohen-Macaulay of dimension $k - 1$.

Proof. Look at the exact sequence:

$$0 \longrightarrow \mathcal{O}_n/I \cap J \longrightarrow \mathcal{O}_n/I \oplus \mathcal{O}_n/J \longrightarrow \mathcal{O}_n/(I + J) \longrightarrow 0.$$

From the assumption $\dim(X \cap Y, o) \leq k - 1$ it follows that $\mathrm{depth}(\mathcal{O}_n/(I + J)) \leq k - 1$. Because (X, o) and (Y, o) are Cohen-Macaulay of dimension k, we know that $\mathrm{depth}(\mathcal{O}_n/I \oplus \mathcal{O}_n/J) = k$. The Depth Lemma gives that

$$\mathrm{depth}(\mathcal{O}_n/(I \cap J)) = \mathrm{depth}(\mathcal{O}_n/(I + J)) + 1,$$

which can be interpreted as the statement of the theorem. □

Theorem 6.5.20 (Auslander-Buchsbaum Formula). *Let $(R,\mathfrak{m})$ be a Noetherian local ring and M be a finitely generated R-module of finite projective dimension. Then:*

$$\mathrm{depth}(M) + \mathrm{pd}_R(M) = \mathrm{depth}(R).$$

Proof. The proof is by induction on $\mathrm{depth}(R)$ and $\mathrm{depth}(R) - \mathrm{depth}(M)$.

Step 1. Suppose first that $\mathrm{depth}(R) = 0$. Then we have to show that M is free, as $\mathrm{pd}_R(M) = \mathrm{depth}(M) = 0$ follows from it. As $\mathrm{depth}(R) = 0$, there exists a nonzero $x \in R$ with $x \cdot \mathfrak{m} = 0$. Now let

$$0 \longrightarrow F_n \xrightarrow{\alpha_n} F_{n-1} \xrightarrow{\alpha_{n-1}} \cdots \xrightarrow{\alpha_1} F_0 \longrightarrow M \longrightarrow 0$$

be a minimal free resolution of M as an R–module. Suppose $n > 0$. We can view F_n as a submodule of F_{n-1} via α_n. Because of the minimality, $F_n \subset \mathfrak{m}F_{n-1}$. Then $xF_n = x\mathfrak{m}F_{n-1} = 0$, hence $F_n = 0$ because F_n is free, a contradiction. It follows that $\mathrm{pd}_R(M) = 0$, hence M is free.

Step 2. Suppose that $\mathrm{depth}(M) = \mathrm{depth}(R)$. Then we can find a maximal regular sequence $f_1,\ldots,f_s$ for both M and R. As M is free if and only M/xM is free for x an nonzerodivisor, it follows that M is free. Hence $\mathrm{pd}_R(M) = 0$.

Step 3. For the induction step we consider two cases.

Case 1. Suppose $\mathrm{depth}(M) > 0$ and $\mathrm{depth}(R) > 0$. This case is done by induction on $\mathrm{depth}(R)$. Take an element $f \in \mathfrak{m}$ which is a nonzerodivisor of M and of R. Take a presentation:

$$0 \longrightarrow K \longrightarrow F \longrightarrow M \longrightarrow 0.$$

From the Snake Lemma:

$$0 \longrightarrow K/f \cdot K \longrightarrow F/f \cdot F \longrightarrow M/f \cdot M \longrightarrow 0$$

is exact. Because the projective dimension stays the same if one divides out a nonzerodivisor, see 6.5.17, and the depth drops by one if one divides out a nonzerodivisor, the theorem follows by induction on $\operatorname{depth}(R)$.

Case 2. Finally, suppose that $\operatorname{depth}(M) = 0$ and $\operatorname{depth}(R) > 0$. The case $\operatorname{pd}_R(M) = 0$ is trivial, as M is then a free R–module. Otherwise, consider a minimal a presentation

$$0 \longrightarrow K \longrightarrow F \longrightarrow M \longrightarrow 0$$

with F free and $K \neq 0$. Then $\operatorname{pd}_R(K) = \operatorname{pd}_R(M) - 1$. Since $\operatorname{depth}(M) = 0$, the inequality in the Depth Lemma is strict, so that $\operatorname{depth}(K) = \operatorname{depth}(M) + 1 = 1$. As $\operatorname{depth}(R) - \operatorname{depth}(K) = \operatorname{depth} *R) - \operatorname{depth}(M) - 1$, the Auslander Buchsbaum Formula holds for K by induction, so $\operatorname{depth}(K) + \operatorname{pd}_R(K) = \operatorname{depth}(R)$. Hence $\operatorname{depth}(M) + \operatorname{pd}_R(M) = \operatorname{depth}(R)$. □

Corollary 6.5.21. *Let* $I \subset \mathbb{C}\{x_1, \ldots, x_n\} = R$ *be an ideal,* $(X, o) = \big(V(I), o\big)$. *Consider a minimal free resolution of* R/I *as* R*–module*

$$0 \longrightarrow F_k \longrightarrow F_{k-1} \longrightarrow \cdots \longrightarrow F_0 = R \longrightarrow R/I \longrightarrow 0.$$

Then $\operatorname{depth}\big((X, o)\big) = n - k$. *In particular,* (X, o) *is Cohen-Macaulay if and only if* $k = n - \dim\big((X, o)\big)$.

This gives a possibility of checking on a computer whether a singularity, defined by polynomials, is Cohen-Macaulay.

Example 6.5.22. Take the union of two planes in $\mathbb{C}^4$ intersecting in a point again. The local ring is $R := \mathbb{C}\{x,y,z,w\}/(xz,xw,yz,yw)$. A free resolution of R as $\mathcal{O} := \mathbb{C}\{x,y,z,w\}$–module is given by

$$0 \longrightarrow \mathcal{O} \xrightarrow{\begin{pmatrix} y \\ -w \\ z \\ -x \end{pmatrix}} \mathcal{O}^4 \xrightarrow{\begin{pmatrix} w & y & 0 & 0 \\ -z & 0 & y & 0 \\ 0 & -x & 0 & w \\ 0 & 0 & -x & -z \end{pmatrix}} \mathcal{O}^4 \xrightarrow{(xz \;\; xw \;\; yz \;\; yw)} \mathcal{O} \longrightarrow R \longrightarrow 0,$$

which shows that R as a $\mathcal{O}$–module has projective dimension three. It follows from the Auslander-Buchsbaum Formula that, as $\mathcal{O}$–module, R has depth one. It follows that R (as R–module) has depth one. We see again that R is not Cohen–Macaulay.

The following corollary of the Auslander-Buchsbaum formula will be used in the proof of the Hilbert-Burch Theorem 6.5.26.

Corollary 6.5.23. *Let* $(R,\mathfrak{m})$ *be a local ring and* M *be an* R*–module. Then*

$$\operatorname{pd}_R(M) \geq \operatorname{depth}\big(\operatorname{Ann}(M), R\big).$$

Proof. We may assume that $\operatorname{pd}_R(M) < \infty$. The proof is by induction on $\operatorname{depth}(M)$. If $\operatorname{depth}(M) = 0$, then, by the Auslander Buchsbaum Formula, $\operatorname{pd}_R(M) = \operatorname{depth}(R) \geq \operatorname{depth}\big(\operatorname{Ann}(M), R\big)$.

We now assume that $\operatorname{depth}(M) > 0$. In particular, from the Auslander-Buchsbaum Formula, it follows that $d := \operatorname{depth}(R) > 0$. Let $0 \longrightarrow F_n \longrightarrow \cdots \longrightarrow F_0 \longrightarrow M \longrightarrow 0$ be a minimal free resolution of M. Choose a regular sequence $x_1,\ldots,x_d$ in R such that $x_{r+1},\ldots,x_d$ is a maximal regular sequence in $\operatorname{Ann}(M)$. Since $\operatorname{depth}(M) > 0$, there exists a nonzerodivisor of M. Thus in particular x_1 is a nonzerodivisor of $R/\operatorname{Ann}(M)$. Hence we have that $r \geq 1$ and, by renumbering, we may suppose x_1 is a nonzerodivisor of M. We obtain, using 1.2.35, the exact sequence

$$0 \longrightarrow F_n/x_1F_n \longrightarrow \cdots \longrightarrow F_0/x_1F_0 \longrightarrow M/x_1M \longrightarrow 0.$$

By induction on $\operatorname{depth}(M)$ we obtain

$$\operatorname{pd}_{R/(x_1)}(M/x_1M) \geq \operatorname{depth}\big(\operatorname{Ann}(M/x_1M),R/(x_1)\big).$$

Now $\operatorname{pd}_R(M) = \operatorname{pd}_{R/(x_1)}(M/x_1M)$ because of 6.5.17 and $\operatorname{depth}\big(\operatorname{Ann}(M/x_1M),R/(x_1)\big) \geq \operatorname{depth}\big(\operatorname{Ann}(M),R\big)$ because of the special choice of $x_1,\ldots,x_d$. □

Finally, in this chapter we will prove the Hilbert-Burch Theorem. The proof needs some preparations. First of all we need a characterization for a complex of type

$$0 \longrightarrow F_n \xrightarrow{\varphi_n} F_{n-1} \xrightarrow{\varphi_{n-1}} \cdots \xrightarrow{\varphi_1} F_0$$

being exact. We start with a lemma.

Lemma 6.5.24. *Let R be a Noetherian local ring and $0 \longrightarrow F_n \xrightarrow{\varphi_n} \cdots \xrightarrow{\varphi_1} F_0$ be a complex*[4] *of finitely generated free R-modules. Let $r_\nu = \sum_{j=\nu}^{n}(-1)^{j-\nu}\operatorname{rank}(F_j)$ for $\nu = 0,\ldots,n$. Then the induced sequence*

$$0 \longrightarrow F_n \otimes_R Q(R) \longrightarrow \cdots \longrightarrow F_0 \otimes_R Q(R) \tag{6.10}$$

is exact if and only if $\operatorname{depth}\big(I_{r_\nu}(\varphi_\nu),R\big) \geq 1$ *for all ν. Moreover, if this is the case, then* $\operatorname{Im}(\varphi_\nu)$ *has* rank r_ν *for all ν.*

Proof.
Step 1. First suppose that (6.10) is exact. To prove that $\operatorname{depth}\big(I_{r_\nu}(\varphi_\nu),R\big) \geq 1$ means to prove that $I_{r_\nu}(\varphi_\nu)$ contains a nonzerodivisor, which means that $I_{r_\nu}(\varphi_\nu)\otimes Q(R) = Q(R)$. Therefore, we may suppose that $R = Q(R)$, and that $0 \longrightarrow F_n \xrightarrow{\varphi_n} \cdots \xrightarrow{\varphi_1} F_0$ is exact, and we have to prove that $I_{r_\nu}(\varphi_\nu) = R$. For $j = n,\ldots,1$ we have exact sequences

$$0 \longrightarrow \operatorname{Im}(\varphi_j) \longrightarrow F_{j-1} \longrightarrow \operatorname{Im}(\varphi_{j-1}).$$

We proceed by descending induction. First of all F_n has rank r_n, and thus, as φ_n is injective, $\operatorname{Im}(\varphi_n)$ has rank r_n. From 1.4.32 it follows that $I_{r_n}(\varphi_n) = R$, and $I_{r_n+1}(\varphi_n) = (0)$. From 1.3.8 it follows that $\operatorname{Im}(\varphi_{n-1})$ has rank r_{n-1}. Again 1.4.32 implies that $I_{r_{n-1}}(\varphi_{n-1}) = R$ and $I_{r_{n-1}+1}(\varphi_{n-1}) = (0)$. The induction step is proved in a completely similar manner.

Step 2. To prove the other direction, assume that $\operatorname{depth}(I_{r_\nu}(\varphi_\nu),R) \geq 1$ for all ν. This condition gives that $I_{r_\nu}(\varphi_\nu)$ contains a nonzerodivisor. As a nonzerodivisor is a unit in $Q(R)$, we get $I_{r_\nu}(\varphi_\nu) \otimes Q(R) = Q(R)$. Therefore, we may assume that $R = Q(R)$ and thus that $I_{r_\nu}(\varphi_\nu) = R$. Under these conditions we have to show exactness, that is, $\operatorname{Im}(\varphi_{\nu+1}) = \operatorname{Ker}(\varphi_\nu)$ for all ν, and that $\operatorname{Im}(\varphi_\nu)$ is free of rank r_ν. By applying 1.3.8 the freeness of $\operatorname{Im}(\varphi_\nu)$ follows if we moreover show that $I_{r_\nu+1}(\varphi_\nu) = (0)$.

[4] Note that there is no "$\longrightarrow 0$" at the right hand side of the complex.

We will proceed by descending induction on ν. So, assume that we have already proved that $I_{r_j+1}(\varphi_j) = (0)$ for $j \geq \nu + 1$, and that the complex

$$0 \longrightarrow F_n \longrightarrow \cdots \longrightarrow F_{\nu+1} \longrightarrow F_\nu$$

is exact. Now consider

$$F_{\nu+1} \longrightarrow F_\nu \longrightarrow \mathrm{Coker}(\varphi_{\nu+1}) \longrightarrow 0.$$

By assumption we have $I_{r_{\nu+1}}(\varphi_{\nu+1}) = R$, and by 1.3.8 it follows that $\mathrm{Im}(\varphi_{\nu+1})$ is *free*, necessarily of rank $r_{\nu+1}$. Moreover Proposition 1.3.8 says that

$$F_\nu = \mathrm{Im}(\varphi_{\nu+1}) \oplus F'_\nu$$

for a *free* module F'_ν. It easily follows from the definition of the r_j that the rank of F'_ν is equal to r_ν. Since $\mathrm{Im}(\varphi_{\nu+1}) \subset \mathrm{Ker}(\varphi_\nu)$ it follows from the definition of Fitting ideals that

$$(R =) I_{r_\nu}(\varphi_\nu) = I_{r_\nu}(\varphi_{\nu|F'_\nu}).$$

Because of size reasons $(\mathrm{rank}(F'_\nu) = r_\nu)$, $I_{r_\nu+1}(\varphi_{\nu|F'_\nu}) = 0$ and therefore $I_{r_\nu+1}(\varphi_\nu) = (0)$. Again from Proposition 1.3.8 it follows that $\mathrm{Im}(\varphi_\nu)$ has rank r_ν, and that $\varphi_{\nu|F'_\nu}$ is injective. This exactly says that $\mathrm{Im}(\varphi_{\nu+1}) = \mathrm{Ker}(\varphi_\nu)$, which is what we had to prove. □

Theorem 6.5.25 (Buchsbaum-Eisenbud). *Let R be a Noetherian local ring, and let*

$$(6.11) \qquad 0 \longrightarrow F_n \xrightarrow{\varphi_n} \cdots \xrightarrow{\varphi_1} F_0$$

be a complex of free R-modules of finite rank. Let $r_\nu = \sum_{j=\nu}^{n}(-1)^{j-\nu}\,\mathrm{rank}(F_j)$ for $\nu = 0,\ldots,n$. Then the complex (6.11) is exact if and only if $\mathrm{depth}(I_{r_\nu}(\varphi_\nu),R) \geq \nu$ *for $\nu = 1,\ldots,n$.*

Proof.
Step 1. Assume the complex is exact. We have to prove that $\mathrm{depth}(I_{r_\nu}(\varphi_\nu),R) \geq \nu$. We use induction on $d = \mathrm{depth}(R)$. The case $d = 0$ is not difficult. The exactness of the complex gives that $\mathrm{pd}_R(\mathrm{Im}(\varphi_\nu)) < \infty$ for $\nu = 1,\ldots,n$. This implies that $\mathrm{pd}_R(\mathrm{Im}(\varphi_\nu)) = 0$ (Auslander-Buchsbaum 6.5.20), that is, $\mathrm{Im}(\varphi_\nu)$ is free and of rank r_ν. By 1.4.32 $I_{r_\nu}(\varphi_\nu)$ contains a nonzerodivisor. As $\mathrm{depth}(R) = 0$, the only nonzerodivisors are the units. Therefore $I_{r_\nu}(\varphi_\nu) = R$ and, consequently, $\mathrm{depth}(I_{r_\nu}(\varphi_\nu),R) = \infty$.

Now assume that $d > 0$. By Exercise 1.3.22 the sequence $0 \longrightarrow F_n \otimes Q(R) \longrightarrow \cdots \longrightarrow F_0 \otimes Q(R)$ is exact. By the previous lemma $\mathrm{depth}(I_{r_\nu}(\varphi_\nu),R) \geq 1$ for all ν. Therefore we can find a nonzerodivisor $x \in \bigcap_{\nu=1}^{n} I_{r_\nu}(\varphi_\nu)$. If x is a unit then $I_{r_\nu}(\varphi_\nu) = R$ for all ν and thus by definition $\mathrm{depth}(I_{r_\nu}(\varphi_\nu),R) = \infty$. Otherwise, $x \in \mathfrak{m}$ is a nonzerodivisor, so that by 1.2.35 we get that

$$0 \longrightarrow F_n/xF_n \longrightarrow \cdots \longrightarrow F_1/xF_1$$

is exact. The induction hypothesis implies that $\mathrm{depth}(I_{r_\nu}(\varphi_\nu \otimes \mathrm{id}_{R/(x)}),R/(x)) \geq \nu - 1$ for $\nu = 2,\ldots,n$. This implies $\mathrm{depth}(I_{r_\nu}(\varphi_\nu),R) \geq \nu$ because of 6.5.5 and the definition of the Fitting ideals, see 1.2.18.

Step 2. For the converse we consider three cases.

Case 1. This is the case that $d = \operatorname{depth}(R) = 0$. Then the only nonzerodivisors in R are units. Thus $R = Q(R)$ and the statement follows from the previous lemma.

Case 2. The case that $d = \operatorname{depth}(R) \geq 2$. This is proved by induction on the dimension of R and on n. The case $n = 0$ is trivial, as obviously the sequence $0 \longrightarrow F_0$ is exact. The case $\dim(R) \leq 1$ follows because then $\operatorname{depth}(R) \leq 1$, hence we are in the first case or in the third case.

Now assume that $\dim(R) > 0$ and $n > 0$. By induction hypothesis we may suppose that

$$0 \longrightarrow F_n \longrightarrow \cdots \longrightarrow F_1$$

is exact. Let $M = \operatorname{Im}(\varphi_2)$ and $K = \operatorname{Ker}(\varphi_1)$. It remains to prove that $K/M = 0$. Let $\mathfrak{p} \neq \mathfrak{m}$ be a prime ideal. By Exercise 1.3.22 the localization of the sequence

$$0 \longrightarrow F_{n\mathfrak{p}} \longrightarrow \cdots \longrightarrow F_{1\mathfrak{p}} \longrightarrow F_{0\mathfrak{p}}$$

is exact. As $\dim(R_\mathfrak{p}) \leq \dim(R) - 1$, it follows from the induction hypothesis with respect to the dimension of the ring that $(K/M)_\mathfrak{p} = 0$ for all prime ideals $\mathfrak{p} \subset R$ different from the maximal ideal $\mathfrak{m}$. By Exercise 1.4.54 this implies that a certain power of $\mathfrak{m}$ annihilates K/M. Thus K/M is a module of dimension zero. Hence $\operatorname{depth}(K/M) = 0$.

We consider the exact sequences

$$0 \longrightarrow \operatorname{Im}(\varphi_\nu) \longrightarrow F_{\nu-1} \longrightarrow \operatorname{Im}(\varphi_{\nu-1}) \longrightarrow 0 \tag{6.12}$$

for $\nu = 2, \ldots, \min\{d,n\}$. Recall that $d = \operatorname{depth}(R)$. Suppose that $n \leq d$. Then $F_n = \operatorname{Im}(\varphi_n)$ is free, and so has depth d, see Exercise 6.5.27, and thus certainly depth $\geq n$. Suppose on the other hand that $j > d$. By assumption $\operatorname{depth}(I_{r_j}(\varphi_j),R) \geq j$. As R has only depth d, it follows that $I_{r_j}(\varphi_j) = R$. Following the first part of the proof of Lemma 6.5.24 we obtain that $\operatorname{Im}(\varphi_j)$ is free of rank r_j for $j \geq d$, In particular, $\operatorname{Im}(\varphi_d)$ has depth at least d. Applying the Depth Lemma 6.5.18 successively to the exact sequences (6.12) gives that $\operatorname{depth}(\operatorname{Im}(\varphi_j)) \geq j$ for $j \geq 2$. As $M = \operatorname{Im}(\varphi_2)$ we get $\operatorname{depth}(M) \geq 2$. Using that F_1 has positive depth, and applying the Depth Lemma to the exact sequence

$$0 \longrightarrow M \longrightarrow F_1 \longrightarrow F_1/M \longrightarrow 0,$$

gives that $\operatorname{depth}(F_1/M) \geq 1$. Note that also $\operatorname{depth}(F_1/K) \geq 1$. This holds because we have an embedding $F_1/K \subset F_0$. Now assume that $K/M \neq 0$. We apply the Depth Lemma to the exact sequence

$$0 \longrightarrow K/M \longrightarrow F_1/M \longrightarrow F_1/K \longrightarrow 0.$$

As both F_1/M and F_1/K have positive depth, it follows that $\operatorname{depth}(K/M) \geq 1$. This is a contradiction, as we have already proved that $\operatorname{depth}(K/M) = 0$. Thus $K/M = 0$, and the sequence is exact.

Case 3. This is the case $d = \operatorname{depth}(R) = 1$. By assumption $\operatorname{depth}(I_{r_2}(\varphi_2),R) \geq 2$. It follows that $I_{r_2}(\varphi_2) = R$. Following the proof of case 2, it follows that $\operatorname{Im}(\varphi_2)$ is free, and

$$F_1 \cong \operatorname{Im}(\varphi_2) \oplus F_1'$$

for a suitable free module F_1' (1.3.8). It remains to prove that the restriction of φ_1 to F_1' is injective. This is trivial if $F_1' = 0$. By 6.5.24 we have that $\operatorname{rank}(F_1') = r_1$. As in the second case we have that $\operatorname{Ker}(\varphi_1|_{F_1'})$ is annihilated by a power of the maximal ideal $\mathfrak{m}$ and thus, if $\operatorname{Ker}(\varphi_1|_{F_1'}) \neq 0$, it follows that $\operatorname{depth}(\operatorname{Ker}(\varphi_1|_{F_1'})) = 0$, and we have an exact sequence

$$0 \longrightarrow \operatorname{Ker}(\varphi_1|_{F_1'}) \longrightarrow F_1' \longrightarrow \operatorname{Im}(\varphi_1) \longrightarrow 0$$

As F_1' has depth one, it follows from the Depth Lemma that $\operatorname{depth}(\operatorname{Ker}(\varphi_1|_{F_1'})) > 0$. This is a contradiction. Therefore, $\varphi_1|F_1'$ is injective and $0 \longrightarrow F_n \longrightarrow \cdots \longrightarrow F_0$ is exact. □

Theorem 6.5.26 (Hilbert-Burch). *Let R be a Noetherian local ring and $I \subset R$ be an ideal.*

(1) Assume that $0 \longrightarrow F_2 \xrightarrow{\alpha_2} F_1 \xrightarrow{\alpha_1} R \longrightarrow R/I \longrightarrow 0$ is a minimal free resolution. Then

(1.1) $\operatorname{rank}(F_1) = \operatorname{rank}(F_2) + 1$,

(1.2) $\operatorname{depth}(I_{\operatorname{rank}(F_2)}(\alpha_2), R) = 2$,

(1.3) $I = aI_{\operatorname{rank}(F_2)}(\alpha_2)$ *for a suitable nonzerodivisor $a \in R$. More precisely the i-th entry of the matrix for α_1 is $(-1)^i a$ times the minor obtained from α_2 by leaving out the i-th row.*

(2) Let α_2 be a $(n-1) \times n$-matrix and suppose that $\operatorname{depth}(I_{n-1}(\alpha_2), R) \geq 2$. *Let $a \in R$ be a nonzerodivisor and define α_1 as in the description of (1.3). Let $I := aI_{n-1}(\alpha_2)$. Then $0 \longrightarrow R^{n-1} \xrightarrow{\alpha_2} R^n \xrightarrow{\alpha_1} R \longrightarrow R/I \longrightarrow 0$ is a free resolution of R/I.*

Proof. The statement (2) is an immediate consequence of 6.5.25, as $r_2 = n-1$, and $r_1 = 1$. We now prove (1.3). Let $\overline{\alpha}_1$ be the map defined by the $1 \times n$-matrix having as i-th entry $(-1)^i$ times the minor obtained from α_2 by deleting the i-th row. Then Laplace expansion gives $\overline{\alpha}_1 \circ \alpha_2 = 0$. On the other hand, by definition, $I_1(\overline{\alpha}_1) = \operatorname{Im}(\overline{\alpha}_1) = I_{\operatorname{rank}(F_2)}(\alpha_2)$. By 6.5.25 we obtain that $\operatorname{depth}(I_{\operatorname{rank}(F_2)}(\alpha_2), R) \geq 2$. Therefore, we can apply again 6.5.25 to the dual complex

$$0 \longrightarrow R \xrightarrow{\overline{\alpha}_1^*} F_1^* \xrightarrow{\alpha_2^*} F_2^*,$$

using that $I_1(\overline{\alpha}_1^*) = I_1(\overline{\alpha}_1) = I_{\operatorname{rank}(F_2)}(\alpha_2)$, to obtain that this complex is exact. Now consider the commutative diagram

$$\begin{array}{ccccccc} & & R & \xrightarrow{\alpha_1^*} & F_1^* & \xrightarrow{\alpha_2^*} & F_2^* \\ & & \downarrow{\scriptstyle\sigma} & & \| & & \| \\ 0 & \longrightarrow & R & \xrightarrow{\overline{\alpha}_1^*} & F_1^* & \xrightarrow{\alpha_2^*} & F_2^* \end{array}$$

The map σ with $\overline{\alpha}_1^* \circ \sigma = \alpha_1^*$ exists because of the exactness of the second row. Because $\sigma \in \operatorname{Hom}_R(R, R)$ there exists an $a \in R$ such that $\sigma(x) = ax$ for all $x \in R$. It remains to prove that a is a nonzerodivisor. So suppose $x \cdot a = 0$ for some $x \in R$. This implies $xI_1(\alpha_1) = 0$. Therefore, $I_1(\alpha_1)$ consists of zerodivisors. This is not possible, as the sequence $0 \longrightarrow F_2 \xrightarrow{\alpha_2} F_1 \xrightarrow{\alpha_1} F_0$ is exact and thus $\operatorname{depth}(I_1(\alpha_1), R) \geq 1$ by Theorem 6.5.25.

To prove (1.1) note that we have now proved that I contains a nonzerodivisor. Hence $I \otimes Q(R) = Q(R)$ and $R/I \otimes Q(R) = 0$. We tensorize the exact sequence with $Q(R)$ and obtain by Exercise 1.3.22 an exact sequence

$$0 \longrightarrow F_2 \otimes_R Q(R) \longrightarrow F_1 \otimes_R Q(R) \longrightarrow Q(R) \longrightarrow 0.$$

This implies $\operatorname{rank}(F_1) = \operatorname{rank}(F_2) + 1$. It remains to prove (1.2). Put $I' = I_{\operatorname{rank}(F_2)}(\alpha_2)$. By dividing by a we get an exact sequence

$$0 \longrightarrow F_2 \xrightarrow{\alpha_2} F_1 \xrightarrow{\frac{\alpha_1}{a}} R \longrightarrow R/I' \longrightarrow 0.$$

Obviously $I' = \operatorname{Ann}(R/I')$, so that from 6.5.23 we obtain $2 \geq \operatorname{pd}_R(R/I') \geq \operatorname{depth}(I',R)$. On the other hand, by 6.5.25, we have $\operatorname{depth}(I',R) \geq 2$. This implies $\operatorname{depth}(I',R) = 2$. □

Exercises

6.5.27. Suppose M is a free R–module. Then $\operatorname{depth}(M) = \operatorname{depth}(R)$.

6.5.28. Let $R \subset S$ be a finite ring extension of local rings, and M be an S–module.

(1) Let $I \subset R$ be be an ideal. Show that $(I \cdot R) \cdot M = (I \cdot S) \cdot M$.

(2) Let $f_1, \ldots, f_s$ be a regular sequence of M as R–module. Show that $f_1, \ldots, f_s$ is a regular sequence of M as S–module.

6.5.29. Let $R \subset S$ be a finite ring extension of local rings, and M be an S–module. Show that $\operatorname{depth}_R(M) = \operatorname{depth}_S(M)$.

6.5.30. Let M,N be R–modules of positive depth. Show that there exists an $f \in R$ which is a nonzerodivisor for both M and N. Generalize this statement to finitely many modules.
(Hint: Use Exercise 1.4.52 and Prime Avoidance 1.1.13.)

6.5.31. The assumption on R to be local in Lemma 6.5.3 is an essential assumption! Show that in the ring $\mathbb{C}[x,y,z]/(xz - z)$ the sequence $x, xy - y$ is a regular sequence, but $xy - y, x$ is not a regular sequence.

6.5.32. Consider the map $f(s,t) = (s,st,t^2,t^3) = (x,y,z,u)$ of Exercise 2.3.18. The image is given by the ideal $I = (yz - ux, y^2 - zx^2, z^2x - uy, z^3 - u^2)$, as shown in that exercise. Let $R = \mathbb{C}\{x,y,z,u\}$. Show that

$$0 \longrightarrow R \xrightarrow{\begin{pmatrix} u \\ -y \\ -z \\ x \end{pmatrix}} R^4 \xrightarrow{\begin{pmatrix} y & u & xz & z^2 \\ -z & 0 & -u & 0 \\ -x & z & -y & u \\ 0 & -x & 0 & -y \end{pmatrix}} R^4 \xrightarrow{(yz-ux,\, y^2-zx^2,\, z^2x-uy,\, z^3-u^2)} R \longrightarrow R/I \longrightarrow 0$$

is a minimal free resolution of I. Conclude by the Auslander-Buchsbaum formula that the germ of the complex space $(V(I),\mathrm{o})$ is not Cohen-Macaulay.

6.5.33. Let $R = \mathbb{C}\{x,y,z\}, I = (xy,xz,yz)$. Compute a free resolution of R/I as R–module. Conclude that $(X,\mathrm{o}) = (V(I),\mathrm{o})$ is Cohen-Macaulay.

6.5.34. Let $R = \mathbb{C}\{x,y,z\}, I = (xy,xz)$. Compute a free resolution of R/I as R–module. Conclude that $(X,\mathrm{o}) = (V(I),\mathrm{o})$ is not Cohen-Macaulay.

6.5.35. Let $R = \mathbb{C}\{x,y\}/(xy)$, and $M = \widetilde{R}$ be the normalization of R. Show that M is not a free R–module. Show by using the Auslander Buchsbaum formula, that M does not have finite projective dimension.

7 Standard Bases

Let an ideal $I \subset \mathbb{C}\{x\}$ be given. In this chapter we will define and study *standard bases* of the ideal I. The main idea here is that one wants to have a "good" representative for any coset of I. In fact, the Weierstraß Division Theorem provides us with the simplest example of a standard basis. Consider $f \in \mathbb{C}\{x\}$ which is regular of order b in x_n. For any $g \in \mathbb{C}\{x\}$ we have a unique representation $g = qf + r$ with r a polynomial of degree less than b. From the uniqueness statement in the Weierstraß Division Theorem, it follows without difficulty that the remainder r only depends on the coset $g + (f)$ of (f). Of course, the remainder r depends on the choice of x_n.

In order to generalize, we would like to have division with remainder through an ideal. The "choice" of x_n is now replaced by the choice of a so-called *monomial ordering* on the monomials of $\mathbb{C}\{x\}$. Having such a monomial ordering, one can talk about the leading term $L(f)$ of f, and the ideal $L(I) = (L(f) : f \in I \setminus \{0\})$ of leading terms of I. A standard basis of I is a set of elements $f_1, \dots, f_m$ of I such that $L(f_1), \dots, L(f_m)$ generate $L(I)$. It is quite trivially true that a standard basis exists. Here one uses only the fact that $\mathbb{C}\{x\}$ is Noetherian. It also turns out that a standard basis of I generates the ideal I. Standard bases of ideals in $\mathbb{C}\{x\}$ generated by polynomials can be computed effectively. The algorithms are explained in the exercises. For example, the computer algebra system SINGULAR can be used for computation (cf. [Singular 2000]).

The first thing one would have is a kind of division with remainder through the f_i, that is, one would like for any $f \in \mathbb{C}\{x\}$ a representation

$$f = \sum_{i=1}^{m} q_i f_i + r \tag{7.1}$$

where no term of r is divisible by the leading term of any f_i. This replaces the condition that r is polynomial of degree less than b in the Weierstraß Division Theorem. In fact, we will give alternative proofs of the Weierstraß Division and Preparation Theorems. In order to control the convergence of the q_i and r, we will define seminorms[1] on $\mathbb{C}\{x\}$ or even $\mathbb{C}[[x]]$ depending on $\rho \in \mathbb{R}^n_+$ defined by $\|f\|_\rho := \sum_\alpha |f_\alpha| \rho^\alpha$. This is also the norm which is used in the books of Grauert and Remmert, see [Grauert-Remmert 1971], and [Grauert-Remmert 1984] to give proofs of the Weierstraß Division and Preparation Theorem. It turns out that if $f_1, \dots, f_m$ is a standard basis, then the r is uniquely determined, and is called the normal form $\mathrm{NF}(f) = \mathrm{NF}(f|I)$ of f. It depends, of course, on the monomial ordering. In particular, it follows that $f \in I$ if and only if $\mathrm{NF}(f) = 0$. This will be used in an essential way in the proof of Grauert's Approximation Theorem in Chapter 8. We will give some applications in Section 7.3.

More generally, there is a theory of standard bases of modules. As it is not more difficult, we will treat the general case from the beginning.

[1] Here semi means that the norm might take infinite value.

7.1 The Division Theorem

Definition 7.1.1. Let $M = \{x^\alpha | \alpha \in \mathbb{N}^n\}$ be the set of monomials in $\mathbb{C}[x]$. A *monomial ordering* is a total ordering on M (respectively $\mathbb{N}^n$) compatible with the semigroup structure, that is, $x^\alpha > x^\beta$ implies $x^{\alpha+\gamma} > x^{\beta+\gamma}$ for all $\gamma \in \mathbb{N}^n$.

We assume furthermore that the orderings considered here are well-orderings, that is, every set of monomials has a minimal element with respect to the ordering. This is equivalent to the property that 1 is the smallest monomial (cf. Exercise 7.1.16).

Examples 7.1.2.

(1) The *lexicographical ordering* $\underset{\text{lex}}{>}$.

$x^\alpha \underset{\text{lex}}{>} x^\beta$ if and only if $\alpha \neq \beta$ and there is an $i \geq 1$ such that $\alpha_j = \beta_j$ for $j < i$ and $\alpha_i > \beta_i$.

(2) The *degree lexicographical ordering* $\underset{\text{deglex}}{>}$.

Let $w = (w_1, \ldots, w_n) \in \mathbb{R}^n$ and $w_i > 0$, $i = 1, \ldots, n$. For $\alpha \in \mathbb{N}^n$ define $|\alpha|_w = \sum_{i=1}^n \alpha_i w_i$ (we will also simply use $|\alpha| = |\alpha|_w$ if w is fixed).

$x^\alpha \underset{\text{deglex}}{>} x^\beta$ if and only if $|\alpha| > |\beta|$ or $|\alpha| = |\beta|$ and $x^\alpha \underset{\text{lex}}{>} x^\beta$.

Definition 7.1.3. Let $>$ be a monomial ordering, and $0 \neq f \in \mathbb{C}\{x\}$ be a power series. Then we may write

$$f = \sum_{i \geq 1} a_i x^{\alpha_i}$$

such that $a_i \neq 0$, $a_i \in \mathbb{C}$ and $x^{\alpha_i} < x^{\alpha_{i+1}}$ for all i. We define

(1) $L(f) = x^{\alpha_1}$, the *leading monomial* of f;

(2) $C(f) = a_1$, the coefficient of the leading monomial;

(3) $\text{Tail}(f) = f - C(f) \cdot L(f) = \sum_{i \geq 2} a_i x^{\alpha_i}$.

Let $I \subset \mathbb{C}\{x\}$ be an ideal. Then $L(I) = (\{L(f) | f \in I, f \neq 0\})$ is called the *leading ideal*. It is the ideal generated by all leading monomials of nonzero elements of I.

Example 7.1.4. Let $f = 2x_1^5 + x_2^6 + x_2^4 + x_1$ and $I = (f)$. With respect to the orderings introduced before (with $w = (1,1)$), we obtain for $L(f)$, $C(f)$, $\text{Tail}(f)$ and $L(I)$ the following (the monomials of $\text{Tail}(f)$ are ordered with respect to the monomial ordering):

	$C(f)$	$L(f)$	$L(I)$	$\text{Tail}(f)$
lex	1	x_2^4	(x_2^4)	$x_2^6 + x_1 + 2x_1^5$
deglex	1	x_1	(x_1)	$x_2^4 + 2x_1^5 + x_2^6$

We also wish to define orderings for the monomials of the free module $\mathbb{C}\{x\}^N$. For a compatible notation we choose as a basis of $\mathbb{C}\{x\}^N$ the vectors $(0, \ldots, 0, 1, 0 \ldots 0)$, and identify them with x_{n+1}^{i-1}. Here the 1 is on the i–th spot, and x_{n+1} is a *new* variable.

Definition 7.1.5. Let $<$ be a monomial ordering on M, the monomials in $\mathbb{C}[x]$. We consider the free module $\mathbb{C}\{x\}^N = \sum_{i=0}^{N-1} \mathbb{C}\{x\}x_{n+1}^i$. Let $M = \{x^\alpha | \alpha \in \mathbb{N}^{n+1}, 0 \leq \alpha_{n+1} \leq N-1\}$ be the set of monomials of $\mathbb{C}\{x\}^N$. A *module ordering* $<_m$ is a total ordering of M compatible with the ordering $<$ on $\mathbb{C}\{x\}$, that is,

(1) $f, f' \in M$ and $f <_m f'$ implies $x^\alpha f <_m x^\alpha f'$.

(2) $f \in M$ and $x^\alpha < x^\beta$ implies $x^\alpha f <_m x^\beta f$.

Exercise 7.1.17 gives several examples for module orderings.

Definition 7.1.6. From now on in this chapter **we use the following ordering.** Let $w = (w_1, \ldots, w_{n+1}) \in \mathbb{R}_+^{n+1}$ be fixed and $<$ be the degree lexicographical ordering. For x^α and $x^\beta \in M$ we define $x^\alpha < x^\beta$ if and only if $|\alpha| < |\beta|$ or if $|\alpha| = |\beta|$ then $\alpha < \beta$ with respect to the lexicographical ordering.
As in Definition 7.1.3 we define for $f \in \mathbb{C}\{x\}^N$ the leading monomial $L(f)$, the leading coefficient $C(f)$, the tail, $\mathrm{Tail}(f)$ and the leading module $L(I)$ for a submodule $I \subset \mathbb{C}\{x\}^N$.

Definition 7.1.7.

(1) Let $m, m' \in M$ be two monomials, $m = x^\alpha$, $m' = x^\beta$. We define $m \mid m'$ if $\alpha_{n+1} = \beta_{n+1}$ and $x^\alpha \mid x^\beta$.

(2) We define an $\mathbb{N}^n$–semimodule structure on $\mathbb{N}^{n+1}$ by putting

$$(\alpha_1, \ldots, \alpha_n) + (\beta_1, \ldots \beta_{n+1}) = (\alpha_1 + \beta_1, \ldots \alpha_n + \beta_n, \beta_{n+1}).$$

So the last component of the semimodule generated by $(\beta_1, \ldots, \beta_{n+1})$ is always equal to β_{n+1}.

Remark 7.1.8. Note that $m \mid m'$ can only occur if m and m' lie in the same summand of $\mathbb{C}\{x\}^N$. In particular, it follows that if $m \mid m'$, then the quotient $\frac{m'}{m}$ is a well-defined element of $\mathbb{C}\{x\}$.

The following result is a generalization of the Weierstraß Division Theorem and, therefore, also called Division Theorem. It is the essential theorem of this section with many consequences. We will need a quite general version, but we will first write down a simpler case.

Theorem 7.1.9 (Grauert's Division Theorem).
Let $f_1, \ldots, f_m$ be elements of $\mathbb{C}\{x\}^N$. Then for every $f \in \mathbb{C}\{x\}^N$ there exist $q_1, \ldots, q_m \in \mathbb{C}\{x\}$ and an element $r \in \mathbb{C}\{x\}^N$ such that

$$f = \sum_{j=1}^{m} q_j f_j + r,$$

having moreover the following two properties.

(1) No monomial of r is divisible by $L(f_j)$ for $j = 1, \ldots, m$;

(2) $L(q_j f_j) \geq L(f)$ for $j = 1, \ldots, m$.[2]

[2] This condition is a kind of minimality condition. Note that in a sum one can always add two terms which cancel. The condition written here avoids this phenomenon.

In order to control the convergence we introduce a seminorm on $\mathbb{C}[[x]]^N$. More precisely let $x = (x_1, \dots, x_n)$ and consider the free $\mathbb{C}[[x]]$–module $\mathbb{C}[[x]]^N$. We choose the basis $\{e_i\}_{i=1,\dots,N}$, $e_i = (0,\dots,0,1,0,\dots,0)$ of $\mathbb{C}[[x]]^N$. We identify it, as in the convergent case, with x_{n+1}^{i-1} for a new variable x_{n+1}, that is, $\mathbb{C}[[x]]^N = \sum_{i=0}^{N-1} \mathbb{C}[[x]]x_{n+1}^i$.

Definition 7.1.10. For $\rho \in \mathbb{R}_+^{n+1}$ and $f = \sum f_\alpha x^\alpha \in \mathbb{C}[[x]]^N$, we define

$$||f||_\rho := \sum_{\alpha \in \mathbb{N}^{n+1}} |f_\alpha| \rho^\alpha \in \mathbb{R}_{\geq 0} \cup \{\infty\}.$$

Here ρ^α by definition is $\prod_{i=1}^{n+1} \rho_i^{\alpha_i}$ for $\alpha = (\alpha_1, \dots, \alpha_{n+1})$.

Definition 7.1.11.

(1) $B_\rho^N = \{f \in \mathbb{C}[[x]]^N : ||f||_\rho < \infty\}$.

(2) $B_\rho = B_\rho^1$.

(3) For $\rho, \overline{\rho} \in \mathbb{R}_+^{n+1}$ we define $\rho \leq \overline{\rho}$ if $\rho_i \leq \overline{\rho}_i$ for $i = 1, \dots, n+1$.

The following proposition is an easy consequence of the results of Section 3.1.

Proposition 7.1.12.

(1) Let $f \in \mathbb{C}[[x]]^N$. Then $f \in \mathbb{C}\{x\}^N \iff$ there exists a $\rho \in \mathbb{R}_+^{n+1}$ with $f \in B_\rho^N$.

(2) If $\rho \leq \overline{\rho}$ then $B_{\overline{\rho}}^N \subset B_\rho^N$.

(3) $\mathbb{C}\{x\}^N = \cup_{\rho \in \mathbb{R}_+^{n+1}} B_\rho^N$.

Proposition 7.1.13. B_ρ^N *is a* Banach B_ρ–module. *This, by definition, means that the norm satisfies the following properties. For all $a \in \mathbb{C}$, $f, g \in B_\rho^N$, $h \in B_\rho$ we have*

(1) $||af||_\rho = |a| ||f||_\rho$;

(2) $||f||_\rho = 0$ if and only if $f = 0$;

(3) $||f + g||_\rho \leq ||f||_\rho + ||g||_\rho$;

(4) $||hf||_\rho \leq ||h||_\rho ||f||_\rho$;

(5) Let $f_1, f_2, f_3, \dots$ be a Cauchy sequence in B_ρ^N. Then the sequence $f_1, f_2, f_3, \dots$ is convergent, that is, there exists an $f \in B_\rho^N$ with $\lim_{k \to \infty} f_k = f$.

Proof. (1), (2) and (3) are obvious.

(4) It suffices to prove statement (4) for the case $N = 1$ as we can view $a \in \mathbb{C}\{x_1, \dots, x_n\}$ as an element of $\mathbb{C}\{x_1, \dots, x_{n+1}\}$. We write $h = \sum_{i=0}^\infty h_i$, and $f = \sum_{j=0}^\infty f_j$ for homogeneous polynomials h_i of degree i and homogeneous polynomials f_j of degree j. Then

$$h \cdot f = \sum_{k=0}^{\infty} \sum_{i+j=k} h_i f_j$$

Note that as h_i and f_j are polynomials, the inequality

$$\|\sum_{i+j=k} h_i f_j\|_\rho \leq \sum_{i+j=k} \|h_i\|_\rho \cdot \|f_j\|_\rho$$

follows easily from the third part of 7.1.13. It follows that

$$\begin{aligned}\|h \cdot f\|_\rho &= \sum_{k=0}^{\infty} \| \sum_{i+j=k} h_i f_j\|_\rho \\ &\leq \sum_{k=0}^{\infty} \sum_{i+j=k} \|h_i\|_\rho \cdot \|f_j\|_\rho \\ &= \Big(\sum_{i=0}^{\infty} \|h_i\|_\rho\Big) \cdot \Big(\sum_{j=0}^{\infty} \|f_j\|_\rho\Big) = \|h\|_\rho \cdot \|f\|_\rho .\end{aligned}$$

(5) We write $f_e = \sum_\alpha f_{e,\alpha} x^\alpha$ for all $e \in \mathbb{N}$. From the definition of the norm we get $\rho^\alpha |f_{e,\alpha} - f_{m,\alpha}| \leq \|f_e - f_m\|_\rho$. It follows that for *fixed* α the $f_{e,\alpha}$ is a Cauchy sequence of complex numbers. As $\mathbb{C}$ is complete we can define

$$f^{(\alpha)} := \lim_{e \to \infty} f_{e,\alpha}; \quad f := \sum_\alpha f^{(\alpha)} x^\alpha \in \mathbb{C}[[x]].$$

We claim that in fact $f \in B_\rho^N$ and $\lim_{e\to\infty} f_e = f$. Let $\varepsilon > 0$ be given. Then there exists a $k_0 = k_0(\varepsilon)$ so that for all $k \geq k_0$ and all $e \geq 0$ we have the inequality

$$\sum_\alpha |f_{k+e,\alpha} - f_{k,\alpha}|\rho^\alpha = \|f_{k+e} - f_k\|_\rho \leq \frac{\varepsilon}{2}. \tag{7.2}$$

This is because $f_1, f_2, \ldots$ is a Cauchy sequence. Furthermore, as $\lim_{e\to\infty} f_{e,\alpha} = f^{(\alpha)}$ we have that for all $s \in \mathbb{N}$ there exists an $e_0 = e_0(s)$ such that

$$\sum_{\alpha:|\alpha|\leq s} \left| f^{(\alpha)} - f_{e+k,\alpha} \right| \rho^\alpha < \frac{\varepsilon}{2} \quad \text{for all } k \geq k_0 \text{ and } e \geq e_0. \tag{7.3}$$

Putting (7.2) and (7.3) together we get for all $k \geq k_0$

$$\sum_{\alpha:|\alpha|\leq s} \left| f^{(\alpha)} - f_{k,\alpha} \right| \rho^\alpha \leq \sum_{\alpha:|\alpha|\leq s} \left| f^{(\alpha)} - f_{e+k,\alpha} \right| \rho^\alpha + \sum_{\alpha:|\alpha|\leq s} |f_{e+k,\alpha} - f_{k,\alpha}|\rho^\alpha < \frac{\varepsilon}{2} + \frac{\varepsilon}{2} = \varepsilon.$$

This holds for all s so that

$$\|f - f_k\|_\rho = \sum_\alpha \left| f^{(\alpha)} - f_{k,\alpha} \right| \rho^\alpha < \varepsilon \quad \text{for all } k \geq k_0. \tag{7.4}$$

As $\|f\|_\rho \leq \|f - f_{k_0}\|_\rho + \|f_{k_0}\|_\rho \leq \varepsilon + \|f_{k_0}\|_\rho < \infty$, it follows $f \in B_\rho^N$. From (7.4) it follows that $\lim_{e\to\infty} f_e = f$. Thus B_ρ^N is complete. □

The following lemma is the basic technical lemma for the proof of the Division Theorem 7.1.9.

Lemma 7.1.14. *Let* $f_1, \ldots, f_m \in \mathbb{C}\{x_1, \ldots, x_n\}^N$. *Then for every* $\varepsilon > 0$ *there exists a* $\rho \in \mathbb{R}_+^{n+1}$ *such that*

$$\|\operatorname{Tail}(f_i)\|_\rho < \varepsilon \|L(f_i)\|_\rho, \quad \text{for } i = 1, \ldots, m.$$

Moreover for $0 < \lambda \leq 1$ *define* $\lambda \circ \rho = (\lambda^{w_1}\rho_1, \ldots, \lambda^{w_{n+1}}\rho_{n+1})$. *Then for all* λ *and* i *we have* $\|\operatorname{Tail}(f_i)\|_{\lambda\circ\rho} < \varepsilon \|L(f_i)\|_{\lambda\circ\rho}$.

Proof. We write $x^{\alpha(i)}$ for the leading term of f_i, and we write $f_i = \sum_\alpha f_\alpha^{(i)} x^\alpha$. For simplicity we will assume that the leading coefficient of f_i is equal to one. Moreover we define

$$f_i^{(1)} = \sum_{\substack{\alpha:|\alpha|=|\alpha(i)| \\ \alpha\neq\alpha(i)}} f_\alpha^{(i)} x^\alpha, \quad f_i^{(2)} = \sum_{\alpha:|\alpha|>|\alpha(i)|} f_\alpha^{(i)} x^\alpha.$$

Obviously $\operatorname{Tail}(f_i) = f_i^{(1)} + f_i^{(2)}$.

It is enough to prove the first part of the lemma, because for $\lambda \le 1$

$$\|\operatorname{Tail}(f_i)\|_{\lambda\circ\rho} \le \lambda^{|\alpha(i)|}\|\operatorname{Tail}(f_i)\|_\rho \quad \text{and} \quad \|L(f_i)\|_{\lambda\circ\rho} = \lambda^{|\alpha(i)|}\|L(f_i)\|_\rho.$$

Step 1. We will choose ρ so small that

$$\|f_i^{(1)}\|_\rho < \frac{\varepsilon}{2}\|x^{\alpha(i)}\|_\rho. \tag{7.5}$$

Let

$$M_i := \{\alpha : |\alpha| = |\alpha^{(i)}|,\ \alpha \neq \alpha^{(i)} \text{ and } f_\alpha^{(i)} \neq 0\}, \quad N_i = \#M_i.$$

The inequality (7.5) will follow from

$$\|f_\alpha^{(i)} x^\alpha\|_\rho < \frac{\varepsilon}{2N_i}\|x^{\alpha(i)}\|_\rho \quad \text{for all } \alpha \in M_i. \tag{7.6}$$

We construct $\rho_1, \ldots, \rho_{n+1}$ by descending induction. We start with $\rho = (\frac{1}{2}, \ldots, \frac{1}{2})$. Write $\alpha = (\alpha_1, \ldots, \alpha_{n+1})$ and $\alpha(i) = (\alpha_1(i), \ldots, \alpha_{n+1}(i))$. The inequality $\alpha > \alpha(i)$ means that there exists a $k = k(i,\alpha)$ such that

$$\alpha_1 = \alpha_1(i), \ldots, \alpha_{k-1} = \alpha_{k-1}(i) \text{ and } \alpha_k > \alpha_k(i). \tag{7.7}$$

Suppose that $\rho_\ell, \ldots, \rho_{n+1}$ are already defined such that (7.6) holds for all α, i with $k(i,\alpha) \ge \ell$. Because of (7.7) the formula (7.6) will still hold after modifying $\rho_1, \ldots, \rho_{\ell-1}$. Let $k \in \{1, \ldots, \ell-1\}$ be the maximum of all $k(i,\alpha)$ for all i and $\alpha \in M_i$, such that $k(i,\alpha) \le \ell - 1$. Consider i and α such that $k = k(i,\alpha)$. We do not modify $\rho_{k+1}, \ldots, \rho_{n+1}$. We can choose ρ_k such that

$$\rho_k^{\alpha_k - \alpha_k(i)} < \frac{\varepsilon}{|f_\alpha^{(i)}|2N_i} \cdot \rho_{k+1}^{\alpha_{k+1}(i)-\alpha_{k+1}} \cdot \ldots \cdot \rho_{n+1}^{\alpha_{n+1}(i)-\alpha_{n+1}}. \tag{7.8}$$

This we can do as $\alpha_k > \alpha_k(i)$. So we may assume that (7.6) holds for all i,α with $k = k(i,\alpha)$. This gives the induction step.

Step 2. Note that by definition of the seminorm we have for all λ with $0 < \lambda \le 1$ the equalities $\|x^{\alpha(i)}\|_{\lambda\circ\rho} = \lambda^{|\alpha(i)|}\|x^{\alpha(i)}\|_\rho$, $\|f_i^{(1)}\|_{\lambda\circ\rho} = \lambda^{|\alpha(i)|}\|f_i^{(1)}\|_\rho$ and the inequality $\|f_i^{(2)}\|_{\lambda\circ\rho} \le \lambda^{|\alpha(i)|+1}\|f_i^{(2)}\|_\rho$. Therefore, if we take

$$\lambda < \frac{\varepsilon\|x^{\alpha(i)}\|_\rho}{2\|f_i^{(2)}\|_\rho} \quad \text{for all } i$$

we get

$$\|\operatorname{Tail}(f_i)\|_{\lambda\circ\rho} \le \|f_i^{(1)}\|_{\lambda\circ\rho} + \|f_i^{(2)}\|_{\lambda\circ\rho} \le \lambda^{|\alpha(i)|}\left(\frac{\varepsilon}{2}\|x^{\alpha(i)}\|_\rho + \lambda\|f_2^{(i)}\|_\rho\right) \le \varepsilon\|L(f_i)\|_{\lambda\circ\rho}.$$

Thus the first part of the lemma holds with $\lambda \circ \rho$. □

We are now finally in the position to prove Grauert's Division Theorem.

Proof of the Division Theorem 7.1.9. Choose an ε with $0 < \varepsilon < 1$, and choose $\rho \in \mathbb{R}_+^{n+1}$ such that the statement of the previous Lemma 7.1.14 holds with ε replaced by ε/m. We may choose ρ so small that moreover $f \in B_\rho^N$. Let the leading term of f_i be x^{α_i} for $i = 1, \dots, m$. Without loss of generality we may assume that the leading coefficients of f_i are equal to one for $i = 1, \dots, m$. Let $\Delta = \langle \alpha_1, \dots, \alpha_m \rangle \subset \mathbb{N}^{n+1}$ be the $\mathbb{N}^n$–semimodule generated by $\alpha_1, \dots, \alpha_m$, see Definition 7.1.7. We inductively define a partition of Δ by

$$\Delta_1 = \langle \alpha_1 \rangle$$
$$\Delta_i = \langle \alpha_i \rangle \cap (\Delta \setminus \langle \alpha_1, \dots, \alpha_{i-1} \rangle).$$

We define maps $q_1, \dots, q_m$ from $\mathbb{C}\{x\}^N$ to $\mathbb{C}\{x\}$ and a map r from $\mathbb{C}\{x\}^N$ to $\mathbb{C}\{x\}^N$ by setting for $g = \sum g_\alpha x^\alpha$:

$$r(g) := \sum_{\alpha \notin \Delta} g_\alpha x^\alpha$$
$$q_i(g) = \frac{1}{x^{\alpha_i}} \sum_{\alpha \in \Delta_i} g_\alpha x^\alpha$$

Note that indeed $q_i(g) \in \mathbb{C}\{x\}$. Furthermore the equality

$$g = \sum_{i=1}^m q_i(g) x^{\alpha_i} + r(g)$$

trivially holds. Moreover the estimates

$$\|r(g)\|_\rho \le \|g\|_\rho \quad \text{and} \quad \|q_i(g)\|_\rho \le \frac{\|g\|_\rho}{\rho^{\alpha_i}} \tag{7.9}$$

hold trivially. We define a sequence of elements $v_0, v_1, v_2, \dots$ in B_ρ^N by

$$v_0 := f$$
$$v_{j+1} := v_j - \sum_{i=1}^m q_i(v_j) f_i - r(v_j) = -\sum_{i=1}^m q_i(v_j) \operatorname{Tail}(f_i).$$

Note that $L(v_{j+1}) > L(v_j)$. The idea of the proof is to define $v = \sum_{j=0}^{\infty} v_j$ and prove that

(1) $f = \sum_{i=1}^m q_i(v) f_i + r(v)$.

(2) $v \in B_\rho^N$.

The first one is easy. As by definition $L(v_{j+1}) > L(v_j)$ the equalities $q_i(v) = \sum_{j=0}^\infty q_i(v_j)$ for all i and $r(v) = \sum_{j=0}^\infty r(v_j)$ hold formally. Now

$$\begin{aligned} f &= \sum_{j=0}^{\infty}(v_j - v_{j+1}) \\ &= \sum_{j=0}^{\infty}(v_j - v_j + \sum_{\ell=1}^{m} q_\ell(v_j) f_\ell + r(v_j)) \\ &= \sum_{\ell=1}^{m}\sum_{j=0}^{\infty} q_\ell(v_j) f_\ell + \sum_{j=0}^{\infty} r(v_j) \\ &= \sum_{\ell=1}^{m} q_\ell(v) f_\ell + r(v). \end{aligned}$$

To show the convergence of v, note that because of 7.1.14 and the estimates (7.9)

$$||v_{j+1}||_\rho = || - \sum_{i=1}^{m} q_i(v_j)\,\mathrm{Tail}(f_i)||_\rho \le \sum_{i=1}^{m} \frac{1}{\rho^{\alpha_i}} ||v_j||_\rho \frac{\varepsilon}{m} \rho^{\alpha_i} = \varepsilon ||v_j||_\rho.$$

This implies that $||v_{j+1}||_\rho \le \varepsilon^{j+1}||f||_\rho$. It now easily follows that $\sum_{j=0}^{k} v_j$ is a Cauchy sequence in B_ρ^N. As B_ρ^N is complete it follows that $v = \sum_{j=0}^{\infty} v_j$ is a well defined element of B_ρ^N. In particular we obtain

$$||v||_\rho \le \frac{1}{1-\varepsilon}||f||_\rho. \tag{7.10}$$

It remains to prove $L(q_i(v)f_i) \ge L(f)$. As $L(v_{j+1}) > L(v_j)$

$$L(q_i(v)f_i) = L(\sum_{j=0}^{\infty} q_i(v_j)f_i) \ge L(q_i(v_0)f_i) = L(q_i(f)f_i) \ge L(f).$$

The last inequality follows from the definition of $q_i(f)$. This proves the theorem. □

Remarks 7.1.15.
(1) Note that we in fact proved something more. Namely, suppose that $0 < \varepsilon < 1$ and $\rho \in \mathbb{R}_+^{n+1}$ are such that Lemma 7.1.14 holds. If f is an element of B_ρ^N, then in the representation $f = \sum_{i=1}^{m} q_i f_i + r$ also $q_i \in B_\rho$ and $r \in B_\rho^N$. In fact we proved the following, see formulas (7.9) and (7.10)

$$\begin{aligned} ||r||_\rho &\le \frac{1}{1-\varepsilon}||f||_\rho \\ ||q_j||_\rho &\le \frac{1}{1-\varepsilon}\frac{||f||_\rho}{||L(f_j)||_\rho} \quad \text{for } j = 1,\ldots,m. \end{aligned}$$

Analyzing the proof of Grauert's Division Theorem carefully, we obtain a more detailed statement, which will be of use later on. Let $s \ge 0$. Suppose moreover that $f_1,\ldots,f_m \in \mathbb{C}\{x_{s+1},\ldots,x_n\}^N$, and let $f \in B_{\overline{\rho}}^N$ for some $\overline{\rho} \in \mathbb{R}_+^{n+1}$. Then for the ρ constructed in the proof of the Division Theorem we can obtain $\rho \le \overline{\rho}$ and $\rho_i = \overline{\rho}_i$ for $i = 1,\ldots,s$. Furthermore, using the estimates of the second part of Lemma 7.1.14, we obtain for all $0 < \lambda < 1$

$$\|r\|_{\lambda\circ\rho} \leq \frac{1}{1-\varepsilon}\|f\|_{\lambda\circ\rho},$$

$$\|q_j\|_{\lambda\circ\rho} \leq \frac{1}{1-\varepsilon}\frac{\|f\|_{\lambda\circ\rho}}{\|L(f_j)\|_{\lambda\circ\rho}} \quad \text{for } j = 1,\dots,m,$$

with $\lambda\circ\rho = (\rho_1,\dots,\rho_s,\lambda^{w_{s+1}}\rho_{s+1},\dots,\lambda^{w_{n+1}}\rho_{n+1})$.

(2) Note that the proof of Grauert's Division Theorem is constructive, in particular we construct the power series r. This construction of r depends on only on the ordering of the $f_1,\dots,f_m$, and of course on the monomial ordering.

Exercises

7.1.16. Prove that a monomial ordering is a well-ordering if and only if 1 is the smallest monomial.

7.1.17 (Examples for module orderings). Let $<$ be a monomial ordering on $\mathbb{C}\{x\}$, $\mathbb{C}\{x\}^N = \sum_{i=0}^{N-1}\mathbb{C}\{x\}x_{n+1}^k$ be the identification as in Definition 7.1.5 and $M = \{x^\alpha | \alpha\in\mathbb{Z}_+^{n+1}, 0\leq\alpha_{n+1}\leq N-1\}$.

Let $w = (w_1,\dots,w_{n+1})\in\mathbb{R}^{n+1}$, $w_{n+1}\neq 0$. For x^α and $x^\beta\in M$ we define $x^\alpha <_m x^\beta$ if and only if

- $|\alpha|_w = \sum_{i=1}^{n+1}\alpha_i w_i < \sum_{i=1}^{n+1}\beta_i w_i = |\beta|_w$ or
- $|\alpha|_w = |\beta|_w$ and $x_1^{\alpha_1}\cdots x_n^{\alpha_n} < x_1^{\beta_1}\cdots x_n^{\beta_n}$.

Prove that $<_m$ is a module ordering. We note that the following module orderings are often used:

(1) degree orderings $w_1 > 0,\dots,w_{n+1} > 0$;
(2) elimination orderings $w_1 = \dots = w_n = 0$.

7.1.18. For x^α and $x^\beta\in M$ we define $x^\alpha <_m x^\beta$ if and only if

- $x_1^{\alpha_1}\cdots x_n^{\alpha_n} < x_1^{\beta_1}\cdots x_n^{\beta_n}$ or
- if $(\alpha_1,\dots,\alpha_n) = (\beta_1,\dots,\beta_n)$ then $\alpha_{n+1} < \beta_{n+1}$.

Prove that $<_m$ is a module ordering.

7.1.19. Formulate and prove Grauert's Division Theorem for the formal power series ring $\mathbb{C}[[x]]$.

7.2 Characterizations and Properties of Standard Bases

Definition 7.2.1. Let $S = \{f_1,\dots,f_m\}\subset\mathbb{C}\{x\}^N$ be an ordered set. For $f\in\mathbb{C}\{x\}^N$ we can, by Grauert's Division Theorem 7.1.9, write $f = \sum q_i f_i + r$, such that no monomial of r is divisible by any of the $L(f_i)$. The r constructed in the proof of Grauert's Division Theorem is uniquely determined with respect to the ordered set S, see 7.1.15. This r is called the *normal form* of f with respect to the (ordered) set S. We will write $r = \mathrm{NF}(f|S)$, and if S is fixed, just $\mathrm{NF}(f)$.

Remark 7.2.2. In general, the normal form depends on the ordering of S. For example, $f = x^2y + xy^2 + y^2$, $f_1 = y^2 - 1$, $f_2 = xy$, $S_1 = \{f_1,f_2\}$ and $S_2 = \{f_2,f_1\}$. Then for the degree lexicographic ordering with $w = (1,1)$ we get

$$\begin{aligned} v &= f + x \\ q_1(v) &= x + 1 \\ q_2(v) &= x \\ r(v) = \mathrm{NF}(f|S_1) &= x + 1 \end{aligned} \qquad \begin{aligned} v &= f + 1 \\ q_1(v) &= x + y \\ q_2(v) &= 1 \\ r(v) = \mathrm{NF}(f|S_2) &= 1. \end{aligned}$$

We will show that if S is a *standard basis* however, then the normal form does not depend on the ordering. First the definition.

Definition 7.2.3. Let $I \subset \mathbb{C}\{x\}^N$ be a submodule. Then $S = \{f_1,\ldots,f_m\} \subset I$ is called a *standard basis* of I if $L(I) = (L(f_1),\ldots,L(f_m))$. A standard basis is called *reduced* if $C(f_i) = 1$ for all i and $L(f_i)$ does not divide any monomial occurring in f_j for $i \neq j$ and occurring in $\mathrm{Tail}(f_i)$.

Example 7.2.4. Let $I = (x^3+y^2,y)$, $w = (1,2)$. Then $\{x^3+y^2,y,y^5\}$ is a standard basis of I in $\mathbb{C}\{x\}$ but not a reduced standard basis of I. The standard basis $\{x^3,y\}$ is reduced.

Remark 7.2.5. Standard bases were introduced by Hironaka (cf. [Hironaka 1964]), Grauert (cf. [Grauert 1972]) for the power series case and, for polynomial rings, by Buchberger (cf. [Buchberger 1965]). In case of polynomial rings, they are also called Gröbner bases. The idea of Gröbner bases is already contained in the work of Gordan, see [Gordan 1899].

We now give some basic results.

Proposition 7.2.6. *Let $I \subset \mathbb{C}\{x\}^N$ be a submodule. Then I has a standard basis.*

Proof. The module $L(I)$ is, by definition, the submodule in $\mathbb{C}\{x\}^N$ generated by the leading monomials of the elements of I. The ring $\mathbb{C}\{x\}$ is Noetherian and, therefore $L(I)$ is generated by finitely many such leading monomials, say $x^{\alpha_1},\ldots,x^{\alpha_m}$. Choose $f_1,\ldots,f_m \in I$ such that $L(f_i) = x^{\alpha_i}$. Then $\{f_1,\ldots,f_m\}$ is a standard basis of I. □

Corollary 7.2.7. *Let $S = \{f_1,\ldots,f_m\}$ be a standard basis of I. Then $I = (f_1,\ldots,f_m)$.*

Proof. For $f \in I$ we have $f = \sum q_i f_i + \mathrm{NF}(f|S)$. This implies that $\mathrm{NF}(f|S) \in I$. If $\mathrm{NF}(f|S) \neq 0$, then $L(\mathrm{NF}(f|S)) \in L(I) = (L(f_1),\ldots,L(f_m))$, which is a contradiction to the definition of $\mathrm{NF}(f|S)$. □

Corollary 7.2.8. *Let $S = \{f_1,\ldots,f_m\}$ and $T = \{g_1,\ldots,g_s\}$ be standard bases of I. Then for all $f \in \mathbb{C}\{x\}^N$ we have* $\mathrm{NF}(f|S) = \mathrm{NF}(f|T)$.

Proof. $f = \sum q_\nu f_\nu + \mathrm{NF}(f|S) = \sum \overline{q_\nu} g_\nu + \mathrm{NF}(f|T)$ implies $\mathrm{NF}(f|S) - \mathrm{NF}(f|T) \in I$. If $\mathrm{NF}(f|S) \neq \mathrm{NF}(f|T)$, then $L(\mathrm{NF}(f|S) - \mathrm{NF}(f|T)) \in L(I)$ is a monomial which appears in $\mathrm{NF}(f|S)$ or in $\mathrm{NF}(f|T)$ which is impossible by definition of NF. □

This result makes the following definition possible.

Definition 7.2.9. Let $I \subset \mathbb{C}\{x\}^N$ be a submodule and S a standard basis of I. Then for any $f \in \mathbb{C}\{x\}^N$ we define $\mathrm{NF}(f|I) = \mathrm{NF}(f|S)$ to be the normal form of f with respect to I.

Corollary 7.2.10. *Let I be a submodule of $\mathbb{C}\{x\}^N$, and $S = \{f_1,\ldots,f_m\}$ be a standard basis of I. Let $f \in \mathbb{C}\{x\}^N$. Then*

(1) $f \in I$ *if and only if* $\mathrm{NF}(f|S) = 0$,

(2) $f \in I$ *if and only if* $\mathrm{NF}(f|I) = 0$,

(3) $f - NF(f|I) \in I$.

Proof. We know that $f = \sum q_\nu f_\nu + \mathrm{NF}(f|S)$ and $I = (f_1, \ldots, f_m)$. Therefore, $\mathrm{NF}(f|S) = 0$ implies $f \in I$. On the other hand, $f \in I$ implies $\mathrm{NF}(f|S) \in I$. If $\mathrm{NF}(f|S) \neq 0$ then $L(\mathrm{NF}(f|S)) \in L(I)$ which is not possible by the definition of NF. □

Corollary 7.2.11. *Let $I \subset \mathbb{C}\{x\}^N$ be a submodule. Then I has a reduced standard basis $S = \{f_1, \ldots, f_m\}$, with the property $L(f_1) < \cdots < L(f_m)$. Such a standard basis is uniquely determined.*

Proof.
Step 1. We first prove the existence. Let $T = \{g_1, \ldots, g_m\}$ be a standard basis of I, and NF the normal form with respect to T. After reordering, and leaving out some of the g_i, we may assume $L(g_1) < \cdots < L(g_m)$. Suppose that $L(g_i)|L(g_j)$ for some $i \neq j$. Then $\{g_1, \ldots, g_{j-1}, g_{j+1}, \ldots, g_m\}$ is also a standard basis for I. Thus we may assume that $L(g_i) \nmid L(g_j)$ for $i \neq j$. Furthermore, dividing g_i by $C(g_i)$ we may assume that $C(g_i) = 1$ for all i. We define

$$f_i := L(g_i) + \mathrm{NF}(\mathrm{Tail}(g_i)|I), \quad i = 1, \ldots, m.$$

We claim that $f_1, \ldots, f_m$ is a reduced standard basis of I. Because $g_i = L(g_i) + \mathrm{Tail}(g_i) = L(g_i) + \sum q_{i\nu} g_\nu + \mathrm{NF}(\mathrm{Tail}(g_i)|T)$, the f_i are in I. Because $L(g_i) = L(f_i)$ we obtain that $S := \{f_1, \ldots, f_m\}$ is a standard basis of I. Furthermore, $L(f_i) \nmid L(f_j)$ for $i \neq j$, and by definition of normal form, no term of $\mathrm{Tail}(f_i)$ is divisible by any of the $L(g_1) = L(f_1), \ldots, L(g_m) = L(f_m)$. Thus $f_1, \ldots, f_m$ is a reduced standard basis.

Step 2. We now show uniqueness. Let $S = \{f_1, \ldots, f_m\}$ and $T = \{g_1, \ldots, g_s\}$ be two reduced standard bases and $L(f_1) < \cdots < L(f_m)$, $L(g_1) < \cdots < L(g_s)$.

We first prove that $s = m$ and $L(f_i) = L(g_i)$. Suppose the converse. Now $L(I) = (L(f_1), \ldots, L(f_m))$ and $L(f_i) \nmid L(f_j)$ for $i \neq j$ (S is a reduced standard basis). Similarly $L(I) = (L(g_1), \ldots, L(g_s))$, $L(g_i) \nmid L(g_j)$ for $i \neq j$. Assume that for some $i \geq 1$ $L(g_j) = L(f_j)$ for $j < i$ and $L(g_i) \neq L(f_i)$. By symmetry we may assume $L(g_i) < L(f_i)$. As they are both standard basis, it follows that $(L(g_1), \ldots, L(g_s)) = (L(f_1), \ldots, L(f_m))$. Hence $L(f_\nu) \mid L(g_i)$ for some ν. There are two cases.

(1) If $\nu \geq i$, then $L(f_\nu) \geq L(f_i)$. Furthermore, because of $L(g_i) < L(f_i)$ we get $L(g_i) < L(f_\nu)$. Thus it is not possible that $L(f_\nu) \mid L(g_i)$, contradiction.

(2) If $\nu < i$, we have, by choice of i, $L(f_\nu) = L(g_\nu)$. But as we have a *reduced* standard basis $L(g_\nu) \nmid L(g_i)$ for $\nu \neq i$. This gives a contradiction because we had $L(f_\nu) \mid L(g_i)$.

So we proved that $s = m$ and $L(g_i) = L(f_i)$ for all i. Now consider $f_i - g_i = \mathrm{Tail}(f_i) - \mathrm{Tail}(g_i) \in I$. If $\mathrm{Tail}(f_i) \neq \mathrm{Tail}(g_i)$ then $L(\mathrm{Tail}(f_i) - \mathrm{Tail}(g_i))$ is a monomial in $L(I)$ and a monomial appearing in $\mathrm{Tail}(f_i)$ or $\mathrm{Tail}(g_i)$, which is a contradiction to the property of both standard bases being reduced. □

We will now give characterization of standard bases which is the theoretical background for Buchberger's algorithm to compute standard bases.

Definition 7.2.12. Let $f, g \in \mathbb{C}\{x\}^N$, with $L(f) = x^\beta$ and $L(g) = x^\gamma$.

(1) We define the least common multiple $\mathrm{lcm}(L(f), L(g))$ to be the monomial x^α with $\alpha = (\max(\beta_1, \gamma_1), \ldots, \max(\beta_n, \gamma_n), \beta_{n+1})$ if $\beta_{n+1} = \gamma_{n+1}$ and to be 0 if $\beta_{n+1} \neq \gamma_{n+1}$.

(2) Let $m = \text{lcm}(L(f),L(g)$. Then we define the *syzygy polynomial* of f and g by

$$\text{spoly}(f,g) = \frac{m}{L(f)} \cdot f - \frac{C(f)}{C(g)} \cdot \frac{m}{L(g)} \cdot g.$$

(Note that in case $m \neq 0$ we have $L(\text{spoly}(f,g)) > m$.)

Examples 7.2.13.

(1) Let $f = 2xy^2 + x^4$, $g = 5x^2y + y^4 \in \mathbb{C}\{x,y\}$ and $w = (1,1)$, then

$$\text{spoly}(f,g) = x^5 - \tfrac{2}{5}y^5.$$

(2) Consider the free $\mathbb{C}\{x_1,x_2\}$–module $\mathbb{C}\{x_1,x_2\}^2 = \mathbb{C}\{x_1,x_2\} \oplus \mathbb{C}\{x_1,x_2\}x_3$, and $w = (1,1,1)$ Let $f = (x_1^2+1,x_2) = 1 + x_1^2 + x_2x_3$ and $g = (x_2,x_1^3) = x_2 + x_1^3x_3$. Then $L(f) = 1$, and $L(g) = x_2$, so that

$$\text{spoly}(f,g) = x_2f - g = (x_1^2x_2,x_2^2 - x_1^3) = x_1^2x_2 + x_2^2x_3 - x_1^3x_3.$$

Definition 7.2.14. Let $f_1,\ldots,f_m \in \mathbb{C}\{x\}^N$. Write $\mathbb{C}[x]^m = \sum_{i=1}^m \mathbb{C}[x] \cdot e_i$, $e_i = (0,\ldots,0,1,0,\ldots,0)$.

(1) A *syzygy on the leading terms* $C(f_1)L(f_1),\ldots,C(f_m)L(f_m)$ is an element $s = \sum_{i=1}^m s_ie_i = (s_1,\ldots,s_m) \in \mathbb{C}[x]^m$ such that

$$s(f_1,\ldots,f_m) := \sum_{i=1}^m s_iC(f_i)L(f_i) = 0.$$

(2) Let $\text{syz}(f_1,\ldots,f_m)$ be the submodule of $\mathbb{C}[x]^m$ of all syzygies of the leading terms of $f_1,\ldots,f_m$.

(3) Write $x^{\alpha(i,j)} = \text{lcm}(L(f_i),L(f_j))$. Then

$$s^{(i,j)} := \frac{x^{\alpha(i,j)}}{L(f_i)}e_i - \frac{C(f_i)}{C(f_j)}\frac{x^{\alpha(i,j)}}{L(f_j)}e_j \in \text{syz}(f_1,\ldots,f_m).$$

(Note that $s^{(i,j)} \cdot (f_1,\ldots,f_m) = \text{spoly}(f_i,f_j)$, where the dot denotes the dot product.)

Lemma 7.2.15. *Consider $f_1,\ldots,f_m \in \mathbb{C}\{x\}^N$, and keep the notations of the previous definition.*

(1) The $\mathbb{C}[x]$–module $\text{syz}(f_1,\ldots,f_m)$ admits a $\mathbb{Z}^{n+1}$–grading, that is, we can write $\text{syz}(f_1,\ldots,f_m) = \oplus_{\nu\in\mathbb{N}^{n+1}}S_\nu$ with

$$S_\nu := \{\sum_{i=1}^m c_ix^{\alpha(i)}e_i \in \text{syz}(f_1,\ldots,f_m) : c_i \in \mathbb{C},\ x^{\alpha(i)}L(f_i) = x^\nu\}.$$

(2) $s^{(i,j)} \in S_{\alpha(i,j)}$.

(3) The set $\{s^{(i,j)}\}$ *generates* $\mathrm{syz}(f_1,\dots,f_m)$.

Proof.
(1) Let $s = \sum_{i=1}^m s_i e_i \in \mathrm{syz}(f_1,\dots,f_m)$ and write $s_i = \sum_{\alpha\in\mathbb{N}^n} c_{i\alpha}x^\alpha \in \mathbb{C}[x]$. For $\nu \in \mathbb{N}^{n+1}$ put

$$s_{i\nu} = \begin{cases} c_{i\beta}x^\beta & \text{if } x^\beta L(f_i) = x^\nu \\ 0 & \text{if } L(f_i) \nmid x^\nu. \end{cases}$$

Define $s_\nu := \sum_{i=1}^m s_{i\nu}e_i$. Then obviously $s_\nu \in S_\nu$ and $s = \sum_\nu s_\nu$.

(2) This is clear.

(3) To prove the third statement, we may assume that $C(f_i) = 1$ for all i. Let $L(f_i) = x^{\alpha_i}$. Because of (1) it is enough to prove the statement for homogeneous $s \in S_\gamma$. We thus write $s = \sum_{i=1}^m c_i x^{\beta_i}$, $c_i \in \mathbb{C}$ and $\beta_i + \alpha_i = \gamma$ for all i with $c_i \neq 0$. Let $\{i_1,\dots,i_k\}$ be the indices such that $c_{i_\nu} \neq 0$. We will prove by induction on k that s is in the module generated by the $s^{(i,j)}$. This is obvious for $k = 0$, as then $s = 0$. If $k \neq 0$, then $k \geq 2$. Note that as $\alpha_{i_j} + \beta_{i_j} = \gamma$, it follows in particular that $x^\gamma \geq \mathrm{lcm}(L(f_{i_1}),L(f_{i_2})) =: x^\alpha$. Thus we can consider

$$s' := s - c_1\frac{x^\gamma}{x^\alpha}s^{(i_1,i_2)}.$$

We have $s' \in \mathrm{syz}(f_1,\dots,f_m)$, and $s' \in \oplus_{\nu=2}^k\mathbb{C}[x]e_{i_\nu}$. By induction, this proves the lemma. □

Theorem 7.2.16 (Buchberger's Criterion). *Let* $I = (f_1,\dots,f_m) \subset \mathbb{C}\{x\}^N$ *be a submodule. The following conditions are equivalent:*

(1) $S = \{f_1,\dots,f_m\}$ *is a standard basis for* I.

(2) $\mathrm{NF}(\mathrm{spoly}(f_i,f_j)|S) = 0$ *for all* i,j.

(3) Let G *be a subset of* $\{s^{(i,j)}\}$ *generating* $\mathrm{syz}(f_1,\dots,f_m)$. *Then* $\mathrm{NF}(\mathrm{spoly}(f_i,f_j)|S) = 0$ *for all* i,j *with* $s^{(i,j)} \in G$.

Proof. The implication (1) $\Longrightarrow$ (2) follows from Corollary 7.2.10, and (2) $\Longrightarrow$ (3) is obvious.

It remains to show (3) $\Longrightarrow$ (1). Without loss of generality, we may assume that $C(f_i) = 1$ for all i. Let $L(f_i) = x^{\alpha_i}$ for $i = 1,\dots,m$. Consider $f \in I$. We have to prove that $L(f)$ is divisible by $L(f_i)$ for some i. As $f \in I$ we can write $f = \sum_{i=1}^m g_i f_i$. To such a representation we associate the monomial $x^\delta = \min\{L(g_i f_i)\}$. Assume that the representation is chosen with maximal x^δ between all possible representations. We will prove that $L(f) = x^\delta$. This would imply $L(f) = L(g_i f_i) = L(g_i)\cdot L(f_i)$ for some i, and thus the theorem. Suppose the converse, that is $L(f) > x^\delta$. We define

$$I_1 := \{i : L(g_i f_i) = x^\delta\}, \quad I_2 := \{i : L(g_i f_i) \neq x^\delta\}.$$

Then we can write

$$f = \sum_{i\in I_1} C(g_i)L(g_i)f_i + \sum_{i\in I_1}\mathrm{Tail}(g_i)f_i + \sum_{i\in I_2} g_i f_i.$$

The assumption $L(f) > x^\delta$ implies $\sum_{i\in I_1} C(g_i)L(g_i)L(f_i) = 0$. Thus we have a homogeneous syzygy on the leading terms of the f_i, in fact in $S_\delta \subset \operatorname{syz}(f_1,\ldots,f_m)$. By assumption on G and Lemma 7.2.15 this syzygy on the leading terms is of type $\sum_{\substack{i,j\in I_1\\ s^{(i,j)}\in G}} c_{ij}x^{\alpha_{ij}}s^{(i,j)} \in S_\delta$ for suitable $c_{ij} \in \mathbb{C}$, and $\alpha_{ij} \in \mathbb{N}^n$. Taking the dot product with $(f_1,\ldots,f_m)$ we obtain

$$\sum_{i\in I_1} C(g_i)L(g_i)f_i = \sum_{\substack{i,j\in I_1\\ s^{(i,j)}\in G}} c_{ij}x^{\alpha_{ij}}\operatorname{spoly}(f_i,f_j).$$

Now consider a nonzero term of the right hand side. Then there exist $i,j \in I_1$, with $s^{(i,j)} \in G$ and $c_{ij}\operatorname{spoly}(f_i,f_j) \neq 0$. As the syzygy on the leading terms is in S_δ we have $x^{\alpha_{ij}}\operatorname{lcm}(L(f_i),L(f_j)) = x^\delta$. It was noted in the second part of Definition 7.2.12 that $L(x^{\alpha_{ij}}\operatorname{spoly}(f_i,f_j)) > x^\delta$.

Now the third assumption implies $\operatorname{NF}(\operatorname{spoly}(f_i,f_j)|S) = 0$ for $s^{(i,j)} \in G$ and, therefore, $\operatorname{spoly}(f_i,f_j) = \sum_{\nu=1}^m q_\nu^{ij}f_\nu$ and $L(q_\nu^{ij}f_\nu) \geq L(\operatorname{spoly}(f_i,f_j))$. Thus in particular we get $x^{\alpha_{ij}}L(q_\nu^{ij}f_\nu) > x^\delta$. We get

$$\begin{aligned}\sum_{i\in I_1} C(g_i)L(g_i)f_i &= \sum_{\substack{i,j\in I_1\\ s^{(i,j)}\in G}} c_{ij}x^{\alpha_{ij}}\sum_{\nu=1}^m q_\nu^{ij}f_\nu\\ &= \sum_{\nu=1}^m\Bigg(\sum_{\substack{i,j\in I_1\\ s^{(i,j)}\in G}} c_{ij}x^{\alpha_{ij}}q_\nu^{ij}\Bigg)f_\nu.\end{aligned}$$

Now we define $\overline{g}_\nu$ by

$$\overline{g}_\nu = \sum_{\substack{i,j\in I_1\\ s^{(i,j)}\in G}} c_{ij}x^{\alpha_{ij}}q_\nu^{ij} + \begin{cases} g_\nu, & \text{for } \nu \in I_2\\ \operatorname{Tail}(g_\nu), & \text{for } \nu\in I_1.\end{cases}$$

Then $f = \sum_{\nu=1}^m \overline{g}_\nu f_\nu$ and $L(\overline{g}_\nu f_\nu) > x^\delta$ for all ν. This is a contradiction to the choice of δ and proves the theorem. □

Example 7.2.17. Let $f_1 = x^4 + x^3y^2$, $f_2 = y^4 + x^2y^3$, $w = (1,1)$. Then $\{f_1,f_2\}$ is a standard basis:

$$\operatorname{spoly}(f_1,f_2) = x^3y^6 - x^6y^3 = -x^2y^3f_1 + x^3y^2f_2.$$

The following lemma gives the possibility to find proper subsets of $\{s^{(i,j)}\}$ which generate $\operatorname{syz}(f_1,\ldots,f_m)$. This is very useful for the application of 7.2.16.

Lemma 7.2.18 (Chain Criterion). *Let $f_1,\ldots,f_m \in \mathbb{C}\{x\}^N$ and assume that a subset G of $\{s^{(i,j)}\}$ generates $\operatorname{syz}(f_1,\ldots,f_m)$. Suppose $s^{(i,j)}$, $s^{(i,k)}$ and $s^{(k,j)}$ are in G and $L(f_i)|\operatorname{lcm}(L(f_j),L(f_k))$. Then $G \setminus \{s^{(j,k)}\}$ generates $\operatorname{syz}(f_1,\ldots,f_m)$.*

Proof. We may assume that $C(f_i) = 1$ for all i, then one computes that

$$s^{(k,j)} = \frac{\operatorname{lcm}(L(f_j),L(f_k))}{\operatorname{lcm}(L(f_i),L(f_j))}s^{(i,j)} - \frac{\operatorname{lcm}(L(f_j),L(f_k))}{\operatorname{lcm}(L(f_i),L(f_k))}s^{(i,k)}.$$

Thus the lemma follows from 7.2.16. □

Exercises

7.2.19 (Product Criterion). Let $f_1, f_2 \in \mathbb{C}\{x\}$, and assume that $\mathrm{lcm}(L(f_1), L(f_2)) = L(f_1) \cdot L(f_2)$. Prove that $\mathrm{NF}(\mathrm{spoly}(f_1,f_2)|\{f_1,f_2\}) = 0$.

7.2.20. Let $<_m$ be an elimination ordering (see 7.1.17 (2)) with $w_1 = \cdots = w_n = 0$, $w_{n+1} = 1$. Let $I \subset \mathbb{C}\{x\}^N$ be a submodule and S a standard basis of I.
Prove that $S \cap \sum_{i=0}^{t} \mathbb{C}\{x\}x_{n+1}^i$ is a standard basis of $I \cap \sum_{i=0}^{t} \mathbb{C}\{x\}x_{n+1}^i$ for all $t \leq N-1$.

7.2.21 (Gröbner bases). In this exercise we will develop the theory of standard bases for polynomial rings rewriting some passages of the Sections 7.1 and 7.2.

- Replace $\mathbb{C}\{x\}$ by $\mathbb{C}[x]$.
- Replace in Definition 7.1.3 $x^{\alpha_i} < x^{\alpha_{i+1}}$ by $x^{\alpha_i} > x^{\alpha_{i+1}}$.
- Rewrite the examples.
- Replace (2) by $L(q_\nu f_\nu) \leq L(f)$, for $\nu = 1, \ldots, m$ in Theorem 7.1.9, and cancel Remark 7.1.15.
- Replace the proof of Theorem 7.1.9 by the following NF–algorithm (normal form algorithm):

  $r = \mathrm{NF}(f|\{f_1, \ldots, f_m\})$
      $r = f$
      $S = \{f_1, \ldots, f_m\}$
      while $r \neq 0$ and $L(f_i) \mid L(r)$ for some i
          choose i minimal with $L(f_i) \mid L(r)$
          $r = r - \frac{C(r)}{C(f_i)} \frac{L(r)}{L(f_i)} f_i$
      if $r \neq 0$
          $r = C(r)L(r) + \mathrm{NF}(\mathrm{Tail}(r)|S)$
      return(r).

  Prove that this algorithm terminates. Use the fact that $<$ is a well-ordering and Definition 7.1.3. Prove that $f = \sum q_\nu f_\nu + r$, $r = \mathrm{NF}(f|\{f_1, \ldots, f_m\})$ for suitable $q_\nu \in \mathbb{C}[x]$ and (1) and (2) of Theorem 7.1.9 are true.
- Check the proofs of 7.2.7, 7.2.8, 7.2.11 and 7.2.10, 7.2.15, and 7.2.16 (here we have to change the definition of $x^\delta = \max\{L(g_i f_i)\}$ to $\min\{L(g_i f_i)\}$ and replace several times $>$ by $<$), and 7.2.18.

7.2.22 (Buchberger's algorithm).
$S = \mathrm{standard}(\{f_1, \ldots, f_m\})$
    $S = \{f_1, \ldots, f_m\}$
    $P = \{(f_i, f_j)\}_{i<j}$
    while $P \neq \emptyset$
    choose $(f,g) \in P$
    $P = P \setminus \{(f,g)\}$
    $h = \mathrm{NF}(\mathrm{spoly}(f,g)|S)$
    if $h \neq 0$
        $P = P \cup \{(f,h)|h \in S\}$
        $S = S \cup \{h\}$
    return(S).

- Prove that the algorithm terminates.
- Use 7.2.16 to prove the S is a standard basis of $(f_1, \ldots, f_m)$.

7.2.23 (Standard bases for formal power series). Standard bases for formal power series can be defined as in 7.2.3.

- Prove that the theory developed in this section holds in the formal case.

- Prove that $\{f_1,\dots,f_m\} \subset \mathbb{C}\{x\}^N$ is a standard basis if and only if $\{f_1,\dots,f_m\} \subset \mathbb{C}[[x]]^N$ is a standard basis.
- Let $\{f_1,\dots,f_m\} \subset \mathbb{C}\{x\}^N$ be a standard basis and $f \in \mathbb{C}\{x\}^N$.
Prove that $\mathrm{NF}(f|\{f_1,\dots,f_m\})$ is the limit of the following sequence:
    (1) $v_0 = f$.
    (2) Suppose v_i is defined. Then $v_{i+1} := v_i - hf_{\pi(i)}$. Here $h = 0$ and $\pi(i) = 1$ if no monomial in v_i is divisible by $L(f_j)$ for all j. Otherwise, let cx^α be the smallest term in v_i such that $L(f_j) \mid x^\alpha$ for some j. Let $\pi(i)$ be minimal with $f_{\pi(i)} \mid x^\alpha$. Then we define $h := c\frac{x^\alpha}{L(f_{\pi(i)})}$.

7.2.24 (Computation of standard bases in the power series ring). Let $<$ be a well ordering on the set of monomials of $\mathbb{C}[x]$. We choose a new variable t and extend this ordering in the following way to $\mathbb{C}[x,t]$:

$$x^\alpha t^a > x^\beta t^b$$

if $\deg(x^\alpha) + a > \deg(x^\beta) + b$ or in case of equality $x^\alpha < x^\beta$. This extension has the following property:
Let $f \in \mathbb{C}[x]$, $f = \sum_{i\geq 1} a_i x^{\alpha_i}$, $a_i \neq 0$ and $x^{\alpha_i} < x^{\alpha_{i+1}}$, that is, the monomials in f are ordered in the sense of a power series ring (cf. Definition 7.1.3). Let $F = f^h = t^{\deg(f)} f\left(\frac{x_1}{t},\dots,\frac{x_n}{t}\right)$ be the homogenization of f, then $F = \sum_{i\geq 1} a_i x^{\alpha_i} t^{\deg(f)-\deg(x^{\alpha_i})}$ and we have $x^{\alpha_i} t^{\deg(f)-\deg(x^{\alpha_i})} > x^{\alpha_{i+1}} t^{\deg(f)-\deg(x^{\alpha_{i+1}})}$, that is, the monomials of F are ordered in the sense of a polynomial ring (cf. Exercise 7.2.21).

Prove the following: *Let $f_1,\dots,f_m \in \mathbb{C}[x]$ and $I = (f_1,\dots,f_m)\mathbb{C}\{x\}$, respectively $I = (f_1,\dots,f_m)\mathbb{C}[[x]]$. Let $<$ be a well-ordering on $\mathbb{C}[x]$ extended as before to $\mathbb{C}[x,t]$. Let $J = (f_1^h,\dots,f_m^h)\mathbb{C}[x,t]$ and $\{G_1,\dots,G_r\}$ be a standard basis of J. Then $\{G_1(t=1),\dots,G_r(t=1)\}$ is standard basis of I.*[3]

7.3 Applications

Proposition 7.3.1. *Let $I \subset \mathbb{C}\{x\}$ such that* $\dim_{\mathbb{C}} \mathbb{C}\{x\}/I < \infty$. *Then* $\dim_{\mathbb{C}} \mathbb{C}\{x\}/I = \dim_{\mathbb{C}} \mathbb{C}\{x\}/L(I)$.

Proof. It is clear that $\{x^\alpha | x^\alpha \notin L(I)\}$ induces a $\mathbb{C}$-basis of $\mathbb{C}\{x\}/L(I)$. We will prove that it also induces a $\mathbb{C}$-basis of $\mathbb{C}\{x\}/I$. Assume $\sum c_\alpha x^\alpha \in I$ for some $c_\alpha \in \mathbb{C}$, not all zero. Let $\beta = \min\{\alpha : c_\alpha \neq 0\}$. Then $x^\beta \in L(I)$, which is a contradiction. This shows that $\{x^\alpha | x^\alpha \notin L(I)\}$ induces a linearly independent set in $\mathbb{C}\{x\}/I$.

To show it is a generating set, let $f \in \mathbb{C}\{x\}$. Modulo I f is equivalent to $\mathrm{NF}(f|I)$, which is a linear combination of the $\{x^\alpha | x^\alpha \notin L(I)\}$. This shows that this set is a basis. □

Now we prove that for $w = (1,\dots,1)$ the Hilbert-Samuel function of an ideal $I \subset \mathbb{C}\{x\}$ coincides with the corresponding function of the leading ideal (cf. Section 4.2).

Corollary 7.3.2. *Let $I \subset \mathbb{C}\{x\}$ be an ideal. Then* $\mathrm{HS}_{\mathbb{C}\{x\}/I} = \mathrm{HS}_{\mathbb{C}\{x\}/L(I)}$.

[3] In practice this way to compute a standard basis is not very efficient. In the computer algebra system SINGULAR for this purpose Mora's algorithm, a modified version of Buchberger's algorithm, see [Mora 1982] is implemented.

Proof.
$$\begin{aligned}\mathrm{HS}_{\mathbb{C}\{x\}/I}(c) &= \dim_{\mathbb{C}}\left(\mathbb{C}\{x\}/I + (x)^c\right)\\ &= \dim_{\mathbb{C}}\left(\mathbb{C}\{x\}/L(I + (x)^c)\right)\end{aligned}$$
by 7.3.1
$$\begin{aligned}&= \dim_{\mathbb{C}}\left(\mathbb{C}\{x\}/L(I) + (x)^c\right)\\ &= \mathrm{HS}_{\mathbb{C}\{x\}/L(I)}(c).\end{aligned}$$
□

We now will prove Lemma 5.1.21 from Chapter 5, which we recall.

Lemma 7.3.3. *Let $\overline{f},\overline{g} \in \mathbb{C}[x,y]$ quasi-homogeneous ($\text{w-deg}(x) = p$, $\text{w-deg}(y) = q$) be such that $(\overline{f},\overline{g})$ is (x,y)-primary. Let $f = \overline{f}+$ terms of higher w-degree, $g = \overline{g}+$ terms of higher w-degree. Then $x^iy^j \in (f,g)$ if $ip + jq > \text{w-ord}(f \cdot g)$.*

Proof. Step 1. We choose $w = (p,q)$ and the ordering with $x < y$. We compute a standard basis for (f,g). As $(\overline{f},\overline{g})$ is in (x,y)–primary, we have that for some c the element x^c is in $(\overline{f},\overline{g})$. It follows that not both $\overline{f},\overline{g}$ are divisible by y. We thus may assume that $L(f) = x^{c_0}$, with $c_0p = \text{w-deg}(\overline{f})$. Now define $g_0 = f$ and $g_1 = \mathrm{NF}(g|\{g_0\})$. Write $L(g_1) = x^{c_1}y^{b_1}$ with $c_0 > c_1$ and $b_1 > b_0 := 0$. We proceed by induction. Assume we have already constructed $\{g_0, \dots, g_{i-1}\}$ with the following properties:

(1) $g_j \in (f,g)$.

(2) $L(g_j) = x^{c_j}y^{b_j}$, $c_j > c_{j+1}$, and $b_{j+1} > b_j$.

(3) $g_j = \mathrm{NF}\left(\mathrm{spoly}(g_{j-2},g_{j-1})|\{g_0, \dots, g_{j-1}\}\right)$.

We then define $g_i := \mathrm{NF}\left(\mathrm{spoly}(g_{i-2},g_{i-1})|\{g_0, \dots, g_{i-1}\}\right)$. We write $L(g_i) = x^{c_i}y^{b_i}$. Then $b_i > b_{i-1}$ because $L\left(\mathrm{spoly}(g_{i-2},g_{i-1})\right) > \mathrm{lcm}(L(g_{i-2},L(g_{i-1})) = x^{c_{i-2}}y^{b_{i-1}}$. Similarly, $c_i < c_{i-1}$. Furthermore we have

$$pc_i + qb_i = \text{w-ord}(\mathrm{spoly}(g_{i-2},g_{i-1})) = pc_{i-2} + qb_{i-1}.$$

This implies

$$pc_{i-1} + qb_i > pc_{i-2} + qb_{i-1}. \tag{7.11}$$

Because the c_i are strictly decreasing and $(\overline{f},\overline{g})$ is (x,y)–primary there is an m such that $L(g_m) = y^{b_m}$. If $g_{m+1} \neq 0$, we let $i \geq 0$ be maximal such that $y^{b_i} \mid L(g_{m+1})$. Then, as g_{m+1} is in normal form with respect to $\{g_0, \dots, g_m\}$, we obtain $i < m$ and $x^{c_i} \nmid L(g_{m+1})$. This implies

$$pc_i + qb_{i+1} > \text{w-deg}(L(g_{m+1})) \geq \text{w-deg}(\mathrm{spoly}(g_{m-1},g_m)) = pc_{m-1} + qb_m,$$

which is a contradiction to formula (7.11), and therefore we get that $g_{m+1} = 0$.

We will now prove that $\{g_0, \dots, g_m\}$ is a standard basis of (f,g). We use Theorem 7.2.16 and Lemma 7.2.18. The property $c_j > c_{j+1}$ and $b_j < b_{j+1}$ implies that $L(g_{j-1}) = x^{c_{j-1}}y^{b_{j-1}} \mid \mathrm{lcm}(L(g_j),L(g_k)) = x^{c_k}y^{b_j}$ for $k \leq j-2$. This implies that already $G = \{s^{(i,i-1)}\}_{i=1,\dots,m}$ generates $\mathrm{syz}(g_0, \dots, g_m)$ (Lemma 7.2.18). By definition $g_i = \mathrm{NF}\left(\mathrm{spoly}(g_{i-2},g_{i-1}) \mid \{g_0, \dots, g_{i-1}\}\right)$ and, therefore, $\mathrm{NF}\left(\mathrm{spoly}(g_{i-2},g_{i-1})|\{g_0, \dots, g_m\}\right) = 0$. The implication (3) $\Longrightarrow$ (1) of Theorem 7.2.16 gives that $\{g_0, \dots, g_m\}$ is a standard basis of (f,g).

Step 2. We will now prove that $\text{w-ord}(f \cdot g) = pc_{m-1} + qb_m$. We already saw that $L(g_m) = y^{b_m}$. Now $\text{w-ord}(g_0) = c_0p$ and

$$\begin{aligned} \text{w-ord}(g_i) &= \text{w-ord}(\text{spoly}(g_{i-2}, g_{i-1})) \\ &= \text{w-ord}(g_{i-1}) + (c_{i-2} - c_{i-1})p \\ &= \text{w-ord}(g_1) + (c_0 - c_{i-1})p. \end{aligned}$$

This implies that

$$\begin{aligned} \text{w-ord}(g_i) &= \text{w-ord}(g_1) + (c_0 - c_{i-1})p \\ &= \text{w-ord}(g_0 \cdot g_1) - c_{i-1}p \\ &= \text{w-ord}(f \cdot g) - c_{i-1}p \end{aligned}$$

For $i = m$ we get $\text{w-ord}(f \cdot g) = pc_{m-1} + qb_m$.

Step 3. Suppose $ip + jq > \text{w-ord}(f \cdot g)$. Because of step 2 we know that $x^i y^j > x^{c_{m-1}} y^{b_m}$.

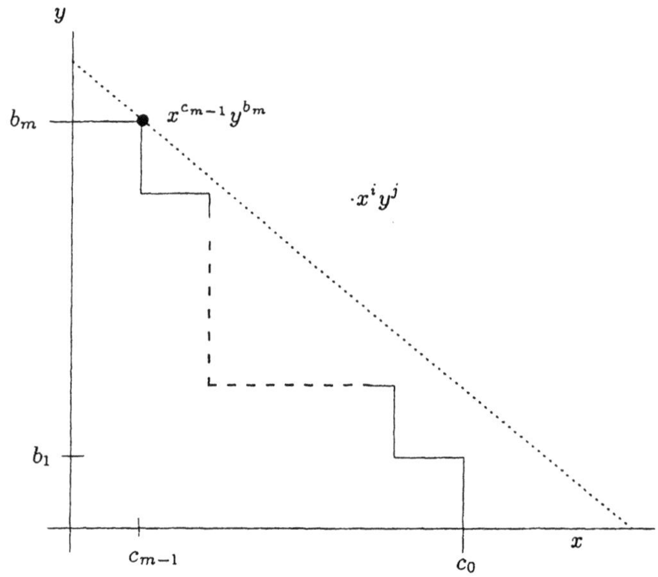

If $x^i y^j \notin (f,g)$ then $L(\text{NF}(x^i y^j | \{g_0, \ldots, g_m\}) \geq x^i y^j > x^{c_{m-1}} y^{b_m}$. Now again we can use (7.11) and the same argument as in step 1 (where we proved that $g_{m+1} = 0$), to obtain a contradiction. □

As a totally different application, we study the ideal of the product of germs of analytic spaces.

Definition 7.3.4. Let $(X, o) \subset (\mathbb{C}^n, o)$ and $(Y, o) \subset (\mathbb{C}^m, o)$ be germs of analytic spaces. Let X and Y be representatives. We define the product $(X \times Y, o)$ to be the germ of $X \times Y$ in 0.

One checks without difficulty that this definition is independent on the choice of the representatives. Let $I \subset \mathbb{C}\{x\}$ be the ideal of (X, o), and $J \subset \mathbb{C}\{y\}$ be the ideal of (Y, o). Then it is a direct check that the product $(X \times Y, o)$ is given as the zero set of $(I + J) \cdot \mathbb{C}\{x,y\}$. That these generate all the functions in the ideal of $(X \times Y, o)$ is not at all trivial.

Theorem 7.3.5. *With the above notation one has*

$$\mathscr{I}(X \times Y, o) = (I + J) \cdot \mathbb{C}\{x,y\}.$$

Proof. The inclusion $\supset$ is easy. For the converse, let $f \in \mathscr{I}(X \times Y, o)$. We choose representatives X of (X, o) and Y of (Y, o) such that f is defined on $X \times Y$ and $f(a,b) = 0$ for all $(a,b) \in X \times Y$. We have to show that $f \in (I + J) \cdot \mathbb{C}\{x,y\}$. Replacing f by $\mathrm{NF}(f \mid J)$, we may assume that no monomial of f is in $L(J)$. We write

$$f = \sum_\alpha f_\alpha y^\alpha, \quad f_\alpha \in \mathbb{C}\{x\}.$$

Therefore, for all α with $f_\alpha \neq 0$ we have that y^α is not $L(J)$. We claim that all f_α are in I, which suffices to prove the theorem (Exercise 7.3.8). Suppose the converse. Then there exists an $a \in X$ with $f_\alpha(a) \neq 0$ for some α. This is because $I = \mathscr{I}(X, o)$. For all $b \in Y$ we have

$$f(a,b) = \sum_\alpha f_\alpha(a) b^\alpha = 0,$$

that is,

$$0 \neq f(a,y) = \sum_\alpha f_\alpha(a) y^\alpha \in \mathscr{I}(Y, o) = J.$$

This is a contradiction, as the leading term of $f(a,y)$ does not lie in $L(J)$. □

This result motivates the following definition.

Definition 7.3.6. Let $A = \mathbb{C}\{x\}/I$ and $B = \mathbb{C}\{y\}/J$ be analytic algebras. Then we define the *analytic tensor product* $A \hat{\otimes} B$ by

$$A \hat{\otimes} B := \mathbb{C}\{x,y\}/(I + J).$$

Thus in case that A and B are reduced, and are the local rings of (X, o) and (Y, o) respectively, the analytic tensor product is the local ring of the product $(X \times Y, o)$.

Remark 7.3.7. We have obvious maps inclusions $i : A \longrightarrow A \hat{\otimes} B$, and $j : B \longrightarrow A \hat{\otimes} B$, which corresponds geometrically to the projection on the factors, if A and B are reduced. On the other hand, we also have a natural surjective map $A \hat{\otimes} B \longrightarrow A$. This corresponds geometrically with the inclusion of one of the factors in the product.

The analytic tensor product satisfies the following universal property. Let C be an analytic algebra, and $\gamma : A \longrightarrow C$ and $\delta : B \longrightarrow C$ be analytic maps. Then there exists a uniquely determined map $\theta : A \hat{\otimes} B \longrightarrow C$ such that $\gamma = \theta \circ i$ and $\delta = \theta \circ j$.

In the case that A and B are reduced, this is easy to prove and follows from the fact that analytic maps $(Z, o) \longrightarrow (X, o)$ and $(Z, o) \longrightarrow (Y, o)$ induce obviously an analytic map $(Z, o) \longrightarrow (X \times Y, o)$.

For the proof of the universal property and more facts on the analytic tensor product we refer to the book Analytische Stellenalgebren, see [Grauert-Remmert 1971].

Exercises

7.3.8. Let $I \subset \mathbb{C}\{x_1,\dots,x_n\}$ be an ideal and $f \in \mathbb{C}\{x_1,\dots,x_n, y_1,\dots,y_m\}$, $f = \sum f_\alpha y^\alpha$ such that $f_\alpha \in I$ for all α. Prove that $f \in I\mathbb{C}\{x_1,\dots,x_n,y_1,\dots,y_m\}$.
(Hint: Use the fact that the normal form with respect to $I\mathbb{C}\{x_1,\dots,x_n,y_1,\dots,y_m\}$ and the degree ordering with $w = (1,\dots,1)$ is additive and maps $(x_1,\dots,y_m)^c$ to $(x_1,\dots,y_m)^c$.)

7.3.9. Let $I \subset \mathbb{C}\{x_0,\dots,x_4\}$ be the ideal generated by the 2–minors of the matrix $\begin{pmatrix} x_0 & x_1 & x_2 & x_3 \\ x_1 & x_2 & x_3 & x_4 \end{pmatrix}$.

(1) Prove that I is a prime ideal.

(2) Prove that $\mathbb{C}\{x_0,\dots,x_4\}/I$ is Cohen-Macaulay.

(Hint: For the first part, consider the map $\varphi : \mathbb{C}\{x_0,\dots,x_4\} \longrightarrow \mathbb{C}\{x_0,x_1\}_{x_0}$ defined by $\varphi(x_2) = \frac{1}{x_0}x_1^2$, $\varphi(x_3) = \frac{1}{x_0^2}x_1^3$ and $\varphi(x_4) = \frac{1}{x_0^3}x_1^4$. Prove that $\mathrm{Ker}(\varphi) = I : (x_0)^\infty$. Then prove that $I = I : (x_0)$. To see this, prove that $(x_1,x_2,x_3)^2 = L(I)$ the leading ideal of I with respect to the degree reverse lexicographical ordering. The six 2–minors of the matrix above are already a standard basis of I with respect to this ordering. For the second part, note that $\mathbb{C}\{x_0,x_4\} \subset \mathbb{C}\{x_0,\dots,x_4\}/I$ is a Noether normalization, and use 6.5.8.)

7.3.10. Use a computer algebra system (for example SINGULAR) to compute a standard basis of the ideal defined in the previous exercise.

8 Approximation Theorems

A canonical construction in local analytic geometry, which is possible without knowing the convergence of the solution, should have a convergent solution. This is always our wish and the underlying philosophy of the Approximation Theorems. We saw already in Chapters 3, 5, 6 and 7 that a lot of canonical formally constructed power series simply (and luckily) converge. Unfortunately, this is not always the case, as we will see in this chapter. This is the reason why we need different Approximation Theorems, adapted to the problem. We start with the famous Approximation Theorem of M. Artin: if a system of analytic equations has a formal solution (which is sometimes really not convergent) then we can find a convergent solution to this system. Even more, we can approximate the formal one by convergent ones up to a given order.

Unfortunately, special properties of the formal solution (as not depending on some variables) do not survive during the approximation process in general. Therefore, for special situations, as needed later in deformation theory, we need special Approximation Theorems as Grauert's Approximation Theorem proved in Section 8.2. In Section 8.3 we will give an overview about general Approximation Theorems and prove some applications.

8.1 Artin's Approximation Theorem

Here it comes:

Theorem 8.1.1 (Artin's Approximation Theorem). *Consider two sets of variables $x = (x_1, \dots, x_n)$ and $y = (y_1, \dots, y_N)$. Let $f_1, \dots, f_m \in \mathbb{C}\{x,y\}$ and $\overline{y} \in (x)\mathbb{C}[[x]]^N$ such that $f_i(\overline{y}) = 0$ for $i = 1, \dots, m$. Let c be a positive integer. Then there exists a $y \in \mathbb{C}\{x\}^N$ such that:*

- *$f_i(y) = 0$ for $i = 1, \dots, m$.*
- *$y \equiv \overline{y}$ modulo $(x)^c$.*

The proof is by induction on n. The case $n = 0$ is trivial. The proof of the induction step is quite long, and will be divided in several lemmas. The first two lemmas will prepare us to prove the Artin Approximation Theorem for a special case. The general case will then be reduced to this special case.

Lemma 8.1.2. *Suppose that Artin's Approximation Theorem holds for $n-1$ variables $x_1, \dots, x_{n-1}$. Consider the following data:*

- *c, a positive integer.*
- *$f_1, \dots, f_m \in \mathbb{C}\{x,y\}$.*
- *A formal Weierstraß polynomial $\overline{P} = x_n^s + \overline{a}_1 x_n^{s-1} + \dots + \overline{a}_s$, $\overline{a}_i \in (x_1, \dots, x_{n-1})\mathbb{C}[[x_1, \dots, x_{n-1}]]$.*

- *Elements:* $\overline{r}_1, \ldots, \overline{r}_N \in \mathbb{C}[[x_1, \ldots, x_{n-1}]][x_n]$ *of degree smaller than* s *in* x_n.

  Suppose that

$$f_i(\overline{r}(x)) \equiv 0 \quad \text{modulo } \overline{P} \text{ for } i = 1, \ldots, m.$$

Then there exists an analytic Weierstraß polynomial P *of degree* s *in* x_n*, and elements* $r_1, \ldots, r_N \in \mathbb{C}\{x_1, \ldots, x_{n-1}\}[x_n]$ *of degree smaller than* s *with respect to* x_n *such that*

$$f_i(r(x)) \equiv 0 \quad \text{modulo } P$$

and $P \equiv \overline{P}$ *and* $\overline{r}_i \equiv r_i$ *modulo* $(x)^c$.

Proof. We may write:

$$\overline{r}_i = \sum_{j=0}^{s-1} (\overline{r}_{ij} + c_{ij}) x_n^j \qquad \overline{r}_{ij} \in (x_1, \ldots, x_{n-1}), \quad c_{ij} \in \mathbb{C}.$$

We consider new variables:

$$A_1, \ldots, A_s; \qquad R_{ij}, 1 \le i \le N, \quad 0 \le j \le s-1$$

and define

$$\widetilde{P} := x_n^s + A_1 x_n^{s-1} + \ldots + A_s$$

$$\widetilde{R}_i := \sum_{j=0}^{s-1} (R_{ij} + c_{ij}) x_n^j \qquad 1 \le i \le N.$$

We do m Weierstraß division with remainder for each f_i and obtain

$$f_i(\widetilde{R}_1, \ldots, \widetilde{R}_N) = Q_i \cdot \widetilde{P} + \sum_{j=0}^{s-1} F_{ij} x_n^j \qquad i = 1, \ldots, m,$$

for some $F_{ij} \in \mathbb{C}\{x_1, \ldots, x_{n-1}, A_1, \ldots, A_s, \{R_{ij}\}\}$. Plugging in $\overline{a}_1, \ldots, \overline{a}_s$ for $A_1, \ldots, A_s$ and, similarly, $\overline{r}_{ij}$ for R_{ij} for $1 \le i \le N$, $0 \le j \le s-1$, we see that we have formal solutions to the equations $F_{ij} = 0$. We may apply the induction hypothesis to find the convergent solutions a_i and the r_{ij}. Then clearly $P = \widetilde{P}(a)$ and $r_i = \sum (r_{ij} + c_{ij}) x_n^j$ satisfy the statement of the lemma. □

Lemma 8.1.3. *Suppose that Artin's Approximation Theorem holds for* $n-1$ *variables* $x_1, \ldots, x_{n-1}$. *Let* $g, f_1, \ldots, f_m \in \mathbb{C}\{x, y\}$, $c \in \mathbb{N}$ *and formal power series* $\overline{y} \in (x)\mathbb{C}[[x]]^N$ *be given such that* $g(\overline{y}) \neq 0$ *and*

$$f_i(\overline{y}) \equiv 0 \qquad \textit{modulo } g(\overline{y}) \qquad i = 1, \ldots, m.$$

Then there exists a $y \in \mathbb{C}\{x\}^N$ *with* $y \equiv \overline{y}$ *modulo* $(x)^c$ *such that* $g(y) \neq 0$ *and*

$$f_i(y) \equiv 0 \qquad \textit{modulo } g(y) \qquad i = 1, \ldots, m.$$

Proof. Consider $g(\overline{y}) \in \mathbb{C}[[x]]$. After a linear coordinate change in the $x_1, \ldots, x_n$, we may assume that $g(\overline{y})$ is regular of order s in x_n. We may enlarge c so that $c > s$. We apply the Weierstraß Preparation Theorem:

$$g(\overline{y}) = \overline{U} \cdot \overline{P},$$

where $\overline{P}$ is a Weierstraß polynomial in x_n of degree s, and $\overline{U}$ is a unit. In particular it follows that $f_i(\overline{y}) \equiv 0$ modulo $\overline{P}$. For $i = 1, \ldots, m$ we do Weierstraß divisions with remainder:

$$\overline{y}_i = \overline{Q}_i \overline{P} + \overline{r}_i,$$

where $\overline{r}_i \in \mathbb{C}[[x_1, \ldots, x_{n-1}]][x_n]$ are polynomials of degree smaller than s in x_n. We will approximate the $\overline{Q}_i, \overline{P}$, and the $\overline{r}_i$. From $\overline{y}_i \equiv \overline{r}_i$ modulo $\overline{P}$ it follows that

$$g(\overline{r}) \equiv g(\overline{y}) \equiv 0 \quad \text{modulo } \overline{P}.$$

The first equivalence follows from the Taylor expansion, the second equivalence holds by construction of $\overline{P}$. Similarly

$$f_i(\overline{r}) \equiv f_i(\overline{y}) \equiv 0 \quad \text{modulo } \overline{P}.$$

By the previous lemma, applied to the functions $g, f_1, \ldots, f_m$, we can find convergent $r_i \in \mathbb{C}\{x\}$, and a convergent Weierstraß polynomial P with $r_i \equiv \overline{r}_i$ and $P \equiv \overline{P}$ modulo $(x)^c$ such that

$$g(r) \equiv 0 \text{ and } f_i(r) \equiv 0 \quad \text{modulo } P \quad \text{for } i = 1, \ldots, m.$$

To define the y_i, take any $Q_i \in \mathbb{C}\{x\}$ such that $Q_i \equiv \overline{Q}_i$ modulo $(x)^c$ and put:

$$y_i := Q_i P + r_i.$$

By construction $y_i \equiv \overline{y}_i$ modulo $(x)^c$. Using Taylor's formula again, as above, we see that

$$g(y) \equiv 0 \text{ and } f_i(y) \equiv 0 \text{ modulo } P \quad \text{for } i = 1, \ldots, m.$$

We are finished if we show that

$$g(y) = U \cdot P \tag{8.1}$$

for a *unit* $U \in \mathbb{C}\{x\}$. It then follows immediately that $f_i(y) \equiv 0$ modulo $g(y)$ for $i = 1, \ldots, m$, and $g(y) \neq 0$ because $P \neq 0$. To show (8.1), observe that by the Weierstraß Division Theorem and the fact that $g(y) \equiv 0$ modulo P there exists a $U \in \mathbb{C}\{x\}$ with $g(y) = U \cdot P$. We only have to show that U is a unit. By assumption, $g(\overline{y})$ is regular of order s in x_n. Because $y \equiv \overline{y}$ modulo x^c and $c > s$ it follows that $g(y)$ is regular of order s in x_n. As P is a Weierstraß polynomial of degree s in x_n, it follows that U is a unit. This finishes the proof of the lemma. □

Lemma 8.1.4. *Suppose that Artin's Approximation Theorem holds for $n-1$ variables $x_1, \ldots, x_{n-1}$. Then Artin's Approximation Theorem for n variables $x_1, \ldots, x_n$ holds under the following assumptions:*

(1) $m \leq N$.

(2) There exists an m–minor Δ of the Jacobian matrix $\left(\frac{\partial f_i}{\partial y_j}\right)$ such that $\Delta(\overline{y}) \neq 0$.

Proof. This we will prove as an application of Newton's Lemma, which itself follows from the Implicit Mapping Theorem. Take any nonzero function $h \in (x)^c$. Then because $f_i(\overline{y}) = 0$, we certainly have:

$$f_i(\overline{y}) \equiv 0 \text{ modulo } \Delta^2(\overline{y})h.$$

We can apply the previous lemma to find a $y' \in \mathbb{C}\{x\}^N$, $y' \equiv \overline{y}$ modulo $(x)^c$, such that

$$f_i(y') \equiv 0 \text{ modulo } \Delta^2(y')h.$$

Putting J to be the ideal generated by the m–minors of the Jacobian matrix we certainly have

$$f_i(y') \equiv 0 \quad \text{modulo } J^2(y')(x)^c,$$

so that Newton's Lemma 3.3.34 applies: there exists a $y \in \mathbb{C}\{x\}^N$, $y \equiv y'$ modulo $J(y')(x)^c$, such that $f_i(y) = 0$ for all $i = 1,\dots,m$. (Note that we needed $m \leq N$ in order to apply Newton's Lemma.) □

Proof of Artin's Approximation Theorem 8.1.1. Define the ideal $\mathfrak{p}$ to be the kernel of the map:

$$\overline{\alpha} : \mathbb{C}\{x,y\} \longrightarrow \mathbb{C}[[x]]$$

defined by $x \mapsto x$, and $y \mapsto \overline{y}(x)$. Note that as we have an injective map $\mathbb{C}\{x,y\}/\mathfrak{p} \hookrightarrow \mathbb{C}[[x]]$, it follows that $\mathfrak{p}$ is a prime ideal. As $f_1,\dots,f_m \in \mathfrak{p}$ it suffices to find a $y(x) \in \mathbb{C}\{x\}^N$ such that $g(x,y(x)) = 0$ for all $g \in \mathfrak{p}$. So we may have assumed $\mathfrak{p} = (f_1,\dots,f_m)$ from the beginning.

Let $\operatorname{ht}(\mathfrak{p}) = k$. By the Jacobian Criterion 4.3.15 applied to the case $I = \mathfrak{p} = \mathfrak{q}$, there exist k variables $z_1,\dots,z_k$ among the x_i and the y_i, such that (after renumbering the f_i)

$$\Delta' := \det\left(\frac{\partial f_i}{\partial z_j}\right)_{1 \leq i,j \leq k} \notin \mathfrak{p}.$$

In particular $\Delta'(x,\overline{y}(x)) \neq 0$. As $0 = f_i(x,\overline{y}(x)) \in \mathbb{C}[[x]]$, it follows that $\frac{\partial f_i(x,\overline{y}(x))}{\partial x_j} = 0$ for all j. From the chain rule it follows that

$$\frac{\partial f_i}{\partial x_j}(x,\overline{y}(x)) = -\sum_k \frac{\partial f_i}{\partial y_k}(x,\overline{y}(x)) \cdot \frac{\partial \overline{y}_k(x)}{\partial x_j}.$$

Hence, $\Delta'(x,\overline{y}(x))$ is in the ideal generated by the k–minors of the matrix

$$\left(\frac{\partial f_i}{\partial y_j}(x,\overline{y}(x))\right)_{\substack{1 \leq i \leq m \\ 1 \leq j \leq N}}.$$

In particular, it follows that one of those minors is nonzero. We may, therefore, apply the previous lemma to find a convergent $y \in \mathbb{C}\{x\}^N$ with $y \equiv \overline{y}$ modulo $(x)^c$ and $f_1(x,y(x)) = \ldots = f_k(x,y(x)) = 0$.

Now $\mathbb{C}\{x,y\}_{\mathfrak{p}}/\mathfrak{p}$ is a regular local ring. (It is even a field.) In the Jacobian Criterion 4.3.15 it was shown that $f_1,\dots,f_k$ generate $\mathfrak{p}\mathbb{C}\{x,y\}_{\mathfrak{p}}$, that is,

$$(f_1,\dots,f_k)\mathbb{C}\{x,y\}_{\mathfrak{p}} = \mathfrak{p}\mathbb{C}\{x,y\}_{\mathfrak{p}}.$$

Because $\mathbb{C}\{x,y\}$ is Noetherian, we can find $h \notin \mathfrak{p}$ such that

$$((f_1,\dots,f_k) : h) = \mathfrak{p}. \qquad (8.2)$$

As $h \notin \mathfrak{p} = \operatorname{Ker}(\overline{\alpha})$ it follows that $h(x,\overline{y}(x)) \neq 0$. By enlarging c we may assume that $h(x,\overline{y}(x)) \notin (x)^c$. Hence $y \equiv \overline{y}$ modulo $(x)^c$ implies that $h(x,y(x)) \notin (x)^c$. In particular $h(x,y(x)) \neq 0$.

For any $g \in \mathfrak{p}$ it follows from formula (8.2) that $hg \in (f_1,\dots,f_k)$. As $h(x,y(x)) \neq 0$ it follows that $g(x,y(x)) = 0$. This is what we had to show. □

Exercises

8.1.5. Deduce from the Artin Approximation Theorem the following somewhat more general version.

Let $f_1,\dots,f_m \in \mathbb{C}\{x,y\}$, and $J \subset \mathbb{C}\{x\}$ be an ideal. Let $\overline{y}(x) \in \mathbb{C}[[x]]^N$ such that $f_i(x,\overline{y}(x)) \in J\cdot\mathbb{C}[[x]]$ for $i = 1,\dots,m$. Then there exists a $y(x) \in \mathbb{C}\{x\}^N$ such that $f_i(x,y(x)) \in J$ for all $i = 1,\dots,m$.

8.1.6. Let the conditions be as in the previous exercise. Let $I \subset \mathbb{C}\{x\}$ be an ideal. Suppose we have a formal solution $\overline{y}(x)$ whose components lie in $I\cdot\mathbb{C}[[x]]$. Then we can find a convergent solution in I. Prove this.

8.1.7. Let $\mathfrak{p} \subset \mathbb{C}\{x\}$ be a prime ideal. Show that $\mathfrak{p}\cdot\mathbb{C}[[x]]$ is a prime ideal.
(Hint: Consider $f,g \in \mathbb{C}[[x]]$ with $f\cdot g \in \mathfrak{p}\cdot\mathbb{C}[[x]]$. Use Exercise 8.1.5 to find for all $c \in \mathbb{N}$ $f_c,g_c \in \mathbb{C}\{x\}$, $f_c \equiv f$ modulo $(x)^c$ and $g_c \equiv g$ modulo $(x)^c$, and $f_c\cdot g_c \in \mathfrak{p}$. Deduce that either $f = \lim_{c\to\infty} f_c \in \mathfrak{p}\cdot\mathbb{C}[[x]]$ or $g = \lim_{c\to\infty} g_c \in \mathfrak{p}\cdot\mathbb{C}[[x]]$.)

8.1.8. Let $\mathfrak{p} \subset \mathbb{C}\{x\}$ be a prime ideal. Suppose that $\mathbb{C}\{x\}/\mathfrak{p}$ is normal. Use Exercise 8.1.5 to see that $\mathbb{C}[[x]]/\mathfrak{p}\cdot\mathbb{C}[[x]]$ is normal.

8.2 Grauert's Approximation Theorem

In this section we prove Grauert's Approximation Theorem. This approximation theorem looks rather complicated, but is very useful. For example, we will use this theorem to prove the existence of a 'semi-universal' deformation of an isolated singularity in Chapter 10.

Definition 8.2.1. We consider *four* sets of variables, $x = (x_1,\dots,x_n)$, $s = (s_1,\dots,s_l)$, $\Phi = (\Phi_1,\dots,\Phi_p)$ and $\Psi = (\Psi_1,\dots,\Psi_q)$. Let $I \subset \mathbb{C}\{s\}$ be an ideal. We can extend I to an ideal $I \subset \mathbb{C}\{x,s\}$. Consider an element $F = (F_1,\dots,F_k) \in \mathbb{C}\{x,s,\Phi,\Psi\}^k$.

(1) An *analytic solution* of the equation $F \equiv 0$ modulo I is a pair $(\phi,\psi) \in \mathbb{C}\{s\}^p \times \mathbb{C}\{x,s\}^q$, that is power series $\phi_1,\dots,\phi_p \in \mathbb{C}\{s\}$, and $\psi_1,\dots,\psi_q \in \mathbb{C}\{x,s\}$ such that

$$F(x,s,\phi(s),\psi(x,s)) \equiv 0 \quad \text{modulo } I\cdot\mathbb{C}\{x,s\}^k,$$

that is, the ϕ and the ψ have the property that

$$F_1(x,s,\phi(s),\psi(x,s)),\dots,F_k(x,s,\phi(s),\psi(x,s)) \in I.$$

(2) A solution of order e of the equation $F \equiv 0$ modulo I is a pair $(\phi,\psi) \in \mathbb{C}[s]^p \times \mathbb{C}\{x\}[s]^q$ such that

$$F(x,s,\phi(s),\psi(x,s)) \equiv 0 \quad \text{modulo } (I + \mathfrak{m}^{e+1}) \cdot \mathbb{C}\{x,s\}^k,$$

where $\mathfrak{m} = (s) = (s_1,\ldots,s_l)$.

We can now formulate the theorem.

Theorem 8.2.2 (Grauert's Approximation Theorem). *Let the notation be as in the previous definition. Let $e_0 \in \mathbb{N}$. Suppose that the system of equations*

$$F \equiv 0 \text{ modulo } I$$

has a solution $(\phi^{(e_0)},\psi^{(e_0)})$ of order e_0. Suppose, moreover, that for all $e \geq e_0$ every solution $(\phi^{(e)},\psi^{(e)})$ of order e with $\phi^{(e)} \equiv \phi^{(e_0)}$ modulo $\mathfrak{m}^{e_0+1}$, and $\psi^{(e)} \equiv \psi^{(e_0)}$ modulo $\mathfrak{m}^{e_0+1}$, extends to a solution $(\phi^{(e)} + \delta^{(e)},\psi^{(e)} + \gamma^{(e)})$ of order $e+1$, with $\delta^{(e)} \in \mathbb{C}[s]^p$ and $\gamma^{(e)} \in \mathbb{C}\{x\}[s]^q$ homogeneous of degree $e+1$ in s. Then the system of equations $F \equiv 0$ modulo I has an analytic solution (ϕ,ψ), with $\phi \equiv \phi^{(e_0)}$ modulo $\mathfrak{m}^{e_0+1}$, and $\psi \equiv \psi^{(e_0)}$ modulo $\mathfrak{m}^{e_0+1}$.

Before giving the proof, we need some preparations. As in Chapter 7, we choose as a basis for the free module $\mathbb{C}\{x,s,\Phi,\Psi\}^k$ the vectors $e_i := (0,\ldots,0,1,0,\ldots,0)$ (the 1 is on the i–th spot) and identify e_i with x_{n+1}^{i-1} by adding a new variable x_{n+1} and enlarging x to $x = (x_1,\ldots,x_{n+1})$.

We choose on $\mathbb{C}\{x,s,\Phi,\Psi\}^k$ the degree lexicographical ordering as in Chapter 7 with weights 1 for all variables. For an element $F \in \mathbb{C}\{x,s,\Phi,\Psi\}^k$ there exists a polyradius $\lambda \in \mathbb{R}_+^{n+1+l+p+q}$ such that $F \in B_\lambda^k$. We will write, in order to distinguish between the variables x,s and ϕ,ψ, the polyradius as $\lambda = (\rho,\tau,\sigma)$, $\rho \in \mathbb{R}_+^{n+1}$, $\tau \in \mathbb{R}_+^l$, $\sigma \in \mathbb{R}_+^{p+q}$.

Proposition 8.2.3. *Let $b_1,\ldots b_t \in \mathbb{C}\{x\}^k$ and $\overline{\rho} \in \mathbb{R}_+^{n+1}$ such that $b_1,\ldots b_t \in B_{\overline{\rho}}^k$ be given. Then for all $\rho \leq \overline{\rho}$ there exists a $K(\rho) \in \mathbb{R}_+$ such that the following holds.*

For all $C \in \mathbb{C}\{x\}[s]^k$, $C = \sum_{|\beta|=e} C_\beta s^\beta$ with $C_\beta \in \mathbb{C}\{x\}$ and for all $\overline{y}_1,\ldots,\overline{y}_t \in \mathbb{C}[s]$ such that

(1) $\overline{y}_j$ is a homogeneous polynomial of degree e in s for $j = 1,\ldots,t$,

(2) $\sum_{j=1}^{t} b_j\overline{y}_j = C$,

there exist $y_j \in \mathbb{C}[s]$, $j = 1,\ldots,t$ such that

(1) y_j is a homogeneous polynomial of degree e in s for $j = 1,\ldots,t$,

(2) $\sum_{j=1}^{t} b_j y_j = C$,

(3) $\|y_j\|_\tau \leq K(\rho)\|C\|_{(\rho,\tau)}$ for all j and all $\rho \leq \overline{\rho}$, $\tau \in \mathbb{R}_+^l$.

The important thing to note here, is that $K(\rho)$ does not depend on C, e and τ. However, it depends (of course) on the b_i and on ρ.

Proof. This is an application of Cramer's rule.

Step 1. We write

$$b_i = \sum_{\alpha\in\mathbb{N}_+^{n+1}} b_{i,\alpha}x^\alpha, \quad C = \sum_{\substack{\alpha\in\mathbb{N}_+^{n+1}\\ |\beta|=e}} C_{\alpha,\beta}x^\alpha s^\beta, \quad \overline{y}_j = \sum_{|\beta|=e} \overline{y}_{j,\beta}s^\beta.$$

The condition then reads

$$\sum_{j=1}^{t} b_{j,\alpha}\overline{y}_{j,\beta} = C_{\alpha,\beta} \text{ for all } \alpha\in\mathbb{N}_+^{n+1} \text{ and } |\beta| = e.$$

This we can view as system of linear equations. More precisely, consider the linear subspace in $\mathbb{C}^t$ spanned by the vectors

$$b^{(\alpha)} := (b_{1,\alpha},\dots,b_{t,\alpha}),\ \alpha\in\mathbb{N}_+^{n+1}.$$

Let r be the dimension of this subspace and choose a basis

$$\{b^{(\alpha_i)} = (b_{1,\alpha_i},\dots,b_{t,\alpha_i}),\ i = 1,\dots,r\}.$$

So for any $\alpha\in\mathbb{N}^{n+1}$, we have an expression $b^{(\alpha)} = \sum_{i=1}^r k_{i,\alpha}b^{(\alpha_i)}$ for some $k_{i,\alpha}\in\mathbb{C}$. By linear algebra, there is an r–minor M of the matrix

$$\left(b_{i,\alpha_j}\right)_{\substack{1\le i\le t\\ 1\le j\le r}}$$

which is nonzero. After renumbering the b_i we may assume that $M = \det(b_{i,\alpha_j})_{1\le i,j\le r}$. We define

(1) for $i = 1,\dots,r$, $y_i := \sum_{|\beta|=e} y_{i,\beta}s^\beta$ with $y_{i,\beta} = \frac{M^{(i,\beta)}}{M}$,

(2) $y_{r+1} := \dots := y_t := 0$.

Here $M^{(i,\beta)}$ is obtained from M by replacing the i–th column of M by the column vector $(C_{\alpha_1,\beta},\dots,C_{\alpha_r,\beta})^t$. We want to show $\sum_{j=1}^t b_{j,\alpha}y_{j,\beta} = C_{\alpha,\beta}$ for all α. By Cramer's rule, we know this is true for $\alpha = \alpha_1,\dots,\alpha_r$. Now

$$C_{\alpha,\beta} = \sum_j b_{j,\alpha}\overline{y}_{j,\beta} = \sum_{i,j} k_{i,\alpha}b_{j,\alpha_i}\overline{y}_{j,\beta} = \sum_i k_{i,\alpha}C_{\alpha_i,\beta} = \sum_{i,j} k_{i,\alpha}b_{j,\alpha_i}y_{j,\beta} = \sum_j b_{j,\alpha}y_{j,\beta},$$

as was to be shown.

Step 2. It remains to define the $K = K(\rho)$ and to show the estimates. We take

$$K = \max_{i,j}\frac{|M_{ij}|}{\rho^{\alpha_j}|M|}.$$

Here M_{ij} is the determinant obtained from M by deleting the i–th row and the j–th column. As $y_{r+1} = \dots = y_t = 0$, we only have to show the estimates for the $y_1,\dots,y_r$. By definition

$$\|y_i\|_\tau = \Big\|\sum_{|\beta|=e}\frac{M^{(i,\beta)}}{M}s^\beta\Big\|_\tau = \sum_{|\beta|=e}\frac{|M^{(i,\beta)}|}{|M|}\tau^\beta.$$

Moreover, by linear algebra we have $M^{(i,\beta)} = \sum_{j=1}^{r} C_{\alpha_j,\beta} M_{ij}$ and, therefore,

$$\frac{|M^{(i,\beta)}|}{|M|} \leq \sum_{j=1}^{r} |C_{\alpha_j,\beta}| \frac{|M_{ij}|}{|M|}.$$

This implies

$$\|y_i\|_\tau \leq \frac{1}{|M|} \sum_{|\beta|=e} \sum_{j=1}^{r} |C_{\alpha_j,\beta}|\, |M_{ij}| \tau^\beta.$$

On the other hand by the definition of the norm

$$\|C\|_{(\rho,\tau)} \geq \sum_{|\beta|=e} \sum_{j=1}^{r} |C_{\alpha_j,\beta}| \tau^\beta \rho^{\alpha_j}.$$

Thus with our choice of K the inequality

$$\|y_i\|_\tau \leq K \cdot \|C\|_{(\rho,\tau)}$$

holds. □

We need the following generalization of this proposition.

Lemma 8.2.4 (Cartan's Lemma). *Let $a_1, \dots, a_r, b_1, \dots, b_t \in \mathbb{C}\{x\}^k$ and $\overline{\rho} \in \mathbb{R}_+^{n+1}$, $\overline{\tau} \in \mathbb{R}_+^l$. Suppose that $a_1, \dots, a_r, b_1, \dots, b_t \in B_{\overline{\rho}}^k$. There exists a $\rho \in \mathbb{R}_+^{n+1}$, $\rho \leq \overline{\rho}$, and $K = K(\rho) \in \mathbb{R}_+$ with the following property.*

For all $\tau \in \mathbb{R}_+^l$ with $\tau \leq \overline{\tau}$ and for all $C \in B_{(\rho,\tau)}^k$, $C = \sum_{|\alpha|=e} C^{(\alpha)} s^\alpha$ with $C^{(\alpha)} \in \mathbb{C}\{x\}$ and for all $\overline{y}_1, \dots, \overline{y}_t \in \mathbb{C}[s]$ and $\overline{z}_1, \dots \overline{z}_r \in \mathbb{C}\{x\}[s]$ such that

(1) $\overline{y}_j$ is a homogeneous polynomial of degree e in s for $j = 1, \dots, t$, and $\overline{z}_i$ is a homogeneous polynomial of degree e in s for $i = 1, \dots, r$ with coefficients in $\mathbb{C}\{x\}$,

(2) $\sum_{i=1}^{r} a_i \overline{z}_i + \sum_{j=1}^{t} b_j \overline{y}_j = C$,

there exist $z_i \in \mathbb{C}\{x\}[s]$, $i = 1, \dots, r$ and $y_j \in \mathbb{C}[s]$, $j = 1, \dots, t$, such that

(1) y_j is a homogeneous polynomial of degree e in s for $j = 1, \dots, t$, and z_i is a homogeneous polynomial of degree e in s for $i = 1, \dots, r$ with coefficients in $\mathbb{C}\{x\}$,

(2) $\sum_{i=1}^{r} a_i z_i + \sum_{j=1}^{t} b_j y_j = C$,

(3) $\|z_i\|_{(\rho,\tau)} \leq K\|C\|_{(\rho,\tau)}$ and $\|y_j\|_\tau \leq K\|C\|_{(\rho,\tau)}$ for all i and j.

The important thing to note here, is that K does not depend on C, e, and τ. However, it depends (of course) on the a_i, the b_i and ρ.

Proof. Step 1. The case $r = 0$ is dealt with in the previous proposition. We consider the normal form NF with respect to the submodule $(a_1, \ldots, a_r)$. Note that as the $\overline{y}_j$ do not depend on x, we have $\mathrm{NF}(\overline{y}_j) = \overline{y}_j$. We take the normal form of $\sum_{i=1}^r a_i \overline{z}_i + \sum_{j=1}^t b_j \overline{y}_j = C$ and get

$$\sum_{j=1}^{t} \mathrm{NF}(b_j)\overline{y}_j = \mathrm{NF}(C). \tag{8.3}$$

We apply the previous proposition. There exists a $K_0(\rho)$ depending on the $\mathrm{NF}(b_j)$ and ρ, and $y_1, \ldots, y_t \in \mathbb{C}[s]$ homogeneous of degree e such that $\sum_{j=1}^t \mathrm{NF}(b_j) y_j = \mathrm{NF}(C)$ and

$$\|y_j\|_\tau \le K_0(\rho) \|\mathrm{NF}(C)\|_{(\rho,\tau)} \text{ for all } \rho \le \overline{\rho}. \tag{8.4}$$

To define the z_i, note that from 7.2.10 it follows that we can write

$$b_j = \sum_{i=1}^{r} \alpha_{ij} a_i + \mathrm{NF}(b_j), \quad C = \sum_{i=1}^{r} \beta_i a_i + \mathrm{NF}(C). \tag{8.5}$$

The precise α_{ij} and β_i we take will be determined in the second step. Plugging (8.5) in $\sum_{i=1}^r a_i \overline{z}_i + \sum_{j=1}^t b_j \overline{y}_j - C = 0$ and using (8.3) we get

$$\sum_{i=1}^{r} (\overline{z}_i + \sum_{j=1}^{t} \alpha_{ij} \overline{y}_j - \beta_i) a_i = 0.$$

Now we define

$$z_i := -\sum_{j=1}^{t} \alpha_{ij} y_j + \beta_i.$$

Obviously,

$$\sum_{i=1}^{r} (z_i + \sum_{j=1}^{t} \alpha_{ij} y_j - \beta_i) a_i = 0.$$

Working backwards, that is, using $\sum_{j=1}^t \mathrm{NF}(b_j) y_j = \mathrm{NF}(C)$ we get

$$\sum_{i=1}^{r} a_i z_i + \sum_{j=1}^{t} b_j y_j = C.$$

Step 2. We still have to tell which α_{ij} and β_i we take that fulfill (8.5), and show the estimates. To do this, we take a standard basis $f_1, \ldots, f_m$ of the submodule $(a_1, \ldots, a_r)$. We can write

$$f_\nu = \sum_{i=1}^{r} h_{\nu i} a_i$$

for some $h_{\nu i} \in \mathbb{C}\{x\}$. Choose $\rho \le \overline{\rho}$ so small that $a_1, \ldots, a_r, b_1, \ldots, b_t, f_1, \ldots, f_m, h_{11}, \ldots, h_{mr} \in B_\rho^k$. We apply Grauert's Division Theorem, see Remark 7.1.15 with $\varepsilon = \frac{1}{2}$. Therefore, by making ρ smaller we get equations

$$b_j = \sum_\nu b_{j\nu} f_\nu + \mathrm{NF}(b_j)$$
$$C = \sum_\nu C_\nu f_\nu + \mathrm{NF}(C),$$

with the following estimates.

(1) $||b_{j\nu}||_\rho \leq 2\frac{||b_j||_\rho}{||L(f_\nu)||_\rho}$;

(2) $||C_\nu||_{(\rho,\tau)} \leq 2\frac{||C||_{(\rho,\tau)}}{||L(f_\nu)||_\rho}$;

(3) $||\,\mathrm{NF}(C)||_{(\rho,\tau)} \leq 2||C||_{(\rho,\tau)}$.

In particular from formula (8.4) we get for $K_0 := K_0(\rho)$

$$||y_j||_\tau \leq K_0||\,\mathrm{NF}(C)||_{(\rho,\tau)} \leq 2K_0||C||_{(\rho,\tau)}, \tag{8.6}$$

so that we have an estimate for the y_i. We define

$$\alpha_{ij} := \sum_\nu b_{j\nu} h_{\nu i}, \quad \beta_i := \sum_\nu h_{\nu i} C_\nu.$$

With these α_{ij} and β_i the equations (8.5) hold. Let $L = 2m \cdot \max_{\nu,i}\left\{\frac{||h_{\nu i}||_\rho}{||L(f_\nu)||_\rho}\right\}$. Then

$$||\alpha_{ij}||_\rho \leq \sum_\nu ||h_{\nu i}||_\rho ||b_{j\nu}||_\rho \leq L||b_j||_\rho,$$
$$||\beta_i||_{(\rho,\tau)} \leq L||C||_{(\rho,\tau)}.$$

From the definition of the z_i and formula (8.6) we get

$$\begin{aligned}
||z_i||_{(\rho,\tau)} &\leq \sum_{j=1}^t ||\alpha_{ij}||_\rho ||y_j||_\tau + ||\beta_i||_{(\rho,\tau)} \\
&\leq \sum_{j=1}^t L||b_j||_\rho ||y_j||_\tau + L||C||_{(\rho,\tau)} \\
&\leq L\max_j\{||b_j||_\rho ||y_j||_\tau\} + L||C||_{(\rho,\tau)} \\
&\leq L\max_j\{2K_0||b_j||_\rho + 1\}||C||_{(\rho,\tau)}.
\end{aligned}$$

Let $K' = L\max_j\{2K_0||b_j||_\rho + 1\}$ and $K = \max\{2K_0, K'\}$. With this K the desired estimates hold. □

We need one more result before we can give the proof of Grauert's Approximation Theorem.

Lemma 8.2.5. *Let $F(x,s,\Theta) \in \mathbb{C}\{x,s,\Theta\}^k$ be a vector of power series in three sets of variables $x = (x_1,\ldots,x_n)$, $s = (s_1,\ldots,s_l)$ and $\Theta = (\Theta_1,\ldots,\Theta_t)$. Let $I \subset \mathbb{C}\{s\}$ be an ideal. Assume that $F \in B^k_{(\rho,\overline{\tau},\sigma)}$ for a suitable polyradius $(\rho,\overline{\tau},\sigma) \in \mathbb{R}^{n+1+l+t}$. Then for all $\varepsilon > 0$, there exists a $K = K(\varepsilon) > 0$ and a $\tau \in \mathbb{R}^l_+$, $\tau \leq \overline{\tau}$, with the following property.*

Let $e \in \mathbb{N}$ and $\theta^{(e)}(x,s) = (\theta_1^{(e)}(x,s),\ldots,\theta_t^{(e)}(x,s))$ such that

(1) the $\theta_i(x,s)$ are polynomials of degree $\leq e$ in s with coefficients in $\mathbb{C}\{x\}$,

(2) the $\theta_i(x,s)$ are in normal form with respect to I,

(3) $\|\theta_\nu^{(e)}(x,s)\|_{(\rho,\lambda\tau)} \leq K$ for $\nu = 1,\ldots,t$ and some λ, $0 < \lambda \leq 1$.

Let $F_e \in \mathbb{C}\{x,s\}^k$ be the degree $e+1$ part in s of the normal form of $F(x,s,\theta^{(e)}(x,s))$ with respect to I, then:

$$\|F_e\|_{(\rho,\lambda\tau)} \leq K\varepsilon.$$

Proof. Let $0 < \alpha < 1$ be given. Then there exists a polyradius $\tau \in \mathbb{R}_+^l$ such that for all $f \in B_{(\rho,\tau)}$

$$\|\operatorname{NF}(f|I)\|_{(\rho,\tau)} \leq \frac{1}{1-\alpha}\|f\|_{(\rho,\tau)}. \tag{8.7}$$

Given α and (ρ,τ), this inequality still holds if we replace τ by $\lambda\tau$ for any λ with $0 < \lambda \leq 1$, see Remark 7.1.15. We develop F:

$$F(x,s,\Theta) = \sum_{\nu \in \mathbb{N}^t} F_\nu(x,s)\Theta^\nu.$$

As this series is convergent, there exists an $A > 0$ such that $\|F_\nu(x,s)\|_{(\rho,\tau)} \leq A^{|\nu|}$. (cf. Exercise 8.2.6). We choose K so small that $AK < 1$. It is an exercise to show that we can choose $K = K(\varepsilon)$ so small that even

$$\sum_{|\nu|\geq 2} (AK)^{|\nu|} = \frac{1}{(1-AK)^t} - 1 - tAK < K\frac{\varepsilon(1-\alpha)}{3},$$

see Exercise 8.2.7. We write:

$$F_\nu(x,s) = F_\nu(x,0) + F'_\nu(x,s) \quad \text{with } F'_\nu \in (s) \quad \text{for } |\nu| \leq 1.$$

We choose δ so small that $\delta < \frac{\varepsilon(1-\alpha)K}{3}$, and $\delta < \frac{\varepsilon(1-\alpha)}{3t}$. By making τ smaller we may assume that $\|F'_\nu(x,s)\|_{(\rho,\tau)} \leq \delta$ for all ν with $|\nu| \leq 1$. Now

$$F(x,s,\theta^{(e)}(x,s)) = F_0(x,0) + F'_0(x,s)$$
$$+ \sum_{|\nu|=1} F_\nu(x,0)\theta^{(e)}(x,s)^\nu + \sum_{|\nu|=1} F'_\nu(x,s)\theta^{(e)}(x,s)^\nu + \sum_{|\nu|\geq 2} F_\nu(x,s)\theta^{(e)}(x,s)^\nu.$$

By assumption the $\theta^{(e)}(x,s)$ are in normal form with respect to I. Moreover the $F_\nu(x,0)$ for all ν with $|\nu| \leq 1$ do not depend on s. It follows that $F_0(x,0) + \sum_{|\nu|=1} F_\nu(x,0)\theta^{(e)}(x,s)^\nu$ is in normal form with respect to I. Furthermore $F_0(x,0) + \sum_{|\nu|=1} F_\nu(x,0)\theta^{(e)}(x,s)^\nu$ does not have terms of degree $e+1$ in the variables s. It follows that F_e, which by definition is the degree $e+1$ part in s of the normal form of $F(x,s,\theta^{(e)}(x,s))$, is equal to the degree $e+1$ part of the normal form of

$$F'_0(x,s) + \sum_{|\nu|=1} F'_\nu(x,s)\theta^{(e)}(x,s)^\nu + \sum_{|\nu|\geq 2} F_\nu(x,s)\theta^{(e)}(x,s)^\nu.$$

Thus, because of (8.7)

$$\begin{aligned}(1-\alpha)\|F_e\|_{(\rho,\lambda\tau)} &\leq \|F_0'(x,s) + \sum_{|\nu|=1} F_\nu'(x,s)\theta^{(e)}(x,s)^\nu + \sum_{|\nu|\geq 2} F_\nu(x,s)\theta^{(e)}(x,s)^\nu\|_{(\rho,\lambda\tau)} \\ &\leq \delta + t\delta K + \sum_{|\nu|\geq 2}(AK)^{|\nu|}.\end{aligned}$$

By choice of K and δ each term on the right hand side is smaller than or equal to $(1-\alpha)\frac{\varepsilon K}{3}$. Hence $\|F_e\|_{(\rho,\lambda\tau)} < \varepsilon K$, as was to be shown. □

Proof of Theorem 8.2.2.
Step 1. We first reduce to the case that $\phi^{(e_0)}$ and $\psi^{(e_0)}$ are in $\mathfrak{m} := (s)$ and $\frac{\partial F}{\partial \Phi_i}|_{s=0}$, $\frac{\partial F}{\partial \Psi_i}|_{s=0} \in \mathbb{C}\{x\}^k$, that is, they do not depend on the Φ and Ψ. Indeed, we can write

$$\phi^{(e_0)} = c + \sum s_\nu(c_\nu + \phi_\nu^{(e_0)\prime}) \qquad \psi^{(e_0)} = d + \sum s_\nu(d_\nu + \psi_\nu^{(e_0)\prime}).$$

Here c,d,c_ν and d_ν are in $\mathbb{C}\{x\}$. Moreover $\phi_\nu^{(e_0)\prime} = \{\phi_{i,\nu}^{(e_0)\prime}\}_{i=1,\ldots,p}$, $\psi_\nu^{(e_0)\prime} = \{\psi_{i,\nu}^{(e_0)\prime}\}_{i=1,\ldots,q}$ with $\phi_{i,\nu}^{(e_0)\prime} \in \mathfrak{m}$, $\psi_{i,\nu}^{(e_0)\prime} \in \mathfrak{m}\cdot\mathbb{C}\{x,s\}$. Define new variables $\Phi' = \{\Phi_{i\nu}'\}$,$\Psi' = \{\Psi_{i\nu}'\}$, and define

$$F'(x,s,\Phi',\Psi') := F(x,s,c+\sum s_\nu(c_\nu+\Phi_\nu'),d+\sum s_\nu(d_\nu+\Psi_\nu')).$$

Now F' has a solution $(\phi^{(e_0)\prime},\psi^{(e_0)\prime})$ of order e_0. Replacing F,$\phi^{(e_0)}$, $\psi^{(e_0)}$ by F',$\phi^{(e_0)\prime}$,$\psi^{(e_0)\prime}$ proves the claim.

Step 2. We can decide whether $F(\phi,\psi) \in I$ by looking at the normal form $\mathrm{NF}(F(\phi,\psi)|I)$. In fact, in 7.2.10 it is proved that:

$$F(\phi,\psi) \in I \iff \mathrm{NF}(F(\phi,\psi)|I) = 0.$$

The normal form $\mathrm{NF}(F(\phi,\psi)|I)$ is a power series, all of whose terms are not in $L(I)$, the ideal of leading terms of I. Moreover, $\mathrm{NF}(F(\phi,\psi)|I)$ is congruent to $F(\phi,\psi)$ modulo I. If (ϕ,ψ) is a solution of $F \equiv 0$ modulo I, then it follows immediately that $(\mathrm{NF}(\phi|I),\mathrm{NF}(\psi|I))$ is also a solution of $F \equiv 0$ modulo I. Similar remarks hold for solutions of order e, as obviously $\sum a_\alpha s^\alpha$ being in normal form implies $\sum_{|\alpha|\leq e} a_\alpha s^\alpha$ is in normal form. So we will and may assume that all our solutions are in normal form.

Step 3. We now choose the polyradius. Let $(\overline{\rho},\overline{\tau},\sigma) \in \mathbb{R}_+^{n+1+l+p+q}$ such that $F \in B^k_{(\overline{\rho},\overline{\tau},\sigma)}$. We consider the elements

$$\frac{\partial F}{\partial \Phi_i}|_{s=0} \in \mathbb{C}\{x\}^k \text{ and } \frac{\partial F}{\partial \Psi_j}|_{s=0} \in \mathbb{C}\{x\}^k, \quad \text{for } i=1,\ldots,p \text{ and } j=1,\ldots,q.$$

By Cartan's Lemma 8.2.4 we can choose a $\rho \in \mathbb{R}_+^{n+1}$, $\rho \leq \overline{\rho}$ and $M = M(\rho) \geq 1$ depending on $\frac{\partial F}{\partial \Psi}|_{s=0}$ and $\frac{\partial F}{\partial \Phi}|_{s=0}$ with the following property. Let $G \in B^k_{(\rho,\overline{\tau})}$, and $\overline{\delta}_1,\ldots,\overline{\delta}_p \in \mathbb{C}[s]$ and $\overline{\gamma}_1,\ldots,\overline{\gamma}_q \in \mathbb{C}\{x\}[s]$ such that

$$G + \sum_{i=0}^{p} \overline{\delta}_i \frac{\partial F}{\partial \Phi_i}|_{s=0} + \sum_{i=0}^{q} \overline{\gamma}_i \frac{\partial F}{\partial \Psi_i}|_{s=0} = 0,$$

and $G,\overline{\delta}_1,\dots,\overline{\delta}_p,\overline{\gamma}_1,\dots,\overline{\gamma}_q$ are homogeneous with respect to s of degree $e+1$ for $e \geq e_0$. Then there exist $\delta_1,\dots,\delta_p \in \mathbb{C}[s]$ and $\gamma_1,\dots,\gamma_q \in \mathbb{C}\{x\}[s]$ satisfying this equation being homogeneous as before, where, moreover,

$$||\delta_i||_\tau,\ ||\gamma_i||_{(\rho,\tau)} \leq M||G||_{(\rho,\tau)} \text{ for all } \tau \leq \overline{\tau}.$$

We take $\varepsilon = \frac{1}{M}$. Let $K = K(\varepsilon)$ and $\tau \leq \overline{\tau}$ as in the previous lemma, applied to $F(x,s,\Phi,\Psi)$ and let $(\phi^{(e_0)},\psi^{(e_0)})$, a solution of order e_0. Because $\phi^{(e_0)} \in \mathfrak{m}$ and $\psi^{(e_0)} \in \mathfrak{m}\mathbb{C}\{x,s\}$ we can choose, by scaling, the polyradius τ so small that, moreover, $||\phi^{(e_0)}||_\tau < K$ and $||\psi^{(e_0)}||_{(\rho,\tau)} < K$.

Step 4. We will for all $e \geq e_0+1$ construct solutions $(\phi^{(e)},\psi^{(e)})$ of order e with $(\phi^{(e)},\psi^{(e)}) \equiv (\phi^{(e-1)},\psi^{(e-1)})$ modulo $\mathfrak{m}^e$ and with the estimates:

$$||\phi_e||_\tau,\quad ||\psi_e||_{(\rho,\tau)} < K.$$

Here ϕ_e and ψ_e are the degree e parts in s of $\phi^{(e)}$ and $\psi^{(e)}$. Let $(\phi^{(e)},\psi^{(e)})$ be a solution of order e, such that $(\phi^{(e)},\psi^{(e)})$ are polynomials in s of degree $\leq e$, in normal form with respect to I and $||\phi_e||_\tau \leq K$, $||\psi_e||_{(\rho,\tau)} \leq K$. Let F_e be the degree $e+1$ part in s of the normal form of $F(x,s,\phi^{(e)},\psi^{(e)})$. By the previous lemma, $||F_e||_{(\rho,\tau)} < \varepsilon K = \frac{K}{M}$. By assumption, there exist $\overline{\delta}^{(e)}$ and $\overline{\gamma}^{(e)}$ homogeneous of degree $e+1$ in s and in normal form with respect to I such that $(\phi^{(e)} + \overline{\delta}^{(e)},\psi^{(e)} + \overline{\gamma}^{(e)})$ is a solution of order $e+1$. By Taylor expansion and taking the degree $e+1$ part of the normal form we obtain

$$F_e + \sum_{i=1}^{p} \frac{\partial F}{\partial \Phi_i}|_{s=0}\overline{\delta}_i^{(e)} + \sum_{i=1}^{q} \frac{\partial F}{\partial \Psi_i}|_{s=0}\overline{\gamma}_i^{(e)} = 0.$$

By choice of M there exist such $\delta_i^{(e)}$ and $\gamma_i^{(e)}$, having moreover the estimates:

$$||\delta_i^{(e)}||_\tau,\ ||\gamma_i^{(e)}||_{(\rho,\tau)} < M||F_e||_{(\rho,\tau)} < M\frac{K}{M} = K.$$

Now $\phi^{(e+1)} := \phi^{(e)} + \delta^{(e)}$ and $\psi^{(e+1)} := \psi^{(e)} + \gamma^{(e)}$. Then $\phi_{e+1} = \delta^{(e)}$ and $\psi_{e+1} = \gamma^{(e)}$, and the desired estimate holds.

Step 5. We now look at the formal solution $\phi = \sum_{e=0}^{\infty} \phi_e$, $\psi = \sum_{e=0}^{\infty} \psi_e$. Then, because ϕ_e and ψ_e are homogenous of degree e with respect to s, and $||\phi_e||_\tau < K$, $||\psi_e||_{(\rho,\tau)} < K$ we obtain

$$||\phi||_{\tau/2} < \sum_{e=0}^{\infty} \left(\tfrac{1}{2}\right)^e \cdot K = 2K,\quad ||\psi||_{(\rho,\tau/2)} < \sum_{e=0}^{\infty} \left(\tfrac{1}{2}\right)^e \cdot K = 2K,$$

showing the convergence of ϕ and ψ. This concludes the proof of Grauert's Approximation Theorem. □

Exercises

8.2.6. Let $f(x,\Theta) = \sum_{\nu \in \mathbb{N}^t} f_\nu \Theta^\nu$. Let (ρ,τ) be a polyradius with $||f||_{(\rho,\tau)} < \infty$. Show that there exists an A with $||f_\nu||_\rho \leq A^{|\nu|}$.

8.2.7. Let $c > 0$, prove that for all x with $|x|$ sufficiently small $\frac{1}{(1-x)^t} < (c+t)x + 1$.
(Hint: Check the monotony behavior of the real function corresponding to the above equality.)

8.3 Some other Approximation Theorems

Now we would like to discuss, without proof, a generalization of Artin's Approximation Theorem 8.1.1:

Theorem 8.3.1. *Let $f_1,\dots,f_m \in \mathbb{C}\{x,y\}$. There is a function $\Theta : \mathbb{N} \to \mathbb{N}$ with the following property.*

Let $f_i(\overline{y}) \equiv 0 \mod (x)^{\Theta(c)}$, $i = 1,\dots,m$, for $\overline{y} \in (x)\mathbb{C}\{x\}^N$, then there exist

$$y \in \mathbb{C}\{x\}^N, \quad y \equiv \overline{y} \mod (x)^c \text{ and}$$
$$f_i(y) = 0, \quad i = 1,\dots,m.$$

The point here is that one does not even have to find a formal solution, but only a solution up to high enough order. The weakness of this theorem however is, that it seems not to be possible in general to compute the function Θ. The theorem is a consequence of Artin's Approximation Theorem and the following theorem, proved in [Pfister-Popescu 1975], [Kurke et al. 1978].

Theorem 8.3.2. *Let K be a field and $f_1,\dots,f_m \in K[[x,y]]$, then there exists a function $\Theta : \mathbb{N} \to \mathbb{N}$ with the following property:*

Let $f_i(\overline{y}) \equiv 0 \mod (x)^{\Theta(c)}$, $i = 1,\dots,m$, for $\overline{y} \in (x)K[[x]]^N$. Then there exists $y \in K[[x]]^N$ such that $y \equiv \overline{y} \mod (x)^c$ and $f_i(y) = 0$, $i = 1,\dots,m$.

Artin's Approximation Theorem 8.1.1 has been generalized in many directions. The most general result is due to Dorin Popescu (cf. [Popescu 1985], [Popescu 1986, 1990]) and has been reproved by Marc Spivakovsky, see [Spivakovsky 1999] and Richard Swan [Swan 1998].

Theorem 8.3.3 (Popescu). *Let A be an excellent Noetherian, and with respect to an ideal $\mathfrak{a}$, Henselian ring. Let $f_1,\dots,f_m \in A[Y]$. Suppose $f_i(\overline{y}) = 0$, $i = 1,\dots,m$, $\overline{y} \in \widehat{A}_{\mathfrak{a}}^N$ (the so-called $\mathfrak{a}$-adic completion). Let $c > 0$ be an integer. Then there exists a $y \in A^N$ such that $y \equiv \overline{y} \mod \mathfrak{a}^c$ and $f_i(y) = 0$, $i = 1,\dots,m$.*

We do not want to explain the notion of excellence. This can be found in Matsumura's book (cf. [Matsumura 1979]). Almost all rings occurring in this book, or more generally in algebraic geometry, are excellent, as, for instance, the rings $K[[x]],\mathbb{C}\{x\},K[x]$ their quotients and their polynomial extensions.

The ring A is Henselian with respect to the ideal $\mathfrak{a}$ if Hensel's Lemma holds with respect to $\mathfrak{a}$ (cf. Corollary 3.3.21, replace there (x) by $\mathfrak{a}$, cf. also [Kurke et al. 1975]).

Another important example of an excellent Henselian local ring is the ring $K\langle x\rangle$ of algebraic power series over a field K, that is $K\langle x\rangle = \{f \in K[[x]],\ F(f) = 0$ for a suitable $F \in K[x,T],\ F \neq 0\}$.

We would like to finish this chapter with some important remarks concerning generalizations of the previous theorem.

Remark 8.3.4. A generalization of Theorem 8.3.3 similar to Theorem 8.3.1 is not possible (cf. a counterexample of Spivakovsky [Spivakovsky 1994]).

Remark 8.3.5. Assume that some of the $\overline{y}_i$ in Theorem 8.1.1 do not depend on some of the x_i. In general, it is not possible to find convergent solutions $y = (y_1,\dots,y_N)$ with the same property. Gabrielov, see [Gabrielov 1971] gave a counterexample of the following type:

$f = aY_1 + bY_2 + cY_3 + d \in \mathbb{C}\{x_1,x_2,x_3, Y_1,Y_2,Y_3\}$, $\overline{y} = (\overline{y}_1,\overline{y}_2,\overline{y}_3) \in \mathbb{C}[[x_1,x_2,x_3]]$ such that $f(\overline{y}) = 0$, $\overline{y}_2,\overline{y}_3 \in \mathbb{C}[[x_1,x_2]]$ and there are no $y = (y_1,y_2,y_3) \in \mathbb{C}\{x_1,x_2,x_3\}$ with $f(y) = 0$ and $y_2,y_3 \in \mathbb{C}\{x_1,x_2\}$.

Remark 8.3.6. In the following special situation the problem of Remark 8.3.5 can be solved.

Let K be a field and $f_1,\dots,f_m \in K\langle x,Y\rangle$, assume $f_i(\overline{y}) = 0$ for $\overline{y} \in (x)K[[x]]$. Assume $y_i \in K[[x_1,\dots,x_{v_i}]]$ $1 \le v_1 \le \dots \le v_N \le n$. Let $c > 0$ be an integer, then there exists a $y \in K\langle x\rangle^N$ such that

$$\begin{aligned} &y \equiv \overline{y} \mod (x)^c \\ &y_i \in K\langle x_1,\dots,x_{v_i}\rangle \\ &f_i(y) = 0, \qquad i = 1,\dots,m. \end{aligned}$$

As we will not explain here, this theorem is a consequence of Popescu's general theorem (cf. [Popescu 1985], [Popescu 1986, 1990]). We obtain the following corollary:

Corollary 8.3.7 (Grauert's Theorem for Algebraic Power Series). *We take the notations of and assumptions of 8.2.1 and 8.2.2. Assume furthermore that $I \subset \mathbb{C}\langle s\rangle$ and $F \in \mathbb{C}\langle x,s,\phi,\psi\rangle^k$. Then the system of equations $F \equiv 0$ modulo I has an algebraic solution (ϕ,ψ) with $\phi \equiv \phi^{(e_0)}$ modulo $\mathfrak{m}^{e_0+1}$, and $\psi \equiv \psi^{(e_0)}$ modulo $\mathfrak{m}^{e_0+1}$.*

Artin's Approximation Theorem and its generalizations have many applications. As an application we give Artin's answer to a question of Grauert.

Proposition 8.3.8. *Let A,B,C be analytic $\mathbb{C}$-algebras and $w : C \longrightarrow B$ a homomorphism.*

(1) Assume that for a surjective homomorphism $\varphi : A \longrightarrow B$ the following diagram is commutative:

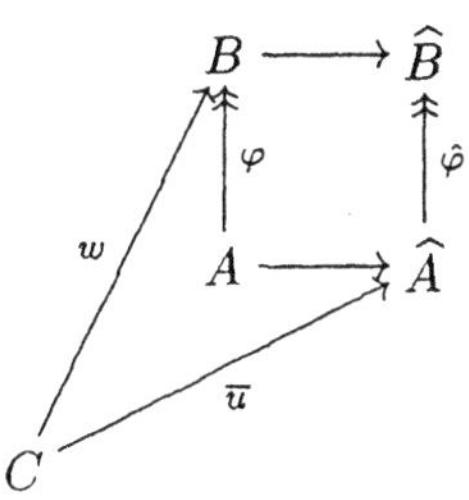

Then, for a given positive integer c, there exists a $u : C \longrightarrow A$ such that $\varphi \circ u = w$ and $u \equiv \overline{u} \mod m^c_{\widehat{A}}$.

(2) Assume the following diagram is commutative:

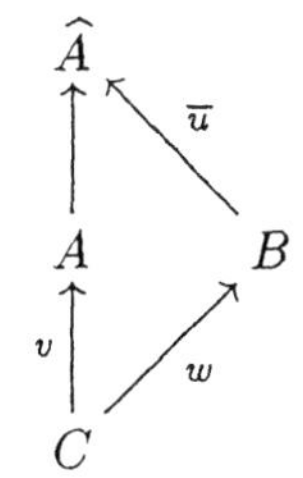

Then for a given positive integer c there exists a $u : B \longrightarrow A$ such that $v = u \circ w$ and $u \equiv \overline{u} \mod m_{\widehat{A}}^c$. If $\hat{\overline{u}} : \widehat{B} \longrightarrow \widehat{A}$ is an isomorphism, then u is also an isomorphism.

Proof. Let $A = \mathbb{C}\{x\}/(f_1,\dots,f_a)$, $B = \mathbb{C}\{y\}/(g_1,\dots,g_b)$, $C = \mathbb{C}\{z\}/(h_1,\dots,h_c)$, $x = (x_1,\dots,x_n)$, $y = (y_1,\dots,y_m)$, $z = (z_1,\dots,z_r)$.

In case of (1), we may assume $y = x$ and $\varphi(x_i \mod (f_1,\dots,f_a)) = x_i \mod (g_1,\dots,g_b)$.

Let $z_i(x) \in \mathbb{C}\{x\}$ be representatives of $w(z_i)$ and $\overline{y}_i(x) \in \mathbb{C}[[x]]$ be representatives of $\overline{u}(z_i)$ (note that $\widehat{A} = \mathbb{C}[[x]]/(f_1,\dots,f_a)$). Then we have, because of the commutativity of the diagram $z_i(x) \equiv \overline{y}_i(x) \mod (g_1,\dots,g_b)$, that is,

$$z_i(x) - \overline{y}_i(x) = \sum_{j=1}^{b} \overline{\alpha}_{ij} g_j, \quad i = 1,\dots,r$$

for suitable $\overline{\alpha}_{ij} \in \mathbb{C}[[x]]$. Furthermore, we have $h_j(\overline{y}(x)) \equiv 0 \mod (f_1,\dots,f_a)$ because of the definition of $\overline{y}$ and the fact that $C = \mathbb{C}\{z\}/(h_1,\dots,h_c)$. This implies $h_i(\overline{y}(x)) = \sum_{j=1}^{a} \overline{\beta}_{ij} f_j$ for suitable $\overline{\beta}_{ij} \in \mathbb{C}[[x]]$.

Now consider the following systems of equations

$$z_i(x) - y_i - \sum_{j=1}^{b} A_{ij} g_j = 0$$

$$h_i(y) - \sum_{j=1}^{a} B_{ij} f_j = 0$$

in the variables $y,\{A_{ij}\},\{B_{ij}\}$. It has the formal solution $(\overline{y},\{\overline{\alpha}_{ij}\},\{\overline{\beta}_{ij}\})$. Using Artin's Approximation Theorem, we obtain for a given c an analytic solution $(y,\{\alpha_{ij}\},\{\beta_{ij}\})$ such that $(y,\{\alpha_{ij}\},\{\beta_{ij}\}) \equiv (\overline{y},\{\overline{\alpha}_{ij}\},\{\overline{\beta}_{ij}\}) \mod (x_1,\dots,x_n)^c$.

We define $u : C \longrightarrow A$ by $u(z_i) = y_i(x) \mod (f_1,\dots,f_a)$. Because of $h_i(y) \equiv 0 \mod (f_1,\dots,f_a)$ it is well-defined and because of $z_i(x) \equiv y_i(x) \mod (g_1,\dots,g_b)$ we have $\varphi \circ u = w$.

To prove (2), let $z_i'(x) \in \mathbb{C}\{x\}$ (respectively $z_i''(y) \in \mathbb{C}\{y\}$) be representatives of $v(z_i)$ (respectively $w(z_i)$). Furthermore, let $\overline{y}_i(x) \in \mathbb{C}[[x]]$ be representatives of $\overline{u}(y_i)$. Similar to case (1), we obtain

$$z_i''(\overline{y}(x)) - z_i'(x) = \sum_{j=1}^{a} \overline{\alpha}_{ij} f_j$$

$$g_i(\overline{y}(x)) = \sum_{j=1}^{a} \overline{\beta}_{ij} f_j$$

for suitable $\overline{\alpha}_{ij},\overline{\beta}_{ij} \in \mathbb{C}[[x]]$. Similar to the first case, the approximation of this formal solution defines the morphism $u : B \longrightarrow A$ with the required properties.

We leave it as an exercise to prove that u is an isomorphism if $\hat{\overline{u}} : \widehat{B} \longrightarrow \widehat{A}$ was already an isomorphism. □

9 Classification of Simple Hypersurface Singularities

This chapter is devoted to the study of hypersurface singularities. There are various equivalence relations one can consider on the set of all functions $f \in \mathbb{C}\{x_1,\ldots,x_n\}$. The two most important ones are right equivalence and contact equivalence. Two functions f and g are called right equivalent if there exists an automorphism φ of $\mathbb{C}\{x_1,\ldots,x_n\}$ such that $\varphi(f) = g$. They are called contact equivalent if there exists an automorphism φ and a unit u in $\mathbb{C}\{x_1,\ldots,x_n\}$ such that $\varphi(f) = u \cdot g$. Until now we considered contact equivalence in this book (without using this name), but in this chapter we will mainly consider right equivalence, because in this case the theorems are somewhat easier to formulate and to prove. (One does not have to worry about the unit.) The aim of this chapter is to give a beginning of the classification of functions up to right equivalence. The first thing is to consider Finite Determinacy Theorems. Let $\mathfrak{m}$ be the maximal ideal of $\mathbb{C}\{x_1,\ldots,x_n\}$. A function f is called (right) k–determined if for all elements g with $f - g \in \mathfrak{m}^{k+1}$ the functions f and g are right equivalent. In particular, a k–determined f is right equivalent to a polynomial of degree at most k. A function is called finitely determined, if there exists a $k \in \mathbb{N}$ such that f is k–determined. It is maybe a little surprising that all functions which have an isolated singularity are finitely determined. The proof, however, is not difficult at all: it is a direct application of Newton's Lemma, see 3.3.34. This result is not strong enough for our purposes, however. We want to find a *small* k such that f is k–determined. So we prove that, if $\mathfrak{m}^{k+1} \subset \mathfrak{m}^2 J(f)$, then f is k–determined. (Here $J(f)$ is the Jacobian ideal, that is, the ideal generated by the partial derivatives of f.) This is what is usually called the Finite Determinacy Theorem, and is more difficult to prove. The idea of the proof is to connect f and g via the path $f + a(g - f)$. So for $a = 0$ we get f, and for $a = 1$ we get g. For all a, we want to prove that there exists a small neighborhood of a in $\mathbb{C}$ such that for all $b \in U$ the functions $f + b(g - f)$ and $f + a(g - f)$ are right equivalent. In order to reach this goal, we need a criterion of local triviality. So, consider an element $F \in \mathbb{C}\{x_1,\ldots,x_n,t\}$ which we view as a family of functions with parameter t. Then we prove that this family is trivial if and only if $\frac{\partial F}{\partial t} \in (x_1,\ldots,x_n)\left(\frac{\partial F}{\partial x_1},\ldots,\frac{\partial F}{\partial x_n}\right)$. In order to prove this, we construct an automorphism in the formal power series ring. The existence of a convergent one then follows from Artin's Approximation Theorem. We will show that this condition holds in the family $f + ag$, so that compactness of [0,1] then completes the proof of the Finite Determinacy Theorem.

As an application of the local triviality we will prove the Mather-Yau Theorem. It says that f and g are contact equivalent, if and only if their Tjurina algebras are isomorphic. We will also prove a corresponding statement for right equivalence. The first reduction in the proof of the Mather-Yau Theorem is to the case that $(f) + J(f) = (g) + J(g)$. Again we connect the functions by a path $f + a(g - f)$. But now one has to be more careful, because it is certainly not true that for all $a \in \mathbb{C}$ the functions $f + a(g - f)$ and f are contact equivalent. For example, the functions f and $-f$ are contact equivalent, but for $a = \frac{1}{2}$ we get that $f + \frac{1}{2}(-f - f) = 0$, which is certainly not contact equivalent to f. But at least we will show that for all $a \in \mathbb{C}$, except for a discrete subset, the functions

f and $f+a(g-f)$ are contact equivalent. As the complement of a discrete subset is still path connected, this is sufficient for the proof of the Mather-Yau Theorem.

In Section 9.2 we start with the classification of functions up to right equivalence. We first define what a simple function is. This is done by putting a topology on $\mathbb{C}\{x_1,\ldots,x_n\}$. A function f is then called simple, if there are only finitely many equivalence classes in a sufficiently small neighborhood of f. The goal of this chapter is to prove that the simple singularities are given by the A-D-E-singularities. We first prove the Splitting Lemma. This allows us to study functions which lie in $\mathfrak{m}^3$. Then, via a sequence of arguments, it is shown that the A-D-E-singularities are simple, and that for nonsimple singularities f we may assume that either $n \geq 3$ and $f \in \mathfrak{m}^3$, or $n=2$ and $f \in \mathfrak{m}^4$, or $n=2$ and $f \in (x,y^2)^3$ or f is nonisolated.

In Section 9.3, we then show that functions with one of these properties are not simple. The idea is to use the Finite Determinacy Theorem, to reduce the study of right equivalence classes to the study of orbits under a group action in $\mathbb{C}^N$ for some $N \gg 0$. In fact, we study the group action on the k-jets, which are the Taylor series of f up to order $k+1$. These orbits are orbits under the action of an algebraic group. This means that all the actions are given by polynomial functions. We then apply some algebraic geometry to prove that orbits of algebraic group actions are locally closed in the Zariski topology. (That is, the intersection of an open and closed subset.) As all orbits contain a smooth point, we can use the group action to show that all points of the orbit are smooth. The next thing to do is to compute the tangent space. We will see that it is equal to $\mathfrak{m}J(f)+\mathfrak{m}^{k+1}/\mathfrak{m}^{k+1}$. In particular, we get the dimension of the orbit of f. We deduce that if $g \notin \mathfrak{m}J(f)$, then in the family $f+t\cdot g$ there are only finitely many t such that $f+tg$ is right equivalent to f. The knowledge of the dimension of the orbits is our main tool for showing that certain functions are not simple.

9.1 Finite Determinacy of Hypersurface Singularities

Definition 9.1.1. Let $f,g \in \mathfrak{m} \subset \mathbb{C}\{x_1,\ldots,x_n\}$, where $\mathfrak{m}$ is, as usual, the maximal ideal.

(1) One says that f is *right equivalent* to g, $f \underset{R}{\sim} g$, if there exists an automorphism φ of $\mathbb{C}\{x_1,\ldots,x_n\}$ such that $\varphi(f)=g$.

(2) One says that f is *contact equivalent* to g, $f \underset{C}{\sim} g$, if there exists an automorphism φ of $\mathbb{C}\{x_1,\ldots,x_n\}$ such that $(\varphi(f))=(g)$, that is, there exists a unit u such that $\varphi(f)=ug$. Thus f and g are contact equivalent exactly if the $\mathbb{C}$-algebras $\mathbb{C}\{x_1,\ldots,x_n\}/(f)$ and $\mathbb{C}\{x_1,\ldots,x_n\}/(g)$ are isomorphic. This is exactly the same as saying that the germs of the analytic hypersurfaces defined by f and g are isomorphic.

In this chapter, we will mainly work with right equivalence. Similar results hold for contact equivalence, but the formulation and proofs will be left to the reader.

Definition 9.1.2. Let $f \in \mathfrak{m} \subset \mathbb{C}\{x_1,\ldots,x_n\}$. Then f is called *k-determined* if all $g \in \mathbb{C}\{x_1,\ldots,x_n\}$ with $f-g \in \mathfrak{m}^{k+1}$ are right equivalent to f. If f is k-determined for some $k \in \mathbb{N}$, then f is called *finitely determined.* In particular, a k-determined f is right equivalent to a polynomial.

The fact that functions which have an isolated singularity are finitely determined follows quite easily from Newton's Lemma 3.3.34, as we will show now.

Theorem 9.1.3. *Let $f \in \mathfrak{m}^2 \subset \mathbb{C}\{x_1,\dots,x_n\}$. Suppose $\mathfrak{m}^{k+1} \subset \mathfrak{m}J(f)^2$. Then f is k-determined. In particular, functions with isolated singularities are finitely determined.*

Proof. Let $g \in \mathfrak{m}$ such that $f - g \in \mathfrak{m}^{k+1}$. Consider the following system of equations, which we want to solve for $y_1,\dots,y_n \in \mathbb{C}\{x_1,\dots,x_n\}$:

$$F(x,y) := f(y) - g(x) = 0.$$

We start with the initial approximation $\overline{y}_i(x) = x_i$. Obviously $\frac{\partial F}{\partial y_i}(x,\overline{y}(x)) = \frac{\partial f}{\partial y_i}(\overline{y}(x)) = \frac{\partial f}{\partial x_i}$. As by assumption $F(x,\overline{y}) = f(x) - g(x) \in \mathfrak{m}^{k+1} \subset \mathfrak{m}J(f)^2$, we can apply Newton's Lemma 3.3.34. There exist $y_1,\dots,y_n \in \mathbb{C}\{x_1,\dots,x_n\}$ with

- $F(x,y) = 0$,
- $y_i(x) \equiv x_i \mod \mathfrak{m}J(f)$. Note that $\mathfrak{m}J(f) \subset \mathfrak{m}^2$.

This says that the y give a coordinate transformation transforming f into g. □

So the fact that germs of holomorphic functions with an isolated singularity are right equivalent to a polynomial is a relatively easy consequence of Newton's Lemma. For our purposes we need, however, a stronger version of the Finite Determinacy Theorem. The statement of the above theorem can be strengthened to the following statement.

Theorem 9.1.4 (Finite Determinacy Theorem). *Let $f \in \mathfrak{m}^2 \subset \mathbb{C}\{x_1,\dots,x_n\}$. Suppose that $\mathfrak{m}^{k+1} \subset \mathfrak{m}^2 J(f)$. Then f is k-determined. In particular, if $k \geq \mu(f) + 1$, then f is k-determined.*

Note that from $\mathfrak{m}^k \subset \mathfrak{m}J(f)$ it follows that $\mathfrak{m}^{k+1} \subset \mathfrak{m}^2 J(f)$. The last condition is somewhat weaker. See Exercise 9.1.13.

The basic idea of the proof of the Finite Determinacy Theorem is to connect f and g via the path $F_t = f + t\cdot(g-f)$, for $0 \leq t \leq 1$. We need a criterion whether for given t,s the element F_t is right equivalent to F_s. The following theorem gives us such a criterion "locally". By this we mean that we give a criterion that F_t is right equivalent to F_0, for small t.

Theorem 9.1.5. *Let $F \in \mathbb{C}\{x_1,\dots,x_n,t\}$ and $c \geq 0$ be an integer. The following conditions are equivalent:*

(1) $\frac{\partial F}{\partial t} \in (x_1,\dots,x_n)^c \left(\frac{\partial F}{\partial x_1},\dots,\frac{\partial F}{\partial x_n}\right)$.

(2) There exists a $\varphi = (\varphi_1,\dots,\varphi_n) \in \mathbb{C}\{x_1,\dots,x_n,t\}^n$ such that

(a) $\varphi_i(x_1,\dots,x_n,0) = x_i$,

(b) $\varphi_i - x_i \in (x_1,\dots,x_n)^c$,

(c) $F(\varphi_1,\dots,\varphi_n,0) = F(x_1,\dots,x_n,t)$.

Corollary 9.1.6. *Suppose that $\frac{\partial F}{\partial t} \in (x_1,\dots,x_n)\left(\frac{\partial F}{\partial x_1},\dots,\frac{\partial F}{\partial x_n}\right)$ for some $F \in \mathbb{C}\{x_1,\dots,x_n,t\}$. Then there exists a small neighborhood U of 0 in $\mathbb{C}$ such that for all fixed $a \in U$ the function $F_a := F(x_1,\dots,x_n,a)$ is right equivalent to F_0.*

Proof. Consider the elements $\varphi_1, \dots, \varphi_n \in \mathbb{C}\{x_1, \dots, x_n, t\}$ which exist according to 9.1.5. There exists a small open neighborhood U of 0 in $\mathbb{C}$, so that for all $a \in U$, the $\varphi_i(x,a)$ are convergent and $\det\left(\frac{\partial \varphi_i(x,a)}{\partial x_j}(0)\right) \neq 0$. Thus for fixed $a \in U$, $(\varphi_1, \dots, \varphi_n)$ is an automorphism of $\mathbb{C}\{x_1, \dots, x_n\}$. Moreover, formula (c) in Theorem 9.1.5 says that it transforms $F(x_1, \dots, x_n, 0)$ into $F(x_1, \dots, x_n, a)$. □

Proof of Theorem 9.1.5.
Step 1. We first prove the easy implication (2) $\Longrightarrow$ (1). As $\varphi_i(x_1, \dots, x_n, 0) = x_i$ it follows easily from the Inverse Function Theorem that $(\varphi_1, \dots, \varphi_n, t)$ is an automorphism of $\mathbb{C}\{x_1, \dots, x_n, t\}$. Let $\psi = (\psi_1, \dots, \psi_n, t)$ be its inverse. Then $\psi_i - x_i \in (x_1, \dots, x_n)^c$, as one easily checks.

By assumption, we have $F(\varphi, 0) = F(x,t)$. Compose this with ψ. Then $F(x,0) = F(\psi,t)$. Hence the derivative of $F(\psi_1, \dots, \psi_n, t)$ with respect to t is zero. Thus

$$\sum_{i=1}^{n} \frac{\partial F}{\partial x_i}(\psi_1, \dots, \psi_n, t) \frac{\partial \psi_i}{\partial t} + \frac{\partial F}{\partial t}(\psi_1, \dots, \psi_n, t) = 0. \tag{9.1}$$

Compose (9.1) with the automorphism $(\varphi_1, \dots, \varphi_n, t)$. Then

$$\sum_{i=1}^{n} \frac{\partial F}{\partial x_i} \cdot \frac{\partial \psi_i}{\partial t}(\varphi_1, \dots, \varphi_n, t) + \frac{\partial F}{\partial t} = 0. \tag{9.2}$$

We already noted that $\psi_i - x_i \in (x_1, \dots, x_n)^c$. But then also $\frac{\partial \psi_i}{\partial t}(\varphi_1, \dots, \varphi_n, t) \in (x_1, \dots, x_n)^c$ follows. So formula (9.2) gives the implication (2) $\Longrightarrow$ (1).

Step 2. We now assume that $\frac{\partial F}{\partial t} \in (x_1, \dots, x_n)^c \left(\frac{\partial F}{\partial x_1}, \dots, \frac{\partial F}{\partial x_n}\right)$. By Artin's Approximation Theorem, or better its (slightly) more general version given in Exercise 8.1.6, it follows that it suffices to find a *formal* $\varphi = (\varphi_1, \dots, \varphi_n) \in \mathbb{C}[[x_1, \dots, x_n, t]]^n$ such that

(1) $\varphi_i(x_1, \dots, x_n, 0) = x_i$,

(2) $\varphi_i - x_i \in (x_1, \dots, x_n)^c$,

(3) $F(\varphi_1, \dots, \varphi_n, 0) = F(x_1, \dots, x_n, t)$.

We now construct those formal $\varphi_1, \dots, \varphi_n$. By assumption, we can write

$$\frac{\partial F}{\partial t} = \sum_{i=1}^{n} \xi_i \frac{\partial F}{\partial x_i}, \quad \xi_i \in (x_1, \dots, x_n)^c.$$

Consider the differential operator $\delta := -\frac{\partial}{\partial t} + \sum_{i=1}^{n} \xi_i \frac{\partial}{\partial x_i}$. Let s be a new variable and consider the map

$$\begin{aligned} \Phi : \mathbb{C}[[x_1, \dots, x_n, t, s]] &\longrightarrow \mathbb{C}[[x_1, \dots, x_n, t, s]] \\ a &\longmapsto \exp(s\delta)(a) := \sum_{k=0}^{\infty} \tfrac{1}{k!} \delta^k(a) s^k \quad \text{for } a \in \mathbb{C}[[x_1, \dots, x_n, t]] \\ s &\longmapsto s. \end{aligned}$$

Put $\Phi_i := \Phi(x_i)$ for $i = 1, \dots, n$, and $\Phi_{n+1} = \Phi(t)$. Note that $\Phi_{n+1} = \Phi(t) = t - s$. Now define $\varphi_i := \Phi_i(x_1, \dots, x_n, t, t)$ for $i = 1, \dots, n$. We have to show that these φ_i satisfy our conditions.

(1) The first condition is easy. Indeed, $\varphi_i(x_1, \dots, x_n, 0)$ by definition is equal to $\delta^0(x_i) = x_i$.

(2) The second condition is also not too difficult. Indeed from the first part we have

$$\Phi(x_i) = x_i + \sum_{k \geq 1} \tfrac{1}{k!} \delta^k(x_i) s^k.$$

From the definition of δ, and the fact that the $\xi_i \in (x_1, \dots, x_n)^c$ it follows immediately that $\Phi(x_i) - x_i \in (x_1, \dots, x_n)^c$. The second condition now follows.

(3) The third condition is the most difficult one. The crucial thing to show is

$$\frac{\partial F(\Phi_1, \dots, \Phi_{n+1})}{\partial s} = 0, \tag{9.3}$$

that is, $F(\Phi_1, \dots, \Phi_n, \Phi_{n+1})$ does not depend on s. Indeed, assuming (9.3) it follows that

$$\begin{aligned} F(\varphi_1, \dots, \varphi_n, 0) = F(\Phi_1, \dots, \Phi_n, \Phi_{n+1})_{s=t} = \\ F(\Phi_1, \dots, \Phi_n, \Phi_{n+1})_{s=0} = F(x_1, \dots, x_n, t). \end{aligned}$$

The first equality holds by definition of the φ_i, the second equality because of (9.3). The third equality is because $\Phi_{i|s=0} = x_i$ for $i \leq n$, and $\Phi_{n+1|s=0} = t$.

It therefore remains to show (9.3). We have the following formulae:

$$\frac{\partial \Phi_i}{\partial s} = \xi_i(\Phi_1, \dots, \Phi_n, \Phi_{n+1}), \tag{9.4}$$

$$\frac{\partial \Phi_{n+1}}{\partial s} = -1, \tag{9.5}$$

$$\delta(F) = 0. \tag{9.6}$$

The second and the third formula follow immediately from the fact that $\Phi_{n+1} = t - s$, and the definition of δ. To prove formula (9.4), consider an arbitrary $a \in \mathbb{C}[[x_1, \dots, x_n, t]]$, and differentiate:

$$\frac{\partial \Phi(a)}{\partial s} = \sum_{k=1}^{\infty} \frac{1}{(k-1)!} \delta^k(a) s^{k-1} = \sum_{k=1}^{\infty} \frac{1}{(k-1)!} \delta^{k-1}(\delta(a)) s^{k-1} = \Phi(\delta(a)).$$

In particular for $i = 1, \dots, n$ we have

$$\frac{\partial \Phi_i}{\partial s} = \Phi(\delta(x_i)) = \Phi(\xi_i) = \xi_i(\Phi_1, \dots, \Phi_n, \Phi_{n+1}).$$

The last equality follows as $\Phi_i = \Phi(x_i)$ for $1 \leq i \leq n$, and $\Phi(t) = \Phi_{n+1}$, by definition. By using the chain rule and the formulae (9.4), (9.5) and (9.6) we get

$$\begin{aligned} \frac{\partial F(\Phi_1, \dots, \Phi_{n+1})}{\partial s} &= \sum_{i=1}^{n} \frac{\partial F}{\partial x_i}(\Phi_1, \dots, \Phi_{n+1}) \frac{\partial \Phi_i}{\partial s} - \frac{\partial F}{\partial t}(\Phi_1, \dots, \Phi_{n+1}) \\ &= \sum_{i=1}^{n} \frac{\partial F}{\partial x_i}(\Phi_1, \dots, \Phi_{n+1}) \xi_i(\Phi_1 \dots, \Phi_{n+1}) - \frac{\partial F}{\partial t}(\Phi_1, \dots, \Phi_{n+1}) \\ &= \delta(F)(\Phi_1, \dots, \Phi_{n+1}) \\ &= 0. \end{aligned}$$

□

Proof of Theorem 9.1.4.
Step 1. Let $g \in \mathfrak{m}^{k+1}$ be given. Let $0 \leq a \leq 1$. Consider

$$F_a(x_1, \dots, x_n, t) := f + (a+t) \cdot g.$$

For all a we will show that there exists an open neighborhood U of a such that for all $a + t \in U$

$$f + (a+t) \cdot g \underset{R}{\sim} f + a \cdot g. \tag{9.7}$$

Because the unit interval [0,1] is compact we can cover the unit interval with finitely many U. As right equivalence is an equivalence relation, it then follows that $f = F_0$ is right equivalent to $F_1 = f + g$.

Step 2. To prove (9.7) we will apply Corollary 9.1.6, and it therefore suffices to prove

$$\frac{\partial F_a}{\partial t} \in \mathfrak{m}^2 \left(\frac{\partial F_a}{\partial x_1}, \dots, \frac{\partial F_a}{\partial x_n} \right). \tag{9.8}$$

Here $\mathfrak{m} = (x_1, \dots, x_n)$. By applying Artin's Approximation Theorem, it suffices to prove this in the formal power series ring $\mathbb{C}[[x_1, \dots, x_n, t]]$.

Step 3. We will first show the following intermediate result.

Let $f \in \mathbb{C}\{x_1, \dots, x_n\}$ and assume $\mathfrak{m}^{k+1} \subset \mathfrak{m}^2 J(f)$. Then $\mathfrak{m}^{k+1} \subset \mathfrak{m}^2 J(f+g)$ for all $g \in \mathfrak{m}^{k+1}$.

In order to prove this, put $F := f + g$. Let $h \in \mathfrak{m}^{k+1}$ be given. We have to find $\overline{\xi}_i \in \mathfrak{m}^2$ such that

$$h = \sum_{i=1}^{n} \overline{\xi}_i \frac{\partial F}{\partial x_i}. \tag{9.9}$$

By applying Artin's Approximation Theorem, it suffices to find $\overline{\xi}_i \in \mathfrak{m}^2 \mathbb{C}[[x_1, \dots, x_n]]$ satisfying (9.9). Given $h \in \mathfrak{m}^{k+1}$, we can by assumption write $h = \sum_{i=1}^{n} \xi_i \frac{\partial f}{\partial x_i}$, $\xi_i \in \mathfrak{m}^2$. But then

$$h = \sum_{i=1}^{n} \xi_i \frac{\partial F}{\partial x_i} - \sum_{i=1}^{n} \xi_i \frac{\partial g}{\partial x_i}. \tag{9.10}$$

As $g \in \mathfrak{m}^{k+1}$ and $\xi_i \in \mathfrak{m}^2$ it follows that $\xi_i \frac{\partial g}{\partial x_i}$ in $\mathfrak{m}^{k+2}$. By assumption $\mathfrak{m}^{k+1} \subset \mathfrak{m}^2 J(f)$, so that we can write

$$\sum_{i=1}^{n} \xi_i \frac{\partial g}{\partial x_i} = \sum_{i=1}^{n} \xi_i^{(1)} \frac{\partial f}{\partial x_i}, \quad \xi_i^{(1)} \in \mathfrak{m}^3. \tag{9.11}$$

(This is the reason we need $\mathfrak{m}^{k+1} \subset \mathfrak{m}^2 J(f)$, as the condition $\mathfrak{m}^{k+1} \subset \mathfrak{m} J(f)$ does not suffice.) Plugging (9.11) in (9.10) we get

$$h = \sum_{i=1}^{n} (\xi_i + \xi_i^{(1)}) \frac{\partial F}{\partial x_i} - \sum_{i=1}^{n} \xi_i^{(1)} \frac{\partial g}{\partial x_i} \quad \text{with} \sum_{i=1}^{n} \xi_i^{(1)} \frac{\partial g}{\partial x_i} \in \mathfrak{m}^{k+3}.$$

Repeating the procedure, we get $\overline{\xi}_i \in \mathfrak{m}^2\mathbb{C}[[x_1,\dots,x_n]]$ satisfying formula (9.9).

Step 4. The rest of the proof runs similar to the third step. Applying the third step to ag we obtain that $\mathfrak{m}^{k+1} \subset \mathfrak{m}^2 J(f+ag)$. Hence

$$g = \sum_{i=1}^{n} \xi_i \frac{\partial(f+ag)}{\partial x_i}, \quad \text{for some } \xi_i \in \mathfrak{m}^2.$$

This implies

$$\frac{\partial F_a}{\partial t} = g = \sum_{i=1}^{n} \xi_i \frac{\partial F_a}{\partial x_i} - t \cdot \sum_{i=1}^{n} \xi_i \frac{\partial g}{\partial x_i}. \tag{9.12}$$

As $g \in \mathfrak{m}^{k+1}$, it follows that $\xi_i \frac{\partial g}{\partial x_i} \in \mathfrak{m}^{k+2}$. Thus we can write

$$\sum_{i=1}^{n} \xi_i \frac{\partial g}{\partial x_i} = \sum_{i=1}^{n} \xi_i^{(1)} \frac{\partial(f+ag)}{\partial x_i}, \quad \xi_i^{(1)} \in \mathfrak{m}^3. \tag{9.13}$$

Plugging in (9.13) in (9.12) we get

$$g = \frac{\partial F_a}{\partial t} = \sum_{i=1}^{n} (\xi_i - t\xi_i^{(1)}) \frac{\partial F_a}{\partial x_i} \quad \mod t^2\mathfrak{m}^{k+3}.$$

Iterating this procedure we obtain

$$\frac{\partial F_a}{\partial t} = \sum_{i=1}^{n} \overline{\xi}_i \frac{\partial F_a}{\partial x_i}, \quad \overline{\xi}_i \in \mathfrak{m}^2\mathbb{C}[[x_1,\dots,x_n,t]].$$

Step 5. For the final statement we apply Exercise 1.3.20. It follows that $\mathfrak{m}^{k-1} \subset J(f)$, from which $\mathfrak{m}^{k+1} \subset \mathfrak{m}^2 J(f)$ follows. So we can apply the first part of the theorem. □

Theorem 9.1.5 can be generalized as follows:

Theorem 9.1.7 (Characterization of Local Analytic Triviality).
Let $f \in \mathbb{C}\{x_1,\dots,x_n,y_1,\dots,y_m\}$ and $c \geq 0$ be an integer. The following conditions are equivalent:

(1) $\frac{\partial f}{\partial y_i} \in (x_1,\dots,x_n)^c \left(\frac{\partial f}{\partial x_1},\dots,\frac{\partial f}{\partial x_n}\right) + (f)$ *for* $i = 1,\dots,m$.

(2) There exist $\varphi_1,\dots,\varphi_n,u \in \mathbb{C}\{x_1,\dots,x_n,y_1,\dots,y_m\}$ *such that*

- $u(x_1,\dots,x_n,0,\dots,0) = 1$
- $\varphi_i(x_1,\dots,x_n,0,\dots,0) = x_i$
- $\varphi_i - x_i \in (x_1,\dots,x_n)^c$
- $f(x_1,\dots,x_n,y_1,\dots,y_m) = u \cdot f(\varphi_1,\dots,\varphi_n,0,\dots,0)$.

If moreover $\frac{\partial f}{\partial y_i} \in (x_1,\dots,x_n)^c \left(\frac{\partial f}{\partial x_1},\dots,\frac{\partial f}{\partial x_n}\right)$ *for all* i, *then we can choose* $u = 1$.

The proof is similar to the proof of 9.1.5 and will be left as Exercise 9.1.15. We now come to the Theorem of Mather and Yau.

Theorem 9.1.8 (Mather-Yau). *Let f and $g \in \mathfrak{m} = (x_1, \ldots, x_n) \subset \mathbb{C}\{x_1, \ldots, x_n\}$. Then the following conditions are equivalent:*

(1) $\mathbb{C}\{x_1, \ldots, x_n\}/(f) \cong \mathbb{C}\{x_1, \ldots, x_n\}/(g)$, *that is, f and g are contact equivalent.*

(2) $\mathbb{C}\{x_1, \ldots, x_n\}/((f) + \mathfrak{m}J(f)) \cong \mathbb{C}\{x_1, \ldots, x_n\}/((g) + \mathfrak{m}J(g))$ *as $\mathbb{C}$-algebras.*

If f and g define isolated singularities these conditions are equivalent to

(3) $\mathbb{C}\{x_1, \ldots, x_n\}/((f) + J(f)) \cong \mathbb{C}\{x_1, \ldots, x_n\}/((g) + J(g))$ *as $\mathbb{C}$-algebras.*

To put it in another way, the function f and g are contact equivalent if and only if the Tjurina algebras of f and g are isomorphic.

Remark 9.1.9. It is not true that functions are right equivalent if and only if their Milnor algebras are isomorphic. Indeed, consider the family of functions $f_t = x^4 + y^5 + t \cdot x^2y^3$. Then one can show, see Exercise 9.1.16:

(1) $f_t \underset{R}{\sim} f_1 \Longleftrightarrow t = a^2b^2,\ a^4 = 1,\ b^5 = 1$.

(2) Consider the automorphism φ_α of $\mathbb{C}\{x,y\}$ defined by $\varphi_\alpha(x) = \alpha^5 x$, and $\varphi_\alpha(y) = \alpha^4 y$, with $\alpha^2 = t^{-1}$. Then $\varphi_\alpha(J(f_t)) = J(f_1)$.

To find the correct theorem for the case of right equivalence, we consider the Milnor algebra $\mathbb{C}\{x_1, \ldots, x_n\}/J(f)$ as $\mathbb{C}\{t\}$-algebra with multiplication $t \cdot h := fh$ for $h \in \mathbb{C}\{x_1, \ldots, x_n\}/J(f)$.

Theorem 9.1.10 (Mather-Yau for Right Equivalence).
Let f and $g \in \mathfrak{m} = (x_1, \ldots, x_n) \subset \mathbb{C}\{x_1, \ldots, x_n\}$. Then the following conditions are equivalent:

(1) $f \underset{R}{\sim} g$.

(2) $\mathbb{C}\{x_1, \ldots, x_n\}/\mathfrak{m}J(f) \cong \mathbb{C}\{x_1, \ldots, x_n\}/\mathfrak{m}J(g)$ *as $\mathbb{C}\{t\}$-algebras.*

If f and g define isolated singularities these conditions are equivalent to

(3) $\mathbb{C}\{x_1, \ldots, x_n\}/J(f) \cong \mathbb{C}\{x_1, \ldots, x_n\}/J(g)$ *as $\mathbb{C}\{t\}$-algebras.*

The proof of this theorem is similar to the proof of 9.1.8 and left as Exercise 9.1.20.

Proof of the Mather-Yau Theorem for Contact Equivalence 9.1.8.
The equivalence (1) $\Longleftrightarrow$ (2) is left as Exercise 9.1.17. The implication (1) $\Longrightarrow$ (3) is easy, and has been proved in 3.4.27. The implication (3) $\Longrightarrow$ (1) is the most difficult case, and will be proved in several steps.

Step 1. We first reduce to the case $(f) + J(f) = (g) + J(g)$. Let $\varphi : \mathbb{C}\{x_1, \ldots, x_n\}/((f) + J(f)) \cong \mathbb{C}\{x_1, \ldots, x_n\}/((g) + J(g))$ be a $\mathbb{C}$-algebra isomorphism. We may assume that $\operatorname{ord}(f), \operatorname{ord}(g) \geq 2$. The isomorphism φ induces an isomorphism between $((f) + J(f) + \mathfrak{m}^2)/\mathfrak{m}^2$ and $((g) + J(g) + \mathfrak{m}^2)/\mathfrak{m}^2$. These are finite-dimensional $\mathbb{C}$-vector spaces. Let k be its dimension, and let $z_1, \ldots, z_n \in \mathbb{C}\{x_1, \ldots, x_n\}$ induce a basis of $\mathfrak{m}/\mathfrak{m}^2$ such that $z_1, \ldots, z_k \in (f) + J(f)$. Let $w_1, \ldots, w_n \in \mathbb{C}\{x_1, \ldots, x_n\}$ induce a basis of $\mathfrak{m}/\mathfrak{m}^2$ such that $w_1, \ldots, w_k \in (g) + J(g)$ and $\varphi([z_{k+i}]) = [w_{k+i}]$ for $i \geq 1$. Define an automorphism ψ on $\mathbb{C}\{x_1, \ldots, x_n\}$ by $\psi(z_i) = w_i$ for all i. This is an automorphism by the Inverse Function Theorem 3.3.7. Then the diagram

$$\begin{array}{ccc}
\mathbb{C}\{x_1,\ldots,x_n\} & \xrightarrow{\psi} & \mathbb{C}\{x_1,\ldots,x_n\} \\
\downarrow & & \downarrow \\
\mathbb{C}\{x_1,\ldots,x_n\}/((f)+J(f)) & \xrightarrow{\varphi} & \mathbb{C}\{x_1,\ldots,x_n\}/((g)+J(g))
\end{array}$$

commutes, that is, $\psi((f)+J(f)) = (g)+J(g)$. Now let $h := \psi(f)$. Then $(h)+J(h) = (g)+J(g)$, because $\psi((f)+J(f)) = ((\psi(f))+J(\psi(f)))$, see 3.4.27. This is what we wanted to show.

Step 2. We may now assume that $(f)+J(f) = (g)+J(g)$. Let $h := f-g$ and $F := g+th \in \mathbb{C}\{x_1,\ldots,x_n,t\}$. So we are connecting f and g by a (complex) path: for $t=0$ we have g, and for $t=1$ we have f. Unlike in the proof of the Finite Determinacy Theorem, it is not true however that for all a, the functions g and $g+ah$ are contact equivalent. In this step we will show that at least for t small the functions g and $g+th$ are contact equivalent. From $(f)+J(f) = (g)+J(g)$, it follows that $(h)+J(h) \subset (g)+J(g)$. Hence

$$\begin{aligned}
\frac{\partial F}{\partial t} = h &= ug + \sum_{i=1}^{n} \xi_i \frac{\partial g}{\partial x_i} \\
&= uF + \sum_{i=1}^{n} \xi_i \frac{\partial F}{\partial x_i} - t\left(uh - \sum_{i=1}^{n} \xi_i \frac{\partial h}{\partial x_i}\right) \\
&= uF + \sum_{i=1}^{n} \xi_i \frac{\partial F}{\partial x_i} + t\left(u^{(1)}g + \sum_{i=1}^{n} \xi_i^{(1)} \frac{\partial g}{\partial x_i}\right)
\end{aligned}$$

Iterating this process we obtain

$$\frac{\partial F}{\partial t} = \overline{u}F + \sum_{i=1}^{n} \overline{\xi}_i \frac{\partial F}{\partial x_i}$$

for suitable $\overline{u},\overline{\xi}_i \in \mathbb{C}[[x_1,\ldots,x_n,t]]$. Using Artin's Approximation Theorem 8.1.1 we obtain

$$\frac{\partial F}{\partial t} \in \left(F, \frac{\partial F}{\partial x_1},\ldots,\frac{\partial F}{\partial x_n}\right).$$

Now we can apply Theorem 9.1.7 to obtain

$$\varphi(F(x_1,\ldots,x_n,0)) = wF(x_1,\ldots,x_n,t)$$

for a suitable map $\varphi = (\varphi_1,\ldots,\varphi_n,t) : \mathbb{C}\{x_1,\ldots,x_n,t\} \longrightarrow \mathbb{C}\{x_1,\ldots,x_n,t\}$ and w a unit in $\mathbb{C}\{x_1,\ldots,x_n,t\}$. Now we want to prove that for small t fixed the map $(\varphi_1,\ldots,\varphi_n,t)$ is an automorphism of $\mathbb{C}\{x_1,\ldots,x_n\}$. This is the place where we need the assumption that f and g define isolated singularities.[1]

At least the map $(\varphi_1,\ldots,\varphi_n,t)$ is an automorphism of $\mathbb{C}\{x_1,\ldots,x_n,t\}$, so has an inverse $(\psi_1,\ldots,\psi_n,t)$, so that

[1] Note that this is not necessary for the proof of (1) $\iff$ (2). Here the fact that for small t fixed the map $(\varphi_1,\ldots,\varphi_n,t)$ is an automorphism follows directly from Theorem 9.1.7 because we have $\frac{\partial F}{\partial t} \in (F) + \mathfrak{m}\left(\frac{\partial F}{\partial x_1},\ldots,\frac{\partial F}{\partial x_n}\right)$.

$$g = F(x_1, \ldots, x_n, 0) = w' F(\psi_1, \ldots, \psi_n, t)$$

for some unit w'. Thus $(V(F), o)$ is as germ of an analytic space isomorphic to $(V(g), o)$. This is the product of an isolated singularity $(V(g), o)$ in $(\mathbb{C}^n, o)$ with the germ of a smooth one-dimensional space $(\mathbb{C}, o)$. The singular locus of $(V(g), o)$ in $(\mathbb{C}^{n+1}, o)$ is a line, so that the singular locus of $(V(F), o)$ is a smooth curve. Now this curve must be given by $x_1 = \cdots = x_n = 0$, as $F = g + t \cdot (f - g)$, and we assumed that the order of both f and g is at least two, and therefore $\left(F, \frac{\partial F}{\partial x_1}, \ldots, \frac{\partial F}{\partial x_n}, \frac{\partial F}{\partial t}\right) \subset (x_1, \ldots, x_n)$. Define $a_i(t) := \psi_i(0, \ldots, 0, t)$. Then $\psi_i - a_i \in (x_1, \ldots, x_n)$, thus for fixed sufficiently small t $(\psi_1 - a_1, \ldots, \psi_n - a_n)$ gives an automorphism of $\mathbb{C}\{x_1, \ldots, x_n\}$. We want to show that $a_i = 0$. For some small representative, the singular point $(0, \ldots, 0, t)$ is mapped via the automorphism $(\psi_1, \ldots, \psi_n, t)$ to a singular point of $V(g)$. As the only singular points of $V(g)$ are given by $x_1 = \cdots = x_n = 0$, it follows that $\psi_i(0, \ldots, 0, t) = 0$. As moreover $\psi_i(x_1, \ldots, x_n, 0) = x_i$, it follows that $\psi_i - x_i \in t \cdot (x_1, \ldots, x_n)$. Thus for fixed small t, the $(\psi_1, \ldots, \psi_n)$, and hence $(\varphi_1, \ldots, \varphi_n)$ gives an automorphism of $\mathbb{C}\{x_1, \ldots, x_n\}$ by the Inverse Function Theorem.

(Note that in this step we only used the inclusion $(h) + J(h) \subset (g) + J(g)$.)

Step 3. We want to redo the argument in step 2 for the function $F := g + ah + th$ for almost all a, that is, for all a except for a discrete subset of $\mathbb{C}$. In order for the argument to work, we need that $(h) + J(h) \subset (g + ah) + J(g + ah)$. This would follow from the statement

$$(9.14) \qquad (g + ah) + J(g + ah) = (g) + J(g)$$

Indeed, if this were true, then $(h) + J(h) \subset (g) + J(g) = (g + ah) + J(g + ah)$ and we can redo the argument of Step 2. Now (9.14) does not hold for all $a \in \mathbb{C}$, but we will show that it holds for almost all $a \in \mathbb{C}$. To prove (9.14) for almost all $a \in \mathbb{C}$, we use the theory of coherent sheaves. Let $U \subset \mathbb{C}^n$ be an open subset such that f and g converge on U and $0 \in U$. Define $G := g + th$, and $J(G) = \left(\frac{\partial G}{\partial x_1}, \ldots, \frac{\partial G}{\partial x_n}\right)$. We have $(G) + J(G) \subset (g) + J(g)$. The ideal sheaves $(G) + J(G)$ and $(g) + J(g)$ in $\mathcal{O}_{U \times \mathbb{C}}$ are coherent sheaves. Therefore, the quotient $\mathcal{M} := ((g) + J(g))/((G) + J(G))$ is a coherent sheaf. Supports of coherent analytic sheaves are analytic sets by 6.2.9, so that $Z := \operatorname{supp}(\mathcal{M})$ is an analytic subset of $U \times \mathbb{C}$. Therefore, the intersection $\widetilde{Z} := Z \cap (\{0\} \times \mathbb{C})$ is an analytic subset of $\{0\} \times \mathbb{C}$. Now $\widetilde{Z}$ is the set of all points $(0, a)$ such that $\mathcal{M}_{(0,a)} \neq (0)$, and therefore includes the set of all points for which the inclusion of stalks $(G) + J(G) \subset (g) + J(g)$ is strict. As an analytic subset of $\{0\} \times \mathbb{C}$, it is either $\{0\} \times \mathbb{C}$ or a discrete set (Identity Theorem 3.1.10). But obviously $(0,0) \notin \widetilde{Z}$, therefore, $\widetilde{Z}$ is discrete. This proves (9.14) for all a except for a discrete subset of $\mathbb{C}$. This discrete subset we call D.

Step 4. Because of $(f) + J(f) = (g) + J(g)$, we see that $1 \notin D$. Choose a path $\gamma : [0,1] \longrightarrow \mathbb{C} \setminus D$ from 0 to 1. For all $a \in \gamma([0,1])$, we have that there exists an open neighborhood U of a such that for all $b \in U$, the function $f + bh$ is contact equivalent to $f + ah$ by Step 3. As $\gamma([0,1])$ is compact, we can cover $\gamma([0,1])$ by a finite number of $U's$. It follows that f and $f + g - f = g$ are contact equivalent. □

Remark 9.1.11. The Morse Lemma follows from the Mather-Yau Theorem. As for $g = x_1^2 + \cdots + x_n^2$ we have $J(g) = (x_1, \ldots, x_n)$ we have that for all f with $J(f) = \mathfrak{m}$ are right equivalent to $x_1^2 + \cdots + x_n^2$. The $\mathbb{C}\{t\}$ module structure can only be trivial, that is, multiplication with t is the zero map. Of course, this is a very difficult proof of the Morse Lemma.

Exercises

9.1.12. Suppose that f and g are right equivalent. Let $k \in \mathbb{N}$. Suppose that f is k–determined. Prove that g is k–determined.

9.1.13. Show by counterexample that the statement $\mathfrak{m}^{k+2} \subset \mathfrak{m}^3 J(f)$ does not imply that f is k–determined in general.

9.1.14. Let R be a commutative ring and $u \in R[[t]]$. Prove that for given $a \in R$ the differential equation

$$\frac{dy}{dt} = u \cdot y$$

has a unique solution $y \in R[[t]]$ with $y(0) = a$.
(Hint: Write $u = \sum\limits_k u_k t^k$, $y = \sum\limits_k y_k t^k$ and compare both sides of the equation.)

9.1.15. Prove Theorem 9.1.7.
(Hint: As an application of Artin's Approximation Theorem, it suffices to find a formal automorphism. Now use the same idea as in Theorem 9.1.5 and induction on the number of parameters. Let $\varphi^{(0)} = (x_1, \ldots, x_n, y_1, \ldots, y_m)$ and $u^{(0)} = 1$. Assume $\varphi^{(k)} = (\varphi_1^{(k)}, \ldots, \varphi_n^{(k)}, y_1, \ldots, y_m)$ and $u^{(k)}$ are defined such that

- $u^{(k)} = 1$ in case of $\frac{\partial f}{\partial y_i} \in (x_1, \ldots, x_n)^c \left(\frac{\partial f}{\partial x_1}, \ldots, \frac{\partial f}{\partial x_n}\right)$
- $u^{(k)}(x_1, \ldots, x_n, y_1, \ldots, y_{m-k}, 0, \ldots, 0) = 1$
- $\varphi_i^{(k)}(x_1, \ldots, x_n, y_1, \ldots, y_{m-k}, 0, \ldots, 0) = x_i$
- $\varphi_i^{(k)} - x_i \in (x_1, \ldots, x_n)^c$.
- $f(\varphi_1^{(k)}, \ldots, \varphi_n^{(k)}, y_1, \ldots, y_{m-k}, 0, \ldots, 0) = u^{(k)} f(x_1, \ldots, x_n, y_1, \ldots, y_m)$.

Choose $\xi_1, \ldots, \xi_n \in (x_1, \ldots, x_n)^c \mathbb{C}\{x_1, \ldots, x_n, y_1, \ldots, y_m\}$ and $v \in \mathbb{C}\{x_1, \ldots, x_n, y_1, \ldots, y_m\}$ such that $\frac{\partial f}{\partial y_{m-k}} = v \cdot f + \sum\limits_{i=1}^{n} \xi_j \frac{\partial f}{\partial x_j}$ (the case $v = 0$ included). Let φ be the automorphism defined by the integral curve of the corresponding vector field and define $\varphi_i := \varphi(x_i)$ $i = 1, \ldots, n$. Note that $\varphi(y_i) = y_i - \delta_{i,m-k} \cdot t$. Then $\frac{\partial \varphi_i}{\partial t} = \xi_i(\varphi_1, \ldots, \varphi_n, y_1, \ldots, y_{m-k-1}, y_{m-k} - t, 0, \ldots, 0)$. Use this property to prove that

$$\frac{\partial f(\varphi_1, \ldots, \varphi_n, y_1, \ldots, y_{m-k-1}, y_{m-k} - t, 0, \ldots, 0)}{\partial t} =$$
$$v(\varphi_1, \ldots, \varphi_n, y_1, \ldots, y_{m-k-1}, y_{m-k} - t, 0, \ldots, 0) \cdot f(\varphi_1, \ldots, \varphi_n, y_1, \ldots, y_{m-k-1}, y_{m-k} - t, 0, \ldots, 0).$$

Use Exercise 9.1.14 to choose $s \in \mathbb{C}[[x_1, \ldots, x_n, y_1, \ldots, y_{m-k}, t]]$ such that

$$\frac{\partial s}{\partial t} = v(\varphi_1, \ldots, \varphi_n, y_1, \ldots, y_{m-k-1}, y_{m-k} - t, 0, \ldots, 0) \cdot s,$$

and $s(t = 0) = 1$. Use Exercise 9.1.14 again to prove that

$$f(\varphi_1, \ldots, \varphi_n, y_1, \ldots, y_{m-k-1}, y_{m-k} - t, 0, \ldots, 0) = s \cdot f(x_1, \ldots, x_n, y_1, \ldots, y_{m-k}, 0, \ldots, 0).$$

Let $\varphi_i := \varphi_i \,|_{t=-y_{m-k}}$ and $s = s \,|_{t=-y_{m-k}}$ then

$$f(\varphi_1, \ldots, \varphi_n, y_1, \ldots, y_{m-k-1}, 0, \ldots, 0) = s \cdot f(x_1, \ldots, x_n, y_1, \ldots, y_{m-k}, 0, \ldots, 0).$$

Use the φ_i, s to define $\varphi_i^{(k+1)}, u^{(k+1)}$.

9.1.16. Prove the two statements in Remark 9.1.9.

9.1.17. Prove the equivalence (1) $\Leftrightarrow$ (2) of the Theorem 9.1.8.

9.1.18. Let g and f in $\mathfrak{m} \subset \mathbb{C}\{x_1,\dots,x_n\}$ be given such that $J(g) \subset \mathfrak{m}^2 J(f)$. Prove that f and g are right equivalent.
(Hint: Follow the steps in the proof of the Finite Determinacy Theorem 9.1.4.)

9.1.19. Let $f \in \mathbb{C}\{x,y\}$ of type

$$f = x^3 + xy^3 + \psi,$$

with $\psi \in (y^6, xy^4, x^2y^3, x^3y^2, x^4)$. Prove that f is right equivalent to $x^3 + xy^3$.

9.1.20. Prove Theorem 9.1.10.
(Hint: Use as in the proof of 9.1.8 the reduction to the case that $J(f) = J(g)$, resp. $\mathfrak{m}J(f) = \mathfrak{m}J(g)$. The $\mathbb{C}\{t\}$–algebra isomorphism implies that $h := f - g \in J(g)$, resp. $h := f - g \in \mathfrak{m}J(g)$. Now just continue as in the proof of 9.1.8.)

9.2 The A-D-E–singularities are simple.

We consider the automorphism group $\mathcal{G}$ of $\mathcal{O}_n = \mathbb{C}\{x_1,\dots,x_n\}$. Let $f \in \mathbb{C}\{x_1,\dots,x_n\}$. The *right equivalence class* of f is exactly the orbit of f under the following group action.[2]

$$\mathcal{G} \times \mathcal{O}_n \longrightarrow \mathcal{O}_n; \quad (G,f) \mapsto G(f).$$

We denote the orbit of f by $\mathcal{G} \cdot f := \{G(f) : G \in \mathcal{G}\}$[3]. The problem with the study of such orbits is, that $\mathcal{O}_n$ as $\mathbb{C}$–vector space is infinite-dimensional! In case of isolated singularities, this is not a severe problem, as we can use the Finite Determinacy Theorem, which allows us to reduce the study of this problem to the finite-dimensional case.

Definition 9.2.1. Let coordinates $x_1,\dots,x_n$ on $(\mathbb{C}^n, o)$ be given, and let $\mathfrak{m}$ be the maximal ideal of $\mathcal{O}_n$. Let $k \geq 1$ be a natural number.

(1) We define the *k-jet space* by $J_k := \mathcal{O}_n/\mathfrak{m}^{k+1}$. Each element in J_k has a representative of type $\sum_{|\nu|\leq k} a_\nu x^\nu$. Note that J_k is a finite-dimensional $\mathbb{C}$–vector space.

(2) Let $f = \sum_\nu a_\nu x^\nu \in \mathcal{O}_n$ be given. The *k – jet* of f is defined by $j^k f = \sum_{|\nu|\leq k} a_\nu x^\nu$.

(3) We can view $\mathcal{G}$ as a subset of $\oplus_{i=1}^n \mathfrak{m}$. Indeed if $G \in \mathcal{G}$ then put $g_i := G(x_i)$. Then G is uniquely determined by the element $(g_1,\dots,g_n)$, see Exercise 3.1.22. We now define

$$\mathcal{G}_k := \{(g_1 \mod \mathfrak{m}^{k+1},\dots,g_n \mod \mathfrak{m}^{k+1}) : (g_1,\dots,g_n) \in \mathcal{G}\}.$$

[2] Group action means that $(G_1 \cdot G_2)(f) = G_1(G_2(f))$.

[3] In case of contact equivalence, we have to use the semi-direct product $\mathcal{G} \rtimes \mathcal{O}_n^*$ of $\mathcal{G}$ with $\mathcal{O}_n^*$. Here $\mathcal{O}_n^*$ is the group of units in $\mathcal{O}_n = \mathbb{C}\{x_1,\dots,x_n\}$. As a set the semi-direct product is just the Cartesian product, but the multiplication is given by $(\varphi_1,w_1)\cdot(\varphi_2,w_2) := (\varphi_1\varphi_2, w_1 \cdot \varphi_1(w_2))$. One needs this strange looking product so that the action of $\mathcal{G} \rtimes \mathcal{O}_n^*$ on $\mathbb{C}\{x_1,\dots,x_n\}$ sending $((\varphi,u),f)$ to $u \cdot \varphi(f)$ is a group action.

(4) We get group actions

$$\mathscr{G}_k \times J_k \longrightarrow J_k.$$

induced by those of $\mathscr{G}$. Here $G \cdot f := j^k(G(f))$. One immediately checks that this is a group action and is well defined, see Exercise 9.2.17.

Example 9.2.2. Let $n = 1$ and $k = 2$. Then $J_2 = \{a_0 + a_1x + a_2x^2 : a_i \in \mathbb{C}\}$, and $G_2 = \{\alpha x + \beta x^2 : \alpha \neq 0\}$. So an element $G = \alpha x + \beta x^2$ acts on $a_0 + a_1x + a_2x^2$ by sending it to $a_0 + a_1(\alpha x + \beta x^2) + a_2(\alpha x + \beta x^2)^2$ which in J_k is equal to $a_0 + \alpha a_1 x + (a_1\beta + a_2\alpha^2)x^2$.

Lemma 9.2.3. *Let coordinates $x_1, \dots, x_n$ on $(\mathbb{C}^n, 0)$ be given. Consider $f \in \mathcal{O}_n$, and $k \in \mathbb{N}$ such that f is k-determined. Consider the group operation of $\mathscr{G}_k$ on J_k, and let $g \in \mathcal{O}_n$ be given. Then*

$$j^k g \in \mathscr{G}_k(j^k f) \iff g \text{ is right equivalent to } f.$$

Proof. The implication $\Longleftarrow$ is obvious. Suppose on the other hand that $j^k g \in \mathscr{G}_k(j^k f)$. Then there exists a $G \in \mathscr{G}$ such that $f \equiv G(g) \mod \mathfrak{m}^{k+1}$. Because f is k–determined, and f and $G(g)$ have the same k–jet, we have that f and $G(g)$ are right equivalent. By transitivity, f and g are right equivalent. □

Definition 9.2.4. On all jet spaces J_k, which are as vector spaces isomorphic to some $\mathbb{C}^N$, we take the usual (Euclidean) $\mathbb{C}$–topology. Consider the projection maps

$$\pi_k : \mathcal{O}_n = \mathbb{C}\{x_1, \dots, x_n\} \longrightarrow J_k = \mathbb{C}\{x_1, \dots, x_n\}/\mathfrak{m}^{k+1}.$$

We define a *topology* on $\mathbb{C}\{x_1, \dots, x_n\}$ as the coarsest topology such that all π_k are continuous.

Thus a basis for the topology for $\mathcal{O}_n$ is given by $\pi_k^{-1}(V)$ for $V \subset J_k$. Informally speaking, a power series is near zero when "some of its coefficients are small", and the coefficients of monomials which are not small are in a high power of the maximal ideal $\mathfrak{m}$. On $\mathscr{G}$ we similarly define a topology. It is the topology induced by the product topology of $\mathfrak{m}^n$. The action of $\mathscr{G}$ on $\mathcal{O}_n$ is continuous. We omit the boring proof, as we do not need the result.

Definition 9.2.5. We consider the action of $\mathscr{G}$ on $\mathcal{O}_n$. Let $f \in \mathfrak{m} \subset \mathcal{O}_n$. Then f is called *simple*, if there exists an open neighborhood U of f, such that the number of orbits which intersect U is finite.

Remark 9.2.6. There is a different way to phrase the definition of simple. Namely, consider families

$$F(x,t) \in \mathbb{C}\{x_1, \dots, x_n, t\}$$

such that $F(x,t) \in (x_1, \dots, x_n)$, and $F(x,0) = f$. Families with this property are also called deformations. Then f is called simple, if there exist finitely many right equivalence classes $G_1, \dots, G_s$ such that for all families $F(x,t)$ as above and t small and fixed the germ $F_t \in \mathbb{C}\{x_1, \dots, x_n\}$ is right equivalent to one of the germs in $G_1, \dots, G_s$. In case that f is finitely determined, the proof that these two definitions are equivalent is left as Exercise 9.2.16.

We will see later on, that simple germs have an isolated singularity. In fact, we have the following Classification Theorem, due to Arnold, whose proof will occupy the remainder of this chapter.

Theorem 9.2.7 (Classification of Simple Singularities). *Suppose that $f \in \mathcal{O}_n$ is simple. Then f is right equivalent to one of the singularities in the following list.*

- $A_k: f(x_1, \ldots, x_n) = x_1^{k+1} + x_2^2 + \ldots + x_n^2, \quad k \geq 0,$[4]
- $D_k: f(x_1, \ldots, x_n) = x_1 x_2^2 + x_1^{k-1} + x_3^2 + \ldots + x_n^2, \quad k \geq 4,$
- $E_6: f(x_1, \ldots, x_n) = x_1^3 + x_2^4 + x_3^2 + \ldots + x_n^2,$
- $E_7: f(x_1, \ldots, x_n) = x_1^3 + x_1 x_2^3 + x_3^2 + \ldots + x_n^2,$
- $E_8: f(x_1, \ldots, x_n) = x_1^3 + x_2^5 + x_3^2 + \ldots + x_n^2.$

We give real pictures in dimension two of A_2, A_3, D_5, D_6, E_6 and E_7.

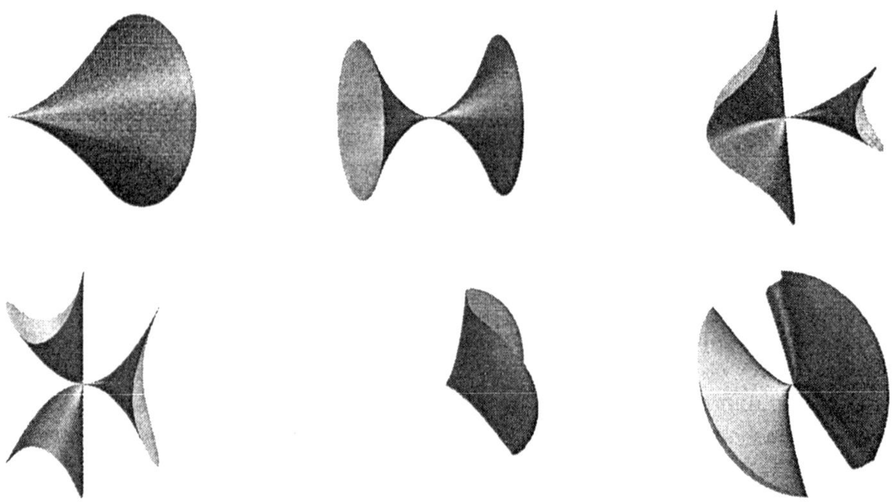

In this section we will show that the A-D-E–singularities of Arnold's list are simple. In the next section we will show that there are no other simple singularities.

We first define the corank of a function. It is the coarsest invariant of right equivalence classes.

Definition 9.2.8. Let $f \in \mathfrak{m}^2 \subset \mathbb{C}\{x_1, \ldots, x_n\}$. The *Hesse matrix* of f is defined by

$$H(f) = \left(\frac{\partial^2 f}{\partial x_i \partial x_j}(0) \right)_{1 \leq i,j \leq n}.$$

The *corank* of f is defined as $\operatorname{Corank}(f) := n - \operatorname{rank}(H(f))$.

The following easy lemma is left as Exercise 9.2.20.

[4] Note that A_0 is the germ of a smooth map.

Lemma 9.2.9. *Let f be right equivalent to g. Then* $\text{Corank}(f) = \text{Corank}(g)$. *Moreover, let F_t be a holomorphic family of functions with $F_t \in \mathfrak{m}^2$ for all t. Then for all t small* $\text{Corank}(F_t) \leq \text{Corank}(F_0)$.

We now prove the Splitting Lemma, which is a generalization of the Morse Lemma.

Lemma 9.2.10 (Splitting Lemma).

(1) Let $f \in \mathbb{C}\{x_1, \dots, x_n\}$, with $f \in \mathfrak{m}^2$ and $\text{Corank}(f)$ *equal to s. Suppose that f is finitely determined. Then f is right equivalent to an element of type*

$$g(x_1, \dots, x_s) + x_{s+1}^2 + \dots + x_n^2.$$

where $g \in \mathbb{C}\{x_1, \dots, x_s\}$ and $g \in \mathfrak{m}^3$.

(2) Let g_1 and $g_2 \in \mathfrak{m}^3 \subset \mathbb{C}\{x_1, \dots, x_s\}$ both have isolated singularities. Then $g_1(x_1, \dots, x_s) + x_{s+1}^2 + \dots + x_n^2$ is right equivalent to $g_2(x_1, \dots, x_s) + x_{s+1}^2 + \dots + x_n^2$ if and only if g_1 is right equivalent to g_2.

(3) Let $f \in \mathbb{C}\{x_1, \dots, x_s\}$ have an isolated singularity. Then $f(x_1, \dots, x_s) + x_{s+1}^2 + \dots + x_n^2$ is simple if and only if f is simple.

Proof.
(1) By following the proof of the Morse Lemma 3.4.30, we may (after renaming coordinates) assume that f is equivalent to a germ of type

$$x_{s+1}^2 + \dots + x_n^2 + \sum_{1 \leq i,j \leq s} x_i x_j H_{ij}. \tag{9.15}$$

As the Corank is s, it follows that none of the H_{ij} for $1 \leq i,j \leq s$ is a unit, hence $H_{ij} \in \mathfrak{m}$. We may rewrite (9.15) as

$$x_{s+1}^2 + \dots + x_n^2 + g(x_1, \dots, x_s) + \sum_{i=s+1}^{n} x_i G_i(x_1, \dots, x_n) =$$

$$g + \sum_{i=s+1}^{n} \left((x_i + \tfrac{1}{2} G_i(x_1, \dots, x_n))^2 - \tfrac{1}{4} G_i(x_1, \dots, x_n)^2 \right).$$

As $H_{ij} \in \mathfrak{m}$, it follows that $G_i \in \mathfrak{m}^2$ for all i, and $g(x_1, \dots, x_s) \in \mathfrak{m}^3$. We can therefore do the following coordinate change:

$$\begin{aligned} x_i' &= x_i & &\text{for } i = 1, \dots, s, \\ x_i' &= x_i + \tfrac{1}{2} G_i(x_1, \dots, x_n) & &\text{for } i = s+1, \dots, n. \end{aligned}$$

After this coordinate change we have that f is right equivalent to a germ of type

$$x_{s+1}^2 + \dots + x_n^2 + g(x_1, \dots, x_s) + \sum_{i=s+1}^{n} x_i G_i(x_1, \dots, x_n). \tag{9.16}$$

with $G_i \in \mathfrak{m}^3$. (Both the g and the G_i may have changed in this process). So we improved the situation. Now we have $G_i \in \mathfrak{m}^3$, which is a stronger statement than $G_i \in \mathfrak{m}^2$. We can continue like this, and we obtain that f is right equivalent to a germ of type (9.16) with $G_i \in \mathfrak{m}^k$. As f is k–determined it follows from Exercise 9.1.12 that (9.16) is also k–determined. Thus (9.16) is right equivalent to its k–jet, and the first part of the theorem follows.

(2) If g_1 and g_2 are right equivalent, then obviously $g_1 + x_{s+1}^2 + \ldots + x_n^2$ and $g_2 + x_{s+1}^2 + \ldots + x_n^2$ are right equivalent. The converse follows without difficulty from the Mather-Yau Theorem for right equivalence, see 9.1.10, but let us give a direct argument. Suppose therefore that $f_1 = g_1(x_1, \ldots, x_s) + x_{s+1}^2 + \ldots + x_n^2$ is right equivalent to $f_2 = g_2(x_1, \ldots, x_s) + x_{s+1}^2 + \ldots + x_n^2$. Hence there exist $\varphi_1, \ldots, \varphi_n \in \mathbb{C}\{x_1, \ldots, x_n\}$, with $\varphi_i = l_i + h_i$, $h_i \in \mathfrak{m}^2$ and l_i linear forms with det $\left(\frac{\partial l_i}{\partial x_j}\right) \neq 0$ such that

$$(9.17) \qquad g_1(\varphi_1, \ldots, \varphi_s) + \sum_{i=s+1}^{n} \varphi_i^2 = g_2(x_1, \ldots, x_s) + \sum_{i=s+1}^{n} x_i^2.$$

Note that $\sum_{i=s+1}^n l_i^2 + \sum_{i=s+1}^n h_i(2l_i + h_i) = \sum_{i=s+1}^n \varphi_i^2$. By assumption, $g_1, g_2 \in \mathfrak{m}^3$, so that

$$(9.18) \qquad \sum_{i=s+1}^{n} l_i^2 = \sum_{i=s+1}^{n} x_i^2.$$

Now suppose we can find $\psi_{s+1}, \ldots, \psi_n \in \mathbb{C}\{x_1, \ldots, x_s\}$ such that

$$(9.19) \qquad (2l_i + h_i)(x_1, \ldots, x_s, \psi_{s+1}, \ldots, \psi_n) = 0 \quad \text{for } i = s+1, \ldots, n.$$

Then we can consider the endomorphism of $\mathbb{C}\{x_1, \ldots, x_s\}$ given by

$$\varphi_1' := \varphi_1(x_1, \ldots, x_s, \psi_{s+1}, \ldots, \psi_n), \ldots, \varphi_s' := \varphi_s(x_1, \ldots, x_s, \psi_{s+1}, \ldots, \psi_n).$$

This is in fact an automorphism. By the Inverse Function Theorem it suffices to show that $(\varphi_1', \ldots, \varphi_s') = (x_1, \ldots, x_s)$. Now $(\varphi_1, \ldots, \varphi_s, l_{s+1} + \frac{1}{2}h_{s+1}, \ldots, l_n + \frac{1}{2}h_n)$ is an automorphism of $\mathbb{C}\{x_1, \ldots, x_n\}$, again by the Inverse Function Theorem. Hence $(\varphi_1, \ldots, \varphi_s, l_{s+1} + \frac{1}{2}h_{s+1}, \ldots, l_n + \frac{1}{2}h_n)$ is the the maximal ideal of $\mathcal{O}_n$. It follows that for all k with $1 \leq k \leq s$ we can write

$$x_k = \sum_{\nu=1}^{s} \xi_\nu \varphi_\nu + \sum_{\nu=s+1}^{n} \xi_\nu (l_\nu + \frac{1}{2}h_\nu), \quad k = 1, \ldots, s.$$

By plugging in ψ_i instead of x_i for $i \geq s+1$ we get by using (9.19)

$$x_k = \sum_{\nu=1}^{s} \xi_\nu(x_1, \ldots, x_s, \psi_{s+1}, \ldots, \psi_n)\varphi_\nu'.$$

Therefore $(\varphi_1', \ldots, \varphi_s') = (x_1, \ldots, x_s)$. Now

$$g_1(\varphi_1', \ldots, \varphi_s') = g_2(x_1, \ldots, x_s)$$

follows from formulae (9.17), (9.18) and (9.19) so that we have an automorphism of $\mathbb{C}\{x_1, \ldots, x_s\}$ transforming g_1 into g_2.

In order to construct the ψ_i, we therefore have to solve the system of equations $F_i = 0$ with

$$F_i := 2l_i + h_i, \quad \text{for } i = s+1, \ldots, n.$$

Now

$$\det\left(\frac{\partial F_i}{\partial x_j}(0)\right)_{i,j\geq s+1} = 2\det\left(\frac{\partial l_i}{\partial x_j}\right)_{i,j\geq s+1}.$$

Therefore, if we show that the right hand side is nonzero, we can apply the Implicit Mapping Theorem to solve this system of equations. The fact that $\det\left(\frac{\partial l_i}{\partial x_j}\right)_{i,j\geq s+1}$ is nonzero, is linear algebra, and treated in Exercise 9.2.21.

(3) First suppose that f is simple. We have to prove that $f+x_{s+1}^2+\ldots+x_n^2$ is simple too. Consider $F_t \in (x_1,\ldots,x_n)\mathbb{C}\{x_1,\ldots,x_n,t\}$ with $F_0 = f + \sum_{i=s+1}^n x_i^2$. Let $k \geq \mu(f)+1$. For t small we have $\mu(F_t) \leq \mu(f)$. Then for all t small, F_t is k–determined by the Finite Determinacy Theorem 9.1.4. We redo the argument of the first step simultaneously for the whole family F_t. For $t = 0$, we get a coordinate change (in fact the identity), so that for all small t we have a coordinate change of $(\mathbb{C}^n, o)$. As F_t is k–determined for all small t we get a family $G_t(x_1,\ldots,x_s)+\sum_{i=s+1}^n x_i^2$, with $G_0 = f$ and $G_t(x_1,\ldots,x_s)+\sum_{i=s+1}^n x_i^2$ is right equivalent to F_t for t small. As f is simple, there exist only finitely many right equivalence classes in which $G_t(x_1,\ldots,x_s)$ can lie. Hence there exist only finitely many equivalence classes in which F_t can lie.

The converse statement is left to the reader. □

Corollary 9.2.11. *Let $f \in \mathfrak{m}^2$ with $\mu(f) < \infty$, and* Corank$(f) = 1$. *Then there exists a $k \geq 2$ such that f is right equivalent to A_k, that is, $x_1^{k+1} + x_2^2 + \ldots + x_n^2$. Moreover, the A_k singularities are simple.*

Proof. By the Splitting Lemma, we only have to consider a function in one variable $f(x) \in \mathbb{C}\{x\}$. If $f(x) = 0$ then $\mu(f) = \infty$. As this is not the case, there exists a unique k with $f(x) = x^{k+1}u$, where u is a unit in $\mathbb{C}\{x\}$. By the Implicit Function Theorem, the function $\sqrt[k+1]{u}$ exists, is holomorphic and a unit in $\mathbb{C}\{x\}$. We do the coordinate change $x' = \sqrt[k+1]{u}x$. In this coordinate $f(x') = x'^{k+1}$. This proves that f is right equivalent to A_k. If F_t is a deformation of f, then $\mu(F_t) \leq \mu(f)$ for all t small. It follows that an A_k singularity only deforms into an A_l singularity with $l \leq k$. In particular, the A_k singularities are simple. □

Now we turn our attention to functions of corank two. By the Splitting Lemma, we only have to consider $f(x,y) \in (x,y)^3 \subset \mathbb{C}\{x,y\}$.

Proposition 9.2.12. *Let $f \in C\{x,y\}$ and $f \in \mathfrak{m}^3$. After a linear coordinate change we may assume that $j^3 f$ is of one of the following types.*

1. *$j^3 f = xy(x+y)$, the zero set of $j^3 f$ consists of three different lines.*
2. *$j^3 f = x^2 y$, two lines coincide.*
3. *$j^3 f = x^3$, the three lines coincide.*
4. *$j^3 f = 0$.*

Suppose F_t is a holomorphic family with $F_0 = f$ and $F_t \in \mathfrak{m}^2$ for all small t. Then for all $t \neq 0$ and small we have for the types (1), (2), resp (3):

(1) *Either* Corank$(F_t) = 0$ *or* 1 *or $j^3 F_t = 0$ consists of three different lines.*

(2) *Either* Corank$(F_t) = 0$ *or* 1 *or $j^3 F_t = 0$ consists of three different lines, or two lines coincide.*

(3) *Either* Corank$(F_t) = 0$ *or* 1 *or* $j^3F_t = 0$ *consists of three different lines, or two lines coincide, or the three lines coincide.*

Proof. Write

$$j^3 f = ax^3 + bx^2y + cxy^2 + dy^3.$$

Case 1. $a = d = 0$. Then $j^3f = xy(bx + cy)$.

(i) If $b = c = 0$ we get *4*.

(ii) Suppose $b = 0$ or $c = 0$, but not both. Without loss of generality, let $c = 0$. Then $j^3f = xy(bx)$. As $b \neq 0$, we can do the coordinate change $y' = by$. Then $j^3f = x^2y'$ which is *2*.

(iii) Both b and c are nonzero. After the coordinate change $x' = bx$, $y' = cy$ we may assume that $b = c$. By putting $x' = \sqrt[3]{b} \cdot x$, $y' = \sqrt[3]{b} \cdot y$ we get $j^3f = x'y'(x' + y')$ which is *1*.

Case 2. Suppose $a \neq 0$ or $d \neq 0$. Without loss of generality, we may assume that $a \neq 0$ and, after the coordinate change $x' = \sqrt[3]{a} \cdot x$, even $a = 1$. We factorize j^3f:

$$j^3f = (x - \alpha y)(x - \beta y)(x - \gamma y).$$

We do the coordinate change $x' = x - \alpha y$, so that we may assume $\alpha = 0$. Hence

$$j^3f = x \cdot (x - \beta y)(x - \gamma y).$$

(i) If $\beta = \gamma = 0$ we get *3*.

(ii) Suppose that $\beta \neq 0$ or $\gamma \neq 0$. Without loss of generality $\beta \neq 0$. Take $y' = x - \beta y$. Then $j^3f = x'y'(x' - \delta y')$ for some δ, and we are back in *Case 1*.

This proves the first part of the proposition. The second part of the proposition follows from the fact that the corank can only drop, see 9.2.9, and the fact that the zeros of the polynomial $j^3(F_t(x,1))$ depend continuously on t. □

Proposition 9.2.13. *Let* $f \in \mathbb{C}\{x,y\}$, $\mu(f) < \infty$.

(1) *Suppose* $j^3f = xy(x + y)$. *Then* f *is right equivalent to* D_4.

(2) *Suppose* $j^3f = x^2y$. *Then there exists a* $k \geq 5$ *such that* f *is right equivalent to* D_k.

(3) *The germs* D_k *for* $k \geq 4$ *are simple.*

Proof. (1) Suppose that $j^3f = xy(x + y)$. It is an exercise to show that from the Finite Determinacy Theorem it follows that $xy(x + y)$ is 3–determined. It follows without difficulty that f is right equivalent to D_4.

(2) Suppose $j^3f = x^2y$. As we suppose $\mu(f) < \infty$, the element f is k–determined for some k. Suppose that we have an $s \geq 4$, such that

$$j^sf = x^2y + ay^s + bxy^{s-1} + x^2\varphi,$$

for a suitable $\varphi \in \mathfrak{m}^{s-2}$, and $a,b \in \mathbb{C}$. This we certainly have for $s = 4$. We do the coordinate transformation $x' = x + \frac{1}{2}by^{s-2}$, $y' = y + \varphi$. Then a simple calculation shows that $j^s f = x'^2 y' + ay'^s$. Renaming x' to x and y' to y we may therefore suppose that $j^s f = x^2 y + ay^s$. There are two cases to consider.

Case 1. We have $a \neq 0$. Do the coordinate change $y' = \sqrt[s]{a} \cdot y$, $x' = \frac{1}{2\sqrt[s]{a}} x$. Then $j^s f = x^2 y + y^s$. As $x^2 y + y^s$ is s–determined (exercise), it follows that f is a D_{s+1} singularity.

Case 2. We have $a = 0$. In this case we write down $j^{s+1} f$. Iterate the above coordinate changes until we come in Case 1, or until $s \geq k$. We will show that $s \geq k$ leads to a contradiction. As f is k–determined it follows that f is right equivalent to $j^s f = x^2 y + ay^s$. If $a = 0$, then f does not have an isolated singularity, contrary to our assumption.

(3) Consider D_4, and a holomorphic family F_t with $F_0 = f = xy(x+y)$. We have that $\mu(F_t) < \infty$. If for fixed t small, the corank of F_t is zero, then F_t has an A_1–singularity by the Morse Lemma. If the corank is one, then F_t is right equivalent to A_k for some k by 9.2.11. As $\mu(F_t) \leq \mu(D_4)$, it follows that $k \leq 4$.[5] Hence D_4 can only deform in finitely many germs. If $\mathrm{Corank}(F_t) = 2$, it follows that $j^3 F_t = 0$ consists of three different lines by 9.2.12. Hence there exist coordinates such that $j^3 F_t = xy(x+y)$ for fixed t. Hence, by what we just proved, $j^3 F_t$ is right equivalent to D_4. This shows that D_4 can deform only in finitely many germs.

The proof that the D_k for $k \geq 5$ are simple, is similar. □

So in view of Proposition 9.2.12, the case to consider now are functions $f \in \mathbb{C}\{x,y\}$ whose 3–jet is x^3.

Proposition 9.2.14. *Let $f \in \mathbb{C}\{x,y\}$ with $\mu(f) < \infty$. Suppose $j^3 f = x^3$. Then either*

(1) f is right equivalent to E_6, E_7 or E_8 or

(2) f is right equivalent to a function which lies in the ideal $(x,y^2)^3 = (x^3, x^2y^2, xy^4, y^6)$.

Moreover, E_6, E_7 and E_8 are simple.

Proof. Suppose $j^3 f = x^3$. We write down the four jet.

$$j^4 f = x^3 + ay^4 + bxy^3 + x^2\varphi,$$

with $a,b \in \mathbb{C}$ and $\varphi \in \mathfrak{m}^2$. We do the coordinate change $x' = x + \frac{1}{3}\varphi$. Hence we may assume

$$j^4 f = x^3 + ay^4 + bxy^3.$$

Case 1. Suppose $a \neq 0$. Then by doing the coordinate change $y' = \sqrt[4]{a} y$ we may assume $a = 1$. After doing the coordinate change $y' = y + \frac{1}{4}bx$ we get $j^4 f = x^3 + y^4 + x^2\psi$, with $\psi \in \mathfrak{m}^2$. By the coordinate change $x' = x + \frac{1}{3}\psi$, we can get rid of ψ to obtain $j^4 f = x^3 + y^4$. As $x^3 + y^4$ is 4-determined it follows that f is right equivalent to E_6. Redoing this proof for a holomorphic family F_t with $F_0 = x^3 + y^4$ shows that E_6 can only deform into singularities of type A_k, D_k or E_6. As $\mu(F_t) \leq \mu(f)$ it follows that E_6 is simple.

[5] It follows from Exercise 9.3.22 that even $k \leq 3$.

Case 2. Suppose $a = 0$, and $b \neq 0$. We may assume then that $b = 1$, by the coordinate change $y' = \sqrt[3]{b}y$. From the Finite Determinacy Theorem it follows that f is 5–determined, but it does not follow from this theorem that f is 4–determined. Indeed, one calculates that $(xy^4,x^2y^3,x^3y^2,x^4) \subset \mathfrak{m}^2 J(f)$, but $y^5 \notin \mathfrak{m}^2 J(f)$. But in fact f is 4–determined. To prove this, we write down the 5–jet of f

$$j^5 f = x^3 + xy^3 + cy^5 + x\varphi,$$

with $c \in \mathbb{C}$ and $\varphi \in \mathfrak{m}^4$. We do the coordinate change $x' = x + cy^2$, and get

$$j^5 f = x'^3 + x'y^3 - 3cx'^2 y^2 + x'\varphi',$$

with $\varphi' \in \mathfrak{m}^4$. Now we do the coordinate change $y' = y - cx'$. Then

$$j^5 f = x'^3 + x'y'^3 - 3cx'^3 y' + \psi,$$

with $\psi \in (x'y'^4,x'^2y'^3,x'^3y'^2,x'^4) \subset \mathfrak{m}^2 J(f)$. Replacing x' by $x'\sqrt[3]{1 - 3cy'}$, and renaming the coordinates we get

$$j^5 f = x^3 + xy^3 + \psi,$$

with $\psi \in (xy^4,x^2y^3,x^3y^2,x^4) \subset \mathfrak{m}^2 J(x^3 + xy^3)$. By Exercise 9.1.19 it follows that f is right equivalent to $x^3 + xy^3$, that is, we have an E_7–singularity. Similarly, one shows that E_7 is simple, and this proof is left to the reader.

Case 3. The final case to consider is $a = b = 0$, that is, $j^4 f = x^3$. So we consider the 5–jet:

$$j^5 f = x^3 + cy^5 + dxy^4 + x^2\varphi,$$

with $c,d \in \mathbb{C}$ and $\varphi \in \mathfrak{m}^3$. Again we can get rid of the $x^2\varphi$ term. The two remaining cases are:

Case 3a. Suppose $c \neq 0$. We can after the coordinate change $y' = \sqrt[5]{c}y$ get that $c = 1$. Now put $y' = y + \frac{1}{5}dx$. Then $j^5 f = x^3 + y'^5 + x^2\psi$, with $\psi \in \mathfrak{m}^3$. By doing the coordinate change $x' = x + \frac{1}{3}\psi$, we can attain $\psi = 0$. Hence $j^5 f = x'^3 + y'^5$. By the Finite Determinacy Theorem, it follows that f is right equivalent to an E_8–singularity. Similarly, it is proved that E_8 is simple.

Case 3b. Consider the case $c = 0$. Then $f \in (x^3,xy^4) + (x,y)^6 \subset (x,y^2)^3$, as was to be proved. □

We now succeeded in our goal of proving that the A-D-E-singularities are simple. In the next section, we will show that these are the only ones. Up to now, we already observed the following

Remark 9.2.15. Suppose $f \in \mathcal{O}_n$ is not simple. Then either

(1) f has a nonisolated singularity, or

(2) $\operatorname{Corank}(f) \geq 3$, or

(3) $\operatorname{Corank}(f) = 2$, and we may suppose $f \in \mathbb{C}\{x,y\}$ with $f \in \mathfrak{m}^4$, or

(4) Corank$(f) = 2$, and we may suppose $f \in \mathbb{C}\{x,y\}$ with $f \in (x,y^2)^3$.

Exercises

9.2.16. For finitely determined germs prove that the two definitions of simple, see Remark 9.2.6, are equivalent.

9.2.17. Show that the group action $\mathscr{G}_k$ on J_k is well-defined and indeed a group action.

9.2.18. Prove the Morse Lemma by using only elementary linear algebra, and applying the Finite Determinacy Theorem.

9.2.19.

(1) Let f be a germ of corank s. Prove that $\mu(f) \geq 2^s$.

(2) Let $f \in \mathfrak{m}^4 \subset \mathbb{C}\{x,y\}$. Show that $\mu(f) \geq 9$.

(3) Let $f \in (y,x^2)^3 \subset \mathbb{C}\{x,y\}$. Prove that $\mu(f) \geq 10$.

Conclude that germs with $\mu(f) < 8$ are simple.

9.2.20. Prove Lemma 9.2.9.

9.2.21. In this exercise we complete proof of the second part of the Splitting Lemma 9.2.10.

Consider matrices B of size $(n-s) \times s$ and C of size $(n-s) \times (n-s)$. Let $A = (a_{ij})_{s+1 \leq i \leq n, 1 \leq j \leq n} := (B,C)$. Let $l_i = \sum_{j=s+1}^{n} a_{ij}x_j$. Suppose that

$$\sum_{i=s+1}^{n} l_i^2 = \sum_{i=s+1}^{n} x_i^2. \qquad (9.20)$$

To complete the proof of the Splitting Lemma it suffices to show $\det(C) = 1$. Prove this. (Hint: Use (9.20) to show that C is orthogonal.)

9.2.22. Let $f \in \mathbb{C}\{x,y\}$ with $j^3 f = xy(x+y)$. Prove, without using the Finite Determinacy Theorem, that f is right equivalent to D_4.

9.3 Orbits

In order to show that functions with one of the properties mentioned in Remark 9.2.15 are not simple, we need to study orbits of group actions more carefully. First a definition.

Definition 9.3.1. An *affine algebraic group* G is an affine variety G, together with regular maps

- $\mu : G \times G \longrightarrow G$,
- $\iota : G \longrightarrow G$,

which satisfy the groups axioms.

Examples 9.3.2.

(1) Consider $G_a(k)$, k a field. As an affine variety, $G \cong k$. Furthermore $\mu(x,y) := x + y$ and $\iota(x) := -x$. This is called the *additive group* .

(2) $G_m(k)$, the *multiplicative group*. As an affine variety it is equal to $k \setminus \{0\}$. Furthermore, $\mu(x,y) := x \cdot y$, and $\iota(x) := x^{-1}$.

(3) Consider the group of $\mathrm{GL}_n(k)$ of invertible $n \times n$ matrices. This is an algebraic group. Indeed it is an affine algebraic set, because it is equal to $k^{n^2} \setminus \{\det(a_{ij}) = 0\}$, so as a complement of an affine hypersurface it is itself affine.

(4) The group $\mathscr{G}_k$ is also an algebraic group because, as in the previous example, it is the complement of an affine hypersurface.

Definition 9.3.3. An affine algebraic group *acts* on an affine variety X, if there exists a *regular map*

$$\varphi : G \times X \longrightarrow X : \quad (g,x) \mapsto g \cdot x$$

such that $(g_1 g_2) \cdot x = g_1 \cdot (g_2 \cdot x)$.

Example 9.3.4. The action of $\mathscr{G}_k$ on J_k.

The main result we need concerning group actions is the following theorem.

Theorem 9.3.5. *Let G be an affine algebraic group, acting on X. Then*

(1) Every orbit $G \cdot x$ is locally closed in the Zariski topology.[6]

(2) Every orbit is smooth.

(3) The boundary of each orbit is a union of orbits of smaller dimension.

The proof will occupy the first part of this section. In fact we need to do some more algebraic geometry, in particular, we need to study so-called *constructible sets.*

Definition 9.3.6. Let X be a topological space, $Z \subset X$ a subset. Z is called *constructible* if it is a finite union of locally closed subsets: $Z = \cup_{i=1}^{n} (U_i \cap F_i)$, with U_i open and F_i closed.

Example 9.3.7. Consider $\mathbb{C}^2$ with the Zariski topology, and let $Z = U_1 \cup F_2$, where U_1 is defined by $U_1 := \{(x,y) : x \neq 0\}$ and $F_2 = \{(x,y) : y = 0\}$.

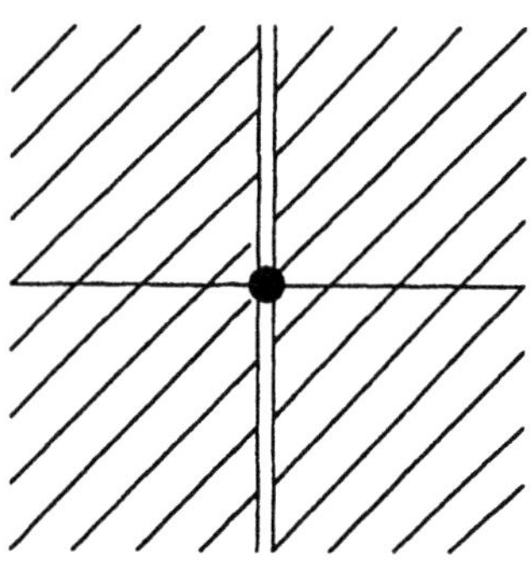

The following set-theoretic lemma is left as Exercise 9.3.20.

[6] Recall that this means that it is the intersection of an open and a closed subset.

Lemma 9.3.8. *Let Z and W be constructible subsets of X. Then $Z \cup W$, $Z \cap W$ and $Z \setminus W$ are constructible. In particular Z^c (the complement) is constructible.*

Lemma 9.3.9. *Let X be an affine variety, and $U \subset X$ be a (Zariski) open subset. Then there exists a finite cover $U = \cup_{i=1}^n U_i$ with U_i affine.*

Example 9.3.10. We have seen in Exercise 6.1.41 that $\mathbb{C}^2 \setminus \{0\}$, as a locally ringed space, is not isomorphic to an affine space. Let $f \in \mathbb{C}[x,y]$. Then $\mathbb{C}^2 \setminus \{f = 0\}$ is affine with coordinate ring $C[x,y,z]/(z \cdot f - 1)$. More generally, the complement of $f = 0$ in an irreducible variety X is affine. Note that we can write

$$\mathbb{C}^2 \setminus \{0\} = (\mathbb{C}^2 \setminus \{y = 0\}) \cup (\mathbb{C}^2 \setminus \{x = 0\}).$$

Proof of Lemma 9.3.9. Consider $Y = X \setminus U$. Then Y is Zariski closed. Consider the ideal $\mathcal{I}(Y) \subset K[X]$. Write $\mathcal{I}(Y) = (f_1, \dots, f_s) \subset K[X]$ and define $U_i := X \setminus V(f_i)$. Then U_i is open and affine. We claim that $U = \cup_{i=1}^n U_i$. As $U_i \subset U$, the inclusion $\cup_{i=1}^n U_i \subset U$ is obvious.

For the other inclusion, suppose we have an element $a \in U \setminus \cup_{i=1}^n U_i$. Then for all i, we have $a \notin U_i$. Therefore $f_i(a) = 0$ for all i. Hence $a \in V(f_1, \dots, f_s) = V(\mathcal{I}(Y)) = Y$. Hence $a \notin U$, a contradiction. □

Theorem 9.3.11 (Chevalley). *Let X,Y be affine, and $\varphi : X \longrightarrow Y$ be a morphism. Then φ maps constructible sets in X to constructible sets in Y.*

Proof. It suffices to prove that a locally closed subset of X is mapped to a constructible set of Y. Let $U \cap F$, with U open and F closed, be a locally closed subset. By Lemma 9.3.9, we can write U as a finite union of open affines. Therefore, it suffices to prove the theorem for the case $X = U$, that is we have to show that the image of a closed subset F of X is constructible. As F now is a finite union of irreducible varieties, it suffices to prove the theorem for the case $X = F$ is irreducible. So we have to show that the image of X is constructible. As $\overline{\varphi(X)}$ is closed, it suffices to show that $\varphi(X)$ is a finite union of locally closed subsets of $\overline{\varphi(X)}$. Thus we may suppose $Y = \overline{\varphi(X)}$, and Y is irreducible. The claim now follows by induction on $\dim(Y)$. The case $\dim(Y) = 0$ is trivial. So assume that the theorem has been proved for all Y' with $Y' \subsetneq Y$. As $\overline{\varphi(X)} = Y$, the image $\varphi(X)$ contains a nonempty open subset U, see Theorem 2.3.12. Consider $Y \setminus U = Z_1 \cup \dots \cup Z_s$. The set $\varphi^{-1}(Z_i)$ is a closed subset of X, and has finitely many irreducible components. Let W_{ij} be such a component. By restriction we get a map $\varphi : W_{ij} \longrightarrow Z_i$. By induction, the image is constructible. Therefore $\varphi(X)$ is the union of U and a constructible set, and therefore constructible itself. □

We can now prove the important Theorem 9.3.5.

Proof of Theorem 9.3.5. (1) Let $x \in X$. Then $G \cdot x$ is constructible. Indeed, $A := G \cdot x$ is the image of the map

$$G \longrightarrow X; \quad g \mapsto g \cdot x.$$

Thus the orbit $A = G \cdot x$ is constructible by the previous Theorem 9.3.11. Write $A = \cup_{i=1}^{s} U_i \cap F_i$. We may suppose that $U_1 \cap F_1$ cannot be removed from this union. Then $F_1 \setminus (F_2 \cup \cdots \cup F_s)$ is an open subset of $F_1 \cup \cdots \cup F_s$. Hence $U := U_1 \cap (F_1 \setminus \left(F_2 \cup \cdots \cup F_s)\right)$ is an open subset of $F_1 \cup \ldots \cup F_s$. As $\overline{A}$ is a closed subset of $F_1 \cup \ldots \cup F_s$, it follows that $U \subset A \subset \overline{A}$ is open in $\overline{A}$. Furthermore, by continuity we have that for all $g \in G$ the set $g^{-1}(U) := \{g^{-1} \cdot x : x \in U\}$ is open in $\overline{A}$. Applying this to $g^{-1} \in G$ we see that $g(U)$ is open in $\overline{A}$. As A is an orbit, it follows that $A = \cup_{g \in G} g(U)$. Thus A is open in $\overline{A}$, or what is the same, A is locally closed.

(2) Let $g \in G$ be fixed. The morphism $g : X \longrightarrow X$ which sends x to $g \cdot x$ is an isomorphism. Now let $x \in X$. Then $G \cdot x$ contains a smooth point. But if $g \cdot x = y$, it follows that y is also a smooth point. Therefore, all points of the orbit are smooth.

(3) Let $A = G \cdot x$. Consider $\overline{A}$, the closure of A. As multiplication with $g \in G$ is continuous, it follows that $g^{-1}(\overline{A})$ is closed. As $g^{-1}(A) = A$, it follows by definition of closure that $g^{-1}(\overline{A}) \supset \overline{A}$. Similarly $g(\overline{A}) \supset \overline{A}$. Now

$$\overline{A} = g \circ g^{-1}(\overline{A}) \supset g(\overline{A}) \supset \overline{A}.$$

Hence we have equality everywhere. Therefore, $\overline{A}$ is G–stable, and $\partial A = \overline{A} \setminus A$ is a union of orbits. As A is open in $\overline{A}$ all the boundary components have smaller dimension. (Use the Krull-dimension.) □

Corollary 9.3.12. *All orbits of the group action of $\mathcal{G}_k$ on J_k are complex submanifolds of J_k.*

The next step is the computation of the (co-)dimension of the orbits. It turns out that this is closely related to the Milnor number.

Theorem 9.3.13. *Let the irreducible algebraic group G act on the vector space $\mathbb{C}^N$. Let $x \in \mathbb{C}^N$. Consider the orbit map*

$$O_x : G \longrightarrow G \cdot x;\ g \mapsto g \cdot x.$$

The differential of the orbit map

$$T_{O_x, id} : T_{G,id} \longrightarrow T_{G \cdot x, x}$$

is surjective. Thus the tangent space to the orbit of x is equal to the image of the differential map at the identity.

Proof. We proved in 4.3.14 that for a dominant map $\varphi : X \longrightarrow Y$ there exists a point $p \in X$ such that $T_{\varphi,p} : T_{X,p} \longrightarrow T_{Y,\varphi(p)}$ is surjective. We apply this to $X = G$, and $Y = G \cdot x$. So there exists a $g \in G$ such that the differential of the orbit map at g is surjective. As we already proved that the orbit is smooth, we only have to show that the rank of the differential of the orbit map is independent of the point $g \in G$. For fixed $h \in G$, consider the automorphism $m_h : G \longrightarrow G$ which sends g to hg. Let $n_h : G \cdot x \longrightarrow G \cdot x$ be the automorphism which sends y to $h \cdot y$. Consider the following commutative diagram of holomorphic maps

$$\begin{array}{ccc} G & \xrightarrow{O_x} & G \cdot x \\ \downarrow m_h & & \downarrow n_h \\ G & \xrightarrow{O_x} & G \cdot x \end{array}$$

If we take derivatives we get a commutative diagram of linear maps

$$\begin{array}{ccc} T_{G,id} & \xrightarrow{T_{O_x,id}} & T_{G\cdot x,x} \\ \downarrow T_{m_h,id} & & \downarrow T_{n_h,x} \\ T_{G,h} & \xrightarrow{T_{O_x,h}} & T_{G\cdot x,h\cdot x}, \end{array}$$

in which the vertical arrows are isomorphisms. It follows that the differential of the orbit map, the horizontal arrows, have constant rank. □

Theorem 9.3.14. *The tangent space to the orbit of $f \in J_k$ under the action of $\mathcal{G}_k$ is equal to $\frac{\mathfrak{m}J(f)+\mathfrak{m}^{k+1}}{\mathfrak{m}^{k+1}}$.*

Proof. As $\mathcal{G}_k$ can be viewed as an open subset of $\oplus_{i=1}^n \mathfrak{m}/\mathfrak{m}^{k+1}$, it follows that the tangent space of $\mathcal{G}_k$ is equal to the tangent space of $\oplus_{i=1}^n \mathfrak{m}/\mathfrak{m}^{k+1}$. The differential of the orbit map

$$\mathcal{G}_k \longrightarrow J_k, \quad h \mapsto f \circ h = h(f),$$

at the identity is given by $(h_1, \dots, h_n) \in \oplus_{i=1}^n \mathfrak{m}/\mathfrak{m}^{k+1} \mapsto h_1 \frac{\partial f}{\partial x_1} + \dots + h_n \frac{\partial f}{\partial x_n}$, because $f(x_1+\varepsilon h_1, \dots, x_n \varepsilon h_n) - f = \varepsilon\big(h_1 \frac{\partial f}{\partial x_1} + \dots + h_n \frac{\partial f}{\partial x_n}\big)$, as a simple calculation shows. (Use Exercise 4.3.22.) □

Remark 9.3.15. Let $f \in \mathfrak{m}^2 \subset \mathcal{O}_n$ have an isolated singularity, and suppose k satisfies $\mathfrak{m}^{k+1} \subset \mathfrak{m}J(f)$. Then the codimension of the orbit of f under $\mathcal{G}_k$ in J_k is equal to $n+\mu(f)$.

Proof. Indeed, applying the previous theorem shows that this codimension is equal to the vector space dimension of $\mathcal{O}_n/\mathfrak{m}J(f)$. Now we have the exact sequence

$$(9.21) \qquad 0 \longrightarrow J(f)/\mathfrak{m}J(f) \longrightarrow \mathcal{O}_n/\mathfrak{m}J(f) \longrightarrow \mathcal{O}_n/J(f) \longrightarrow 0.$$

The vector space dimension of $J(f)/\mathfrak{m}J(f)$ is by Nakayama's Lemma the minimal number of generators of $J(f)$. As f has an isolated singularity, $V(J(f)) = \{0\}$, or better, $J(f)$ is an $\mathfrak{m}$–primary ideal. As the (Chevalley) dimension of $(\mathbb{C}^n, o)$ is n, it follows that number of generators of $J(f)$ is at least n. But then the number of generators of $J(f)$ is exactly n, because $\frac{\partial f}{\partial x_1}, \dots, \frac{\partial f}{\partial x_n}$ are n generators of $J(f)$. The statement therefore follows from the exact sequence (9.21). □

Theorem 9.3.16. *Let $f \in \mathfrak{m}^2 \subset \mathbb{C}\{x_1, \dots, x_n\}$, with finite Milnor number. Let $g \notin \mathfrak{m} \cdot J(f)$. Then there exist only finitely many $t \in \mathbb{C}$ such that $f + t \cdot g$ is right equivalent to f.*

Proof. Choose k such that $\mathfrak{m}^{k+1} \subset \mathfrak{m}J(f)$. Hence g does not lie in the tangent space to the orbit through f in J_k under the action of $\mathcal{G}_k$. This implies that the line $f+tg$ in the jet space J_k does not lie entirely in the orbit of f. Therefore, the line $f + tg$ in J_k intersects the orbit in only finitely many points, as the intersection of the orbit with the line is a constructible subspace of the line. This is exactly the statement of the theorem. □

Proposition 9.3.17. *Let $f \in \mathbb{C}\{x_1, \dots, x_n\}$ be simple. Then f has an isolated singularity.*

Proof. Suppose not. Then the Milnor number $\mu(f)$ is not finite. As $\mathfrak{m}J(f) \subset J(f)$, it follows that in particular $\dim_{\mathbb{C}}\left(\mathbb{C}\{x\}/\mathfrak{m}J(f)\right) = \infty$, that is, for all $k > 0$, there exists a $g_k \in \mathfrak{m}^k \setminus \mathfrak{m}J(f) + \mathfrak{m}^{k+1}$, as otherwise $\mathfrak{m}^k \subset \mathfrak{m}J(f)$ by Nakayama. Hence for all sufficiently small $t \neq 0$ the germ $f_k := f + tg_k$ is not right equivalent to f. This is because otherwise, g_k had to be in the tangent space of the orbit of f under $\mathscr{G}_k$, but by assumption it is not. Thus even f_k and f are not in the same orbit under $\mathscr{G}_k$. Moreover for $k < s$ also f_k is not right equivalent to f_s. Indeed otherwise the classes of f_s and f_k in the jet space J_k would be $\mathscr{G}_k$ equivalent. But they are not, as in J_k the class of f_s is equal to the class of f. Hence there are infinitely many nonequivalent function germs in any small neighborhood of f, and therefore f is not simple. □

Examples 9.3.18. (1) Let a complex number a with $a^2 \neq 4$ be given. Then

$$f_a(x,y) = x^4 + y^4 + ax^2y^2$$

has an isolated singularity, hence a finite Milnor number. One calculates that $\mathfrak{m}^5 \subset \mathfrak{m}^2 J(f_a)$. Hence f_a is 4–determined. Now $J(f_a)$ has two generators, and $\mathfrak{m}J(f_a)$ has 4 generators. Therefore, not all of the degree four monomials in x,y can be in $\mathfrak{m}J(f_a)$. Indeed, $\mathfrak{m}J(f_a)$ looks like,

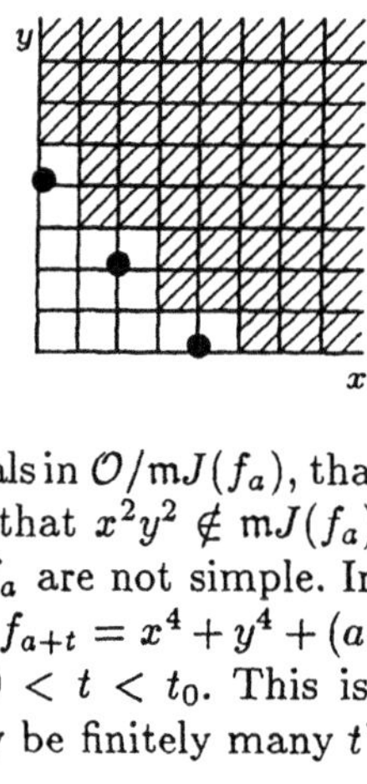

Here dots denote nonzero monomials in $\mathcal{O}/\mathfrak{m}J(f_a)$, that are however equal up to a nonzero constant modulo $\mathfrak{m}J(f_a)$. We see that $x^2y^2 \notin \mathfrak{m}J(f_a)$. As this holds for all $a \in \mathbb{C}$ with $a^2 \neq 4$, it follows that all germs f_a are not simple. Indeed, suppose the converse. Then there exists a (small) t_0 such that $f_{a+t} = x^4 + y^4 + (a+t)x^2y^2$ are mutually right equivalent for infinitely many t with $0 < t < t_0$. This is impossible as for t small x^2y^2 is not in $\mathfrak{m}J(f_{a+t})$, so there can only be finitely many t's in the family f_{a+t} that are right equivalent.

(2) Consider the family $f_a := x^3 + y^6 + axy^4$, such that a is a complex number with $4a^3 + 27 \neq 0$. The zero set of f_a consists of three (different) parabolas. One calculates that f_a then has an isolated singularity. Moreover, one shows that the monomial $xy^4 \notin \mathfrak{m}J(f_a)$, see Exercise 9.3.21 so that all germs f_a are not simple as in the first example.

(3) Consider the family $f_a := x^3 + y^3 + z^3 + axyz$ with $a^3 + 27 \neq 0$. Then f_a has an isolated singularity, and $xyz \notin \mathfrak{m}J(f_a)$, see Exercise 9.3.21. Thus again f_a is not simple.

We now can reach our goal of proving that there are no other simple singularities except for the A-D-E–singularities.

Theorem 9.3.19. *Germs $f \in \mathcal{O}_n$ with one of the following properties are not simple.*

(1) Nonisolated singularities.

(2) $n = 2$ *and* $f \in \mathfrak{m}^4$.

(3) $n = 2$ *and* $f \in (x,y^2)^3$.

(4) f *with* $\mathrm{Corank}(f) \geq 3$.

In particular, Arnold's Classification Theorem of Simple Singularities 9.2.7 holds.

Proof. It was shown in the previous section, see 9.2.15 that either f is right equivalent to an A-D-E–singularity, or right equivalent to a germ with one of the above listed properties. So if we show that singularities with one of the above properties are not simple, Arnold's Classification Theorem follows.

(1) The fact that nonisolated singularities are not simple was proved in 9.3.17.

(2) We calculate modulo $\mathfrak{m}^5$. If f is simple, then $j^4 f \in J_4$ has the property that in a neighborhood U of $j^4 f$ there are only finitely many orbits under the action of $\mathscr{G}_4$. We look at the tangent space at the orbit. As $f \in \mathfrak{m}^4$, the tangent space $\mathfrak{m}J(j^4 f) + \mathfrak{m}^5/\mathfrak{m}^5$ is a subset of $\mathfrak{m}^4/\mathfrak{m}^5$. This tangent space has at most dimension 4, as $\mathfrak{m}J(j^4 f)$ has four generators which all lie in $\mathfrak{m}^4$. It follows that the orbit of $j^4 f$ has dimension at most 4. This holds for all $f \in \mathfrak{m}^4$. Note that $\mathfrak{m}^4/\mathfrak{m}^5$ is $\mathscr{G}_4$–stable, and has dimension 5. As a finite union of subvarieties of dimension four never can fill up an open subset of a vector space of dimension 5, it follows that all neighborhoods of $j^4 f$ intersect infinitely many orbits. In particular f is not simple.

(4) The argument is similar, but now we calculate modulo $\mathfrak{m}^4$. We know that $\mathfrak{m}J(f)/\mathfrak{m}^4$ has dimension at most n^2, but $\mathfrak{m}^3/\mathfrak{m}^4$ has dimension $\binom{n+2}{3}$, which is bigger.

(3) In this case the argument is somewhat more subtle. We calculate modulo $\mathfrak{m}^7$, so we consider the action of $\mathscr{G}_6$ on the jet space J_6. We take the weighted order by defining the weight of x to be equal to two, and the weight of y to be equal to one. So all terms of elements in $(x,y^2)^3$ have weighted order at least six. Let $f \in (x,y^2)^3$. The four generators $x\frac{\partial f}{\partial x}, y\frac{\partial f}{\partial x}, x\frac{\partial f}{\partial y}, y\frac{\partial f}{\partial y}$ of $\mathfrak{m}J(f)$ have weighted order at least six, five, seven and six respectively. So we have at most three linearly independent elements in $\mathfrak{m}J(f)$ of weighted order six, namely those given by $x\frac{\partial f}{\partial x}$, $y^2\frac{\partial f}{\partial x}$, and $y\frac{\partial f}{\partial y}$. There are however four of them in $(x,y^2)^3$. We draw those weighted order six elements for the example $f = x^3 + y^6$. The box with the cross means that that monomial is not in $\mathfrak{m}J(f)$.

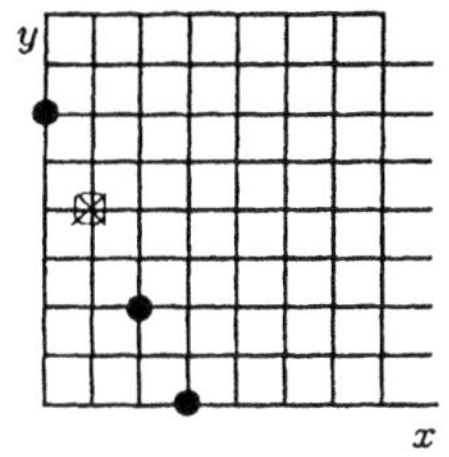

A counting argument shows that $(x,y^2)^3 + \mathfrak{m}^7/\mathfrak{m}^7$ is a vector space of dimension sixteen. It follows that the orbit $\mathscr{G}_6 \cdot f$ has dimension at most sixteen, as its tangent space has at most dimension 16. However, $\mathscr{G}_6 \cdot f$ does not contain the whole of $(y,x^2)^3$ modulo $\mathfrak{m}^7$, as otherwise the tangent space of $\mathscr{G}_6 \cdot f$ would contain $(y,x^2)^3$ modulo $\mathfrak{m}^7$. So the intersection of $\mathscr{G}_6 \cdot f$ with $(y,x^2)^3$ modulo $\mathfrak{m}^7$ has dimension at most fifteen. It now follows as in the other cases that f is not simple. □

Exercises

9.3.20. Prove Lemma 9.3.8.

9.3.21.

(1) Consider the family $f_a := x^3 + y^6 + axy^4$ for $a \in \mathbb{C}$. Show that f_a has an isolated singularity exactly when $4a^3 + 27 \neq 0$. Show that $xy^4 \notin \mathfrak{m}J(f_a)$.

(2) Do the same for the family $f_a = x^3 + y^3 + z^3 + axyz$.

9.3.22. Let F_t be a holomorphic family. Suppose that for all sufficiently small nonzero t,s the germs F_t and F_s are right equivalent. Suppose that F_0 is not right equivalent to F_t for t sufficiently small and nonzero. Prove that $\mu(F_t) < \mu(F_0)$.
(Hint: Use the third part of Theorem 9.3.5 and Remark 9.3.15.)

10 Deformations of Singularities

In the final chapter of this book, we study deformations of germs of complex spaces. The ultimate goal is to prove the existence of a semi-universal deformation of (X, o), in case it has an isolated singularity. As the proof of this theorem is quite involved, we will first treat some special cases. So, in Section 10.1, we will consider isolated *hypersurface* singularities (X, o) given by $f = 0$. A deformation $(X_S, \mathrm{o}) \longrightarrow (S, \mathrm{o})$ over some germ of a complex space (S, o) is simply given by $F_S = 0$ for some element $F_S \in \mathbb{C}\{x\}\widehat{\otimes}\mathcal{O}_{S,\mathrm{o}}$, with $F_S(x,0) = f$. There is a natural notion of isomorphism of deformations. Moreover, there is the notion of induced family, also called pull-back. If $(T, \mathrm{o}) \xrightarrow{\varphi} (S, \mathrm{o})$ is a holomorphic map, we get a map $\varphi : \mathbb{C}\{x\}\widehat{\otimes}\mathcal{O}_{S,\mathrm{o}} \longrightarrow \mathbb{C}\{x\}\widehat{\otimes}\mathcal{O}_{T,\mathrm{o}}$, which we denote by the same name. Then $\varphi(F_S)$ is an element of $\mathbb{C}\{x\}\widehat{\otimes}\mathcal{O}_{T,\mathrm{o}}$, which gives a deformation $(X_T, \mathrm{o}) \longrightarrow (T, \mathrm{o})$ of (X, o). A very important special case is that $(T, \mathrm{o}) \hookrightarrow (S, \mathrm{o})$, in which case one just restricts the family to the subspace (T, o). A versal deformation $(X_S, \mathrm{o}) \longrightarrow (S, \mathrm{o})$ is a deformation from which every other deformation $(X_T, \mathrm{o}) \longrightarrow (T, \mathrm{o})$ can be induced. If, moreover, the map on tangent spaces $(\mathfrak{m}_{T,\mathrm{o}}/\mathfrak{m}_{T,\mathrm{o}}^2)^* \longrightarrow (\mathfrak{m}_{S,\mathrm{o}}/\mathfrak{m}_{S,\mathrm{o}}^2)^*$ is uniquely determined, it is called semi-universal. We will show in Remark 10.1.10 , that it is too much to ask for uniqueness of the map $(T, \mathrm{o}) \longrightarrow (S, \mathrm{o})$. The main tool in the proof of the existence of a semi-universal deformation of an isolated hypersurface singularity is Grauert's Approximation Theorem.

In Section 10.2 we will consider deformations of germs of complex spaces, defined by say $f_1 = \cdots = f_k = 0$. In general, defining a deformation by simply deforming $f_1, \ldots, f_k$ to $F_{1S}, \ldots, F_{kS}$ as in the hypersurface case, does not lead to a good notion. This is because the dimension might change, or components might disappear. In case (S, o) is smooth and one-dimensional, so that $\mathcal{O}_{S,\mathrm{o}} = \mathbb{C}\{s\}$, we will show in Example 10.2.1 that it is a good idea to take the condition that s is a nonzerodivisor of $\mathbb{C}\{x\}\widehat{\otimes}\mathcal{O}_{S,\mathrm{o}}/(F_{1S}, \ldots, F_{kS})$. We will show that this property is equivalent to the fact that one can lift all the relations $(r_1, \ldots, r_k)$, that is $f_1 r_1 + \ldots + f_k r_k = 0$, to a relation between the $F_{1S}, \ldots, F_{kS}$. This is called flatness, and can be defined for general $\mathcal{O}_{S,\mathrm{o}}$, for example, $\mathbb{C}[\varepsilon]/(\varepsilon^2)$, the local ring of the "double point" $(\mathbb{T}, \mathrm{o})$. As ε is always a zerodivisor, the flatness definition by the lifting property of the relations makes more sense. We then classify all deformations of (X, o) over the double point $(\mathbb{T}, \mathrm{o})$. These deformations are classified by the so-called $T^1_{X,\mathrm{o}}$, which is a finite-dimensional vector space in case (X, o) has an isolated singularity. If one has a deformation over a germ of a complex space (S, o), one can define the Kodaira-Spencer map. To every element of the tangent space $(\mathfrak{m}_{S,\mathrm{o}}/\mathfrak{m}_{S,\mathrm{o}}^2)^*$, one assigns a deformation over $(\mathbb{T}, \mathrm{o})$, hence an element of $T^1_{X,\mathrm{o}}$. We then prove the existence of a versal deformation, again by applying Grauert's Approximation Theorem, in case there exists a deformation over a smooth space (S, o) such that the Kodaira-Spencer map is surjective. This case applies, for example, to isolated complete intersection singularities, see Exercise 10.2.33, and to isolated Cohen-Macaulay codimension two singularities. An essential ingredient is the Hilbert-Burch Theorem.

In the final section, we will prove the existence of a semi-universal deformation in a more general setting, namely under the assumption that $T^1_{X,\mathrm{o}}$ is a finite-dimensional vector space. The basic idea is to do this order by order. Let (X, o) be defined by $f \in$

$\mathbb{C}\{x\}^k$. On first order, one does the following. Take a basis $g_1, \ldots, g_\tau$ of $T^1_{X,0}$, and take a new ring $\mathbb{C}[[s_1, \ldots, s_\tau]]$ with maximal ideal $\mathfrak{m} = (s_1, \ldots, s_\tau)$. We then have a flat deformation

$$f + s_1 g_1 + \ldots, s_\tau g_\tau \tag{10.1}$$

over the space with local ring $\mathbb{C}[[s_1, \ldots, s_\tau]]/\mathfrak{m}^2$. In general, it is not possible to lift this family to a flat deformation over $\mathbb{C}[[s_1, \ldots, s_\tau]]/\mathfrak{m}^3$. We will show however, that there is a minimal I with $\mathfrak{m}^3 \subset I \subset \mathfrak{m}^2$ such that there exists a deformation over $\mathbb{C}[[s_1, \ldots, s_\tau]]/I$ lifting (10.1). In general, suppose we have a deformation of $(X, 0)$ over a space with local ring $\mathcal{O}/J$. We will show that there exists a minimal I with $\mathfrak{m}J \subset I \subset J$, such that there exists an extension of the deformation over the space with local ring $\mathcal{O}/I$. The existence of this minimal I will be shown by obstruction theory. We will construct an element in $T^2_{X,0} \otimes I$ for all I with $\mathfrak{m}J \subset I \subset J$, $T^2_{X,0}$ a vector space, the so called obstruction space. It is zero if and only if the family can be lifted over the space with local ring $\mathcal{O}/I$. Then we will successively extend the family in a maximal way order by order. This leads to a formal deformation. The fact that it is formally semi-universal is not very difficult. In fact, by applying the ideas of Section 10.2, one can show that it is semi-universal, by applying Grauert's Approximation Theorem, in case the formal deformation is in fact convergent. The final subtle point is to prove that there exists a convergent deformation which is formally semi-universal. Also in this proof Grauert's Approximation Theorem is used. We moreover prove the following criterion for (formal) semi-universality. Suppose a formal deformation of $(X, 0)$ over a formal algebra $\mathbb{C}[[s]]/J$ is given. Consider the obstruction element in $T^2_{X,0} \otimes J/\mathfrak{m}J$, which leads to the *obstruction map* $(T^2_{X,0})^* \longrightarrow J/\mathfrak{m}J$. Then the formal deformation is semi-universal if and only if the Kodaira-Spencer map is an isomorphism, and the obstruction map is surjective.

10.1 Deformations of Hypersurface Singularities

We make precise in the following definition the notion of deformations of hypersurface singularities.

Definition 10.1.1. Let $(X, 0) \subset (\mathbb{C}^n, 0)$ be a hypersurface singularity, defined by an $f \in \mathbb{C}\{x_1, \ldots, x_n\}$, and $(S, 0)$ be a germ of a complex space. A *deformation* $(X_S, 0)$ is a germ of a complex space in $(\mathbb{C}^n \times S, 0)$ defined by an element $F_{S,0} \in \mathbb{C}\{x_1, \ldots, x_n\}\hat{\otimes}\mathcal{O}_{S,0}$ such that $F_{S,0}(x, 0) = f$. Here $F_{S,0}(x, 0)$ is the image of $F_{S,0}$ under the natural surjective map $\mathbb{C}\{x_1, \ldots, x_n\}\hat{\otimes}\mathcal{O}_{S,0} \longrightarrow \mathbb{C}\{x_1, \ldots, x_n\}$.

We now show a deformation of the E_7-singularity.

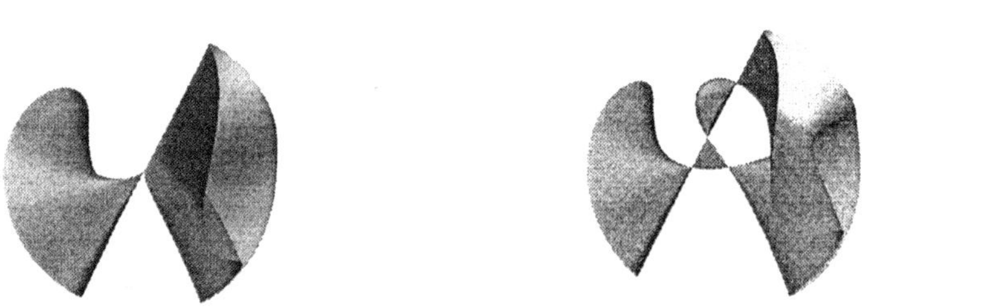

Example 10.1.2. Let f and g be elements in $\mathbb{C}\{x_1,\ldots,x_n\}$. Then $F_{S,o} := f + s \cdot g \in \mathbb{C}\{x_1,\ldots,x_n,s\} = \mathbb{C}\{x_1,\ldots,x_n\}\hat{\otimes}\mathbb{C}\{s\}$ gives a deformation of $(X,o) := \big(V(f),o\big)$.

Definition 10.1.3. Let (X_S,o) be a deformation of (X,o) over (S,o), defined by $F_{S,o}$. Let $(T,o) \longrightarrow (S,o)$ be an analytic map between germs of complex spaces. Then we have an induced map $\mathcal{O}_{S,o} \longrightarrow \mathcal{O}_{T,o}$ on the ring level, and hence a map:

$$\mathbb{C}\{x_1,\ldots,x_n\}\hat{\otimes}\mathcal{O}_{S,o} \longrightarrow \mathbb{C}\{x_1,\ldots,x_n\}\hat{\otimes}\mathcal{O}_{T,o}.$$

The image of $F_{S,o}$ under this map we denote by $F_{T,o}$. Then $F_{T,o}$ defines a deformation (X_T,o). This deformation we call the *pull-back* of (X_S,o) over $(T,o) \longrightarrow (S,o)$, or alternatively the deformation induced by (X_S,o) via the map $(T,o) \longrightarrow (S,o)$. Often it is denoted by $(X \times_S T,o)$. One also says that in this case the diagram

$$\begin{array}{ccc} (X \times_S T,o) & \longrightarrow & (X_S,o) \\ \downarrow & & \downarrow \\ (T,o) & \longrightarrow & (S,o) \end{array} \tag{10.2}$$

is Cartesian.

One of the most important examples is the case of an inclusion $(T,o) \hookrightarrow (S,o)$. In this case, we just restrict the family over (S,o) to a family over the subspace (T,o).

Definition 10.1.4. Let (X_S,o) and (X'_S,o) be two deformations of (X,o) over (S,o), defined by $F_{S,o}$ and $G_{S,o}$ in $\mathbb{C}\{x_1,\ldots,x_n\}\hat{\otimes}\mathcal{O}_{S,o}$ respectively. A *morphism of deformations* $\Phi : (X_S,o) \longrightarrow (X'_S,o)$ is given by an analytic map

$$\Phi : \mathbb{C}\{x_1,\ldots,x_n\}\hat{\otimes}\mathcal{O}_{S,o} \longrightarrow \mathbb{C}\{x_1,\ldots,x_n\}\hat{\otimes}\mathcal{O}_{S,o}$$

such that

(1) it transforms fibers into fibers, that is the diagram

$$\begin{array}{ccc} \mathbb{C}\{x_1,\ldots,x_n\}\hat{\otimes}\mathcal{O}_{S,o} & \xrightarrow{\Phi} & \mathbb{C}\{x_1,\ldots,x_n\}\hat{\otimes}\mathcal{O}_{S,o} \\ \downarrow & & \downarrow \\ \mathcal{O}_{S,o} & = & \mathcal{O}_{S,o} \end{array}$$

is commutative,

(2) $(\Phi(F_{S,o})) = (G_{S,o})$,

(3) $\Phi_{|s=0} = Id$.

We now classify the deformations of a hypersurface singularity over the double point $(\mathbb{T},o)$ which is the germ of the complex space consisting of just one point, but with local ring $\mathbb{C}[\varepsilon]/(\varepsilon^2)$.

Theorem 10.1.5. *Let (X,o) be a hypersurface singularity defined by $f \in \mathbb{C}\{x_1,\ldots,x_n\}$. Then two deformations given by $f + \varepsilon g_1$ and $f + \varepsilon g_2$ over $(\mathbb{T},o)$ are isomorphic if and only if $g_1 - g_2 \in (f) + J(f)$. In particular, the isomorphism classes of deformations of (X,o) over $(\mathbb{T},o)$ are in one-one correspondence with the elements of the Tjurina algebra $\mathbb{C}\{x\}/(f,J(f))$ of f.*

Proof. Consider an automorphism Φ of $(\mathbb{C}^n \times \mathbb{T}, o)$ over $(\mathbb{T}, o)$. It looks like $\Phi(x_i) = x_i + \varepsilon h_i$, and $\Phi(\varepsilon) = \varepsilon$. Furthermore, consider a unit u in $\mathbb{C}\{x\}\widehat{\otimes}\mathcal{O}_{T,o}$ which is congruent to 1 modulo ε. So $u = 1 + \varepsilon u_1$. If the deformations $f + \varepsilon g_1$ and $f + \varepsilon g_2$ are isomorphic we are able to find h_i and u_1 such that

$$(1 + \varepsilon u_1)\Phi(f + \varepsilon g_1) = f + \varepsilon g_2.$$

From Taylor's formula and the fact that $\varepsilon^2 = 0$ it follows that the left hand side is equal to

$$f + \varepsilon \left(\sum_{i=1}^{n} h_i \frac{\partial f}{\partial x_i} + u_1 f + g_1 \right).$$

This yields $g_1 - g_2 \in (f) + J(f)$ as claimed. The converse is proved by reversing the steps. □

Definition 10.1.6. Let a deformation (X_S, o) of (X, o) over (S, o) be given.

(1) The deformation (X_S, o) is called *versal* if for any deformation (X_T, o) of (X, o) over (T, o) there exists a map $(T, o) \longrightarrow (S, o)$ such that (X_T, o) is isomorphic to the pull-back of (X_S, o) over $(T, o) \longrightarrow (S, o)$.

(2) The deformation (X_S, o) of (X, o) is called *semi-universal* if the induced map on tangent spaces $(\mathfrak{m}_{T,o}/\mathfrak{m}_{T,o}^2)^* \longrightarrow (\mathfrak{m}_{S,o}/\mathfrak{m}_{S,o}^2)^*$ is uniquely determined. (Here * denotes the $\mathbb{C}$-dual.)

Theorem 10.1.7. *Let $f \in \mathbb{C}\{x_1, \dots, x_n\}$. Suppose that the Tjurina number $\tau(f) = \dim_{\mathbb{C}} \mathbb{C}\{x_1, \dots, x_n\}/(f, \frac{\partial f}{\partial x_1}, \dots \frac{\partial f}{\partial x_n})$ is finite, that is, suppose that $(X, o) := (V(f), o)$ has an isolated singularity. Let $g_1, \dots, g_p \in \mathbb{C}\{x_1, \dots, x_n\}$ be such that they represent generators of the vector space $\mathbb{C}\{x_1, \dots, x_n\}/(f, \frac{\partial f}{\partial x_1}, \dots \frac{\partial f}{\partial x_n})$. Then*

$$F_{S,o}(x_1, \dots, x_n) := f + \sum_{i=1}^{p} s_i \cdot g_i \in \mathbb{C}\{x_1, \dots, x_n, s_1, \dots, s_p\}$$

gives a versal deformation (X_S, o) of (X, o) over $(S, o) = (\mathbb{C}^p, o)$. If moreover $g_1, \dots, g_p$ represent a basis, then the deformation is semi-universal.

Before giving the proof of this theorem, we give some examples.

Example 10.1.8. We take the function $f = x^3 \in \mathbb{C}\{x\}$. A basis for the Tjurina algebra is given by the classes of $\{1, x\}$. Therefore, by the theorem, a versal deformation is given by $F(x, s_1, s_2) = x^3 + s_1 x + s_2$. As an example, we look at the deformation given by $G(x,t) = x^3 + tx^2$ of $(V(f), o)$. The germ of the complex space defined by $G(x,t)$ looks like

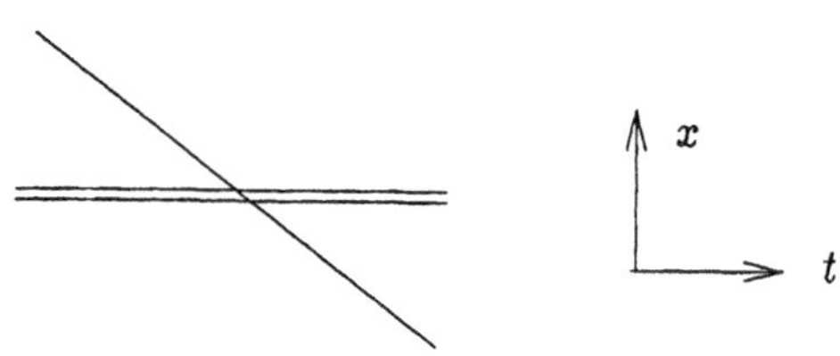

Versality says that we can find a map $(T,o) \longrightarrow (S,o)$, that is, dually a map of rings

$$\mathbb{C}\{s_1,s_2\} \longrightarrow \mathbb{C}\{t\},$$

or to put it in another way, two power series $s_1(t),s_2(t)$ such that

$$F(x,s_1(t),s_2(t)) = uG(\Phi(x,t))$$

for some unit $u \in \mathbb{C}\{x,t\}$ and some automorphism Φ of $\mathbb{C}\{x,t\}$. It turns out that we can take the unit u to be one. So this means that we have to find an automorphism $\Phi : (\mathbb{C} \times T,o) \longrightarrow (\mathbb{C} \times T,o)$ with

- $\Phi = (\phi(x,t),t)$,
- $\phi(x,0) = x$,
- $F(x,s_1(t),s_2(t)) = G(\phi(x,t),t)$.

To find this ϕ we rewrite

$$x^3 + tx^2 = (x + \tfrac{1}{3}t)^3 - \tfrac{1}{3}t^2x - \tfrac{1}{27}t^3.$$

With $\phi(x,t) = x - \frac{1}{3}t$ we get

$$G(\phi(x,t),t) = x^3 - \tfrac{1}{3}t^2x + \tfrac{2}{27}t^3.$$

We can therefore take

$$s_1(t) = -\tfrac{1}{3}t^2;\ s_2(t) = \tfrac{2}{27}t^3.$$

therefore the map $(T,o) \longrightarrow (S,o)$ is given by the following "picture".

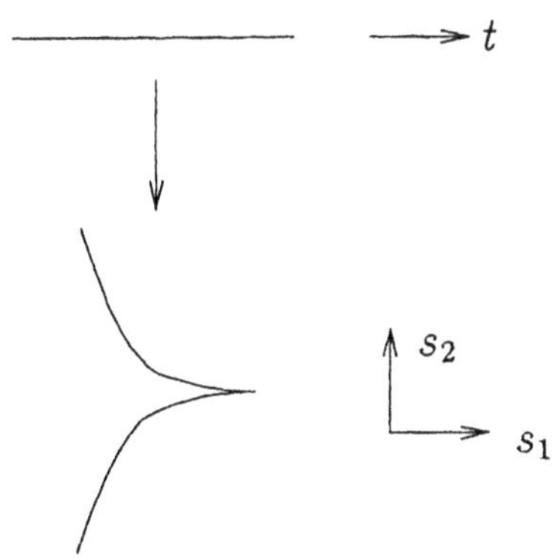

Example 10.1.9. Let $(X,o) = \big(V(f),o\big)$ be a germ of an isolated singularity. We take $G = f + \varepsilon \cdot h \in \mathbb{C}\{x_1,\dots,x_n,\varepsilon\}/(\varepsilon^2)$ for some $h \in \mathbb{C}\{x_1,\dots,x_n\}$. This defines a deformation $(X_{\mathbb{T}},o)$ of (X,o) over the double point $(\mathbb{T},o)$. As $\mathbb{C}\{x_1,\dots,x_n\}/(f,\frac{\partial f}{\partial x_1},\dots,\frac{\partial f}{\partial x_n})$ is a finite-dimensional vector space we can write

$$h = b_0 \cdot f + \sum_{i=1}^{\tau} b_i g_i + \sum_{i=1}^{n} h_i \frac{\partial f}{\partial x_i} \tag{10.3}$$

with $b_i \in \mathbb{C}$ for $i \geq 1$, $b_0, h_i \in \mathbb{C}\{x_1, \dots, x_n\}$ and g_i as in the statement of theorem 10.1.7. We define the automorphism Φ of $(\mathbb{C}^n \times \mathbb{T}, o)$ by

$$\Phi(x_i) = x_i - h_i \cdot \varepsilon, \quad \Phi(\varepsilon) = \varepsilon.$$

By using Taylor's formula (10.3) and the fact that $\varepsilon^2 = 0$ we get

$$\begin{aligned} G(\Phi(x,\varepsilon)) &= f(x_1 - h_1\varepsilon, \dots, x_n - h_n\varepsilon) + \varepsilon h \\ &= f(x_1, \dots, x_n) - \sum_{i=1}^{n} \varepsilon h_i \frac{\partial f}{\partial x_i} + \varepsilon h. \end{aligned}$$

Using formula (10.3) we get

$$f(x_1, \dots, x_n) - \sum_{i=1}^{n} \varepsilon h_i \frac{\partial f}{\partial x_i} + \varepsilon h = (1 + \varepsilon b_0) f(x_1, \dots, x_n) + \varepsilon \sum_{i=1}^{\tau} b_i g_i.$$

This shows that the deformation $(X_{\mathbb{T}}, o) \longrightarrow (\mathbb{T}, o)$ is isomorphic to the pull back of the semi-universal family via the map $(\mathbb{T}, o) \longrightarrow (S, o)$ given by

$$\begin{aligned} \mathbb{C}\{s_1 \dots, s_\tau\} &\longrightarrow \mathbb{C}\{\varepsilon\}/(\varepsilon^2) \\ s_i &\mapsto \varepsilon b_i. \end{aligned}$$

Remark 10.1.10. The map $(T, o) \longrightarrow (S, o)$ in the definition of a semi-universal deformation in general is not uniquely determined. For example, let $f = x^2$. A semi-universal deformation is given by $x^2 + s$. We claim that the deformations of $(V(f), o)$ given by $x^2 + t$ and $x^2 + t + t^2$ are isomorphic. Indeed with $\Phi(x) = x\sqrt{1+t}$, $\Phi(t) = t$ and $u = 1 + t$ we get

$$\Phi(x^2 + t + t^2) = (x\sqrt{1+t})^2 + t + t^2 = x^2 + tx^2 + t + t^2 = u(x^2 + t).$$

Thus the deformation $x^2 + t$ is induced by the map given by $s(t) = t$, but also induced by the map $s(t) = t + t^2$.

Proof of Theorem 10.1.7. The proof uses Grauert's Approximation Theorem[1] 8.2.2 and the ideas of the last example. Let (X_T, o) be a deformation of (X, o) over (T, o), defined by $G(x,t) \in \mathbb{C}\{x_1, \dots, x_n\}\widehat{\otimes}\mathcal{O}_{T,o}$. We have $\mathcal{O}_{T,o} = \mathbb{C}\{t_1, \dots, t_q\}/J$ for some ideal J. We have to find the following.

- A map $(T, o) \longrightarrow (S, o)$ or dually a map $\mathcal{O}_{S,o} \longrightarrow \mathcal{O}_{T,o}$, which in turn means that we have to find p power series $s(t) = (s_1(t_1, \dots, t_q), \dots, s_p(t_1, \dots, t_q))$ with $s \equiv 0 \mod t$.
- An automorphism Φ of $\mathbb{C}\{x\}\widehat{\otimes}\mathcal{O}_{T,o}$ with $\Phi(x)_{|t=0} = x$ and $\Phi(t) = t$. We put $\Phi(x_i) = \phi_i$, and $\phi = (\phi_1, \dots, \phi_n)$.
- a unit $u \in \mathbb{C}\{x\}\widehat{\otimes}\mathcal{O}_{T,o}$ with $u \equiv 1 \mod \mathfrak{m}_{T,o}$

[1] It is possible for the hypersurface case to prove the theorem without using Grauert's Approximation Theorem. For more complicated cases Grauert's Theorem is a necessary tool. We use it in this proof for didactical reasons.

such that

$$G(\Phi(x,t)) = u \cdot F(x,s(t)). \tag{10.4}$$

By taking a representative of $G(x,t)$ in $\mathbb{C}\{x,t\}$ which defines also a deformation of (X,o), we may assume that $J = (0)$. We view (10.4) as a system of equations, which we have to solve for Φ, u and s. Here Φ and u may depend on x and t, but the p functions $s_1,\dots,s_p$ are only allowed to depend on $t_1,\dots,t_q$. We show that equation (10.4) has a solution by applying Grauert's Approximation Theorem 8.2.2. We obviously have a solution of (10.4) of order 0. To apply this theorem, we have to show that every solution (ϕ,u,s) of (10.4) of order i in t can be extended to a solution of (10.4) of order $i+1$. Having this solution (ϕ,u,s) of order i, we look at the homogeneous part of *degree* $i+1$ of $G(\phi(x,t)) - uF(x,s(t))$ and write it as

$$\sum_{|\nu|=i+1} t^\nu h_\nu$$

with $h_\nu \in \mathbb{C}\{x_1,\dots,x_n\}$. Just as in the second example we can write for all ν;

$$h_\nu = b_{0\nu} f + \sum_{j=1}^{p} b_{j\nu} g_j + \sum_{j=1}^{n} h_{\nu j} \frac{\partial f}{\partial x_j}$$

with $b_{j\nu} \in \mathbb{C}$, $j \geq 1$, and $b_{0\nu}, h_{\nu j} \in \mathbb{C}\{x_1,\dots,x_n\}$. We define

$$\begin{aligned} \phi'_j &= \phi_j - \sum_{|\nu|=i+1} t^\nu h_{\nu j} \qquad j = 1,\dots,n \\ s'_j &= s_j + \sum_{|\nu|=i+1} t^\nu b_{j\nu} \qquad j = 1,\dots,p, \\ u' &= u + \sum_{|\nu|=i+1} t^\nu b_{0\nu} \end{aligned}$$

As $G \equiv f$ modulo t, it follows from Taylor's formula that $(\phi'_1,\dots,\phi'_n,u',s'_1,\dots,s'_p)$ is a solution of (10.4) of order $i+1$. (Work this out!) This proves the fact that the given family is versal. The fact that the map is uniquely determined on the level on tangent spaces, follows easily from the classification of isomorphism classes of deformations over the double point $(\mathbb{T},o)$, see 10.1.5. □

10.2 Deformations of Singularities

We now consider deformations of germs of complex spaces. So consider a singularity (X,o) defined by the equations

$$f_1(x) = \ldots = f_k(x) = 0,$$

for $f_i \in \mathbb{C}\{x_1,\dots,x_n\} = \mathbb{C}\{x\}$. Naively, one probably would like to say that a deformation of (X,o) over a space (S,o) is given by the zero set of functions,

$$F_1(x,s) = \ldots = F_k(x,s) = 0$$

with $F_i \in \mathbb{C}\{x\}\widehat{\otimes}\mathcal{O}_{S,o}$ and $F_i(x,0) = f_i(x)$ for $i = 1, \dots k$. However, this kind of deformation is too general, because a "general fiber" of $(X_S, o) \longrightarrow (S, o)$ does not sufficiently reflect the properties of (X, o). To see this, look at the following example.

Example 10.2.1. Let $(X_S, o) \longrightarrow (S, o)$ be an analytic map. Here (X_S, o) is defined by

$$\begin{array}{ccccc} F_1 & = & xy - s & = & 0 \\ F_2 & = & xz - s & = & 0 \\ F_3 & = & yz - s & = & 0 \end{array}$$

Although this is a space defined in 4-space, we would like to imagine the situation by the following picture:

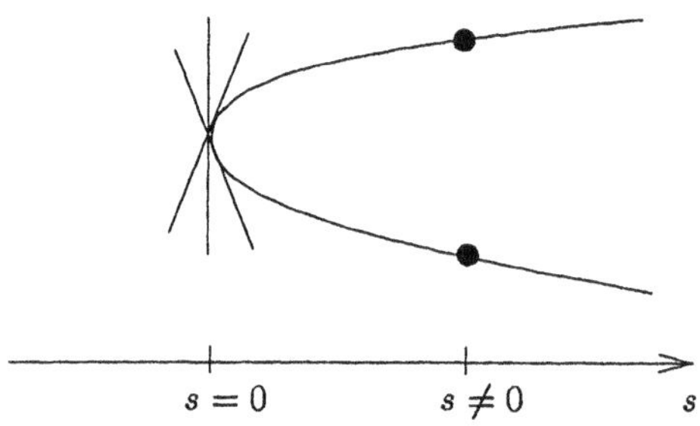

The problem here is that s is zerodivisor of $\mathbb{C}\{x,y,z,s\}/(F_1,F_2,F_3)$. The total space (X_S, o) has four components, of which three are in the fiber $s = 0$. In fact, a primary decomposition of (F_1,F_2,F_3) is (cf. 1.4.12):

$$(F_1,F_2,F_3) = (x,y,s) \cap (x,z,s) \cap (y,z,s) \cap (x - y, y - z, x^2 - s).$$

We therefore have a decomposition $(X_S, o) = (X_0, o) \cup (X_1, o)$, where (X_0, o) is the union of three coordinate axes, and (X_1, o) is the parabola. Take a function f, vanishing on (X_1, o), but not on (X_0, o). Then obviously $s \cdot f = 0$ as elements of $\mathcal{O}_{X_S,o}$, but both s and f are nonzero in $\mathcal{O}_{X_S,o}$. This expresses the fact that s is a zerodivisor. In fact it can be proved by using the Active Lemma 4.1.10, see Exercise 10.2.28, that

Lemma 10.2.2. *s is a zerodivisor $\Longleftrightarrow$ there exists a component of (X_S, o) in the fiber above zero.*

The above example hopefully makes clear to the reader that for a "nice" one parameter family (with parameter s), one should impose the condition that s is a nonzerodivisor. This we call flatness:

Definition 10.2.3. A $\mathbb{C}\{s\}$-module M is called *flat* if and only if s is a nonzerodivisor of M.

For example a *finitely generated* $\mathbb{C}\{s\}$-module M is flat if and only if M is free, see 1.3.9.

It turns out that it is not so easy at all to construct (nontrivial) one parameter *flat* deformations of singularities. Therefore, it is a natural idea to construct deformations by "power-series expansion". That is, construct deformations over $(\mathbb{T}, o)$, then try to lift over the germ of the complex space with local ring $\mathbb{C}[s]/(s^3)$, etc. But then one has the problem of defining flatness of a $\mathbb{C}[s]/(s^2)$-module, as s is always a zerodivisor of $\mathbb{C}[s]/(s^2)$. So we need to give a different definition of flatness of a $\mathbb{C}\{s\}$-module, which gives a good generalization to rings with nilpotent elements. The following lemma is of importance for solving this problem.

Lemma 10.2.4. *Let $(S,o) = (\mathbb{C},o)$, and consider a deformation of (X,o) as above. Then the deformation is flat (that is, s is a nonzerodivisor) if and only if for every relation between the f_i:*

$$f_1 r_1 + \ldots + f_k r_k = 0,$$

we can find a lift *$R_i(x,s)$, that is, $R_i(x,0) = r_i(x)$ and*

$$F_1 R_1 + \ldots + F_k R_k = 0.$$

Proof. Suppose that s is a nonzerodivisor, and take a relation $\sum f_i r_i = 0$. Take *any lift* $R'_i(x,s)$ of the r_i, and look at $F_1 R'_1 + \ldots + F_k R'_k$. This might not be zero, but we know it is if we plug in $s = 0$. This shows that this expression is divisible by s:

$$F_1 R'_1 + \ldots + F_k R'_k = s\Phi.$$

This says that $s\Phi = 0$ as element of $\mathcal{O}_{X_S,o}$. As s is a nonzerodivisor it follows that $\Phi = 0$ as element of $\mathcal{O}_{X_S,o}$. Hence we can write $\Phi = \sum \alpha_i F_i$ for some α_i. Now put $R_i = R'_i - s\alpha_i$. It follows that $F_1 R_1 + \ldots + F_k R_k = 0$, so we found a lift of the r_i.

On the other hand, suppose that we can lift every relation. We need to show that s is a nonzerodivisor. So suppose that $s\Phi = 0 \in \mathcal{O}_{X_S,o}$, that is

$$s\Phi = F_1 R_1 + \ldots + F_k R_k.$$

Putting $s = 0$ we get a relation $\sum f_i r_i = 0$, which by assumption can be lifted to a relation $\sum F_i R'_i = 0$. Then

$$s\Phi = \sum F_i (R_i - R'_i).$$

As both R_i and R'_i are lifts of the r_i, it follows that $R_i - R'_i$ is divisible by s. Hence the power-series $\frac{R_i - R'_i}{s}$ exists. Because in the power-series ring s is a nonzerodivisor it follows that

$$\Phi = \sum F_i \frac{R_i - R'_i}{s},$$

expressing the fact that $\Phi = 0$ as element of $\mathcal{O}_{X_S,o}$. This is what we had to show. □

This result motivates the following definition.

Definition 10.2.5. Let $(X,o) \subset (\mathbb{C}^n,o)$ be a germ of a complex space. Let $\mathcal{I}(X,o) = (f_1,\ldots,f_k)$. Let (S,o) be a further germ of a complex space. Let $(X_S,o) \subset (\mathbb{C}^n \times S,o)$ be defined by $F_1,\ldots,F_k$, $F_i \in \mathbb{C}\{x\}\widehat{\otimes}\mathcal{O}_{S,o}$. Then $(X_S,o) \longrightarrow (S,o)$ is called a *(flat)*[2] *deformation* of (X,o) over (S,o) if

(1) $F_i(x,0) = f_i(x)$ for $i = 1,\ldots k$,

(2) for every relation $r = (r_1,\ldots,r_k) \in \mathbb{C}\{x\}^k$ between the f_i, that is $f \cdot r := \sum_{i=1}^k f_i r_i = 0$, there exists a lift $R = (R_1,\ldots,R_k) \in (\mathbb{C}\{x\}\widehat{\otimes}\mathcal{O}_{S,o})^k$, that is $R_i(x,0) = r_i(x)$, which is a relation between the F_i. This means that $F \cdot R := \sum_{i=1}^k F_i R_i = 0$.[3]

[2] Later on we will often write deformation instead of flat deformation.

[3] This is, in fact, equivalent to saying that $\mathbb{C}\{x\}\widehat{\otimes}\mathcal{O}_{S,o}/(F_1,\ldots,F_k)$ is a flat $\mathcal{O}_{S,o}$-module in the general set-up of flatness, see [Matsumura 1979], Theorem 49.

A priori the definition depends on the choice of the generators of (X_S, o) (and of (X, o)), but it is not so difficult to show that it in fact does not. See Exercise 10.2.29.

Proposition 10.2.6. *Let (X, o) be a germ of a complex space, defined by the ideal $I = (f_1, \dots, f_k)$ in $\mathbb{C}\{x\}$. Let $g_1, \dots, g_k$ be further elements in $\mathbb{C}\{x\} =: \mathcal{O}$. Consider $(\mathbb{T}, o)$, the double point. Then $f_1+\varepsilon g_1, \dots, f_k+\varepsilon g_k$ defines a (flat) deformation $(X_{\mathbb{T}}, o) \longrightarrow (\mathbb{T}, o)$ if and only if*

$$f_1 \mapsto g_1, \dots, f_k \mapsto g_k$$

gives a well-defined element of $\operatorname{Hom}_{\mathcal{O}}(I, \mathcal{O}_{X,o}) =: N_{X,o}$.[4] *The module $N_{X,o}$ is called the* normal module *of (X, o).*

Proof. Suppose first that $f + \varepsilon g := (f_1 + \varepsilon g_1, \dots, f_k + \varepsilon g_k)$ defines a (flat) deformation $(X_{\mathbb{T}}, o) \longrightarrow (\mathbb{T}, o)$. We have to show that the map $f_i \longrightarrow g_i$ induces a well defined map in $\operatorname{Hom}_{\mathcal{O}}(I, \mathcal{O}_{X,o})$. Thus an element $h = \sum r_i f_i$ is sent to $\sum r_i g_i$. In order to show that this is well-defined, we have to show that if $\sum r_i f_i = 0$, it follows that $\sum r_i g_i \in I$. But $\sum r_i f_i = 0$ means that $r = (r_1, \dots, r_k)$ is a relation, which by assumption can be lifted to a relation $r + \varepsilon s$ between $f + \varepsilon g$, hence

$$0 = (f + \varepsilon g) \cdot (r + \varepsilon s) = f \cdot r + \varepsilon(f \cdot s + r \cdot g) = \varepsilon(f \cdot s + r \cdot g)$$

because $\varepsilon^2 = 0$. Hence $r \cdot g \in (f_1, \dots, f_k)$ as was to be shown. The converse is proved by reversing the arguments. □

Lemma 10.2.7. *Let (X, o) be a germ of a complex space. Then (flat) deformations $(X_{\mathbb{T}}, o) \longrightarrow (\mathbb{T}, o)$ are in one-one correspondence with the elements of the normal module $N_{X,o}$.*

Proof. We apply the previous proposition.

Step 1. Suppose that $g_1, \dots, g_k$ and $g'_1, \dots, g'_k$ give the same element in $N_{X,o}$. We have to show that the ideals $(f_1 + \varepsilon g_1, \dots, f_k + \varepsilon g_k)$ and $(f_1 + \varepsilon g'_1, \dots, f_k + \varepsilon g'_k)$ are equal. For this it suffices, by symmetry, to show one inclusion. Now, by assumption, $g_i - g'_i \in I$, that is, there exist $a_{ij} \in \mathbb{C}\{x\}$ such that $g_i - g'_i = \sum_j a_{ij} f_j$. It then follows that, as $\varepsilon^2 = 0$

$$f_i + \varepsilon g_i = f_i + \varepsilon g'_i + \varepsilon \sum_j a_{ij} f_j = (f_i + \varepsilon g'_i) + \varepsilon \sum_j a_{ij}(f_j + \varepsilon g'_j),$$

This gives the inclusion

$$(f_1 + \varepsilon g_1, \dots, f_k + \varepsilon g_k) \subset (f_1 + \varepsilon g'_1, \dots, f_k + \varepsilon g'_k).$$

Step 2. The fact that equality of the ideals $(f_1 + \varepsilon g_1, \dots, f_k + \varepsilon g_k)$ and $(f_1 + \varepsilon g'_1, \dots, f_k + \varepsilon g'_k)$ implies that $g_1, \dots, g_k$ and $g'_1, \dots, g'_k$ give the same element in $N_{X,o} = \operatorname{Hom}_{\mathcal{O}}(I, \mathcal{O}_X)$ is proved similarly. □

Example 10.2.8. We consider the germ of the complex space (X, o) defined by the 2–minors of the matrix[5]

[4] Note that $N_{X,o}$ is not an invariant of the isomorphism class of (X, o), as it might depend on the embedding.

[5] Readers knowing some algebraic geometry will recognize that this singularity is the cone over the rational normal curve of degree four.

$$\begin{pmatrix} x_0 & x_1 & x_2 & x_3 \\ x_1 & x_2 & x_3 & x_4 \end{pmatrix}.$$

We will denote by f_{ij} the minor obtained by taking the i–th and j–th column. So for example $f_{13} = x_0x_3 - x_1x_2$. It has been shown in Exercise 7.3.9 that the ideal I defined by the minors is a prime ideal. Thus (X, o) is irreducible. We determine the normal module $N_{X,o}$.

Step 1. We have to determine the relations between the f_{ij}. We get eight relations in the following way: take three columns, double one of the rows, and compute the determinant. As two rows are the same, the determinant is zero, so that we indeed get a relation. We get relations r_{ijk} if we double the first row, and s_{ijk} if we double the second row. As an example, we compute r_{123}. So we take the first three columns, and put the first row on top. We get the following matrix

$$\begin{pmatrix} x_0 & x_1 & x_2 \\ x_0 & x_1 & x_2 \\ x_1 & x_2 & x_3 \end{pmatrix},$$

and hence the relation $x_0f_{23} - x_1f_{13} + x_2f_{12} = 0$. We claim that all relations are generated by the relations constructed in this way. We will only indicate how to prove this, leaving the reader to fill in the details. First of all, we have that $\mathbb{C}\{x_0,x_4\} \subset \mathcal{O}_{X,o}$ is a Noether normalization. Because $\mathcal{O}_{X,o}$ is a free $\mathbb{C}\{x_0,x_4\}$–module, see Exercise 7.3.9, it follows that x_0,x_4 is a regular sequence in $\mathcal{O}_{X,o}$. Thus all relations between the elements

$$\begin{aligned} -x_1^2 &= f_{12}(0,x_1,x_2,x_3,0), \\ -x_1x_2 &= f_{13}(0,x_1,x_2,x_3,0), \\ -x_1x_3 &= f_{14}(0,x_1,x_2,x_3,0), \\ -x_2^2 &= f_{23}(0,x_1,x_2,x_3,0) \\ -x_2x_3 &= f_{24}(0,x_1,x_2,x_3,0), \\ -x_3^2 &= f_{34}(0,x_1,x_2,x_3,0) \end{aligned}$$

can be lifted to relations between the f_{ij}. The relations between these monomials are easily determined, and there are indeed eight of them.

Step 2. We can easily generate 8 elements $n_1, \ldots, n_8$ of the normal module $N_{X,o}$. These we get by putting an extra ε at one of the eight spots of the matrix. The fact that the relations can be lifted, follows because we can lift the relations r_{ijk} and s_{ijk} by again taking three columns and double a row! The $n_1, \ldots, n_4$ correspond to the four entries in the top row, and the $n_5, \ldots, n_8$ correspond to the four entries in the bottom row. To give an example we determine n_1. We get the following matrix

$$\begin{pmatrix} x_0+\varepsilon & x_1 & x_2 & x_3 \\ x_1 & x_2 & x_3 & x_4 \end{pmatrix}.$$

We take the first minor, and get $x_0x_2 + \varepsilon x_2 - x_1^2$. Thus $f_{12} \to x_2$ etc. It turns out that the normal module is generated by *nine* elements, namely by $n_1, \ldots, n_8$ and one further element n_9. We put these in the following table. The rows give the elements of $N_{X,o}$ by giving the value on f_{ij} in the columns.

	f_{12}	f_{23}	f_{34}	f_{13}	f_{24}	f_{14}
n_1	x_2	0	0	x_3	0	x_4
n_2	$-x_1$	x_3	0	0	x_4	0
n_3	0	$-x_2$	x_4	$-x_1$	0	0
n_4	0	0	$-x_3$	0	$-x_2$	$-x_1$
n_5	$-x_1$	0	0	$-x_2$	0	$-x_3$
n_6	x_0	$-x_2$	0	0	$-x_3$	0
n_7	0	x_1	$-x_3$	x_0	0	0
n_8	0	0	x_2	0	x_1	x_0
n_9	x_0	$-x_2$	x_4	0	0	x_2

The fact that n_9 is also an element of the normal module is a direct check. We want to show that $n_1, \ldots, n_9$ generate the normal module. To show this, we first note that an element in $N_{X,o}$ is determined by the values on f_{12},f_{23} and f_{34}, that is, by the first three columns. To see this, look for example at the relation r_{123}. It says that $x_0 f_{23} - x_1 f_{13} + x_2 f_{12} = 0$. So if $\varphi \in N_{X,o}$, it follows that

$$x_1\varphi(f_{13}) = x_0\varphi(f_{23}) + x_2\varphi(f_{12}),$$

so we can compute $x_1\varphi(f_{13})$. As x_1 is a nonzerodivisor of $\mathcal{O}_{X,o}$ we can compute $\varphi(f_{13})$. Similarly one sees that one can compute $\varphi(f_{24})$ and $\varphi(f_{14})$.

By doing column operations, we see that we have normal module elements φ of the following type:

(1) $\varphi(f_{12}) = x_i$, $\varphi(f_{23}) = 0$, $\varphi(f_{34}) = 0$, for $i = 0,1,2$;

(2) $\varphi(f_{12}) = 0$, $\varphi(f_{23}) = x_i$, $\varphi(f_{34}) = 0$, for $i = 1,2,3$;

(3) $\varphi(f_{12}) = 0$, $\varphi(f_{23}) = 0$, $\varphi(f_{34}) = x_i$, for $i = 2,3,4$.

Take an element φ of $N_{X,o}$. We claim that $\varphi(f_{23}) \in (x_1,x_2,x_3)$. So calculate modulo (x_1,x_2,x_3). The ideal I modulo (x_1,x_2,x_3) is generated by one element, namely x_0x_4. We look at

$$0 = \varphi(r_{123}) = x_0\varphi(f_{23}) - x_1\varphi(f_{13}) + x_2\varphi(f_{12}).$$

Modulo (x_1,x_2,x_3) we get $x_0\varphi(f_{23}) = 0$. Hence a representative (also called $\varphi(f_{23})$) of $\varphi(f_{23})$ in $\mathbb{C}\{x\}$ modulo (x_1,x_2,x_3) is divisible by x_4. Similarly, using the relation s_{234}, we get that $\varphi(f_{23})$ is divisible by x_0. Thus $\varphi(f_{23})$ is divisible by x_0x_4, which is modulo I equal to x_1x_3.

It is somewhat more difficult, but similar, to show that $\varphi(f_{12}) = 0 \mod (x_0,x_1,x_2)$, and $\varphi(f_{34}) = 0 \mod (x_2,x_3,x_4)$. Altogether this shows that $N_{X,o}$ has the nine generators written down above.

Definition 10.2.9. Let two deformations $(X_S, o) \longrightarrow (S, o)$ and $(X'_S, o) \longrightarrow (S, o)$ of (X, o) over (S, o) be given. A *morphism of* $(X_S, o) \longrightarrow (S, o)$ to $(X'_S, o) \longrightarrow (S, o)$ is given by a map $\Phi : (X_S, o) \longrightarrow (X'_S, o)$ such that:

(1) Φ restricts to the identity map on (X, o).

(2) Φ preserves fibers, that is, the diagram

$$\begin{array}{ccc}(X_S,o) & \xrightarrow{\Phi} & (X'_S,o)\\ \downarrow & & \downarrow\\ (S,o) & \xrightarrow{\text{Id}} & (S,o)\end{array}$$

is commutative.

Theorem 10.2.10. *Let $\Phi : (X_S, o) \longrightarrow (X'_S, o)$ be a morphism of deformations of (X, o) over (S, o). Then Φ is an isomorphism.*

Proof. This is a consequence of the Inverse Function Theorem. A morphism of deformations of (X, o) induces the identity on (X, o), and preserve fibers. Therefore, Φ^* is induced by a morphism which, by abuse of notation, we also call Φ^*

$$\begin{array}{rcl}\mathbb{C}\{x_1,\dots,x_n\}\hat{\otimes}\mathcal{O}_{S,o} & \longrightarrow & \mathbb{C}\{x_1,\dots,x_n\}\hat{\otimes}\mathcal{O}_{S,o},\\ x_i & \mapsto & x_i+h_i, \quad i=1,\dots n,\\ s\in\mathcal{O}_{S,o} & \mapsto & s.\end{array}$$

where $h_i \in \mathfrak{m}_{S,o} \cdot \mathbb{C}\{x\}\hat{\otimes}\mathcal{O}_{S,o}$, $\mathfrak{m}_{S,o}$ the maximal ideal of $\mathcal{O}_{S,o}$. To prove that Φ^* is an isomorphism, we may suppose that $\mathcal{O}_{S,o} = \mathbb{C}\{s\}$. Therefore, the determinant of the Jacobian matrix of Φ^* is equal to one. Hence the inverse of Φ^*, and hence of Φ exists, by the Inverse Function Theorem. □

Remark 10.2.11. Let $(X, o) \subset (\mathbb{C}^n, o)$ be a singularity, given by $f = (f_1, \dots, f_k)$. Consider deformations $(X_S, o) \longrightarrow (S, o)$ and $(X'_S, o) \longrightarrow (S, o)$ over (S, o). Suppose (X_S, o) is defined by $F_S = (F_1, \dots, F_k)$, and (X'_S, o) is defined by $F'_S = (F'_1, \dots, F'_k)$, both lifting $(f_1, \dots, f_k)$. To give an isomorphism $\Phi : (X_S, o) \longrightarrow (X'_S, o)$ means that one has to find:

- Elements $\Phi_1, \dots, \Phi_n \in \mathbb{C}\{x_1, \dots, x_n\}\hat{\otimes}\mathcal{O}_{S,o}$ with $\Phi_i(x, 0) = x_i$.
- A $k \times k$–matrix Λ with entries in $\mathbb{C}\{x_1, \dots, x_n\}\hat{\otimes}\mathcal{O}_{S,o}$ such that $\Lambda_{|s=0} = \text{Id}$.

satisfying

$$F'_S = F_S(\Phi_1, \dots, \Phi_n) \circ \Lambda \in (\mathbb{C}\{x\}\hat{\otimes}\mathcal{O}_{S,o})^k.$$

(Here $\circ$ denotes matrix multiplication.) The elements Φ_i give the relative isomorphism, inducing the identity on (X, o), whereas the matrix Λ changes generators for the ideal of (X_S, o).

Now we want to classify deformations of (X, o) over the double point $(\mathbb{T}, o)$. Suppose that the ideal of (X, o) is equal to $I \subset \mathcal{O}_n$. Consider Θ_n, the free $\mathcal{O}_n$–module of all derivations of $\mathcal{O}_n$. Elements θ of Θ_n look like $\theta = h_1 \frac{\partial}{\partial x_1} + \dots + h_n \frac{\partial}{\partial x_n}$ for $h_i \in \mathcal{O}_n$. Note that we have a map

$$\begin{aligned}\alpha : \Theta_n &\longrightarrow N_{X,o} = \text{Hom}_{\mathcal{O}_n}(I, \mathcal{O}_{X,o})\\ \theta &\mapsto (f \mapsto \theta(f)).\end{aligned}$$

So we can define:

Definition 10.2.12. $T^1_{X,o} := \text{Coker}(\alpha)$.[6]

[6] It is, in fact true that $T^1_{X,o}$ is an invariant of the isomorphism class of (X, o), see Exercise 10.2.26.

Theorem 10.2.13. *Let $(X,o) \subset (\mathbb{C}^n,o)$ be a germ of a complex space. Isomorphism classes of deformations of (X,o) over $(\mathbb{T},o)$ are in one-one correspondence with elements of $T^1_{X,o}$.*

Proof. Suppose (X,o) is defined by $f_1,\dots,f_k$. Consider two deformations $(X_{\mathbb{T}},o) \longrightarrow (\mathbb{T},o)$ and $(X'_{\mathbb{T}},o) \longrightarrow (\mathbb{T},o)$ given by $f_1+\varepsilon g_1,\dots,f_k+\varepsilon g_k$ and $f_1+\varepsilon g'_1,\dots,f_k+\varepsilon g'_k$ respectively. A morphism of deformations between $(X_{\mathbb{T}},o)$ and $(X'_{\mathbb{T}},o)$ is induced by a map $\mathbb{C}\{x\}\otimes\mathbb{C}[\varepsilon]/(\varepsilon^2) \longrightarrow \mathbb{C}\{x\}\otimes\mathbb{C}[\varepsilon]/(\varepsilon^2)$ with $(x_1,\dots,x_n,\varepsilon) \mapsto (x_1+\varepsilon h_1,\dots,x_n+\varepsilon h_n,\varepsilon)$ for some $h_1,\dots h_n \in \mathbb{C}\{x\}$. Then $(X'_{\mathbb{T}},o)$ is given by the zero set of

$$f_i(x_1+\varepsilon h_1,\dots,x_n+\varepsilon h_n)+\varepsilon g_i = f_i + \varepsilon\left(\sum_{j=1}^{n} h_j\frac{\partial f_i}{\partial x_j} + g_i\right) \quad \text{for } i=1,\dots k.$$

This holds because $\varepsilon^2=0$. Hence we see that the map $I \longrightarrow \mathcal{O}_{X,o}$ given by $f_i \mapsto g_i - g'_i$ is in the image of Θ_n. Similarly, one shows that elements of Θ_n induce isomorphisms. This is what we needed to show. □

Examples 10.2.14.
(1) Let (X,o) be a germ of an isolated hypersurface, defined by $f\in\mathbb{C}\{x\}$. Then $T^1_{X,o} = \mathbb{C}\{x\}/(f,J(f))$. This follows from Example 10.1.9.

(2) We take the singularity of Example 10.2.8. Recall that the normal module elements are determined by their values on f_{12}, f_{23} and f_{34}. A direct calculation shows that $\frac{\partial}{\partial x_0} = n_1$, $\frac{\partial}{\partial x_1} = n_2+n_5$, $\frac{\partial}{\partial x_2} = n_3+n_6$, $\frac{\partial}{\partial x_3} = n_7+n_4$, $\frac{\partial}{\partial x_4} = n_8$. Thus $T^1_{X,o}$ is generated by four element n_2, n_3, n_4, and n_9. We claim that $(x_0,x_1,x_2,x_3,x_4)T^1_{X,o} = 0$. Thus we have to do twenty checks!! The reader hopefully forgives us for just doing one of those. We will show that the class of x_0n_4 is the zero element in $T^1_{X,o}$.We look at the values on f_{12}, f_{13} and f_{34}. They are 0, 0 and $-x_0x_3$ which modulo I is equal to 0, 0 and $-x_2x_1$. Thus it is equal to $-x_1n_8 = -x_1\frac{\partial}{\partial x_4}$, hence the trivial element in $T^1_{X,o}$. Altogether we see that $\dim_{\mathbb{C}}(T^1_{X,o}) = 4$, with a basis given by the classes of the normal module elements n_2,n_3,n_4 and n_9.

Theorem 10.2.15. *Let (X,o) be a germ of a complex space with an isolated singularity. Then $T^1_{X,o}$ is a finite-dimensional $\mathbb{C}$-vector space.*

Proof. Let $X\subset U$ be a small representative of (X,o) such that 0 is the only singular point of X. Consider the sheaf $\mathcal{T}^1_X := \mathcal{H}om_{\mathcal{O}_U}(\mathcal{I},\mathcal{O}_X)/\Theta_U$. Here Θ_U is the sheaf of derivations on U, $\mathcal{I}$ is the sheaf of ideals defining X. The map $\Theta_U \longrightarrow \mathcal{H}om_{\mathcal{O}_U}(\mathcal{I},\mathcal{O}_X)$ is defined as in the definition of $T^1_{X,o}$. By the basic coherence theorems $\mathcal{T}^1_X$ is a coherent sheaf on X, whose stalks in $p\in X$ is equal to $T^1_{X,p}$. As X has only one singularity, these are zero for $p\neq 0$, by Exercise 10.2.27. Hence $\mathcal{T}^1_X$ is a coherent sheaf concentrated in the point 0, and therefore the set of global sections $\Gamma(\mathcal{T}^1_X,U) = T^1_{X,o}$ is a finite-dimensional $\mathbb{C}$-vector space by Theorem 6.2.12. □

The converse of this theorem does not hold, see Exercise 10.2.30. We now come to the definition of induced family and of versal deformations.

Definition 10.2.16. Let (X,o) be a germ of a complex space.

(1) Suppose $(X_S, o) \longrightarrow (S, o)$ is a deformation of (X, o), defined by $F_1, \ldots, F_k \in \mathbb{C}\{x\}\hat{\otimes}\mathcal{O}_{S,o}$. Let $p : (T, o) \longrightarrow (S, o)$ be an analytic map. We define the *pull-back*, or the *induced deformation*

$$(X \times_S T, o) \longrightarrow (T, o)$$

as follows. The map p induces a map $p^* : \mathcal{O}_{S,o} \longrightarrow \mathcal{O}_{T,o}$, which in turns induces a map $p^* : \mathbb{C}\{x\}\hat{\otimes}\mathcal{O}_{S,o} \longrightarrow \mathbb{C}\{x\}\hat{\otimes}\mathcal{O}_{T,o}$. Then (X_T, o) is given by $p^*(F_1), \ldots, p^*(F_k)$. It is proved in Exercise 10.2.32 that the map $(X \times_S T, o) \longrightarrow (T, o)$ is indeed flat.

(2) A deformation $(X_S, o) \longrightarrow (S, o)$ is called *versal* if for every deformation $(X_T, o) \longrightarrow (T, o)$ there exists a map $p : (T, o) \longrightarrow (S, o)$ such that the deformations $(X_T, o) \longrightarrow (T, o)$ and $(X \times_S T, o) \longrightarrow (T, o)$ are isomorphic. The versal deformation is called *semi-universal* if moreover the map on tangent spaces is uniquely determined.

Definition 10.2.17. Let $(X_S, o) \longrightarrow (S, o)$ be a deformation of (X, o). Suppose (X, o) is defined by $f \in \mathbb{C}\{x\}^k$, and that the deformation is given by $F \in \left(\mathbb{C}\{x\}\hat{\otimes}\mathcal{O}_{S,o}\right)^k$. Suppose $\mathfrak{m}_{S,o}$ has a minimal system of generators $s_1, \ldots, s_p$. Write $F = f + \sum s_i g_i$. We denote by $[g_i]$ the class of $g_i \mod \mathfrak{m}_{S,o}$ in $T^1_{X,o}$. Then the map

$$\begin{array}{rcl} T_{S,o} = \left(\mathfrak{m}_{S,o}/\mathfrak{m}^2_{S,o}\right)^* & \longrightarrow & T^1_{X,o} \\ \sum\limits_i a_i s_i^* & \mapsto & \sum\limits_i a_i [g_i] \end{array}$$

is called the *Kodaira-Spencer map*. It is easily shown that the definition of the Kodaira-Spencer map is independent of the choices made.

The main goal in this chapter is to prove that for (X, o) with an isolated singularity, a semi-universal deformation exists. In general, the proof of this theorem needs many additional preparations, but we proved it already in the first section for hypersurface singularities. In the remainder of this chapter, we will prove a special case of this theorem, which includes the case of hypersurface singularities and complete intersection singularities (see Exercise 10.2.33). Moreover we treat the case of Cohen-Macaulay singularities of codimension two by applying the Hilbert-Burch Theorem 6.5.26.

Theorem 10.2.18. *Let (X, o) be a germ of a complex space, defined by $f \in \mathbb{C}\{x\}^k$. Suppose $\dim_{\mathbb{C}}(T^1_{X,o}) < \infty$. Let $(S, o) = (\mathbb{C}^\tau, o)$. Suppose that*

(1) there exists a (flat) deformation $(X_S, o) \longrightarrow (S, o)$ defined by[7]

$$F_S(x,s) := f + s_1 g_1 + \ldots + s_\tau g_\tau \in \left(\mathbb{C}\{x\} \otimes \mathbb{C}\{s_1, \ldots, s_\tau\}\right)^k,$$

(2) the Kodaira-Spencer map of $(X_S, o) \longrightarrow (S, o)$ is surjective.

Then $(X_S, o) \longrightarrow (S, o)$ is a versal deformation of (X, o). If moreover the Kodaira-Spencer map is an isomorphism, then $(X_S, o) \longrightarrow (S, o)$ is a semi-universal deformation.

For the proof of this theorem, we need the following preparations, which we will also need in the next section.

[7] Note that not necessarily $g_i \in \mathbb{C}\{x\}$.

Definition 10.2.19. For a local Artinian $\mathbb{C}$-algebra A we denote by $(\mathbb{A}, o)$ the germ of the complex space consisting of one point o, but with local ring A.

Definition 10.2.20. Let A and B be local $\mathbb{C}$-algebras, $\mathfrak{m}$ the maximal ideal of B. Suppose $\varphi : B \longrightarrow A$ is a surjective morphism of $\mathbb{C}$-algebras. Then

(1) φ is called a *small surjection* if $\mathfrak{m} \cdot \operatorname{Ker}(\varphi) = 0$,

(2) φ is called a *simple surjection* if $\operatorname{Ker}(\varphi)$ is a one-dimensional $\mathbb{C}$-vector space.

Example 10.2.21. The surjection $\mathbb{C}[\varepsilon]/(\varepsilon^3) \longrightarrow \mathbb{C}[\varepsilon]/(\varepsilon^2)$ is a simple surjection. The surjection $\mathbb{C}[\varepsilon,\eta]/(\varepsilon,\eta)^2 \longrightarrow \mathbb{C}[\varepsilon,\eta]/(\varepsilon,\eta)$ is small, but not simple. The surjection φ : $\mathbb{C}[\varepsilon,\eta]/(\varepsilon^2,\eta^2) \longrightarrow \mathbb{C}[\varepsilon,\eta]/(\varepsilon,\eta)$ is not small, as $\varepsilon\eta$ is not zero in $\mathbb{C}[\varepsilon,\eta]/(\varepsilon^2,\eta^2)$.

Lemma 10.2.22. *Let (X, o) be a germ of a complex space, with $f \in \mathbb{C}\{x\}^k$ generators of the ideal of (X, o). Let $\varphi : B \longrightarrow A$ be a small surjection of local Artinian $\mathbb{C}$-algebras. Consider a deformation $(X_{\mathbb{B}}, o) \longrightarrow (\mathbb{B}, o)$, defined by $F_B \in \left(\mathbb{C}\{x\}\widehat{\otimes}B\right)^k$. Let $(X_{\mathbb{A}}, o) \longrightarrow (\mathbb{A}, o)$ be the restriction of $(X_{\mathbb{B}}, o) \longrightarrow (\mathbb{B}, o)$ to $(\mathbb{A}, o)$, defined by the image F_A of F_B in $\left(\mathbb{C}\{x\}\widehat{\otimes}A\right)^k$. Let r_A be a relation between the F_A, that is, $F_A \cdot r_A = 0$. Then there exists a lift r_B of r_A with $F_B \cdot r_B = 0$.*

Proof. The proof is by induction on the vector space dimension $\dim_{\mathbb{C}}(A)$. The case $\dim_{\mathbb{C}}(A) = 1$ is easy. In this case $A = \mathbb{C}$, and the deformation $(X_{\mathbb{A}}, o)$ is equal to (X, o). By assumption on flatness, every relation $r = r_A$ can be lifted to a relation r_B between the F_B.

Now assume $\dim_{\mathbb{C}}(A) > 1$. We can always find $0 \neq \alpha \in A$ with $\mathfrak{m}_A \cdot \alpha = 0$. Consider $A' := A/(\alpha)$. By induction we can find a lift r'_B of r'_A with $r'_B \cdot F_B = 0$. Moreover, take any lift $\widetilde{r}_B$ of r_A. As both $\varphi(\widetilde{r}_B)$ and $\varphi(r'_B)$ project to r'_A, it follows that $\varphi(\widetilde{r}_B) - \varphi(r'_B) = h \cdot \alpha$ for some $h \in \mathbb{C}\{x\}^k$. Since both $\varphi(\widetilde{r}_B)$ and $\varphi(r'_B)$ are relations between the F_A we get $h \cdot F_A \cdot \alpha = 0$. Moreover $F_A \cdot \alpha = f \cdot \alpha$, because $\mathfrak{m}_A \cdot \alpha = 0$. Thus h is a relation between the f, and can by flatness be lifted to a relation h_B between the F_B. Now define $r_B := r'_B + \alpha' h_B$ for any α' with $\varphi(\alpha') = \alpha$. Then r_B is a relation, and lifts r_A. This proves the lemma. □

Lemma 10.2.23. *Consider:*

(1) A singularity (X, o) defined by $f \in \mathbb{C}\{x\}^k$.

(2) $\varphi : B \longrightarrow A$ a small *surjection of local Artinian $\mathbb{C}$-algebras. Let $\alpha_1, \dots ,\alpha_p$ be a $\mathbb{C}$-basis of* $\operatorname{Ker}(\varphi)$,

(3) A deformation $(X_{\mathbb{A}}, o) \longrightarrow (\mathbb{A}, o)$, defined by F_A and two lifts $(X_{\mathbb{B}}, o) \longrightarrow (\mathbb{B}, o)$, and $(X'_{\mathbb{B}}, o) \longrightarrow (\mathbb{B}, o)$ of $(X_{\mathbb{A}}, o) \longrightarrow (\mathbb{A}, o)$, defined by F_B and F'_B respectively, both being lifts of F_A.

Then there exist elements $n_1, \dots ,n_p \in N_{X,o}$ with $F_B - F'_B = \alpha_1 n_1 + \dots + \alpha_p n_p$.

Conversely, let F_B be given. Then $F_B + \alpha_1 n_1 + \dots + \alpha_p n_p$ also defines a deformation of (X, o) over $(\mathbb{B}, o)$ for all $n_1, \dots ,n_p \in N_{X,o}$.

Proof. We can write:

$$F_B = F_A + \sum \alpha_i g_i; \quad F'_B = F_A + \sum \alpha_i g'_i.$$

Let r be a relation, that is, $r \cdot f = 0$. We will see that $r \cdot (g_i - g'_i) \in (f_1, \ldots, f_k)$, that is, $g_i - g'_i$ defines an element of $N_{X,0}$. Consider lifts R_B and R'_B of r, which are relations between the F_B and the F'_B respectively. By Lemma 10.2.22, we may assume that the restrictions of R_B and R'_B to $\mathbb{C}\{x\}^k \otimes A$ are equal. Therefore we can write:

$$R_B = R_A + \sum c_i \cdot \alpha_i; \quad R'_B = R_A + \sum c'_i \cdot \alpha_i.$$

for complex numbers c_i and c'_i. As $B \longrightarrow A$ is a small extension, it follows that $\mathfrak{m}\alpha_i = 0$ for all i. Hence

$$0 = R_B \cdot F_B = R_A \cdot F_A + \sum (r \cdot g_i)\alpha_i + \sum (c_i \cdot f)\alpha_i.$$

Similarly

$$0 = R'_B \cdot F'_B = R_A \cdot F_A + \sum (r \cdot g'_i)\alpha_i + \sum (c'_i \cdot f)\alpha_i.$$

By subtracting and looking at the coefficients of α_i we see that for all $i = 1, \ldots, p$:

$$r \cdot (g_i - g'_i) \in (f_1, \ldots, f_k).$$

Therefore, $n_i := g_i - g'_i$ is an element of $N_{X,0}$. The converse statement is much simpler, and left as Exercise 10.2.34. □

Proof of Theorem 10.2.18. Let $\mathcal{O}_{T,0} = \mathbb{C}\{t\}/J$, and $(X_T, 0) \longrightarrow (T, 0)$ be a deformation of $(X, 0)$, given by equations $F_T(x,t) = 0$. Versality means that we have to find a map $(T, 0) \longrightarrow (S, 0)$ such that the deformation $(X \times_S T, 0) \longrightarrow (T, 0)$ is isomorphic to $(X_T, 0) \longrightarrow (T, 0)$. Dually the map $(T, 0) \longrightarrow (S, 0)$ is given by an analytic map $\mathcal{O}_{S,0} \longrightarrow \mathcal{O}_{T,0}$. Thus we have to find τ power series $s_1(t), \ldots, s_\tau(t) \in \mathbb{C}\{t\}$ to define this map. The space $(X \times_S T, 0)$ is given by the equations $F_S(x,s(t)) = 0$. To say that this deformation is isomorphic to $F_T(x,t) = 0$ means that we can find an automorphism Φ of $\mathbb{C}\{x,t\}$ with $\Phi(t) = t$, $\Phi(x)_{|t=0} = x$, and a matrix $\Lambda \in \mathbb{C}\{x,t\}$ with $\Lambda_{|t=0}$ the identity matrix such that

$$F_S(x,s(t)) \equiv (F_T \circ \Phi)(x,t) \circ \Lambda \mod J \cdot \mathbb{C}\{x,t\}^k, \tag{10.5}$$

with $\mathcal{O}_{T,0} = \mathbb{C}\{t\}/J$. Let $\Phi(x_i) = \phi_i$ and $\phi = (\phi_1, \ldots, \phi_n)$.

We want to apply Grauert's Approximation Theorem 8.2.2 in order to find a solution to the equation (10.5). We have to solve for ϕ, Λ and s, but the s is only allowed to depend on t, and not on x. The equation (10.5) has a solution of order zero, as it then reads $f = f$. Thus we have to show that every solution of (10.5) of order e with respect to t extends to a solution of (10.5) of order $e + 1$. Therefore, suppose (ϕ,Λ,s) is a solution of (10.5) of order e. Thus

$$F_S(x,s(t)) \equiv F_T(\phi,t) \circ \Lambda \mod (J + (t)^{e+1}) \cdot \mathbb{C}\{x,t\}^k. \tag{10.6}$$

Here (t) is the maximal ideal of $\mathbb{C}\{t\}$. We will change the order $e + 1$ terms in t of s,ϕ and Λ in such a way that (10.6) holds modulo $(J + (t)^{e+2})$.

Consider the small surjection

$$\sigma : B := \mathbb{C}\{t\}/(J + (t)^{e+2}) \longrightarrow \mathbb{C}\{t\}/(J + (t)^{e+1}) =: A.$$

Take a basis $\alpha_1, \dots, \alpha_p$ of $\operatorname{Ker}(\sigma)$. We can factorize σ into p simple surjections σ_i, for $i = 1, \dots, p$. Thus, using induction, we may suppose that $\sigma : B \longrightarrow A$ is a small surjection, the kernel of σ is of dimension one, and $\operatorname{Ker}(\sigma)$ is generated by a monomial α of degree $e + 1$.

Step 1. Change of $s(t)$: We apply 10.2.23 to $\sigma : B \longrightarrow A$, $F_B = F_S(x,s(t))$ and $F'_B = F_T(\phi,t) \circ \Lambda$. The $s(t)$, $\phi(x,t)$ and $\Lambda(x,t)$ are given by induction. It follows that

$$(10.7) \qquad F_S(x,s(t)) - F_T(\phi,t) \circ \Lambda = n \cdot \alpha$$

for an element $n \in \mathbb{C}\{x\}^k$ such that $f \mapsto n$ is an element of the normal module $N_{X,o}$. By assumption the Kodaira-Spencer map is surjective, that is, the elements $a_i : f \longrightarrow g_i$ mod (s) for $i = 1, \dots, \tau$ project to generators of $T^1_{X,o}$. Thus as elements of $T^1_{X,o}$, we can write

$$n = b_1 a_1 + \ldots + b_\tau a_\tau$$

for suitable $b_i \in \mathbb{C}$. Now we can redefine the $s(t)$:

$$s_1(t) := s_1(t) - b_1\alpha, \ \dots, \ s_\tau(t) := s_\tau(t) - b_\tau\alpha.$$

Step 2. Change of $\phi(x,t)$. After the change of $s(t)$ we have just done, we have reached the stage that the $n = (n_1, \dots, n_k) \in N_{X,o}$ in (10.7) defines a trivial element in $T^1_{X,o}$. Then, obviously, $n\Lambda^{-1}_{|t=0}$ is a trivial element too. Therefore, there exist $c_1, \dots, c_n \in \mathbb{C}\{x\}$ such that $\theta(f) = n\Lambda^{-1}_{|t=0}$, with $\theta := c_1\frac{\partial}{\partial x_1} + \ldots + c_n\frac{\partial}{\partial x_n}$. We redefine the automorphism $(\phi_1, \dots, \phi_n, t)$ in the following way.

$$\phi_1 := \phi_1 - c_1\alpha, \ \dots, \ \phi_n := \phi_n - c_n\alpha.$$

It follows from Taylor's expansion that after this redefinition, the n defines the trivial element of $N_{X,o}$, that is, n_i is the zero element in $\mathcal{O}_{X,o}$ for $i = 1, \dots, k$.

Step 3. Change of Λ. After these redefinitions the elements n in (10.7) have entries n_i which all give the zero element in $\mathcal{O}_{X,o} = \mathbb{C}\{x\}/(f_1, \dots, f_k)$. Thus we can find elements $c_{ji} \in \mathbb{C}\{x\}$ such that $n_i = \sum_j c_{ji} f_j$. Redefine the matrix Λ by putting $\Lambda := \Lambda - (c_{ji})\alpha$. After this redefinition, we reached the stage that $n = 0 \in \mathbb{C}\{x\}^k$. This is what we had to show. □

To finish this section, we will prove the existence of a versal deformation for $(X, o) \subset (\mathbb{C}^n, o)$ that have an isolated singularity, are Cohen-Macaulay, and have dimension $n - 2$. The Auslander-Buchsbaum formula 6.5.20 implies that the projective dimension of $\mathcal{O}_{X,o}$ is equal to two. The main ingredient is now the Hilbert-Burch Theorem 6.5.26 and Theorem 10.2.18. Indeed, by the Hilbert-Burch Theorem, one has the following free $R := \mathcal{O}_n$- resolution of $\mathcal{O}_{X,o} = R/I$, where $I = \mathcal{I}(X, o)$:

$$0 \longrightarrow F_2 \xrightarrow{\alpha_2} F_1 \xrightarrow{\alpha_1} R \longrightarrow R/I \longrightarrow 0.$$

Here α_2 is an $(t - 1) \times t$ matrix for some t. Generators $f_1, \dots, f_t$ for the ideal I are obtained by taking the $(t-1)$-minors of the matrix α_2. Equivalently, one has the following resolution of the ideal I:

$$0 \longrightarrow F_2 \xrightarrow{\alpha_2} F_1 \xrightarrow{\alpha_1} I \longrightarrow 0.$$

Proposition 10.2.24. *With the above notation, we get deformations of (X,o) over any germ of a complex space (S,o) in the following way. Let E be a matrix with entries in $\mathcal{O}_n\hat{\otimes}\mathfrak{m}_{S,o}$. Consider the matrix α_2+E, and consider $I_{t-1}(\alpha_2+E)$. This is an ideal in $\mathcal{O}_n\hat{\otimes}\mathcal{O}_{S,o}$, and defines a space (X_S,o) such that $(X_S,o)\longrightarrow(S,o)$ is a flat deformation of (X,o). Moreover, all deformations of (X,o) are obtained this way.*

Proof. Step 1. Consider α_1+E', the $1\times t$ matrix where the i–th entry equals $(-1)^i$ times the minor obtained from α_2+E by leaving out the i–th row. Then the sequence

$$(\mathcal{O}_n\hat{\otimes}\mathcal{O}_{S,o})^{t-1}\xrightarrow{\alpha_2+E}(\mathcal{O}_n\hat{\otimes}\mathcal{O}_{S,o})^{t}\xrightarrow{\alpha_1+E'}I_{t-1}(\alpha_2+E)\longrightarrow 0$$

is certainly a complex, that is, $(\alpha_1+E')\circ(\alpha_2+E)=0$. This implies that we can lift all the relations between the minors of α_2, which are generators of the ideal I. So we get indeed a flat deformation.

Step 2. Conversely, suppose we have a deformation (X_S,o) of (X,o) over (S,o). The flatness says that we can lift the relations. Hence we get a complex

$$(\mathcal{O}_n\hat{\otimes}\mathcal{O}_{S,o})^{t-1}\xrightarrow{\alpha_2'}(\mathcal{O}_n\hat{\otimes}\mathcal{O}_{S,o})^{t}\xrightarrow{\alpha_1'}\mathcal{O}_n\hat{\otimes}\mathcal{O}_{S,o}\longrightarrow\mathcal{O}_{X_S,o}\longrightarrow 0. \tag{10.8}$$

where $\alpha_1'=\alpha_1$ modulo $\mathfrak{m}_{S,o}$ and $\alpha_2'=\alpha_2$ modulo $\mathfrak{m}_{S,o}$. We leave it as Exercise 10.2.37 to prove that the sequence (10.8) is exact, that is, the kernel of α_1' is generated by the lifts of the set of $t-1$ generators of the kernel of α. We claim that the map α_2' is in fact injective. To show this suppose $\alpha_2'(a_1,\dots,a_{t-1})=0$. Calculating modulo $\mathfrak{m}_{S,o}$, we see that $(a_1,\dots a_{t-1})$ is zero modulo $\mathfrak{m}_{S,o}$. This is because α_2 is injective. Take a basis $b_1,\dots,b_p$ of $\mathfrak{m}_{S,o}/\mathfrak{m}_{S,o}^2$, and write $(a_1,\dots,a_{t-1})=\sum(a_{i1},\dots,a_{it-1})b_i$ modulo $\mathfrak{m}_{S,o}^2$ for $a_{ij}\in\mathcal{O}_n$. From $\alpha_2'(a_1,\dots,a_{t-1})=0$ it follows by calculating modulo $\mathfrak{m}_{S,o}^2$ that $a_{ij}=0$. Thus $(a_1,\dots a_{t-1})$ is zero modulo $\mathfrak{m}_{S,o}^2$. Going on like this, we see that $(a_1,\dots a_{t-1})$ is zero modulo $\mathfrak{m}_{S,o}^k$ for all k, hence 0 by Krull's Intersection Theorem. So we get an exact sequence

$$0\longrightarrow(\mathcal{O}_n\hat{\otimes}\mathcal{O}_{S,o})^{t-1}\xrightarrow{\alpha_2'}(\mathcal{O}_n\hat{\otimes}\mathcal{O}_{S,o})^{t}\xrightarrow{\alpha_1'}\mathcal{O}_n\hat{\otimes}\mathcal{O}_{S,o}\longrightarrow\mathcal{O}_{X_S,o}\longrightarrow 0.$$

Let $I=\mathscr{I}(X_S,o)\subset\mathcal{O}_n\hat{\otimes}\mathcal{O}_{S,o}$. It follows from the Hilbert-Burch Theorem that $I=aI_{t-1}(\alpha_2')$ for some nonzerodivisor a. Calculating modulo $\mathfrak{m}_{S,o}$ we see that $a\equiv 1 \mod \mathfrak{m}_{S,o}$. Thus a is a unit in $\mathcal{O}_n\hat{\otimes}\mathcal{O}_{S,o}$. Then we change the minors of α_2' to a^{-1} times the minors, to get the desired generators of the ideal I. □

Theorem 10.2.25. *Let (X,o) be a Cohen-Macaulay codimension two isolated singularity, whose ideal is given by the $(t-1)$–minors of a matrix A with entries in $\mathcal{O}_n$. Let $\tau=\dim_{\mathbb{C}}(T^1_{X,o})$. Consider matrices α_i such that the deformations given by the $(t-1)$–minors of a matrix $A+\varepsilon\alpha_i$ form a basis of the vector space of isomorphism classes of deformations over $(\mathbb{T},o)$. Then the ideal given by the $(t-1)$–minors of the matrix $A+\sum_{i=1}^{\tau}s_i\alpha_i$, which has entries in $\mathcal{O}_n\hat{\otimes}\mathbb{C}\{s_1,\dots,s_\tau\}$ is the ideal of a space (X_S,o) which gives a semi-universal deformation of (X,o) over $(S,o)\cong(\mathbb{C}^\tau,o)$.*

Proof. The theorem follows from Proposition 10.2.24 and Theorem 10.2.18. □

Exercises

10.2.26.

(1) Let $\phi : \mathbb{C}\{x\} \longrightarrow \mathbb{C}\{x\} =: \mathcal{O}$ be an automorphism, and $I \subset \mathcal{O}$ be an ideal. Prove that $\lambda \mapsto \varphi\lambda\varphi^{-1}$ induces an isomorphism $\mathrm{Hom}_{\mathcal{O}}(I,\mathcal{O}/I) \longrightarrow \mathrm{Hom}_{\mathcal{O}}(\varphi(I),\mathcal{O}/\varphi(I))$.

(2) Let $(X,o) \cong (Y,o)$ be germs of complex spaces. Prove that $T^1_{X,o} \cong T^1_{Y,o}$.

10.2.27. Let (X,o) be a germ of a smooth space. Show that $T^1_{X,o} = 0$.

10.2.28. Prove Lemma 10.2.2.

10.2.29. Prove that the definition of flat deformation is independent of the choice of generators of (X_S,o).

10.2.30.

(1) Consider the germ of the complex space (X,o) defined by the ideal $I = (xz,xw,yz,yw) \subset \mathbb{C}\{x,y,z,w\}$. Thus $V(I)$ is the union of two planes in $\mathbb{C}^4$ intersecting in a point. Show that $T^1_{X,o} = 0$. Singularities with this property are called rigid.

(2) Give an example of a nonisolated singularity (X,o) with finite-dimensional $T^1_{X,o}$.

10.2.31. Consider the curve singularity $\big(V(xy,xz,yz),o\big)$. Compute a semi-universal deformation of this curve singularity.

10.2.32. Prove that for a (flat) deformation $(X_S,o) \longrightarrow (S,o)$, and an analytic map $p : (T,o) \longrightarrow (S,o)$ the map $(X \times_S T,o) \longrightarrow (T,o)$ is flat.

10.2.33. Let (X,o) be a complete intersection with isolated singularity. Describe a versal deformation of (X,o).

10.2.34. Finish the proof of Lemma 10.2.23.

10.2.35. Consider $t(t-1)$ variables x_{ij} for $1 \le i \le t-1$ and $1 \le j \le t$. Consider the generic determinantal variety $(X,o) \subset (\mathbb{C}^{t(t-1)},o)$ defined by the $((t-1)$–minors of the matrix (x_{ij}). Prove that (X,o) is rigid.

10.2.36. Use a computer algebra system (for example SINGULAR) to compute Example 10.2.8 (Hint: Try in SINGULAR the following:

```
ring r =0,(x(0..4)),(c,dp);
matrix m[2][4] = x(0),x(1),x(2),x(3),x(1),x(2),x(3),x(4);
ideal i = minor(m,2);
module s = syz(i);
qring q = std(i);
module s = imap(r,s);
syz(transpose(s));)
```

10.2.37. Prove that the sequence (10.8) is exact.
(Hint: Use the same method as in the proof of the injectivity of α_2'.)

10.3 Existence of a Semi-Universal Deformation

In this section we will finally give the proof that for an isolated singularity a versal deformation exists. We need many preparations.

Definition 10.3.1. Let (X, o) be a germ of a complex space, A a local Artinian $\mathbb{C}$-algebra with maximal ideal $\mathfrak{m}_A$. Suppose $\mathfrak{m}_A^{e+1} = 0$. Then a deformation $(X_{\mathbb{A}}, o) \longrightarrow (\mathbb{A}, o)$ is called a *deformation of order e*.

Definition 10.3.2. Let (X, o) be a germ of a complex space with local ring $\mathcal{O}_{X,o} = \mathbb{C}\{x\}/(f_1, \dots, f_k)$. Let $A = \mathbb{C}[[s]]/J$ be a local formal $\mathbb{C}$-algebra in the variables $s_1, \dots, s_p$, with maximal ideal $\mathfrak{m}_A$. By abuse of notation, we denote by J also the ideal $J \cdot \mathbb{C}\{x\}[[s]]$.

(1) A *formal deformation* (X_A, o) of (X, o) over A is given by a quotient of $\mathbb{C}\{x\}[[s]]/J$ by an ideal $(F_1, \dots F_k)$ such that

 (a) $F_i(x,0) = f_i$ for $i = 1, \dots, k$.

 (b) For every relation $r \in \mathbb{C}\{x\}^k$, that is $r \cdot f = 0$, there exists $R \in \mathbb{C}\{x\}[[s]]^k$ such that $R \cdot F = 0$ as elements of $\mathbb{C}\{x\}[[s]]/J$ and $R(x,0) = r$.

(2) For all $e \in \mathbb{N}$, consider the local Artinian $\mathbb{C}$-algebra A_e defined by $A_e = \mathbb{C}[[s]]/(J + \mathfrak{m}_A^{e+1})$. Note that a formal deformation induces a deformation $(X_{\mathbb{A}_e}, o) \longrightarrow (\mathbb{A}_e, o)$ of order e for all e.

(3) A formal deformation (X_A, o) is called *versal* if for all local *Artinian* $\mathbb{C}$-algebras B, and all deformations $(X_{\mathbb{B}}, o) \longrightarrow (\mathbb{B}, o)$ there exists a homomorphism $A_e \longrightarrow B$ for some $e \gg 0$ and an isomorphism of deformations $(X_{\mathbb{A}_e} \times_{\mathbb{A}_e} \mathbb{B}, o) \cong (X_{\mathbb{B}}, o)$. It is called *semi-universal*, if the induced homomorphism on the tangent spaces $(\mathfrak{m}_A/\mathfrak{m}_A^2)^* \longrightarrow (\mathfrak{m}_B/\mathfrak{m}_B^2)^*$ is uniquely determined.

Remark 10.3.3. Let (X_A, o) and (X_B, o) be two formal semi-universal deformations of a germ of a complex space (X, o). Then the formal analytic algebras A and B are isomorphic.

To see this, note that, because (X_A, o) is semi-universal, we have a unique map $\alpha : A_1 \longrightarrow B_1$, inducing $(X_{\mathbb{B}_1}, o) \longrightarrow (\mathbb{B}_1, o)$. Similarly, we get a map $\beta : B_1 \longrightarrow A_1$. By semi-universality, the composition $\beta \circ \alpha$ is the identity on A_1, and $\alpha \circ \beta$ is the identity on B_1. Hence A_1 and B_1 are isomorphic. Now let $e \in \mathbb{N}$ be given and consider $(X_{\mathbb{B}_e}, o) \longrightarrow (\mathbb{B}_e, o)$. By semi-universality, for some q there exists a map $A_q \longrightarrow B_e$, so that $(X_{\mathbb{B}_e}, o) \longrightarrow (\mathbb{B}_e, o)$ is induced by $(X_{\mathbb{A}_q}, o) \longrightarrow (\mathbb{A}_q, o)$. As $\mathfrak{m}_{B_e}^{e+1} = 0$ by definition, we can take $q = e$. Thus we get a map $\phi : A_e \longrightarrow B_e$. Conversely, we get a map $\psi : B_e \longrightarrow A_e$. The composition $\psi \circ \phi : A_e \longrightarrow A_e$ is such that the pull-back of the deformation $(X_{\mathbb{A}_e}, o) \longrightarrow (\mathbb{A}_e, o)$ is isomorphic to $(X_{\mathbb{A}_e}, o) \longrightarrow (\mathbb{A}_e, o)$. By semi-universality, this map is uniquely defined up to first order, and is the identity modulo the square of the maximal ideal. By the Inverse Function Theorem[8], the map $\psi \circ \phi$ is an isomorphism, and similarly $\phi \circ \psi$ is an isomorphism. Note that in particular $\dim_{\mathbb{C}}(A_e) = \dim_{\mathbb{C}}(B_e)$ for all $e \in \mathbb{N}$.

Remark 10.3.4. Suppose that (X_A, o) is a formal semi-universal deformation of (X, o). Then the Kodaira-Spencer map $(\mathfrak{m}_A/\mathfrak{m}_A^2)^* \longrightarrow T^1_{X,o}$ is an isomorphism.

Namely, to construct the inverse mapping, consider an element of $T^1_{X,o}$, that is, a deformation over the space defined by the local ring $\mathbb{C}[[\varepsilon]]/\varepsilon^2$. Because the formal deformation is semi-universal, we get a unique map $A_1 \longrightarrow \mathbb{C}[[\varepsilon]]/\varepsilon^2$. By looking at the coefficients of ε of the values of all elements in $\mathfrak{m}_A/\mathfrak{m}_A^2$ we get an element of $(\mathfrak{m}_A/\mathfrak{m}_A)^*$. One checks that this gives the inverse of the Kodaira-Spencer map.

[8] This theorem holds in the formal case too, just ignore the convergence.

Thus, as already said, we first want to show the existence of a formal semi-universal deformation. The idea is to do this order by order, that is, define deformations $(X_{\mathbb{A}_e}, o) \to (\mathbb{A}_e, o)$ of order e for all $e \in \mathbb{N}$. For $e = 1$, we define a deformation over $(\mathbb{A}_1, o)$, with $A_1 = \mathbb{C}[[s_1, \ldots, s_\tau]]/\mathfrak{m}^2$, and τ is the dimension of $T^1_{X,o}$. Given A_e, it might be quite obvious that we would like to have a deformation of order $e+1$ over an as big as possible space $\mathbb{A}_{e+1}$. In order to see that such a space exists, we will develop obstruction theory. To formulate this, we need to define a further $\mathcal{O}_{X,o}$-module, called $T^2_{X,o}$.

Definition 10.3.5. Let a singularity (X, o), with local ring $\mathcal{O}_{X,o} = \mathcal{O}_n/I$ be given. For simplicity, we write $\mathcal{O}_n = \mathcal{O}$. Take generators $f_1, \ldots, f_k$ of I. We consider the map $\mathscr{F} := \mathcal{O}^k \longrightarrow \mathcal{O}$, which sends the $e_i = (0, \ldots, 0,1,0, \ldots, 0)$ to f_i. We let $\mathscr{R}$ be the kernel of this map. Thus, we have an exact sequence

$$0 \longrightarrow \mathscr{R} \longrightarrow \mathscr{F} \longrightarrow \mathcal{O} \longrightarrow \mathcal{O}_{X,o} \longrightarrow 0.$$

Let $\mathscr{R}_0 \subset \mathscr{R}$ be the submodule generated by the so-called Koszul relations $f_i e_j - f_j e_i \in \mathscr{R}$. Note that $f_i e_j - f_j e_i$ is indeed in $\mathscr{R}$, as obviously $f_i \cdot f_j - f_j \cdot f_i = 0$. Now we can define

$$T^2_{X,o} := \operatorname{Hom}_{\mathcal{O}}(\mathscr{R}/\mathscr{R}_0, \mathcal{O}_{X,o}) / \operatorname{Hom}_{\mathcal{O}}(\mathscr{F}, \mathcal{O}_{X,o}).$$

Lemma 10.3.6. *$T^2_{X,o}$ is a well-defined $\mathcal{O}_{X,o}$-module.*

Proof. The fact that it is an $\mathcal{O}_{X,o}$-module is easy, as we are looking at homomorphisms to $\mathcal{O}_{X,o}$. Thus if we multiply a homomorphism with something in the ideal of (X, o), it will certainly become the zero homomorphism. As $\mathscr{R} \subset \mathscr{F}$ we get dually a map

$$\operatorname{Hom}_{\mathcal{O}}(\mathscr{F}, \mathcal{O}_{X,o}) \longrightarrow \operatorname{Hom}_{\mathcal{O}}(\mathscr{R}, \mathcal{O}_{X,o})$$

Now $\mathscr{R}_0$ is generated by elements of type $f_i e_j - f_j e_i$. If $\varphi \in \operatorname{Hom}(\mathscr{F}, \mathcal{O}_{X,o})$, then this element is sent to $f_i \varphi(e_j) - f_j \varphi(e_i) = 0 \in \mathcal{O}_{X,o}$. This shows that we get an induced map

$$\operatorname{Hom}_{\mathcal{O}}(\mathscr{F}, \mathcal{O}_{X,o}) \longrightarrow \operatorname{Hom}_{\mathcal{O}}(\mathscr{R}/\mathscr{R}_0, \mathcal{O}_{X,o}).$$

It remains to prove that the definition is independent of the choice of the generators $f_1, \ldots, f_k$ of the ideal of $\mathcal{O}_{X,o}$. This is left as Exercise 10.3.25. □

Theorem 10.3.7. *Let (X, o) be a germ of a complex space with an isolated singularity. Then $T^2_{X,o}$ is a finite-dimensional $\mathbb{C}$-vector space.*

The proof is left as Exercise 10.3.26. The main theorem on obstruction theory is the following.

Theorem 10.3.8. *Let (X, o) be a germ of a complex space. Consider a local Artinian (resp. formal, resp. analytic) $\mathbb{C}$-algebra A, and a small surjection $B \longrightarrow A$, with kernel J. This means that we have an exact sequence*

$$0 \longrightarrow J \longrightarrow B \longrightarrow A \longrightarrow 0.$$

and $\mathfrak{m} \cdot J = 0$, $\mathfrak{m}$ the maximal ideal of B. Therefore, J is a vector space over $\mathbb{C} = B/\mathfrak{m}$. Suppose we have a deformation $(X_{\mathbb{A}}, o) \longrightarrow (\mathbb{A}, o)$ of (X, o). For a short notation we write ξ for this deformation. Then we have the following statements.

(1) *There exists a well-defined* obstruction element

$$\operatorname{ob}(\xi)_{B\to A} \in T^2_{X,0} \otimes J.$$

This obstruction element $\operatorname{ob}(\xi)_{B\to A}$ *is zero, if and only if there exists a flat deformation of* $(X_{\mathbb{B}}, 0) \longrightarrow (\mathbb{B}, 0)$ *extending the given deformation* ξ*,*

(2) *Suppose we have a commutative diagram*

$$\begin{array}{ccccccccc} 0 & \longrightarrow & J & \longrightarrow & B & \longrightarrow & A & \longrightarrow & 0 \\ & & \downarrow & & \downarrow & & \downarrow & & \\ 0 & \longrightarrow & J' & \longrightarrow & B' & \longrightarrow & A' & \longrightarrow & 0 \end{array}$$

such that the rows are small extensions. Let ξ' *be the deformation over* A' *induced by* ξ*. Let* $\beta : T^2_{X,0} \otimes J \longrightarrow T^2_{X,0} \otimes J'$ *be the induced map. Then*

$$\operatorname{ob}(\xi')_{B'\to A'} = \beta\big(\operatorname{ob}(\xi)_{B\to A}\big).$$

Proof. We give the proof for local Artinian $\mathbb{C}$-algebras, the remaining cases we leave as Exercise 10.3.22. We first construct the obstruction element $\operatorname{ob}(\xi)_{B\to A}$. Suppose the deformation ξ, that is $(X_{\mathbb{A}}, 0) \longrightarrow (\mathbb{A}, 0)$, is given by the elements $F_A \in \big(\mathbb{C}\{x\} \otimes A\big)^k$. Let $r \in \mathscr{R}$ be a relation. By flatness we get a relation r_A between the F_A. Now take *arbitrary* lifts $F_B \in \mathbb{C}\{x\} \otimes B$ of F_A and a lift r_B of r_A. Then we consider

$$r_B \cdot F_B \in \mathcal{O}_{X,0} \otimes B.$$

As r_B is a lift of r_A, and F_B is a lift of F_A it follows that the image of $r_B \cdot F_B$ in $\mathbb{C}\{x\} \otimes A$ is zero. Thus we have in fact

$$r_B \cdot F_B \in \mathcal{O}_{X,0} \otimes J.$$

We claim that the map $r \mapsto r_B \cdot F_B$ defines an element in $\operatorname{Hom}_{\mathcal{O}}(\mathscr{R}/\mathscr{R}_0, \mathcal{O}_{X,0} \otimes J) = \operatorname{Hom}_{\mathcal{O}}(\mathscr{R}/\mathscr{R}_0, \mathcal{O}_{X,0}) \otimes J$,[9] which induces the obstruction element $\operatorname{ob}(\xi)_{B\to A} \in T^2_{X,0} \otimes J$.

Step 1. We first show that the element $r_B \cdot F_B$ depends on r and on F_B, but not on the particular lift r_B. In particular it is independent of the lift r_A. So, take another lift r'_A of r, with $r'_A \cdot F_A = 0$, and a lift r'_B of r'_A. First we show that we may assume that the restrictions r_A of r_B and r'_A of r'_B to A are equal. The idea of the proof is the same as in the proof of Lemma 10.2.22, so it is done by induction on $\dim_{\mathbb{C}}(A)$. This part of the proof does not use the fact that $B \longrightarrow A$ is a small extension. The case $\dim_{\mathbb{C}}(A) = 1$ is our assumption. To do the induction step, take an element $0 \neq \alpha \in A$ with $\mathfrak{m}_A \cdot \alpha = 0$. We may suppose by induction that the image of r_B and r'_B in $A/(\alpha)$ are equal. So we can write $r_A - r'_A = h \cdot \alpha$ for some $h \in \mathbb{C}\{x\}^k$. By assumption $r_A \cdot F_A = r'_A \cdot F_A = 0$. Hence $h \cdot F_A\alpha = 0$. Now $\mathfrak{m}_A\alpha = 0$, and $F_A \equiv f$ modulo $\mathfrak{m}_A$, so that $h \cdot f = 0$. Thus h is a relation, which can be lifted to a relation h_A between the F_A. We lift h_A to h_B, and change r'_B to $r''_B := r'_B - h_B\alpha'$, for a suitable α' which maps to α. Then r''_B is also a lift of r. Moreover $r'_B \cdot F_B$ is equal to $r''_B \cdot F_B$, and the restrictions of r_B and r''_B to A are equal.

[9] This equality holds because the tensor product is taken over $\mathbb{C}$ and, therefore, all modules are free.

Thus we now reached the stage that $r_B - r'_B$ has entries in $\mathbb{C}\{x\} \otimes J$, J being the kernel of the map $B \longrightarrow A$. As $\mathfrak{m}_B J = 0$ by assumption, and $F_B \equiv f$ modulo $\mathfrak{m}_B$ it follows that $(r_B - r'_B) \cdot F_B = (r_B - r'_B) \cdot f$, which, as the f_i are zero in $\mathcal{O}_{X,o}$, is zero in $\mathcal{O}_{X,o} \otimes J$. This shows that $r_B \cdot F_B$ is independent of the lift r_B of r.

Therefore the map $r \mapsto r_B \cdot F_B$ gives a well-defined element in $\mathrm{Hom}_{\mathcal{O}}(\mathscr{R}, \mathcal{O}_{X,o}) \otimes J$. This element might depend on the choice of the lift F_B.

Step 2. The restriction of the homomorphism to the submodule $\mathscr{R}_0$ generated by the Koszul relations is zero. Indeed, we can take as lift of $r = f_i e_j - f_j e_i$ simply the element $r_B := F_{Bi} e_j - F_{Bj} e_i$. Then $r_B \cdot F_B = 0$.

Step 3. We now show that the class of $r_B \cdot F_B$ gives a well-defined element in $T^2_{X,o} \otimes J$. Take another lift F'_B of F_A. Then $F_B - F'_B \in C\{x\}^k \otimes J$. We look at $r_B \cdot (F_B - F'_B)$. As $\mathfrak{m}_B J = 0$, and $r_B \equiv r$ modulo $\mathfrak{m}_B$, it follows that this is equal to $r \cdot (F_B - F'_B)$. It follows that the map $r \mapsto r_B \cdot (F_B - F'_B)$ is equal to the map $r \mapsto r \cdot (F_B - F'_B) \in \mathrm{Hom}_{\mathcal{O}}(\mathscr{F}, \mathcal{O}_{X,o}) \otimes J$. This is what we had to show.

Step 4. Suppose that an extension of ξ to a deformation $(X_{\mathbb{B}}, o) \longrightarrow (\mathbb{B}, o)$ exists. Then we can find F_B and for every $r \in \mathscr{R}$ an element r_B such that $r_B \cdot F_B = 0 \in \mathbb{C}\{x\} \otimes J$. Thus, certainly, the obstruction element is zero.

Step 5. On the other hand, suppose that $\mathrm{ob}(\xi)_{B \to A} = 0$. Take an arbitrary lift $F'_B = (F'_{B1}, \dots, F'_{Bk})$ of F_A. Because $\mathrm{ob}(\xi)_{B \to A} = 0$, the element in $\mathrm{Hom}_{\mathcal{O}}(\mathscr{R}/\mathscr{R}_0, \mathcal{O}_{X,o}) \otimes J$ given by

$$r \mapsto r_B \cdot F'_B, \quad r_B \text{ any lift of } r,$$

is, in fact, in $\mathrm{Hom}_{\mathcal{O}}(\mathscr{F}, \mathcal{O}_{X,o}) \otimes J$. Hence, there exist $h_1, \dots, h_k \in \mathbb{C}\{x\} \otimes J$ such that the map is given by

$$r = (r_1, \dots, r_k) \mapsto \sum_{i=1}^{k} r_i h_i.$$

Now we define

$$F_B = (F_{B1}, \dots, F_{Bk}) := (F'_{B1}, \dots, F'_{Bk}) - (h_1, \dots, h_k).$$

By construction, the element in $\mathrm{Hom}_{\mathcal{O}}(\mathscr{R}, \mathcal{O}_{X,o}) \otimes J = \mathrm{Hom}_{\mathcal{O}}(\mathscr{R}, \mathcal{O}_{X,o} \otimes J)$, given by

$$r \mapsto r_B \cdot F_B, \quad r_B \text{ any lift of } r,$$

is the zero map. We claim that the F_B defines a deformation which extends F_A. To show this, for every relation $r \in \mathscr{R}$ a lift r'_B such that $r'_B \cdot F_B = 0$ has to be found. Now by construction $r_B \cdot F_B = 0$ as element of $\mathcal{O}_{X,o} \otimes J$. Therefore, there exist $a_1, \dots, a_k$ in J with $r_B \cdot F_B = a_1 f_1 + \ldots + a_k f_k$. Put $a = (a_1, \dots, a_k)$. As $\mathfrak{m}_B J = 0$, and $F_B \equiv f$ modulo $\mathfrak{m}_B$, it follows that $a \cdot F_B = a \cdot f$. Now define $r'_B := r_B - a$. Then r'_B is a lift of r with $r'_B \cdot F_B = 0$. As we can do this for all r, it follows that F_B gives a (flat) deformation of (X, o).

Step 6. The second statement of the theorem is immediate from the construction of the obstruction element. □

Corollary 10.3.9. *Let (X, o) be a germ of a complex space, and $A = \mathcal{O}/J$ a local Artinian (resp. formal, resp. analytic) $\mathbb{C}$-algebra, and $A' := \mathcal{O}/\mathfrak{m}J$. Let ξ be a deformation $(X_{\mathbb{A}}, o) \longrightarrow (\mathbb{A}, o)$, and consider the small extension*

$$0 \longrightarrow J/\mathfrak{m}J \longrightarrow \mathcal{O}/\mathfrak{m}J \longrightarrow \mathcal{O}/J \longrightarrow 0,$$

and let $\mathrm{ob}(\xi)_{A' \to A} \in T^2_{X,o} \otimes (J/\mathfrak{m}J)$ be the obstruction element. By Exercise 1.2.46 this obstruction element gives a map which we, by abuse of notation, give the same name:

$$\mathrm{ob}(\xi)_{A' \to A} : \left(T^2_{X,o}\right)^* \longrightarrow J/\mathfrak{m}J.$$

Let $I \mod \mathfrak{m}J$ be the image of the map $\mathrm{ob}(\xi)_{A' \to A}$. Then I is the unique minimal ideal (with respect to inclusion) with the properties:

(1) $\mathfrak{m}J \subset I \subset J$

(2) Let $B = \mathcal{O}/I$. Then there exists a deformation $(X_{\mathbb{B}}, o) \longrightarrow (\mathbb{B}, o)$ of (X, o) inducing the deformation ξ.

The small surjection $B \longrightarrow A$ we call a maximal extension with respect to the deformation $(X_{\mathbb{A}}, o) \longrightarrow (\mathbb{A}, o)$.

Proof. We can write

$$\mathrm{ob}(\xi)_{A' \longrightarrow A} = \alpha_1 \otimes h_1 + \ldots + \alpha_p \otimes h_p,$$

for some linear independent elements $\alpha_1, \ldots, \alpha_p$ of $T^2_{X,o}$ and $h_1, \ldots h_p$ of J. Let $I = (h_1, \ldots, h_p) + \mathfrak{m}J$. To show that a deformation $(X_{\mathbb{B}}, o) \longrightarrow (\mathbb{B}, o)$ exists, we consider the local Artinian $\mathbb{C}$-algebra $\mathcal{O}/I$. Then $\mathrm{ob}(\xi)_{\mathcal{O}/I \to A}$ is equal to $\alpha_1 \otimes h_1 + \ldots + \alpha_p \otimes h_p$, but this time considered as element of $T^2_{X,o} \otimes (J/I)$, thus $\mathrm{ob}(\xi)_{\mathcal{O}/I \to A}$ is zero. (See the second part of Theorem 10.3.8.) Hence we forced the obstruction element to be zero, and therefore there exists an extension of ξ over the Artinian $\mathbb{C}$-algebra $\mathcal{O}/I$. To show that I is minimal and unique with respect to the two properties, let K be another ideal satisfying $\mathfrak{m}J \subset K \subset J$, such that there exists an extension of ξ over $\mathcal{O}/K$. The obstruction element is $\alpha_1 \otimes h_1 + \ldots + \alpha_p \otimes h_p$, and therefore is zero considered as element of J/K. Hence $I = (h_1, \ldots, h_p) + \mathfrak{m}J \subset K$, which is what we had to prove. □

Lemma 10.3.10. *Let $A = \mathbb{C}[[s]]/J$ be a formal algebra, (X_A, o) be a formal deformation of (X, o) over A, and $(X_{\mathbb{A}_e}, o)$ be the induced deformation of order e for all e. Consider the small surjection*

$$\mathbb{C}[[s]]/J + \mathfrak{m}^{e+1} \longrightarrow \mathbb{C}[[s]]/J + \mathfrak{m}^{e}. \tag{10.9}$$

Assume that

(1) $\mathbb{C}[[s]]/J + \mathfrak{m}^e \longrightarrow \mathbb{C}[[s]]/J + \mathfrak{m}^{e-1}$ is a maximal extension for the deformation $(X_{\mathbb{A}_{e-1}}, o) \longrightarrow (\mathbb{A}_{e-1}, o)$,

(2) $\dim_{\mathbb{C}}\left(\frac{J+\mathfrak{m}^{e+1}}{\mathfrak{m}J+\mathfrak{m}^{e+1}}\right) = \dim_{\mathbb{C}}\left(\frac{J+\mathfrak{m}^{e}}{\mathfrak{m}J+\mathfrak{m}^{e}}\right)$.

Then (10.9) is a maximal extension for the deformation $(X_{\mathbb{A}_e}, o)$.

Proof. If $\mathbb{C}[[s]]/J + \mathfrak{m}^{e+1}$ is not maximal, then there exists an ideal $I \subsetneq J + \mathfrak{m}^{e+1}$ with $\mathfrak{m}J + \mathfrak{m}^{e+1} \subset I \subset J + \mathfrak{m}^e$, and such that the deformation extends to a deformation over $\mathbb{C}[[s]]/I$. Then the inclusion $I + \mathfrak{m}^e \subset J + \mathfrak{m}^e$ is in fact an equality because of the first condition. Therefore, the minimal number of generators of I modulo $\mathfrak{m}^e$ is equal to the minimal number of generators of J modulo $\mathfrak{m}^e$, that is, $\dim_{\mathbb{C}}\left(\frac{I+\mathfrak{m}^e}{\mathfrak{m}I+\mathfrak{m}^e}\right) = \dim_{\mathbb{C}}\left(\frac{J+\mathfrak{m}^e}{\mathfrak{m}J+\mathfrak{m}^e}\right)$. As $I \subsetneq J + \mathfrak{m}^{e+1}$, it follows that there is a minimal set of generators of J modulo $\mathfrak{m}^{e+1}$, which has an element of degree e. In formula's

$$\dim_{\mathbb{C}}\left(\frac{J+\mathfrak{m}^e}{\mathfrak{m}J+\mathfrak{m}^e}\right) = \dim_{\mathbb{C}}\left(\frac{I+\mathfrak{m}^e}{\mathfrak{m}I+\mathfrak{m}^e}\right) \le \dim_{\mathbb{C}}\left(\frac{I}{\mathfrak{m}I+\mathfrak{m}^{e+1}}\right) =$$
$$\dim_{\mathbb{C}}\left(\frac{I}{\mathfrak{m}J+\mathfrak{m}^{e+1}}\right) < \dim_{\mathbb{C}}\left(\frac{J+\mathfrak{m}^{e+1}}{\mathfrak{m}J+\mathfrak{m}^{e+1}}\right).$$

But this is in contradiction to the second condition. □

It is time for an example.

Example 10.3.11. We consider the germ of the complex space $(X, 0)$ defined by the 2–minors of the matrix

$$\begin{pmatrix} x_0 & x_1 & x_2 & x_3 \\ x_1 & x_2 & x_3 & x_4 \end{pmatrix}.$$

The modules $N_{X,0}$ and $T^1_{X,0}$ have been described in 10.2.8 and 10.2.14. Thus we have the following deformation ξ over $\mathbb{C}[[s_1,s_2,s_3,t]]/\mathfrak{m}^2 =: A$, with $\mathfrak{m} = (s_1,s_2,s_3,t)$:

$$\begin{aligned}
F_{12} &= f_{12} - s_1x_1 + tx_0 \\
F_{23} &= f_{23} + s_1x_3 - s_2x_2 - tx_2 \\
F_{34} &= f_{34} + s_2x_4 - s_3x_3 + tx_4 \\
F_{13} &= f_{13} - s_2x_1 \\
F_{24} &= f_{24} + s_1x_4 - s_3x_2 \\
F_{14} &= f_{14} - s_3x_1 + tx_2
\end{aligned}$$

The f_{ij} are the 2–minors of the above matrix as in Example 10.2.8. The F_{ij} give a (flat) deformation over $\mathbb{C}[[s_1,s_2,s_3,t]]/(s_1,s_2,s_3,t)^2$, so we must be able to lift the relations. Now we give the lift R_{ijk} and S_{ijk} of the relations r_{ijk} and s_{ijk}.

$$\begin{aligned}
R_{123} &= x_0F_{23} - (x_1 + s_1)F_{13} + (x_2 + s_2)F_{12} \\
R_{124} &= x_0F_{24} - (x_1 + s_1)F_{14} + (x_3 + s_3)F_{12} - tF_{13} \\
R_{134} &= x_0F_{34} - (x_2 + s_2)F_{14} + (x_3 + s_3)F_{13} - tF_{14} - tF_{23} \\
R_{234} &= (x_1 + s_1)F_{34} - (x_2 + s_2 + t)F_{24} + (x_3 + s_3)F_{23} \\
S_{123} &= x_1F_{23} - (x_2 + t)F_{13} + x_3F_{12} \\
S_{234} &= x_2F_{34} - x_3F_{24} + x_4F_{23} \\
S_{124} &= x_1F_{24} - (x_2 + t)F_{14} + x_4F_{12} - tF_{23} \\
S_{134} &= x_1F_{34} - x_3F_{14} + x_4F_{13} - tF_{24}
\end{aligned}$$

These R_{ijk} and S_{ijk} are all zero modulo $(s_1,s_2,s_3,t)^2$ expressing the fact that we have a (flat) deformation. We obtain the obstruction element by looking at the quadratic terms in s and t, which one gets by simply expanding the terms:

$$
\begin{aligned}
r_{123} \mapsto R_{123} &= s_2tx_0 \\
r_{124} \mapsto R_{124} &= s_3tx_0 - s_1tx_2 + s_2tx_1 \\
r_{134} \mapsto R_{134} &= s_3tx_1 - s_1tx_3 \\
r_{234} \mapsto R_{234} &= 0 \\
s_{123} \mapsto S_{123} &= s_2tx_1 \\
s_{124} \mapsto S_{124} &= s_3tx_1 - s_1tx_3 + s_2tx_2 \\
s_{134} \mapsto S_{134} &= -s_1tx_4 + s_3tx_2 \\
s_{234} \mapsto S_{234} &= 0
\end{aligned}
$$

This map of $\mathscr{R} \longrightarrow \mathcal{O}_{X,o} \otimes J$ gives the element $\mathrm{ob}(\xi)_{A' \to A}$ in $T^2_{X,o} \otimes J$, where $A' = \mathbb{C}[[s_1,s_2,s_3,t]]/\mathfrak{m}^3$ and $J = \mathfrak{m}^2/\mathfrak{m}^3$. Now, according to Corollary 10.3.9, we need a basis of $T^2_{X,o}$ to describe the map $\mathrm{ob}(\xi)_{A' \to A} : \left(T^2_{X,o}\right)^* \longrightarrow \mathfrak{m}^2/\mathfrak{m}^3$, and compute I, the image of this map. We will see that we already computed part of a basis, which is enough for our purpose. We have in particular that the elements α_1, α_2 and α_3 given in the first three columns of the following table, (obtained by looking at the coefficients of s_1t, s_2t and s_3t) give three elements in $\mathrm{Hom}_{\mathcal{O}}(\mathscr{R}/\mathscr{R}_0, \mathcal{O}_{X,o})$. The last six columns correspond to the images of the relations with respect to the canonical basis of $\mathrm{Hom}_{\mathcal{O}}(\mathscr{F}, \mathcal{O}_{X,o})$. We will see that α_1, α_2 and α_3 are linearly independent modulo $\mathrm{Hom}_{\mathcal{O}}(\mathscr{F}, \mathcal{O}_{X,o})$, and therefore induce linear independent elements of $T^2_{X,o}$.

(10.10)

$r_{123} \mapsto$	0	x_0	0	x_2	x_0	0	$-x_1$	0	0
$r_{124} \mapsto$	$-x_2$	x_1	x_0	x_3	0	0	0	x_0	$-x_1$
$r_{134} \mapsto$	$-x_3$	0	x_1	0	0	x_0	x_3	0	$-x_2$
$r_{234} \mapsto$	0	0	0	0	x_3	x_1	0	$-x_2$	0
$s_{123} \mapsto$	0	x_1	0	x_3	x_1	0	$-x_2$	0	0
$s_{124} \mapsto$	$-x_3$	x_2	x_1	x_4	0	0	0	x_1	$-x_2$
$s_{134} \mapsto$	$-x_4$	0	x_2	0	0	x_1	x_4	0	$-x_3$
$s_{234} \mapsto$	0	0	0	0	x_4	x_2	0	$-x_3$	0

To see whether these elements are linearly independent in $T^2_{X,o}$ we have to look at the "trivial" elements in $T^2_{X,o}$, that is, elements of $\mathrm{Hom}(\mathscr{F}, \mathcal{O}_{X,o})$ which are given by the final six columns of the above table (10.10). These six columns are the duals of f_{12}, f_{23}, f_{34}, f_{13}, f_{24}, and f_{14} respectively. Thus, for example, the first row and the final six columns says that r_{123} is the relation $x_2f_{12} + x_0f_{23} + 0f_{34} - x_1f_{13} + 0f_{24} + 0f_{14} = 0$. An elementary but boring check now shows that α_1, α_2 and α_3 represent linearly independent elements in $T^2_{X,o}$. They can therefore be considered as part of a basis of $T^2_{X,o}$ and $\mathrm{ob}(\xi)_{A' \to A} = \alpha_1 \otimes s_1t + \alpha_2 \otimes s_2t + \alpha_3 \otimes s_3t$. Hence the minimal ideal I of the corollary is given by the ideal $(s_1t, s_2t, s_3t) + (s_1, s_2, s_3, t)^3$. Note, however, that the family considered above, in fact, defines a deformation over $\mathbb{C}\{s_1,s_2,s_3,t\}/(s_1t,s_2t,s_3t)$, as all the $R_{ijk} \cdot F$ and $S_{ijk} \cdot F$ vanish modulo (s_1t,s_2t,s_3t) and not only modulo $(s_1t,s_2t,s_3t) + (s_1,s_2,s_3,t)^3$.

Theorem 10.3.12. *Let $A = \mathbb{C}[[s]]/J$ be a formal local $\mathbb{C}$-algebra, and (X, o) be a germ of a complex space. Let (X_A, o) be a formal deformation of (X, o). For all k put $A_k := A/\mathfrak{m}^{k+1}$ and $J_k = J + \mathfrak{m}^{k+1}$. Suppose that*

(1) the Kodaira-Spencer map of (X_A, o) is surjective (resp. an isomorphism),

(2) for all $e \geq 2$, the small extension $A_{e+1} \longrightarrow A_e$ is a maximal extension with respect to the deformation $(X_{\mathbb{A}_e}, o) \longrightarrow (\mathbb{A}_e, o)$.

Then (X_A, o) is a formal versal (resp. semi-universal) deformation.

Example 10.3.13. The family written down in the previous example is formally versal. It is a deformation over $A = \mathbb{C}[[s_1,s_2,s_3,t]]/(s_1t,s_2t,s_3t)$. Details are left as an exercise.

The basic idea of the proof is as the proof of Theorem 10.2.18, but there is a subtle difficulty popping up, which is taken care of by Proposition 10.3.15. In its proof we need the following lemma.

Lemma 10.3.14. *Suppose that $\varphi : B \longrightarrow A$ is a simple surjection, and let $\psi : C \longrightarrow B$ be a map of local Artinian $\mathbb{C}$-algebras such that*

(1) $\varphi \circ \psi : C \longrightarrow A$ is surjective,

(2) φ does not have a section.

Then ψ is surjective.

Proof. Suppose the converse. Let $J = \mathrm{Ker}(\psi)$. Then we have a injective map $\psi : C/J \longrightarrow B$. Moreover $\varphi \circ \psi : C/J \longrightarrow A$ is surjective. As $\psi : C/J \longrightarrow B$ is not surjective, and $\dim_{\mathbb{C}}(B) = \dim_{\mathbb{C}}(A) + 1$, it follows that $\dim_{\mathbb{C}}(C/J) = \dim_{\mathbb{C}}(A)$. In Exercise 10.3.23 you will prove that a surjective algebra map of Artinian $\mathbb{C}$-algebras of the same (vector space) dimension is an isomorphism. Thus $\varphi \circ \psi$ is an isomorphism. Denote the inverse by $\beta : A \longrightarrow C/J$. Then the map $\psi \circ \beta$ is a section of φ. This is a contradiction. □

Proposition 10.3.15. *Let local Artinian $\mathbb{C}$-algebras $A = \mathbb{C}[[s_1, \dots, s_p]]/I$, and $A' = \mathbb{C}[[s_1, \dots, s_p]]/J$ be given, for some ideals I, J with $\mathfrak{m}J \subset I \subset J$. Let two further Artinian $\mathbb{C}$-algebras B and B' be given. Let $\psi : A \longrightarrow A'$ be the canonical surjection. Let $\xi : (X_{\mathbb{A}'}, o) \longrightarrow (\mathbb{A}', o)$ be a deformation of (X, o). Suppose that*

(1) the small extension $A \longrightarrow A'$ is a maximal extension with respect to the deformation $\xi : (X_{\mathbb{A}'}, o) \longrightarrow (\mathbb{A}', o)$;

(2) we have a simple surjection $\pi : B \longrightarrow B'$;

(3) we have a map $\phi' : A' \longrightarrow B'$ and a deformation $(X_{\mathbb{B}}, o) \longrightarrow (\mathbb{B}, o)$ inducing Cartesian diagrams

$$\begin{array}{ccc} (X_{\mathbb{B}'}, o) & \longrightarrow & (X_{\mathbb{A}'}, o) \\ \downarrow & & \downarrow \\ (\mathbb{B}', o) & \xrightarrow{\phi'} & (\mathbb{A}', o) \end{array} \qquad \begin{array}{ccc} (X_{\mathbb{B}'}, o) & \longrightarrow & (X_{\mathbb{B}}, o) \\ \downarrow & & \downarrow \\ (\mathbb{B}', o) & \longrightarrow & (\mathbb{B}, o) \end{array}$$

Then there exists a map $\alpha : A \longrightarrow B$ making the following diagram commutative:

$$\begin{array}{ccc} A & \xrightarrow{\alpha} & B \\ \downarrow\psi & & \downarrow\pi \\ A' & \xrightarrow{\phi'} & B'. \end{array} \tag{10.11}$$

Proof. We consider the so-called fibered product

$$A' \times_{B'} B := \{(a,b) \in A' \times B : \phi'(a) = \pi(b)\}.$$

Defining addition and multiplication component wise $A' \times_{B'} B$ is a local $\mathbb{C}$-algebra. We have a natural projection $pr_1 : A' \times_{B'} B \longrightarrow A'$ which sends (a,b) to a. Note that pr_1 is also a simple surjection. Indeed, $pr_1(a,b) = 0$ implies that $a = 0$, and thus, by definition $\pi(b) = 0$. Thus b is in the kernel of π, which by assumption is one-dimensional. It is immediate from the definition that making the diagram (10.11) commutative is the same as making the diagram

$$\begin{array}{ccc} & & A' \times_{B'} B \\ & \overset{\phi}{\nearrow} & \downarrow{\scriptstyle pr_1} \\ A & \xrightarrow{\psi} & A' \end{array}$$

commutative.

Case 1. Suppose that pr_1 has a section, that is, there exists a map $s : A' \longrightarrow A' \times_{B'} B$ with $pr_1 \circ s$ the identity. This is the easy case, as we then can define $\phi := s \circ \psi$.

Case 2. Suppose that pr_1 does not have a section.

Step 1. We will show that $A' \times_{B'} B$ is a quotient of $\mathbb{C}[[s_1, \dots .s_p]]$.[10] We first construct a map ω making the diagram

$$\begin{array}{ccc} \mathbb{C}[[s_1, \dots .s_p]] & \xrightarrow{\omega} & A' \times_{B'} B \\ \downarrow & & \downarrow{\scriptstyle pr_1} \\ A & \xrightarrow{\psi} & A' \end{array}$$

commutative. Indeed, by assumption A' is a quotient of $\mathbb{C}[[s_1, \dots .s_p]]$. Consider the composition

$$\beta : \mathbb{C}[|s_1, \dots ,s_p]] \longrightarrow A' \xrightarrow{\phi'} B'.$$

As $\pi : B \longrightarrow B'$ is surjective, we can find $b_i \in B$ with $\pi(b_i) = \beta(s_i)$ for all $i = 1, \dots ,p$. Now define

$$\omega(s_i) = (s_i, b_i) \in A' \times_{B'} B.$$

As $\mathbb{C}[[s_1, \dots ,s_p]]$ is a formal free algebra, we can extend ω to an algebra map. Applying Lemma 10.3.14 to the case $C = \mathbb{C}[[s_1, \dots ,s_p]]/\operatorname{Ker}(\omega)$, $\psi = \omega, B = A' \times_{B'} B$, $A = A'$ and $\varphi = pr_1$, we obtain $A' \times_{B'} B \cong \mathbb{C}[[s_1, \dots ,s_p]]/\operatorname{Ker}(\omega)$.

Step 2. Write $\mathcal{O} = \mathbb{C}[[s_1, \dots ,s_p]]$. Put $K := \operatorname{Ker}(\omega)$. Thus we get

$$A' \times_{B'} B \cong \mathcal{O}/K \twoheadrightarrow \mathcal{O}/J = A'.$$

Recall that $A = \mathcal{O}/I$. We will show that ω induces a map $\phi : A \longrightarrow A' \times_{B'} B$. To show this we have to show that $I \subset K$. Now $\mathcal{O}/K \twoheadrightarrow \mathcal{O}/J$ is a simple surjection, thus in particular small by Exercise 10.3.24. Hence $\mathfrak{m}J \subset K \subset J$. As by assumption $A \longrightarrow A'$ is a maximal extension with respect to $(X_{\mathbb{A}'}, o) \longrightarrow (\mathbb{A}', o)$, we have $I \subset K$, if there exists a deformation over $\mathcal{O}/K$ inducing the deformation ξ. This we show in Step 3.

[10] Note that both A and A' are quotients of the same $\mathbb{C}[[s_1, \dots .s_p]]$.

Step 3. We now write $D = A' \times_{B'} B$ for short. To complete the proof, it therefore suffices to show that there exists a deformation

$$(X_{\mathbb{D}}, o) \longrightarrow (\mathbb{D}, o)$$

inducing ξ. Suppose ξ is given by functions $F_{A'} \in \left(\mathbb{C}\{x\}\hat{\otimes}A'\right)^k$. Moreover, by assumption we have deformations $(X_{\mathbb{B}}, o)$ given by say $F_B \in \left(\mathbb{C}\{x\}\hat{\otimes}B\right)^k$, and $(X_{\mathbb{B}'}, o)$ given by say $F_{B'} \in \left(\mathbb{C}\{x\}\hat{\otimes}B'\right)^k$. As the deformation over $(\mathbb{A}', o)$ and $(\mathbb{B}, o)$ induces the same deformation over B', it can be easily seen that we may assume that

$$\phi'(F_{A'}) = F_{B'} = \pi(F_B).$$

Thus the pair $(F_{A'}, F_B)$ defines an element in $\left(\mathbb{C}\{x\}\hat{\otimes}D\right)^k$. To show that it defines a (flat) deformation, we have to lift relations. So let $r \in \mathbb{C}\{x\}^k$ with $r \cdot f = 0$ be an arbitrary relation. By flatness we can find $R_{A'} \in \mathbb{C}\{x\}^k \otimes A'$ with $R_{A'} \cdot F_{A'} = 0$. Let $R_{B'}$ be the image of $R_{A'}$ under the natural map $\mathbb{C}\{x\}^k \otimes A' \longrightarrow \mathbb{C}\{x\}^k \otimes B'$. By Lemma 10.2.22 we can find an $R_B \in \mathbb{C}\{x\}^k \otimes B$ lifting $R_{B'}$ with $R_B \cdot F_B = 0$. The pair $(R_{A'}, R_B)$ therefore is an element of $\mathbb{C}\{x\}^k\hat{\otimes}D = \mathbb{C}\{x\}^k\hat{\otimes}(A' \times_{B'} B)$, and $(R_{A'}, R_B) \cdot (F_{A'}, F_B) = 0$. Hence the relation r lifts. This finishes the proof of the proposition. □

Proof of Theorem 10.3.12. Consider a local Artinian $\mathbb{C}$-algebra B, and suppose that we have a deformation $(X_{\mathbb{B}}, o) \longrightarrow (\mathbb{B}, o)$ of (X, o). Let $A_e := A/\mathfrak{m}^{e+1} = \mathbb{C}[[s]]/(J + \mathfrak{m}^{e+1})$. We have an induced deformation $(X_{\mathbb{A}_e}, o) \longrightarrow (\mathbb{A}_e, o)$. Consider the following small surjection

$$(10.12) \qquad 0 \longrightarrow J + \mathfrak{m}^{e+1}/(\mathfrak{m}J + \mathfrak{m}^{e+2}) \longrightarrow \mathbb{C}[[s]]/(\mathfrak{m}J + \mathfrak{m}^{e+2}) \longrightarrow A_e \longrightarrow 0$$

By assumption, for all e the ideal $J \mod \mathfrak{m}^{e+2}$ is minimal among those ideals for which there exists an extension of $(X_{\mathbb{A}_e}, o) \longrightarrow (\mathbb{A}_e, o)$ over some small extension of A_e.

To show the versality, we have to show that there exists an $e \in \mathbb{N}$ and a morphism of $\mathbb{C}$-algebras $A_e \longrightarrow B$ such that the deformation $(X_{\mathbb{B}}, o) \longrightarrow (\mathbb{B}, o)$ is isomorphic to $(X_{\mathbb{A}_e} \times_{\mathbb{A}_e} \mathbb{B}, o) \longrightarrow (\mathbb{B}, o)$. This we will prove by induction on the vector space dimension $\dim_{\mathbb{C}}(B)$. We suppose that $X_{\mathbb{A}_e}$ is defined by elements $F^{(e)}(x,s) = (F_1^{(e)}(x,s), \ldots, F_k^{(e)}(x,s)) \in \mathbb{C}\{x\}^k \otimes A_e = \mathbb{C}\{x\}^k \otimes \mathbb{C}[[s]]/J_e$. (Recall that $J_e = J + \mathfrak{m}^{e+1}$.)

Step 1. Consider the induced deformation over $B/\mathfrak{m}_B^2$. As the Kodaira Spencer map is surjective, we can find a map $A_1 \longrightarrow B$ inducing the deformation over $B/\mathfrak{m}_B^2$. This map is uniquely determined in case the Kodaira-Spencer map is an isomorphism.

Step 2. Now take a nonzero element $\alpha \in B$, such that $\mathfrak{m}_B\alpha = 0$. Put $B' = B/(\alpha)$, and denote by $\pi : B \longrightarrow B'$ the canonical projection. Then $\dim_{\mathbb{C}}(B') = \dim_{\mathbb{C}}(B) - 1$, and therefore, by induction, we have for suitable $e \in \mathbb{N}$ a morphism of $\mathbb{C}$-algebras

$$A_e \xrightarrow{\sigma'} B'$$

that induces a deformation which is isomorphic to the given deformation $(X_{\mathbb{B}'}, o) \longrightarrow (\mathbb{B}', o)$, induced by $(X_{\mathbb{B}}, o) \longrightarrow (\mathbb{B}, o)$ via the projection π. By Proposition 10.3.15, there exists a map $\sigma : A_{e+1} \longrightarrow B$, making the diagram

$$\begin{array}{ccc} A_{e+1} & \xrightarrow{\sigma} & B \\ \downarrow & & \downarrow \pi \\ A_e & \xrightarrow{\sigma'} & B' \end{array}$$

commutative.

Step 3. Now that we have our map σ we get two different deformations over $(\mathbb{B}, o)$: The given deformation $(X_{\mathbb{B}}, o) \longrightarrow (\mathbb{B}, o)$, and the deformation which is induced by σ. By assumption the restriction of the two deformations to $(\mathbb{B}', o)$ are isomorphic. Hence there exist

(1) a $k \times k$ matrix Λ' with entries in $\mathbb{C}\{x\} \otimes B'$, lifting the identity matrix and

(2) a B'– automorphism Φ' defined by $\phi'_1 = \Phi'(x_1), \ldots, \phi'_n = \Phi'(x_n) \in \mathbb{C}\{x\} \otimes B'$ lifting the identity,

such that

$$F^{(e)}(x, \sigma'(s_1), \ldots, \sigma'(s_p)) = F'_B(\phi'_1, \ldots, \phi'_n, t) \circ \Lambda'.$$

We now take any lift Λ of Λ', and $\phi_1, \ldots, \phi_n$ of $\phi'_1, \ldots, \phi'_n$. We will change σ, Λ, and $\phi_1, \ldots, \phi_n$ in such a way that:

$$F^{(e+1)}(x, \sigma(s_1), \ldots, \sigma(s_p)) = F_B(\phi_1, \ldots, \phi_n, t) \circ \Lambda.$$

This change of σ, ϕ_i, and Λ can be done exactly as in the proof of 10.2.18. □

Theorem 10.3.16 (Schlessinger). *Let (X, o) be a germ of a complex space. Suppose that $T^1_{X,o}$ is a finite-dimensional vector space, of dimension p, and $a_1, \ldots a_p$ are elements of $N_{X,o}$ projecting to a basis of $T^1_{X,o}$. The following construction gives a formal semi-universal deformation of (X, o).*

(1) Consider new variables $s_1, \ldots, s_p$, and consider $A := \mathbb{C}[[s_1, \ldots, s_p]]$, with maximal ideal $\mathfrak{m} = (s_1, \ldots, s_p)$. We set $A_1 := A/\mathfrak{m}^2$ and define a first order deformation:

$$(X_{\mathbb{A}_1}, o) \longrightarrow (\mathbb{A}_1, o),$$

by taking $F^{(1)}(x, s) = f + \sum_{i=1}^{p} s_i a_i \in \mathbb{C}\{x\}^k \otimes A_1$.

(2) Now suppose the deformation $(X_{\mathbb{A}_e}, o) \longrightarrow (\mathbb{A}_e, o)$, for $A_e = \mathcal{O}/J_e$ has already been constructed. We apply 10.3.9 to find an ideal J_{e+1}, minimal with respect to the properties mentioned in that corollary. Then define $A_{e+1} := \mathcal{O}/J_{e+1}$. So by 10.3.9, we can find $(X_{\mathbb{A}_{e+1}}, o) \longrightarrow (\mathbb{A}_{e+1}, o)$ defined by $F^{(e)}(x, s)$. Continuing like this we get formal power series $F \in \mathbb{C}\{x\}[[s]]/J$, which defines a (flat) deformation over $J = \cap_e J_e$.

Moreover, all formal semi-universal deformations of (X, o) are obtained by this construction.

Lemma 10.3.17. *Consider the ideals J and J_e constructed in the statement of Theorem 10.3.16. Then $J_e = J + \mathfrak{m}^{e+1}$.*

Proof. As $J_{e+k} \subset J_e$ for all k by construction, and $J = \cap_e J_e$, it suffices to show $J_{e+1} + \mathfrak{m}^{e+1} = J_e$ and iterate this equation. This is proved by induction on e. The case $e = 1$ is trivial, as $J_1 = \mathfrak{m}^2$. So assume we proved $J_e + \mathfrak{m}^e = J_{e-1}$. As $\mathfrak{m}^{e+1} \subset \mathfrak{m}J_{e-1} \subset J_e \subset J_{e-1}$, and $J_{e+1} \subset J_e$ by construction, the inclusion $J_{e+1} + \mathfrak{m}^{e+1} \subset J_e$ follows. To show the other inclusion, observe that there is a deformation of order e over the space with local ring $\mathcal{O}/J_{e+1} + \mathfrak{m}^{e+1}$ (lifting the deformation $(X_{\mathbb{A}_{e-1}}, o) \longrightarrow (\mathbb{A}_{e-1}, o)$) by simply restricting. Hence, by minimality of J_e we only have to show

$$\mathfrak{m}J_{e-1} \subset J_{e+1} + \mathfrak{m}^{e+1} \subset J_{e-1}.$$

The inclusion $J_{e+1} + \mathfrak{m}^{e+1} \subset J_{e-1}$ is easy, as $J_{e+1} \subset J_e \subset J_{e-1}$, and $\mathfrak{m}^{e+1} \subset \mathfrak{m}^e \subset J_{e-1}$ by construction. For the other inclusion, we use $\mathfrak{m}J_{e-1} = \mathfrak{m}J_e + \mathfrak{m}^{e+1} \subset J_{e+1} + \mathfrak{m}^{e+1}$. The equality follows from the induction hypothesis, the inclusion holds because $\mathfrak{m}J_e \subset J_{e+1}$ by definition. □

Proof of Theorem 10.3.16. The fact that the constructed deformation is semi-universal is an immediate consequence of Theorem 10.3.12, Lemma 10.3.17 and Corollary 10.3.9. It remains to show that all formal versal deformations come from a construction as in Theorem 10.3.16. This we do by contradiction. Thus let (X_A, o) be a formal semi-universal deformation. Now certainly, because of the versality, the Kodaira-Spencer map is surjective. Furthermore $\dim_{\mathbb{C}}(\mathfrak{m}_A/\mathfrak{m}_A^2)$ is equal to the dimension of $T^1_{X,o}$, because the deformation is semi-universal. This implies that the Kodaira-Spencer map is an isomorphism. Now suppose that for some e the small extension $A_{e+1} \longrightarrow A_e$ is not maximal with respect to the deformation $(X_{\mathbb{A}_e}, o) \longrightarrow (\mathbb{A}_e, o)$. We can follow the construction of Theorem 10.3.16 to find a formal semi-universal deformation over A' with $A'_e = A_e$, and $A'_{e+1} \twoheadrightarrow A_{e+1}$. But we have formal semi-universal deformations over both A and A' it follows that $\dim_{\mathbb{C}}(A'_{e+1}) = \dim_{\mathbb{C}}(A_{e+1})$, see Remark 10.3.3. Thus A_{e+1} was maximal after all, proving the claim. □

Remark 10.3.18. Note that in fact we proved something stronger. Indeed, under the assumptions of Theorem 10.3.16, suppose we have a deformation $(X_B, o) \longrightarrow (\mathbb{B}, o)$ of order q. We have an induced deformation of order $q-1$ over the space defined by the Artinian ring $B_{q-1} := B + \mathfrak{m}^q/\mathfrak{m}^q$. Suppose we have a map $\varphi : A_e \longrightarrow B_{q-1}$ inducing this deformation. Then there exists a lift $\widetilde{\phi} : A_f \longrightarrow B$ for some $f \geq e$, that is, we have a commutative diagram

$$\begin{array}{ccc} A_f & \xrightarrow{\widetilde{\phi}} & B \\ \downarrow & & \downarrow \\ A_e & \xrightarrow{\phi} & B_{q-1}, \end{array}$$

such that $\widetilde{\phi}$ induces the given deformation $(X_{\mathbb{B}}, o) \longrightarrow (\mathbb{B}, o)$. This property of the constructed versal deformation is of great use in the proof of the existence of a convergent semi-universal deformation.

Theorem 10.3.19. *Let (X_S, o) and (S, o) be germs of complex spaces, and suppose $(X_S, o) \longrightarrow (S, o)$ is formally semi-universal. Then (X_S, o) is a semi-universal deformation.*

Proof. Let a deformation $(X_T,\mathrm{o}) \longrightarrow (T,\mathrm{o})$ be given. Suppose that this deformation is given by functions $F_T(x,t) \in \left(\mathbb{C}\{x\}\hat{\otimes}\mathcal{O}_{T,\mathrm{o}}\right)^k$. We have to find a map $(T,\mathrm{o}) \longrightarrow (S,\mathrm{o})$ such that the deformation $(X_T,\mathrm{o}) \longrightarrow (T,\mathrm{o})$ is isomorphic to $(X_S \times_S T,\mathrm{o}) \longrightarrow (T,\mathrm{o})$. Write $\mathcal{O}_{S,\mathrm{o}} = \mathbb{C}\{s\}/J$, and $\mathcal{O}_{T,\mathrm{o}} = \mathbb{C}\{t\}/J'$. To give a map $(T,\mathrm{o}) \longrightarrow (S,\mathrm{o})$ means that dually we have to find a map $\phi : \mathbb{C}\{s\} \longrightarrow \mathbb{C}\{t\}$ with $\phi(J) \subset J'$. The induced deformation $(X_S \times_S T,\mathrm{o}) \longrightarrow (T,\mathrm{o})$ is given by $F_S(x,\phi(s)) \in \mathbb{C}\{x,t\}^k$. This has to be isomorphic to the given deformation defined by F_T. Therefore, we have to find an automorphism $\Phi : \mathbb{C}\{x,t\}^n \longrightarrow \mathbb{C}\{x,t\}^n$, and a matrix $\Lambda \in \mathbb{C}\{x,t\}^k$ such that

$$F_S(x,\phi(s)) = (F_T \circ \Phi)(x,t) \circ \Lambda \mod J' \cdot \mathbb{C}\{x,t\}. \tag{10.13}$$

This is a system of system of equations we want to solve for ϕ, Φ and Λ. We want to apply Grauert's Approximation Theorem, and we therefore have to show that every solution of (10.13) of order e can be lifted to a solution of order $e+1$. By assumption, (X_S,o) is formally semi-universal, hence is given by a sequence of maximal extensions by Theorem 10.3.16. Note that Remark 10.3.18 says that every solution of order e can be lifted to a solution of order $e+1$. Thus we may indeed apply Grauert's Approximation Theorem. □

Corollary 10.3.20. *Let (X,o) be a germ of a complex space, and let (X_A,o) be a (formal) deformation over $A = \mathbb{C}[[s]]/J$. Then (X_A,o) is (formally) semi-universal if and only if*

(1) The Kodaira-Spencer map is an isomorphism,

(2) The obstruction map $\left(T^2_{X,\mathrm{o}}\right)^* \longrightarrow J/\mathfrak{m}J$ is surjective.

A similar statement holds for convergent deformations.

Proof. From Theorem 10.3.19, it suffices to consider the formal case.

Step 1. First suppose that the Kodaira-Spencer map is an isomorphism, and that the obstruction map is surjective. Consider the small extensions

$$\begin{array}{ccccc}
J/\mathfrak{m}J & \longrightarrow & \mathbb{C}[[s]]/\mathfrak{m}J & \longrightarrow & \mathbb{C}[[s]]/J \\
\downarrow & & \downarrow & & \downarrow \\
(J+\mathfrak{m}^{e+1})/(\mathfrak{m}J+\mathfrak{m}^{e+2}) & \longrightarrow & \mathbb{C}[[s]]/\mathfrak{m}J+\mathfrak{m}^{e+2} & \longrightarrow & \mathbb{C}[[s]]/J+\mathfrak{m}^{e+1}.
\end{array}$$

We obtain, using the second part of Theorem 10.3.8, for the corresponding obstruction map $\left(T^2_{X,\mathrm{o}}\right)^* \longrightarrow J+\mathfrak{m}^{e+1}/\mathfrak{m}J+\mathfrak{m}^{e+2}$ a commutative diagram

$$\begin{array}{ccc}
\left(T^2_{X,\mathrm{o}}\right)^* & \longrightarrow & J/\mathfrak{m}J \\
& \searrow & \downarrow \\
& & J+\mathfrak{m}^{e+1}/\mathfrak{m}J+\mathfrak{m}^{e+2}.
\end{array}$$

As by the second assumption, the obstruction map $\left(T^2_{X,\mathrm{o}}\right)^* \longrightarrow J/\mathfrak{m}J$ is surjective, the image of the obstruction map of the second extension is thus $(J+\mathfrak{m}^{e+2})/(\mathfrak{m}J+\mathfrak{m}^{e+2})$. By Corollary 10.3.9 we see that for all e the extension $\mathbb{C}[[s]]/J+\mathfrak{m}^{e+2} = A_{e+1} \longrightarrow A_e = \mathbb{C}[[s]]/J+\mathfrak{m}^{e+1}$ is maximal with respect to the deformation $(X_{\mathbb{A}_e},\mathrm{o}) \longrightarrow (\mathbb{A}_e,\mathrm{o})$. Thus we have a formal versal deformation by Theorem 10.3.12.

Step 2. Now suppose (X_A, o) is formally semi-universal. Consider the formal deformation constructed in Theorem 10.3.16. By construction, for all e, the obstruction element associated to the extension

$$0 \longrightarrow (J + \mathfrak{m}^{e+1})/(\mathfrak{m}J + \mathfrak{m}^{e+2}) \longrightarrow \mathbb{C}[[s]]/\mathfrak{m}J + \mathfrak{m}^{e+2} \longrightarrow \mathbb{C}[[s]]/J + \mathfrak{m}^{e+1} \longrightarrow 0,$$

maps surjectively onto $J_{e+1}/\mathfrak{m}J_{e+1} = J + \mathfrak{m}^{e+2}/\mathfrak{m}J + \mathfrak{m}^{e+2}$. Now for some $e \gg 0$ the minimal number of generators of $J_{e+1} \mod \mathfrak{m}^{e+2}$ is equal to the minimal number of generators of J by Exercise 1.3.19. Thus the obstruction map is surjective. The Kodaira-Spencer map is an isomorphism by definition of the first order deformation. □

Now we come to the proof of the main theorem.

Theorem 10.3.21 (Grauert). *Let (X, o) be a germ of a complex space. Suppose that $\dim_{\mathbb{C}}(T^1_{X,o}) < \infty$. Then there exists an analytic semi-universal deformation of (X, o).*

Proof. The previous theorem says that it remains to give a convergent deformation which is formally semi-universal. Let (X, o) be defined by $f = (f_1, \ldots, f_k) \in \mathbb{C}\{x\}^k$, and let $r_1, \ldots, r_t \in \mathbb{C}\{x\}^k$ generate the module of relations between the $f_1, \ldots, f_k$. Take a formal semi-universal deformation of (X, o) over $A = \mathbb{C}[[s]]/J$, and let $\mathfrak{m} = (s)$. Let e be a natural number, such that the minimal number of generators of J is equal to the minimal number of generators of J modulo $\mathfrak{m}^{e+1}$. As $\mathbb{C}[[s]]$ is Noetherian local ring, such an e exists according to Exercise 1.3.19. Let $g_1, \ldots, g_p$ be the generators of $J + \mathfrak{m}^{e+1}/\mathfrak{m}^{e+1}$. We may assume that the g_i are polynomials in s. In order to apply Grauert's Approximation Theorem, we will have to interpret the flatness condition in terms of equations. Now flatness means that we can find lifts $F, R_1, \ldots, R_t \in \mathbb{C}\{x, s\}$ of $f, r_1, \ldots, r_t$ and vectors $M_1, \ldots, M_t \in \mathbb{C}\{x, s\}^p$ with

$$R_i \cdot F - g \cdot M_i = 0 \text{ for } i = 1, \ldots t. \tag{10.14}$$

We want to apply Grauert's Approximation Theorem 8.2.2 to this set of equations. Our formal semi-universal deformation of (X, o) over $A = \mathbb{C}[[s]]/J$ gives solutions modulo $\mathfrak{m}^{q+1}$ of (10.14), $F^{(q)}$, $R_i^{(q)}$, $g^{(q)}$, and $M_i^{(q)}$ for all $q \geq e$, lifting those for $q = e$. Now suppose $F^{(q)'}$, $R_i^{(q)'}$, $g^{(q)'}$ and $M_i^{(q)'}$ is a solution modulo $\mathfrak{m}^{q+1}$ of (10.14) lifting $F^{(e)}$, $R_i^{(e)}$, $g^{(e)}$, and $M_i^{(e)}$. In order to apply Grauert's Approximation Theorem, we have to prove that the $F^{(q)'}$, $R_i^{(q)'}$, $g^{(q)'}$ and $M_i^{(q)'}$ can be lifted to solutions modulo $\mathfrak{m}^{q+2}$. The ideal $J'_q = (g_1^{(q)'}, \ldots, g_p^{(q)'})$ has p generators, just as many as J_e. From Lemma 10.3.10 it follows that the surjection $\mathbb{C}[[s]]/J'_q \longrightarrow \mathbb{C}[[s]]/J_e$ is induced by a sequence of maximal small extensions. Therefore J'_q can by 10.3.16 be extended to an ideal J', so that there exists a formal semi-universal deformation over $\mathbb{C}[[s]]/J'$. Hence from Remark 10.3.3 it follows that the formal algebras $\mathbb{C}[[s]]/J'$ and $\mathbb{C}[[s]]/J$ are isomorphic, and thus that $\mathbb{C}[[s]]/J'_q$ and $\mathbb{C}[[s]]/J_q$ are isomorphic for all q. Thus for dimension reasons the minimal ideal I with

(1) $\mathfrak{m}J'_{q+1} \subset I \subset J'_q$ and,

(2) there exists a lift of the deformation over $(\mathbb{A}'_q, o)$ to one over $(\mathbb{A}'_{q+1}, o)$ with $A'_{q+1} = \mathbb{C}[[s]]/I$,

is the ideal J'_{q+1}, as this is also the case for J_{q+1}. Thus there does indeed exists a lift $F^{(q+1)'}$, $R_i^{(q+1)'}$, $g^{(q+1)'}$ and $M_i^{(q+1)'}$ of $F^{(q)'}$, $R_i^{(q)'}$, $g^{(q)'}$ and $M_i^{(q)'}$. Thus Grauert's Approximation Theorem 8.2.2 applies, and the theorem is proved. □

Exercises

10.3.22. Prove Theorem 10.3.8 for the formal and the analytic case.
(Hint: Use Exercise 1.3.19 to show that there exists an $n \in \mathbb{N}$ with $\mathcal{O}_{X,o} \otimes J \cong \mathcal{O}_{X,o} \otimes (J + \mathfrak{m}^n/\mathfrak{m}^n)$, and use the theorem already proved for the Artinian case.)

10.3.23. Let $\varphi : B \longrightarrow A$ be a surjective map of Artinian $\mathbb{C}$-algebras. Suppose that $\dim_{\mathbb{C}}(A) = \dim_{\mathbb{C}}(B)$. Show that φ is an isomorphism.

10.3.24. Prove that a simple surjection of Artinian $\mathbb{C}$-algebras is small.

10.3.25. Show that the definition of of $T^2_{X,o}$ is independent of the choice of generators of the ideal of (X, o).

10.3.26. Prove Theorem 10.3.7.
(Hint: Use similar arguments as in the corresponding proof for $T^1_{X,o}$.)

10.3.27. Prove Theorem 10.3.21 for the algebraic case: Let (X, o) be an isolated singularity, defined by algebraic power series. Suppose $\dim_{\mathbb{C}}(T^1_{X,o}) < \infty$. Then there exists a semi-universal deformation of (X, o) which is defined by algebraic power series.
(Hint: Use Corollary 8.3.7.)

Bibliography

[Artin 1968] Artin, M.: *On the solution of analytic equations.* Inv. math. **5**, 277–291.

[Brieskorn-Knörrer 1986] Brieskorn, E., Knörrer, H.: *Plane Algebraic Curves.* Birkhäuser, Basel, Boston, Stuttgart.

[Buchberger 1965] Buchberger, B.: *Ein Algorithmus zum Auffinden der Basiselemente des Restklassenringes nach einem nulldimensionalen Polynomideal.* Dissertation Innsbruck.

[Eisenbud 1995] Eisenbud, D.: *Commutative Algebra with a View Toward Algebraic Geometry.* GTM **150**. Springer-Verlag, Berlin etc..

[Gabrielov 1971] Gabrielov, A.M.: *The formal relation between analytic functions.* Funktional Anal i Priložen **5**, no. 4, 64–65.

[Gordan 1899] Gordan, P.: *Neuer Beweis des Hilbert'schen Satzes über homogene Funktionen.* Nachrichten König. Ges. der Wiss. zu Göttingen, 240–242.

[Grauert 1972] Grauert, H.: *Über die Deformation isolierter Singularitäten analytischer Mengen.* Inv. Math **15**, 171–198.

[Grauert-Remmert 1971] Grauert, H., Remmert, R.: *Analytische Stellenalgebren.* Grundl. **176**. Springer Verlag, Berlin etc.

[Grauert-Remmert 1984] Grauert, H., Remmert, R.: *Coherent Analytic Sheaves.* Grundl. **265**. Springer Verlag, Berlin etc.

[Gunning-Rossi 1965] Gunning, R., Rossi, H.: *Analytic Functions of Several Complex Variables.* Prentice-Hall, Englewood Cliffs, NJ.

[Hironaka 1964] Hironaka, H.: *Resolution of Singularities of an Algebraic Variety over a Field of Characteristic Zero.* Ann. of Math. **79**, 109–326.

[Kurke et al. 1975] Kurke, H., Pfister, G., Roczen, M.: *Henselsche Ringe und algebraische Geometrie.* VEB Deutscher Verlag der Wissenschaften, Berlin.

[Kurke et al. 1978] Kurke, H., Pfister, G., Popescu, D.; Roczen, M.; Mostowski, T.: *Die Approximationseigenschaft lokaler Ringe.* Springer Lecture Notes 634.

[Lang 1965] Lang, S.: *Algebra.* Addison-Wesley, Reading, Massachusetts.

[Matsumura 1979] Matsumura, H.: *Commutative Algebra.* W.A. Benjamin, New York.

[Mora 1982] Mora, T.: *An Algorithm to Compute the Equations of Tangent Cones.* Proc. EUROCAL 82, Lecture Notes in Comp. Sci. **144**, 158–165.

[Popescu 1985] Popescu, D.: *General Néron Desingularization.* Nagoya Math. J. **100**, 97–126.

[Popescu 1986, 1990] Popescu, D.: *General Néron Desingularization and Approximation.* Nagoya Math. J. **104**, 85–115; **118**, 45–53.

[Pfister-Popescu 1975] Pfister, G., Popescu, D.: *Die strenge Approximationseigenschaft lokaler Ringe.* Inv. Math. **30** (82), 145–147.

[Schlessinger 1968] Schlessinger, M.; *Functors of Artin rings.* Trans. Am. Math. Soc. **130**, 208–222.

[Singular 2000] Greuel, G.-M., Pfister, G., Schönemann,H.: *The Computeralgebra System* SINGULAR, http://www.singular.mathematik.uni-kl.de.

[Spivakovsky 1994] Spivakovsky, M.: *Non-existence of the Artin Function for Henselian Pairs.* Math. Ann. **299**, 727–729.

[Swan 1998] Swan, R.: *Neron-Popescu Desingularization*, in Algebra and Geometry, Ed. Ming-chang Kang, International Press, Cambridge, 135–192.

[Spivakovsky 1999] Spivakovsky, M.: *A new proof of D. Popescu's Theorem on smoothing of ring homomorphisms.* J. of AMS **294**, 381–444.

[Van der Waerden 1971] Van der Waerden, B.L.: *Moderne Algebra*, eighth ed. Springer Verlag, Berlin etc.

Index

U

W

Z

GPSR Compliance

The European Union's (EU) General Product Safety Regulation (GPSR) is a set of rules that requires consumer products to be safe and our obligations to ensure this.

If you have any concerns about our products, you can contact us on ProductSafety@springernature.com

In case Publisher is established outside the EU, the EU authorized representative is:

Springer Nature Customer Service Center GmbH
Europaplatz 3
69115 Heidelberg, Germany

Batch number: 08567405

Printed by Printforce, the Netherlands